DYNAMICS OF METEOROLOGY AND CLIMATE

Richard S. Scorer
Emeritus Professor and Honorary Senior Research Fellow
in Mathematics and Environmental Technology,
Imperial College of Science, Technology and Medicine, London

JOHN WILEY & SONS
Chichester • New York • Weinheim • Brisbane • Singapore • Toronto

PRAXIS

Published in association with
PRAXIS PUBLISHING
Chichester

The White House,
Eastergate, Chichester,
West Sussex, PO20 6UR, England

Based on *Environmental Aerodynamics* first published in 1978.
This book published in 1997 by
John Wiley & Sons Ltd
in association with Praxis Publishing Ltd

Wiley Editorial Offices

John Wiley & Sons Ltd, Baffins Lane,
Chichester, West Sussex, PO19 1UD, England

John Wiley & Sons, Inc., 605 Third Avenue,
New York, NY 10158-0012, USA

Wiley-VCH Verlag GmbH, Pappelallee 3,
D-69469 Weinheim, Germany

Jacaranda Wiley Ltd, G.P.O. 33 Park Road, Milton,
Queensland 4001, Australia

John Wiley & Sons (Asia) Pte Ltd, 2 Clementi Loop #02-01,
Jin Xing Distripark, Singapore 12981

John Wiley & Sons (Canada) Ltd, 22 Worcester Road,
Rexdale, Ontario, M9W 1L1, Canada

Library of Congress Cataloguing-in-Publication Data
Scorer, R. S. (Richard Segar), 1919–
Dynamics of Meteorology and Climate / Richard Scorer.
p. cm.
Based on Environmental Aerodynamics. Chichester, England : E.
Horwood : New York : Halsted Press, 1978.
"Published in association with Praxis Publishing, Chichester."
Includes bibliographical references and index.
ISBN 0-471-96815-3 (Cloth : alk. paper). —ISBN 0-471-96816-1 (pbk. : alk. paper)
1. Mesometeorology. 2. Dynamic meteorology. 3. Fluid mechanics. 4. Air—Pollution.
I. Scorer, R. S. (Richard Segar), 1919– Environmental aerodynamics. II Title.
QC883.5.S358 1997 96-28413
551.5--dc20 CIP
A catalogue record for this book is available from the British Library

ISBN 0-471-96815-3 Cloth ISBN 0-471-96816-1 Paperback
Typeset by Heather FitzGibbon, Christchurch, Dorset
Printed and bound in Great Britain by MPG Books Ltd, Bodmin, Cornwall

DYNAMICS OF METEOROLOGY AND CLIMATE

WILEY-PRAXIS SERIES IN ATMOSPHERIC PHYSICS

Series Editor: John Mason, B.Sc., Ph.D.

This series reflects the major developments which are occurring in studies of the physical and dynamic processes which take place within the Earth's atmosphere. With increasing concern over important environmental issues such as ozone depletion, global warming and dispersal of pollutants, this is a vibrant research area.

Written by international authors chosen for their expertise and high reputations in all aspects of atmospheric physics and climatology, the titles in this series provide a forum for the presentation and dissemination of new ideas and scientific results in this important field.

These books will appeal to: professional meteorologists; atmospheric physicists; climatologists; civil, aeronautical, mechanical and chemical engineers; applied mathematicians; aerobiologists; architects, environmental scientists and ecologists; together with postgraduate and undergraduate students in these fields and, in some cases, pre-university college students and non-scientists with a keen interest in environmental issues.

For further details of the books listed below and ordering information, why not visit the Praxis Web Site at http://www.praxis-publishing.co.uk

PHYSICS OF THE UPPER POLAR ATMOSPHERE
Asgeir Brekke, Professor in Physics, The Auroral Observatory, Institute of Mathematical and Physical Science, University of Tromsø, Tromsø, Norway

APPLICATIONS OF WEATHER RADAR SYSTEMS: **A Guide to Uses of Radar Data in Meteorology and Hydrology**
Second edition
Christopher G. Collier, Professor of Environmental Remote Sensing, Telford Research Institute, University of Salford, UK

DYNAMICS OF METEOROLOGY AND CLIMATE
Richard Scorer, Emeritus Professor and Honorary Senior Research Fellow in Mathematics and Environmental Technology, Imperial College of Science, Technology and Medicine, London, UK

Forthcoming titles

PHYSICS OF THE ENVIRONMENT AND CLIMATE
Gérard Guyot, Research Director, Laboratory of Bioclimatology of INRA, Avignon, France

OPTICS OF LIGHT SCATTERING MEDIA: **Problems and Solutions**
Alex A. Kokhanovsky, Institute of Physics, Academy of Sciences, Minsk, Belarus

DYNAMIC ANALYSIS OF CLIMATE: **Atmospheric Circulation, Perturbations, Climatic Evolution**
Marcel Leroux, Professor of University, Université Jean Moulin, Lyon, France

To Margaret

Satellites give us a link between the great outdoors and the world conjured up by the weather chart, which will be on roughly the same scale. Although it clearly reveals the weather systems there is a choice to be made, as in every photographic printing process, how the shading represents the intensities seen by the satellite. In all the many satellite pictures in this book the whole range of visible grades may be used or the grades which can be printed will be used to emphasise a particular part of the spectrum leaving the grades outside that range unused.

In this picture the big crack in the fast-ice which is very warm compared with the ice around it (which is shown black so that everything colder is shown with the same blackness) and the jet of warm air coming from the estuary of Kangerdlugssuaq on the SE coast of Greenland appears less black because it is less cold. In the waveband (Channel 3) used here some scattered sunshine is also seen particularly in the bright patch near the end of the crack in the ice: this is where the warm water has made some steaming fog with droplets which are small compared with the wavelength. This image is made with an enlargement (of about 18 times) to show the individual pixels which are 1.1 km square immediately beneath the satellite at the left-hand edge of this frame, becoming stretched more as we go to the right side, away from the direct vertical. It is close to local midday at 1529 UT on 20.02.90 when the ice is near to its maximum extent.

Table of contents

Part 2 Turbulent phenomena, clouds and dispersion

The colour plate section is positioned between pages 352 and 353

Foreword

The atmosphere and the oceans provide an extraordinarily rich natural environment in which evolution, governed by the fundamental principles of fluid dynamics, proceeds. This dynamical evolution is subject to a multitude of complicating factors—nonuniform irradiation by the sun, tidal forces of moon and sun, Coriolis forces associated with the Earth's rotation, boundary constraints associated with the irregular geometry of the continents and of the Earth's solid surface—and is inevitably complex in detail and uncertain in both local and global predictability. Yet certain fundamental processes, namely those associated with vorticity, with gravity and inertial waves, and with the intrinsic instabilities of fluid flow and the turbulence to which these instabilities give rise, provide the key to an understanding of this evolution, and must lie at the heart of any serious attempt to approach the major problems of meteorology and climate. To provide an essential understanding of these fundamental processes, and to knit them together into a coherent view of our planetary atmosphere–ocean system is a vast undertaking of immense challenge, and one that is of the greatest importance in trying to assess man's impact on this system, whether at the local or the global level.

Dick Scorer has brought a lifetime of study and reflection to bear upon these problems, and provides in this book an exceptionally wide-ranging discussion both of the fundamental dynamical principles, and of their meteorological manifestations, illustrated by a wide range of excellent photographs, including many from satellites. The treatment builds on previous editions, and includes a completely new section on short-term weather forecasting and long-term climatic change. The discussion, at times provocative, will not fail to stimulate. This volume provides a refreshing and individual approach to the subject which can be warmly recommended both to hardened professionals, and to undergraduate and graduate students entering upon a wonderfully rewarding field of study.

KEITH MOFFATT
Isaac Newton Institute, Cambridge

Author's preface to the previous edition

Natural Aerodynamics, the predecessor to this book, was written twenty years ago at a time when air pollution, gliding, and other environmental applications of fluid mechanics were of rapidly growing interest. Since then there has developed even greater interest in mesoscale meteorology, which I like to call out-of-doors hydrodynamics, not only because of its important applications to aviation but because we have begun to regard the environment as something to live in, along with other species, and not merely something crudely to exploit. The physics of the environment has had a fair share of the high school syllabuses: by comparison the beauties of its mechanics have been neglected and this new book will, I hope, enable university teachers to take it much more in their stride.

We are all concerned with man-sized, town-sized, and country-sized phenomena of fluid motion. The medium looks very different from that treated in textbooks on fluid mechanics because they give so little attention to the forces of buoyancy and rotation which are dominant out-of-doors. Because of this the motion has been regarded by engineers and theoreticians as too complicated for them, and accordingly rather avoided. Nevertheless fascinating laboratory experiments and theoretical treatments of a fairly simple kind do exist and there is no reason why they should not be part of standard courses.

The basis of our subject in conventional fluid dynamics is described in the first two chapters of this book. They form a commentary on fluid mechanics which could be specially useful to engineers and physicists who are beginning to grapple with environmental problems. Chapter 3 is a discussion of fluid rotation giving a simpler treatment of secondary flow. This makes it accessible and useful to many who might otherwise have been content with qualitative arguments about pressure and accelerations which often give explanations which are difficult to use with confidence or for prediction.

Chapter 4 discusses many of the subtleties of dynamical meteorology which originate in the earth's rotation. I hope this survey of wind systems will be helpful to engineers, geographers, and anyone interested in weather.

Chapters 5 and 6 are necessarily somewhat more mathematical because they deal with precisely defined phenomena. Both topics—waves and billows—have increased in importance for aviation, oceanography, and many branches of physical meteorology and deserve serious mathematics, although even now a great deal remains in the realm of physical inference.

Turbulence is treated in Chapter 7 from the standpoint of mechanism, unlike the more conventional statistical and phenomenological approaches which are less useful out-of-doors. This leads to the next two chapters which are about turbulence as we find it in nature, and contain some remarkably simple but effective methods of analysis, and to Chapter 10 which applies these methods to dispersion of pollution. Chapter 11 surveys mechanical behaviour in which the condensation of clouds is essential. It may also remind us that although the earth is a red and ultramarine planet with a blue atmosphere, it is half covered with white clouds which dominate the motion.

Although the last chapter is short, one may hope that it deals with the part of environmental aerodynamics most ripe for exciting developments in the near future. If this book can inspire a few who might otherwise have thought of themselves as engineers or physicists to enter the biological realm, it will have achieved one of its main aims.

Chapters 3, 5, and 6, with suitable extracts from Chapters 1 and 2, are the heart of one lecture course and Chapters 7–10 of another which have been given in various forms over the last two decades. They contain what does not appear to be assembled elsewhere. Chapters 1, 2, and 4 may be specially useful to atmospheric physicists, who may also find in Chapter 11 many topics which have received little treatment elsewhere. Thus it is to be hoped that the book will be found useful background reading for many environmental specialisms. Some geographers with a feel for the phenomena may find a new way into mathematics.

Courses with an environmental emphasis are being introduced in many disciplines and many participants will have difficulty in extracting for themselves the appropriate parts of more conventional books on fluid mechanics. I hope that they may be helped by this new book to find insight into the joys of the great out doors.

I acknowledge gratefully the privilege of having many research students to work with me: there is reference to many of their works, and for me their cooperation has always been a source of inspiration. I should also say that my indebtedness to Professors F. H. Ludlam, C.-S. Yih and R. R. Long is greater than my references to their writings may imply.

R. S. Scorer
Imperial College, London
February 1977

Author's preface to this edition

The earliest edition of this book was published in 1958 as *Natural Aerodynamics*, and was mainly descriptive about topics like vorticity and partly turbulent flow. It was intended to make people aware of the fluid mechanics going on in the environment around them. The next edition, *Environmental Aerodynamics* (1978), was far more mathematical and was based on two lecture courses, on 'The Flow of Stratified Fluids' and 'Partly Turbulent Flow', and was published also in Russian. This latest edition contains a variously edited version of all that was in the second with some additions, but faces the great issue of possible climate change—not by discussing what it might change to, but how it has changed in the recent past and what changed effects determine it now or might dominate in the near future.

By the *recent past* I do no mean since meteorological observations began about a century ago, but recent in terms of the earth's history. By recent I mean the extent of remote sensing backwards to events of the last 200 000 years. Of these the last 100 000, the last 10 000 and the last 1000 have been filled up in extraordinary detail by interpretation of cores in the deep ice, in the ocean bed, in river deltas, and even in peat in places where humanity has not been too busy.

We can now think about our place in the evolutionary processes, and many simplistic ideas have to be abandoned in the context of global warming. We need global models and computers of the greatest available capacity to bring all the physics and mechanics together. We have to include the oceans and recognise that many mechanisms have to be represented by parameters because in reality they are dominated by scales of motion too small to be described in the grid of points that a computer must use in its finite difference techniques to solve the differential equations. And this is where we come up against the two major blockages to progress: the discovery of valid parameters and the chaos that the numerical solutions quickly get into.

Even if we soon know enough to persuade us that drastic measures will be, indeed are, advisable, indeed necessary, there is the problem of getting action that goes against all the opportunist tactics our race has learnt in the course of evolving from small numbers to becoming top species. The development of the two blockages may seem to devalue the mathematical methods which were so exquisitely devised in the nineteenth century and improved in the twentieth century. We can no longer imagine an exact answer to be progressively approached: the art of interpreting what we can get is enhanced in value. While the remote sensors expand into the detail of the present, and the past, we have to retain the physical understanding to which mathematics provides a marvellous short-cut.

Thus I hope the reader will sympathise with my need to have the whole book in mind when writing each paragraph, and will quickly discover the advantage of making a first approach through the index. There are many mixtures of principles, approximations and rough guesses which make the learning about nature most enjoyable. I like Jeffries's definitions of *approximate* and *rough*: A first idea is *rough*, and by gradual improvement it becomes more *approximate* to our ultimate hope. Approximation means *getting closer*, not making a rough guess, although that may sometimes be all that is both necessary and desired.

I am much obliged to the journals *Nature* and *Science* and those of their authors who gave permission to reproduce diagrams. Those reproduced from *Nature* remain the copyright of Macmillan Magazines Limited. Their papers have helped greatly in telling the history of past climate. Thanks are also due to the authors A. G. Smith, D. G. Smith and B. M. Funnell and their publisher Cambridge University Press for the use of three of their World Maps reminding us of the importance of continental drift which has been probably the main factor in causing the very long-term changes in the Earth's changing climate. There are also other authors whose works have educated me and they are acknowledged where appropriate.

The demands and questions from Series Editor John Mason and Clive Horwood of Praxis Publishing have served to mould this book into a form fit for this series on Atmospheric Physics. The multiplicity of pictures is necessary in order to be sure the reader knows what is in mind all the time. I am also deeply indebted to the Satellite Receiving Station in the University of Dundee and to NOAA, for the pictures which have greatly advanced our understanding.

Finally, without the continuous encouragement and tolerance of my wife this work could not have been completed and I am grateful indeed.

RICHARD SCORER
Wimbledon, 1997

List of illustrations, plates and tables

CHAPTER 3

CHAPTER 4

CHAPTER 5

CHAPTER 6

CHAPTER 7

CHAPTER 8

CHAPTER 9

CHAPTER 10

CHAPTER 11

CHAPTER 12

CHAPTER 13

CHAPTER 14

CHAPTER 15

CHAPTER 16

CHAPTER 17

CHAPTER 19

CHAPTER 20

COLOUR PLATES (*positioned between pages 352 and 353*)

TABLES

Part 1
Fundamental theory, vorticity, waves and instability

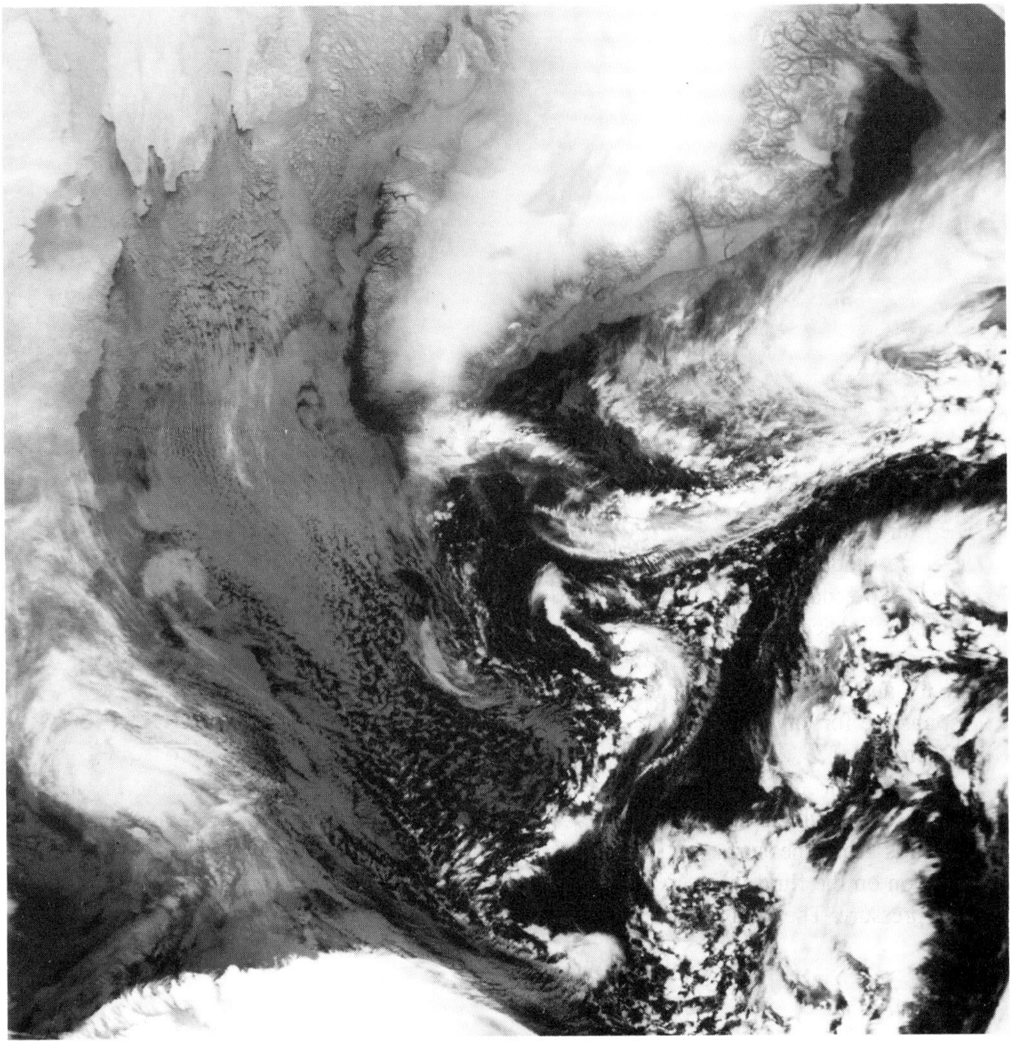

The North Atlantic is an area of continuous interplay between warm and cold. Warm means salty water; cold means fresh meltwater, and this is a Channel 4 picture (waveband 10.5–11.5 μm). The location of the scene in the frontispiece on p. vi is easily found. The coldest objects are pure white and the warmest jet black, the intensity of the received signal being portrayed in photographic negative. The scale is reduced by a factor of 12 so that this picture represents an area about 3000 km square.

The hinterland of the Kangerdlugssuaq estuary and the highest peaks of Greenland is evident, and the clouds over the middle of the snow plateau are warmer than the snow below. The warm current up the SW coast of Greenland is clearly free of ice and cloud. It is that area which has supported most of the human population throughout several centuries. The low cloud is a dull grey, warmer than the higher cloud but colder than the sea, which is black.

Most of the other satellite pictures of clouds etc. have a scale three times this one, but their scale can be checked by reference to visible bits of coastline which can be identified. The warm jet from the Kangerdlugssuaq estuary is only rarely seen, and interpretation of the interplay of the cold and warm water must be inferred as long as salinity measurements *in situ* remain unavailable.

1

Fundamental equations

1.1 IRROTATIONAL MOTION AND REAL FLUIDS

There are several theorems in the classical treatment of irrotational motion of incompressible fluids which have great significance for the study of real fluids. These are well described in many good textbooks: here we shall discuss them rather briefly, surveying their significance in a less orthodox manner.

In an inviscid fluid it is assumed that the only forces acting are body forces such as gravity and the fluid pressure which acts perpendicularly to any surface. Viscosity is responsible for any forces which may act along a surface. In this context a surface is a mathematical concept, and does not refer only to the outer boundary of the fluid. To calculate the force due to the pressure we consider the total force on the fluid inside a surface S, of which **dS** is an element of area dS, the vector being directed normally outwards from the surface. If **l** is a unit vector in a fixed direction, the force in this direction on the fluid inside S due to the pressure on the element is $-p\mathbf{dS}.\mathbf{l}$, p being the fluid pressure. The total force on the surface is therefore

$$-\int_{\text{surf}} \mathbf{l}.\,p\mathbf{dS} = -\int_{\text{vol}} \text{div}(\mathbf{l}p)\,\mathrm{d}\tau = -\mathbf{l}.\int_{\text{vol}} \mathbf{grad}\,p\,\mathrm{d}\tau \qquad (1.1.1)$$

in which $\mathrm{d}\tau$ is an element of volume. A dot denotes a scalar product. The integrals are over the whole surface or throughout the enclosed volume. Here we have used the divergence theorem, namely that the outward flux from the inside to the outside of the surface which encloses it is equal to the divergence of the vector from all the elements of volume enclosed. Thus for any vector field **A**, $\int_{\text{vol}} \text{div}\,\mathbf{A}\,\mathrm{d}\tau = \int_{\text{surf}} \mathbf{A}.\mathbf{dS}$, and the identity that $\text{div}(\mathbf{l}p) = \mathbf{l}.\mathbf{grad}\,p + p\,\text{div}\,\mathbf{l}$, the last term of which is zero because **l** is a constant vector.

The expression (1.1.1) is the component in the direction **l** and therefore the total vector force on the enclosed fluid is

$$-\int_{\text{surf}} p\,\mathbf{dS} = -\int_{\text{vol}} \mathbf{grad}\,p\,\mathrm{d}\tau \qquad (1.1.2)$$

which shows that the force per unit volume at a point is $-\mathbf{grad}\,p$.

We take Newton's law to be absolutely accurate for all purposes in terrestrial fluid mechanics if only because no measurements are practicable for detecting any errors. It states that the rate of change of momentum in the direction **l** is equal to the total force in that direction. Accordingly, if ρ is the density of the fluid (mass per unit volume) and **v** is the velocity of the fluid, the law, in a field of gravity, is

$$\int_{\text{vol}} \frac{\mathrm{D}}{\mathrm{D}t}(\mathbf{v}.\mathbf{l})\, \rho \mathrm{d}\tau = -\int_{\text{surf}} \mathbf{l}.p\mathrm{d}\mathbf{S} + \int_{\text{vol}} \mathbf{l}.\mathbf{g}\rho \mathrm{d}\tau . \qquad (1.1.3)$$

In this, $\mathrm{D}/\mathrm{D}t$ is the rate of change following the same fluid particles, often called the **material rate of change**, or the **Lagrangian derivative**, and $\rho \mathrm{d}\tau$ is the element of mass envisaged in the integral. The volume and the surface are thought of as defined by material particles which therefore move with the fluid, and (1.1.3) equates the rate of change of momentum in the direction **l** produced by the forces in that direction. We have presented the **body force** as if it were **gravity** only, which is equal to $\mathbf{g}\rho$ per unit volume. There are body forces, such as electromagnetic ones, which are not necessarily proportional to the mass, but in this book the only body force we shall be concerned with is gravity, and later Coriolis forces in Chapter 4. Expression (1.1.3) is the **l**-component of

$$\int_{\text{vol}} \frac{\mathrm{D}\mathbf{v}}{\mathrm{D}t} \rho \mathrm{d}\tau = -\int_{\text{vol}} \mathbf{grad}\, p\, \mathrm{d}\tau + \int_{\text{vol}} \mathbf{g}\rho\, \mathrm{d}\tau$$

and since the volume is unspecified, the integrands must be equal, and so

$$\boxed{\frac{\mathrm{D}\mathbf{v}}{\mathrm{d}t} = -\frac{1}{\rho}\, \mathbf{grad}\, p + \mathbf{g}.} \qquad (1.1.4)$$

From this general form of the equation of motion we deduce the hydrostatic equation. The acceleration is zero, and gravity acts in the negative direction of the gradient of the gravitational potential gz and so (1.1.4) is $\mathbf{grad}\, p = \rho\, \mathbf{grad}(-gz)$, or

$$\boxed{\frac{\partial p}{\partial z} = -g\rho.} \qquad (1.1.5)$$

We shall always take the z-**coordinate** as being measured **vertically upwards** for any situation in which we are concerned with gravity.

In the next section we have set out several mathematical identities of which we shall make use. The identity (1.2.7) may be written in the form

$$\mathrm{div}\,(\phi\, \mathbf{grad}\, \phi) = \phi\, \mathrm{div}\, \mathbf{grad}\, \phi + \mathbf{grad}\, \phi . \mathbf{grad}\, \phi \qquad (1.1.6)$$

and this may be integrated, using the divergence theorem, to give

$$\int_{\text{surf}} \phi\, \mathbf{grad}\, \phi . \mathbf{dS} = \int_{\text{vol}} \text{div}(\phi\, \mathbf{grad}\, \phi)\, \mathrm{d}\tau$$

$$= \int_{\text{vol}} \phi \nabla^2 \phi\, \mathrm{d}\tau + \int_{\text{vol}} (\mathbf{grad}\, \phi)^2\, \mathrm{d}\tau. \tag{1.1.7}$$

In order to study the conditions which have to be imposed, or specified, for a field of fluid motion to be determined, we imagine that there are two fields possible in a given situation and denote their velocities by $\mathbf{v}$ and $\mathbf{v}'$. If $\mathbf{u} = \mathbf{v} - \mathbf{v}'$ and if div $\mathbf{v}$ and **curl** $\mathbf{v}$ are both specified at all points in the region of fluid under consideration, so that they are equal to div $\mathbf{v}'$ and **curl** $\mathbf{v}'$ respectively, we must have that

$$\text{div}\, \mathbf{u} = 0 \quad \text{and} \quad \mathbf{curl}\, \mathbf{u} = \mathbf{0} \tag{1.1.8}$$

throughout the region. The second of these equations means that there exists a potential ϕ such that $\mathbf{u} = \mathbf{grad}\, \phi$, and so the first then tells us that

$$\boxed{\text{div}\, \mathbf{u} = \text{div}\, \mathbf{grad}\, \phi = \nabla^2 \phi = 0} \tag{1.1.9}$$

at all points of the region.

The assumption that div $\mathbf{v}$ and **curl** $\mathbf{v}$ are specified everywhere thus gets rid of the first term on the right in (1.1.7). If the left-hand side were made zero by some conditions imposed at the boundary surface $\mathbf{S}$, the term $\int_{\text{vol}} (\mathbf{grad}\, \phi)^2\, \mathrm{d}\tau$ would have to be zero, which means that **grad** ϕ would have to be of zero magnitude everywhere so that $\mathbf{v}$ would equal $\mathbf{v}'$. We can therefore set about finding what boundary conditions would achieve this result of making the velocity field unique. We shall discuss why it is useful to assume that div $\mathbf{v}$ and **curl** $\mathbf{v}$ should be zero later.

The integral $\int_{\text{surf}} \phi\, \mathbf{grad}\, \phi . \mathbf{dS} = \int_{\text{surf}} \phi \mathbf{u} . \mathbf{dS}$ can obviously be made zero if the component of $\mathbf{u}$ in the direction of $\mathbf{dS}$, that is normal to the surface, is zero; and this is equivalent to saying that the normal component of $\mathbf{v}$ is specified on the boundary of the fluid. Alternatively, if the boundary consists of a single surface and the tangential component of the velocity $\mathbf{v}$ is everywhere specified on it, the tangential component of $\mathbf{v}'$ is the same, and that of $\mathbf{u}$ is zero. **grad** ϕ therefore has zero component along the surface so that ϕ is constant on the surface. If the surface consisted of more than one part, ϕ could have different but constant values on the separate parts. In the case of one part

$$\int_{\text{surf}} \phi\, \mathbf{grad}\, \phi . \mathbf{dS} = \phi \int_{\text{surf}} \mathbf{grad}\, \phi . \mathbf{dS} = \phi \int_{\text{vol}} \text{div}\, \mathbf{grad}\, \phi\, \mathrm{d}\tau = 0$$

using first the divergence theorem and then (1.1.9). The result is thus as required and we may state **Kelvin's Uniqueness Theorem** as follows:

If div **v and curl v are specified at all points of a region and either the normal component of v is specified at all points of the boundary, or, when the boundary consists of a single surface, the tangential component of v is specified at all points of it, v is unique.**

If the normal component of $\mathbf{v}$ is specified at all points on the boundary but div $\mathbf{v}$ and **curl** $\mathbf{v}$ are not specified, we may discuss the energy of the motion to see what the effect is. When div $\mathbf{v} = 0$ the fluid is **incompressible**, and when **curl** $\mathbf{v} = 0$ everywhere, the motion has zero vorticity and is described as **irrotational**. Let T be the kinetic energy of the motion $\mathbf{v}$ in the case in which div $\mathbf{v}$ and **curl v** are restricted to be zero everywhere, and let $\mathbf{v}'$ be the actual motion which has kinetic energy T'. If in this case we let $\mathbf{u} = \mathbf{v}' - \mathbf{v}$, and if we assume that the density is uniform throughout the fluid and is taken as unity, we have

$$0 \leq \int_{\text{vol}} u^2 \, \mathrm{d}\tau = \int_{\text{vol}} (v'^2 - v^2) \, \mathrm{d}\tau - 2\int_{\text{vol}} \mathbf{v}.\mathbf{u} \, \mathrm{d}\tau$$

$$= 2T' - 2T - 2\int_{\text{vol}} \mathbf{grad}\, \phi.\mathbf{u} \, \mathrm{d}\tau$$

where we have written $\mathbf{v} = \mathbf{grad}\, \phi$ because $\mathbf{curl\, v} = \mathbf{0}$ everywhere. Irrotational flow is therefore called **potential** flow, and the velocity is the gradient of the potential, ϕ. Therefore

$$T' - T \geq \int_{\text{vol}} \mathbf{grad}\, \phi.\mathbf{u} \, \mathrm{d}\tau = \int_{\text{vol}} \text{div}(\phi \mathbf{u}) \, \mathrm{d}\tau - \int_{\text{vol}} \phi \, \text{div}\, \mathbf{u} \, \mathrm{d}\tau$$

$$= \int_{\text{surf}} \phi \mathbf{u}.\mathbf{dS} - \int_{\text{vol}} \phi \, \text{div}\, \mathbf{u} \, \mathrm{d}\tau$$

$$= -\int_{\text{vol}} \phi \, \text{div}\, \mathbf{v}' \, \mathrm{d}\tau \tag{1.1.10}$$

because the component of $\mathbf{u}$ normal to S is zero since $\mathbf{v}'$ and $\mathbf{v}$ have the same normal component on S. If div $\mathbf{v}'$ is zero everywhere then the right side of (1.1.10) is zero and $T' \geq T$. The difference between $\mathbf{v}$ and $\mathbf{v}'$ is that $\mathbf{v}'$ may have vorticity: if it does not then **curl** $\mathbf{v}'$ is zero and the uniqueness theorem proves that it is the same as $\mathbf{v}$. If it is different from $\mathbf{v}$ somewhere, where **curl** $\mathbf{v}'$ is not zero, then u^2 is positive somewhere and $T' > T$. We can now state **Kelvin's Minimal Theorem** as follows:

Of all incompressible motions (having div $\mathbf{v} = 0$ **everywhere) of which the normal component is specified on the boundary, the unique irrotational motion has less kinetic energy than any other motion satisfying the same boundary conditions.**

This does not hold for a case in which the density is not uniform because it appears as a variable in the integrand and we cannot arrive at (1.1.10). Thus we could have two fluids with identical velocities everywhere, but one having greater density where the velocity was greatest.

If, on the other hand, div $\mathbf{v}'$ is not zero, so that the fluid is compressible, div $\mathbf{v}'$ may take positive or negative values and the integral (1.1.10) may be positive or negative, except in the case in which ϕ is not variable and $\mathbf{v}$ is zero, for in that case the irrotational motion is zero and no other motion can have less kinetic energy. The treatment of the integral in (1.1.10) is further complicated by the fact that the density is not uniform in a compressible fluid. The important point is that if there are pressure waves being propagated through the fluid, the kinetic energy might be more or it might be less than in the case of the incompressible irrotational motion satisfying the same boundary conditions. In this book we shall only very occasionally be concerned with effects of this kind in compressible fluids (see section 2.1 and Chapter 7), and in most cases div $\mathbf{v} = 0$, but by contrast we shall be very much concerned with rotational motion in which $\mathbf{curl\, v} \neq \mathbf{0}$.

If we can demonstrate that an inviscid incompressible fluid cannot acquire vorticity, then any motion set up by moving the boundaries must be irrotational (because the initial state of rest is irrotational) and must have the least possible kinetic energy satisfying those boundary conditions.

First we imagine that the motion is set up by a sudden jerking of the boundary. We shall see in section 1.4 that the effect of viscosity is to conduct vorticity through the fluid from the boundary in the way heat is conducted from a hot surface. This means that if there is no vorticity in the fluid initially it will take a finite time to produce any by viscosity. When the boundary is jerked suddenly, the motion set up is irrotational except for a very thin layer at the boundary. This has the effect of allowing the fluid to slip along the boundary while being subjected to a specified normal component of velocity.

During the infinitesimal time Δt at very large pressure P is created with very large gradients inside the fluid. A finite 'impulsive pressure' Π, equal to $P\Delta t$, is generated and sets up a finite velocity $\mathbf{v}$ during the time Δt. The change in momentum is then equal to the impulse on the surface enclosing the fluid, and so, considering the component in a direction $\mathbf{l}$, we find

$$\int_{\text{vol}} \mathbf{l}.\mathbf{v}\rho \, \mathrm{d}\tau = -\int_{\text{surf}} \mathbf{l}\,\Pi.\mathbf{dS}$$

$$= -\int_{\text{vol}} \text{div}\,(\mathbf{l}\,\Pi)\,\mathrm{d}\tau = -\mathbf{l}.\int_{\text{vol}} \mathbf{grad}\,\Pi\,\mathrm{d}\tau \qquad (1.1.11)$$

and since this holds for any surface marked out in the fluid, and for any direction $\mathbf{l}$, we must have

$$\rho\mathbf{v} = -\,\mathbf{grad}\,\Pi,$$

or

$$\mathbf{v} = -\frac{1}{\rho}\,\mathbf{grad}\,\Pi \qquad (1.1.12)$$

and this is equal to **grad** ϕ only if the **curl** of it is zero. Using (1.2.8) we see that this would be the case only if

$$\mathbf{grad}\,\Pi \times \mathbf{grad}\,\frac{1}{\rho} = \mathbf{0} \qquad (1.1.13)$$

everywhere. This could not be the case unless the surfaces Π = constant happened to be the same as the surfaces ρ = constant; and this would only be the case if it were specially contrived to be so. The motion would therefore be irrotational if ρ were uniform, in which case the velocity and impulsive pressure would be related by

$$\mathbf{v} = \mathbf{grad}\,\phi \quad \text{and} \quad \phi = -\Pi/\rho \qquad (1.1.14)$$

and motion set up by jerking the boundaries suddenly is therefore irrotational.

We come finally to **Kelvin's Circulation Theorem**. This applies to an inviscid fluid under conservative body forces when the fluid is well mixed. This means that the forces have a potential, of which gravity $g = -$ **grad** gz is the most important example, and the fluid is **homentropic**. Being homentropic means the condition of the fluid would not be altered if two particles were interchanged. The two simplest cases of this are a fluid of uniform density that is incompressible and a compressible fluid in adiabatic motion with all its elements having the same potential temperatures, or entropy. The **circulation** around a circuit always consisting of the same fluid particles is $\oint \mathbf{v}.\mathbf{ds}$, and it changes with time at a rate

$$\frac{\mathrm{D}}{\mathrm{D}t}\oint \mathbf{v}.\mathbf{ds} = \frac{\mathrm{D}\mathbf{v}}{\mathrm{d}t}.\mathbf{ds} + \oint \mathbf{v}.\frac{\mathrm{D}}{\mathrm{D}t}\,\mathbf{ds} \tag{1.1.15}$$

where **ds** is the vector element of arc of the circuit; and the ring on the integral sign indicating that it is a line integral round the (closed) circuit. Assuming that **v** has no discontinuities in the fluid we have that $\frac{\mathrm{D}}{\mathrm{D}t}\,\mathbf{ds} = \mathbf{dv}$, which is the change in fluid velocity along the infinitesimal arc **ds**, and so the last term in (1.1.15) is $\oint \mathbf{v}.\mathbf{dv} = \oint \mathrm{d}(\frac{1}{2}v^2) = 0$ because **v** is continuous. The first term on the right of (1.1.15) may, by (1.1.4), be written

$$\oint \frac{\mathrm{D}\mathbf{v}}{\mathrm{D}t}.\mathbf{ds} = \oint \mathbf{g}\,.\mathbf{ds} - \oint \frac{1}{\rho}\,\mathbf{grad}\,p.\mathbf{ds} \tag{1.1.16}$$

of which the first term on the right is equal to $\oint \mathbf{grad}\,(gz).\mathbf{ds} = 0$ because the integral of any gradient vector is zero round a closed circuit, which is the same as saying that **g** is a **conservative** force field.

The last term of (1.1.16) may be written as

$$\oint \frac{1}{\rho}\,\mathbf{grad}\,p.\mathbf{ds} = \frac{1}{\rho}\oint \mathbf{grad}\,p.\mathbf{ds} = 0$$

in the case where ρ is uniform, or

$$= \oint \mathbf{grad}\left(\int \frac{\mathrm{d}p}{\rho}\right).\mathbf{ds} = 0$$

in the case when the fluid is **homentropic**, meaning that for all particles the density is the same function of the pressure. Thus we may write $\frac{1}{\rho}\mathbf{grad}\,p = \mathbf{grad}\,e$, or $\frac{1}{\rho}\mathrm{d}p = \mathrm{d}e$, or $e = \frac{\mathrm{d}p}{\rho}$. In consequence (1.1.15) shows that $\oint \mathbf{v}.\mathbf{ds}$ **does not alter with time provided that the circuit is composed always of the same particles. This is the same as saying that the same number of vortex lines always passes through a given material circuit, so that the vortex lines move with the fluid**. The vortex lines are a system of lines having the vorticity, **curl v**, directed along the lines, with magnitude equal to the number of lines crossing unit area.

If the fluid is neither homentropic nor of uniform density

$$\frac{\mathrm{D}}{\mathrm{D}t}\oint \mathbf{v}\,.\,\mathbf{ds} = -\oint \frac{1}{\rho}\mathbf{grad}\,p\,.\,\mathbf{ds} = \int_{\text{surf}} \mathbf{curl}\left(\frac{1}{\rho}\ \mathbf{grad}\,p\right).\,\mathbf{dS} \tag{1.1.17}$$

where the surface of which **dS** is the element is any of which the circuit is the boundary, or edge. By (1.2.8) this is

$$\boxed{\frac{\mathrm{D}}{\mathrm{D}t}\oint \mathbf{v}\,.\,\mathbf{ds} = -\int_{\text{surf}} \mathbf{grad}\ \frac{1}{\rho} \times\ \mathbf{grad}\,p\,.\,\mathbf{dS}} \tag{1.1.18}$$

which is known as **Bjerknes's circulation theorem**. It is related to (1.1.13), and states that the **pressure field in an inviscid fluid can only create circulation when the density is not constant on the surfaces of constant pressure** (i.e. when the isopycnic (density) and isobaric (pressure) surfaces are not the same). This effect is represented by the term $\mathbf{R} \times \mathbf{f}$ in (1.4.1).

To summarise, therefore, we see that when motion is generated by movement of the boundaries it is irrotational if the fluid is inviscid and of uniform density. Irrotational motion has less kinetic energy than any other possible motion compatible with the same boundary conditions; so that the addition of vorticity always adds energy, and it is clear why turbulence which is a sink of energy is rotational. If the fluid is compressible the boundaries can be moved without setting the whole fluid in motion immediately, so that it can have less energy than an incompressible fluid. On the other hand there may be pressure waves added to the motion, which thus contains kinetic energy even if the boundaries are not moving, and so a compressible fluid can have more or less kinetic energy than an incompressible fluid satisfying the same boundary conditions even when both motions are irrotational.

The motion is determinate only if div **v** and **curl v** are specified in some way at all points of the fluid. **This means that turbulence, which is a state that is too complicated for curl v to be so specified, is indeterminate, and can only be treated by means of additional assumptions or information** (see Chapter 9).

1.2 NOTATION, COORDINATE SYSTEMS, IDENTITIES

When we express vectors in terms of their components we use rectangular Cartesian coordinates (x, y, z) with z **vertically upwards** unless otherwise specified. We may use coordinates (x_1, x_2, x_3) when convenient. Thus gravity is

$$\mathbf{g} = (0, 0, -g) = -\ \mathbf{grad}\ (gz) = -\nabla(gz) \tag{1.2.1}$$

The total fluid **velocity** is written as

$$\mathbf{v} = (u, v, w), \qquad \text{or} \qquad = (v_1, v_2, v_3) \tag{1.2.2}$$

and its ith component as v_i. The **vorticity** is

$$\boldsymbol{\omega} = \ \mathbf{curl\ v}\ = \nabla \times \mathbf{v} = (\xi, \eta, \zeta) = (\omega_1, \omega_2, \omega_3) \tag{1.2.3}$$

and $\times$ is used to denote the **vector product**. The **vector operator** ∇ is equal to $\left(\frac{\partial}{\partial x}, \frac{\partial}{\partial y}, \frac{\partial}{\partial z}\right)$.

$$\omega_i = (\mathbf{curl\ v})_i = \epsilon_{ijk}\frac{\partial v_k}{\partial x_j} = \frac{\partial v_k}{\partial x_j} - \frac{\partial v_j}{\partial x_k}.$$

The divergence of velocity is (using the repeated suffix summation convection)

$$\text{div } \mathbf{v} = \nabla.\mathbf{v} = \frac{\partial u}{\partial x} + \frac{\partial v}{\partial y} + \frac{\partial w}{\partial z} = \frac{\partial v_i}{\partial x_i}, \tag{1.2.4}$$

and the material rate of change of the quantity σ associated with a fluid element is

$$\frac{\mathrm{D}\sigma}{\mathrm{D}t} = q\frac{\mathrm{D}\sigma}{\mathrm{D}s} = \frac{\partial \sigma}{\partial t} + \mathbf{v}.\mathbf{grad}\ \sigma = \frac{\partial \sigma}{\partial t} + v_i\frac{\partial \sigma}{\partial x_i} \tag{1.2.5}$$

where t is time, s is **distance** along a particle trajectory, and q is the **speed** which is the magnitude of $\mathbf{v}$. σ may represent any quantity such as density, temperature or concentration of a particular component of pollution or composition.

The following identities are noted for reference but are not proved here (a proof would be obtained by expressing all the vectors as components in 3 dimensions):

$$\frac{\mathrm{D}\mathbf{v}}{\mathrm{D}t} = \frac{\partial \mathbf{v}}{\partial t} + \mathbf{grad}\left(\frac{1}{2}q^2\right) - \mathbf{v}\times\mathbf{curl\ v}. \tag{1.2.6}$$

$$\text{div}(\sigma\mathbf{v}) = \sigma\,\text{div } \mathbf{v} + \mathbf{v}.\mathbf{grad}\ \sigma \tag{1.2.7}$$

$$\mathbf{curl}\ (\sigma\mathbf{v}) = \sigma\,\mathbf{curl\ v} + \mathbf{grad}\ \sigma\times\mathbf{v} \tag{1.2.8}$$

$$\mathbf{curl\ curl\ v} = \mathbf{grad}\ \text{div } \mathbf{v} - \nabla^2\mathbf{v} \tag{1.2.9}$$

where

$$\nabla^2\mathbf{v} = (\text{div } \mathbf{grad})\mathbf{v}.$$

$$\text{div }(\mathbf{v}\times\boldsymbol{\omega}) = \boldsymbol{\omega}.\mathbf{curl\ v} - \mathbf{v}.\mathbf{curl}\ \boldsymbol{\omega} \tag{1.2.10}$$

$$\mathbf{curl}\ (\mathbf{v}\times\boldsymbol{\omega}) = \mathbf{v}\ \text{div } \boldsymbol{\omega} - \ \text{div } \mathbf{v} + (\boldsymbol{\omega}.\mathbf{grad})\mathbf{v} - (\mathbf{v}.\mathbf{grad})\boldsymbol{\omega} \tag{1.2.11}$$

These last two identities are true for any two vector fields $\mathbf{v}$ and $\boldsymbol{\omega}$. Although we are mostly concerned with the cases in which $\boldsymbol{\omega} = \mathbf{curl\ v}$, and also for any scalar σ and the vector $\boldsymbol{\omega} = \mathbf{curl\ v}$,

$$\text{div } \boldsymbol{\omega} = 0 \quad \text{and} \quad \mathbf{curl\ grad}\ \sigma = \mathbf{0} \ \text{ always} \tag{1.2.12}$$

and if **curl v** = 0, there exists a scalar, ϕ, such that $\mathbf{v} = \mathbf{grad}\ \phi$ and ϕ is a potential.

1.3 THE NAVIER–STOKES EQUATIONS

The main purpose of this section is to make clear the assumptions on which the **Navier–Stokes** equations are based. They are the most widely used equations in fluid mechanics and it is important not to use them incorrectly in the complex situations which occur in nature.

The most **fundamental assumption**, without which the equations could not be derived, is that the **stresses** (forces) in a fluid **are related linearly** and **instantaneously** to the **rate of strain** (distortion). Thus twice the rate of strain produces twice the force, and the fluid has no memory of previous motion: the forces come into existence or vanish immediately when the distorting motion begins or stops. Viscous forces are due to the motion of the molecules and so this assumption implies that the molecular velocities are large compared with the differences of fluid velocity in different parts of the fluid, and this certainly will do for ordinary motions of the atmosphere, and for speeds of the fluid less than the speed of sound.

The rate of distortion at a point in the fluid can be represented as the rate of closing up of the angles between three planes of particles which are mutually orthogonal at the moment in question. The quantity $\partial w/\partial y$ is a measure of the rate of rotation about the x-axis of a line of particles on the y-axis. When added to $\partial v/\partial z$, which is the rate of rotation about the x-axis of particles on the z-axis, the sum is the rate of closing up of the angle between lines of particles momentarily located on the axes (see Fig. 1.3.i). This quantity is one of the components of rate of strain. A distortion is also produced if the liquid is stretched along the y-axis or compressed along the z-axis, a rectangular shape remaining rectangular, although the diagonals of a square or rectangle may close up or move apart. The magnitudes of these distortions are represented by the quantities $\partial v/\partial y$ and $\partial w/\partial z$ (Fig. 1.3.ii). There are thus six quantities required to define the rate of distortion and which may therefore enter the equations, namely

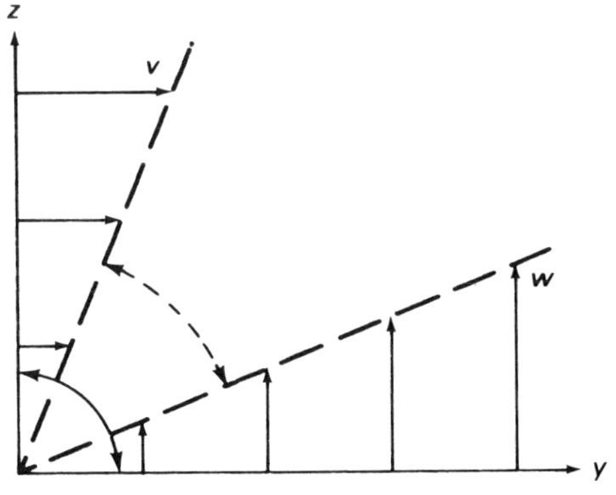

Fig. 1.3.i Rate of fluid distortion measured by gradients of velocity.

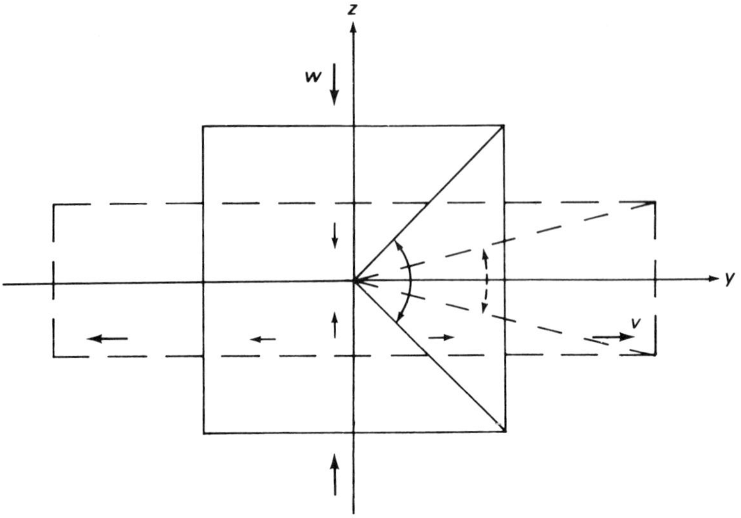

Fig. 1.3.ii Distortion by stretching (v) or compression (w).

$$e_{xx} = 2\frac{\partial u}{\partial x}, \quad e_{yy} = 2\frac{\partial v}{\partial y}, \quad e_{zz} = 2\frac{\partial w}{\partial z}$$

$$e_{yz} = e_{zy} = \frac{\partial w}{\partial y} + \frac{\partial v}{\partial z}$$

$$e_{zx} = e_{xz} = \frac{\partial u}{\partial z} + \frac{\partial w}{\partial x}$$

$$e_{xy} = e_{yz} = \frac{\partial v}{\partial x} + \frac{\partial u}{\partial y} \qquad (1.3.1)$$

The **vorticity** measures the local angular velocity of the fluid and may be thought of as being twice its value. The rotation around the x-axis is obtained by reversing the sign of $\partial v/\partial z$ in e_{yz}, thereby adding the rates of rotation of two lines of particles momentarily at right angles to each other. Thus the **components of vorticity** are

$$\xi = \frac{\partial w}{\partial y} - \frac{\partial v}{\partial z}, \quad \eta = \frac{\partial u}{\partial z} - \frac{\partial w}{\partial x}, \quad \zeta = \frac{\partial v}{\partial x} - \frac{\partial u}{\partial y}. \qquad (1.3.2)$$

The distortion of Fig. 1.3.ii has no total rotation; also rotations may be accompanied by no distortions. A rigid, or 'solid', rotation Ω produces a velocity in a particle at position $\mathbf{x}$ measured from a fixed origin and equal to

$$\mathbf{v} = \mathbf{\Omega} \times \mathbf{x}$$

or

$$v_i = \Omega_j x_k - \Omega_k x_j. \qquad (1.3.3)$$

For **rigid material** all e_{ij}s $(i \neq j)$ in (1.3.1) are zero and $\partial w/\partial y = -\partial v/\partial z$, and so

$$\xi = 2\frac{\partial w}{\partial y}, \quad \eta = 2\frac{\partial u}{\partial z}, \quad \zeta = 2\frac{\partial v}{\partial x}. \tag{1.3.4}$$

Since $\partial u/\partial x = 0$ in rigid material, the velocity of a particle relative to the origin of coordinates has x-component given by (1.3.3) to be

$$u = y\frac{\partial u}{\partial y} + z\frac{\partial u}{\partial z} = \frac{1}{2}(y\zeta - z\eta) \tag{1.3.5}$$

and comparing this with (1.3.3) we see that the velocity is the same as that produced by a rigid rotation equal to $\frac{1}{2}\boldsymbol{\omega}$.

There are nine derivatives of the velocity [set out in (1.3.1)] which define a 3×3 matrix whose members a_{ij} are defined by

$$a_{ij} = \partial u_i/\partial x_j \qquad i, j = 1, 2, 3,$$

and this may be expressed as the sum of the symmetric matrix

$$e_{ij} = \tfrac{1}{2}(a_{ij} + a_{ji})$$

and the antisymmetric matrix

$$\sigma_{ij} = \tfrac{1}{2}(a_{ij} - a_{ji})$$

Since the diagonal members of σ_{ij} are zero, the vector $\sigma_{ij}x_j$ is the same as that obtained as the vector product of $\boldsymbol{\Omega}$ and $\mathbf{x}$, where $\Omega_i = \sigma_{jk}$. Thus σ_{ij} represents a solid rotation superimposed in the total motion on the distortion represented by e_{ij}. The solid rotation causes no internal stresses because it does not distort the fluid, and so only the component e_{ij} of the total motion a_{ij} is related to the stresses.

By a suitable transformation of coordinates representing a pure rotational displacement, the matrix e_{ij} may be reduced to a diagonal matrix. This is equivalent to saying that a quadric surface has three principal axes on which the distance from the origin is stationary under rotation, or that the distortion of the fluid elements can be fully described as the sum of extensions or compressions along the three principal axes. This is a property of second-degree surfaces in space and is nothing to do with any physical property of the fluid. We shall make use of this fact in a moment.

In specifying the rate of strain and rotation of the fluid relative to the particle at the point by the nine derivatives of the velocity components, we have placed no restrictions on the kind of motion we are discussing other than that it is continuous and differentiable. But if we assume that the stresses produced at a point depend on these quantities only, we imply that they are determined by something on a very small scale within the fluid, and not by events taking place a finite distance away. Thus the rates of strain are presumed not to change significantly across the region in which the events determining the strain are taking place. This is equivalent to neglecting any dependence on space derivatives of the velocity higher than the first. This assumption might be wrong if the first derivatives varied significantly over the mean free path of the molecules. The flow around a meteor in the high atmosphere may be incorrectly represented because the meteor may be smaller than the mean free path at great heights.

For the equations we are going to derive to be valid, the quantities e_{ij} do not have to be small although in assuming them to be linearly related to the stress we must make similar assumptions about the derivatives of the stress and assume that one component does not influence the relationship between others.

We will also assume that the fluid is isotropic, which means that it has no directional properties. Since we shall be studying stratified fluids we must assume that the shearing motion and the density gradients are too small to affect the linear relationship. This is reasonable if the stresses are produced on the molecular scale because, in a distance many times greater than the mean free path, there is no significant stratification. Where a very hot jet enters a cold environment the temperature gradients may be very large indeed, but here it is assumed that the equations remain valid by allowing for variations of viscosity with temperature without any local anisotropy. The cases in which these assumptions break down are those in which high precision is impossible, because the motion is so turbulent as to make direct application of the equations out of the question.

For the isotropic case some economy of argument is obtained by using the principal axes of rate of strain, denoted by $OXYZ$. For these

$$e_{YZ} = e_{ZX} = e_{XY} = 0 \tag{1.3.6}$$

The rotational displacement effecting the transformation from $Oxyz$ to $OXYZ$ is specified by its three components, which are chosen so that the equations (1.3.6) are satisfied. The result is that, relative to the principal axes, the rate of strain is expressed completely in terms of stretching or compressing motion along the axes. According to the assumptions, then, the stress p_{ij} are given by equations of the form

$$p_{XX} = -p + \tfrac{1}{2}\lambda\,(e_{XX} + e_{YY} + e_{ZZ}) + \mu e_{XX} \tag{1.3.7}$$

for it must depend on e_{YY} in the same ways as on e_{ZZ} because of isotropy. The symbol p_{ij} represents the component in the i-direction of the total stress per unit area across a surface perpendicular to the j-direction. For the principal axes of stress

$$p_{YZ} = p_{ZX} = p_{XY} \tag{1.3.8}$$

and the axes must, from the assumption of isotropy, coincide with the principal axes of rate of strain.

Thus only two coefficients of viscosity are required, λ to define the effect of divergence, for $e_{ii} = 2 \operatorname{div} \mathbf{v}$, and the other to define the effect of stretching. The coefficient $-p$ appears in (1.3.7) because we know that there is a stress, which we call the **normal pressure**, when there is no motion. The negative sign indicates, by convention, taken from elasticity theory, that a stress is positive when it is a tension.

If the axes $Oxyz$ have directions denoted by the unit vectors **l**, **m**, **n** relative to the principal axes $OXYZ$ which vary from point to point (i.e. **l** is a unit vector along Ox) and if **V** is the fluid velocity referred to $OXYZ$, then

$$\frac{1}{2}e_{xx} = \frac{\partial u}{\partial x} = (\mathbf{l}.\mathbf{grad})(\mathbf{l}.\mathbf{V})$$

$$= \left(l_1\frac{\partial}{\partial X} + l_2\frac{\partial}{\partial Y} + l_3\frac{\partial}{\partial Z}\right)(l_1U + l_2V + l_3W)$$

$$= l_1^2\frac{\partial U}{\partial X} + l_2^2\frac{\partial V}{\partial Y} + l_3^2\frac{\partial W}{\partial Z}$$

$$= \frac{1}{2}\left(l_1^2 e_{XX} + l_2^2 e_{YY} + l_3^2 e_{ZZ}\right) \quad (1.3.9)$$

because

$$\left.\begin{aligned} e_{YZ} &= \frac{\partial V}{\partial Z} + \frac{\partial W}{\partial Y} = 0 \\ \text{and} \quad e_{XX} &= 2\frac{\partial U}{\partial X}, \text{ etc.} \end{aligned}\right\} \quad (1.3.10)$$

according to the definition of the principal axes. Because the axes are orthogonal

$$l_1^2 + l_2^2 + l_3^2 = 1 \quad (1.3.11)$$

and so

$$2 \operatorname{div} \mathbf{v} = e_{xx} + e_{yy} + e_{zz} = e_{XX} + e_{YY} + e_{ZZ} \quad (1.3.12)$$

and

$$e_{yz} = \frac{\partial w}{\partial y} + \frac{\partial v}{\partial z} = (\mathbf{m}.\mathbf{grad})\ (\mathbf{n}.\mathbf{V}) + (\mathbf{n}.\mathbf{grad})\ (\mathbf{m}.\mathbf{V})$$

$$= m_1n_1e_{XX} + m_2n_2e_{YY} + m_3n_3e_{ZZ} \quad (1.3.13)$$

because of the relationships (1.3.10). There are, of course, two similar equations for e_{zx} and e_{xy}.

In obtaining all these relationships we put the space derivatives of the l's, m's and n's equal to zero. This is because, although they vary from point to point with the variations of the principal axes, we are concerned with derivatives of quantities relative to two sets of axes defined at the moment at a point.

In Fig. 1.3.iii we show a rectangular box of sides δx, δy, δz, with one corner at the origin. All the stresses with a non-zero moment about Ox are indicated, and shown acting at the mid-points of the appropriate faces of the box. The magnitudes of the moments are calculated as if the forces were uniform over the face on which they act and as if they varied uniformly. Thus the higher-order terms in δx etc. which involve higher derivatives than the first are neglected, according to the fundamental principle of calculus. The two forces indicated by the double-thickness arrows contribute together a moment about Ox equal to

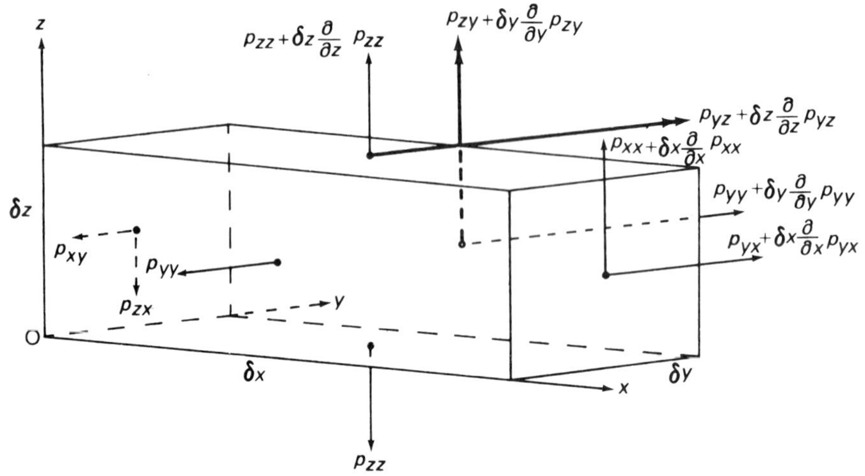

Fig. 1.3.iii Stresses on a box of fluid of dimensions δx, δy, δz, which have a moment about Ox.

$$-p_{yz}\,.\,\delta x\delta y\,.\,\delta z+p_{zy}\,.\,\delta y\delta z\,.\,\delta x \tag{1.3.14}$$

and all the remaining terms in the moment are of higher order. These terms are the products of the forces, the areas on which they act, and their distance from Ox. The sum of all the moments is equal to the moment of inertia multiplied by the angular acceleration of the box. The moment of inertia is of the order of the mass times the square of a linear dimension of the box, and is therefore of the order of $(\delta x)^5$. Since the angular acceleration is necessarily finite and the quantity (1.3.14) is of the order of $(\delta x)^3$, the quantity must be zero; otherwise the acceleration would have to tend to infinity as the size of the box considered tended to zero. Therefore

$$p_{yz}=p_{zy} \tag{1.3.15}$$

and similarly

$$p_{zx}=p_{xz},\ p_{xy}=p_{yx}.$$

The same result is true for the stresses in an elastic solid as a consequence of the same argument.

To relate the stresses to those expressed in terms of the principal axes we consider the forces on a small tetrahedron formed by the origin O and points A, B, C along the axes OXYZ where the plane ABC is perpendicular to the direction Ox. Since the forces are proportional to the areas on which they act, i.e. to the square of the dimension, while the momentum is proportional to the mass, i.e. to the cube of the dimension, the resultant forces must be zero, or otherwise an infinitesimal tetrahedron would experience an arbitrarily large acceleration. Thus the components in the direction Ox are in equilibrium, and so

$$p_{xx}\text{ABC}=l_1p_{XX}\text{OBC}+l_2p_{YY}\text{OCA}+l_3p_{ZZ}\text{OAB}$$

and so since OBC = l_1ABC,

$$p_{xx} = l_1^2 p_{XX} + l_2^2 p_{YY} + l_3^2 p_{ZZ} \tag{1.3.16}$$

and in view of (1.3.11)

$$p_{xx} + p_{yy} + p_{zz} = p_{XX} + p_{YY} + p_{ZZ}. \tag{1.3.17}$$

This last result means that what we may call the **mean pressure**, namely

$$p_m = -\tfrac{1}{2}(p_{xx} + p_{yy} + p_{zz}), \tag{1.3.18}$$

is the same in any set of coordinates and has meaning without reference to the coordinate system.

The forces on the tetrahedron in Fig. 1.3.iv in the direction of Oz are also in equilibrium, and so

$$p_{xz}\text{ABC} = n_1 p_{XX}\text{OBC} + n_2 p_{YY}\text{OCA} + n_3 p_{ZZ}\text{OAB}$$

i.e.

$$p_{xz} = l_1 n_1 p_{XX} + l_2 n_2 p_{YY} + l_3 n_3 p_{ZZ} \tag{1.3.19}$$

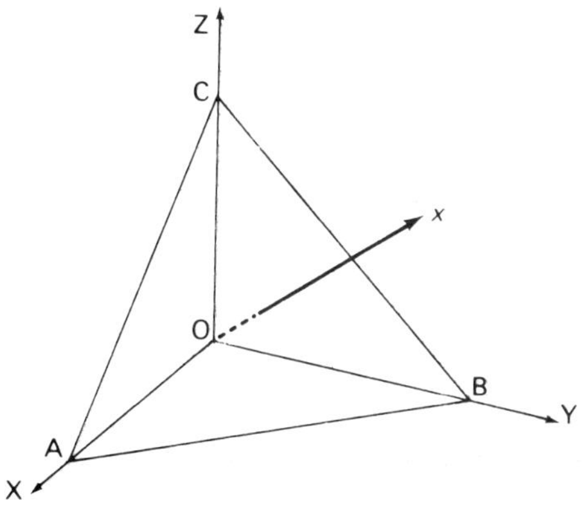

Fig. 1.3.iv Ox is perpendicular to plane ABC.

The relation (1.3.7) between the stresses and the rate of strain may be written

$$p_{XX} = -p + \lambda \operatorname{div} \mathbf{v} + \mu e_{XX} \tag{1.3.20}$$

and from this, and similar expressions for p_{YY} and p_{ZZ}, we may now obtain the relationship between the stress and the rate of strain in terms of the original coordinates Oxyz. Equations (1.3.16), (1.3.20) and (1.3.9) give

$$\boxed{\begin{aligned} p_{xx} &= (-p+\lambda \operatorname{div} \mathbf{v})\left(l_1^2+l_2^2+l_3^2\right)+\mu\left(l_1 e_{XX}+l_2 e_{YY}+l_3 e_{ZZ}\right) \\ &= -p+\lambda \operatorname{div} \mathbf{v}+\mu e_{xx}. \end{aligned}} \tag{1.3.21}$$

Since the axes are rectangular, $l_i n_i = 0$ and so substituting (1.3.20) into (1.3.19) we find that, using (1.3.13),

$$\boxed{p_{xz} = \mu e_{xz}.} \tag{1.3.22}$$

The two equations ((1.3.21) and (1.3.22)) relate the stresses p_{ij} to the rates of strain e_{ij} through two coefficients of viscosity which are physical properties of the fluid and have the dimensions L^2T^{-1}. The mean pressure is now seen to be [from (1.3.18)]

$$\begin{aligned} p_{\mathrm{m}} &= -\tfrac{1}{3}\left(p_{xx}+p_{yy}+p_{zz}\right) \\ &= -\tfrac{1}{3}\left[-3p+3\lambda \operatorname{div} \mathbf{v}+\mu(e_{xx}+e_{yy}+e_{zz})\right] \\ &= p-\left(\lambda+\tfrac{2}{3}\mu\right) \operatorname{div} \mathbf{v}. \end{aligned} \tag{1.3.23}$$

The quantity p has already been called the **normal pressure** and it has been assumed to be the same for the components of stress in each of the equations (1.3.7) because of the isotropic property of the fluid. When the divergence of velocity is zero and the density is not varying, $p_{\mathrm{m}} = p$. We do not have any practicable means of detecting any difference between p_{m} and p in any situation in which div **v** was not zero unless the coefficient $\lambda+\frac{2}{3}\mu$ were large. Although $p_{\mathrm{m}} = p$ when there is no motion, it does not follow that they are necessarily always the same, and p has not been defined except as the constant in (1.3.7). We now have a definition of it in (1.3.23), and some authors make the error of logic in arguing that p is always the same as p_{m} because it could not otherwise be defined in such a way as to indicate how it might be measured, except as the same quantity as p_{m}. We depend either on experiment or on additional theory to determine the value of $\lambda+\frac{2}{3}\mu$. Kinetic theory indicates that when the molecules occupy only a negligibly small fraction of the whole volume, $\lambda+\frac{2}{3}\mu$ is zero, and although such an argument is not valid for a liquid, this does not matter because then div **v** is very small. No measurement has been devised which gives any magnitude to $\lambda+\frac{2}{3}\mu$, and the success of the assumption that it is zero indicates at least that it has a much smaller magnitude than μ. It is therefore usual to pursue the theory on the assumption that it is zero and that the stress and rate of strain are related by

$$\left.\begin{aligned} p_{xx} &= -p-\tfrac{2}{3}\mu \operatorname{div} \mathbf{v} + \mu e_{xx} \\ p_{yz} &= \mu e_{yz}, \end{aligned}\right\} \tag{1.3.24}$$

or

$$p_{ij} = -\delta_{ij}\left(p + \tfrac{2}{3}\mu \operatorname{div} \mathbf{v}\right) + \mu e_{ij}. \tag{1.3.25}$$

We may now obtain the equation of motion for a fluid by equating the rate of change of momentum of unit volume to the sum of the forces on unit volume. Thus

$$\rho \frac{\mathrm{D}v_i}{\mathrm{D}t} = \rho F_i + \frac{\partial}{\partial x_j} p_{ij} \tag{1.3.26}$$

in which **F** is a body force proportional to mass. In our case $\mathbf{F} = \mathbf{g}$.

From this we have by insertion of (1.3.25)

$$\rho \frac{\mathrm{D}v_i}{\mathrm{D}t} = \rho F_i - \frac{\partial p}{\partial x_j} - \frac{2}{3}\frac{\partial}{\partial x_i}(\mu \operatorname{div} \mathbf{v}) + \frac{\partial}{\partial x_j}(\mu e_{ij}). \tag{1.3.27}$$

In the case in which μ is the same at all points of the fluid, and is therefore a constant in the last two terms, we may write $\nu = \mu/\rho$ and get

$$\frac{\mathrm{D}v_i}{\mathrm{D}t} = F_i - \frac{1}{\rho}\frac{\partial p}{\partial x_i} - \frac{2}{3}\nu\frac{\partial}{\partial x_i}\frac{\partial v_j}{\partial x_j} + \nu\frac{\partial}{\partial x_j}\left(\frac{\partial v_i}{\partial x_j} + \frac{\partial v_j}{\partial x_i}\right) \tag{1.3.28}$$

and ν is called the **kinematic viscosity** because it relates to the motion. μ is called the **dynamic viscosity** and may be defined as the force across unit area in motion with shear of unit magnitude acting in the direction of the shear. **In the case of plane flow in which v_i varies in the j-direction the force is $\mu \frac{\partial v_i}{\partial x_j}$ in the i-direction**. If this force varies in the j direction it produces on unit volume a force in the i-direction equal to $\frac{\partial}{\partial x_j}\mu\frac{\partial v_i}{\partial x_j}$, or $\mu\frac{\partial^2 v_i}{\partial x_j^2}$ when μ is a constant in space. This is the most important effect of viscosity on the motion and it is due to the variation of shear stress from one point of the fluid to another. **The force is zero if the velocity gradients are linear.**

In turbulent flow the eddies are often thought of as producing an effect similar to that of the molecular viscosity but of greater magnitude (see section 9.3). However, if the equation is to be applied on the assumption that μ represents an eddy viscosity, the form (1.3.27) rather than (1.3.28) must be used because usually the coefficient varies from one point to another in the fluid. In particular, the horizontal flow of air over the ground has a level of turbulence and a velocity gradient which both decrease with height, and in some cases the variation of one, and in some cases that of the other, is the more important.

The application of the theory to turbulence also has to be made with great care because turbulence is not usually isotropic but has directional proportions, and so the assumptions on which the equation is based on fail. This is not serious when the velocity is everywhere in one direction (say x) and varies only in another (say z, i.e. vertically).

The equation (1.3.28) can be written in vector form as follows:

$$\boxed{\frac{\mathrm{D}\mathbf{v}}{\mathrm{D}t} = \mathbf{F} - \frac{1}{\rho}\,\mathbf{grad}\,p + \nu\nabla^2\mathbf{v} + \frac{1}{3}\nu\,\mathbf{grad}\,(\mathrm{div}\,\mathbf{v}),} \tag{1.3.29}$$

which is the **Navier–Stokes equation. It is valid for the motion of a fluid with uniform isotropic viscous properties in which the stress is instantaneously and linearly related to the rate of strain.** Without the viscous terms it is the classical equation for an inviscid fluid, namely (1.1.4). The most important viscous terms in practice are those already mentioned, namely those producing a shearing stress, such as $\nu\frac{\partial^2 v}{\partial x^2}$ and $\nu\frac{\partial^2 w}{\partial x^2}$, which are the forces acting on unit volume of fluid in the y- and z-directions due to non-uniform gradients in the x-direction of the y- and z-components of velocity. **These forces act so as to straighten out velocity profiles, and vanish when the profiles are linear.**

The final term in (1.3.29) does not concern us much in this book because we consider scarcely any situations in which div **v** is not rather small and, still less, has a significant gradient in a case where ν is not negligible or its turbulent equivalent is known with any precision.

1.4 THE VORTICITY EQUATION

It will be recalled that Kelvin's minimal theorem (page 6) showed that of all the possible flow patterns which conform to given rigid boundaries of the fluid, the irrotational pattern possesses the least kinetic energy. Moreover this irrotational pattern is unique, and it is the pattern created by a sudden jerking of the boundaries into the motion. If the fluid possesses some vorticity it possesses more kinetic energy, yet the transport of fluid is not increased by that extra vorticity. The irrotational flow can be thought of as transporting the fluid and the **vorticity as useless additional kinetic energy**, often in what we call turbulent form, which the fluid may or may not possess for some cause. Thus if the water in a bucket is stirred with a spoon there must be some vorticity present if it is still moving when the spoon is withdrawn, because the irrotational motion is zero when the bucket is at rest.

This argument applies strictly only to incompressible fluids of uniform density. When the density varies, the motion set up by jerking the boundaries contains vorticity, and if the fluid is compressible, waves are transmitted through it. While they are being transmitted there are internal motions taking place which cannot be discovered from knowledge of how the boundaries are moving at the moment in question. We shall not be concerned much with the transmission of sound waves which may be irrotational, but turn our attention to the mechanisms which create vorticity and the phenomena which accompany it. The complicated motion set up in a milk bottle by holding it in the hand and rotating the wrist back and forth, or the motion of the air in a car which turns a corner sharply, starts by being irrotational but soon becomes very confused and the milk or air becomes mixed up. The **mixing is achieved only by generating vorticity**.

The water in a conical flask (or for that matter any fairly steep-sided vessel) can be sloshed round by moving the flask quickly in small horizontal circles sliding the base on the table without rotation, and the result is to induce rotation in the whole mass of fluid.

A wine decanter is very suitable for this experiment. According to Kelvin's circulation theorem, rotation cannot be imparted to the fluid by moving the boundaries, yet there is the rotation to be seen. Such a rapid movement of a flask is a favourite method employed by chemists doing titrations to mix up the liquid when they wish to disperse one more drop of reagent throughout the whole volume. In this case the production of vorticity is achieved by causing the free surface to fold over so that discontinuities of velocity are introduced into the bulk of the fluid.

Vorticity is thus associated with turbulence, overturning, and the production of kinetic energy which does not transport the whole fluid effectively. In machinery it is often wasteful to produce vorticity; on the other hand it may be desirable to generate mixing. For example, the combustion chamber in a gas turbine engine must be designed to produce very good quick mixing between air and fuel. It is produced in the flow past obstacles, and also in buoyant convection when a fluid is heated at the bottom or cooled at the top, or when salt is dissolved into the fluid at the top. The vorticity equation, which we shall now derive, describes the way in which vorticity, denoted by **curl v** or $\boldsymbol{\omega}$, can be generated.

The equation of motion (1.3.29) may be written

$$\rho \frac{\mathrm{D}\mathbf{v}}{\mathrm{D}t} = \rho F - \mathbf{grad}\, p + \mu\left(\nabla^2 \mathbf{v} + \frac{1}{3}\mathbf{grad}\ \mathrm{div}\ \mathbf{v}\right),$$

and if we consider the case in which μ is a constant (otherwise we would have to take the form (1.3.27)), then by taking the **curl** we eliminate the terms in p and div **v** which occur in gradients by (1.2.12). Thus

$$\mathbf{curl}\left(\rho \frac{\mathrm{D}\mathbf{v}}{\mathrm{D}t}\right) = \mathbf{curl}\ (\rho \mathbf{g}) + \mu\ \mathbf{curl}\ \nabla^2 \mathbf{v}$$

in the case where the only body force is gravity. By using the identities (1.2.6), (1.2.8), (1.2.9) and (1.2.12) this is found to lead to

$$\rho\ \mathbf{curl}\left(\frac{\partial \mathbf{v}}{\partial t} - \mathbf{v} \times \boldsymbol{\omega}\right) + \mathbf{grad}\ \rho \times \mathbf{f} = \mathbf{grad}\ \rho \times \mathbf{g} - \mu\ \mathbf{curl}\ \mathbf{curl}\ \boldsymbol{\omega}$$

where we have denoted the **fluid acceleration**, $\mathrm{D}\mathbf{v}/\mathrm{D}t$, by **f**. The order of the operators **curl** and $\partial/\partial t$ is interchangeable and so, using (1.2.11),

$$\frac{\partial \boldsymbol{\omega}}{\partial t} + \boldsymbol{\omega}\ \mathrm{div}\ \mathbf{v} - (\boldsymbol{\omega}.\mathbf{grad})\mathbf{v} + (\mathbf{v}.\mathbf{grad})\boldsymbol{\omega} = \mathbf{grad}\ \ln \rho \times (\mathbf{g} - \mathbf{f}) - \nu\ \mathbf{curl}\ \mathbf{curl}\ \boldsymbol{\omega},$$

from which using (1.2.9) we obtain

$$\boxed{\frac{\mathrm{D}\boldsymbol{\omega}}{\mathrm{D}t} = -\boldsymbol{\omega}\ \mathrm{div}\ \mathbf{v} + (\boldsymbol{\omega}.\mathbf{grad})\mathbf{v} + \mathbf{R} \times (\mathbf{g} - \mathbf{f}) + \nu \nabla^2 \boldsymbol{\omega}.} \qquad (1.4.1)$$

Here we have introduced the symbol **R**, of which the magnitude is often written as β, to denote $\frac{1}{\rho}\,\mathbf{grad}\,\rho = \mathbf{grad}\,\ln\rho$, which is a representation of the density field. Equation (1.4.1) is the **vorticity equation** and the four terms on the right describe mechanisms by which the vorticity of a parcel of fluid is changed: the left side is equal to this rate of change. We now discuss these mechanisms.

The meaning of the first term on the right is best made clear by transferring it to the left. By means of the **equation of continuity**, namely

$$\boxed{\frac{\mathrm{D}\rho}{\mathrm{D}t} = -\rho\,\mathrm{div}\,\mathbf{v},} \tag{1.4.2}$$

which asserts that the rate of increase of density is equal to the inflow per unit volume multiplied by the density, i.e. to the inflow of mass per unit volume, we then obtain

$$\frac{\mathrm{D}}{\mathrm{D}t}\frac{\boldsymbol{\omega}}{\rho} = \left(\frac{\boldsymbol{\omega}}{\rho}\,.\,\mathbf{grad}\right)\mathbf{v} + \frac{1}{\rho}\mathbf{R}\times(\mathbf{g}-\mathbf{f}) + \frac{\nu}{\rho}\nabla^2\boldsymbol{\omega}, \tag{1.4.3}$$

which shows that the second term on the right of (1.4.1) represents a mechanism operating on ω when div **v** is zero which is the same as that operating on ω/ρ when div **v** is not zero. However, our objective is to isolate the terms on the right and elucidate the different vorticity-producing mechanisms which they represent. **In the adiabatic motion of a well mixed fluid** ρ is proportional to $p^{1/\gamma}$ and so **R**, which is in the direction of **grad** ρ, is also in the direction of **grad** p. If we neglect the effect of viscosity represented by the last term in (1.4.3) for the moment, then the equation of motion is (1.1.4), namely

$$\frac{1}{\rho}\,\mathbf{grad}\,p = \mathbf{g} - \mathbf{f} \tag{1.4.4}$$

in which case $\mathbf{R}\times(\mathbf{g}-\mathbf{f}) = \mathbf{0}$, because $\mathbf{grad}\,p \times \mathbf{grad}\,\rho$ vanishes.

The consequence of this is that we may discuss the meaning of the separate terms in (1.4.1) ignoring the term in div **v** but remembering that when div **v** is not zero the conclusion applies to $\boldsymbol{\omega}/\rho$ instead of to $\boldsymbol{\omega}$, provided that the motion is adiabatic and the fluid well mixed, so that the density is constant along the isobars.

The clue to the meaning of the term $(\boldsymbol{\omega}\,.\,\mathbf{grad})\,\mathrm{v}$ has already been given in the discussion following (1.1.15), where it was shown that if the fluid is inviscid and either homentropic (well mixed and adiabatic) or of uniform constant density, the number of vortex lines through a circuit composed always of the same particles is constant. **Thus the vortex lines move with fluid**.

Let it be supposed that we have a set of lines marked out in the fluid and that they represent a vector field **a** which can be deduced from the lines by their direction and the number crossing unit area. Alternatively the lines may be marked along their length so that the magnitude of **a** is represented by the distance between the marks. Thus where the lines are far apart the marks are at short intervals, indicating a small magnitude. This is achieved by marking vortex tubes so as to indicate constant increments of volume along the tube (Fig. 1.4.i). Only if the vector **a** is non-divergent can it be marked out in this way.

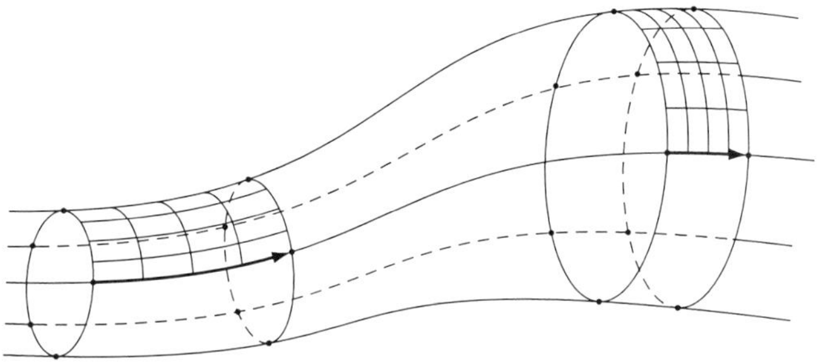

Fig. 1.4.i Relation between vortex lines, vortex tubes and the vorticity vector.

If the motion is also non-divergent, the volume occupied by a given set of particles remains constant, and so the same relationships between the set of lines and the vector field defined by them will persist as the lines are carried by the motion. If the vector **a** is now regarded as the material segment AB of one of the lines, in time Δt its two ends will be displaced by the velocity field distances $\mathbf{v}\Delta t$ and $(\mathbf{v} + \delta\mathbf{v})\Delta t$ where $\delta\mathbf{v}$ is the difference between the velocities of the particles at the two ends of the vector **a** (see Fig. 1.4.ii). Thus

$$\delta\mathbf{v} = (\mathbf{a} \cdot \mathbf{grad})\,\mathbf{v} \tag{1.4.5}$$

to the first order in the small quantity **a**. At the end of the interval Δt, the material vector has become $\mathbf{a} + \dfrac{\mathrm{D}\mathbf{a}}{\mathrm{D}t}\Delta t$ and so its change is equal to each of the quantities $\dfrac{\mathrm{D}\mathbf{a}}{\mathrm{D}t}\Delta t$ and $(\mathbf{a} \cdot \mathbf{grad})\,\mathbf{v}\Delta t$. Therefore

$$\boxed{\frac{\mathrm{D}\mathbf{a}}{\mathrm{D}t} = (\mathbf{a} \cdot \mathbf{grad})\,\mathbf{v}} \tag{1.4.6}$$

is the equation giving the rate of change, due to the motion, of a vector field **a** defined by material lines in the fluid which are carried by the motion. This is the form taken by (1.4.1) for $\boldsymbol{\omega}$ when the fluid is incompressible, of uniform density, and inviscid. For a homentropic fluid the term in **R** in (1.4.3) vanishes and the vorticity must be deduced from the vortex lines carried by the fluid either by the number crossing unit area, as before, or by the length of an intercept of a tube of unit mass, instead of unit volume.

The x-component of $-\boldsymbol{\omega}\,\mathrm{div}\,\mathbf{v} + (\boldsymbol{\omega} \cdot \mathbf{grad})\,\mathbf{v}$, which are the first two terms on the right of equation (1.4.1), is

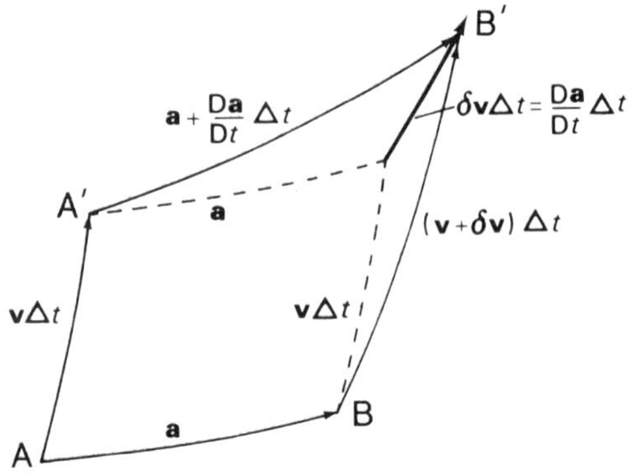

Fig. 1.4.ii Advection of the vorticity vector in which AB moves to A′B′.

$$-\xi\left(\frac{\partial v}{\partial y}+\frac{\partial w}{\partial z}\right)+\eta\frac{\partial u}{\partial y}+\zeta\frac{\partial u}{\partial z} \tag{1.4.7}$$

which is therefore a measure of the rate of change of ξ due to the two terms. The first term states that if there is already some ξ, the x-component of vorticity, more will be created if $\partial v/\partial y+\partial w/\partial z$ is negative. This simply refers to the fact that if the fluid is expanded or contracted in planes perpendicular to the x-axis, the moment of inertia about the x-direction is accordingly increased or decreased, and according to the conservation of momentum the rotation about the x-direction is correspondingly decreased or increased. This pattern of distortion is rather special and unlikely to occur by itself: it is more probable that the divergence is zero or nearly so, in which case, since div **v**, which is equal to $\frac{\partial u}{\partial x}+\frac{\partial v}{\partial y}+\frac{\partial w}{\partial z}$, is zero, the component becomes

$$\xi\frac{\partial u}{\partial x} \tag{1.4.8}$$

which represents the fact that if the fluid is being stretched in the x-direction it is being contracted in planes normal to it. For this reason it is customary to think of **the vortex lines as being intensified by stretching along their length**.

The second and third terms in (1.4.7) are of similar structure, and we therefore consider only

$$\eta\frac{\partial u}{\partial y} \tag{1.4.9}$$

which indicates that if there is alreayd a non-zero y-component of vorticity, η, and the x-velocity, u, varies in the y-direction so as to rotate lines of particles situated along the y-axis, they will acquire a component in the x-direction, and so some ξ will be created (see Fig. 1.4.iii).

The creation of vorticity by the density gradients is represented in equation (1.4.1) by the term

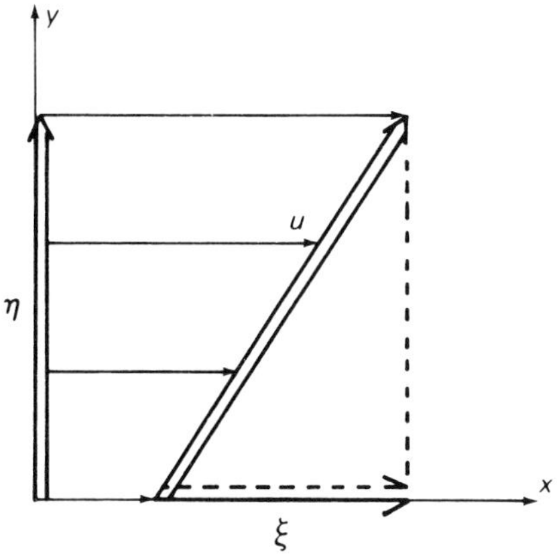

Fig. 1.4.iii Rotation of vortex lines by the fluid motion.

$$\frac{\mathrm{D}\boldsymbol{\omega}}{\mathrm{D}t} = \mathbf{R} \times (\mathbf{g} - \mathbf{f}) \tag{1.4.10}$$

This shows that the negative acceleration, **f**, has a similar effect to gravity, the inertial and gravitational masses of a piece of matter being the same. Vorticity, or rotation, is created so as to rotate the fluid until the surfaces of constant density are normal to the vector **g** – **f**. When the fluid is at rest, **f** is zero, and we have the familiar result that rotation of the surfaces of constant density is induced when they are not horizontal, so as to bring them into the horizontal.

R is a vector perpendicular to the surfaces of constant density, and the axis about which the vorticity is generated is horizontal and in the surface, perpendicular to both **g** and **R**. If the fluid undergoes an acceleration, as for instance when it passes round a bend in a pipe, the centrifugal forces are added to the weight. If the acceleration is along the direction of motion, as in a vehicle accelerated in a straight line, the fluid behaves as if **g** were in the direction taken by a weight hung on a thread in the vehicle. Water in a partly filled vessel on the vehicle sloshes to the back as the vehicle accelerates forward. In a centrifuge or similarly rotating system, $\mathbf{g} \ll \mathbf{f}$, and hence the surfaces of constant density tend to become cylinders whose axis is the axis of rotation.

The creation of vorticity by viscosity is represented by the equation

$$\boxed{\frac{\mathrm{D}\boldsymbol{\omega}}{\mathrm{D}t} = \nu\nabla^2\boldsymbol{\omega}} \tag{1.4.11}$$

when the viscosity is uniform. This is a mechanism which acts on each component of vorticity separately: thus the x-component is

$$\frac{\mathrm{D}\xi}{\mathrm{D}t} = \nu\nabla^2\xi \tag{1.4.12}$$

and this is the same as the equation representing the conduction of heat by a thermal conductivity ν when ξ is the temperature. The effect is to redistribute the vorticity so that its gradient becomes more nearly linear. The usual ultimate effect is to make the vorticity more nearly uniform so that the fluid is made more nearly to rotate as a solid. Where the gradient of ξ increases in the x-direction the flux of ξ conducted decreases (increases negatively) in the x-direction, and so the value of ξ increases.

If ν is not constant, (1.4.12) is correct if the variation in ν is due solely to variations in ρ. If ρ is a constant and μ is variable, the derivation must be made from equation (1.3.27): this is not simple enough to write in vector form and operate upon as we did in the derivation of (1.4.1), even when div $\mathbf{v} = 0$ everywhere. This is important when stresses are produced by eddies because they can vary very substantially across the fluid. Because of its complexity, this situation is best discussed by simplifying the equations as far as possible for the special case under consideration, and this is discussed in Chapter 9.

1.5 THE ISOPYCNIC VORTICITY THEOREM

In the case of an inviscid fluid

$$\frac{D\boldsymbol{\omega}}{Dt} = -\boldsymbol{\omega}\,\mathrm{div}\,\mathbf{v} + (\boldsymbol{\omega}\,.\,\mathbf{grad})\mathbf{v} + \mathbf{R}\times(\mathbf{g}-\mathbf{f}). \qquad (1.5.1)$$

The first term on the right is parallel to the vorticity. The second term represents the production of changes in the vorticity by moving the vortex lines with the fluid. The third term represents the creation of new vorticity perpendicular to **R**, that is in the surfaces of constant ρ. Consequently, **if at one moment the vortex lines lie in surfaces of constant density**, called isopycnic surfaces, because all new vorticity is created in those surfaces or moves materially with them, **they will remain in them**.

Only four circumstances can alter this situation. Viscosity might produce significant amounts of vorticity not lying in the surfaces or, for some reason, the vorticity might not initially lie in them. The free surface might become folded over so as to produce velocity discontinuities which then diffuse by viscosity, or the density might be altered by heating or some other agency.

The case of special interest to us in the atmosphere or ocean is that in which both the flow and the isopycnic surfaces are horizontal. If the vorticity is also horizontal initially, it will be found to be in the same surfaces at points downstream. This is a common situation in the atmosphere, for although there is obviously a vertical component of vorticity due to the earth's rotation which represents a turning of the order of one rotation per day, it is usually very small compared with the horizontal component due to the change of wind strength or direction with height.

In the atmosphere away from the bottom boundary the vorticity is, for practical purposes, entirely produced by the action of gravity on density gradients. That means that new vorticity is always horizontal and in the contours of the isopycnic surfaces. The acceleration of the fluid in the ocean and atmosphere is generated as a result of small departures of the density from horizontal uniformity, and is always small compared with **g**, and so the term $\mathbf{R}\times\mathbf{f}$ is usually negligible.

It will be seen in Chapter 5 that the mathematical complexity of waves in the atmosphere is so considerable that almost no progress has been made except in the case to which this theorem is applicable, and even then only in the two-dimensional case. The theorem shows that in spite of the apparent speciality of this case it is nevertheless relevant to a great many real situations.

It should be noted that airflow over a horizontal surface acquires vorticity due to the drag of the ground, but this does not invalidate the application of the theorem because that vorticity is horizontal and lies in the stream surfaces.

1.6 ADIABATIC CHANGES IN THE ATMOSPHERE: THE POTENTIAL TEMPERATURE

While some changes in the air are of a very varied nature owing to conduction, convection, radiation, evaporation and condensation, many are **adiabatic**, meaning 'without passage (of heat) across (the boundary)'. When applied to a parcel of gas, an adiabatic change is one in which there is no transfer of heat by radiation or conduction into or out of the parcel. However, the energy may be transferred by mechanical work being done on the boundary to compress the gas, causing its temperature to rise, or cooling it when the gas is allowed to expand.

If $\mathrm{d}Q$ is the energy change in a parcel of unit mass at temperature T,

$$\mathrm{d}Q = c_{\mathrm{V}}\,\mathrm{d}T + Ap\,\mathrm{d}V \tag{1.6.1}$$

where c_{V} is the specific heat of the gas at constant volume, A is the mechanical equivalent of heat and $p\,\mathrm{d}V$ is the work done by the gas in expanding its volume by $\mathrm{d}V$ at pressure p. Using the calculus notation we neglect terms which are of the second order or higher in the changes represented by the symbol d: thus, for example, the pressure might change slightly during the expansion $\mathrm{d}V$.

In a change at strictly constant pressure we could write

$$\mathrm{d}Q = c_{\mathrm{p}}\,\mathrm{d}T \tag{1.6.2}$$

where c_{p} is the specific heat at constant pressure, and would include any work done, and heat lost, by any expansion $\mathrm{d}V$.

The gas laws for a perfect gas, which is a very good approximation for changes in the atmosphere, where the volume occupied by the material of the molecules is a very small fraction of the total volume, are expressed by

$$pV = RT, \quad \text{or} \quad p = R\rho T \tag{1.6.3}$$

where R is the gas cosntant (note that this quantity R is not related to the density gradient vector $\mathbf{R}$. This physical quantity in the gas law equation is not used except in this section) and ρ is the density, so that in a change

$$p\,\mathrm{d}V = R\,\mathrm{d}T - V\,\mathrm{d}p = R\,\mathrm{d}T - \frac{1}{\rho}\,\mathrm{d}p \tag{1.6.4}$$

and for the case of constant pressure, we can equate the value of $\mathrm{d}Q$ in equations (1.6.1) and (1.6.2) with $\mathrm{d}p = 0$, to give

$$c_v\, dT + AR\, dT = c_p\, dT$$

or

$$A = (c_p - c_v)/R \tag{1.6.5}$$

which simply expresses the difference between the two specific heats in terms of A, the mechanical equivalent of heat. We introduce the symbol γ for their ratio

$$c_p/c_v = \gamma = 1.403, \tag{1.6.6}$$

very approximately for the atmosphere, which is composed, predominantly of diatomic molecules. It has a different value for mon- or tri-atomic gases.

When $dQ = 0$, and the change is adiabatic, and the change in temperature is due to the change in density when work is done on the boundary of the parcel, to change its volume, and so (1.6.1) becomes

$$0 = c_v\, dT + (c_p - c_v)p\, dV/R$$

$$= dT + (\gamma - 1)(dT - V\, dp) \qquad \text{(by (1.6.4)}$$

and using (1.6.3) to eliminate V we get, with $V = \dfrac{1}{\rho}$,

$$\frac{dT}{T} = \frac{\gamma - 1}{\gamma}\frac{dp}{p} \quad \text{or} \quad \frac{dp}{p} = \gamma\frac{d\rho}{\rho} \tag{1.6.7}$$

which means that

$$\frac{T}{T_o} = \left(\frac{p}{p_o}\right)^{\frac{\gamma-1}{\gamma}} = \left(\frac{\rho}{\rho_o}\right)^{\gamma-1}$$

where the suffix o indicates the starting value before the change. These results may be written as

$$\frac{\rho}{\rho_o} = \left(\frac{T}{T_o}\right)^{\frac{1}{\gamma-1}};$$

$$\frac{p}{p_o} = \left(\frac{\rho}{\rho_o}\right)^{\gamma}$$

or

$$\frac{Dp}{Dt} = c\left(\frac{D\rho}{Dt}\right) \tag{1.6.9}$$

where $c = \sqrt{\gamma RT}$ is the velocity of sound (see section 2.1).

We **define the potential temperature θ, and its reciprocal τ,** by

$$\boxed{\theta = \frac{1}{\tau} = T\left(\frac{p_1}{p}\right)^{\frac{\gamma-1}{\gamma}}} \tag{1.6.10}$$

where p_1 is a reference pressure, usually taken as 1000 mb or roughly mean sea level pressure, in practical meteorology. The '**modified pressure**' ϖ is defined by

$$\varpi = \frac{\gamma R}{\gamma - 1}\left(\frac{p}{p_1}\right)^{\frac{\gamma-1}{\gamma}} \tag{1.6.11}$$

and in this way we obtain the identity

$$\boxed{\frac{1}{\rho}\ \mathbf{grad}\, p = \frac{1}{\tau}\ \mathbf{grad}\ \varpi.} \tag{1.6.12}$$

Consequently, since molecular viscosity is of negligible importance in atmospheric motion except on a very small scale such as around raindrops, cloud droplets, leaves, and so on, the equation of motion (1.3.29) becomes

$$\boxed{\frac{\mathrm{D}\mathbf{v}}{\mathrm{D}t} = \mathbf{g} - \frac{1}{\tau}\ \mathbf{grad}\ \varpi.} \tag{1.6.13}$$

The strategic purpose of this is to identify the equation for adiabatic motion with the classical case of incompressible motion in which a parcel of fluid does not undergo changes of density. In our case a parcel retains the same potential temperature, and when we take the curl of the equation of motion the quantity ϖ disappears as does p in the classical case, and we simply replace ρ by τ in the vorticity equation. (See note after equation (1.4.1)). The vorticity equation (1.4.1) is

$$\boxed{\frac{\mathrm{D}\boldsymbol{\omega}}{\mathrm{D}t} = -\ \boldsymbol{\omega}\ \mathrm{div}\ \mathbf{v} + (\boldsymbol{\omega}\,.\,\mathbf{grad})\mathbf{v} + \mathbf{R}\times(\mathbf{g}-\mathbf{f})} \tag{1.6.14}$$

where now

$$\boxed{\mathbf{R} = \frac{1}{\tau}\ \mathbf{grad}\ \tau.}$$

In considering the motion of the air, which is compressible, we can take over very many of the deductions made for an incompressible fluid, including those concerning the vorticity, simply by replacing ρ by τ. In those from which the pressure is not eliminated we may find it necessary to replace p by ϖ. The method is not useful when there are significant non-adiabatic mechanisms operating. In the cases mentioned above in which viscosity is important we are not usually concerned about compressibility, and so this transformation is not called for anyway. **This simple exchange of symbols takes care of the fact that in vertical motion the air changes its temperature because of changes of pressue due to compression by the air's own weight**. However a modification of the equation of continuity is also required: this is discussed in Chapter 5, equation (5.7.6).

1.7 BERNOULLI'S EQUATION FOR AN INVISCID FLUID IN STEADY MOTION; BERNOULLI SURFACES

For an inviscid fluid the equation of **steady motion** is, by (1.2.6),

$$\mathbf{grad}\ \frac{1}{2}q^2 - \mathbf{v}\times\boldsymbol{\omega} = -\mathbf{grad}\ gz - \frac{1}{\rho}\ \mathbf{grad}\ p \tag{1.7.1}$$

the z-axis being vertical. If we now put

$$\frac{1}{\rho}\ \mathbf{grad}\ p = \ \mathbf{grad}\ \frac{p}{\rho} - p\ \mathbf{grad}\ \frac{1}{\rho}$$

or

$$\frac{1}{\tau}\ \mathbf{grad}\ \varpi = \ \mathbf{grad}\ \frac{\varpi}{\tau} - \varpi\ \mathbf{grad}\ \frac{1}{\tau} \tag{1.7.2}$$

then

$$\mathbf{grad}\left(\frac{1}{2}q^2 + \frac{\varpi}{\tau} + gz\right) - \varpi\ \mathbf{grad}\ \frac{1}{\tau} = \mathbf{v}\times\boldsymbol{\omega} \tag{1.7.3}$$

with a corresponding equation using p and ρ, which will be taken as read for the rest of this section.

If initially (i.e. somewhere upstream) τ is constant along the streamlines and vortex lines, which means that there exist surfaces containing streamlines and vortex lines which are also isentropic (or, in the case of an incompressible fluid, if the density is constant on streamlines and vortex lines), and this is usually true of horizontal airflow over level ground, and $\mathbf{v}$ and $\boldsymbol{\omega}$ are horizontal vectors and $\mathbf{grad}\ \tau$ is a vertical vector, then $\mathbf{v}.\mathbf{grad}\ \frac{1}{\tau}$ and $\boldsymbol{\omega}.\mathbf{grad}\ \frac{1}{\tau}$ both vanish. Writing

$$\boxed{gH = \frac{1}{2}q^2 + \frac{\varpi}{\tau} + gz} \tag{1.7.4}$$

from (1.7.3) we have that

$$\boxed{\mathbf{v}\,.\,\mathbf{grad}\,(gH) = \boldsymbol{\omega}\,.\,\mathbf{grad}\,(gH) = 0} \tag{1.7.5}$$

The consequence of (1.7.5) is that gH **is constant in the directions of the vorticity and the velocity and is therefore constant in the surfaces containing these vectors, which we call the Bernoulli surfaces.**

The assumption about the isentropic surfaces was seen in section 1.5 to be relevant to the atmosphere (and for the same reasons to the ocean). In cases where the isopycnic theory is not applicable so that the vortex lines do not lie in the isentropic surfaces, and never will, the motion is almost certainly unsteady and we cannot obtain any useful result about the constancy of gH.

Bernoulli's theorem is only valid for reversible flows because it is essentially a statement about conservation of energy, and there can be no such result for viscous flows. The classical applications are for a fluid of uniform potential temperature or density for which $\mathbf{grad}\ \tau = \mathbf{grad}\ \rho = \mathbf{0}$. Then

$$\mathbf{grad}\,(gH) = \mathbf{v} \times \boldsymbol{\omega} \tag{1.7.6}$$

and so gH is constant along vortex lines and streamlines. Since it is not necessarily constant throughout the fluid, Bernoulli surfaces always exist in this case, and each streamline defines a surface, namely the one containing all the vortex lines through it, or vice versa. **Any streamline intersecting one of the vortex lines through a given streamline must intersect them all**, unless there is no vorticity and gH is constant throughout the fluid.

For a well mixed fluid, τ is uniform and so the pressure is expressible in terms of ρ [see discussion following (1.1.16)], and in this special case

$$\frac{1}{\rho}\,\mathbf{grad}\,p = \mathbf{grad}\int\frac{\mathrm{d}p}{\rho} \tag{1.7.7}$$

and so

$$\boxed{gH = \frac{1}{2}q^2 + \int\frac{\mathrm{d}p}{\rho} + gz = \text{ constant on a Bernoulli surface.}} \tag{1.7.8}$$

This is the well known case found in elementary textbooks which is much **used in theory of compressible flow** if it is unstratified. **If the motion is also irrotational** then the right side of (1.7.6) vanishes because $\boldsymbol{\omega} = \mathbf{0}$ and gH **is constant throughout the fluid**. In addition, if the flow is unsteady but irrotational we can write $\mathbf{v} = \mathbf{grad}\ \phi$ and the equation of motion leads to

$$gH = \frac{1}{2}q^2 + \int\frac{\mathrm{d}p}{\rho} + gz + \frac{\partial\phi}{\partial t} = \text{ constant} \tag{1.7.9}$$

throughout the fluid. Since the essential feature of stratified or buoyant fluids is that their motion has a significant vorticity, these cases are of little relevance here.

1.8 THE STATIC STABILITY

The potential temperature defined in (1.6.10) may be written

$$\theta = \frac{p}{R\rho}\left(\frac{p_1}{p}\right)^{\frac{\gamma-1}{\gamma}} = k_0^{1/\gamma}\frac{p_1^{\frac{\gamma-1}{\gamma}}}{R} \tag{1.8.1}$$

The static stability is the restoring force due to buoyancy on a unit mass of fluid displaced adiabatically a unit distance vertically. The stability is positive if the force is of opposite sign to the displacement. In an incompressible fluid it is equal to $-\frac{g}{\rho}\frac{\partial \rho}{\partial z}$ because $\delta\rho$ is the density anomaly and is equal to $\delta z \frac{\partial \rho}{\partial z}$. Because in a gas the coefficient of expansion is $1/T$, the temperature anomaly may be used in the compressible case, and this is the potential temperature anomaly. Thus the **static stability** is

$$g\beta = -\frac{g}{\tau}\frac{\partial \tau}{\partial z} = \frac{g}{\theta}\frac{\partial \theta}{\partial z}, \tag{1.8.2}$$

the derivatives being obtained in the undisturbed static state. β is positive when the potential temperature increases upwards.

The **temperature gradient**, in which the restoring force is zero and the displaced parcel is undergoing adiabatic changes, is the one in which the parcel is indistinguishable from its surroundings, and is called the **adiabatic lapse rate**. It is obtained by combining equations (1.6.1) and (1.6.2) with the hydrostatic equation (1.1.5). By eliminating p and ρ and differentiating to obtain $\partial T/\partial z$ we find

$$\frac{\partial T}{\partial z} = -\frac{(\gamma-1)g}{\gamma R} = \Gamma, \tag{1.8.3}$$

which is a constant determined by gravity and the physical properties of air. For the earth's atmosphere at sea level it is 9.86°C per km, and for practical purposes may be taken as 1°C per 100 m throughout the atmosphere.

Because the **adiabatic lapse rate** is independent of temperature it would mean that if the atmosphere were in a state of neutral stability with a temperature of 300 K at the ground it would be at the absolute zero at a height of 30 km. Radiative equilibrium between the earth at 300 K and outer space at 0 K would be at $[\frac{1}{2}(300)^4]^{\frac{1}{4}}$ K, according to a simple application of **Stefan's law**, and this is about 250 K, so that it may be deduced that the atmosphere is highly stratified, the stratosphere having a temperature perhaps as high as 250 K at 30 km.

The rate at which a parcel of air is cooled adiabatically if it is displaced in stratified surroundings is very nearly the same as in the neutral, or well mixed, case. The difference arises from the fact that for (1.8.3) ρ is not the density of the parcel but of the surroundings. But in practice this is of no importance because of other errors. Thus a parcel cannot be displaced very fast in a stratified environment without the temperature difference causing mixing except when waves are the cause of the displacement, in which case the pressure variations due to the waves are of small magnitude. For practical purposes, therefore, in any surroundings we shall not be misled by the assumption that the temperature of an unmixed parcel falls at the rate of 1°C per 100 m when it is displaced vertically.

We shall need later to calculate the density of a displaced parcel. If ρ_0 is its original density, then in adiabatic displacements.

$$\frac{\rho}{\rho_0} = \left(\frac{T}{T_0}\right)^{\frac{1}{\gamma-1}} \text{ exactly} \tag{1.8.4}$$

and

$$T = T_0 + \Gamma\zeta \text{ approximately} \tag{1.8.5}$$

for a vertical displacement ζ. Therefore

$$\begin{aligned}\frac{\rho}{\rho_0} &= \left(1 + \frac{\Gamma}{T}\zeta\right)^{\frac{1}{\gamma-1}} \simeq 1 + \frac{\Gamma}{(\gamma-1)T}\zeta \\ &= 1 - \frac{g}{\gamma RT}\zeta \\ &= 1 - \frac{g}{c^2}\zeta \end{aligned} \tag{1.8.6}$$

where

$$c = (\gamma RT)^{1/2}$$

is the velocity of sound = 316 m s^{-1} approx. The approximate formula (1.8.6) is good as long as

$$0.3\frac{g}{c^2}\zeta \ll 1, \tag{1.8.8}$$

this being the ratio of the first neglected term to the last one taken in arriving at (1.8.6). The condition (1.8.8) requires that the displacement be small compared with 30 km, and is certainly valid for distances of the order of 3 km, which is greater than those in most practical cases to which we may wish to apply (1.8.6).

The customary concept of static instability is that the buoyance force is in the direction of the displacement, thereby causing an acceleration in that direction. This is the **displaced parcel theory**. But we shall find it useful to think in terms of a **rotational displacement**. When there is no motion the vorticity equation is

$$\frac{\mathrm{D}\boldsymbol{\omega}}{\mathrm{D}t} = \mathbf{R} \times \mathbf{g}, \quad \text{or} \quad \frac{\mathrm{D}\xi}{\mathrm{D}t} = -\,\mathbf{g}\,\beta \sin\phi \tag{1.8.9}$$

where $-\beta$ is the magnitude of **R** defined for equation (1.6.14) and where ξ is the x-component of vorticity, assumed horizontal, and ϕ is the angular displacement of the parcel (see Fig. 1.8.i) about the x-axis of a local tube of fluid. The tube has angular velocity $\frac{1}{2}\xi$, which means that

$$\frac{D\phi}{Dt} = \tfrac{1}{2}\xi, \quad \text{and} \quad \frac{D\xi}{Dt} = 2\frac{D^2\phi}{Dt^2}. \tag{1.8.10}$$

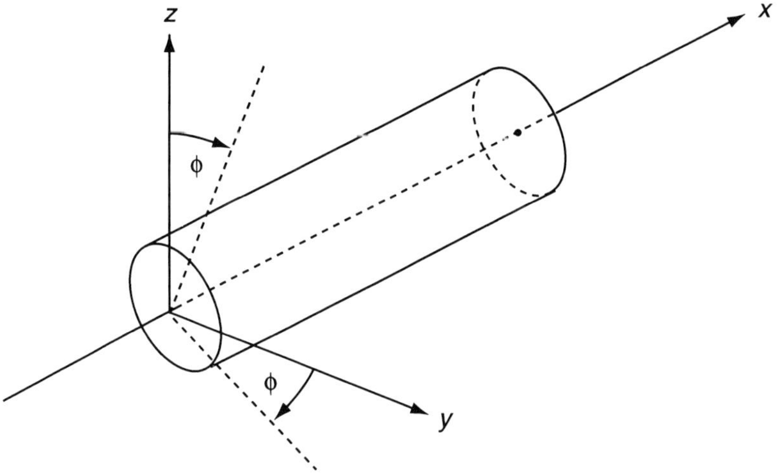

Fig. 1.8.i The angular displacement ϕ of a cylindrical parcel with a horizontal axis.

In this discussion β denotes the magnitude of **R** and is equal to $-\beta = \dfrac{1}{\tau}\dfrac{\partial\tau}{\partial z}$ in the initial equilibrium state, and so (1.8.9) becomes

$$\frac{D^2\phi}{Dt^2} = \frac{g}{2\tau}\frac{\partial\tau}{\partial z}\sin\phi, \quad = -\frac{1}{2}g\beta\phi \text{ for small } \phi. \tag{1.8.11}$$

This indicates instability, with ϕ growing exponentially with time, if β is negative and the potential temperature decreases with height. This is, of course, the same as the condition derived by the parcel displacement argument.

The interest in this form of the criterion for local static instability is that vorticity is generated by it, and it cannot occur without generating vorticity. The consequence is that mixing takes place and the motion becomes turbulent.

If a parcel of fluid is displaced upwards (Fig. 1.8.ii) we can think of this as being the result of vorticity being generated in horizontal rings around it. A parcel which is penetrating through the surrounding fluid (Fig. 1.8.iii) has motion which can be thought of as being due to the vortex rings which have been generated around it in accordance with (1.8.9). The motion does not remain simple but soon becomes a tangle of small-scale vortices which cause the mixing (see Chapter 10): this is because once the surfaces of constant potential temperature become disturbed, the fluid is everywhere subject to the generation of vorticity, and irregularities grow anywhere.

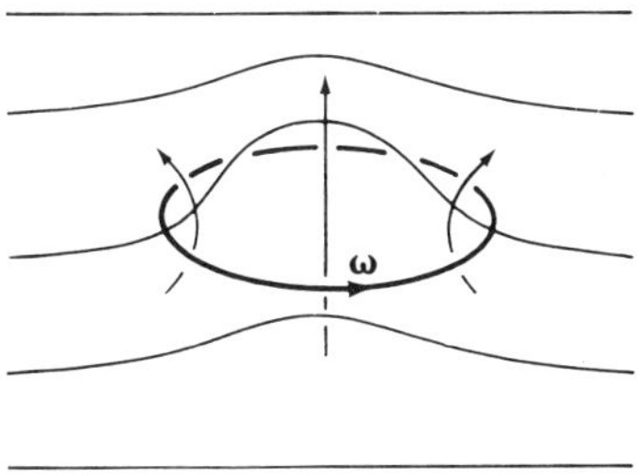

Fig. 1.8.ii Generation of vortex rings by the vertical displacement of a stratified fluid.

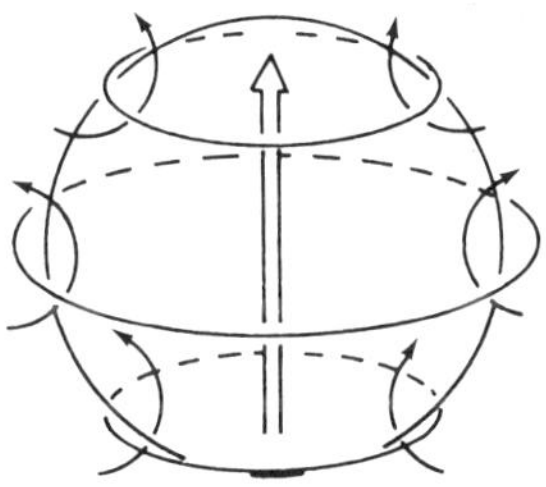

Fig. 1.8.iii Vortex rings generated by static instability transporting a parcel through its surroundings.

2

Various phenomena of fluid flow

2.1 THE IMPORTANCE OF SOUND PROPAGATION

Our environment is full of evidence that sounds are transmitted as pressure waves through the air. Speech, traffic noise, bangs and squeaks, music, the resonance of the bathroom and the rattling of windows in loose frames are all transmitted without our feeling the pressure waves except through our ears.

The transmission is not instantaneous, as appears to be the case with light. We see someone hammering in a stake a hundred yards away, and the bang of the hammer arrives a third of a second after the appearance of the impact that obviously caused it. And if we stand some distance away from a large flat bit of wall we can hear the echo arrive back to us when we call: indeed we can arrange to time the delay by emitting bangs at regular intervals to coincide with the arrival of the echo and count the time for N bangs. Thus if there were 60 bangs in exactly a minute, the sound must have travelled there and back in one second. By careful measurement of the distance we could easily measure the speed of sound with considerable accuracy.

A typical speed would be about 316 m s^{-1}. A lightning flash may be followed after, say, 3 s by the first rumble of thunder indicating that it was roughly 1 km away at its nearest point, but we have to remember that the lightning flash may be several times that length so that the arrival of the noise may occupy up to around 10 s, which one might have regarded as being due to 'echoing around the clouds'—for otherwise how would an instantaneous flash cause a long rumble?

A musical note of moderate pitch, such as the 100 Hz 'mains hum' close to G (an octave and a half below middle C, which is 261.6 Hz), would have a wavelength of 3.16 m and would resonate very well in a bathroom of that width. Louder and deeper bangs could be composed of wavelengths around 10 m. The high-pitched whistle which a dog can hear but people cannot has a frequency of about 30 000 Hz and a wavelength near to 1 cm.

We can often see dust particles 'floating' in the air, particularly in sunbeams, but we cannot usually see any oscillations as noises pass through the air. This indicates that the displacement of the air particles is very small compared with the wavelength, which means that we can expect considerable accuracy from a linearised theory and that trains of waves travelling through the same air in different directions at the same time do not interfere with each other significantly.

A source of sound sends out waves which are in the shape of expanding spheres, so that they very soon become like plane waves passing any observation point. A good model therefore is a plane wave, which can be analysed like a wave travelling along a cylindrical tube of unit cross-section with no friction on the wall, all the displacements being parallel to the axis of the tube, which we will assume to be aligned in the x-direction (Fig. 2.1.i).

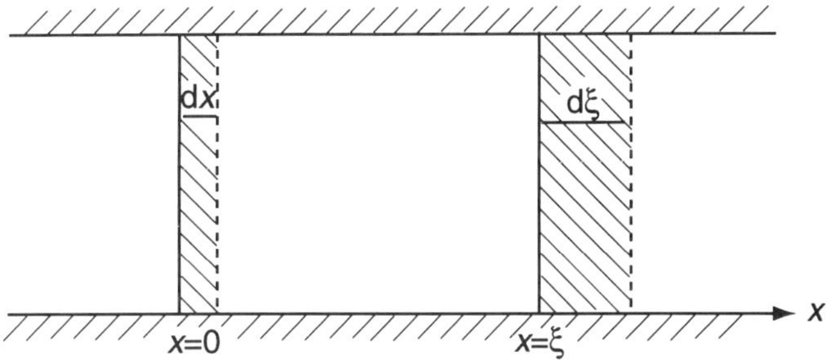

Fig. 2.1.i Displacements of air particles at a wave travelling along a tube with no friction, like a bit of a plane wave perpendicular to the direction of propagation.

A plane sound wave in the yz-plane moves in the x-direction and we focus our attention on the air at $x = 0$ which is displaced to $x = \xi$ by the wave, the fluid being otherwise at rest. We have three equations for the pressure, p, the density, ρ, and the fluid velocity, $\mathbf{v} = (u, 0, 0)$. We assume **adiabatic changes** (section 1.6) on the grounds that the waves produce very small and very rapidly changing gradients of temperature with negligible conversion of the energy to heat and loss by radiation of conduction. The pressure and density are related by (1.6.7), i.e.

$$\frac{\mathrm{d}p}{p} = \gamma \frac{\mathrm{d}\rho}{\rho}$$

$$\frac{1}{p}\frac{\partial p}{\partial x} = \frac{\gamma}{\rho}\frac{\partial \rho}{\partial x} \tag{2.1.1}$$

and so the pressure gradient is

$$\frac{\partial p}{\partial x} = \frac{\gamma p}{\rho}\frac{\partial \rho}{\partial x} = \gamma RT \frac{\partial \rho}{\partial x} \tag{2.1.2}$$

and the **equation of continuity** states that the fluid within the block between $x = 0$ and $x = \mathrm{d}x$ is the same material as in its displaced position between $x = \xi$ and $x = \xi + \frac{\partial \xi}{\partial x}\mathrm{d}x$; therefore to the first order in $\partial\xi/\partial x$

$$\rho_0 \, \mathrm{d}x = \rho\left(1 + \frac{\partial \xi}{\partial x}\right) \mathrm{d}x \tag{2.1.3}$$

and

$$\rho = \rho_0\left(1 - \frac{\partial \xi}{\partial x}\right),$$

and so the density gradient is

$$\frac{\partial \rho}{\partial x} = -\rho_0 \frac{\partial^2 \xi}{\partial x^2}; \tag{2.1.4}$$

and thirdly, in the **equation of motion** in which we neglect the advective term $u\frac{\partial}{\partial x}$ because the fluid velocity is very small, the acceleration is $\partial^2\xi/\partial t^2$, and so, writing

$$c^2 = \gamma RT \tag{2.1.5}$$

and $\rho = \rho_0$ to the first order in the perturbation we have, by (2.1.2), (2.1.4) and (2.1.5),

$$\begin{aligned}\frac{\partial^2 \xi}{\partial t^2} &= -\frac{1}{\rho}\frac{\partial p}{\partial x} = \frac{c^2}{\rho}\frac{\partial \rho}{\partial x} \\ &= c^2 \frac{\partial^2 \xi}{\partial x^2},\end{aligned} \tag{2.1.6}$$

which is the wave equation. Solutions are of the form

$$\xi = f(x - ct) + g(x + ct) \tag{2.1.7}$$

where f and g are arbitrary functions representing the form of the wave moving in the + or – direction of x, and travelling a distance c in unit time: $x - ct$ takes the same value at $x = 0$ and at $x = \pm ct$ at a later value of t for the functions f and g respectively. Thus the wave form moves with speed c, which is dependent on the gas constant R, the ratio of the specific heats γ, and the temperature T, but not separately on the pressure or density. Thus the speed is independent of the wavelength, and we say that the process of **sound transmission is not dispersive**. But we shall find, in Chapters 5–7, that when the waves are very long and have speed dependent on gravity in a stratified air mass or are affected by the earth's rotation, they are dispersive.

The Mach number, M, is the ratio of the fluid speed to the speed of sound, so that

$$M = U/c$$

and the generation (2.1.6) is valid for Mach numbers small compared with unity.

2.2 EXPLOSION AND OTHER EXPANSION WAVES

When heat is applied to a body of air in a concentrated manner its pressure is increased because the velocity of its molecules is increased. The pressure in the surrounding air is unaltered at first but the heated air expands until its pressure becomes the same, and it pushes the air outwards. In an extreme case when heat is applied suddenly or gas is evolved from solid or liquid very quickly we call it an explosion. Some fireworks are designed to create a big bang, and a shock wave travels outwards causing an outward displacement as it passes so that air moves away from the explosion with a jump.

The atom bomb exploded in 1945 created an expansion of about 1 km^3, and sent out a spherical shock wave, and this produced an outward displacement of the air of about 1.6 m in a small fraction of a second at a distance of 10 km. The sudden rise in pressure was enough to break windows and to topple walls which it hit at right angles. If the same total expansion had taken place over about ten minutes the gradual displacement of 1.6 m would have been almost impossible to detect among other slight movements of the air that continually surround us. Nevertheless that displacement would have occurred and the same heat would have produced as big a convection current into the stratosphere as was observed after the atom bomb explosion.

The important point is that a pressure wave travels outwards from the point of application of heat with the speed of sound and becomes imperceptible very quickly, except that it causes the density of the heated air to decrease, and the hot air begins, slowly by comparison with the speed of sound, to rise owing to its buoyancy.

At the site of the heating it appears that the effect is simply to decrease the density. The outward displacement nevertheless travels at a finite speed of about 1140 km per hour. Therefore if it occurs in the middle of a large continent it cannot contribute to the creation of a sea breeze at the coast until at least an hour later.

An explosion at the ground sets up a hemispherically expanding wave (Fig. 2.2.i) of which the top reaches a level where the density is about a tenth of an atmosphere in about a minute, and the lower part of the wave has become an almost vertical circular wall of radius nearly 20 km expanding horizontally. Most of the mass of the atmosphere is below the top of that wall, and the effect of the stratification of the air on the propagation of the wave will be studied in Chapter 7.

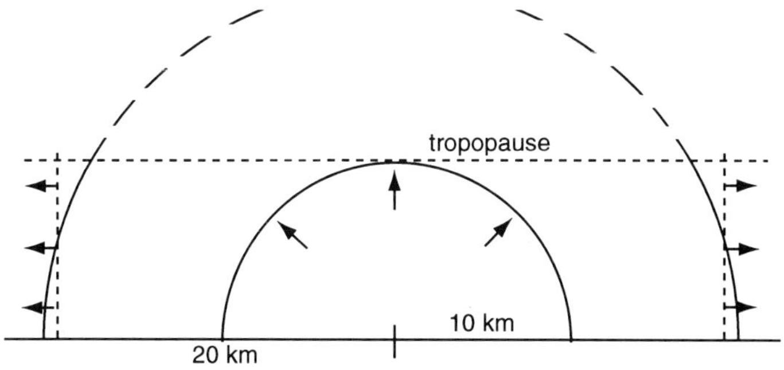

Fig. 2.2.i At a distance comparable with the depth of the atmosphere the shock wave due to an explosion becomes like a vertical wave front as seen in the troposphere.

2.3 PRACTICAL APPROXIMATIONS

The most important consequence of this is that for short distance from boundaries like buildings and small hills, but more so within laboratories and machines as long as the speeds of the airflow are small compared with the speed of sound, **the air behaves like an incompressible fluid**.

Most air has been at rest at some time recently, and therefore possessed no vorticity, except that due to the rotation of the earth. In many circumstances it is set in motion by the pressure applied by its surroundings being changed, and this can only impart vorticity to it if the fluid has variations of density. So if it is of uniform density, or of uniform potential temperature in cases where the vertical pressure gradient is important, it can only acquire vorticity by viscosity conducting it from a rigid boundary. With this in mind we now give a description of a variety of fluid flow patterns, but starting with the assumption that the fluid **is inviscid, irrotational, incompressible and also, in the simplest cases, in steady motion**.

Under these restrictions the medium is often described as a **perfect fluid**, but except for the simplicity of the mathematics in which it is much like electric and magnetic fields and some other cases like flow through porous media, a perfect fluid would be more aptly described as a dull and, certainly (sometimes embarrassingly), useless fluid.

2.4 POTENTIAL FLOW

Without vorticity **curl v = 0**, and there exists a **potential** ϕ which is defined by

$$\phi_P = \int_0^P \mathbf{v}\,.\,\mathbf{ds}. \tag{2.4.1}$$

This value is independent of the path of integration because the circulation around any closed circuit in a potential flow field is zero. Thus

$$\mathbf{v} = \mathbf{grad}\ \phi. \tag{2.4.2}$$

In many problems if the integral defining the potential were calculated from a different origin, not O, that would be equivalent to adding a constant to the value everywhere, but this would not affect the velocity because the constant has no gradient. In some physical problems it is more convenient to define the potential so that the vector points down the gradient, but in fluid mechanics it has become conventional to have it pointing in the positive direction.

The equation of continuity for an incompressible fluid is simply a statement that there is no flow out of any small volume, meaning that the divergence of the velocity is zero. Thus div $\mathbf{v} = 0$, which becomes, with (2.4.2),

$$\nabla^2\phi = 0. \tag{2.4.3}$$

When it is required to be known, the pressure is computed from Bernoulli's equation (1.7.8) when a formula has been found which satisfies (2.4.3) and complies with the boundary conditions, which usually means having no component of velocity across any rigid boundary. We now describe some interesting special cases:

The potential

$$\phi = -mr, \tag{2.4.4}$$

in which r is distance from the origin of coordinates, describes the flow radially outwards from a source at the origin with velocity m/r^2 (Fig. 2.4.i). The total outward flux of volume, which is called the strength of the source, is $4\pi m$. If m is negative it is called a sink, and the flow is inwards. A flow which is similar in planes all passing through the same axis is best described by spherical polar coordinates (r, θ, λ) and is independent of λ. The coordinate x is measured in the direction of $\theta = 0$. The potential

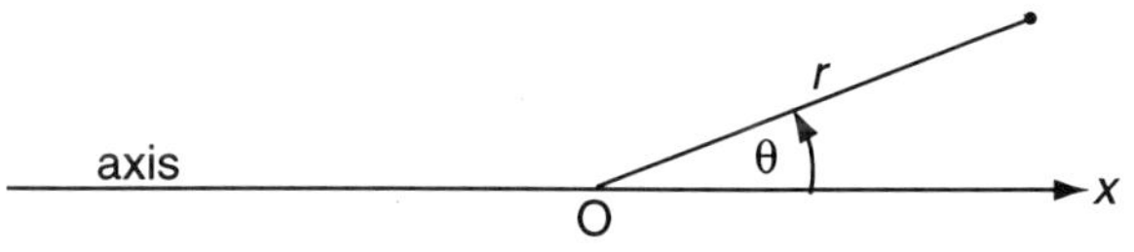

Fig. 2.4.i Coordinates used for the potential of the flow from a point source.

$$\phi = Ux = Ur\cos\theta \tag{2.4.5}$$

represents a uniform stream of speed U in the x-direction.

The flow due to a doublet of strength μ is represented by

$$\phi = -\mu\cos\theta/r^2\,. \tag{2.4.6}$$

This is made evident by noting that the left-hand side of

$$\partial x\frac{\partial}{\partial x}\frac{m}{r} = -\,m\partial x\cos\theta/r^2 \tag{2.4.7}$$

represents the potential of a sink (m/r) together with a source ($-m/r$) placed a distance ∂x away from it in the x-direction. The moment of the doublet so formed is $\mu = m\partial x$. If we now add the doublet to the uniform stream, since they are both proportional to cos θ, we obtain

$$\phi = \left(Ur - \frac{\mu}{r^2}\right)\cos\theta \tag{2.4.8}$$

so that the radial component of velocity is

$$\frac{\partial\phi}{\partial r} = \left(U + \frac{2\mu}{r^3}\right)\cos\theta \tag{2.4.9}$$

and this is zero on $r = a$ if

$$\mu = -\,Ua^3/2 \tag{2.4.10}$$

This, therefore, is the condition for the radial velocity to be zero on the sphere $r = a$, and so

$$\phi = U\cos\theta\left(r + \frac{a^3}{2r^2}\right) \tag{2.4.11}$$

is the potential of the steady flow of the uniform stream around the sphere of radius a, fixed at the origin. The flow pattern is shown in Fig. 2.4.ii(a). At larger r in all directions the flow is a uniform stream U in the x-direction (parallel to $\theta = 0$).

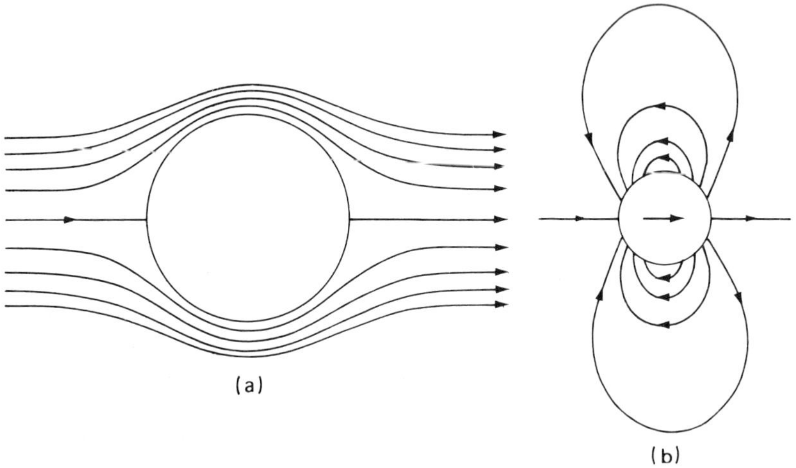

Fig. 2.4.ii (a) Streamlines of flow around a sphere shown in a plane through the axis of the flow. (b) Lines of flow due to a sphere moving through otherwise stagnant fluid.

2.5 VIRTUAL MASS OF AN ACCELERATED BODY

The pressure at any point on the sphere when it is fixed in a stream U is obtained from (1.7.8) and this gives for $\partial/\partial t = 0$, when the velocity is constant,

$$p = p_0 - \tfrac{1}{8}\rho U^2(5 - 9\cos^2\theta). \tag{2.5.1}$$

The pressure on a sphere moving unsteadily in otherwise stationary fluid is obtained by using (1.7.9) and (2.4.11) and subtracting the potential of uniform flow from the potential of the sphere in the uniform stream to give

$$\phi = Ua^3\cos\theta/2r^2. \tag{2.5.2}$$

The flow lines are plotted in Fig. 2.4.ii(b). We calculate $\partial/\partial t$ in (1.7.9) by noting that in Fig. 2.4.i we have $x = r\cos\theta$:

$$\frac{\partial\theta}{\partial t} = U\sin\theta/r, \qquad \frac{\partial r}{\partial t} = -U\cos\theta \tag{2.5.3}$$

at the fixed point (r, θ) for a coordinate system moving with the sphere with speed U. Then after putting $r = a$ we obtain the pressure on the moving sphere in the form

$$p = p_0 \tfrac{1}{8}\rho U^2(5 - 9\cos^2\theta) + \tfrac{1}{2}\rho\dot{U}a\cos\theta. \tag{2.5.4}$$

In (2.5.1) and (2.5.4), p_0 is the pressure at an infinite distance in the stream or in the fluid, where it is at rest, respectively. The pressure produces no force on the sphere if the motion is steady because it is the same on the upstream and downstream sides. This is true of any body in a stream of 'perfect' fluid. But when the sphere has an acceleration $\dot{U}$ through a fluid otherwise at rest the total force, F, is obtained by integrating the last term of (2.5.4) over the surface. Thus

$$F = -\int_0^\pi p\cos\theta \,.\, 2\pi a^2 \sin\theta\, \mathrm{d}\theta = \frac{2}{3}\pi\rho a^3\dot{U} = \frac{1}{2}M'\dot{U} \tag{2.5.5}$$

where M' happens in this case to be the mass of fluid displaced by the sphere. The quantity $\tfrac{1}{2}M'$ is called the **virtual mass** of the sphere because it represents the apparent mass of the fluid which has to be set in motion when the sphere is accelerated. When a sphere is started in motion from rest the motion is initially irrotational (see equation (1.1.14)), and except for the inertia of the substance of the sphere itself is well represented by the formula (2.5.5), but viscous forces soon change the flow as described below.

Many other shapes of body may be represented by a distribution of sources and sinks but not necessarily any shape. The virtual mass of other shapes may be quite differently related to the mass of fluid displaced. Thus a thin flat plate has zero virtual mass if accelerated from rest in its own plane; but if accelerated in a plane perpendicular to itself it requires a large mass of fluid to be set in motion. In accelerated or steady irrotational flow in general there may be a couple due to the pressure on the surface, but this is zero for the sphere because of its axial symmetry.

Classical fluid mechanics is concerned with the application of the above methods to a variety of contrived problems which are well described at length in several excellent textbooks. But in problems of the great outdoors they are seldom satisfactory, because the air is not a perfect fluid, but more importantly because the air motion is seldom irrotational: gravity continually creates vorticity through the density gradients, and turbulent eddies are found in the wake of almost all solid bodies.

2.6 FREE STREAMLINES

The theory of free streamlines is well described in many books (e.g. Lamb, Arts 73–78). A surface is assumed to exist which satisfies the following conditions: that it streams away from a sharp edge of a body, the pressure is constant on the free streamline, and the fluid on the other side of the streamline is stagnant (or it might be air providing a constant pressure on a water surface). The fluid on the free streamline has a constant speed and the streamline is taken to be its boundary. If there were a stagnant fluid on only one side of the free streamline it would be a section of a vortex sheet and would be unstable (see

Chapter 8), and would become unsteady downstream, but the theory is a beginning to the study of wakes provided by classical hydrodynamics.

Examples are displayed in Figs 2.6.i & ii. The vena contracta, (c), is fairly realistic for a jet of water emerging from an orifice, and the upper half may be taken to illustrate the flow on a horizontal base from the bottom of a submerged sluice gate, and in this case gravity is obviously important. The velocity approaches a steady value far downstream. To be suitable for this treatment the flow is probably required to be two-dimensional or strictly axisymmetric. The potential flow in a corner of angle α is given by

$$\phi = Ur^{2\pi/\alpha}\cos\frac{2\pi\theta}{\alpha}. \qquad (2.6.1)$$

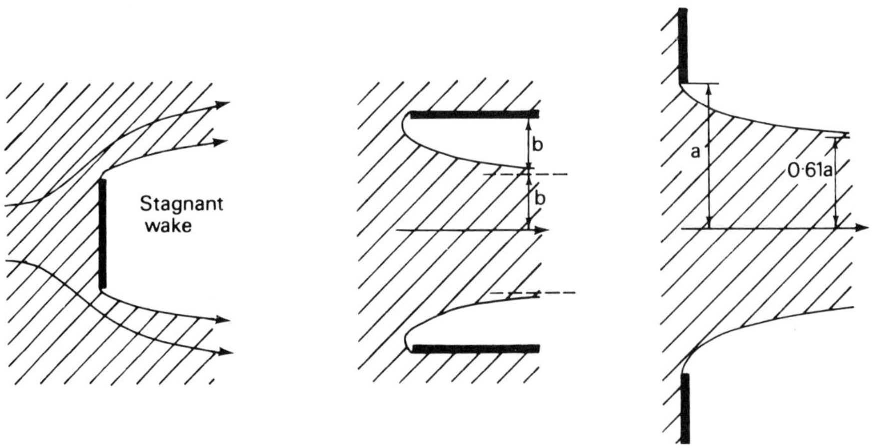

Fig. 2.6.i Free streamlines behind a flat plate, flow into a slit from horizontal boundaries (Borda's mouthpiece), and out of a slit in a vertical wall (vena contracta).

When one of the planes is rigid at $\theta = 0$ and the other is the interface with a fluid of different density in which there is no motion and the pressure is hydrostatic, the pressure along the free boundary must vary linearly like gz. In steady flow, therefore, $(\partial\phi/\partial r)^2$ must increase linearly along $\theta = \alpha$, according to Bernoulli's equation because the pressure is the same on the two sides of the interface. Therefore

$$r^{\left(\frac{2\pi}{\alpha}-1\right)} \propto r, \qquad \text{or} \qquad \alpha = 2/3.$$

On the basis of this argument the flow of a light fluid over a wedge of heavier stagnant fluid would turn through 60° (Fig. 2.6.ii(a)). Hot fluid from a chimney into stagnant cool air would converge initially at this angle as shown in (b). The conclusion applies only close to the corner and is not relevant if friction is retarding the emerging fluid at the chimney wall, or if the exterior fluid is not stagnant.

The argument shows, however, that the flow cannot be irrotational on both sides of an interface dividing two fluids resting on a plane (a) unless one is stagnant and the other

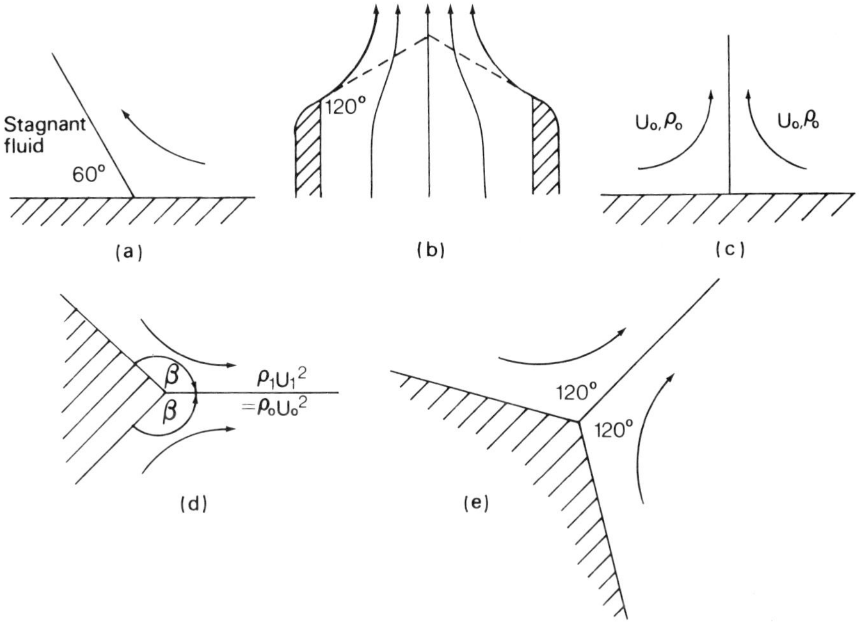

Fig. 2.6.ii (a) Flow of a light fluid up a wedge of stagnant heavy fluid. (b) Emergence of hot gas from a chimney in stagnant air. (c) Front between two fluids perpendicular to boundary. (d) Junction of irrotational streams at a horizontal interface. (e) Junction of irrotational streams at inclined interface.

turns through exactly 60°. Alternatively the two fluids must have the same density and speed and the interface must be normal to the boundary as in (c). Alternatively, the interface may be horizontal, the two angles the same and ρU^2 the same on the two sides as in (d), or both may turn through an angle of 60° as in (e).

2.7 THE SURFACE OF ROTATING LIQUIDS

The shape of the free surface of a liquid steadily rotating about a vertical axis may be calculated if the tangential (circumferential) velocity, v, is given. This is because the pressure within the fluid having no vertical acceleration is given by the hydrostatic equation

$$p = g\rho h + p_0 \qquad (2.7.1)$$

where h is the height of the free surface above the level at which pressure, p_0, is taken to be uniform. The fluid acceleration is v^2/r, r being the distance from the vertical axis, and this is equal to $\frac{1}{\rho}\partial p/\partial r$ according to the equation of motion.

Thus for a fluid in solid rotation Ω

$$\frac{\partial p}{\partial r} = \rho \frac{(r\Omega)^2}{r} \tag{2.7.2}$$

and so

$$h = \frac{\Omega^2}{2g} r^2 \tag{2.7.3}$$

and the surface is a paraboloid of revolution (Fig 2.7.i). h is the height above the centre point where the pressure is p_0.

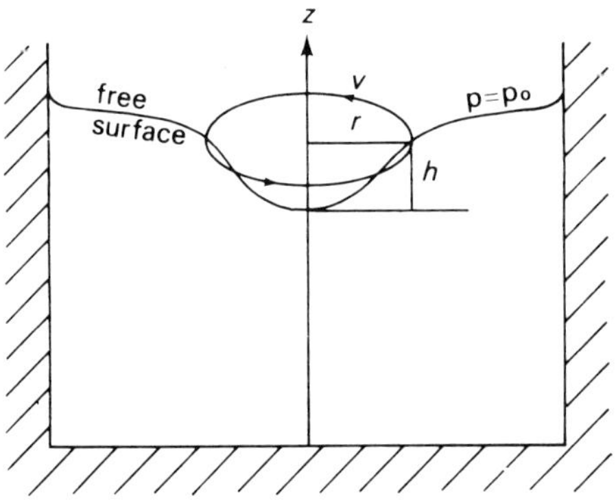

Fig. 2.7.i Coordinates and shape of surface of rotating fluid.

For a fluid in irrotational flow the circulation around the axis is the same at all radii. Consequently, $2\pi r v = 2\pi k = \text{const.}$ (independent of r) and so $v = k/r$, $\phi = k \log r$. By the same argument in this case

$$h = h_0 - \frac{k^2}{4gr^4}, \tag{2.7.4}$$

which is the shape of the 'bath plug vortex', and h_0 is approached asymptotically at large depth. This is also the shape of a funnel cloud in a tornado if the air is well mixed and has a uniform absolute humidity, and the cloud base is a surface of constant pressure. (See also section 11.10.)

If the tea in a teacup is set in rotation, the free surface tends to become paraboloidal near the centre because viscosity tends to make the rotation close to solid, while further out the circulation is made nearly uniform (i.e. without vertical component of vorticity; see Chapter 3) and so the curvature out there is convex in vertical sections through the axis, as in the bath plug. At the wall, surface tension makes it curve upwards if the solid boundary is wetted.

2.8 LAMINAR VISCOUS FLOW

(1) Flow between coaxial rotating cylinders

The stress due to viscosity in a fluid in simple shearing along streamlines with no stretching of the fluid along the streamlines is given by the second of the equations (1.3.24). This simply states that the shear stress is proportional to the viscosity and the shear.

If a fluid of viscosity μ is set in rotation about a vertical axis with tangential velocity v equal to k/r, which is irrotational because it has potential $\phi = k \log r$, the moment of the momentum of a cylindrical shell of given mass and radius r is independent of r. The moment of the tangential stress between such cylinders is also seen to be independent of r, for it is proportional to $\mu 2\pi r . r\, dv/dr$. Such a velocity distribution is therefore set up around a rigid rotating cylinder in a large coaxial vessel.

Between two rigid cylinders of radius r_1 and r_2 rotating with angular velocities Ω_1 and Ω_2, a steady velocity is obtained by adding a solid rotation, which produces no tangential stress within the fluid, to an irrotational motion. Thus, provided the motion is not unstable (see, for example, section 8.6),

$$\begin{aligned} v &= r\Omega + \frac{k}{r} \\ &= r_1\Omega_1 \quad \text{or} \quad r_2\Omega_2 \text{ at } r_1 \text{ or } r_2 \text{ respectively} \end{aligned} \tag{2.8.1}$$

where

$$k = \frac{r_1^2 r_2^2(\Omega_1 - \Omega_2)}{r_2^2 - r_1^2}, \qquad \Omega = \frac{r_1^2\Omega_1 - r_2^2\Omega_2}{r_1^2 - r_2^2} \tag{2.8.2}$$

(2) Flow in pipes of circular cross-section

The steady flow down a uniform circular pipe will produce a shearing stress, on a cylinder of fluid of radius r coaxial with the pipe (see Fig. 2.8.i), which is equal to the difference between the pressure forces on the two ends. Thus

$$-\pi r^2 l \frac{\partial p}{\partial x} = \mu 2\pi r l \frac{\partial v}{\partial r}$$

or (2.8.3)

$$v = v_0 - \frac{r^2}{4\mu}\frac{\partial p}{\partial x}$$

where v_0 is the velocity on the axis $r = 0$. The profile of velocity is therefore parabolic.

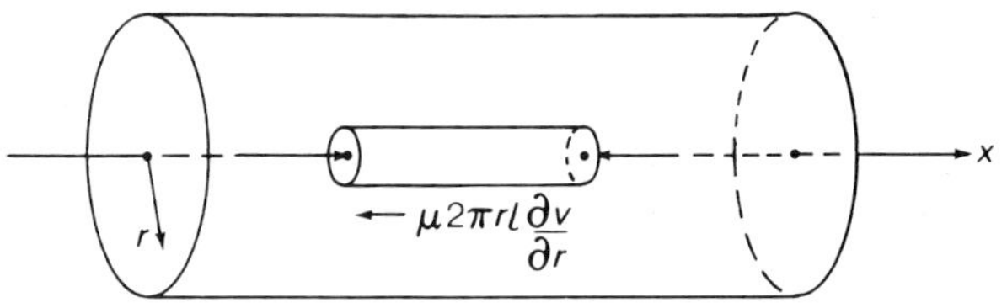

Fig. 2.8.i Viscous force on a cylinder of fluid in laminar viscous flow in a circular pipe.

(3) The Hele–Shaw cell

By similar argument the steady two-dimensional flow between two fixed parallel planes, $z = \text{constant}$, is also seen to be parabolic, with the formula

$$v = v_0 - \frac{z^2}{\mu}\frac{\partial p}{\partial x}. \tag{2.8.4}$$

In this latter case, if the flow is zero on the walls, the velocity in the centre of a gap of depth $2h$ between them is

$$v = -\frac{h^2}{\mu}\frac{\partial p}{\partial x} \tag{2.8.5}$$

and if the gap is narrow so that the viscous flow is very slow, the inertia forces represented by the acceleration along or perpendicular to the streamlines in the equation of motion can be neglected so that equation (2.8.5) may be taken to hold in the central plane, $z = h$, and indeed at every level between the planes in the form

$$v = -k\frac{\partial p}{\partial x}$$

or

$$v = \mathbf{grad}\,(-kp). \tag{2.8.6}$$

Since it has a velocity potential the flow is irrotational, and has the same pattern throughout the depth, although of magnitude varying across the depth. Since irrotational flow is unique with given boundaries, the irrotational flow around two-dimensional obstacles may be reproduced between parallel plates with the appropriate obstacles inserted in the gap. This apparatus is called a **Hele–Shaw cell**, and the streamlines may be made visible by introducing streamers of dye (see Fig. 2.8.ii).

(4) Three-dimensional cases

The same principle is embodied in **Darcy's law** for the flow through a rigid porous medium, namely

$$v = -\sigma\,\mathbf{grad}\,p, \tag{2.8.7}$$

which is valid in three dimensions. The flow is assumed to be governed by this law on the grounds that the flow is everywhere down the pressure gradient, proportional to the porosity σ, and slow enough for inertia forces to be neglected. This is the same as **Ohm's law** for the electric current in a conducting medium.

Taylor's electrolytic tank embodies the same idea. The equipotentials in a conducting liquid between two conducting plate electrodes, one at each end of a tank, may be mapped out using a probe on an insulated wire lead, and they would represent the potential flow (of a perfect fluid) around obstacles made of insulating material placed in the tank. Taylor's original application was intended to reveal the flow of air over humps on the ground by making models of the topography out of paraffin wax; but the analogy

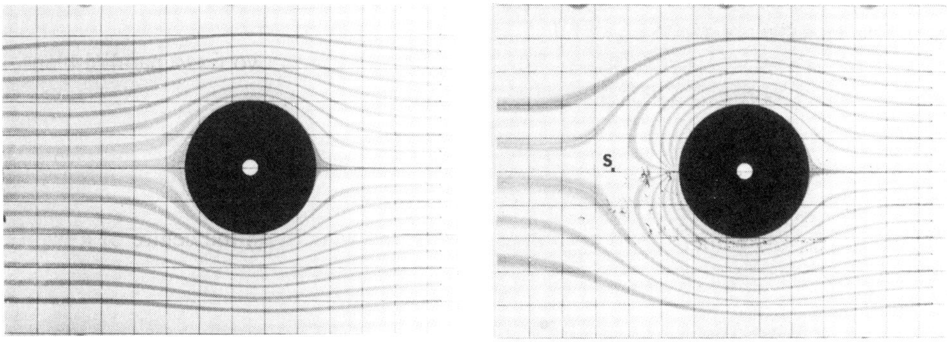

(a) (b)

Fig. 2.8.ii Flow in a Hele-Shaw cell with an obstacle in the narrow gap between two glass plates. The streamlines, marked by dye, show plane irrotational flow. (a) In the left picture the flow passes around a circular obstacle. The dye on the streamlines passing through the low-velocity region is spread sideways more than elsewhere. The flow can be seen to be from the right. (b) With suction applied at a sink close to the rear stagnation point of (a) taking 0.6 of the flow. Equally the pattern represents a flow from the left towards a source and a circular cylinder, or flow round a long obstacle delineated by the streamlines from the stagnation point, S. This is the situation envisaged in equations (2.6.1) and Fig. 3.7.iii).

in this case is poor because there is usually an increase of wind with height so that the real flow is not irrotational.

The condition at a rigid boundary of a fluid is that if the fluid is viscous, the velocity there is zero. The flow around small bubbles of air in water might be thought to follow the pattern of irrotational flow because the shearing stress in the liquid at an interface with a gas is negligible and the boundary condition for an inviscid fluid might be thought to be applicable. However, although the surface tension tends to make the surface spherical, impurities introduce resistance to motion on the surface. Because the pressure is greatest, according to (2.5.1), at the front and rear stagnation points on a sphere, the spherical shape due to surface tension is slightly flattened. See Fig. 2.13.iv(b).

One other important factor makes the flow of a viscous fluid near to a free surface different from that of an inviscid fluid, even though there is no shearing stress at the boundary and therefore no restraint on the tangential velocity. This is that the fluid elements are distorted in curved irrotational flow. The same flow in a viscous fluid would therefore have shearing stresses in it, and the flow must be different at the free boundary because of this. For example, the flow near to the surface of a path plug vortex in viscous fluid cannot be irrotational with velocity proportional to k/r because it would require a shearing stress to maintain it. Secondary flows are therefore produced by the reduced speed at the surface and there is a flow down the surface into the drain as well as close to the bottom of the bath (see Chapter 3).

The dissipation of energy by viscous forces in an incompressible fluid may be thought of in terms of the following equation:

(1) rate at which surface tractions do work + (2) rate of working of normal forces on the surface = (3) rate of increase of kinetic energy = (4) rate at which energy is dissipated internally.

In the case of steady irrotational flow, only the first and last two terms are not zero and the rate of dissipation is equal to the input by the surface tractions. Surface tractions are therefore necessary to maintain irrotational flow in a viscous fluid, particularly if there is any curvature because curvature distorts the shape of fluid elements.

If there are no surface tractions and no fixed boundaries the rate of dissipation is proportional to the square of the vorticity, and is equal to $\frac{1}{2}\mu\omega^2$ per unit volume, ω being the magnitude of **curl v**. This is the rate at which turbulent energy is dissipated (see Chapter 9).

2.9 BOUNDARY LAYERS

It was shown by equation (1.4.12) that each component of vorticity is separately conducted through the fluid by viscosity. If a boundary layer of a viscous fluid is suddenly set in tangential motion, the fluid at the surface is carried with it and vorticity is created in the fluid close to it, and is therefore gradually conducted into the main body of fluid.

The distance, δ, at which the fluid is moving subsequently with, say, 5% of the velocity at the boundary is determined by the kinematic viscosity ν (dimensions L^2T^{-1}) and the time t since the boundary was originally jerked into motion, and not by other factors. Therefore, to have the correct dimensions, δ, can only take the following mathematical form:

$$\delta \propto (\nu t)^{\frac{1}{2}}. \qquad (2.9.1)$$

Since the same argument applies to the distance at which any multiple of the velocity is achieved, the mathematical form of the velocity profile remains constant and its dimension expands in the same way as $t^{\frac{1}{2}}$.

By rough analogy, the boundary layer, which is the layer within distance δ of the boundary, grows on a flat plate which moves through the fluid, the time for growth being represented by distance x from the leading edge of the plate divided by the speed V of the plate through the fluid. Fig. 2.9.i shows how the velocity profile changes downstream on a plate fixed in a stream. If we think of a column of fluid ABCD moving downstream with the layer growing in it, $t = x/V$, and so

$$\delta \propto x^{\frac{1}{2}}\nu^{\frac{1}{2}}U^{-\frac{1}{2}}. \qquad (2.9.2)$$

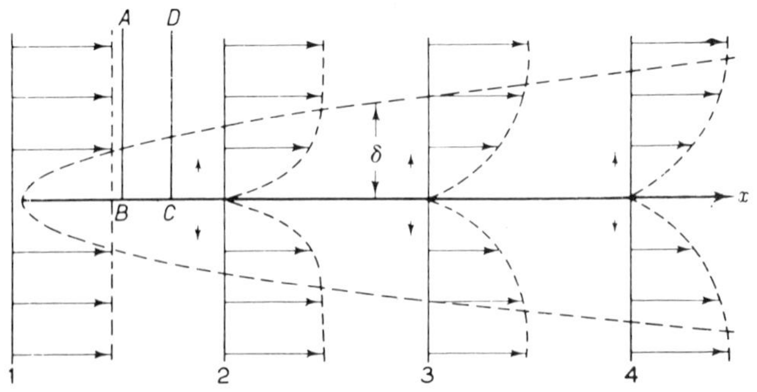

Fig. 2.9.i Change in velocity profile in a stream flowing across a flat plate.

At the trailing edge of the plate the vorticity continues to diffuse outwards, but the vorticity of opposite signs on the two sides now begins to cancel out and the curvature of the velocity profile is reversed. In Fig. 2.9.ii the shaded area represents the deficit of momentum flux suffered by the stream, which is equal to the drag force on the plate.

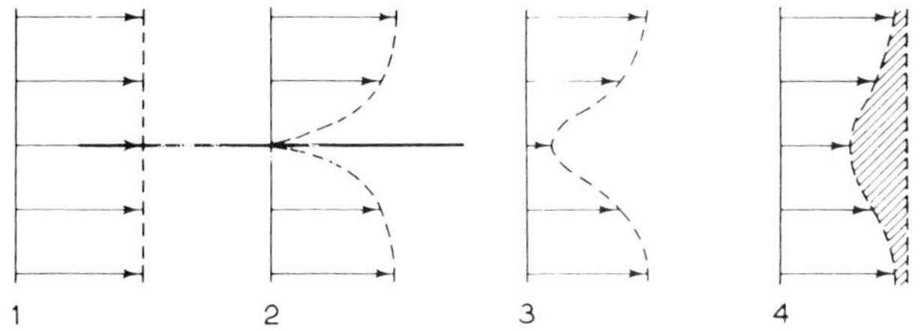

Fig. 2.9.ii Spread of the velocity deficit behind a flat plate.

Although the shaded area remains constant downstream it is spread over an everwidening wake. In practice the flow in the wake is not steady unless the **Reynolds number** (see section 2.14) is small.

Boundary layers grow in the downstream direction on all solid surfaces in viscous fluids unless the fluid is accelerated. In Fig. 2.9.i we see that the volume flux in the boundary layer is less than in the same depth of fluid at a point upstream. Therefore continuity of flow requires that there shall be a small velocity away from the surface. This means that the flat plate acts, as far as the fluid outside the boundary layer is concerned, like a body whose thickness increases downstream. If the fluid is being accelerated by motion down a pressure gradient, for example into a converging channel or over the front face of an obstacle where the flow moves away from the forward stagnation point which divides the flow, the thickness of the boundary layer may be reduced. On the other hand, in a diverging channel where the fluid is moving towards higher pressure, the boundary layer is thickened by the deceleration, the shear stress is reduced and the fluid is decelerated more quickly than in a uniform channel.

The pressure at a point downstream may be equal at a certain point to the stagnation pressure of the slow-moving fluid close to the wall, and in that case the flow is reversed beyond that point (Fig. 2.9.iii) and separation takes place. The separation occurs where the forward viscous drag due to the layers away from the boundary is less than the

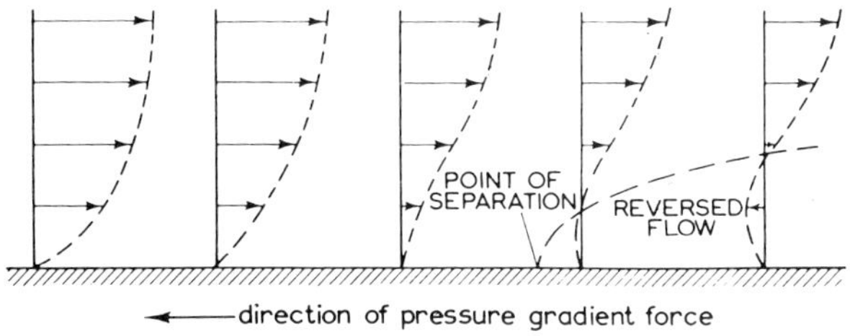

Fig. 2.9.iii Separation and reversal of boundary layer flow in an adverse pressure gradient (e.g. in diverging flow).

backward force due to the reversed pressure gradient. There is an assumption made in boundary layer theory that, because the acceleration perpendicular to the wall is very small, there is negligible pressure gradient in that direction. The isobars are therefore assumed to extend into the boundary layer from the main flow beyond (see Fig. 2.9.iv).

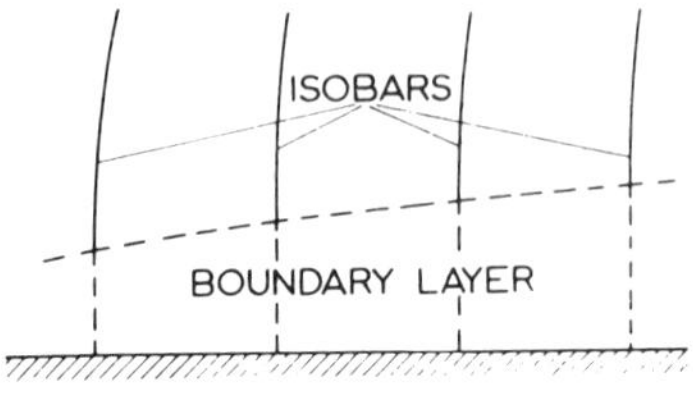

Fig. 2.9.iv External pressure gradient projected through a boundary layer.

The separation of the flow in a diverging channel such as is shown in the left diagram of Fig. 2.9.v can be prevented by causing the surface to move in the direction of the flow at the point of separation. Thus a roller might be installed where the curvature is large (Fig. 2.9.v, right).

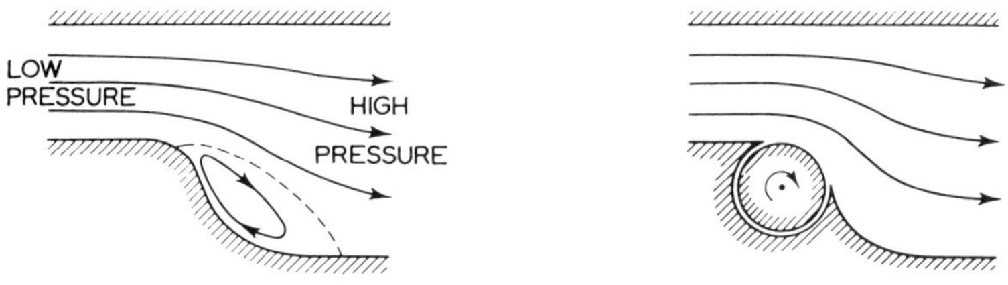

Fig. 2.9.v Left: separation in a diverging channel. Right: separation prevented by motion of the boundary near the point of separation.

Another method of preventing boundary layer separation is to introduce a jet of air in the direction of motion where it would otherwise separate. This is called the **Coanda effect**, and is illustrated for the case of a jet flap on a wing in Fig. 2.9.vi. The flow becomes turbulent when a wing is stalled by separation of the flow from its upper surface. The lift is greatly reduced because the region of low pressure in the high-speed flow over the upper surface is lost.

The Coanda effect can be illustrated by blowing a jet of air over a corner of a solid object. If the jet is made to impinge on the wall close to the corner, the boundary layer formed is too shallow at the corner to cause separation. If it is lifted slightly, so that slower-moving fluid lies between the velocity maximum and the boundary, separation may occur (Fig. 2.9.vii).

This last phenomenon is also illustrated if the jet is of water. The pressure on the outer surface of the thin layer is uniform, being equal to the atmospheric pressure, and so any adverse pressure gradient is reduced in magnitude at the solid surface. If an upturned pan is held under a smooth-flowing kitchen tap, the water flows over the corner at the edge of

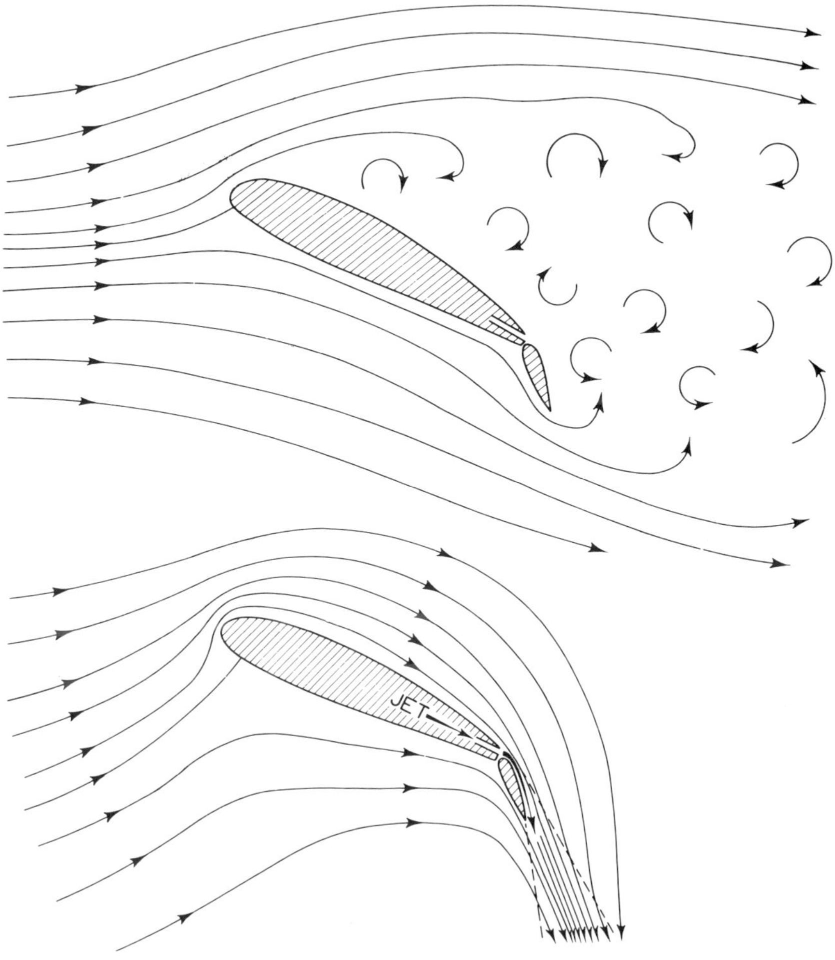

Fig. 2.9.vi The Coanda effect, or jet flap, in which separation from the surface of a wing and stalling are prevented by blowing air over the upper surface of a flap.

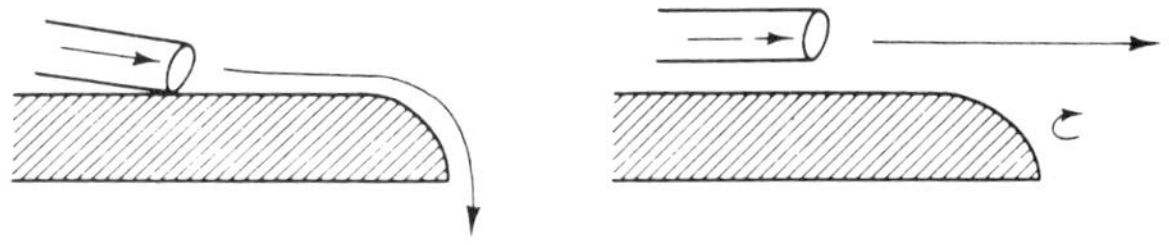

Fig. 2.9.vii A jet blown over an edge close to the surface will adhere to the corner but will separate if slower-moving fluid lies between it and the rigid surface.

the base and down the sides as if it were stuck to them, instead of escaping horizontally as might be expected at a sharp corner.

The velocity close to the boundary may be increased on the upper surface of a wing by placing vortex generators upstream of the separation point. These are small aerofoils whose tip vortices bring faster-moving air from outside the boundary layer close to the surface so as to prevent separation there (Fig. 2.9.viii).

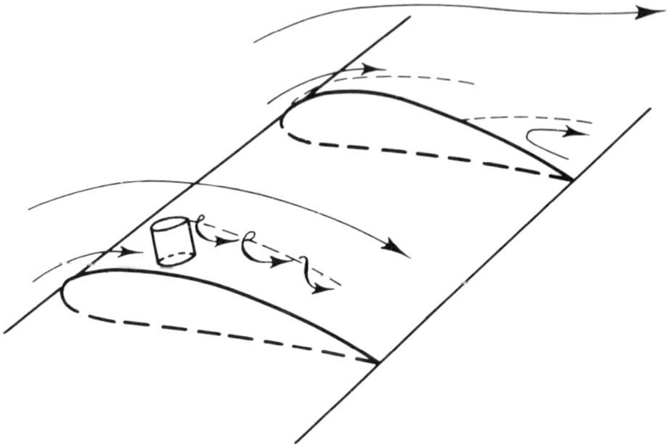

Fig. 2.9.viii Vortex generators prevent separation by bringing faster-moving fluid from outside the boundary layer close to the surface.

Another device to avoid separation is the **slotted wing**, which is a slot carrying air from the lower surface where the pressure is higher into the boundary layer on the upper surface directed so as to act like the Coanda jet. This is also used on the leading edge of a wing on approach to landing when the airspeed is reduced and the angle of attack of the wing is increased (see Fig. 2.9.xi). (See also the slotted wingtips of soaring birds of prey in Chapter 15).

If the air on a sloping hillside is warmed by sunshine heating the ground, the buoyant air flows up the slope. This causes separation of the flow coming down the slope (Fig. 2.9.ix). On the other hand, if the ground cools at night, the air cooled by proximity to it is drawn by gravity down the slope and separation is stopped. (See also section 5.18.)

Another technique for preventing separation is to remove the slow-moving fluid by suction through a slot or porous surface, but this may require an additional source of power to make it effective.

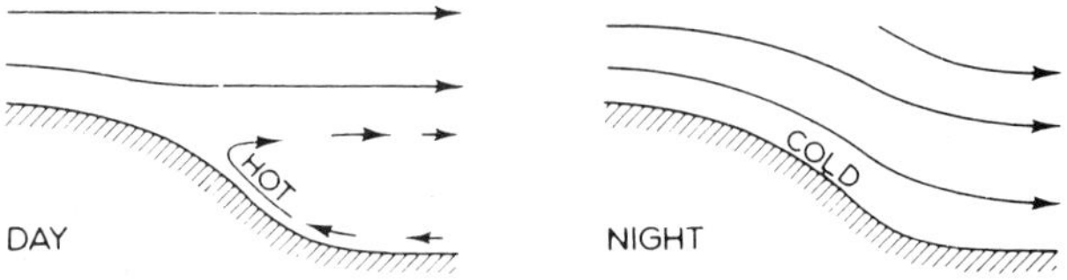

Fig. 2.9.ix The downslope flow of air on a cooled surface can prevent the separation of flow from a hillside and alter the main flow above the cooled layer.

As we shall see in Chapter 9, turbulence can sometimes be very effective in spreading the momentum more uniformly across an airstream and in this respect it acts like a greatly increased viscosity, and quickly decreases the thickness of the boundary layer (Fig. 2.9.x).

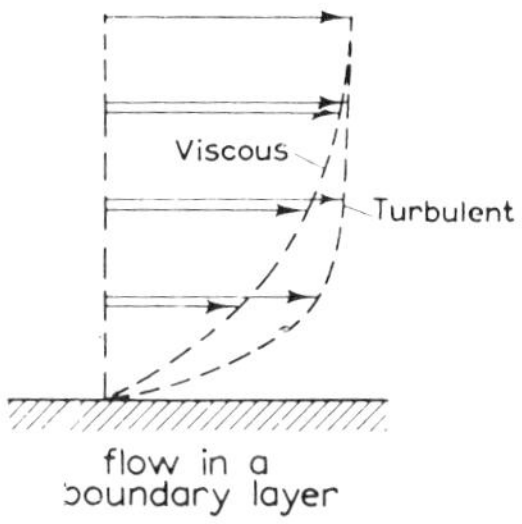

Fig. 2.9.x The presence of turbulence reduces the thickness of the boundary layer by introducing shear stresses in the main flow which are absent close to the boundary where eddies are reduced in size.

The objective is to create turbulence only in the places where it produces a desired effect as it does in the wakes of the vortex generators described above. A small sharp-edged ridge, a sort of trip-wire, perhaps about 1 cm in height on an aircraft wing, may cause the boundary layer to be reduced in thickness by turbulence on a wing so as to avoid separation (Fig. 2.9.xi). It is known that the velocity profile of the flow in tubes, which we have already described as being parabolic when it is laminar, is flattened when the flow becomes turbulent and makes the shear layer on the wall thinner. But this also increases the viscous drag of getting through the pipe because the shear at the wall is made much greater; and this may be a greater disadvantage.

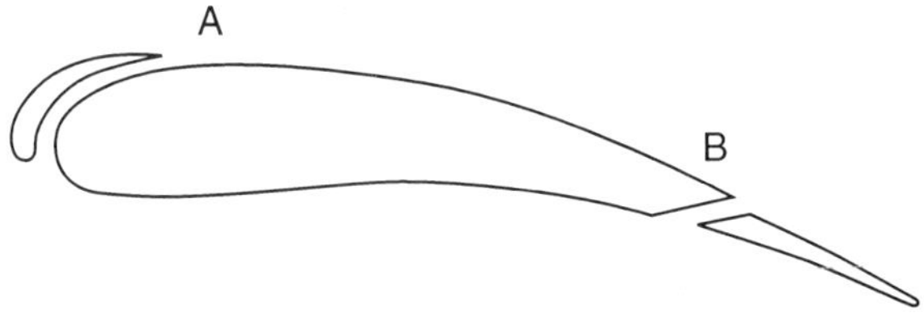

Fig. 2.9.xi Slot on leading edge and on trailing area of wing to prevent separation at A and B.

The famous Hurricane aircraft was renowned for its aerial versatility, which was due to the effectiveness of its control surfaces. As it happened there was a sharp-edged ridge around the rear of the propeller spinner which was, according to my informant, made to prevent oil from the propeller gears flowing back over the fuselage and windshield. It could have given the additional advantage of creating turbulence where the airflow would have separated from the fuselage during tight manoeuvres, and improved the effectiveness of the control surfaces particularly on the tailplane and rudder (Fig. 2.9.xii).

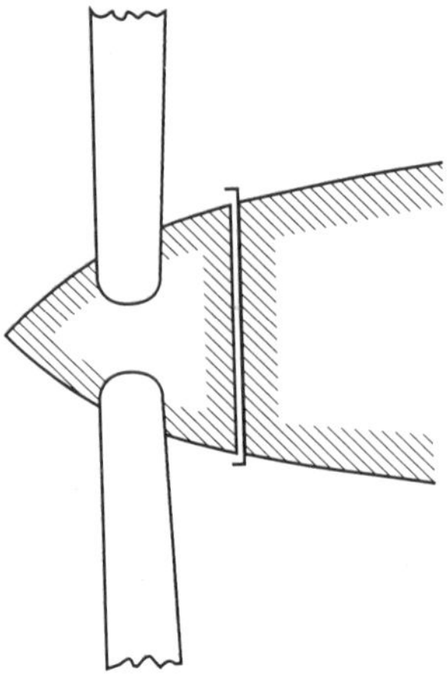

Fig. 2.9.xii Section of propeller spinner on Hurricane aircraft.

Although turbulence increases the effective flux of momentum across the streamlines, the eddies are reduced in size close to the wall and there may be a **laminar sub-layer** which acts as a lubricant; but because the shear stress further from the wall is increased by the turbulence, the pressure gradient down the pipe required to drive the flow is still greater when the flow is turbulent than when it is steady and laminar.

In the bend of a channel, separation is less likely to occur on the inside wall because the pressure there is reduced by the centrifugal forces required by the curved flow (see section 3.2). On the outside of the bend there is a higher-pressure region and the boundary layer increases as it approaches this region. Fig. 2.9.xiii shows how this can be avoided by using a **cascade of aerofoils**, and no large-scale eddies are formed when there is a large number of small aerofoils; but the relative positions of a line of particles across

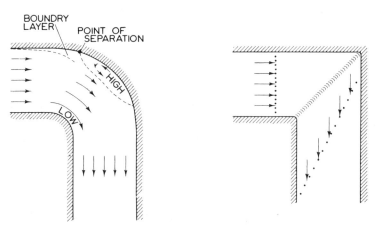

Fig. 2.9.xiii The separation in a channel bend occurs on the approach to the high-pressure region on the outside. It can be prevented by introducing a 'cascade of aerofoils' into a square corner.

the flow are changed by the corner. The problems caused by the use of this device to carry the air leaving from the working section of a wind tunnel back to the entry end were studied by Squire, who did not wish to ignore the centrifugal forces (see section 3.3).

2.10 REYNOLDS NUMBER

The Reynolds number is non-dimensional, and is the ratio of the **inertia forces** $\left(\text{mass} \times \text{acceleration}, \rho \times \frac{\mathrm{D}\mathbf{v}}{\mathrm{D}t}\right)$ to the **viscous forces** (viscosity × shear, $\mu \times \nabla^2 \mathbf{v}$). Each of these quantities is composed of a speed V, a length of d, and a kinematic viscosity, ν. These are composed from the factors which determine the flow; and the density appears in both mass and viscosity, and so when we use the kinematic viscosity, ν, we find the only non-dimensional number which can be composed from them is:

$$\mathrm{Re} = \frac{\rho U \div [d \div U]}{\mu d^{-1} . d^{-1} U} = \frac{Ud}{\nu} \tag{2.10.1}$$

where U and d are a representative speed and length appropriate to the situation. Alternatively, Re may be regarded as the ratio of a pressure force, represented by $\frac{1}{2}\rho U^2$ through Bernoulli's equation, to a viscous force per unit area represented by viscosity × velocity gradient (i.e. $\mu U/d$).

At small Reynolds numbers the inertia forces become negligible; at large values the viscous forces are negligible, but it should be noted that this is usually because the flow has become turbulent and inertia forces in the form of eddies have become dominant (see Chapter 9). After turbulence has set in, the flow pattern is not much altered by further increases in Re because the velocities in the eddies are proportional to the mean flow which created them, and so the ratio of their inertia forces to those of the mean flow is not altered by an increase in speed, until it approaches the speed of sound.

At small Reynolds number the flow around a small particle is determined by the viscosity, μ, and the particle dimension, a. The drag force, D, on it when its velocity is U must have the mathematical form

$$D \propto \mu a U \tag{2.10.2}$$

because no other combination of μ, a and U generates a quantity with the dimensions of force. The constant of proportionality was first shown by Stokes to be 6π, and is applicable to a solid sphere moving slowly through a liquid. If it moves under gravity and the densities of the sphere and the liquid differ by $\Delta\rho$, the terminal velocity is given by

$$6\pi\mu a U = \tfrac{4}{3}\pi^3 g \Delta\rho. \tag{2.10.3}$$

For reasons given previously (towards the end of section 2.8) this formula is not necessarily valid for liquid spheres or bubbles of gas because they allow slip velocities at their surface. However, bubbles very often do obey this equation because of impurities which inhibit motion along the surface.

It has just been remarked that the formula (2.10.3) with the constant of proportionality equal to 6 was valid according to Stokes's analysis, only when the inertia terms in equation (1.3.2), such as $\mathbf{v}.\mathbf{grad}\,\mathbf{v}$, were small compared with the viscous terms like $\nu\nabla^2\mathbf{v}$, for by this assumption the equation was made linear in $\mathbf{v}$. This implies a very small Reynolds number indeed, indeed small compared with unity! This may be valid for ball bearings sinking slowly through very thick oil, but is scarcely valid for raindrops, although it has to be admitted that the formula may still actually not be misleading. The reason is that the flow in less viscous fluids may be unstable with separation of the flow from the surface or at least unsteady motions may occur in the wake, which make the calculation of the energy dissipated by viscosity according to Stokes's analysis, which is for laminar flow, to be in considerable error as a basis for calculating the drag. Thus it could be that the assumption that the inertia terms are negligible is correct but that the dissipation is actually greater than is calculated from the assumption that the flow is steady. At this point the discussion must be postponed until the subject of turbulent wakes has been considered. The transition to turbulence in the wakes of spheres has only been considered in the case of spheres much larger than the small particles in the air because of the difficulty of making measurements in the latter case.

The difference between high and low Reynolds number drag is well illustrated by a parachute and a dandelion seed (Fig. 2.10.i). A parachute has a large Reynolds number and is a device for creating intense eddying motion in its wake as it descends through the air. The eddies are generated as the flow separates from the perimeter and from the edge of the hole in its centre, and the drag force, produced by the higher pressure on its lower side which is advancing through the air, may be thought of as the force required to

Fig. 2.10.i This picture of dandelion seeds before being blown off the stalk of the flower shows their hairy nature. The individual fibres have a very low Reynolds number because of their small dimension and collectively cause a viscous force to produce a very slow sinking through the air, and thereby maximise their transport in the wind.

produce the eddying motion. All air brakes on aircraft work on this principle: they are called spoilers, because they spoil the laminar airflow over the wing surface and generate eddying energy.

The dandelion seed, by contrast, illustrates the principle that the flow is dominated by viscosity at low Reynolds number. The very small branches present a very large surface over which the air must flow and exert the viscous drag due to the shearing stress at the surface. Eddies are not left behind, but fluid is dragged downwards with the seed. Because the seed sinks through the air the drag force is always being applied to new air, and so it leaves behind it a steadily lengthening trial of downward moving air. This gradually communicates its downward momentum to a widening volume by viscosity. The energy is dissipated as heat.

2.11 PATHS OF SMALL PARTICLES: EFFICIENCY OF CATCH

When a small particle is 'embedded in' a fluid, which means that its fall speed (terminal velocity) is small compared with the velocity of the stream past some larger obstacle, its acceleration is approximately that of the particles of the fluid. Thus soap bubbles do not collide with a house because they are carried along very approximately by the fluid in which they are embedded. However, they may well collide with a tree because the curvature of the trajectories of the air particles is greater than that in the flow round a house. This causes the bubble, which is heavier than the air it displaces, to travel in a more curved path than the air. Very tiny bubbles have smaller terminal velocities than big ones, and may therefore not collide with the tree trunk but only with small branches as the wind caries them along. A soap bubble itself has a good efficiency of catch of obstacles of smaller dimension such as twigs of a tree or blades of grass.

Smoke particles, which have fall speeds of the order of less than one millimetre a second, collide with fine fibres and are therefore filtered out of the air by a cloth, but are not captured on buildings. Indeed, the blackening of buildings is very much affected by the small eddies on sharp corners and protuberances. These produce flows with large curvature which can cause smoke particles to collide with the surface. The broadscale flow around the building does not cause particles to collide with it unless they have a large fall speed: thus heavy raindrops hit the window panes but small drizzle and fog droplets do not. On the other hand, drizzle and fog are very effectively captured on tiny branches and especially on the fine needles of pine trees. Because of this, the rain water falling to thc ground is greater under pine trees, where they are at a high enough altitude frequently to be in cloud, than it is in the open in the same neighbourhood. Contrary to popular belief, trees do not attract rain and they certainly increase the evaporation to the air of water drawn from some considerable depth in the soil: however, in some areas their contribution to ground water through the capture of cloud particles may be very important. They may also cause soil to become more spongy and hold back the moisture from draining out of the land.

The flow on the upstream side of an obstacle is very approximately potential flow. The paths of particles are indicated in Fig. 2.11.ii. Because they have an effectively negative mass, bubbles of gas in a liquid cannot collide with an obstacle because their track is always more curved than the liquid particle trajectories.

Fig. 2.11.i(a) Condensation direct from vapour to ice: crystals growing on a twig at temperatures below the frost point. (Photo M. Koldovsky.)

Fig. 2.11.i(b) Condensation onto twigs when there is a prevailing drift of the supersaturated air from the right. (Photo M. Koldovsky.)

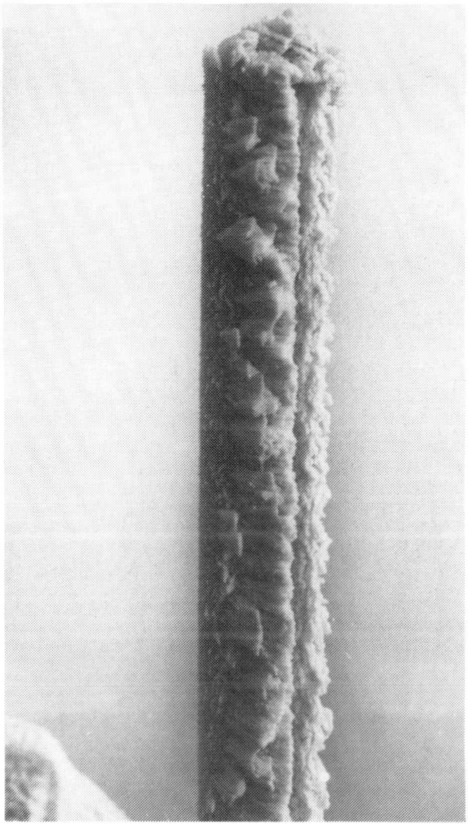

Fig. 2.11.i(c) The condensation and freezing and deposition of supercoded fog (or cloud) droplets on a post with a prevailing drift of air from the right. The droplets freeze onto the point of impact. (Photo M. Koldovsky.)

If the particle to be captured has a significant size it may be captured even if its centre is not on a collision path. Small cloud or drizzle droplets may be caught in the wake of a raindrop even though they do not collide with the advancing lower surface, because the airflow in the wake is unsteady (see Chapter 10) and may frequently bring cloud particles close to the upper surface of the drop.

The **efficiency of catch** of a body moving through a cloud of particles is defined as the proportion of the particles lying in the cylinder swept out by the geometrical outline of the body which actualy collide with and are captured by that body.

The efficiency of catch of bubbles of gas of different sizes in water is usually very small, but if the body were a wire of smaller dimension than the bubbles it might capture some. The efficiency of catch of a raindrop falling through a cloud depends very much on its size relative to the cloud droplets. If they are very small it is low, but for almost drizzle size it may approach unity. In those circumstances where there is a considerable capture through the wake of the raindrop, it could exceed unity.

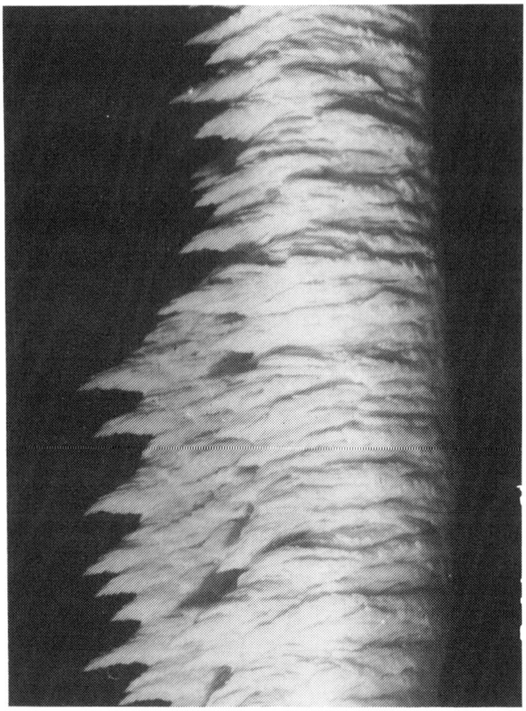

Fig. 2.11.i(d) Deposition of supercooled droplets from cloud onto a steel wire. The droplets, being liquid, spread over the surface of the deposit before becoming ice. This depends on the temperature being only 2° or 3°C below freezing. (Photo M. Koldovsky.)

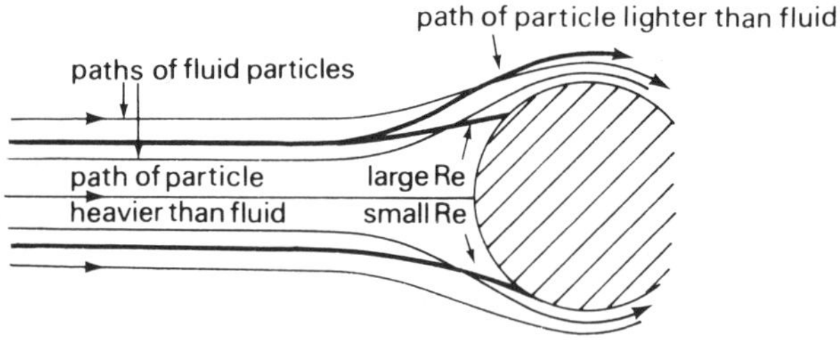

Fig. 2.11.ii Paths of particles approaching an obstacle. Particles lighter than the fluid always avoid the obstacle, but heavier particles may collide with it.

The efficiency of collision is not always the same as the efficiency of catch and retention, because it depends on the mechanism of attachment. If a hailstone falls through a cloud of supercooled cloud droplets they stick to it if they freeze on impact. If they do not freeze on impact they may nevertheless wet the surface and be caught that way. On the other hand if they are too plentiful the latent heat of fusion may be enough to raise the temperature of the whole surface of the hailstone to 0°C, and the unsteady motion may cause splashing of water from the surface. Likewise, when a raindrop reaches about 5 mm in diameter the surface tension is no longer sufficient to hold the surface spherical in the presence of the aerodynamic forces, and the raindrop will usually break up if it exceeds this size.

In the early days of the spraying of crops with pesticide, the particles were released from an aircraft and the downward motion of the air in the aircraft downwash together with the gravitational settling of them under their own weight in still air were used to deposit the particles on the crop. Thus large particles were more effectively caught on the crop, but mainly on the upper surfaces only. More recently, smaller spray particles have been used and the spray released in a cross-wind: the particles are then captured on the crop as they are carried through it by the wind. This technique is easier to employ because it is not necessary to wait for the air to be calm. It is much less wasteful of pesticide because the large particles used to ensure settling often represent an overdose.

The most efficient way to dose a pest which has an airborne phase is to use a fine spray which sinks very slowly through the air but is composed of particles large enough to be caught on the insect as it flies through the spray cloud (see section 14.7).

The shapes of obstacles enlarged by capture of material from a fluid stream can vary enormously. Thus a hailstone may be spherical, although often it has a conical face, which indicates that it has achieved a preferential orientation as it falls. When super-cooled droplets are deposited by freezing onto wires, the growth may be Y shaped, perhaps because sideways oscillation of the wire in the wind produces a stagnation point which oscillates across the front of the wire causing two points of maximum collection. The growth is uneven because a slight protuberance grows more rapidly.

In light winds the growth of ice on wires and twigs may be more by condensation (collision of molecules) than by ordinary collision of particles. The rate of sublimation depends on the gradient of concentration of the deposited material, and this will be greater at the tips of small protuberances than on parts of the surface with a larger geometrical dimension. Thus hoar frost often consists of spiky growths, and ice crystals falling through air which is supersaturated with water vapour below 0°C grow with dendritic shapes (Fig. 2.11.iii) at the tips of which the boundary layer is thinner and therefore the concentration gradient greater. (See next page.)

An interesting aspect of falling ice crystal growth is its extraordinary symmetry. As soon as one dendritic (branch) grows larger than the others it tends to be dragged back by viscosity to the rear of the crystal, exposing other dendrites at the front to thinner boundary layers and greater concentration gradients. The growth is also determined by the crystal structure, which imposes hexagonal symmetry.

2.12 DRAG COEFFICIENT: C_D

At large Reynolds number the drag force on a body in a stream U is caused by the creation of kinetic energy in eddies whose velocities are proportional to the stream speed and whose size is determined by the geometry of the body. The drag D is then given by

$$D = C_D \tfrac{1}{2} \rho U^2 d^2, \tag{2.12.1}$$

Fig. 2.11.iii Dendritic shapes of ice crystals growing in supersaturated air as they fall. The growth on a protuberance causes the orientation of the falling crystal to expose the opposite point preferentially forward so that the growth is symmetrical. The plates fall down a helical path, rotating as they fall.

where d^2 here represents the frontal area of the body. The Mach number M is U/c where c is the velocity of sound. As U approaches c, the flow over a body is changed because some parts of it become supersonic and the pressure distribution is completely altered by the appearance of shock waves. The pattern of separation is also changed, and a large part of the drag when $U > c$, and $M > 1$ is due to the production of shock waves which travel out as a cone through the air like the wave pattern of a ship on water. The drag is then proportional to a higher power of U, or to put it another way, C_D is decreased as M is increased through unity.

At low Re, i.e. in the Stokes regime, the drag is proportional to U (see, for example, equation (2.10.2)). Thus according to the definition of C_D in (2.12.1), C_D must be increased at low Reynolds numbers. Table 2.12.i indicates the drag coefficient of spheres at different Re, and Table 2.12.ii indicates more specifically the fall speed of spheres with the density of water falling through air. Liquid spheres disintegrate when the diameter exceeds about 5 mm. For most common objects of laboratory size, except those of deliberate aerodynamic design such as propeller blades and aircraft wings, the relationship (2.12.1) holds over the range of velocities commonly experienced, and C_D is in the neighbourhood of unity for $50 < \mathrm{Re} < 10^5$. However, the upper limit varies according to the size of the body because that determines the value of Re at the corresponding Mach number.

It needs to be emphasised that at large Re and small M the degree of turbulence in the fluid stream and the nature of the solid surface upstream of the point of separation can significantly alter the thickness of the boundary layer and also alter the point of separation, as described in section 2.9. When the separation point is moved back, the wake size is considerably reduced and so not only the drag coefficient but also the drag itself may

Table 2.12.i. Drag coefficient of spheres

Re	C_D
0.002	10 000
0.005	5 000
0.01	2 400
0.05	480
0.1	235
0.4	64
0.8	34
1	27
2	15
5	7
10	4
20	2.6
50	1.6
100	1.1

$$\mathrm{Re} = \frac{2aw_t}{\nu}$$

$$a = \text{radius}$$

$$w_t = \text{fall speed}$$

$$\nu_a = \mu_a/\rho_a \ \text{(for air)} = 0.15\ \mathrm{cm^2\ s^{-1}} \text{ at } 20°\mathrm{C}$$

$$\nu_w = 0.015\ \mathrm{cm^2\ s^{-1}} \text{ (for water) at } 5°\mathrm{C}$$

$$C_D \tfrac{1}{2}\rho_a w_t^2 \pi a^2 = \tfrac{4}{3}\pi a^3 g(\rho_w - \rho_a)$$

$$\rho_{a,w} = \text{density (air, water)}$$

be reduced by increasing the velocity (and Re). In the great outdoors almost any situation may arise because the large Reynolds numbers are common at low wind speeds.

2.13 SEPARATION AND IRREVERSIBILITY

All viscous flow is irreversible because it converts work into heat by the action of the viscous stresses. On the other hand, most of the energy which makes motion irreversible in practice is generated as eddies. Although the eddies are converted into heat eventually by viscosity, it is the eddy-making mechanism which is the primary cause of the irreversibility, and eddies are most often the result of separation of the flow in cases when buoyancy forces are not important.

Table 2.12.ii. Fall speed of water drops in air

Diameter. mm		Fall speed m s^{-1}
0.001		0.0003
0.01		0.03
0.1		0.27
0.2	Cloud	0.72
0.3		1.2
0.4		1.6
0.5		2.1
0.6		2.5
0.7		2.9
0.8		3.3
0.9	Drizzle	3.7
1.0		4.0
1.2		4.6
1.4		5.2
1.6		5.7
1.8		6.1
2.0	Rain	6.5
2.2		6.9
2.4		7.3
2.6		7.6
2.8		7.8
3.0		8.1
3.2	Begins break-up	8.3
3.6		8.6
4.0		8.8
4.6		9.0
5.2		9.1
5.8	Break-up certain	9.2

The two most important examples of eddy-making are wakes and jets. The wake of an obstacle in a fluid stream is initially bounded by the surface of separation, which streams away from the line of separation of a body. In the case of a sphere (Fig. 2.13.i) the separation begins not far from the rear stagnation point S′, but it moves forwards as the flow develops, and for large Reynolds number settles down along a line at about 80° from the forward stagnation point S. In front of this line the flow outside the boundary layer is almost indistinguishable from the calculated potential flow. Behind this line, eddies are unsteady and grow in size when they are shed from the body and are carried downstream. The two-dimensional flow behind a cylinder usually takes the form of a **vortex street** (Fig. 2.13.ii, often named after **Von Karman**) in which eddies of opposite rotation are

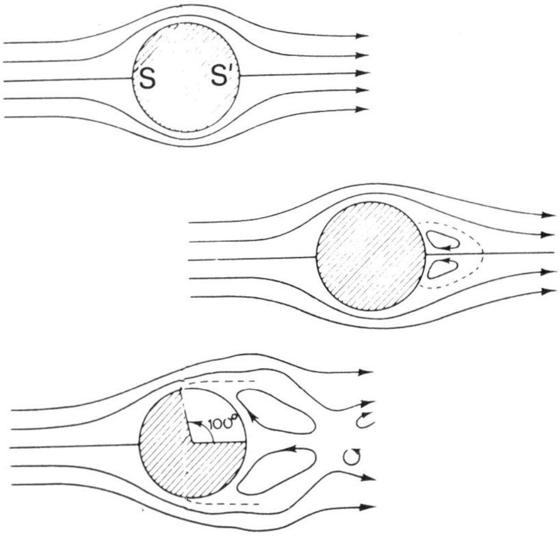

Fig. 2.13.i Initially (or at very low Re) a sphere has no wake, but after a short time the motion separates near the rear stagnation point and the separation point moves forward until at large speeds it is at about 100° from it. The wake is then very turbulent. The point of separation may be moved back from the 100° position by turbulence in the airstream or roughness on the sphere causing turbulence in the boundary layer.

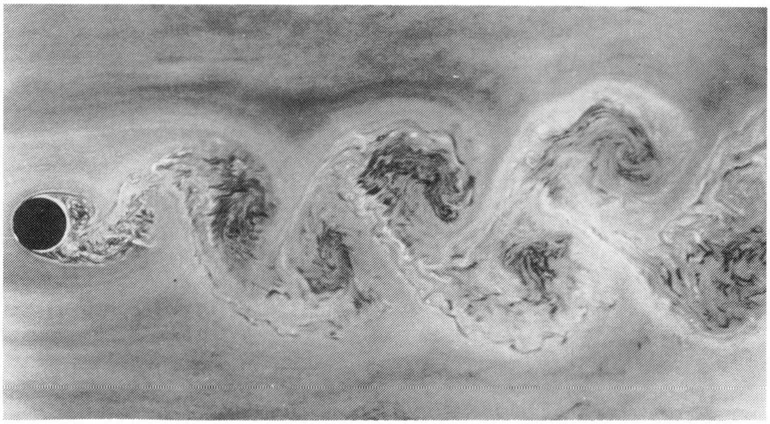

Fig. 2.13.ii Photo of the wake of a circular cylinder made visible by particles floating on the surface of a water channel. In this case the 'particles' were tiny bubbles of froth generated by introducing some detergent and allowing a small amount of air to be sucked into the recirculating pump.

shed alternately: behind a sphere the flow is more chaotic but it sometimes takes the form of a vortex with a helical core.

Behind a cylinder of almost any shape the flow tends to be in the form of a vortex sheet if the flow settles down. It is then periodic, and this causes wires to oscillate in the wind. The frequency, σ, of oscillation (often called the **Strouhal** frequency) is found to be related to the velocity U of the stream and the obstacle diameter d by

$$\sigma = S\ U/d, \tag{2.13.1}$$

S being the **Strouhal number**, which is equal to about 0.2 for circular cylinders.

Many experiments performed in wind tunnels at large Reynolds number produce results which are characteristic of the tunnel. This is because of their individual levels of turbulence and fan frequencies, which affect the separation properties of objects, and this particularly applies to artificially smooth laboratory objects such as spheres and circular cylinders. Ordinary objects found in nature usually have rough surfaces and protuberances and are subjected to winds with a considerable **level of turbulence** (e.g. 0.2, which means that the fluctuating component of the velocity is around a fifth of the wind speed). It should also be remembered that some authors include a factor 2π in the definition of the Strouhal number (2.13.1), while others define it as the reciprocal of the number given there. The island of Jan Mayen (Fig. 2.13.iii) has a Strouhal number of the order of 0.2, even though the Reynolds number is around 10^{10}, and has an irregular shape.

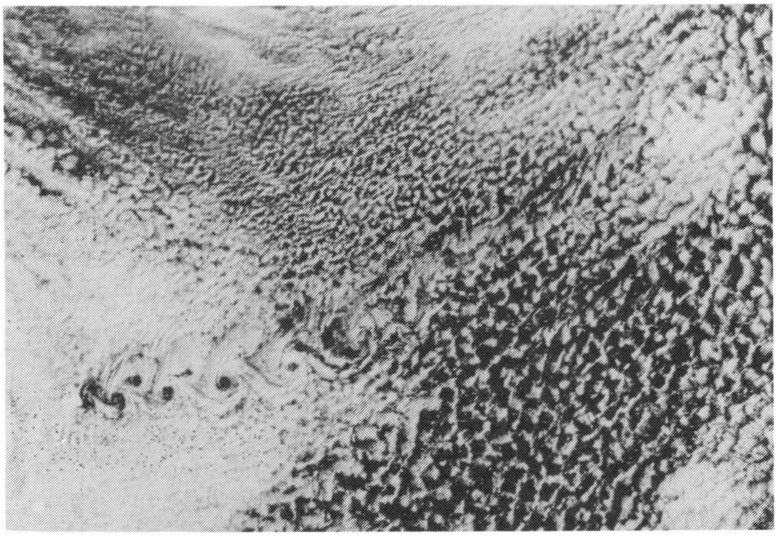

Fig. 2.13.iii A satellite (NOAA 4) photograph of the vortex street in the westerly airflow downstream of Jan Mayen, which has two peaks, 2277 m 10 km wide and 769 m 40 km to the SW. The pattern is made visible by the clouds at 1500–2000 m, which lie beneath a stable layer which causes the flow to be nearly two-dimensional when the mountains on the island reach higher than the clouds. The cyclonic vortices are stronger. [Photograph by courtesy of University of Dundee.]

Most important in practice as far as separation is concerned are salient edges. These are edges from which free streamlines may be generated, and the vortex sheets so produced roll up into eddies which are the beginning of the vortex street formation. Although vortex streets are considered theoretically as if they were composed of two parallel rows of alternate vortices, in practice the vortex spacing usually grows in size as they proceed downstream, and only after about 10 more have been shed does the spacing between neighbours become nearly constant. One special kind of salient edge is the strake, which is a helical protuberance often seen on chimneys. This ensures that the vortices are not shed at the same moment along the whole length of the chimney, and thereby avoids the oscillating force due to the shedding causing the chimney as a whole to vibrate.

The vortex motion produces a flow towards the obstacle in between the two lines of vortices. This means that the vortices do not travel as fast as the windspeed; but soon their rotation decreases, and by the time about five pairs have been shed they are travelling as fast as the wind. It can usually be seen that their spacing increases until that speed has been attained. They have then become **fossil vortices** with the pattern retained for a further few spacings, in the clouds. Vigorous convection will soon disperse the pattern.

Typical **salient edges** from which separation takes place are the corner of a building, the apex of a roof, the top of a cliff, or wall on a hill. Corners on vehicles have the same effect. They ensure that the flow behind the obstacle is turbulent and produces a drag coefficient of the order of unity. Wakes and separation lines are often made visible by smoke, as illustrated in Fig. 2.13.iv.

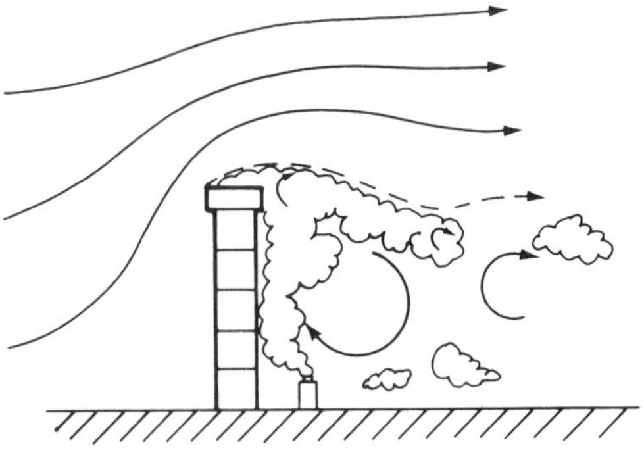

Fig. 2.13.iv The separation on a wall often occurs at the upwind edge of the top. Smoke from a generator behind the wall is carried forward to this line and then is either recirculated in a standing eddy behind the wall or carried away intermittently with moving eddies.

Jets are produced when fluid emerges from a hole in a wall or from a nozzle. The fluid on the high-pressure size of a hole approaches it very much according to a classical theory in which $\mathbf{v} = \mathbf{grad}\,\phi$ and $\nabla^2\phi = 0$. On the low-pressure side (Fig. 2.13.v) the flow is quite different and consists of a turbulent jet (see sections 9.12 and 10.9). There is often a naivety in the explanation given of the mechanism of entrainment manifested in turbulent jets: it is said in many school textbooks that the ambient fluid flows into the jet because, according to Bernoulli's equation, the large velocity in it is accompanied by a low pressure. This is misleading because in order to set up the jet a large excess of pressure has to be set up on the other side of the wall. When the instability on the boundary of the jet mixes some of the ambient fluid into it and carries the fluid with it, by

(a) (b)

Fig. 2.13.v (a) The release of a pilot balloon at the Col des Aravis. (b) After the release the pressure distribution on the baloon surface flattens it. The balloon expands greatly as it rises, and is therefore very flabby at low altitudes and changes shape easily in the pressure field of the airflow.

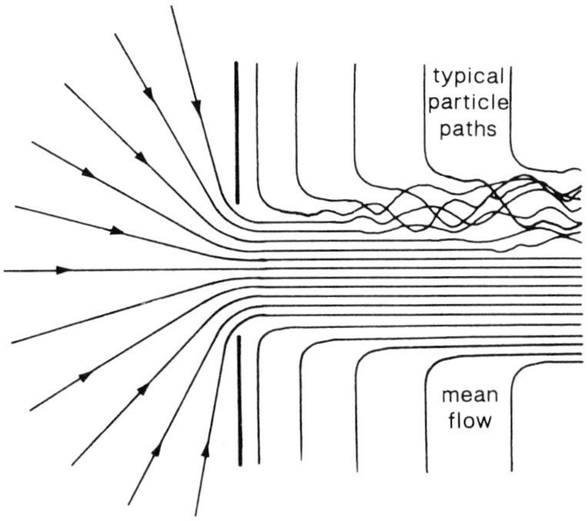

Fig. 2.13.vi The flow into a sink or on the upstream side of an obstacle is usually close to potential flow, whereas the flow from an orifice, which is the wake of the obstacle containing the orifice, is turbulent, irreversible and mixes with and entrains the surrounding.

diluting its momentum, a low-pressure region is created from where the fluid is entrained. If a laminar constant velocity jet could be set up, the pressure on its boundary would take the ambient value but Bernoulli's equation would still be inapplicable. This is because the fluid inside and outside the jet would have different values of the Bernoulli constant, gH, defined in (1.7.9) since there would be a vortex sheet between them. Thus when the flow becomes unsteady and turbulent, Bernoulli's equation is no longer valid. (See section 1.7.)

If flow were reversible it would be most inconvenient for animals, because, unless they were moving or there was a wind, they would breathe in at each breath the air breathed out from the previous one. This would be most unfortunate in sleep! However, the outward breath is a jet and the air breathed in comes from all directions as into a sink. Life is utterly dependent on the processing of fluids within the body, and would be impossible if irreversible motions were not dominant.

In a jet the momentum flux is the same at all distances along it, therefore R^2W^2 is constant, the cross-sectional area being represented by R^2, and the axial component of velocity W being proportional to the volume flux and also to the 'density of momentum'. If W is to decrease then R must increase in proportion, and so the volume flux, which is proportional to R^2W, increases. This means that the jet can only slow down by carrying a greater volume of the surrounding air with it, which means that there must be entrainment, and this is initiated by the instability of the vortex sheet. The low pressure is not produced by the high velocity in the jet but only by its mixing with the ambient fluid. Thus a water jet can cause low pressure by entraining air, a fact made use of in the vacuum pump attached to the tap in the chemistry laboratory.

2.14 NON-DIMENSIONAL NUMBERS

A useful way of thinking of the enormous variety of phenomena in fluid flow is to divide them into ranges of the various important non-dimensional numbers. We observe a transition from laminar flow in which inertia forces are negligible (Stokes regime) at low Reynolds number to the high Reynolds number regime, where the viscous forces are negligible and the flow has become fully turbulent downstream of an obstacle. The other important non-dimensional numbers may also be at the extreme ends of their range. Thus the Mach number U/c has been assumed to be small throughout, so that effects of compressibility may be ignored.

The most important other numbers which occur in our study are concerned with buoyancy. The **Froude** number is the ratio of buoyancy or gravity forces to inertia forces, or the ratio of gravitational potential energy to kinetic energy. In studies of buoyancy effects in the following chapters this number is in the region of unity; otherwise it is taken as zero. On the other hand, it is not usually very large. Thus we may often make what is called the **Boussinesq approximation**, which is appropriate when density differences are not large. There is a **buoyancy number**

$$B = \Delta\rho/\rho, \tag{2.14.1}$$

which is relevant to convection caused by density anomalies $\Delta\rho$. In most cases $B \ll 1$. The accelerations, according to Archimedes's principle, are gB, which is therefore small compared with g. Boussinesq's approximation is to neglect density anomalies except when combined with g. This is equivalent to neglecting the fluid acceleration $\mathbf{f}$ in comparison with $\mathbf{g}$ in the vorticity equation (1.4.1) on the grounds that $\mathbf{f}$ is of the order of $\mathbf{g}B$, but not, of course, in (1.4.4), which is a version of the equation of motion.

The buoyancy manifests itself in a stably stratified fluid through the non-dimensional number

$$\beta h \approx -\frac{h}{\rho}\,\text{grad}\,\rho \approx -h\frac{\partial\rho}{\rho\delta z} \tag{2.14.2}$$

in which h is a typical height displacement in the situation. This is, therefore, the value of B if $\Delta\rho$ is taken as the magnitude of density differences. β is a measure of the density stratification. The Froude number may be expressed in various forms such as

$$\mathrm{F} = \frac{g\beta h^2}{U^2} \quad \text{or} \quad \frac{gBh}{U^2} \tag{2.14.3}$$

In cases where vorticity is important we find the **Richardson number**, Ri, to be very relevant. But it differs from the other non-dimensional numbers in that it usually varies across the field and has only a local value because it is a ratio of gradients. Thus

$$\mathrm{Ri} = \frac{g\beta}{\left(\dfrac{\partial u}{\partial z}\right)^2} \tag{2.14.4}$$

represents the ratio of stabilising effect of the density stratification to the destabilising effect of velocity gradients. We shall see in Chapter 8 that simple stability criteria based on a local number are not to be had, and this reminds us that the first form of F given in

(2.14.3) contains the gradient β, and is a local number. Many of the solutions to problems containing velocity and density gradients as given are therefore particular and not general at all, for F and Ri are infinitely variable across the field of flow. However, in situations in which Re is the relevant number, the flow field is determined by the viscosity before the problem is presented to us.

Vorticity and density stratification therefore have to be studied on the basis of the question, 'what are the significant phenomena?' rather than 'what are the significant non-dimensional numbers?', for there are too many of them. The Richardson number may be regarded as the ratio of the potential energy of a particle displaced in a staple stratification to its kinetic energy, which is very much a local number and not a number representative of the flow pattern. Thus we may often find the number expressed as

$$\mathrm{Ri} = \frac{g\beta h^2}{U^2}.$$

It has been regarded as providing a criterion for whether an airstream of speed U and stratification β will pass over a hill of height h. But that is not valid because in a stratified stream the vertical displacements alter the pressure in the streamlines below. (For further discussion, see section 6.15.)

3

Secondary vorticity

3.1 DEFINITIONS

The **primary flow** in a fluid stream is the motion which transports the fluid. It consists, in steady flow, of the motion along the streamlines. This might seem to be the total flow, but motion in planes normal to the streamlines may exist, namely rotation around them or shearing across them. This is called the **secondary flow**, and a measure of it is the magnitude of the **streamwise component of vorticity**, which is called the **secondary vorticity**. It is important because it represents extra energy, which is often wasteful, has scouring or other effects, and can lead to mixing. We shall discuss the mechanisms whereby it is generated, which are of particular interest because the vorticity generated at a boundary by viscosity is perpendicular to the direction of flow over the boundary. It occurs in pipes and rivers and in the corners of the cross-section of channels. In all these cases secondary vorticity occurs as a result of a component of vorticity existing normal to streamlines being rotated usually as a result of curvature of the primary flow.

The distance measured along the path of a particle is denoted by s and the fluid speed by q. The curvature of the path is κ, a is the tangential component of $\mathbf{f}$, the acceleration, and τ is the torsion of the path defined below and not to be confused with the quantity defined for the same symbol in equation (1.6.10).

We use a system of moving axes, as employed in elementary differential geometry. The unit vector directed along the tangent to the particle path is denoted by $\mathbf{t}$, which is called the **unit tangent**. By definition

$$\mathbf{t} = \mathbf{v}/q. \tag{3.1.1}$$

Since $\mathbf{t}\,.\,\mathbf{t} = 1$, by differentiation with respect to s we obtain

$$\mathbf{t}\,.\,\frac{\partial \mathbf{t}}{\partial s} = 0, \tag{3.1.2}$$

which means that the change in $\mathbf{t}$ is always in a direction normal to $\mathbf{t}$, because it is of fixed magnitude. $\mathbf{t}$ and $\partial\mathbf{t}/\partial s$ define a plane, called the **osculating plane**, which is the plane of closest contact with the particle path. This may be thought of as defined by three successive points on the curve or two successive tangents, just as a tangent is defined by two successive points. The unit vector in the direction of $\partial\mathbf{t}/\partial s$ is denoted by **n** and is called the **unit principal normal**. The curvature, κ, is given by

$$\boxed{\frac{\partial\mathbf{t}}{\partial s} = \kappa\mathbf{n}} \tag{3.1.3}$$

In the case of circular motion it is seen that κ^{-1} is the radius and so this is the radius of curvature in other cases. The magnitude of the curvature is

$$\kappa = \mathbf{n}.\frac{\partial\mathbf{t}}{\partial s} = \kappa\left|\frac{\partial\mathbf{t}}{\partial s}\right| \tag{3.1.4}$$

Since $\mathbf{t}.\mathbf{n} = 0$,

$$\mathbf{n}.\frac{\partial\mathbf{t}}{\partial s} = -\mathbf{t}.\frac{\partial\mathbf{n}}{\partial s} = \kappa. \tag{3.1.5}$$

The **unit binormal**, denoted by **b**, is defined by

$$\mathbf{b} = \mathbf{t}\times\mathbf{n} \tag{3.1.6}$$

and with **n** defines the plane normal to the curve. **t**, **n**, **b** form a right-handed system of axes (Fig. 3.1.i).

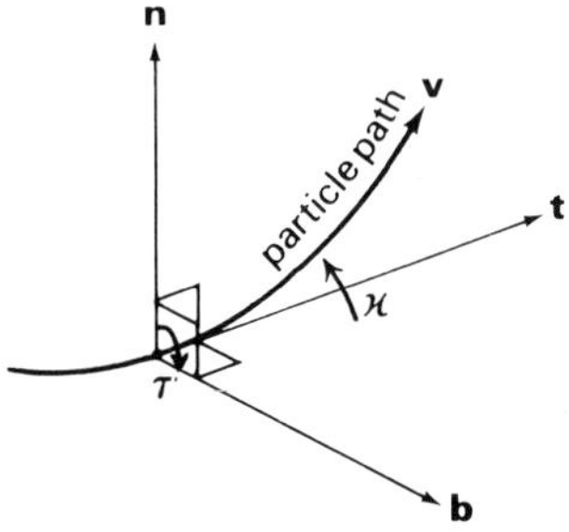

Fig. 3.1.i The unit tangent, normal and binormal to a particle path.

As we proceed along the trajectory, **t** is tipping up at rate κ in the direction of **n**. At the same time **n** and **b** are rotating around **t** at rate τ, which is a measure of the torsion. It is the rate at which the oscilating plane rotates around **t** as the point moves along the curve. For a plane curve, $\tau = 0$.

As **t** tips in the direction of **n**, **n** tips backwards in the direction of $-\mathbf{t}$ at rate κ, and rotates towards **b** at the rate τ.† **b** rotates in the direction of $-\mathbf{n}$ at rate τ, and is in a fixed direction normal to the plane if the curve is plane. The acceleration of a particle has components in the direction of **t** and **n** only. We now give mathematical expression to these ideas.

Since **n** is a unit vector, $\partial\mathbf{n}/\partial s$ has no component along **n** and so

$$\boxed{\frac{\partial\mathbf{n}}{\partial s}=-\kappa\mathbf{t}+\tau\mathbf{b}.} \tag{3.1.7}$$

The coefficient of **t** must be κ so as to conform with (3.1.5) when the scalr product of (3.1.7) with **t** is taken, and (3.1.7) may be taken as defining τ. Since $\mathbf{t}.\mathbf{b}=0$

$$\mathbf{t}.\frac{\partial\mathbf{b}}{\partial s}=-\mathbf{b}.\frac{\partial\mathbf{t}}{\partial s}=0 \quad \text{by (3.1.3)}$$

and so $\partial\mathbf{b}/\partial s$ has a component only in the **n** direction of magnitude τ. This may be obtained by differentiating $\mathbf{b}.\mathbf{n}=0$, and using (3.1.3). Thus

$$\boxed{\frac{\partial\mathbf{b}}{\partial s}=-\tau\mathbf{n}.} \tag{3.1.8}$$

Equations (3.1.3), (3.1.7), and (3.1.8) are called **Frenet's formulae**.

Since the acceleration is in the plane of **t** and **n**, in steady flow by (1.2.6),

$$\mathbf{f}=\frac{\mathrm{D}\mathbf{v}}{\mathrm{D}t}=(\mathbf{v}.\mathbf{grad})\mathbf{v}=\mathbf{grad}\ \frac{1}{2}q^2-\mathbf{v}\times\boldsymbol{\omega} \tag{3.1.9}$$

$$=(\mathbf{f}.\mathbf{t})\mathbf{t}+(\mathbf{f}.\mathbf{n})\mathbf{n} \tag{3.1.10}$$

because **f** is in the plane of **t** and **n**,

$$=q(\mathbf{t}.\mathbf{grad})q\mathbf{t} \quad \text{by (3.1.1)}$$

$$=q^2(\mathbf{t}.\mathbf{grad})\mathbf{t}+q\,\mathbf{t}(\mathbf{t}.\mathbf{grad}\ q) \tag{3.1.11}$$

$$=\kappa q^2\mathbf{n}+a\mathbf{t} \tag{3.1.12}$$

because $\mathbf{t}.\mathbf{grad}=\partial/\partial s$, and $a=\mathbf{v}.\mathbf{grad}\ q$ is the tangential acceleration. Thus, the components of the acceleration are

† τ is used in this chapter to mean torsion; but it means the reciprocal of potential temperature in Chapter 1, see (1.6.10), which is the main use in this book. Its use here is conventional in differential geometry. The same symbol has other uses in Chapter 9, see (9.3.7) or 9.9 where it is a time interval.

$$\mathbf{f}\,.\,\mathbf{t} = \mathbf{t}\,.\,\mathbf{grad}\ \tfrac{1}{2}q^2 = a, \tag{3.1.13}$$

$$\mathbf{f}\,.\,\mathbf{n} = \kappa q^2 = \text{ the centrifugal force on unit mass,} \tag{3.1.14}$$

$$\mathbf{f}\,.\,\mathbf{b} = 0 \text{ by definition of } \mathbf{b} \tag{3.1.15}$$

$$\mathbf{f} \times \mathbf{t} = (\mathbf{f}\,.\,\mathbf{n})(\mathbf{n} \times \mathbf{t}) = \kappa q^2 \mathbf{b} \tag{3.1.16}$$

$$\begin{aligned} \boldsymbol{\omega}\,.\,\mathbf{f} &= \boldsymbol{\omega}\,.\,\mathbf{grad}\,\tfrac{1}{2}q^2 \quad \text{by (3.1.9)} \\ &= \boldsymbol{\omega}\,.\,\mathbf{grad}\left(\tfrac{1}{2}\mathbf{v}\,.\,\mathbf{v}\right) \\ &= \mathbf{v}\,.\,(\boldsymbol{\omega}\,.\,\mathbf{grad})\,\mathbf{v}. \end{aligned} \tag{3.1.17}$$

The secondary vorticity is

$$\omega_s = \boldsymbol{\omega}\,.\,\mathbf{t} = \frac{1}{q}\boldsymbol{\omega}\,.\,\mathbf{v} \tag{3.1.18}$$

and represents twice the rotation around the direction $\mathbf{t}$ in unit time. In studying secondary vorticity a better representation is often the amount of rotation in unit distance, especially if we relate it to the geometry of a pipe shape or channel which causes curvature, or position in relation to a given obstacle. Thus the relevant quantity is ω_s/q.

The above formulas will be used in subsequent analysis.

3.2 THE SECONDARY VORTICITY EQUATION

Although it might seem to reduce the generality of the discussion, we shall consider only steady flow. It must be recognized that secondary flow loses its precision of definition in unsteady flow, and motion in three dimensions is complicated enough. We shall show that instabilities can be studied even in steady flow, because if small disturbances can be shown to grow in the downstream direction that is equivalent to instability following the fluid. The fact that the disturbance is not propagated upstream simply means that we have fixed our coordinates to the upstream end of the disturbance, and our main interest in instability is how a steady stream may become turbulent.

We take the vorticity equation in the form (1.4.1) for steady motion:

$$(\mathbf{v}\,.\,\mathbf{grad})\,\boldsymbol{\omega} = -\boldsymbol{\omega}\ \text{div}\ \mathbf{v} + (\boldsymbol{\omega}\,.\,\mathbf{grad})\,\mathbf{v} + \mathbf{R} \times (\mathbf{g} - \mathbf{f}) - \nu\nabla^2\boldsymbol{\omega}. \tag{3.2.1}$$

Then, using the appropriate results of the last section, we have .

$$(\mathbf{v}\,.\,\mathbf{grad})\frac{\boldsymbol{\omega}\,.\,\mathbf{t}}{q} = \boldsymbol{\omega}\,.\left[\frac{1}{q}(\mathbf{v}\,.\,\mathbf{grad})\mathbf{t} + \mathbf{t}\left(\mathbf{v}\,.\,\mathbf{grad}\ \frac{1}{q}\right)\right] + \frac{\mathbf{t}}{q}\,.\,(\mathbf{v}\,.\,\mathbf{grad})\boldsymbol{\omega}$$

(3.2.1)

$$= \boldsymbol{\omega}\,.\left[\kappa\mathbf{n} - \frac{1}{q^3}\mathbf{t}\left(\mathbf{v}\,.\,\mathbf{grad}\frac{1}{2}q^2\right)\right] + \frac{\mathbf{t}}{q}\,.\,\left[-\boldsymbol{\omega}\ \mathrm{div}\ \mathbf{v} + (\boldsymbol{\omega}\,.\,\mathbf{grad})\mathbf{v} + \mathbf{R}\times(\mathbf{g}-\mathbf{f}) + \nu\nabla^2\boldsymbol{\omega}\right]$$

(3.1.13) (3.1.7)

$$= \boldsymbol{\omega}\,.\left[\kappa\mathbf{n} - \frac{1}{q^2}\mathbf{t} + (\mathbf{f}\,.\,\mathbf{t})\right] - \frac{\boldsymbol{\omega}\,.\,\mathbf{t}}{q}\ \mathrm{div}\ \mathbf{v} + \frac{1}{q^2}\boldsymbol{\omega}\,.\,\mathbf{f} + \frac{\mathbf{t}}{q}\,.\,\left[\mathbf{R}\times(\mathbf{g}-\mathbf{f}) + \nu\nabla^2\boldsymbol{\omega}\right]$$

(3.1.10)

$$= \boldsymbol{\omega}\,.\left[\kappa\mathbf{n} + \frac{1}{q^2}\mathbf{n}(\mathbf{f}\,.\,\mathbf{n})\right] - \frac{\boldsymbol{\omega}\,.\,\mathbf{t}}{q}\ \mathrm{div}\ \mathbf{v} - \frac{1}{q}\mathbf{R}\,.\,\mathbf{f}\times\mathbf{t} + \frac{\mathbf{t}}{q}\mathbf{R}\times\mathbf{g}\,.\,\mathbf{f} + \nu\frac{\mathbf{t}}{q}\,.\,\nabla^2\boldsymbol{\omega}$$

(3.1.14) (3.1.16)

$$= 2\kappa\boldsymbol{\omega}\,.\,\mathbf{n} + \kappa q\mathbf{R}\,.\,\mathbf{b} + \frac{1}{q}\mathbf{R}\times\mathbf{g}\,.\,\mathbf{t} - \frac{\boldsymbol{\omega}\,.\,\mathbf{t}}{q}\ \mathrm{div}\ \mathbf{v} + \nu\frac{1}{q}\,.\,\nabla^2\boldsymbol{\omega}. \tag{3.2.2}$$

The term in this equation in div **v** on the right has the same physical meaning as in the original vorticity equation and is of very small importance. Likewise the viscous term which has come through unscathed has no new significance. We shall give no further attention to these two terms but will concentrate on the effects of curvature and gravity. For these purposes the equation is strictly for an **incompressible inviscid fluid**, and takes the form

$$\boxed{\mathbf{v}\,.\,\mathbf{grad}\frac{\omega_s}{q} = 2\kappa\boldsymbol{\omega}\,.\,\mathbf{n} + \kappa q\mathbf{R}\,.\,\mathbf{b} + \frac{1}{q}\mathbf{R}\times\mathbf{g}\,.\,\mathbf{t}.} \tag{3.2.3}$$

Since

$$\boldsymbol{\omega}\,.\,\mathbf{n} = \boldsymbol{\omega}\,.\,\mathbf{b}\times\mathbf{t} = \mathbf{t}\times\boldsymbol{\omega}\,.\,\mathbf{b} = \frac{1}{q}(\mathbf{v}\times\boldsymbol{\omega})\,.\,\mathbf{b}$$

$$= \frac{1}{q}\left(\mathbf{grad}\ \frac{1}{2}q^2 - \mathbf{f}\right).\,\mathbf{b} = \frac{1}{q}\ \mathbf{grad}\ \frac{1}{2}q^2\,.\,\mathbf{b} \tag{3.2.4}$$

we have that

$$2\kappa\boldsymbol{\omega}\,.\,\mathbf{n} + \kappa q\mathbf{R}\,.\,\mathbf{b} = \kappa q\left[\frac{2}{q^2\rho}\rho\,\mathbf{grad}\,\frac{1}{2}q^2 + \frac{2}{q^2\rho}\frac{1}{2}q^2\,\mathbf{grad}\,\rho\right].\,\mathbf{b}$$

$$= \kappa q\,\mathbf{grad}\,\ln\left(\frac{1}{2}\rho q^2\right).\,\mathbf{b}$$

if for the moment we ignore the gravity term, as would be justifed when $\mathbf{f} \gg \mathbf{g}$.

For example this could be the case in a curved pipe, then (3.2.2) becomes

$$\mathbf{v}\,.\,\mathbf{grad}\,\frac{\omega_s}{q} = \kappa q\,(\mathbf{grad}\,L)\,.\,\mathbf{b} \tag{3.2.5}$$

where

$$L = \ln\left(\tfrac{1}{2}\rho q^2\right). \tag{3.2.6}$$

In the case where there are no density gradients we can use (3.2.4) to represent $\boldsymbol{\omega}\,.\,\mathbf{n}$ in (3.2.3), but in gernal we can see that the gradients of ρ and gradients of q have a similar effect in producing secondary vorticity when there is curvature. If the flow is horizontally stratified the term in $\mathbf{g}$ vanishes initially (upstream) and so the application of (3.2.5) is not restricted to the case in which $\mathbf{f} \ll \mathbf{g}$, it is simply required the $\mathbf{R} = \mathbf{0}$. In thinking of the forces acting on the fluid we see that the inertial effects are represented by $\mathbf{grad}\,L$.

Returning to (3.2.3) we notice that, in the derivation, the term $2\kappa\boldsymbol{\omega}\,.\,\mathbf{n}$ originated from two different parts of the vorticity equation. First, it comes from $(\mathbf{v}\,.\,\mathbf{grad})\mathbf{t}$. This means that as the direction of $\mathbf{t}$ changes along the path of the particle, vorticity which was directed along the principal normal now finds itself having a component along the tangent, so creating secondary vorticity. The second half comes from the term $(\boldsymbol{\omega}\,.\,\mathbf{grad})\mathbf{v}$, which refers to the advection of the vortex lines with the fluid. On the inside of the bend the velocity is greater than on the outside, so that a line of particles AB which was initially perpendicular to $\mathbf{t}$ is rotated to A′B′ so as no longer to be normal to the streamlines (Fig. 3.2.i). Thus $\boldsymbol{\omega}$ acquires a component along $\mathbf{t}$ if it had a component along $\mathbf{n}$ initially.

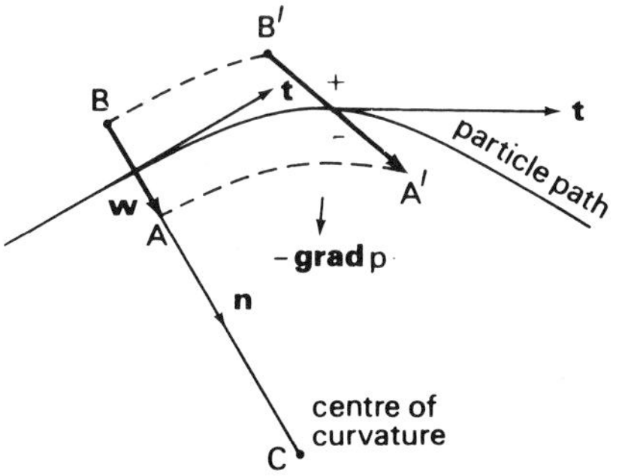

Fig. 3.2.i Streamwise vorticity is created out of vortex lines across the stream by the rotation of the direction of the stream and by the consequent rotation of the vortex lines in the opposite direction.

The acceleration of the flow on the inside of a bend is a property of irrotational flow. It can be thought of as arising because, in order to pass round the bend, the fluid must be subjected to a pressure gradient force towards the centre of curvature. The particles on the inside therefore move towards lower pressure and are accelerated, while those on the outside are decelerated. Alternatively we may consider the case of irrotational motion: since there is no vorticity in the plane of the bend, the rotation of two lines of particles at right angles to one another must be equal and opposite. Thus the tangent, represented by a line of particles along the streamline, rotates in the opposite direction to a line along the normal. Consequently, a component of vorticity along the normal, which introduces no vorticity into the plane of the motion (which may therefore be considered irrotational), is rotated towards the tangent and the angle between them closed up at twice the rotation of either. From this we see why the two contributions are equal.

The effect of the secondary vorticity is seen by considering the case of horizontal flow passing round a bend in a horizontal plane. The vector **g** is downwards, in the direction of **b**. This is, to a first approximation, in a fixed direction, although the swirl introduced by the secondary vorticity may cause the streamlines not to be plane. In Fig. 3.2.ii we have illustrated the case of flow entering the bend with velocity increasing upwards, and having vorticity in the direction of –**n**. The Bernoulli surfaces containing the streamlines and vortex lines are initially horizontal planes parallel to the plane of the motion. The **Bernoulli vector**, **B**, defined as the normal to the Bernoulli surface at a point, is given by

$$\mathbf{B} = \mathbf{t} \times \boldsymbol{\omega}. \tag{3.2.7}$$

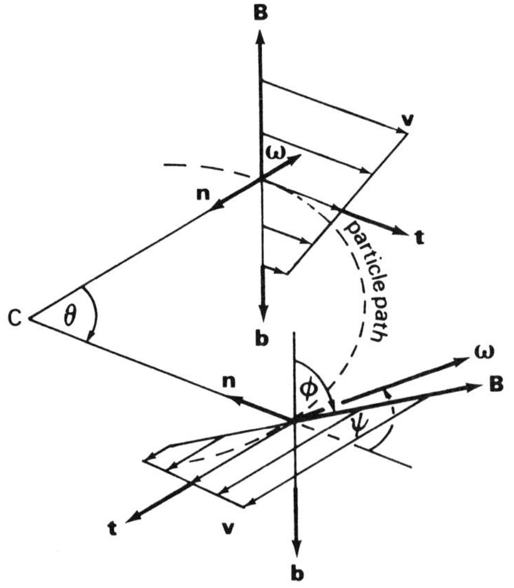

Fig. 3.2.ii The Bernoulli vector, which moves with the fluid, is rotated around the streamline through angle ϕ when the flow turns through angle θ, because of the streamwise component of $\boldsymbol{\omega}$ which is produced by the curvature.

The effect of the bend is to rotate the vector **B** around **t** through an angle ϕ while the vorticity vector $\boldsymbol{\omega}$ is rotated in the plane of the motion through an angle ψ relative to **t** and **n**. θ is the angle of the bend. The fast-moving fluid is displaced towards the outside of the bend.

The rotation of the vector **B** will be the consequence of the creation of secondary vorticity when the flow is confined in a circular pipe, as in Hawthorne's experiments to be described shortly. But in a wide, shallow river the flow near the middle will be almost horizontal. This results in the displacement of the faster-moving surface fluid towards the outside of the bend only producing vertical motion near the sides of the river, so that in the middle **B** will remain vertical.

The meaning of the term $\kappa q\mathbf{R}.\mathbf{b}$ Is now evident. Its effect is to bring the densest fluid to the outside of the bend: it represents the effect of density gradients in L in (3.2.5), and a density gradient in the direction of **b** produces secondary flow in the same sense as a vorticity vector in the direction of **n** (which is a gradient of speed in the direction of **b**).

The effect of gravity is to modify this effect by tending to move the densest fluid to the bottom of the stream. In this case it is represented by $\frac{1}{q}\,\mathbf{r}\times\mathbf{t}.g\mathbf{b}$ because **g** is parallel to **b**: this is equal to $-\frac{g}{q}\mathbf{R}.\mathbf{n}$, which tends to rotate the isopycnic surfaces until they contain **b**, which is horizontal in this case.

3.3 SQUIRE'S FORMULA

If the fluid is inviscid and of uniform density ($\mathbf{R} = \mathbf{0}$) and in steady motion the secondary voriticity equation is simply

$$\frac{\mathrm{D}}{\mathrm{D}t}\frac{\boldsymbol{\omega}.\mathbf{t}}{q} = 2\kappa\boldsymbol{\omega}.\mathbf{n} \tag{3.3.1}$$

or

$$q\frac{\partial}{\partial s}\frac{\omega_\mathrm{s}}{q} = 2\kappa\omega_\mathrm{n} \tag{3.3.2}$$

where subscripts 's' and 'n' refer to components along the streamline and normal. For a small change in stream direction $\mathrm{d}\theta$, since $\mathrm{d}\theta = \kappa\,\mathrm{d}s$,

$$\mathrm{d}\frac{\omega_\mathrm{s}}{q} = 2\frac{\omega_\mathrm{n}}{q}\kappa\,\mathrm{d}s = 2\frac{\omega_\mathrm{n}}{q}\,\mathrm{d}\theta.$$

Approximately, and perhaps exactly, the speed may remain constant as the streamline turns through $\mathrm{d}\theta$, and Squire's version of the formula is [1]

$$\boxed{\mathrm{d}\omega_\mathrm{s} = 2\omega_\mathrm{n}\,\mathrm{d}\theta,} \tag{3.3.3}$$

which expresses the way in which vorticity normal to the flow becomes streawise, the factor 2 being of particular significance as explained in the previous section.

The particular application with which Squire was concerned was the passage of boundary layer air through a cascade of aerofoils in the corner of a closed-circuit wind tunnel. The flow is made to turn through 90° in a horizontal plane by passing between the aerofoils shown in horizontal section in Fig. 3.3.i. In this way the flow is guided round the corner as a set of separate flat layers of air and the consequences of the large velocity gradients are avoided because the bend for each particle is short. A marked line of particles before the bend becomes broken into a series of pieces each of which is rotated, and emerges as a nearly continuous line with only very small-scale turbulence, of the order of the thickness of the aerofoils.

At the top and the bottom of the tunnel, however, there is a boundary layer with vorticity directed across the stream (Fig. 3.3.ii). In each bend this causes secondary flow to be generated with a component of velocity across the floor and ceiling towards the inside of the bend, with a small outward flow in the body of the tunnel to compensate. Squire's formula shows that the amount of secondary flow emerging from the cascade is independent of the dimensions of the cascade and dependent mainly on the angle turned through, but its duration is less in a short bend, and the up- and down-motions on the aerofoils quickly cancel one another after leaving the aerofoil.

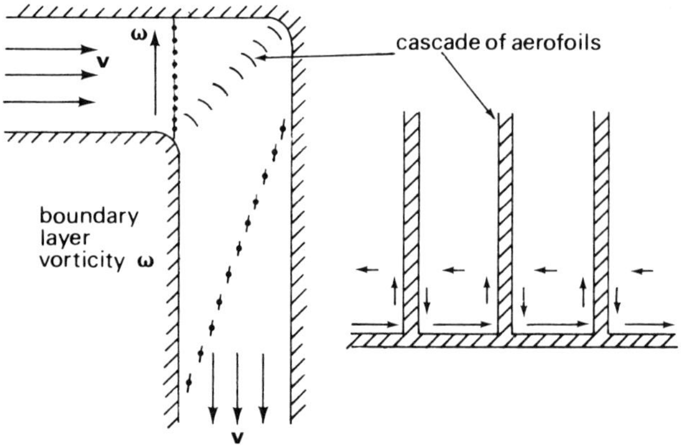

Fig. 3.3.i In a wind tunnel the boundary layer vorticity is across the stream. In traversing a bend through a cascade of aerofoils a line of pacticles becomes broken into several sections forming a diagonal line, and in each section of the cascade, secondary flow occurs. But this produces a small displacement compared with that generated in the longer bend that would be necessary without the cascade.

Fig. 3.3.ii The flow between the cascades is mainly observable as boundary layer flow towards the inside of the bend in each section of the cascade. The diagram shows the secondary motion near the bottom looking downstream in the direction of the stream.

3.4 FLOW IN RIVER BENDS

The water close to the bed of a river is retarded by the frictional drag of the bed and by the drag due to the wakes of obstacles and protrusions on the bed. There is also drag due to solid particles carried with the stream which are intermittently slowed down or brought to rest on the bed. Since rivers are usually much wider than they are deep we can think of them as having slow-moving fluid at the bottom with the main body of the water above moving faster (Fig. 3.4.i).

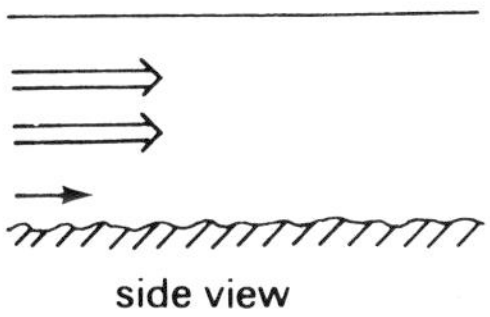

Fig. 3.4.i Velocity profile in a river.

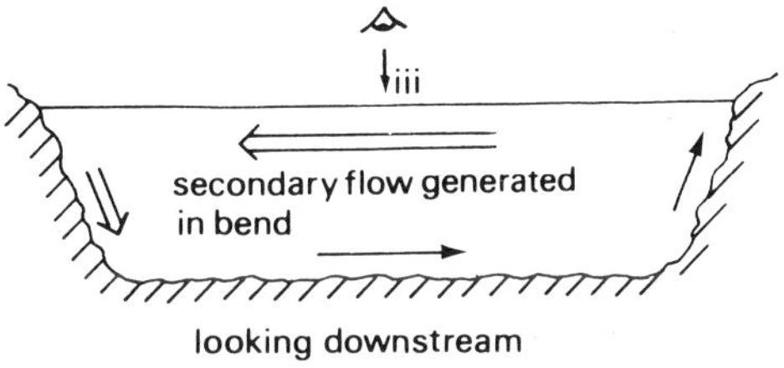

Fig. 3.4.ii Secondary motion generated in a bend to the right.

After the bend the vorticity vector which was initially directed across the stream will lie at an angle to it and a component ω_s will be generated. The slow-moving bottom water will be turned through a much greater angle than the faster-moving surface water. We may think of them both being subjected to the same pressure gradient towards the centre of curvature which is determined by the slope of the free surface with the vertical acceleration being negligible, so that the hydrostatic equation applies throughout the depth.

The effect is that fast-moving water is moved to the outside bank (Figs 3.4.ii and 3.4.iii). This causes people to observe that the theoretical idea that the flow is decelerated on the outside and accelerated on the inside is not realised, but the flow is three-dimensional and not plane. The outer bank is therefore eroded, while slow-moving fluid from which silt is readily deposited rises on the inside. The channel is therefore moved bodily outwards on a bend, which causes meandering to be increased with time (Fig. 3.4.iv).

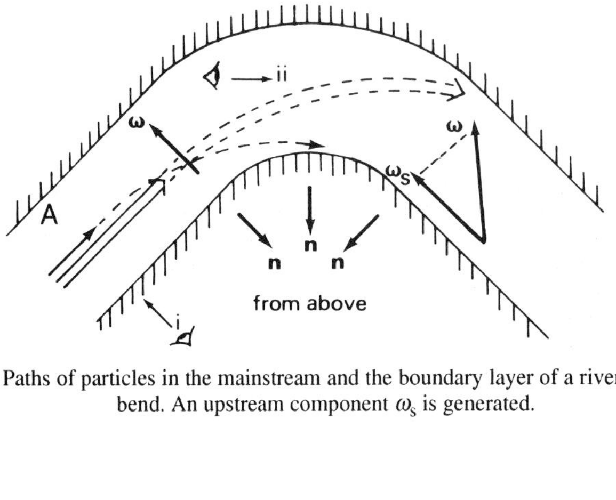

Fig. 3.4.iii Paths of particles in the mainstream and the boundary layer of a river in a right-hand bend. An upstream component ω_s is generated.

Fig. 3.4.iv A photograph taken from position A above the left wall of the channel, so that the vertical wall on the inside of the bend can be seen. The streaks are originated by blobs of white poster paint on the floor of the channel, and are seen to flow up to the base and then up the inside wall of the bend, showing how the slow boundary layer flow on the base is accelerated to greater curvature to the right than the main flow, which itself determines the low pressure on the inside of the bend.

We have argued on the assumption that all the vorticity we are concerned with is in the fluid as it enters the region of consideration, and that no more is produced by the walls in that region. In practice new vorticity is continously being produced. If this did not happen, the secondary flow in a second bend in the opposite direction to the first would be very much influenced by (a) how far the stream had been rotated, so that the Bernoulli surfaces had taken up new positions, (b) by the magnitude of the secondary vorticity and the angular momentum associated with it on leaving the first bend. In rivers the vorticity is generated fairly rapidly in the bottom fluid, so that in practice the flow in a second bend may be very little influenced by what happened in the first. (See Fig. 3.4.v.)

In other situations, as soon as the velocity field has been significantly altered by the secondary flow, the kind of first approximation we have been discussing is no longer applicable except in a qualitative way. In many cases the geometry of the situation is so complicated that there is no question of a detailed calculation anyway. The growth of meanders† often leads to the formation of the ox-bow lakes of geography book fame. (See Figs 3.4.vi & xi.) Any formulation of the erosion process requires assumptions which are not true to real situations, but are nevertheless interestting. (See Fig. 3.4.viii.)

One of the most remarkable cases of secondary flow occurs where a side channel draws water from a straight-flowing river. The flow on the bed can be thought of as determined by the drag of the flow above together with the acceleration due to the pressure field imposed by the shape of the free surface. This pressure field is determined by the flow of the bulk of the fluid above the boundary layer on the bed, and this is

† Named after the river Menderes in Caria, Asia Minor; named *Meander* in Latin, *Maiandros* in Greek, and described by Xenophon ca 375 BC.

Fig. 3.4.v Flow through an S-bend in a channel with paint streaks showing the direction of bottom water flow.

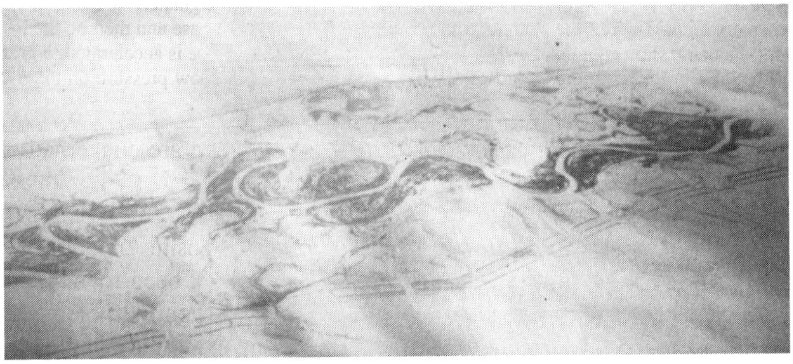

Fig. 3.4.vi An aerial view of the river Ural where it has several meanders.

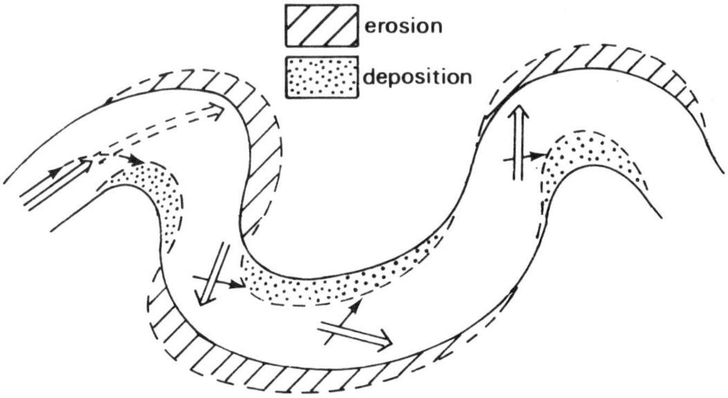

Fig. 3.4.vii Because of the erosion of the outer bank of a bend by the faster-moving component of the stream, and the deposition of silt on the inside where the slow-moving bottom water rises, the meandering of a river is increased with time.

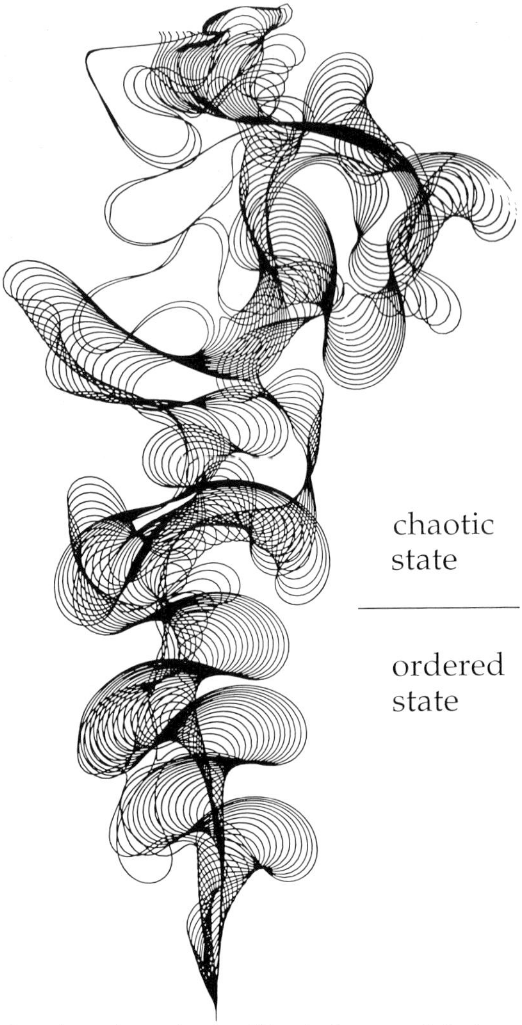

Fig. 3.4.viii By using a time series describing the incremental displacement of a river channel (in the artificial case of a uniformly erodable surface) Stolum [2] illustrated the possibilities waiting to be realised when a river flows across a suitable surface of mobile. The initial spatiotemporal evolution of a simulated river starting with a nearly straight line, shown at intervals of 200 time steps for the first 5000 iterations. Both an ordered and a chaotic state occur, with the transition between them initiated by a cutoff cluster. The ordered state in the lower part of the figure has reached a mature stage in which the train of bends is still highly symmetrical around the original axis, while at the same time each bend is growing into an asymmetrical shape (Kinoshita shape). The chaotic state seen in the upper half of the figure was initiated by a cutoff cluster occurring in the ordered state.

indicated in the diagram for a case in which a side channel draws off one third of the flow of the main stream. The pressure field then consists of a small pressure gradient in the downstream direction, which balances the drag, together with larger gradients which produce the decelerations in the main stream and the curved flow into the side channel. The main flow is of almost constant depth and is approximately non-divergent, and can

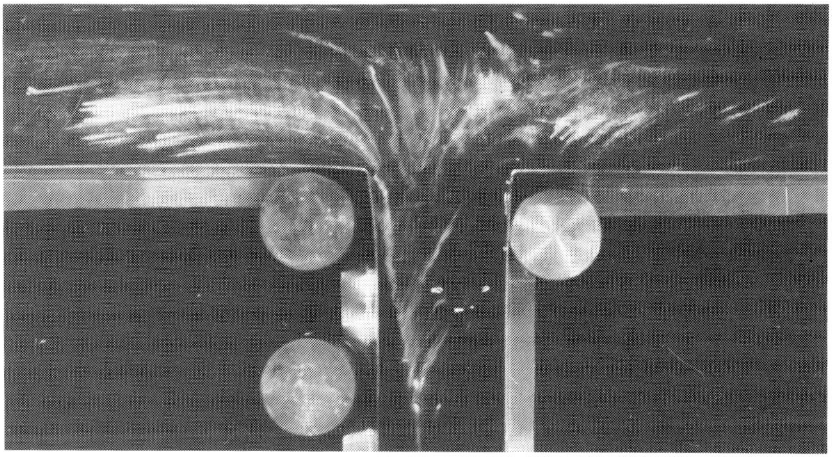

Fig. 3.4.ix Bottom (boundary layer) flow shown by white paint streaks in a water channel from which about one third of the flow is taken off into a side channel at right angles. The pattern is analysed in Fig. 3.4.x.

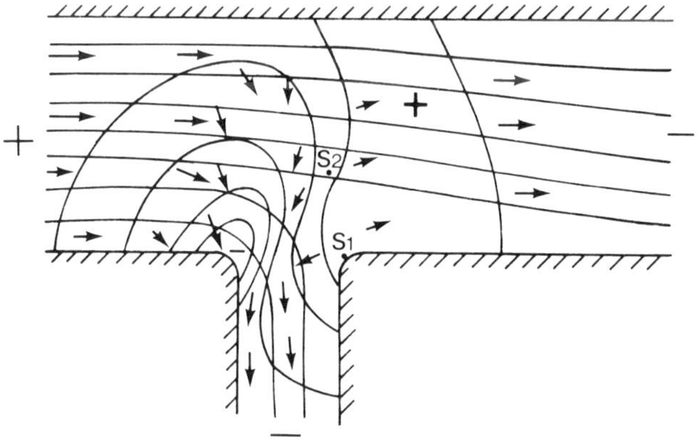

Fig. 3.4.x The contours indicate the topography (+ and –) of the free surface and the streamlines of the main flow where a side channel draws off about a third of the flow of a straight stream. The arrows indicate the bottom flow and S_1 and S_2 the stagnation points in it.

therefore be represented by streamlines whose spacing is inversely proportional to the speed (Fig. 3.4.x). The flow at the bottom, indicated by short arrows in the diagram, is not non-divergent, and the stagnation points in particular are places of downflow. Their precise position is very much dependent upon the details of the case. Low and high pressure are indicated by negative and positive signs respectively.

The most important features are the rising pressure into which the mainstream flow is decelerated and the low pressure on the inside of the curved flow into the side stream. The consequence is that at least two thirds of the bottom flow goes into the side channel. Configurations can be designed to make all the bottom flow go that way. A simple explanation often given for the collapse of Mesopotamian civilisation is that those in charge of the irrigation, on which the agriculture and the civilisation depended, did not understand secondary flow. Consequently the channels were continuously blocked with silt because they insisted on drawing them off at right angles from the main stream. Every effort to dredge them led to the raising of the average level of the irrigated areas so that eventually they could not be irrigated by gravity, and pumping on the scale require become impossible.

Fig. 3.4.xi The Murray River in New South Wales, Australia, with traces of previous meanders in its flood plain.

There is some residual secondary flow in the side channel and so the maximum accumulation is on the upstream side of it. There is one stagnation point, S_1, beneath the pressue maximum where the main flow stagnates on the downstream corner of the sidestream entrance. There is another, S_2, at the point where the pressure gradient force and the force due to the drag of the upper layers are equal and opposite. At this point the main flow line crosses the isobar at right angles.

To avoid drawing silt into a channel it should be taken from the outside of a bend in the main stream (Fig. 3.4.xii) where the bottom flow is away from the outside bank. Or the amount of silt taken from the main channel can be reduced to less than its proportionate amount by building a wall which causes the side channel flow to be decelerated as it enters, before the flow is curved into the side channel direction. In this way the bottom flow in the main stream is directed away from the side stream entrance to a slight extent (Fig. 3.4.xii).

We now turn to discuss the residual secondary flow remaining downstream of a bend in the flow.

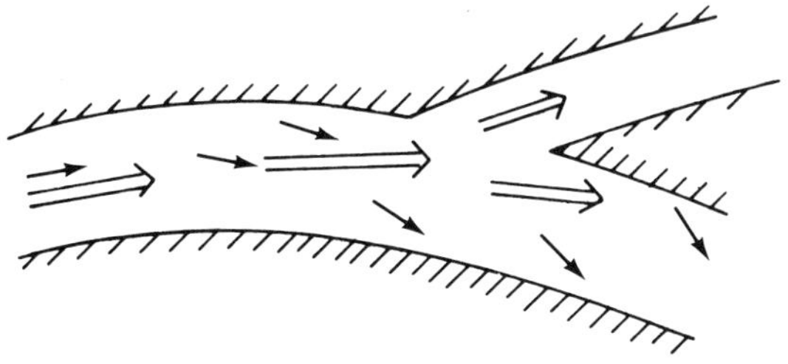

Fig. 3.4.xii By drawing water from the outside of a bend in the main channel almost no bottom flow and silt enter the side channel.

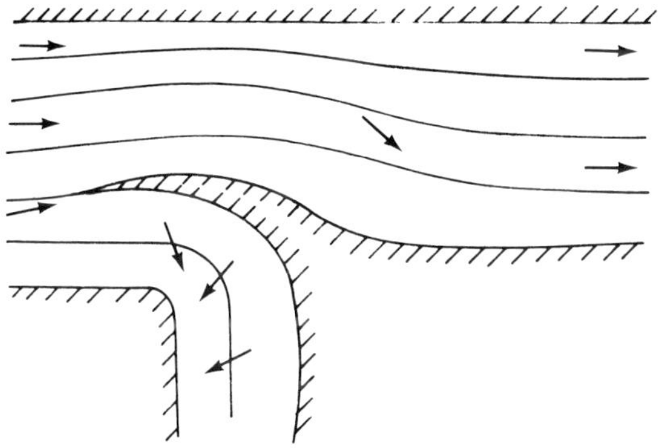

Fig. 3.4.xiii The silt carried into a side channel can be reduced if the flow into it is decelerated before entering the bend.

3.5 OSCILLATIONS IN A PIPE BEND

In this discussion we ignore the effect of density gradients except to remark upon their effect in a qualitative way at the end. Gravity may therefore be ignored, and for convenience we speek of the motion taking place in a horizontal bend (Fig. 3.5.i) of a pipe of circular cross-section. The basic equation (3.2.5) may then be written

$$q \frac{\partial}{\partial s} \frac{\omega_s}{q} = \kappa q \, \mathbf{grad} \left(\ln \frac{1}{2} \rho q^2 \right) . \mathbf{b}. \tag{3.5.1}$$

In the notation of section 3.2 illustrated in Fig. 3.5.ii, ϕ is the angle through which the Bernoulli surfaces have been rotated around the streamline. If, for the purposes of discussing the main mechanism under study, we contrive a case in which the Bernoulli surfaces have the same rate of secondary rotation with distance downstream across the whole section of the pipe, we may write

$$\omega_s = 2 \frac{\mathrm{D}\phi}{\mathrm{D}t} = 2q \frac{\partial \phi}{\partial s}. \tag{3.5.2}$$

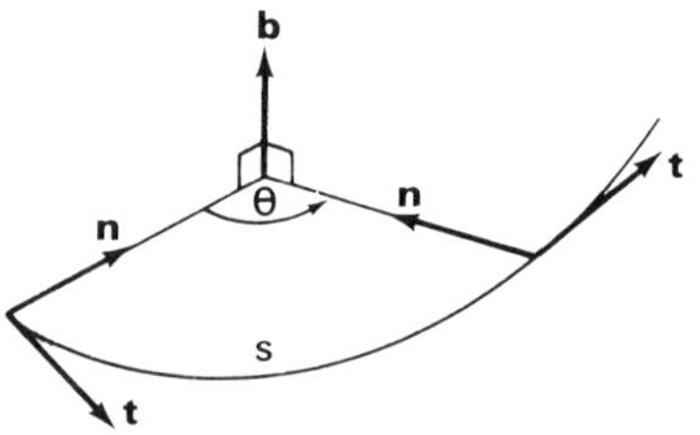

Fig. 3.5.i Coordinates for the motion around a horizontal bend with **b** vertical.

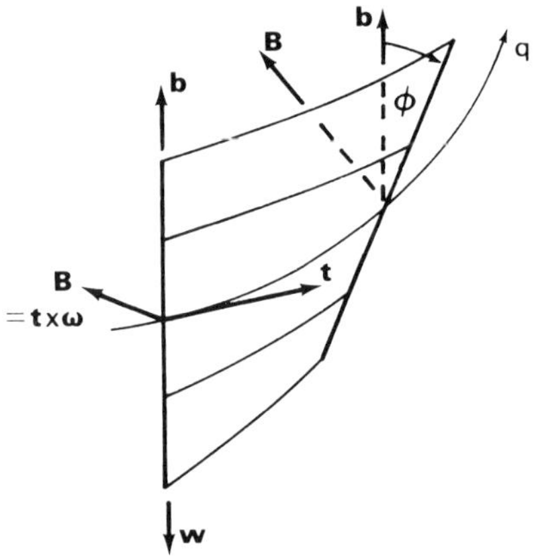

Fig. 3.5.ii The rotation of the Bernoulli surface, to which the vector **B** is perpendicular, through angle ϕ as it traverses a bend. $\phi = 0$ in the equilibrium position in which **B** is paralell to **n**.

If we also assume that all the flow has the same curvature as the centre line of the pipe (which is equivalent to assuming both that the diameter of the pipe is small compared with the radius of curvature of the bend and that change in curvature due to the helical motion of the particles is small compared with the curvature) then **b** may be regarded as a constant vertical vector and κ a constant scalar. The significance of this assumption is illustrated in Fig. 3.5.iii. If $\phi = 0$ when the Bernoulli surfaces contain vertical generators with the fast-moving fluid on the outside of the bend, (3.5.1) becomes

$$2q\frac{\partial^2\phi}{\partial s^2} = \kappa q\,\mathbf{grad}\left(\ln\frac{1}{2}\rho q^2\right)\sin\phi, \tag{3.5.3}$$

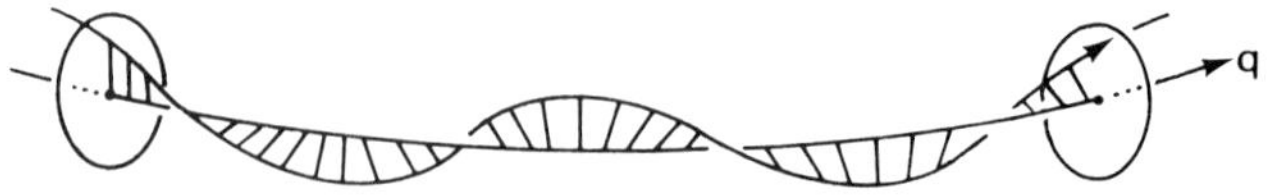

Fig. 3.5.iii When there is secondary motion the paths of the particles are helical. The assumptions made in the simple theory assume that the helix is of small radius and has long pitch compared with the radius of the bend. The effects of curvature are calculated as if the particles passed along the axis of the pipe.

and if θ is the angle turned through by the central (undisturbed) streamline

$$\kappa\, \mathrm{d}s = \mathrm{d}\theta$$

and so

$$\frac{\partial^2 \phi}{\partial \theta^2} = \frac{1}{2\kappa}\, \mathbf{grad}\left(\ln \frac{1}{2}\rho q^2\right) \sin\phi \tag{3.5.4}$$

where the gradient is measured across the Bernoulli surfaces in the negative direction of **B**, which is parallel to **n** when $\phi = 0$.

This equation represents a system which executes stable oscillations when the gradient of q is negative, and the period of oscillation in θ for small values of ϕ is (by analogy with the pendulum equation)

$$2\pi\left(-2\kappa/\mathbf{grad}\ \ln \tfrac{1}{2}\rho q^2\right)^{1/2}. \tag{3.5.5}$$

The unstable situation in which q increases inwards ($\phi > 90$) is discussed in section 3.8.

In a real case it is unlikely that the period indicated by (3.5.5) will be the same for all streamlines. For it to be the same, if y is the coordinate in the direction of **B**, it would be required that

$$\frac{\partial}{\partial y}\frac{1}{\kappa}\frac{\partial}{\partial y}\ln\frac{1}{2}\rho q^2 = \text{const (indep. of } y) \tag{3.5.6}$$

or, ignoring the variations in κ,

$$\rho q^2 \propto e^{-\alpha y} \tag{3.5.7}$$

where α is a constant and the zero of y may be put at any convenient distance. With any other form of profile a more complicated motion results.

Hawthorne [3] demonstrated the reality of these secondary oscillations by blowing air through a pipe of circular cross-section of radius a. The stream was given a velocity gradient in the vertical direction and passed horizontally round a bend. A straight section ended at station 0, where the position of the Bernoulli surfaces was identified in the cross-section by measuring the contours of constant stagnation pressure p_0 where

$$p_0 = \rho g H = \tfrac{1}{2}\rho q^2 + p + \rho g z,$$

or, since in this case the effect of gravity can be ignored and the hydrostatic equation subtracted out, we may write, with $\rho g z$ added to p (or subtracted from p_0),

$$p_0 = p + \tfrac{1}{2}\rho q^2. \tag{3.5.8}$$

These contours were measured across the open end of the pipe by a pitot tube; consequently there might be some very slight errors due to the influence of the pressure field in the diverging and turbulent flow outside. A section of curved pipe was then attached, and the position of the Bernoulli surfaces measured in the same way at the new open end. In order to determine the magnitude of the secondary vorticity at this position, which we call station 1, a section of straight tube of length b was attached and the position of the Bernoulli surfaces measured at its downstream open end, station 1b.

The expected behaviour is indicated in Fig. 3.5.iv, where the Bernoulli vector **B**, perpendicular to the surfaces, is used to denote their orientation. The line AB is the intersection of the central Bernoulli surface with plane of the cross-section, and is normal to **B.**

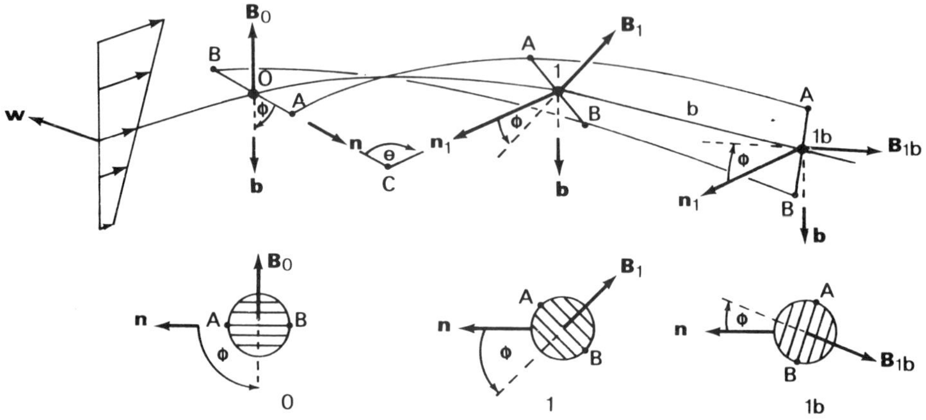

Fig. 3.5.iv On the left are shown the velocity profile and vorticity vector in the stream aproaching station 0, which is at the beginning of the pipe bend. At station 0 the various vectors and the position of the line of particles AB which lie on a Bernoulli surface are shown at the beginning of the bend and at station 1 at the end of the bend. Finally, at station 1b they are shown at the end of a straight section of pipe of length b. Below the diagrams show the positions of the surfaces in the pipe cross-section as seen looking upstream.

ϕ is the angle between the plane of the bend given by **n** and the negative direction of **B**. The change in ϕ in the length b represents the secondary motion persisting as it existed at station 1. The cross-sections are viewed facing upstream.

The straight section b was then removed and another section of the bend added, to increase θ, and the same procedure repeated. The diagrams in Fig. 3.5.v show the results obtained by Hawthorne. The pipe had an internal diameter of 6 in and the radius of the bend was 30 in. Measurements were taken at intervals of 30° of bend and the linear extension tube had length b equal to 25 in. The upper diagrams show the configuration of the stagnation pressure at each value of θ, and the lower ones the configuration at the end of the straight extension. The dynamic pressures are measured in inches of water, and the maximum shows a slow decline due to friction as the length of pipe is increased, while the minimum is increased.

In Hawthorne's experiment the velocity profile was linear, with zero velocity at one extreme. Thus the condition (3.5.7) was not satisfied and the different Bernoulli surfaces rotated through different angles. Nevertheless the results expected from the simple theory are clearly apparent and may be summarised in Fig. 3.5.vi, which indicates the angle through which the Bernoulli vector is rotated. The oscillations appear as considerably

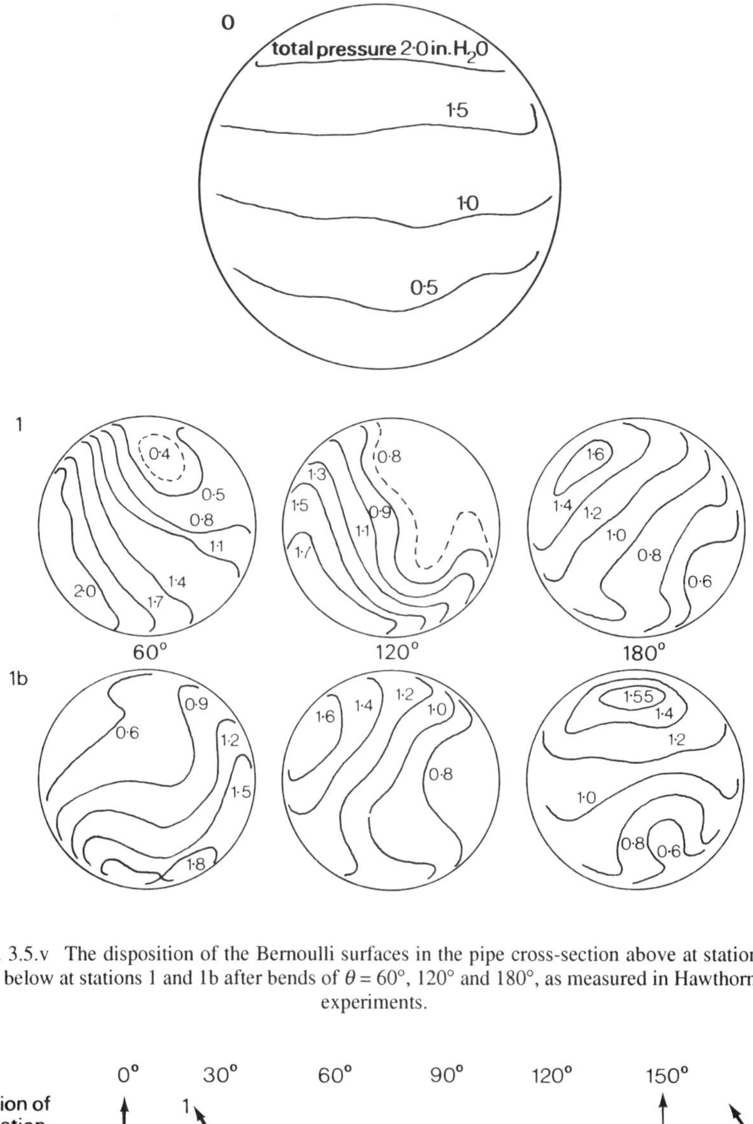

Fig. 3.5.v The disposition of the Bernoulli surfaces in the pipe cross-section above at station 0, and below at stations 1 and 1b after bends of $\theta = 60°$, 120° and 180°, as measured in Hawthorne's experiments.

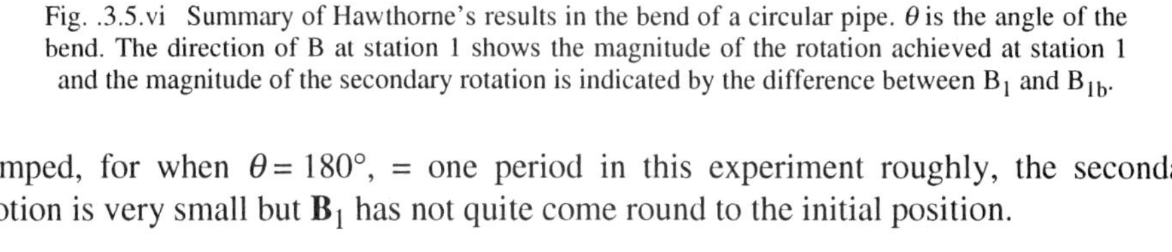

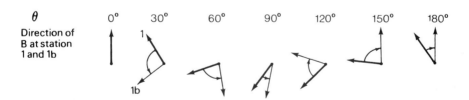

Fig. .3.5.vi Summary of Hawthorne's results in the bend of a circular pipe. θ is the angle of the bend. The direction of B at station 1 shows the magnitude of the rotation achieved at station 1 and the magnitude of the secondary rotation is indicated by the difference between B_1 and B_{1b}.

damped, for when $\theta = 180°$, = one period in this experiment roughly, the secondary motion is very small but $\mathbf{B}_1$ has not quite come round to the initial position.

In another experiment, Hawthorne used a pipe of internal diameter $6\frac{3}{8}$ in with a radius curvature of the central line of $9\frac{9}{16}$ in. He turned the flow through 90° and then added increasing lengths of straight pipe to study the fate of the Bernoulli surfaces as friction and other influences began to operate. The ultimate state would, of course, be with the maximum velocity in the middle but his results, shown in Fig. 3.5.vii, clearly indicate the tendency of the low stagnation pressure fluid to move towards the middle. This result will be discussed further in section 3.8.

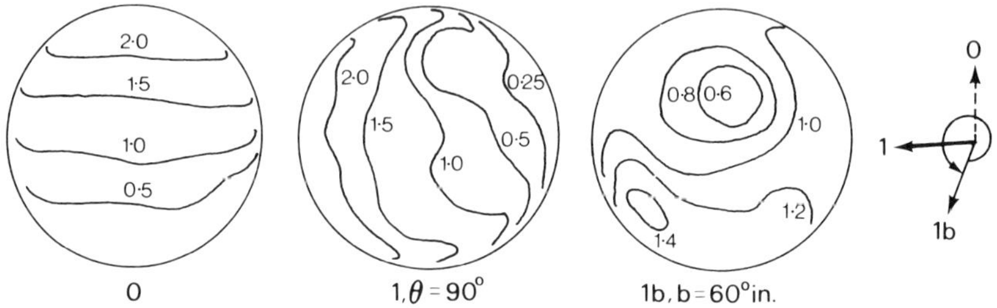

Fig. 3.5.vii The flow with Bernoulli surfaces depicted at station 0 was turned through 90° and the distribution was measured at increasing lengths of straight pipe. At $b = 60$ in the tendency for the low stagnation fluid to occupy the centre of the channel is clear.

The flow in a ventilating duct of rectangular cross-section was also studied by Hawthorne, where the tendency of the high stagnation pressure fluid to move to the outside of the bend, with oscillations around that position, was clearly revealed.

Dean [4] studied the case of a very long bend, as in a closely coiled pipe, where an equilibrium between the inertial and viscous forces is established. The high-velocity fluid is produced by the pressure gradient in the middle of the pipe, and low-velocity fluid at the walls by friction. The high-speed fluid moves to the outside of the bend, and is decelerated by the friction, while the low-speed fluid moving inwards to take its place is accelerated. The velocity configurations arrived at are shown in Fig. 3.5.viii. These are taken from a paper by McConologue and Srivastava [5], who extended Dean's work. The contours show the downstream velocity and the streamlines of the secondary flow. In the case shown in the upper-half diagram, the flux down the pipe was about one quarter of that in the case shown in the lower half, and the secondary flow was about one-seventh. Reference to the original paper should be made for details.

The pressure drop in a pipe bend is a combination of the loss due to friction and that due to the setting up of the secondary flow. If the flow emerges from the bend just at the extreme point of an oscillation when the secondary flow is zero, the loss will be less than if, for that bend, the secondary flow has large magnitude at that point. The loss in some pipe bends of 90° can be nearly doubled by the generation of secondary circulation. Indeed, it has long been known that this loss, expressed as a proportionate increase in the loss that would be experienced due to friction in a straight pipe of the same length as the bend, does not increase monotonically with the angle θ, and correspondingly with the length of the bend. Such a result is to be expected according to this simple theory, and the graph in Fig. 3.5.ix shows the kind of relationship to be expected with a given size of pipe and radius of bend. If a given θ is chosen and κa varied, a similar periodic variation would be found. (κ = curvature of bend, a = radius of pipe cross-section.)

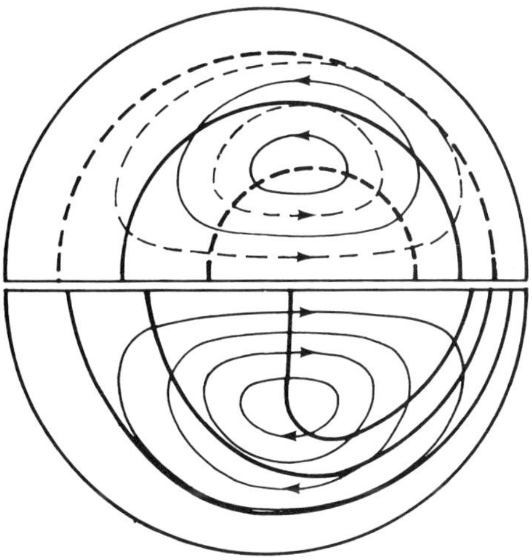

Fig. 3.5.viii Steady-state flow of a viscous fluid in a long pipe bend shown in cross-section. The upper and lower halves of this diagram represent different cases, the flux down the pipe being about four times greater in the lower half. The contours show the Bernoulli surfaces, and the streamlines the secondary flow, according to calculation by McConologue and Srivastava.

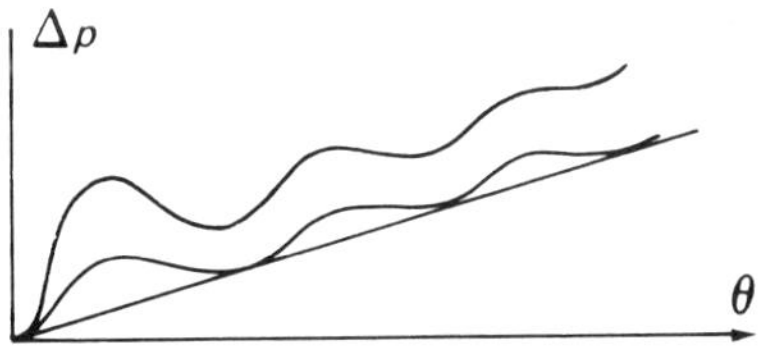

Fig. 3.5.ix The pressure drop to be expected in a pipe bend increases linearly with the length (angle) of the bend but in addition a further loss occurs because the secondary flow set up. The middle curve shows how this would vary if the oscillations had the same period over the whole cross-section, but in practice the loss is more like the upper curve.

The straight line indicates the pressure drop, Δp, in the same length of straight pipe, and the lower curve the minimum loss in a bend of angle θ. The upper curve shows the loss to be expected if all the Bernoulli surfaces did not rotate together, so that at no point in the bend is the secondary vorticity zero everywhere across the section. This problem was first studied experimentally by Eustice [6].

3.6 SECONDARY MOTION BEHIND OBSTACLES: HAWTHORNE'S HOT HUMP

The flow behind obstacles is usually turbulent, and so it is not realistic to pretend to much detail in a theory. The very simple theory which follows is intended really to give an idea of the secondary vorticity which has been created when the flow reaches the widest point of the obstacle. Beyond there, separation can be a dominant factor in determining the flow, and this varies very much with the obstacle shape.

To discuss general principles, we need to have a simple representation of an obstacle. This is done by placing a doublet in a stream, the strength being related to the size of the obstacle. A source may be added to represent the inflow in the wake of the obstacle towards it relative to the main stream. A more complicated obstacle could be represented by a system of doublets, or even a distribution of sources and sinks separated by a finite distance but producing closed circulations around which the main stream flows. Its wake could be represented by a system of sources rather than one alone, but that is not a profitable line of discussion unless we have a particular pressing problem on hand.

A uniform stream is represented by a stream function increasing linearly in the perpendicular direction. A simple two-dimensional obstacle in it could be represented by the classical stream function.

$$\psi = Uy + m\theta + \frac{\mu}{r}\sin\theta \tag{3.6.1}$$

where (x, y) and (r, θ) are rectangular and polar coordinates in the plane normal to the (cylindrical) obstacle, which in this case is a circular cylinder of radius $(\mu/U)^{1/2}$. The wake size is determined by the source strength, m. The formula (3.6.1) enables us to calculate the position of any streamline, and in principle we could do the same with more complicated formula representing a more complicated case. The situation is illustrated in Fig. 3.6.i and Fig. 2.8.ii(b).

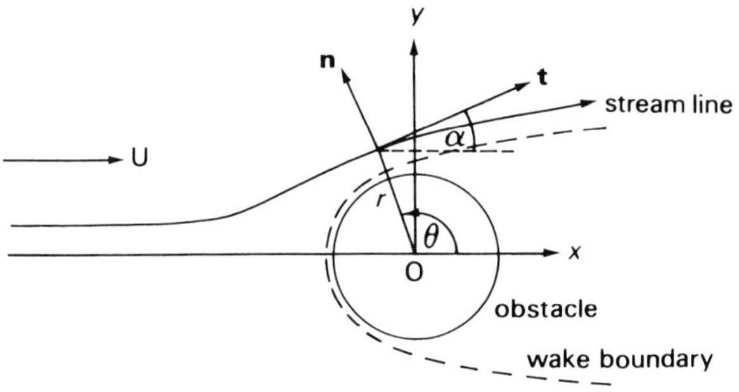

Fig. 3.6.i Coordinates for discussion of the flow around an obstacle with a wake. See Fig. 2.8.ii(b) and imagine the wake boundary to be defined by the streamline through S which is taken to be the forward stagnation point for the case of flow from left to right (which is simply reversing the flow of the experiment and making the sink into a source).

We now imagine that the actual flow past the cylinder varies in the direction of the z-axis. Thus we might have a velocity stratification which would be accompanied by a corresponding variation in the values of m and μ so as to retain the same geometry of the flow, together with a density stratification, represented by

$$\frac{\partial U}{\partial z} = \lambda U, \quad \frac{\partial m}{\partial z} = \lambda m, \quad \frac{\partial \mu}{\partial z} = \lambda \mu, \quad \frac{\partial \rho}{\partial z} = -\beta\rho \tag{3.6.2}$$

The gradient of L in (3.2.6) is in the direction of z and may be calculated at any point of the flow. Since it is proportional to ρq^2 the gradient of its logarithm will be proportional to $\beta\lambda^2$. Equation (3.2.5) or (3.5.1) which neglects the effects of gravity is now

$$q\frac{\partial}{\partial s}\frac{\omega_s}{q} = \kappa q\frac{\partial}{\partial z}\ln\frac{1}{2}\rho q^2 \tag{3.6.3}$$

which, when integrated with respect to s along the streamline, is

$$\frac{\omega_s}{q} = \int_{x=-\infty}\frac{\partial\alpha}{\partial s}\frac{\partial}{\partial z}\ln\frac{1}{2}\rho q^2\,\mathrm{d}s$$

$$= \int_{x=-\infty}\frac{\partial}{\partial z}\ln\frac{1}{2}\rho q^2\mathrm{d}\alpha. \tag{3.6.4}$$

This is an easy quantity to evaluate in principle but very laborious in practice by numerical step-by-step integration. The formula looks simple enough but depends on following the variables along the particular streamline from far upstream in order to evaluate the secondary vorticity at a point in the field. However it is a relatively easy calculation to program for a computer, and the secondary vorticity is thereby obtained in one of the relevant planes, say the plane normal to the stream through the obstacle at its widest point.

The configuration of the Bernoulli surfaces is not readily obtained from the distribution of ω_s because, except when it represents a solid rotation of a section of flow, it cannot be assumed to be precisely equal to $q\partial\phi/\partial s$, as in the last section. To map out these surfaces it would be necessary to obtain the field of secondary vorticity in each plane of a series of steps in the x-direction to calculate the value of div $\mathbf{v}$ in each plane, and then by integration obtain the velocity distribution in each plane. From this the increment of displacement of the Bernoulli surfaces could be calculated. The calculation in this exercise would be an order of magnitude greater than needed to obtain the first approximation to the value of ω_s alone, and if it were undertaken it would be just as well to proceed from knowledge of the displaced positions to calculate a second approximation to the value of **grad** $(\ln\frac{1}{2}\rho q^2)$ before proceeding to the next step. However, such an exercise is scarcely justified in view of the other errors inherent in the crudity of the original approximation used, (3.6.1), to obtain the streamlines.

The problem requires specification of the three non-dimensional numbers

$$\frac{U}{\lambda m},\quad \frac{U}{\lambda^2\mu},\quad \frac{\lambda}{\beta} \tag{3.6.5}$$

so that there is a three-dimensional family of cases even for the simple case of a cylinder in a flow with exponential functions of z for velocity and density in the undisturbed stream. In any real case there must be ends to the cylinder and the functions are not exponential, and so only very broad results are likely to emerge from any study if they are to be simple enough to comprehend, unless the calculation is undertaken because of a special interest in some definite configuration.

An axisymmetric obstacle can be represented using Stokes's stream function, thus

$$\psi = \frac{1}{2}Ur^2\sin\theta + m\cos\theta + \frac{\mu}{r^2}\sin^2\theta. \tag{3.6.6}$$

This case is more complex, and a decision has to be made as to the definition of the streamlines to be used in the calculation of (3.6.4). Hawthorne and Martin (1955) performed this calculation for the flow over a hemisphere, assuming the flow to be approximately the same as the potential flow round a sphere, each Bernoulli surface being horizontal far upstream. They also included the effect of gravity and retained the term $\frac{1}{q}\mathbf{R}\times\mathbf{g}.\mathbf{t}$ from (3.2.3) but, because variation in ρ was negligible in comparison with variations in q^2 and $g \gg f$, they wrote $\frac{1}{2}q^2$ for $\frac{1}{2}\rho q^2$ in L. They expressed their calculations in the form of maps of the magnitude of ω_s/q far downstream. However, as we have already pointed out it is more realistic to express them as values in the plane where the obstacle is widest. Indeed they carried out an experiment in which the rotation was made visible by trails of smoke issuing from holes on the hemisphere. A fairing was placed in the wake of the hemisphere to reduce separation and turbulence behind it so that the secondary vorticity would be visible, and the trails were seen to be helical, showing a very complicated overturning of the Bernoulli surfaces.

The series of experiments described by Hawthorne and Martin show the flow over Hawthorne's Hot Hump. This consisted of a hemisphere placed on a boundary. The viscous drag of the boundary upstream produced a velocity gradient and the surface was heated (or cooled) to produce a density gradient by conduction. When the hot boundry was at the bottom of the fluid the density gradient was unstable, but it was stable when the hemisphere was placed on the roof of the wind tunnel. The sense of the secondary vorticity was observed by the helical motion of smoke trails downstream of the median plane.

With forward shear and a cold surface on the bottom of the tunnel the two effects are as shown in Fig. 3.6.ii.

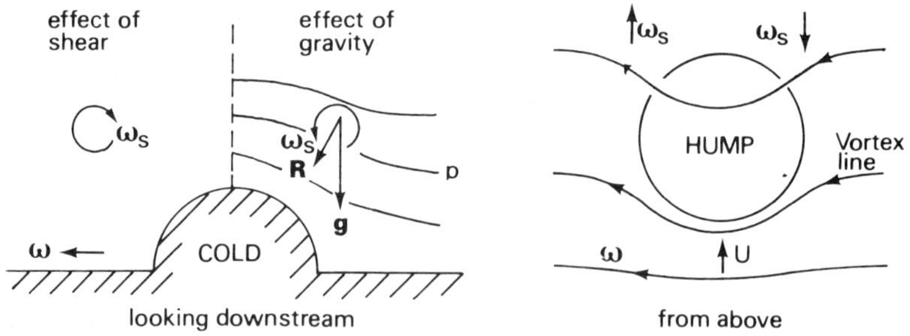

Fig. 3.6.ii The sense of the secondary vorticity produced in the flow over a hemispherical hump with a stable gravitational stratification and forward shear due to friction. The right-hand figure shows the distortion of the initial vortex lines.

In the flow over the hemisphere the vortex lines, which are horizontal and lying across the stream far upstream, are held back by the obstacle and are therefore stretched and rotated so that, when seen from above as in the right-hand figure, they point backwards on the right and forwards on the left of the obstacle. This introduces a secondary component which produces downward motion behind the obstacle, as indicated in the left half of the left diagram (Fig. 3.6.ii). The effect of gravity on the density gradient is to produce a motion tending to restore the isopycnic surfaces to the horizontal because they are humped up over the obstacle. This brings **R** back into coincidence with **g**.

Over a hot hump the secondary vorticity generated by gravity is in the opposite direction. Close to the floor where the vertical displacement is negligible the effect is very small, whereas the effect of the holding back of the vortex lines is a maximum. Consequently there is on each side a surface below which the vorticity was dominantly produced from the shear and above which it was dominantly produced by gravity. There is therefore an inflow towards the axis of the flow along these surfaces. Both effects fall to zero on the vertical plane of the flow.

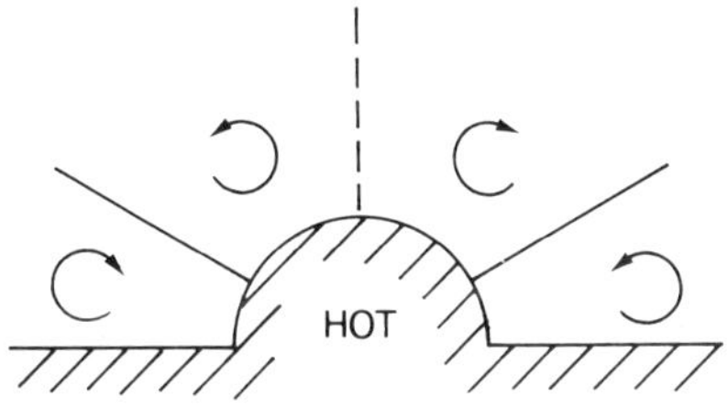

Fig. 3.6.iii The secondary circulation set up over a hot hump viewed looking in the downstream direction.

No account is taken in this theory of the possibility of the propagation of gravity waves through the fluid, and we shall see in Chapter 5 that this method is not appropriate for that problem in spite of Hawthorne's suggestion that it might be. In the gravitationally stable case of the cold hump there would be a wave pattern similar to that produced by a ship on the surface of water (see section 6.6). In the other, unstable, case the presence of the hump would set off buoyant convection. This is discussed in Chapters 11, 12 and 14.

An obvious conclusion from these considerations is that when the density gradients are small and can be neglected, the secondary motion generated by an obstacle will produce a downward flow behind it towards the ground. This is the case on many occasions when the low-level air is stirred up by motion over rough ground, since there is usually shear generated by ground drag.

The downflow which is commonly seen in exhaust motion behind a car is quite different and is the result of the velocity and aerodynamic design of the car, and ensures downflow on the rear window which keeps it clean. Racing cars have fins on the back to avoid that downflow and produce a downforce. Downflow produces lift!

3.7 SECONDARY FLOW IN FRONT OF OBSTACLES

The distortion of the vortex lines by the presence of an obstacle in a stream with shear is responsible for some interesting phenomena on the upstream side of the obstacle. This is well illustrated by the case of a cylindrical pillar standing up from the bed of a river. The main body of the flow is fairly uniform and so the flow on the upstream side of the cylinder is very similar to the potential flow pattern with a line of stagnation points down the front. On this line the pressure is a maximum for the neighbourhood and increases away from the bottom with increasing wind strength, and this feature of the pressure distribution is communicated to the boundary layer on the bed of the stream because the vertical accelerations are small compared with gravity. Consequently we would expect a flow outwards from the bottom of the column (Fig. 3.7.ii) in the upstream direction close to the bed. An alternative argument leading to this conclusion is that the vortex lines of the oncoming flow are all wraped around the obstacle (Fig. 3.7.i) and they are stretched so as to intensify the vorticity, which is the same as saying that they are continuously accumulated by this wrapping around the obstacle at its foot (Fig. 3.7.i).

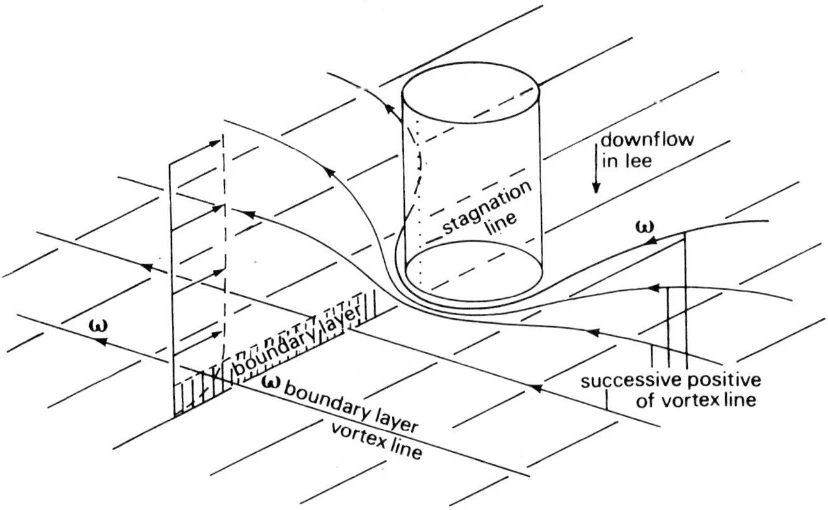

Fig. 3.7.i The vortex lines in the boundary layer are wrapped round an obstacle protruding from the boundary and produce downflow along the front stagnation line, outflow at the base of the obstacle on the upstream side, and downflow behind it.

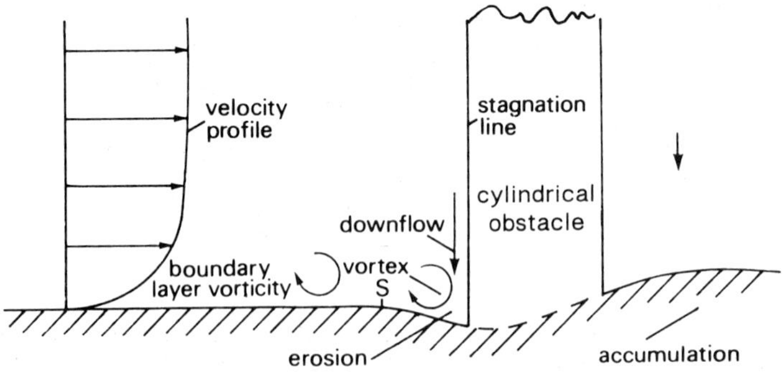

Fig. 3.7.ii The central plane of Fig. 3.7.i, showing where the bed of the flow is eroded at the foot of an obstacle.

The outflow at the base occurs all round the front of the obstacle (Fig. 3.7.iii) as far as the point where the flow separates from the sides. In the wake, two influences are at work. A general downflow is induced by the secondary vorticity; but close to the floor the flow may be into the wake. This is because the vortices in the wake, which may or may not be moving downstream, have low pressure in their centres towards which there is an inflow in the boundary layer. The consequence is that there is a an accumulation in the wake of some of the material eroded from the bed at the foot of the pillar.

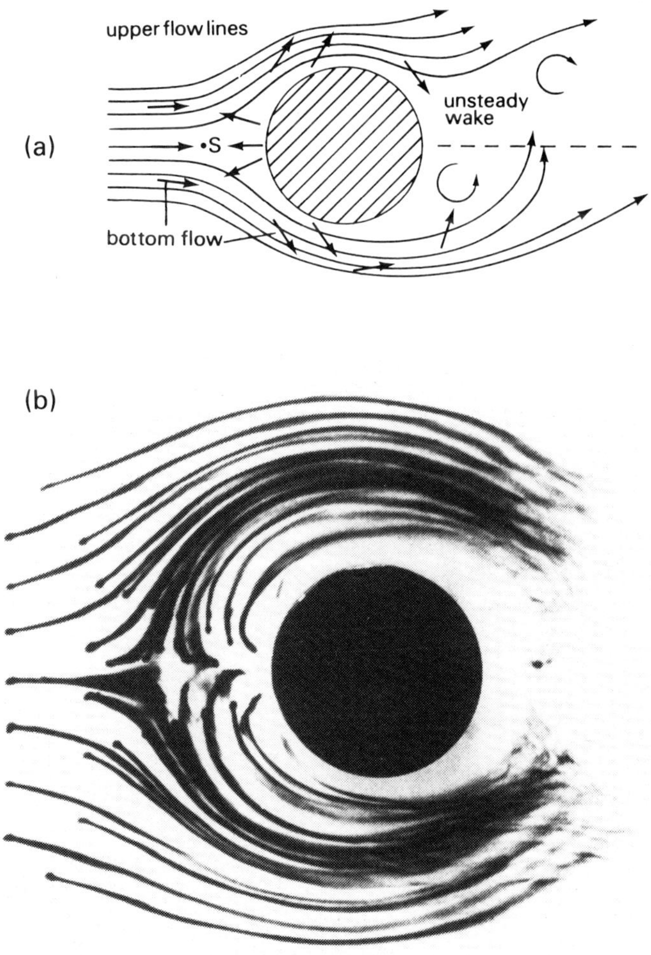

Fig. 3.7.iii (a) Main flow, shown by flow lines, and flow in the boundary layer close to the surface around a cylindrical pillar standing in a stream, shown by hevy arrows. (b) Photograph (in negative) showing streamers of water paint drawn out along the streamlines at the bottom boundary in a water channel. The paint begins to leave the boundary in the unsteady part of the flow. Outside the boundary layer the flow upstream is very similar to potential flow (see Fig. 2.13.ii) and contains a vortex street in the wake of the cylinder.

Fig. 3.7.iii(c) The flow on the bottom of the channel is seen surrounding the stagnation point S (in (a)) with the flow coming from the right. A slow flow rises from around S and can be seen in an arc held off the cylinder at a higher level. This picture is a short time exposure.

The erosion at the foot has often been the cause of the undermining of bridge piers. The most effective way to avoid this is to build the pier with a sharp upstream edge which cuts the stream into two without any significant region of stagnation. The vortex lines are cut by, rather than wrapped around, the obstacle, and as a consequence no significant secondary flow is set up (Fig. 3.7.vi(a)).

Such a design can only be used with success if the stream direction is known for certain to be more or less unvarying. In the case of a wide, shallow (or braided) river, the

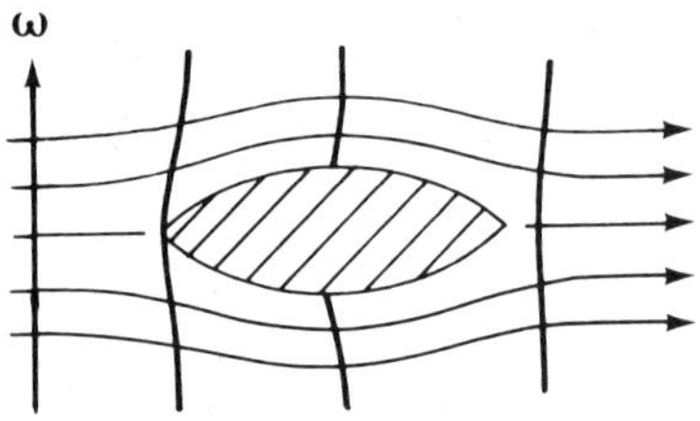

Fig. 3.7.iv (a) Vortex lines cut by a sharp upstream edge of an obstacle produce little secondary flow.

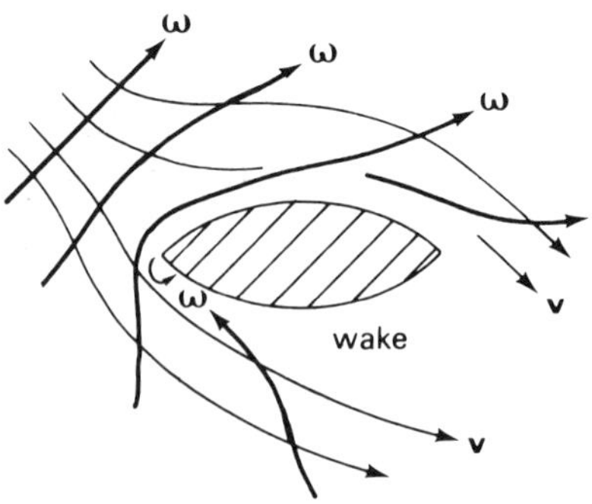

Fig. 3.7.iv (b) If the direction of flow becomes oblique to a sharp-edged pier, the vortex lines may be wrapped around it and cause disastrous erosion at its base.

direction of flow relative to a bridge pier may change after or during a period of flood and the erosion in that case, due to the wrapping of vortex lines around one side, may be disastrous (Fig. 3.7vi(b)).

In a wind stream past an obstacle such as a large cooling tower or tall building, the shear may extend to the top of the building and beyond, although it may be more intense in a shallow boundary layer close to the ground. In that case the downflow induced by the wrapping of vortex lines is present above the boundary layer, and this may cause pollution emitted from the top to be drawn down in the wake. This phenomenon must not be confused with the downwash due to vortices shed from the obstacle (see Chapter 12).

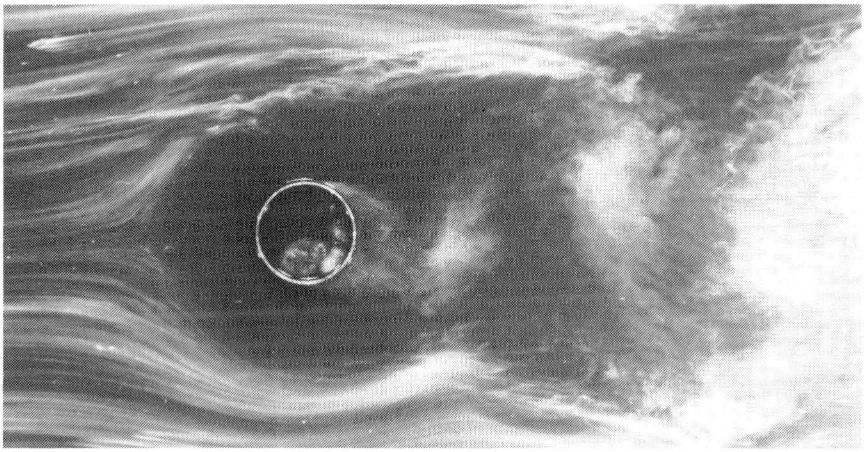

Fig. 3.7.v(a) Instantaneous bottom flow past a circular cylinder when a vortex street is shed. The flow at the surface is shown in Fig. 2.13.ii on page 67.

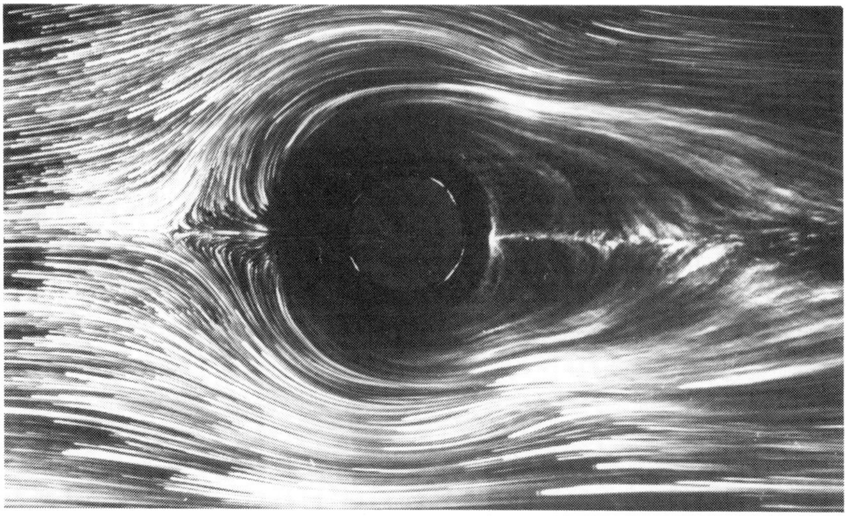

Fig. 3.7.v(b) Long 'time exposure' of the paint tracks made after the vortex street flow has been drained off. The flow is from the left.

If the surface of the water in an experimental channel is made to have a large surface tension (e.g. by deposition of dust) the motion of the surface water is retarded by attachment to the walls and to the obstacle, then the retardation produces a mirror image of the vorticity in the bottom boundary layer. The same effect may be produced by a wind blowing the surface water in the direction opposite to the flow (see Fig. 3.7.vi).

Fig. 3.7.vi The outflow from a bridge pier in the river Shannon. The water flow is towards the pier from the right. The wind is blowing from the left, so that the vorticity in the top layer of water is a vector directed into the picture, and this becomes draped around the obstacle, causing an outflow from it at the surface, and an upwelling close to the pier.

3.8 MOTION IN A TEACUP

The configuration of the vortex lines in a teacup stirred so as to give it vorticity about a vertical axis is very complex. A circulation in vertical planes, with horizontal circles as vortex lines, develops in the boundary layer at the bottom. This causes a circulation of the main flow with a slow outward flow above the boundary layer (Fig. 3.8.i). This in turn produces a vortex with a vertical axis with the same circulation at all radii and a decrease of tangential velocity almost like r^{-1}. Near the axis of symmetry the viscosity prevents the large shears that this velocity distribution would require. Consequently the rotation becomes more nearly solid, with tangential velocity varying like r, and the surface concave near the middle (see section 2.5).

The inflow in the boundary layer at the bottom produces the slow upward and outward flow in the main body of the fluid and also causes tea leaves to accumulate in the middle of the bottom of the cup even though they are heavier than water and therefore might be expected to be centrifuged to the outside.

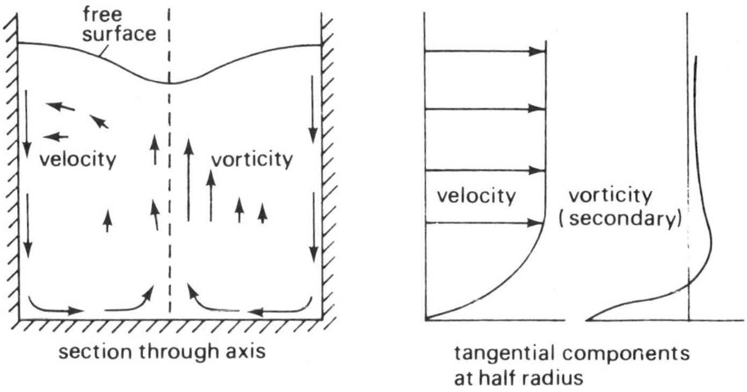

Fig. 3.8.i The velocity and vorticity components in a vertical plane through the axis of a cup with fluid rotating in it, and the velocity and vorticity profiles of the flow.

On the outer vertical wall the flow is centrifugally unstable (see section 3.10) and the fluid whose tangential velocity has been reduced there descends the wall and feeds the inflow in the bottom boundary layer with fluid of smaller circulation than the main body of fluid.

If a cup of liquid is rotated about a vertical axis of symmetry with the fluid initially at rest, a secondary flow is set up, with the fluid in the bottom boundary layer being centrifuged outwards and forced up the outer wall. This is different from the reverse of the flow just described in a few details. The flow on the outer wall is initially turbulent because of the large shear there by the instability is gradually suppressed by the stable stratification due to the increase of circulation outwards (see section 8.6).

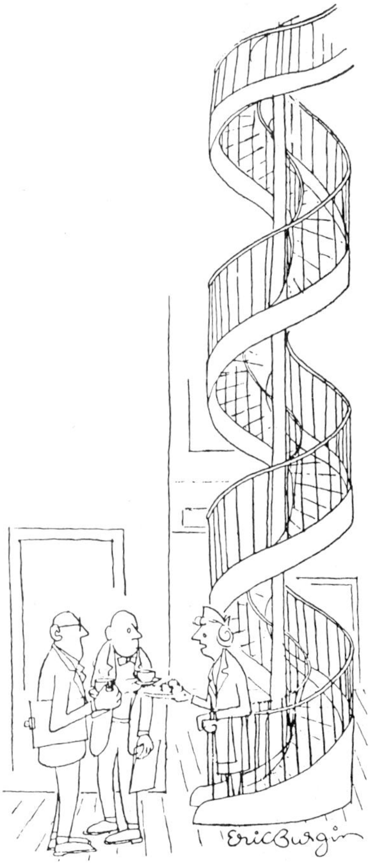

Fig. 3.8.ii 'You'll find it's already stirred.' (Reproduced by permission of *Punch*.)

3.9 FLOW IN THE CORNERS OF STRAIGHT CHANNELS

It is observed that there is a secondary circulation with flow from the centre of a channel towards a corner, and outwards from the corner along the walls. This is most noticeable in channels of triangular section, having acute-angled corners.

A ready explanation is provided as follows: the vortex lines generated in the boundary layer of a uniform stream are parallel to the walls and perpendicular to the stream (Fig. 3.9.i). In a corner the vortex lines turn to follow the general shape of the section, but since the flow in a corner is retarded more than near the centre of the wall, the vortex lines are held back there. This produces a streamwise component of vorticity with a secondary flow as described. This explanation is likely to be correct, but it must be emphasized that it does not amount to a proof that this kind of secondary flow must occur. This is because, on the face of it, the vortex lines in the corner could well coincide with the lines of constant velocity so that they would not actually be distorted.

A corollary of this argument is that the flow along a convex edge will generate a secondary flow with flow towards the edge along the solid surface and away from it elsewhere because the vortex lines will be carried forward by the main stream more rapidly near the edge (Fig. 3.9.ii). This phenomenon is not known to be common.

It is also possible that, if it is observed, it might be due to the upflow towards the edge induced in corners below.

There are two principles to guide a verbal argument in this kind of problem:

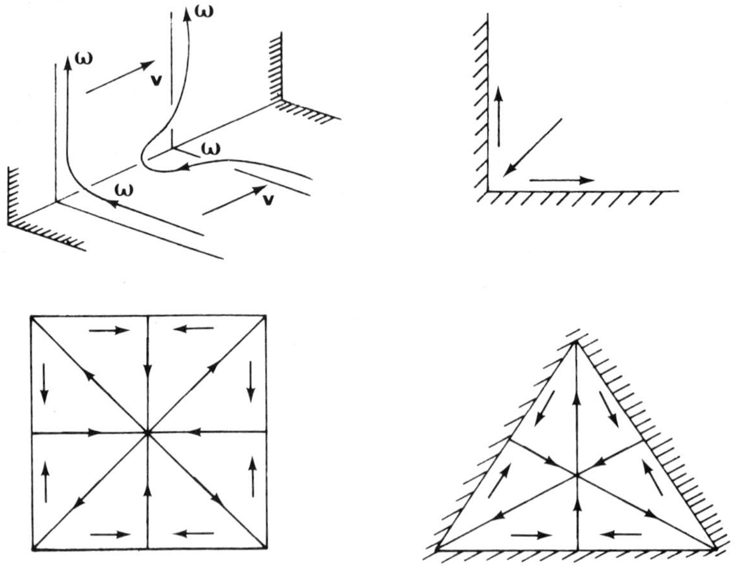

Fig. 3.9.i Distortion of the boundary layer vortex lines in a corner of a channel, and the resultant secondary flow in a cross-section of a square and of a triangular channel.

Fig. 3.9.ii Distortion of vortex lines in flow along an angular edge, and the consequent secondary flow.

(i) The vortex lines are carried along by the motion of the fluid, but it is necessary to remember that vorticity is being created in a boundary layer and is carried (each component separately) by the viscosity down the gradient (of vorticity).

(ii) The flow field associated with a vortex line (which cannot end in the fluid and must either be a closed curve or end on a boundary) is such that at any point the velocity is in the direction of the maximum rate of change of the solid angle subtended by the vortex line at the point. The l ines of flow produced by the vortex line are perpendicular to lines along which the solid angle subtended by the line is constant. Thus a circular vortex line is associated with a flow which is perpendicular to all circles coaxial with the vortex line. If such a flow does not occur in the presence of a vortex line it must be because other lines also exist in the space of the fluid.

The argument, based on qualitative inference, given above, is very weak, and is given because it is easy to remember. But any vortex line moving under its own influence is retarded where the curvature is large because of the opposing influences of the nearby elements of vorticity, whereas in places where the curvature is small it is moving under the influence of the rest of the vortex line and all the elements contribute in the same direction. If a closed vortex line is bulged outwards in the neighbourhood of one point on it, the solid angle subtended by it at points in the neighbourhood of a point on the bulge is reduced. Consequently the bulge will move so as to rotate around its unbulged position, but will not be unstable. The influence will spread out from the bulge, so that, in effect, disturbances of the position of a steady vortex line are dispersive.

4

The rotating earth

4.1 FORMAL MATHEMATICAL TREATMENT

When Ruskin wrote 'These questions about movement lead partly far away into high mathematics' he was expressing a cry of anguish which has often been echoed throughout the world of geographers and others who are very much concerned with what goes on in the great outdoors. Engineers and navigators learn the mathematical rules because they have an instinctive and habitual faith in their correctness; but many of them even then can only give fallacious physical explanations of the formulas they use. Mathematicians tend not to care about physical explanations. In all these categories of people there are exceptions, of which the reader is no doubt one; and for them we shall try to produce both mathematical and physical arguments which explain the phenomena we observe. We shall consider the mathematical argument first because that will establish the precise formula which requires physical explanation.

We shall not consider in detail the effects of the gravitational fields of the sun and moon because they only produce small effects, which scarcely influence the motion in the troposphere as much as the normal errors of measurement. For a similar reason we do not expect the effects of the earth's rotation to have effects in times which are small when compared with a day, and this is demonstrated in detail in section 5.8 for the case of gravity waves.

Newton's law of motion applies to motion relative to a frame of reference which is not accelerating. The motion of our practical frame of reference (Fig. 4.1.i), which is fixed to the surface of the earth at the point 0, which we occupy, consists of a rotation around the axis of the earth once a day, with an angular velocity denoted by Ω. To this is added a motion in an easterly direction of speed $\Omega R \sin\theta$, where R is the radius of the earth and θ is the co-latitude, and an acceleration towards the axis of the earth of magnitude $\Omega^2 R \sin\theta$. Of these the rotation and the acceleration will affect the motion relative to a frame of reference which as no acceleration. The mathematical statement of this situation is as follows.

A vector (see Fig. 4.1.ii) which has three components in the directions of axes fixed at C, such as a position vector, a displacement, or a time derivative of one of these, and which has magnitude $\mathbf{r}$ and rate of change $d\mathbf{r}/dt$ relative to the axes, has an additional

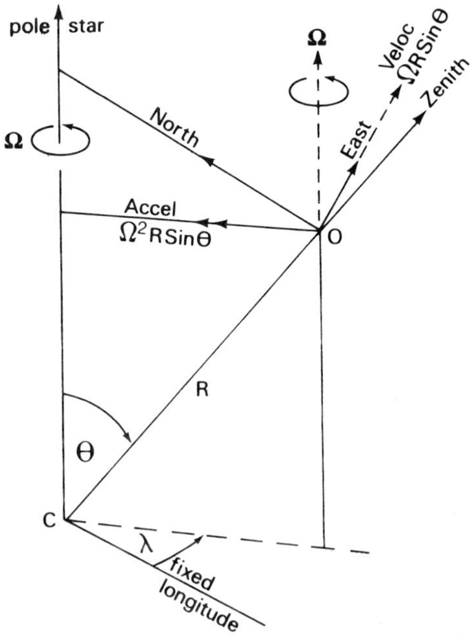

Fig. 4.1.i The velocity, acceleration and rotation of a coordinate system OENZ at a point on the earth's surface relative to a non-rotating frame at the centre of the earth. Acceleration is denoted by a double arrow.

rate of change $-\boldsymbol{\omega} \times \mathbf{r}$ relative to a set of axes, denoted by a subscript 1, which has the same origin and is rotating relative to the fixed set with angular velocity $\boldsymbol{\omega}$. Therefore

$$\left(\frac{\mathrm{d}\mathbf{r}}{\mathrm{d}t}\right)_1 = \frac{\mathrm{d}\mathbf{r}}{\mathrm{d}t} - \boldsymbol{\omega} \times \mathbf{r}. \tag{4.1.1}$$

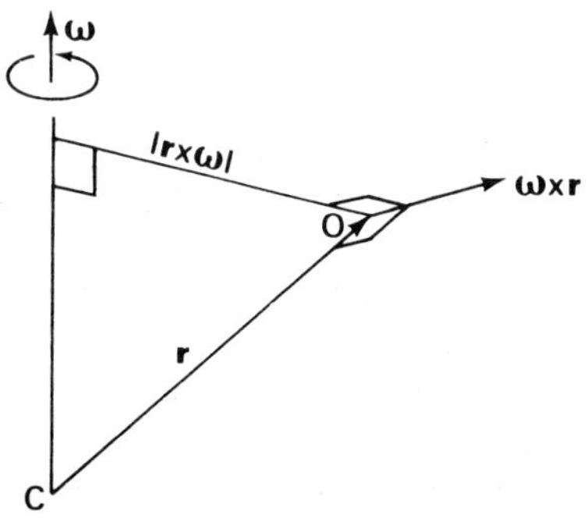

Fig. 4.1.ii Diagram illustrating the basis of the argument for equation (4.1.1).

This is simply a statement of the fact that a point whose position vector is $\mathbf{r}$ in both coordinate frames at the instant in question and which is fixed in the non-rotating frame has velocity $-\boldsymbol{\omega}\times\mathbf{r}$ in the rotating one, because $\boldsymbol{\omega}\times\mathbf{r}$ is a vector perpendicular to $\boldsymbol{\omega}$ and $\mathbf{r}$ and of magnitude equal to $\boldsymbol{\omega}\times$ (the projection of $\mathbf{r}$ on the normal to $\boldsymbol{\omega}$) (see Fig. 4.1.ii). Therefore

$$\frac{d\mathbf{r}}{dt}=\left(\frac{d\mathbf{r}}{dt}\right)_1+\boldsymbol{\omega}\times\mathbf{r}=\left[\left(\frac{d}{dt}\right)_1+\boldsymbol{\omega}\times\right]\mathbf{r}. \tag{4.1.2}$$

Evidently this equation relates the time derivatives of a vector relative to the two frames, and we may apply it a second time, or apply it directly to $d\mathbf{r}/dt$ instead of to $\mathbf{r}$. Thus

$$\begin{aligned}\frac{d}{dt}\frac{d}{dt}\mathbf{r}&=\left[\left(\frac{d}{dt}\right)_1+\boldsymbol{\omega}\times\right]\left[\left(\frac{d}{dt}\right)_1+\boldsymbol{\omega}\times\right]\mathbf{r}\\&=\left(\frac{d^2\mathbf{r}}{dt^2}\right)_1+\left(\frac{d}{dt}\right)_1\boldsymbol{\omega}\times\mathbf{r}+\boldsymbol{\omega}\times\left(\frac{d\mathbf{r}}{dt}\right)_1+\boldsymbol{\omega}\times(\boldsymbol{\omega}\times\mathbf{r})\\&=\left(\frac{d^2\mathbf{r}}{dt^2}\right)_1+2\boldsymbol{\omega}\times\left(\frac{d\mathbf{r}}{dt}\right)_1+\boldsymbol{\omega}\times(\boldsymbol{\omega}\times\mathbf{r})\end{aligned} \tag{4.1.3}$$

because $\boldsymbol{\omega}$ is a constant; or

$$\frac{d\mathbf{V}}{dt}=\left(\frac{d\mathbf{v}_1}{dt}\right)_1+2\boldsymbol{\omega}\times\mathbf{v}_1+\boldsymbol{\omega}\times(\boldsymbol{\omega}\times\mathbf{r}) \tag{4.1.4}$$

where $\mathbf{V}$ is the velocity relative to the fixed frame, and $\mathbf{v}_1$ relative to the rotating frame. $\boldsymbol{\omega}$ is assumed to be a constant.

This gives the acceleration 'in space' in terms of the position, velocity and acceleration relative to a rotating frame with origin at the centre of the earth, C. If we now transfer to a rotating frame with origin at the point 0 on the earth's surface we are at, the vorticity $\mathbf{v}_1$ is the same as the velocity relative to our local frame fixed to the earth, which we call $\mathbf{v}$. Therefore

$$\frac{d\mathbf{V}}{dt}=\left(\frac{d\mathbf{v}}{dt}\right)_0+2\boldsymbol{\omega}\times\mathbf{v}+\boldsymbol{\omega}\times(\boldsymbol{\omega}\times\mathbf{r}) \tag{4.1.6}$$

where subscript 0 indicates the value in the frame of reference with which we are moving at 0, and $\mathbf{r}$ is the radius vector from the centre of the earth.

An important particular case of acceleration is that due to gravity. The weight vector of a body of unit mass as normally measured is the mass multiplied by '**apparent gravity**. This is defined as the acceleration of a freely falling particle, instantaneously at rest, relative to our frame moving with the earth. If we call this quantity $\mathbf{g}^*$, then since $\mathbf{g}$ is the actual acceleration we have that, since $\mathbf{v}=\mathbf{o}$

$$\mathbf{g}=\mathbf{g}^*+\boldsymbol{\omega}\times(\boldsymbol{\omega}\times\mathbf{r}). \tag{4.1.6}$$

This relates the attraction of the earth towards its centre C to the apparent vertical, down which $\mathbf{g}^*$ is directed, the difference being due to the centrifugal force represented by $\boldsymbol{\omega} \times (\boldsymbol{\omega} \times \mathbf{r})$, whose magnitude is $\Omega^2 R \sin\theta$ as we have already remarked. The earth is an oblate spheroid and the vertical does not point to the centre of the earth except at the poles and on the equator (see Fig. 4.1.iii).

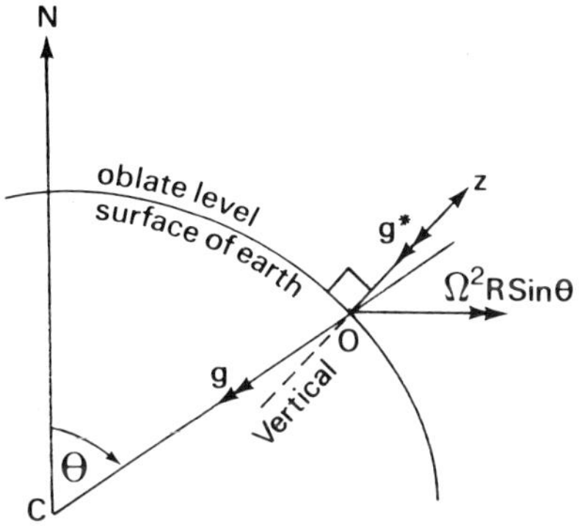

Fig. 4.1.iii Apparent gravity is perpendicular to the surface of the earth, which is an oblate sphere. The force of gravity is towards the centre of the earth, and the difference is the centrifugal force of the earth's rotation.

The equation of motion for fluid relative to our local axes fixed to the earth is therefore derived from the basic equation

$$\frac{d\mathbf{V}}{dt} = \mathbf{F} + \mathbf{g}, \tag{4.1.7}$$

where $\mathbf{F}$ represents *all* the forces other than gravity, and is

$$\left(\frac{d\mathbf{v}}{dt}\right)_0 + 2\boldsymbol{\omega} \times \mathbf{v} = \mathbf{F} + \mathbf{g}^*. \tag{4.1.8}$$

From now on we shall write this in the conventional form

$$\boxed{\frac{D\mathbf{v}}{Dt} = -\frac{1}{\rho}\,\mathbf{grad}\,p + \mathbf{v} \times \mathbf{f} + \mathbf{g} + \mathbf{F}.} \tag{4.1.9}$$

$\mathbf{F}$ now includes any viscous or other forces *not* due to the pressure gradient or gravity, and $\mathbf{g}$ is identified as is more usual with observed, or 'apparent', gravity, namely

$$\mathbf{g} = (0,\ 0,\ -g). \tag{4.1.10}$$

The quantity

$$\mathbf{f} = 2\boldsymbol{\Omega} = (0, 2\Omega\sin\theta, 2\Omega\cos\theta) = (0, f_2, f_3) \tag{4.1.11}$$

Fig. 4.1.iv Practical demonstration of the rotation of the earth carried out by Foucault at the Panthéon in Paris.

is called the **Coriolis parameter**, the components being in the eastward, northward and upward vertical directions.

The symbol f is conventionally used for the Coriolis parameter in discussions of our topic. The vector **f** is used elsewhere, particularly in Chapter 3, to represent the acceleration of the fluid, but there is no cause for confusion. Thus f usually means f_3 because the effect of f_2 is very small. In equation (4.1.9) f_2 is multiplied by the vertical velocity only, and this is usually small compared with the wind.

4.2 PHYSICAL INTERPRETATION

The quantity $\mathbf{v} \times \mathbf{f}$ appears to have been added to the forces on a fluid which produce its acceleration. It is usually called the 'deviating force', or 'Coriolis force' after the French engineer who died in 1843. The horizontal component of this force, which is due to the vertical component of **f**, is an acceleration to the right of magnitude vf_3, where v is the horizontal speed relative to the ground.

In this chapter the discussion will view the fluid FROM ABOVE in the NORTHERN HEMISPHERE. To interpret it for the southern hemisphere it may be viewed as if from below, i.e. from inside the earth, or left and right may be interchanged. CYCLONIC means COUNTERCLOCKWISE WHEN VIEWED FROM ABOVE in the northern hemisphere, but clockwise in the southern hemisphere. Clockwise is a northern hemisphere convention: clocks turn with the sun, which is opposite the earth's rotation. CYCLONIC means rotating WITH THE EARTH.

Considering therefore the vertical component of the earth's rotation, it causes the ground to rotate cyclonically beneath us. If we travel from O to P at a uniform speed v in time t (Fig. 4.2.i) the rotation of the earth will have caused our end point also to have traversed the circular arc P_0P along which it has travelled from its original position P_0, and we shall in space have traversed the circular arc from O to P. Since the vertical component of the earth's rotation is $\Omega \cos \theta$, the angle P_0OP is $t\Omega \cos \theta$ and the arc P_0P traversed by P is $vt \times \Omega \cos \theta$. In time t with acceleration a from rest a particle traverses a distance $\frac{1}{2}at^2$. The acceleration in this case therefore is $2v\,\Omega \cos \theta$.

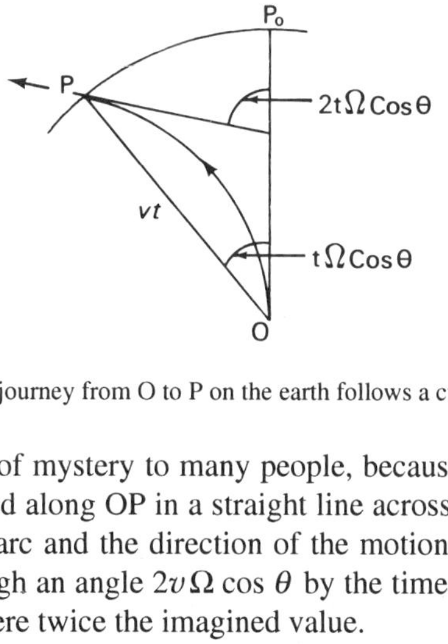

Fig. 4.2.i A straight journey from O to P on the earth follows a curved path in space.

The factor 2 is a source of mystery to many people, because it appears that in getting to P one has merely travelled along OP in a straight line across the ground. However, the track is Along the circular arc and the direction of the motion 'in space' (i.e. relative to fixed axis) has turned through an angle $2v\Omega \cos \theta$ by the time it arrives at P: it therefore behaves as if the rotation were twice the imagined value.

The horizontal component of the earth's rotation is $\Omega \sin \theta$ and is around a horizontal axis pointing northward. The eastern horizon is sinking and the western horizon is rising (Fig. 4.2.ii), so that the weight of a particle is decreased or increased according as it moves eastward or westward by an amount $2v\Omega \sin \theta$ per unit mass. With a typical velocity of 10 m s^{-1} and Ω equal to $2\pi/(24 \times 60 \times 60)$ rad s^{-1}, this acceleration is of the order of 10^{-3} m s^{-2}, which is about $10^{-4} \times$ gravity. It is therefore not productive of any discernible effects because small changes in temperature produce larger changes in the weight of the air. If the whole depth of the atmosphere were moving at 10 m s^{-1} the pressure at the ground would be changed by about 0.1 mb, which is about the weight of

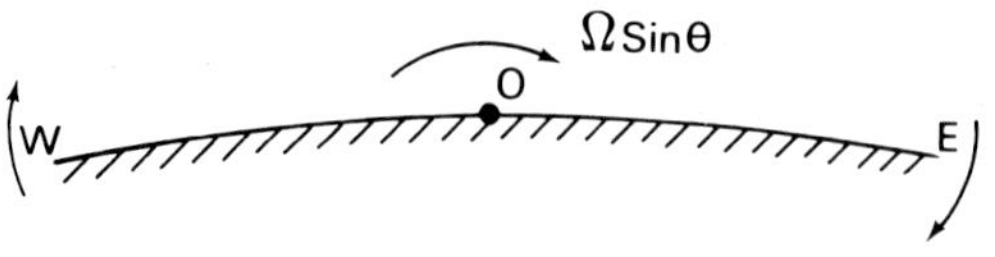

Fig. 4.2.ii The horizontal component of the earth's rotation causes the western horizon to tip upwards

the bottom metre of air. The pressure at the ground varies normally by a few mb in the time between successive weather charts and across the area represented on the chart, and the velocity does not change enough to affect the last significant figure reported in a pressure measurement, namely 0.1 mb.

Ferrel's law (after the American meteorologist, d. 1891) states that the air is deviated to the right as a result of the earth's rotation; Buys-Ballot's law (after the Dutch meteorologist, d. 1890) states that if the motion is straight there is lower pressure on the left. These are merely particular applications of the principle, half a century after the death of Coriolis.

4.3 THE RELATIONSHIP BETWEEN PRESSURE GRADIENT AND WIND

There can be no general relationship between pressure gradient and wind because it is related to the acceleration. Decelerating air is most often moving towards higher pressure. It is not correct to say that but for the earth's rotation the air would move towards low pressure; however, we may say, that but for friction it would move equally often towards high pressure. The fact that it moves nearly along the isobars is a measure of the small influence of friction on the motion, because it means that on the whole the pressure field is not being required to do much work on the air.

The facts that the pressure is the easiest parameter of the air to measure accurately and that the air motion is nearly horizontal makes it easy to get a useful relationship. As a first approximation the vertical component of (4.19) is

$$\boxed{0 = -\frac{1}{\rho}\frac{\partial p}{\partial z} - g} \tag{4.3.1}$$

because the vertical acceleration is small compared with g the vertical velocity being small, and the deviating and frictional forces are also negligible. This is the well known **hydrostatic equation**, and deviations from it in the atmosphere due to vertical acceleration are usually negligible except in waves of the kind described in Chapter 5, and in some violent storm downdraughts or funnel clouds.

There are several types of horizontal motion, according to the relative importance of the terms in the horizontal component of (4.1.9), namely

$$\frac{\mathrm{D}\mathbf{v}_\mathrm{h}}{\mathrm{D}t} = -\frac{1}{\rho}\mathbf{grad}_\mathrm{h}p + \mathbf{v}_\mathrm{h} \times \mathbf{f} + \mathbf{F}_\mathrm{h} \tag{4.3.2}$$

the horizontal component of the velocity **v**, of **F** and of the gradient of p being given a subscript h, and the vertical component of **f** being taken.

(1) The geostrophic wind

Very often the air travels large distances with only a small change in speed and direction during a day. The acceleration is nearly zero. Also the frictional force is small compared with the remaining terms and all that is then left of (4.3.2) is

$$0 = -\frac{1}{\rho}\frac{\partial p}{\partial x} + vf_3 + 0 \tag{4.3.3}$$

in which the x direction is taken across the direction of motion and v is the speed. The wind is then in **geostrophic** (earth-turning) **balance** and

$$\boxed{v_G = \frac{1}{\rho}\frac{\partial p}{\partial x},} \tag{4.3.4}$$

where f is written for f_3 and v_G is the geostrophic wind speed. It is inversely proportional to the spacing of the isobars, with low pressure on the left.

(2) The ageostrophic wind

This is the departure of the actual wind from geostrophic, denoted by v_A. Thus

$$\mathbf{v} = \mathbf{v}_G + \mathbf{v}_A. \tag{4.3.5}$$

In the absence of friction, therefore, but with a significant acceleration, we find, on subtracting (4.3.3) from the appropriate component of (4.3.2), that

$$\boxed{\frac{D\mathbf{v}}{Dt} = \mathbf{v}_A \times \mathbf{f}.} \tag{4.3.6}$$

The acceleration is thus proportion to the ageostrophic wind and directed at right angles to the left of it.

(3) Antitriptic winds

These were defined by Jeffreys to be those in which the frictional force balanced the pressure gradient force. The concept is only useful for winds close to the ground where the frictional force is opposed to the velocity. The motion is then directed towards low pressure. If we imagine that the drag is proportional to kv^2 we would obtain

$$kv^2 = -\frac{1}{\rho}\frac{\partial p}{\partial y} \tag{4.3.7}$$

where the y-direction is the direction of motion and v the speed.

A wind in a ravine or canyon, a sea breeze, katabatic or anabatic wind, usually has a significant antitriptic component.

(4) Shearing stresses

If the wind varies with height, from causes to be discussed below, and there is some mixing of superposed layers, momentum is transferred to the slower-moving layer.

In the case of drag at the surface, work must be done to overcome it and so there is a component of velocity towards low pressure (Fig. 4.3.i). On the other hand the friction at the surface may be small, as it often is over the sea, while the wind speed increases with height enough for the lower layers to be dragged forward by the upper layers, whose direction may also differ. The wind then has a component towards high pressure (Fig. 4.3.ii). It is not correct to say that because of friction the surface wind always has a component towards low pressure: it may have one towards high pressure, and if it has one towards low pressure this could be on account of a retarding stress due to layers above moving more slowly or in a different direction. In the trade winds the flow across the isobars is primarily due to the slower-moving upper layers, a change in direction with height, or an equatorward drift. The matter is discussed further in section 4.11.

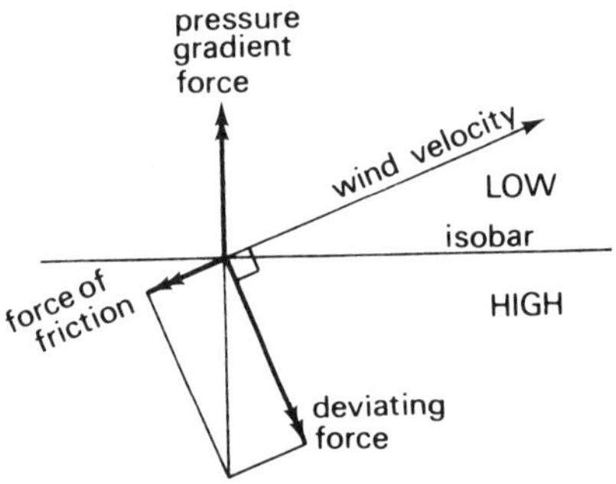

Fig. 4.3.i Deviation of the wind from the isobar on account of surface friction.

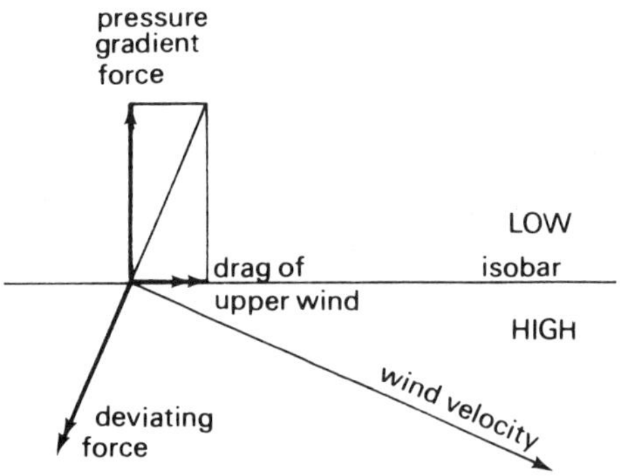

Fig. 4.3.ii Deviation of the wind from the isobar due to drag of upper wind.

(5) The gradient wind: curved flow

When the motion is circular we may readily represent the acceleration by the quantity v^2/r, where r is the radius of curvature of the air path. If the only other forces are the pressure gradient force and deviating force, we have the following equation, balancing the radial acceleration with the radial forces:

$$\boxed{\mp\frac{v^2}{r} = \mp\frac{1}{\rho}\frac{\partial p}{\partial r} + vf.} \tag{4.3.8}$$

The upper sign is appropriate to cyclonic motion in which the pressure gradient force is radially inwards and the deviating force outwards. They are reversed for an anticyclone (Fig. 4.3.iii).

The **gradient wind equation** (4.3.8) is a good approximation to reality when the flow is curved without much tangential acceleration.

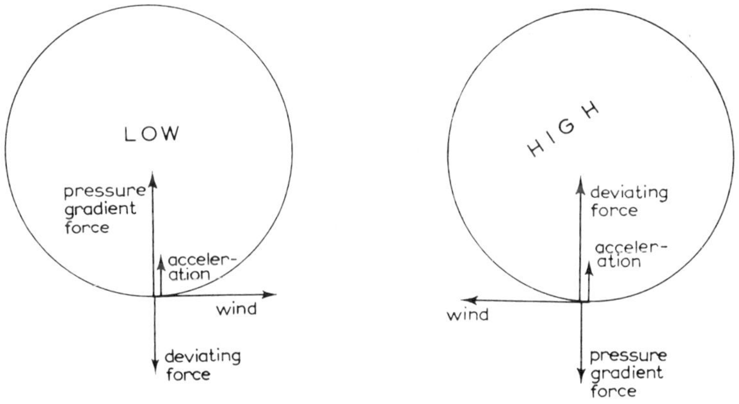

Fig. 4.3.iii Balance of forces in curved frictionless flow.

(6) Changing pressure field: isallobaric wind

The isallobaric wind is that component of the wind due to the changing pressure field. If, during the change, the wind is roughly geostrophic we may differentiate the geostrophic wind equation with respect to time and obtain

$$0 = -\frac{\partial}{\partial t}\left(\frac{1}{\rho}\ \mathbf{grad}\, p\right) + \frac{\partial}{\partial t}(\mathbf{v} + \mathbf{f}) \tag{4.3.9}$$

or

$$\mathbf{f} \times \frac{\partial \mathbf{v}}{\partial t} = -\frac{1}{\rho}\ \mathbf{grad}\ \frac{\partial p}{\partial t}. \tag{4.3.10}$$

If we now assume that there is very little variation in the acceleration from place to place we may equate $\mathrm{D}v/\mathrm{D}t$ to $\partial v/\partial t$. This is by no means always justified in practical cases, but pursuing this approximation we have

$$\mathbf{f} \times \frac{D\mathbf{v}}{Dt} = -\frac{1}{\rho}\,\mathbf{grad}\,\frac{\partial p}{\partial t} = \mathbf{f} \times \mathbf{v}_i \times \mathbf{f} \qquad (4.3.11)$$

and find that in this case the isallobaric wind is the ageostrophic wind given by (4.3.6). The **isallobaric wind** is therefore directed at right angles to the isallobars (lines of constant $\partial p/\partial t$) towards the region of greatest rate of fall of pressure. It has magnitude given by

$$\boxed{\mathbf{v}_i = \frac{1}{\rho f^2}\frac{\partial \dot{p}}{\partial x}} \qquad (4.3.12)$$

because $\mathbf{v}_i$ and $\mathbf{f}$ are approximately perpendicular.

This classical piece of theory is a very unsatisfactory explanation of why the wind blows across the isobars towards a region of maximum pressure fall when the gradient is increased very rapidly. For if the fall is rapid, geostrophic flow does not have time to be approximately balanced and the inflow merely represents this fact. (4.3.12) is then quite inaccurate because in a small fraction of a day the Coriolis effect cannot be dominant. Reality is more complex.

(7) The geostrophic acceleration: confluence and diffluence

When the flow is almost straight and the pressure distribution nearly steady so that curvature and isallobaric effects can be neglected, the ageostrophic component may be deduced from the rate of change of the geostrophic wind following the particles along the streamlines, which are approximately the isobars.

Thus the air flows across the isobars towards lower pressure where it is being accelerated and towards higher pressure where it is being decelerated. This is illustrated in Fig. 4.3.iv, and is common at the entrance and exit of a jet stream.

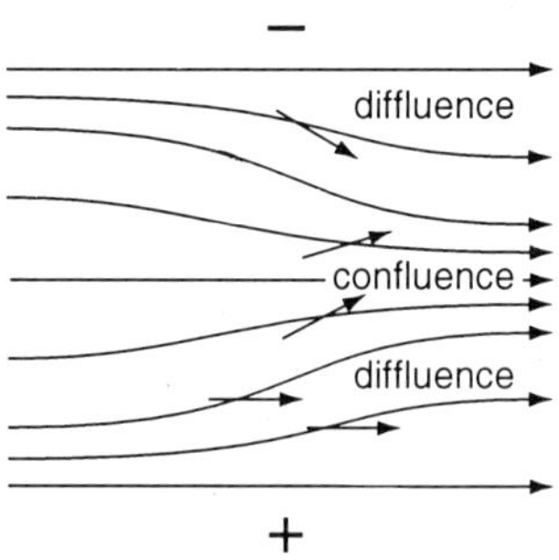

Fig. 4.3.iv Ageostrophic motion due to confluence and diffluence.

If the flow is geostrophic in these cases it would also be non-divergent, which means it would have no vertical component of velocity; but when a vertical velocity occurs it would be called convergent or divergent in the horizontal flow field. The ageostrophic flow (across the isobars) would be given approximately by

$$v_A = \frac{1}{f}(\mathbf{v}_G \cdot \mathbf{grad})\mathbf{v}_G, \tag{4.3.13}$$

the direction being across the isobars towards low pressure for an acceleration.

4.4 MOTION IN THE UPPER AIR: ISOBARIC CONTOURS

It is not convenient to draw maps of the pressure distribution except at the surface where it is corrected to sea level. For the upper air we draw contours of an isobaric surface. Fig. 4.4.i shows a vertical cross-section of an isobaric surface which is at height h at A. If Δp is the difference between the pressure at A and at B, both at height h, since the pressure is the same at A as at A′ a height Δh above B, we must have that $\Delta p = g\rho\Delta h$. Therefore

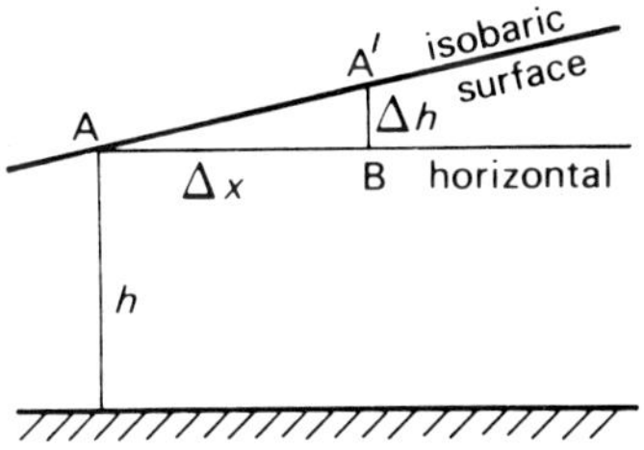

Fig. 4.4.i Relationship between isobars at constant height and contours of an isobaric surface.

$$\frac{1}{\rho}\frac{\partial p}{\partial x} = g\frac{\partial h}{\partial x} \tag{4.4.1}$$

an isobar at height h may be represented by the contour of the isobaric surface through the point. The **geostrophic wind equation** (4.3.4) may be written

$$\boxed{\mathbf{v}_G = \frac{g}{f}\frac{\partial h}{\partial x}} \tag{4.4.2}$$

and a map of contours of an isobaric surface looks very like a map of isobars at a constant height. The advantage of the former is that the factor g/f is independent of the density, and therefore of height, and the same scale factor may be used for contour maps at any height to relate the geostrophic wind to the spacing of the contour lines.

4.5 THE THERMAL WIND

If there is a horizontal gradient of temperature the horizontal pressure gradient must change with altitude because two neighbouring columns of air do not have the same weight. Fig. 4.5.i represents a vertical cross-section of the air perpendicular to the

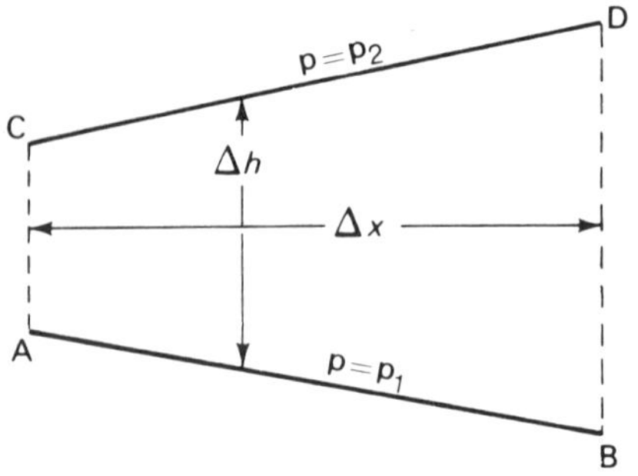

Fig. 4.5.i Relationship between thermal wind and isobaric thickness.

horizontal isotherms in constant pressure surfaces p_1 and p_2. Since (4.4.2) holds at the two levels denoted by subscripts 1 and 2, the difference in geostrophic wind Δv_G is given by

$$\Delta v_G = \frac{g}{f} \Delta \frac{\partial h}{\partial x} = \frac{g}{f} \frac{\partial}{\partial x} \Delta h \tag{4.5.1}$$

where Δh is the thickness of the layer between the two isobaric surfaces. Since the height of a column of air of given horizontal section and given weight is proportional to temperature, the thickness is proportional to temperature, at a given pressure. This means that

$$\frac{\partial}{\partial x} \Delta h = \frac{\Delta h}{T} \left(\frac{\partial T}{\partial x} \right)_{p=\text{const}}, \tag{4.5.2}$$

T being the temperature (in K). Therefore

$$\frac{\partial v_G}{\partial z} \simeq \frac{\Delta v_G}{\partial h} = \frac{g}{fT} \left(\frac{\partial T}{\partial x} \right)_{p=\text{const}}. \tag{4.5.3}$$

At a level where the isobaric surfaces are horizontal, the vertical gradient of geostrophic wind blows along the isotherms with low temperature on the left, and is therefore called the **thermal wind**. Where the isobaric surfaces are not horizontal a correction is required, and so in practice the isotherms are not drawn on charts, but instead contours of the thickness between two isobaric surfaces are drawn, and these represent the difference between the winds at the two levels. This is equivalent to using (4.5.1) rather than (4.5.3). The name 'thermal wind' persists because the wind shear is attributable to horizontal temperature gradients.

4.6 ESTIMATING THE VERTICAL VELOCITY

The problem facing a weather forecaster is as follows: all cloud formation and rain are due to upward motion of the air, and the disappearance of cloud is due to subsidence. But the vertical velocity of the air over a large area is too difficult to measure because of its small magnitude (10^{-3} up to 1 m s^{-1}). Furthermore, because of the sparsity of observations and their lack of accuracy, it is not possible to measure the horizontal wind so that the horizontal gradients of u and v can reveal w by means of the simplified version of the equation of continuity, namely

$$-w = \int \left(\frac{\partial u}{\partial x} + \frac{\partial v}{\partial y} \right) dz. \tag{4.6.1}$$

To add to the difficulty, if we were to measure the pressure accurately, and were to assume the horizontal wind to be geostrophic to determine its value, we would find it to be non-divergent, and leading to zero vertical velocity. Mathematically we note that the components of the geostrophic wind are

$$u_G = -\frac{1}{\rho f}\frac{\partial p}{\partial y}, \quad v_G = \frac{1}{\rho f}\frac{\partial p}{\partial x} \tag{4.6.2}$$

so that

$$\frac{\partial u_G}{\partial x} + \frac{\partial v_G}{\partial y} = 0 \tag{4.6.3}$$

except for horizontal gradients of ρf, which are relatively too small to lead to a significant value of w; and in any case, variations in ρf are not a significant cause, and are not directly related to vertical motions.

In other words, vertical motion is associated with the ageostrophic component of the wind. We are not concerned here with the mathematical techniques used in numerical weather forecasting, but with understanding the basic mechanisms whereby upward motion can be associated with measurable and, preferably, easily observable phenomena. For some people these ideas are very helpful in understanding what can be seen in the sky and on a weather chart an in making one's own forecast or making the best use of public ones.

When vertical motion occurs there must be a stretching of vertical columns of fluid because the vertical velocity is zero at the ground. Almost the whole atmosphere possesses vorticity because it is rotating with the earth, and even in anticyclones the absolute vorticity is positive, i.e. in the same direction as the earth's rotation, even though it is less than that of the earth. Consequently vertical stretching of the vortex lines intensifies the vorticity, most signifi-cantly near the ground.

At some **high altitude** the vertical column of fluid must be subject to vertical compression and horizontal divergence of the flow, and this results in a decrease in the vertical component of vorticity, probably to a value less than that of the earth. In consequence, relative to the earth, the circulation **appears anticyclonic**.

Near the bottom of the atmosphere, therefore, there must be an increase of cyclonic vorticity when air rises. This means an increase in the pressure gradient, and therefore a fall in pressure at the centre beneath the rising air. This is the logic behind the close association of falling pressure and rising air, which in turn is associated with cloud and rain. Thus it is not so much low pressure as falling pressure that is associated with rising air: low pressure has been thought of as the cause of rain and the association has sometimes been explained by asserting that friction causes the air to flow in towards low pressure in the manner we have discussed in Chapter 3. It should be pointed out that the rain is associated mainly with troughs of low pressure in the map of the isobars, and that behind a moving trough the air is very often subsiding and clear of cloud, and that the vertical motion is mainly due to the lifting of warm air by the often subsiding cold air, and has little to do with friction. It is true that friction is a cause of inflow, but this often has the effect of causing the low pressure centre to fill up because the friction decreases the circulation. Thus tropical cyclones tend to die out (fill up) over land because the sea is a source of heat and moisture and has reduced surface friction.

In practice, therefore, a forecaster estimates the amount of vertical motion as much by keeping a watch on the vorticity of the geostrophic wind as by any other means. This can be determined directly from the pressure field and is equal to $\frac{1}{\rho f}\nabla^2 p$. This vorticity is increased by convergence at low levels, and low-pressure centres are formed or deepened by convergence. The amount of convergence that is produced in this way far exceeds that due to friction in those latitudes where the earth's rotation is important. Close to the equator, where the vertical component of the earth's rotation is negligible, large gradients of pressure are not generated.

4.7 SUTCLIFFE'S DEVELOPMENT THEORY

This theory is not explicitly used today because the predictions made with modern computer models are usually better. The computer gives answers but does not 'explain' details of how it obtains its answers. In making the models, the forecasters have decided what factors to include and how to interpret the models' output in terms of, weather. Undoubtedly, Sutcliffe's theory played a large part in deciding how to include many of the factors in the model. This point is discussed further in Chapter 16 in connection with climate change.

Development is the name given to the formation of low-pressure centres and the vertical motion and weather phenomena which accompany it. We concentrate our attention on the horizontal acceleration field, which is the cause of all the ageostrophic motion in the absence of friction.

If subscripts 1 and 2 denote the wind vector at two levels, $\mathbf{grad}_h$ denotes the horizontal component of the gradient, and a prime denotes the difference between the two levels, then

$$\mathbf{v}' = \mathbf{v}_2 - \mathbf{v}_1 \tag{4.7.1}$$

and this is related to the thickness pattern (see equation (4.5.1) referred to in section 4.5. The acceleration at the upper level is

$$\frac{D\mathbf{v}_2}{Dt} = \frac{\partial \mathbf{v}_2}{\partial t} + (\mathbf{v}_2 \,.\, \mathbf{grad}_\mathrm{h})\mathbf{v}_2$$

$$= \frac{\partial \mathbf{v}_2}{\partial t} + (\mathbf{v}_1 \,.\, \mathbf{grad}_\mathrm{h})\mathbf{v}_1 + (\mathbf{v}' \,.\, \mathbf{grad}_\mathrm{h})\mathbf{v}_1 + (\mathbf{v}_2 \,.\, \mathbf{grad}_\mathrm{h})\mathbf{v}'. \qquad (4.7.2)$$

Consequently

$$\frac{D\mathbf{v}_2}{Dt} - \frac{D\mathbf{v}_1}{Dt} = (\mathbf{v}' \,.\, \mathbf{grad}_\mathrm{h})\mathbf{v}_1 + \left[\frac{\partial}{\partial t} + \mathbf{v}_2 \,.\, \mathbf{grad}_\mathrm{h}\right]\mathbf{v}'. \qquad (4.7.3)$$

According to (4.3.6) the ageostrophic motion is entirely produced by the acceleration. Since it is also responsible for the convergence or divergence of the horizontal component of the motion it is responsible for the vertical motion. Upward motion is associated with convergence below and divergence above, and also with increasing cyclonic circulation and falling pressure in the lower layers. Equation (4.7.3) gives the difference between the accelerations above and below and expresses this difference, which determines the amount of cyclogenesis, in terms of the wind field below ($\mathbf{v}_1$), the advection by the thermal wind ($\mathbf{v}' \,.\, \mathbf{grad}_\mathrm{h}$), and the rate of change of the thermal wind field ($\mathbf{v}'$) as seen at the upper level (where $\mathbf{v} = \mathbf{v}_2$). We now consider the various mechanisms involved.

(1) Thermal steering of the surface pressure field

As a practical approximation to the wind field we may take the geostrophic wind, and for $(\mathbf{v}' \,.\, \mathbf{grad}_\mathrm{h})\mathbf{v}_1$ we write $(\mathbf{v}'_\mathrm{G} \,.\, \mathbf{grad}_\mathrm{h})\mathbf{v}_\mathrm{G}$. This is a vector in the direction in which the surface isobars are turned as we move along with the thermal wind, i.e. along the isotherms of the thickness pattern with cold air on the left. This moves the surface pressure pattern in the direction of the thermal wind, and is called **thermal steering**.

To exemplify the argument leading to this conclusion we imagine a closed cyclonic circulation in the surface isobars, with straight isotherms of the thickness pattern across the top of it (Fig. 4.7.i). The cold air is indicated to lie to the north so that this is a simplified version of a typical mid-latitude situation of a cyclone in the westerlies. The low-level geostrophic wind $\mathbf{v}_{\mathrm{G}1}$ is circulating around the depression, equally called a low, or cyclone. As we pass along the direction of the thermal wind from A to B, C to E, or F to G, the surface geostrophic wind changes by the addition of a vector directed northwards (i.e. along FA, or GB). This is the contribution of the thermal steering term (first on the right-hand side) to the relative acceleration given by (4.7.3). According to (4.3.6) the ageostrophic wind is directed at right angles to this to the left, namely westwards (to the left in Fig. 4.7.i(a)). This means that the upper air is undergoing a displacement as shown in the middle of the lower diagram, namely to the left relative to the lower layers, and this implies upward motion with cyclonic development at the surface ahead of the depression and anticyclonic development as a consequence of downward motion behind it. Thus the depression is moved in the direction of the thermal wind.

A second example is provided by a pattern of diffluence, as shown in Fig. 4.7.ii. Here the geostrophic wind is decreasing in the downstream direction, which is also taken to be the direction of the thermal wind; and so the relative acceleration is a vector directed in

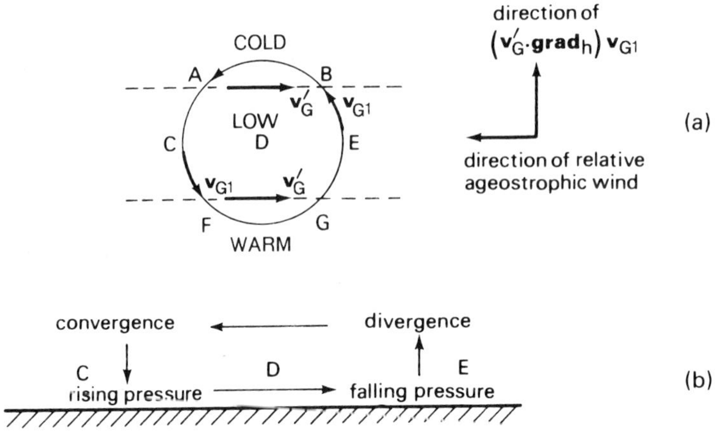

Fig. 4.7.i Mechanism of movement of a low steered by the thermal wind: (a) represents the pressure map; (b) is the vertical cross-section of (a).

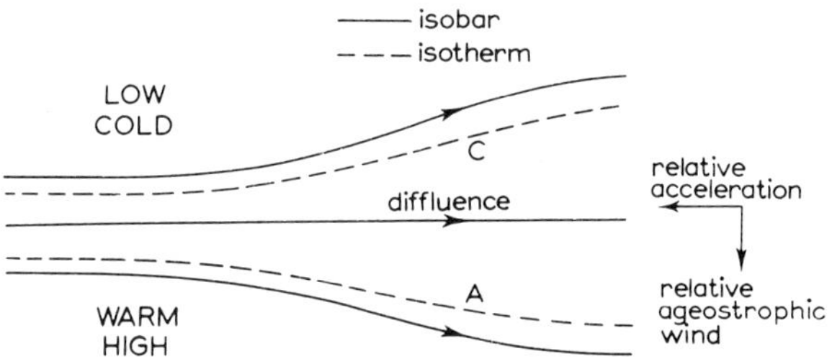

Fig. 4.7.ii Movement of a diffluence pattern in the direction of the thermal wind.

the upstream direction, opposite to the wind in this case. The relative ageostrophic wind is therefore at right angles to this, indicating low-level convergence on the left and divergence on the right. Therefore there is cyclogenesis (C), or falling pressure, on the left and anticyclogenesis (A), or rising pressure, on the right, so that the pressure gradient is increased, and the pressure pattern is moved in the direction of the thermal wind. It is interesting to compare this with the ageostrophic motion indicated in Fig. 4.3.iv, which is deduced without reference to the upper air and the thermal wind pattern, and indicates a quite different effect.

Equation (4.7.3) could equally be written with the subscripts 1 and 2 interchanged and the sign of $\mathbf{v}'$ changed. Then we would be looking at how the thermal wind advected the upper air pressure field. But similar conclusions would not be valid because there is not necessarily the same relationship between cyclogenesis and vertical motion as exists at the surface where there is a rigid horizontal boundary to it. One could argue that of course the pressure pattern is moved along with the upper wind because it is due to the weight of the upper air: but that would be erroneous because the pressure field at the surface is not actually due to the upper air which moves along with the thermal wind. The isotherms in the case illustrated in Fig. 4.7.i do not indicate warm air over the cyclone. In fact 'cold lows' are common in the atmosphere, hurricanes being the extreme case. The low pressure is produced by warm air in the stratosphere, and the low pressure usually indicates that the tropopause is lower than average: that warm air is not moved along with the cyclone, but that a wave travels along the tropopause and is induced by the cyclone below.

Thus although we have indicated by two examples how the pressure pattern is moved along by the thermal wind we have only indicated the argument that can be used in any particular case. To prove this result of Sutcliffe's in a general way would require a formal representation of the surface boundary condition and some other restraint on the motion in the neighbourhood which would make cyclogenesis the necessary consequence of the ageostrophic motion deduced from the acceleration, because obviously we have assumed that the compensating vertical motions occur in a particular way. Nevertheless this may be regarded as a theorem for the circumstances supposed, which are like the real atmosphere.

This is a superb example of deducing all the aspects of the motion that are important from the geostrophic wind field which cannot usually produce them. It represents a very considerable breakthrough in dynamical thinking.

(2) The formation of lows by heating

The term $\partial \mathbf{v}'/\partial t$ on the right-hand side of (4.7.3) represents the effect of local changes in the temperature field. We illustrate this by imagining the air to be warmed over a region, such as Britain or Iberia on a summer's day. The thermal wind will blow at high level with warm air on the right, that is anticyclonically around the heated region if there is initially no pressure gradient across the region (Fig. 4.7.iii): as it develops the local rate of change is in that same direction, and by equation (4.7.3) this is also the direction of the relative acceleration. The ageostrophic component is at right angles to this to the left, that is outwards from the heat region. This produces divergence aloft and convergence below,

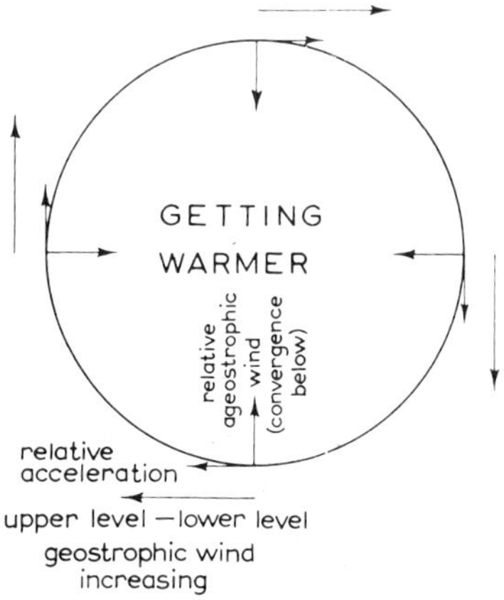

Fig. 4.7.iii The generation of a surface low by an anticyclonic acceleration aloft around warmed air.

and therefore upward motion in the middle, an increase in low-level vorticity, and the establishment of a heat low. The heating of course produces an expansion wave which produces an outflow and a fall of low-level pressure, as described in section 2.1.

We conclude that heating produces centres of low pressure. This may seem rather an obvious result to a reader well schooled in the idea that hot air rises after producing low pressure beneath it because it is less dense. He is probably familiar with this in the case of the establishment of sea breezes and similar phenomena; but in many such cases the earth's rotation is negligible and friction important, and the winds may even be antitriptic. In the present case the earth's rotation is the dominant influence and friction negligible. It illustrates the principle that the earth's rotation is a cause of the development of low pressure by dynamical mechanisms. The hurricane is the extreme example which develops because there is a heat source, but the lighter air that has been warmed by the heat source is not the initial cause of the low pressure at the centre.

The low pressure is intensified by the inflow at the bottom, because it accompanies the vertical stretching of the vortex lines, and the increase in the circulation. Indeed the Coriolis force, which makes the inflowing air deviate to the right, may be said to produce the low pressure. By contrast, if the low pressure is produced in the absence of the earth's rotation, the inflow causes the low to fill up, and friction retards this filling a little. In our case the effect of friction is to facilitate inflow without increasing the circulation, thereby hastening the filling up of the low.

The message about the mechanisms which operates in this case is that there is no straight sequence of cause and effect, but that many changes take place together and in concert. This sea breeze case is an obvious one to test the principle.

(3) The intensification of temperature gradients

A horizontal temperature gradient may be intensified at an already existing zone of large temperature gradient by the wind blowing across the isotherms towards a region where it blows along them, for example in a frontal zone. In Fig. 4.7.iv the flow depicted in the left diagram will bring warm and cold air closer together and intensify the temperature gradient by advection in the upper air, represented by the term $(\mathbf{v}_2 \cdot \mathbf{grad}_h)\mathbf{v}'$ in (4.7.3). The thermal wind, which is along the isotherms, is being accentuated by advection and so the relative ageostrophic wind is at right angles to the vector $\mathbf{v}'$ which is being generated,

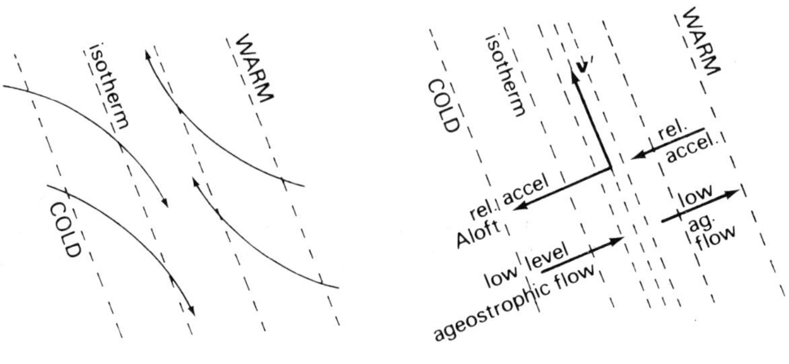

Fig. 4.7.iv Intensification of thermal gradient.

producing a flow away from the zone of increasing temperature gradient aloft and motion towards it below. The zone is therefore a region of falling pressure. In this case it is a trough of low pressure, not a centre, and the development is typical of a cold front.

The same result can be deduced for a warm front, which is illustrated in Fig. 4.7.v.

When a cold front approaches an anticyclone, and the thermal wind increases, the pressure falls ahead of it and rises behind it so as to displace the front through the anticyclone, which reforms behind it. The warm front approaching a cold air mass may produce the opposite result, depicted in Fig. 4.7.v, where the divergence ahead of the advancing warm is accompanied by rising pressure, and the front is halted. In general the intensification of a temperature gradient produces rising pressure in the cold air, and falling pressure in the warm air, but it must be remembered that it is only one of the mechanisms we have been concerned with and any actual case might produce a more complex result.

It cannot be too much emphasised that whatever part friction may play in atmospheric motion it is not one of the major causes of significant development on the scale of cyclones and anticyclones. The intensification of fronts is probably one of the most important, and surprising, consequences of the effect of the earth's rotation.

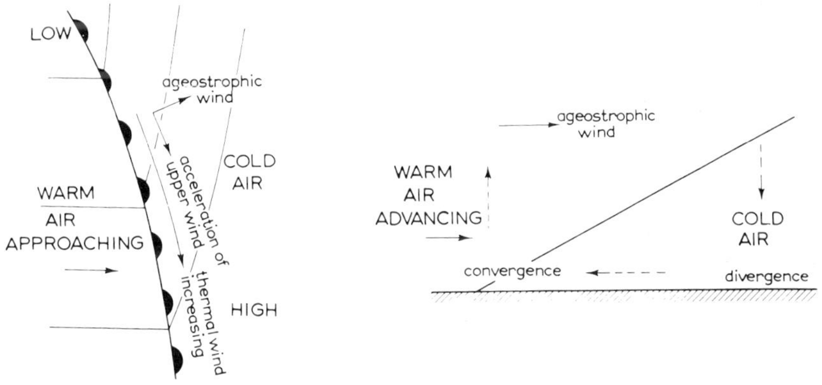

Fig. 4.7.v The left diagram shows the surface isobars and the surface position of a warm front. On the right are shown the consequences in a vertical section of the ageostrophic wind caused by the approach of the warm air.

4.8 FRONTS AS DISCONTINUITIES

The first useful attempt to analyse the mechanics of fronts was made by Margules, who showed that there could be a discontinuity of wind across a stationary or moving front with a fixed configuration. On either side of a sloping discontinuity of temperature (and therefore of density) the geostrophic wind is given by

$$v_G = \frac{1}{\rho f}\frac{\partial p}{\partial x} \tag{4.8.1}$$

for the y-component of the wind. Fig. 4.8.i shows a vertical section normal to the wind component, with the plane of the front sloping upwards at angle α. The pressure gradient just below the discontinuity $\Delta\rho$ exceeds that just above the front by an amount $\frac{\partial p}{\partial x}(g\Delta\rho\,.\,\Delta z)$ and ρ is replaced by $\rho+\Delta\rho$. Neglecting the higher-order terms in the discontinuity we note that, with temperature T,

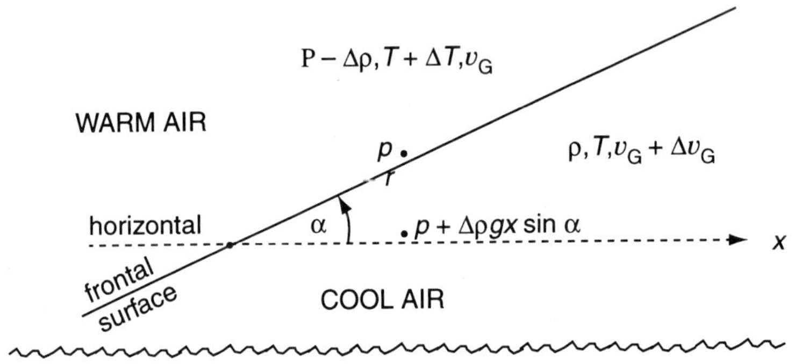

Fig. 4.8.i A discontinuity of density $\Delta\rho$ inclined at an angle α to the horizontal.

$$\frac{1}{\rho+\Delta\rho}=\frac{1}{\rho}\left(1+\frac{\Delta\rho}{\rho}\right)^{-1}=\frac{1}{\rho}\left(1-\frac{\Delta\rho}{\rho}\right)=\frac{1}{\rho}\left(1+\frac{\Delta T}{T}\right) \tag{4.8.2}$$

and so below the discontinuity, where $\Delta z = \alpha x$ when $\alpha = 10^{-2}$, say,

$$\begin{aligned} v_G+\Delta v_G &= \frac{1}{(\rho+\Delta\rho)f}\frac{\partial}{\partial x}(b+g\Delta\rho x\alpha) \\ &= v_G\left(1+\frac{\partial T}{T}\right)-\frac{g}{f}\frac{\Delta T}{T}\alpha \end{aligned} \tag{4.8.3}$$

where we have neglected the term in $\Delta T\Delta\rho$. The Boussinesq approximation (which states that the effect of density variations on the inertia, represented by $\frac{\Delta T}{T}v_G$, is small compared with the final term in (4.8.3) which contains g) may be tested by noting that in CGS units v is of the order of 10^{+3} (i.e. 10 m s^{-1}), g is about 10^3 and f^{-1} is about 3×10^4, so that g/vf is also of the order of 3×10^4. Thus we may write that

$$\nabla\Delta v_G = +\frac{\Delta T}{T}\frac{g\alpha}{f} \tag{4.8.4}$$

is the change in the geostrophic wind in passing up through the frontal surface is given by this positive quantity in (4.8.4).

We have been guided by Sutcliffe's ideas to see that the ageostrophic motion is the cause of development. The addition of any uniform horizontal velocity to the whole of

Fig. 4.8.i does not alter the conclusion; but the geostrophic wind cannot be accompanied by a vertical velocity because it is not horizontally divergent, or convergent, because

$$\operatorname{div} v_G = \frac{\partial u_G}{\partial x} + \frac{\partial v_G}{\partial y} = \frac{1}{\rho f}\left(-\frac{\partial^2 p}{\partial x \partial y} + \frac{\partial^2 p}{\partial y \partial x} \right) = 0. \tag{4.8.5}$$

Fronts had long been observed by seamen as fairly sharp discontinuities of wind direction and strength and temperature, and they had long noticed the variation of wind with height, indicating convergence into a rainy region low down and divergence out of it at high levels indicated by the spreading out of high clouds.

These ideas were consolidated in the inter-war years by close examination of weather charts over land areas where routine observations were adequate for that purpose, and the ideas became known as the **Norwegian frontal theory of cyclones**. The incorporation of it into mathematical models in more recent years is discussed in Chapter 16.

Fronts are not normally stationary, and the wind is not normally zero, so that development of some sort is to be expected. The geostrophic wind is discontinuous at a front, and it is likely that so is the ageostrophic wind, and there may be up- or downslope winds in either air mass. In Fig. 4.8.ii motions which typically occur in the vertical plane normal to a front are shown. If air from A and B comes into juxtaposition at C, the temperature gradient can be sharpened, with the consequence that the cold air advances undercutting the warm air according to the mechanism described in section 4.7(3).

Fig. 4.8.ii Intensification of temperature contrast at a front.

The effect of the motion on the largest scale causes stretching of the regions of large gradient (of temperature, humidity or whatever), and the intensification of the gradient there. Motion on every scale causes intensification on smaller scales, leading ultimately to the transfer on the molecular scale which actually does reduce the gradient. It is not correct that on the average, the eddies of one scale effect a transfer which reduces gradients on larger scales unless these are defined by averaging over relatively large distances which actually includes an increasing number of shorter but sharper steps in concentration. The discrimination of scales of turbulence is considered in Chapter 9.

In some cases the ageostrophic motion leads to a decrease in the temperature gradient, as often happens at a cold front of type (a), called a katafront, in which the cold air is warmed by subsidence. The consequence may be low-level ageostrophic flow away from the front and sinking motion in the warm air also. In that case the cloud systems, which were produced when fronts caused cyclogenesis, are rapidly dispersed.

Fig. 4.8.iii depicts the motion in three dimensions at a warm front of a typical cyclone, where the motion is like that in Fig. 4.8.ii(a). The cold front is a similar motion in reverse, and the tracks of different parts of the warm and cold air masses are shown in Fig. 4.8.iv.

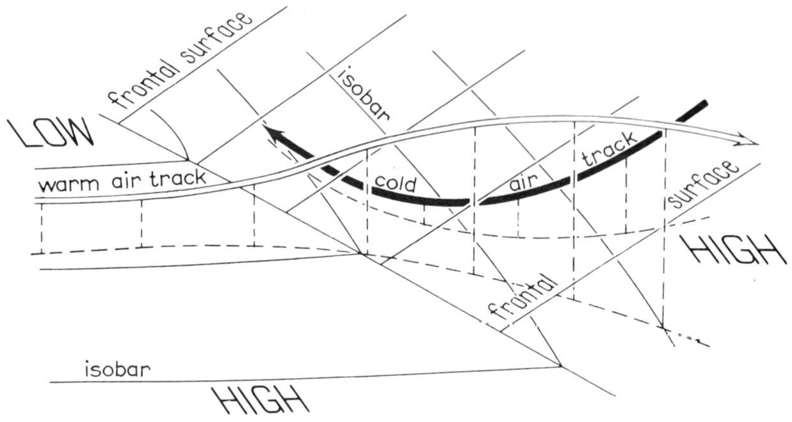

Fig. 4.8.iii The isobars at ground level, the frontal surface at a warm front, and the typical tracks of warm air above the surface and cold air ahead of and below it.

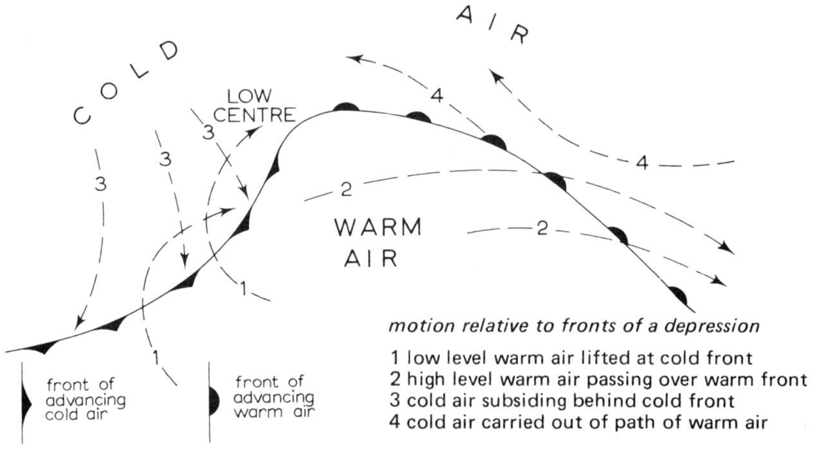

Fig. 4.8.iv Typical trajectories of air in the warm sector of a cyclone. The position of the fronts is shown as their position at ground level. The warm sector is larger at greater height.

4.9 JET STREAMS

The warm air above a front is subject to the thermal wind and therefore usually blows almost along the front with the cold air on the left. A typical cross-section of a frontal zone is shown in Fig. 4.9.i, in which the vertical scale is exaggerated. It shows the order of magnitude of the wind maximum and its horizontal and vertical extent, but obviously individual cases differ considerably. The jet stream is the name given to the wind maximum. Such concentrations of velocity are easily reproduced in rotating masses of fluid in which temperature gradients are generated. In Fig. 4.9.i the front is depicted as a zone of large gradient rather than a discontinuity, and this is nearer to reality, but the effects are the same.

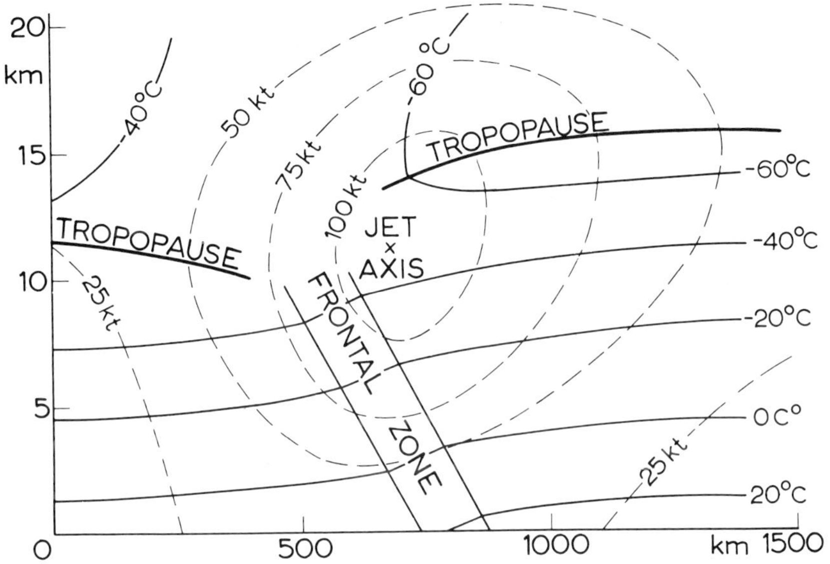

Fig. 4.9.i Typical dimensions and simplified structure of a jet stream showing isotherms in the section and wind speed blowing across it (into the paper).

Above a jet stream the temperature gradient is reversed, usually in the stratosphere, and the wind decreases with height. The strength of a jet stream varies along a front, and has a maximum to the west of the centre of a cyclone which is active on a wave of a front. Fig. 4.9.ii shows a typical distribution of the jet stream maximum in relation to the wave, which is moving much more slowly, perhaps at 30 to 50 kt. The warm air in the jet

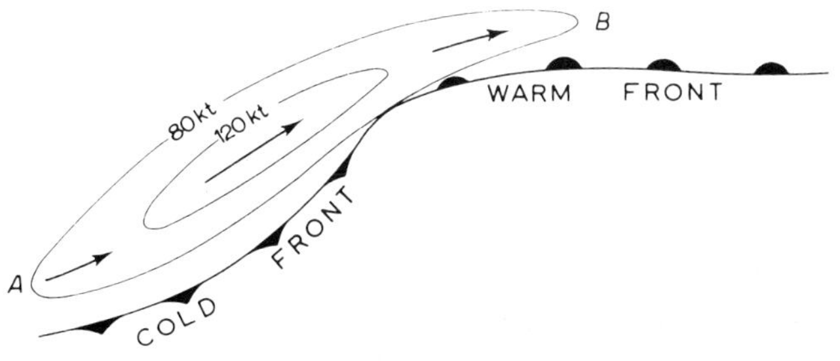

Fig. 4.9.ii Typical position and speed distribution of jet stream maximum in relation to surface position of fronts in a warm sector cyclone. This diagram is best interpreted with Figs 4.8.iv, 4.8.v and 4.9.i in mind.

stream therefore overtakes the cyclone, and is accelerated into the maximum at the jet entrance at A, and decelerated as it emerges at the jet exit at B.

The energy required to produce the acceleration is derived from a **direct circulation** in which the cold air sinks and the warm air rises over it, thereby decreasing the potential energy of the air. The motion in this circulation is ageostrophic, and is largely accounted for by the mechanism of the acceleration of the geostrophic wind described in section 4.7(3). At a jet exit the deceleration is effected by an indirect circulation in which the warm air sinks and moves ageostrophically towards higher pressure: kinetic energy is converted into potential energy. The jet maximum, or *axis* of the jet, is usually about 1 km higher at the entrance than at the exit because the maximum is further downwind at the lower levels.

In spite of their name, depressions are the object of the most delightful mechanisms in the atmosphere. Their general anatomy, with the direction of motion of the warm air revealed by cirrus and altocumulus clouds, was known to mariners long before aviators could explore them, before balloon soundings were in regular use, and before scientists had made any significant analysis of them and made numerical models of them.

The condensation of clouds increases the time scale of depressions because rising warm air may be continuously warmed by the condensation and release of rain. Thus there are probably no cyclones of comparable scale on Mars. Venus and Jupiter have deeper atmospheres with clouds of quite different chemical composition. The readiness of water to form rain and its very large latent heat probably make the earth's atmosphere beautiful in a way which is rather rare in the universe. Certainly no other planet in the solar system has such rapidly varying and partly clouded skies.

4.10 HURRICANES

In sections 9.8 and 9.10, mechanism whereby vortices may be generated as a result of convection clouds are described. The part played by clouds discussed in more detail in section 14.10. The staring point is in an air mass possessing some absolute rotation as a result of being on the rotating earth. The feature of hurricanes which makes them different from larger cyclones is that they depend on a continuous supply of heat to maintain the circulation. This heat source is the rain, whose latent heat of condensation is given to the air in which the rain condenses. This heat source exists mainly because the sea level pressure at the centre is lower than in the surrounding air: thus the air temperature is lower by between 3° and 5°C at the centre because the air expands adiabatically as it moves inwards. In consequence much more violent convection is supported there.

Because of the strong winds and rough sea, the air spiralling inwards at the sea surface is subject to considerable retardation due to friction. In the absence of friction the tangential velocity (relative to axes fixed in space) would increase like r^{-1}, r being distance from the centre: instead, it is observed to increase inwards roughly like $r^{-1/2}$. Near the centre the rotation of axes fixed to the earth is a small contribution which can be neglected. The air does not penetrate to the very centre of the storm but only to the outer edge of the eye, which is a region free of rain clouds and is composed of air sucked down from the stratosphere by the low pressure in the vortex. The stratospheric air in the eye is like the air in a bath plug vortex, and is very warm owing to its descent from the stratosphere. It is very dry also because the stratosphere is very dry but near the surface it mixes with other air and sometimes contains cloud. Some stratospheric air is mixed with the cloud wall surrounding the eye, and by evaporating the clouds it is cooled. Therefore there is a violent cascade of cold air surrounding the eye down the inside of the wall of cloud (see Fig. 14.11.ii).

The incoming air rises in the ring of clouds around the eye and spreads outwards beneath the tropopause. Because there is little friction in this high level outflow the tangential velocity relative to fixed axes decreases like r^{-1} and at three or four times the radius of the eye the rotation of the spreading canopy of the cloud is anticyclonic relative to the earth. This is discussed further in section 14.11.

In the large horizontal gradient of wind speed due to the rapid increase in tangential velocity as the air approaches the centre, the convection clouds are wound up into spiral lines. The convection is intensified by the ageostrophic motion producing inward, upward motion and cloud formation in the sea towards which the centre is moving. The sky clears more rapidly in the quarter away from which the centre is moving.

If a hurricane crosses the coast and goes inland it rapidly declines in intensity because the heat source of the sea is discontinued. It is often alleged that it is the increase in surface friction that is the more significant cause of the filling up of the low-pressure centre; but this is belied by the fact that if a cold front is drawn into the circulation of a hurricane as it crosses from sea to land it may continue longer before the pressure rises in the centre. This illustrates the important fact that it is the action of lifting the warm air (either by the cold air behind the cold front or by the surrounding air which has not been similarly warmed by the sea) that causes both the convergence and the intense rotation. The low pressure is produced by the lowering of the tropopause in the eye so as to provide warmer air overhead.

Hurricanes may become mid-latitude cyclones if they move to a high enough latitude (say 37°) and are regenerated by the inflow of a cold front. The sea is not warm enough to provide the heat source needed to maintain a hurricane except in low latitudes in the autumn. The heat flows from the sea largely as latent heat in the form of evaporated water; the heat is released when the rain falls out.

Hurricanes are called typhoons in the Pacific and Indian Oceans. The general name is tropical cyclone. They occur in the Arabian Sea, in the Bay of Bengal, in the Western Pacific and in the Atlantic oceans in low latitudes. Occasionally they cross the Malay Peninsular into the Indian Ocean and migrate towards Bengal. They are known in Northern Australia and the China Sea. Their movement is from the East in the Trade Winds at first but they curve along the east side of the continents, sometimes entering the Westerlies of the Eastern US or Japan. They occasionally cross Central America into the Eastern Pacific from the Caribbean Sea, and pass from the Arabian Sea into southern Arabia, where they have been responsible for outbreaks of plagues of desert locusts (see Chapter 15).

Figs 14.11.i and 14.11.ii show a typical cross-section of a mature hurricane and a radar picture of one near the coast of Florida. The dimensions of a hurricane can vary from case to case over a wide range, but typically the circulation and cloud systems extend over 100–300 km diameter; the eye may be from 5 to 50 km wide; and the speed of its centre from very slow to 15–20 kt.

Apart from the strong wind, the most severe effect is the raising of a tidal wave if the occurrence of a natural high tide coincides with a strong onshore wind in the right-hand advancing quarter of the storm. If the speed of the storm is such as to raise a wave in its centre of about 1 m above normal sea level, this wave is increased in height as it moves into shallow coastal water, and ships of up to 5000 tons have been carried as much as 400 m inland on occasion.

4.11 NUMERICAL MODELLING OF CYCLONIC DEVELOPMENT

Many ideas such as those described in this chapter have become much less used in forecasting because they have been replaced by numerical models. Of particular importance are global models because it is recognised that the whole atmosphere–ocean system is continuously evolving as one, and particular events can seldom be attributed to a particular single cause. Thus thermal steering is much better treated by an analysis of a much larger part of the system than is implied by the application of Sutcliffe's methods.

The use of global computer models (GCMs) requires continuous use because there are routines required for the assimilation of new observations to avoid the false introduction (and, of course, the false removal) of gravity wave motions which arise because of the incorrect matching of pressure fields with wind fields. If a good and correct match is not achieved when new pressure, temperature and wind observations are received, forecasts are likely to become unreliable, for although the wind is approximately geostrophic and may be deduced from a good chart, developments are all associated with the ageostrophic component.

Numerical modelling has become a new part of the science which is learnt only with the use of large computers. The results may be compared with observations which have been archived in detail, and the use of computers has opened up a new field of experimentation in which new values of parameters that are used to represent physical processes such as convection and radiation, which depend on cloud amount and which is one of the most difficult but obvious and apparently simple factors to predict.

4.12 THE EKMAN SPIRAL

We consider the situation in which the air velocity is horizontal and constant in direction at each height over a large area. Thus there are no accelerations and the equation of motion (4.3.2) is then

$$0 = -\frac{1}{\rho}\,\mathbf{grad}\ p + \mathbf{v} \times \mathbf{f} + \frac{\partial}{\partial z} K \frac{\partial v}{\partial z}. \tag{4.12.1}$$

We have expressed the stress, represented in (1.3.25), by the pressure p and an eddy viscosity, K, multiplying a velocity gradient according to the conventional mathematical treatment to represent the Reynolds stress, which is described in section 9.3, as if it were a viscous stress. The effect of variations in ρ are negligible except in so far as they produce a thermal wind which will be represented by the vertical variation of the horizontal pressure gradient. The pressure gradient may be represented by the geostrophic wind (u_G, v_G) according to (4.3.4) and so the two components of (4.12.1) are

$$(u - u_G)f = \frac{\partial}{\partial z} K \frac{\partial v}{\partial z},$$

$$(v - v_G)f = -\frac{\partial}{\partial z} K \frac{\partial u}{\partial z}. \tag{4.12.2}$$

G. I. Taylor applied Ekman's original theory, concerning wind-driven motions in the ocean, to the atmosphere. That work is well summarised by Brunt (1939), p. 253. If we take the classical assumptions, that there is no thermal wind so that (u_G, v_G) is a constant, and that K is independent of z, the solution of (4.12.2) is readily seen to be

$$u + \mathrm{i}v - (u_G + \mathrm{i}v_G) = C\,\mathrm{e}^{-(1+\mathrm{i})mz + i\gamma} \tag{4.12.3}$$

in which C and γ are real constants determined by the boundary condition at the bottom of the atmosphere. Here

$$m^2 = f/2K \tag{4.12.4}$$

with m taken to be positive. The negative real exponent is chosen in (4.12.3) on the grounds that the wind becomes equal to the geostrophic wind at sufficient height. Taylor's theory was designed to obtain an estimate of the magnitude of K, at a time when its order was not known, by measuring the deflection of the surface wind from the isobars. He applied the boundary condition that the stress was in the direction of the wind at the surface. This is expressed by saying that $u + \mathrm{i}v$ is parallel to $\partial u/\partial z + \mathrm{i}\partial v/\partial z$ at $z = 0$. The reader is referred to Brunt for details.

Equation (4.12.3) shows that the horizontal wind vector approaches the geostrophic value with increasing z along a spiral. In the ocean a positive exponent is used for wind-driven motions which fade out downwards, and Ekman assumed the ocean surface to be level so that the geostrophic flow was zero. This is simply to say that the pressure gradient in the air is cancelled by an immeasurably small tilt of the ocean surface. In that case the equations (4.12.2) become

$$U + \mathrm{i}V = -\mathrm{i}\frac{K}{f}\frac{\partial^2}{\partial z^2}(U + \mathrm{i}V) \tag{4.12.5}$$

of which the solution is

$$U + \mathrm{i}V = \mathrm{e}^{(1+\mathrm{i})mz}(U_0 + \mathrm{i}V_0) \tag{4.12.6}$$

where K and m now have values appropriate to the water and $U_0 + \mathrm{i}V_0$ is the current at the surface. The velocity vector thus follows out a spiral of magnitude decreasing with depth.

The total momentum of the wind-driven motion, which is in the direction of what may be called the 'surface ocean current', is perpendicular to the stress applied by the wind. Assuming the stress to be along the wind direction, the current is perpendicular to the wind itself. The surface stress is equal and opposite to the Coriolis force on the current. An important consequence is that an anticyclone generates convergence of the surface ocean flow, even though there may well be a significant outflow of the air near the surface. This mechanism is illustrated in Fig. 4.12.i and is the classical explanation of the Sargasso Sea under the Bermuda high-pressure area. For the effect to be significant the anticyclone must be present in the same area for long periods of time, so that the corresponding geostrophic flow in the ocean, which is comparatively slow, can produce a noticeable effect.

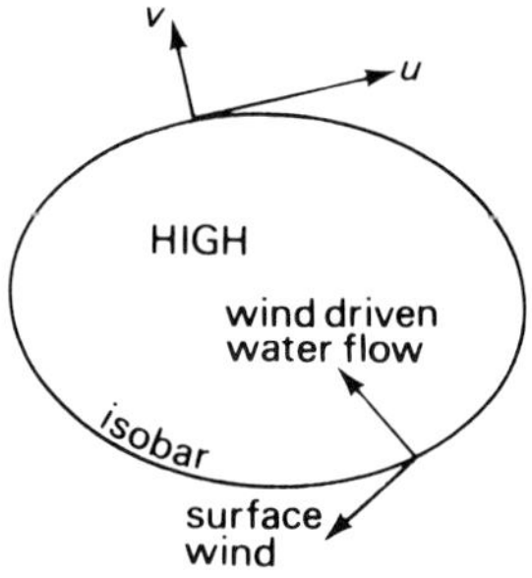

Fig. 4.12.i u is the wind component along the isobar, and v the outward-flowing component from an anticyclone; the surface water is driven in towards the centre of the high.

The cross-isobar flow of the wind due to drag at the surface may be estimated in the case of no thermal wind ($u_G + iv_G = \text{const}$) by integrating equations (4.12.2) from the surface up to infinite z. Thus if we choose a coordinate system such that $v_G = 0$, and estimate the total transverse flow v, we have

$$\int_0^\infty v \, dz = -\frac{1}{f}\left[K\frac{\partial u}{\partial z}\right]_0^\infty = \tau_{xz}/\rho f, \tag{4.12.7}$$

where τ_{xz} is the stress at the surface in the x-direction. This result does not depend on any assumptions about K. Equation (4.12.7) suggests that the surface stress could be determined by observations of v, but attempts to do this by Sheppard and others revealed the importance of the thermal winds. The winds did not follow out a spiral according to (4.12.3), but revealed that the geostrophic wind varied very significantly with height. This was so even in the area of the Windward Islands, where horizontal homogeneity had been expected of the large geographical extent of the Trade Winds.

The Ekman spiral in the ocean and the Taylor spiral in the atmosphere are of considerable interest because of the influence of the earth's rotation on motion systems, but the details of any actual case are much more complicated, and no precise conclusions of the simple theory can be made use of quantitatively.

5

Waves in a stratified fluid

5.1 THE PURPOSE OF A THEORY: ASSUMPTIONS, APPROXIMATIONS, NOTATIONS

Obviously we try to solve as many of our technical problems by mathematics as we can: in every case we have to make assumptions and approximations in order to make the analysis tractable, but we must insist that the most important aspects of the mechanisms of reality are retained in our model.

There are many original papers using very sophisticated mathematics designed to resolve certain mathematical issues, and we are lucky if they take us nearer to matching our theory to the observational (out of doors) and experimental (laboratory) facts known. For example, one way to determine the precise nature of a steady flow pattern is to imagine the flow started up from rest and see what pattern the motion tends towards after a long time: this method has been used to discover what the flow is like at great heights and close to an obstacle. There do not exist, nor are there likely to exist, any observations from nature which would confirm that theory. Moreover the theory cannot be used in any discriminating way in practice to help understand observations, nor does it make any useful difference to predictions we might make of the flow over any actual mountain because other factors dominate any particular case. Yet the issue of the flow at great height has been given more attention by theorists than almost any other. Indeed many regard that aspect of the problem as solved, whereas in practice the real atmosphere never behaves like the models used; in fact almost all actual cases have features such as a level at which the wind speed is zero, which make them insoluble by the mathematics available today.

The most developed theories are for waves on deep water, and they have advanced our understanding of breaking waves and hydraulic jumps, but the ideas do not work as well for a stratified fluid with no well defined upper surface. The motion pattern of river flow over an obstacle has not been very significantly advanced by the excellent exlanations of waves at the free surface.

One very great difficulty in practice is that actual mountain shapes are very complicated, and the airflow very variable. It therefore seems at first sight desirable to develop a theory by which we can calculate the flow over any given shape in any given stream. However, to do this for the two-dimensional case of a simple stream is difficult enough, and most mountains are not two-dimensional while even the best methods are not applicable in the three-dimensional case. Consequently they are not really useful in the sense that they can be employed to predict the airflow tomorrow, or even today. However many of the mechanisms which dominate what happens in actual cases can be understood by studying special theoretical cases in which the appropriate mechanisms are dominant.

If more complicated calculations for everyday use could be done we would stil get very little further than we would without them. This is because the differences between reality and theory lie in the considerable unsteadiness of actual flows, and in the fact that the flow often separates from the surface, particularly in the lee of a steep obstacle where the predictions would be most useful if they could be made. Buoyancy produced by clouds and changes in the wind and temperature structure during the day, together with a variety of mathematically quite intractable situations leading to turbulence of various kinds, all militate against quantitative precision.

In such circumstances it is gratifying that a simple theory does enable us to gain considerable insight into the mechanisms, and it is these simpler mathematical aspects which will occupy our attention. It must also be appreciated that great precision in prediction is not necessarily of much use nor worth obtaining. We go through life managing most of the time by having a large safety factor which is scarcely ever invoked, and we have other things to do and enjoy besides giving attention to the quantitative aspects of our experiences which will be soon over. The observational and computational effort needed to make accurate calculations about all we do and experience would simply be wasted, and for that reason much that is theoretically possible will never be done. The final criterion is whether we can understand the world we live in, and find our way through it, not whether we can predict it and control it.

Except in those sections where it is stated otherwise, we shall be studying steady, adiabatic or incompressible, two-dimensional ($\partial/\partial y = 0$), inviscid flow of a stream that is disturbed from straight horizontal flow over level ground by passage over an obstacle. The undisturbed, or initial, state is represented by the subscript 0. Thus u_0 is the velocity possessed by a particle far upstreamm along the streamline. In the adiabatic (or incompressible) case τ (or ρ) is constant along a streamline and equal to τ_0 (or ρ_0). In this chapter τ is $1/\theta$, the reciprocal of the potential temperature, as in section 1.6.

The velocity and vorticity are assumed to have the following forms:

$$\mathbf{v} = (u, v, w),\ \mathbf{v}_0 = (u_0, v_0, 0);\ \boldsymbol{\omega} = (\xi, \eta, \zeta),\ \boldsymbol{\omega}_0 = \left(-\frac{\partial v_0}{\partial z_0}, \frac{\partial u_0}{\partial z_0}, 0\right). \qquad (5.1.1)$$

In motion which does not vary in the y-direction there are no pressure gradients or other forces (except the Coriolis force in specified cases) in that direction and so in the oncoming stream

$$v = v_0(z), \qquad (5.1.2)$$

which means that **each Bernoulli surface has its own velocity in the y-direction**. Since this does not change at all, it has no effect on other components of the motion. Thus v_0 will not appear in our equations representing the dynamics, and $v_0(z)$ may be added to any result we obtain when $\partial/\partial y = 0$. Likewise ξ and ζ, the vorticity components, will not appear in the analysis.

We shall analyse wave motions using approximations only when they are unavoidable, in order to discover how valid our conclusions are. The result is that the analysis is not similar to the classical analysis of small perturbation theory and our results are valid for some very large disturbances from horizontal flow.

The continuity of mass flow between two Bernoulli surfaces requires that (see Fig. 5.1.i) for two-dimensional motion

$$\rho_0 u_0 \, \mathrm{d}z_0 = \rho u \, \mathrm{d}z = -\rho w \, \mathrm{d}x, \tag{5.1.3}$$

which can be written as follows:

$$u = \frac{\rho_0}{\rho} u_0 \frac{\partial z_0}{\partial z}, \quad w = -\frac{\rho_0}{\rho} u_0 \frac{\partial z_0}{\partial x}. \tag{5.1.4}$$

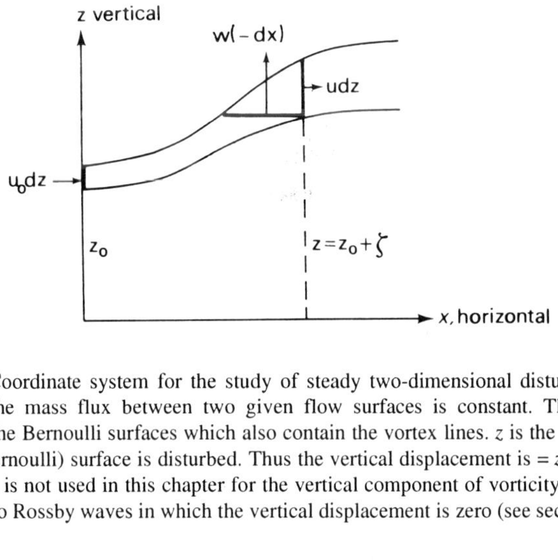

Fig. 5.1.i Coordinate system for the study of steady two-dimensional disturbances for which $\partial/\partial y = 0$. The mass flux between two given flow surfaces is constant. The streamlines are sections of the Bernoulli surfaces which also contain the vortex lines. z is the height from which a stream (Bernoulli) surface is disturbed. Thus the vertical displacement is $= z - z_0$ (see equation (5.4.6)), and is not used in this chapter for the vertical component of vorticity (except in leading up to Rossby waves in which the vertical displacement is zero (see section 7.2)).

Thus

$$\mathbf{grad}\, z_0 = \left(\frac{\partial z_0}{\partial x}, 0, \frac{\partial z_0}{\partial z}\right) = \frac{\rho}{\rho_0 u_0}(-w, 0, u) = \frac{\rho}{\rho_0 u_0} \mathbf{v} \times \mathbf{j}, \tag{5.1.5}$$

where

$$\mathbf{j} = (0,\ 1,\ 0). \tag{5.1.6}$$

If n is length measured normal to the Bernoulli surfaces, on which z_0 is constant,

$$\frac{\partial}{\partial z_0} = \frac{\partial n}{\partial z_0}\frac{\partial}{\partial n} = \frac{1}{\text{grad}\, z_0} \frac{\mathbf{grad}\, z_0 \,.\, \mathbf{grad}}{\text{grad}\, z_0},$$

where

$$\text{grad } z_0 = |\mathbf{grad}\ z_0|, \text{ and so}$$

$$\frac{\partial}{\partial z_0} = \frac{1}{(\text{grad } z_0)^2}\ \mathbf{grad}\ z_0 \,.\, \mathbf{grad} \tag{5.1.7}$$

$$= \frac{\rho_0 u_0}{\rho q_1^2} \mathbf{v} \times \mathbf{j} \,.\, \mathbf{grad}. \tag{5.1.8}$$

The form (5.1.7) is applicable in any steady motion; while (5.1.8) is appropriate to our two-dimensional case, in which $v = 0$, and the speed is given by

$$q_1^2 = u^2 + w^2, \tag{5.1.9}$$

and $\text{grad } q_1^2 = \text{grad } q^2$, ignoring the velocity in the y-direction.

5.2 THE FLOW EQUATION FOR STEADY MOTION

For inviscid steady motion the vorticity equation (1.4.1) takes the form

$$\frac{\mathrm{D}\boldsymbol{\omega}}{\mathrm{D}t} = -\boldsymbol{\omega} \text{ div } \mathbf{v} + (\boldsymbol{\omega} \,.\, \mathbf{grad})\mathbf{v} + \mathbf{R} \times (\mathbf{g} - \mathbf{f}), \tag{5.2.1}$$

and the equation of continuity for steady motion is

$$\rho \text{ div } \mathbf{v} = -\mathbf{v} \,.\, \mathbf{grad}\ \rho. \tag{5.2.2}$$

In the kind of two-dimensional motion under consideration the y-component of velocity, v, has a gradient only in the direction normal to the flow surfaces. These surfaces contain $\boldsymbol{\omega}$, and so the second term on the right of (5.2.1) has no y-component. Using (5.2.2) and (1.2.6) the y-component of (5.2.1) in steady motion is

$$\mathbf{v} \,.\, \mathbf{grad}\ \eta - \frac{\eta}{\rho} \mathbf{v} \,.\, \mathbf{grad}\ \rho = \left[\mathbf{R} \times \left(\mathbf{g} - \mathbf{grad}\ \tfrac{1}{2} q^2 + \mathbf{v} \times \boldsymbol{\omega}\right)\right]_y. \tag{5.2.3}$$

But

$$\mathbf{R} = \frac{1}{\tau} \frac{\mathrm{d}\tau}{\mathrm{d}z_0}\ \mathbf{grad}\ z_0 = -\beta\ \mathbf{grad}\ z_0, \tag{5.2.4}$$

and this is parallel to $\mathbf{v} \times \boldsymbol{\omega}$. (There is an exactly similar equation for the incompressible case with ρ written for τ.) Since $\mathbf{v} \,.\, \mathbf{grad}\ \rho_0 = 0$,

$$\mathbf{v} \,.\, \mathbf{grad}\left(\frac{\rho_0}{\rho}\eta\right) = \frac{\eta \rho_0}{\rho^2}\ \mathbf{v} \,.\, \mathbf{grad}\ \rho + \frac{\rho_0}{\rho}\ \mathbf{v} \,.\, \mathbf{grad}\ \eta$$

$$= \left[\frac{\rho_0}{\rho} \beta\ \mathbf{grad}\ z_0 \times\ \mathbf{grad}\left(gz + \frac{1}{2} q^2\right)\right]_y$$

$$=\left[\frac{\beta}{u_0}(\mathbf{v}\times\mathbf{j})\times\mathbf{grad}\left(\left(gz+\frac{1}{2}q^2\right)\right]_y \qquad \text{by (5.1.5)}$$

$$=\frac{\beta}{u_0}\mathbf{v}.\mathbf{grad}\left(gz+\frac{1}{2}q^2\right)$$

$$=\mathbf{v}.\mathbf{grad}\left[\frac{\beta}{u_0}\left(gz+\frac{1}{2}q^2\right)\right] \tag{5.2.5}$$

because β/u_0 is constant along a streamline and can therefore be taken inside the operator $\mathbf{v}.\mathbf{grad}$. Consequently

$$\boxed{\frac{\rho_0}{\rho}\eta-\frac{\beta}{u_0}\left(gz+\frac{1}{2}q^2\right)=\text{ constant along a streamline.}} \tag{5.2.6}$$

In the incompressible case, $\rho_0=\rho$ and so the equation is slightly simpler.

The valure of the constant in (5.2.6) may readily be obtained by an alternative derivation (see (5.2.12)), as follows:

As in (1.7.4)

$$gH=\frac{1}{2}q^2+\frac{\varpi}{\tau}+gz=\text{ constant along a streamline,} \tag{5.2.7}$$

and the equation of motion (1.7.3) is

$$\mathbf{grad}\,(gH)=\varpi\,\mathbf{grad}\,\frac{1}{\tau}+\mathbf{v}\times\boldsymbol{\omega}. \tag{5.2.8}$$

where H was defined in (1.7.4).

By (5.1.7) and (5.1.8)

$$\frac{\mathrm{d}}{\mathrm{d}z_0}(gH)=\frac{1}{(\mathbf{grad}\,z_0)^2}\,\mathbf{grad}\,z_0\,.\,\mathbf{grad}\,(gH)$$

$$=\frac{\varpi}{\tau}\tau\frac{\mathrm{d}}{\mathrm{d}z_0}\frac{1}{\tau}+\frac{\rho_0u_0}{\rho q^2}\mathbf{v}\times\mathbf{j}.\mathbf{v}\times\boldsymbol{\omega}$$

$$=\frac{\varpi}{\tau}\beta+\frac{\rho_0u_0}{\rho}\eta \tag{5.2.9}$$

because the component of $\mathbf{v}\times\boldsymbol{\omega}$ which is normal to $\mathbf{v}$ and to $\mathbf{j}$ is $\mathbf{v}\times\eta\mathbf{j}$, and so

$$\mathbf{v}\times\mathbf{j}.\mathbf{v}\times\eta\mathbf{j}=q^2\eta. \tag{5.2.10}$$

Thus, substituting for $\varpi\tau$ from (5.2.7) into (5.2.9), which eliminates the pressure is therefore equivalent to using the vorticity equation, we have

$$\frac{\mathrm{d}}{\mathrm{d}z_0}(gH) = \beta\left(gH - \frac{1}{2}q^2 - gz\right) + \frac{\rho_0 u_0}{\rho}\eta,$$

and so

$$\frac{\rho_0}{\rho}\eta - \frac{\beta}{u_0}\left(\frac{1}{2}q^2 + gz\right) = \frac{1}{u_0}\left[\frac{\mathrm{d}}{\mathrm{d}z_0}(gH) - \beta gH\right] = \frac{1}{u_0\tau}\frac{1}{\mathrm{d}z_0}(gH\tau), \tag{5.2.11}$$

which is constant on a streamline because it is completely determined by values far upstream ($\beta = \beta_0$, $\tau = \tau_0$). The corresponding equation for the incompressible case has $\rho = \rho_0$ on the left and ρ for τ on the right; thus it is

$$\eta - \frac{\beta}{u_0}\left(\frac{1}{2}q^2 + gz)\right) = \frac{1}{u_0\rho}\frac{\mathrm{d}}{\mathrm{d}z_0}(\rho gH). \tag{5.2.12}$$

In (5.2.11), H and τ [and in (5.2.12) H and ρ] are constant along a streamline.

We have thus obtained a value for the constant in (5.2.6). These results depend fundamentally on the isopycnic vorticity theorem (1.5) and are only true if somewhere in the flow the vorticity vector lies in the isopycnic surfaces; otherwise we could not write $\eta\mathbf{j}$ for $\boldsymbol{\omega}$ to obtain (5.2.10).

The result (5.2.6) is possible only in two-dimensional motion because only in that case is the velocity field defined by the spacing of the Bernoulli surfaces, which also defines the density field.

5.3 THE FLOW EQUATION FOR MOTION IN THREE DIMENSIONS

Instead of (5.2.9) we have

$$\frac{\mathrm{d}}{\mathrm{d}z_0}(gH) = \frac{\mathrm{d}}{\mathrm{d}z_0}\frac{1}{\tau} + \frac{1}{(\mathrm{grad}\ z_0)^2}\ \mathbf{grad}\ z_0 \,.\, \mathbf{v}\times\boldsymbol{\omega}$$

$$= \beta\left(gH - \frac{1}{2}q^2 - gz\right) + \frac{|\mathbf{v}\times\boldsymbol{\omega}|}{|\mathbf{grad}\ z_0|} \tag{5.3.1}$$

because $\mathbf{v}\times\boldsymbol{\omega}$ is parallel to $\mathbf{grad}\ z_0$ everywhere if it is somewhere on each streamline, by the theorem of (1.5). Whence

$$\frac{|\mathbf{v}\times\boldsymbol{\omega}|}{|\mathbf{grad}\ z_0|} - \beta\left(\frac{1}{2}q^2 + gz\right) = \frac{1}{\tau}\frac{\mathrm{d}}{\mathrm{d}z_0}(gH\tau)$$

$$= \text{constant on a stream (Bernoulli) surface.} \tag{5.3.2}$$

This form of the equation is deceptively simple. To use it, the equation of continity must be used simultaneously, which is very difficult to do, and it is unlikely that any attempt to do this will be made. This is because of the difficulty of specifying boundary conditions for a medium capable of transmitting waves similar to the motion under study. This difficulty is best appreciated when it is compared with the procedure for solving the two-dimensional case.

5.4 THE WAVE EQUATION FOR AN INCOMPRESSIBLE STRATIFIED STREAM IN STEADY TWO-DIMENSIONAL MOTION

In this case $\rho = \rho_0$ and we may use (5.1.4) to express the velocity and vorticity in terms of the field of z_0 as follows:

$$\eta = \frac{\partial u}{\partial z} - \frac{\partial w}{\partial x} = \frac{\partial}{\partial z}\left(u_0 \frac{\partial z_0}{\partial z}\right) + \frac{\partial}{\partial x}\left(u_0 \frac{\partial z_0}{\partial x}\right)$$

$$= \frac{\mathrm{d}u_0}{\mathrm{d}z_0}(\nabla z_0)^2 + u_0 \nabla^2 z_0. \tag{5.4.1}$$

because

$$\frac{\partial u_0}{\partial z} = \frac{\partial u_0}{\partial z_0}\frac{\partial z_0}{\partial z} \quad \text{and} \quad \frac{\partial u_0}{\partial x} = \frac{\partial u_0}{\partial z_0}\frac{\partial z_0}{\partial x}$$

$$q^2 = u_0^2 (\nabla z_0)^2, \tag{5.4.2}$$

and so (5.2.6) becomes

$$\frac{\eta}{u_0} - \frac{\beta}{u_0^2}\left(\frac{1}{2}q^2 + gz\right) = \text{ constant along a streamline}$$

$$= \nabla^2 z_0 + \frac{1}{u_0}\frac{\mathrm{d}u_0}{\mathrm{d}z_0}(\nabla z_0)^2 - \frac{\beta}{u_0^2}\left[\frac{1}{2}u_0^2(\nabla z_0)^2 + gz\right]$$

$$= \nabla^2 z_0 + \left(\alpha - \frac{1}{2}\beta\right)(\nabla z_0)^2 - \frac{g\beta z}{u_0^2}. \tag{5.4.3}$$

Here,

$$\beta = \frac{1}{\tau}\frac{\partial \tau}{\partial z_0}$$

$$\alpha = \frac{1}{u_0}\frac{\mathrm{d}u_0}{\mathrm{d}z_0} \tag{5.4.4}$$

represent the density stratification and the vorticity in the undisturbed stream, so that

$$\alpha - \frac{1}{2}\beta = \frac{\mathrm{d}}{\mathrm{d}z_0}\left(\ln u_0 + \frac{1}{2}\ln \rho\right) = \frac{\mathrm{d}}{\mathrm{d}z_0}\ln\left(\rho^{1/2}u_0\right) = S \tag{5.4.5}$$

represents the total effect of stratification of the stream due to shear and density gradients together.

More insight is gained by expressing this equation in terms of the vertical displacement of the fluid, ζ, given by

$$\boxed{z_0 = z - \zeta.} \tag{5.4.6}$$

Then

$$\nabla z_0 = \nabla z - \nabla\zeta = \left(-\frac{\partial\zeta}{\partial x},\ 1-\frac{\partial\zeta}{\partial z}\right)$$

$$(\nabla z_0)^2 = 1 - 2\frac{\partial\zeta}{\partial z} + (\nabla\zeta)^2$$

$$\nabla^2 z_0 = -\nabla^2\zeta, \tag{5.4.7}$$

and (5.4.3) becomes

$$\nabla^2\zeta - \left(\alpha - \frac{1}{2}\beta\right)\left[1 - 2\frac{\partial\zeta}{\partial z} + (\nabla\zeta)^2\right] + \frac{g\beta}{u_0^2}\zeta = \text{const on a streamline.} \tag{5.4.8}$$

In particular

$$\boxed{\nabla^2\zeta - \left(\alpha - \frac{1}{2}\beta\right)\left[2\frac{\partial\zeta}{\partial z} - (\nabla\zeta)^2\right] + \frac{g\beta}{u_0^2}\zeta = 0} \tag{5.4.9}$$

in the case where the undisturbed stream is horizontal with ζ and its gradients equal to zero, the constant in (5.4.8) then being $-\alpha + \frac{1}{2}\beta$. Lamb (see §235 Eq. 12) gave this equation in linearised form with $\alpha = 0$.

5.5 PRACTICAL APPROXIMATIONS

In motion which is essentially determined by the action of gravity on the density stratification, the first and last terms in (5.4.9) are necessarily important. For example, we could have a case in which $\alpha = \frac{1}{2}\beta$, in which waves are obviously possible. We now consider the importance of the term in $\frac{1}{2}\beta$, noting that there could obviously be cases in which α is large enough to be important because its magnitude can be made as large as we choose by having a large enough velocity gradient in the oncoming stream.

When a streamline has a vertical tangent, $\partial\zeta/\partial z$ is equal to unity and $(\nabla\zeta)^2$ is the same order of magnitude as $\zeta\nabla\zeta^2$. If the wavelength is $2\pi/k$ then comparing the first and last terms in (5.4.9), which are both involved in an essential way for gravity waves, we see that k^2 is the same order of magnitude as $g\beta/u_0^2$. The term in $\frac{1}{2}\beta$ is comparable only if β, or $\frac{1}{2}\beta\zeta^2k^2$, is comparable with $g\beta\zeta/u_0^2$. This would require u_0^2 to be comparable with $g\zeta$, which is unlikely for waves of large enough amplitude for the flow to be nearly vertical somewhere. Or, comparing the first term in (5.4.9) with $\beta(\nabla\zeta)^2$, we see that the second is only important if $\beta\zeta$ is comparable with unity. This means that the displacement would have to be comparable with the height in which the density (or potential temperature) is decreased (increased) by a factor e^{-1}, often called a scale height. In an actual case, $\beta\zeta$ is roughly equal to $\Delta\rho/\rho$, and the density anomaly is usually a small fraction of the density.

In the atmospheric cases, β is of the order of 10^{-2} km^{-1}, and so for the term in β to be important the displacement would have to be of the order of 100 km.

In neglecting this term we are making the **Boussinesq approximation** (see section 2.12) because the terms in β not containing g originate in the term $\mathbf{R} \times \mathbf{f}$ in the vorticity equation. As a result we are saying that the accelerations, since they are produced by buoyancy forces created by density anomalies in which $\Delta\rho \ll \rho$, are small compared with gravity. The density variations are taken to have a small effect on the inertial mass of the fluid, but a significant effect through gravity.

Another approach is to assume that the coefficients in (5.4.9) are constants and that the displacements are small enough for the non-linear term $(\nabla\zeta)^2$ to be negligible. In that case we may write

$$\xi = e^{Sz}\zeta \tag{5.5.1}$$

where

$$S = \alpha - \tfrac{1}{2}\beta, \tag{5.5.2}$$

and equation (5.4.9) then becomes

$$\nabla^2\xi + \left(\frac{g\beta}{u_0^2} - S^2\right)\xi = 0. \tag{5.5.3}$$

When $\alpha = 0$ the term in β without g is important only if $\frac{1}{4}\beta$ is comparable with g/u_0^2. This condition is never fulfilled in practice because the scale height is always much more than the height from which free fall under gravity produces a speed u_0.

In practical cases velocity gradients, represented by α, can be much more important than density gradients, represented by β, except that density discontinuities can be realised and kept sharp much more easily than discontinuities of velocity. Density discontinuities are treated in section 5.10.

If S is neglected in (5.5.3) it might as well be neglected in (5.5.1), which is equivalent to neglecting β in (5.4.9). The substitution (5.5.1) has sometimes been used in order to obtain an equation of the form (5.5.3), which is easier to handle than (5.4.9). This is only possible if S is a constant throughout the whole depth or layer of the medium under consideration. If a solution of (5.4.9) is sought in such a case, of the form $\exp\, i(Sz + kx)$, it is found that for a range of values of k there is no solution which tends to zero for large z. This is on account of the term in $\beta\partial\zeta/\partial z$ which makes even the oscillatory solutions proportional to $\exp\frac{1}{2}\beta z$. As a result, if waves occur in an atmosphere in which $\alpha = 0$ and β and u_0 are the same at all heights, the amplitude must become large at great height provided the waves last long enough for a steady solution to be relevant. Certainly long waves produced by the semidiurnal tide do have a very large amplitude at heights around 80 km, but these are discussed later in section 7.4. If α exceeds $\frac{1}{2}\beta$, the difficulty disappears. The meaning of this is that the pressure fluctuations necessary for waves low down in the atmosphere can only occur if the wave amplitude in the layers of very low density at high altitude is very large. However a large increase in velocity with height reduces the amplitude because fast-moving fluid is equivalent to dense fluid: this is evident from equation (5.4.5), where variations in speed can produce variations in $\rho^{1/2}u_0$ as large as or even larger than those due to variations in density.

The following section shows that the term in β is not a significant diffulty in comparison with the effects of compressibility in the atmosphere.

5.6 THE FLOW EQUATION FOR A COMPRESSIBLE ATMOSPHERE

If we use the full Equation of Continuity (5.1.4), which contains density variations, in the place of (5.4.1) and (5.4.2) we have, for the compressible case

$$\eta = \frac{\rho_0}{\rho}\left[\frac{\mathrm{d}u_0}{\mathrm{d}z_0}(\nabla z_0)^2 + u_0\nabla^2 z_0\right] + u_0\left[\frac{\partial z_0}{\partial z}\frac{\partial}{\partial z} + \frac{\partial z_0}{\partial x}\frac{\partial}{\partial x}\right]\frac{\rho_0}{\rho} \tag{5.6.1}$$

$$q^2 = \frac{\rho_0^2}{\rho^2}u_0^2(\nabla z_0)^2 \tag{5.6.2}$$

from which it follows from (5.2.6) that

$$\left(\frac{\rho_0}{\rho}\right)^2\left[\frac{\mathrm{d}u_0}{\mathrm{d}z_0}(\nabla z_0)^2 + u_0\nabla^2 z_0\right] + \frac{\rho_0}{\rho}u_0\left[\frac{\partial z_0}{\partial z}\frac{\partial}{\partial z} + \frac{\partial z_0}{\partial x}\frac{\partial}{\partial x}\right]\frac{\rho_0}{\rho}$$

$$-\frac{\beta_0}{u_0}\left[\frac{1}{2}\frac{\rho_0^2}{\rho_2}u_0^2(\nabla z_0)^2 + gz\right] = \text{constant along a streamline.} \tag{5.6.3}$$

We now use equations (5.4.6) and (5.4.7) and assume adiabatic displacements so that we may use (1.8.6) in the form

$$\frac{\rho_0}{\rho} = 1 + \frac{g}{c^2}\zeta, \tag{5.6.4}$$

which is valid for displacements small compared with 30 km in the atmosphere, and obtain

$$-\left(1+2\frac{g}{c^2}\zeta\right)\nabla^2\zeta + \left(\alpha - \frac{1}{2}\beta\right)\left(1+2\frac{g}{c^2}\zeta\right)\left[(\nabla\zeta)^2 - 2\frac{\partial\zeta}{\partial z}\right]$$

$$+\frac{g}{c^2}\left(1+\frac{g}{c^2}\zeta\right)\left[\left(1-\frac{\partial\zeta}{\partial z}\right)\frac{\partial\zeta}{\partial z} - \left(\frac{\partial\zeta}{\partial x}\right)^2\right] - \frac{g\beta}{u_0^2}\zeta = 0. \tag{5.6.5}$$

The left-hand side of this equation is equal to a constant along each streamline, and this constant has been put equal to zero on the assumption that somewhere in the stream the flow is undisturbed and horizontal so that ζ and its first two derivatives are zero there. We have also neglected terms in the square of $g\zeta/c^2$.

In the atmosphere, β and g/c^2 are of the order of 10^{-7} and $10^{-6}\,\mathrm{cm}^{-1}$ respectively, so that even when $\alpha = 0$ the term in β without g is negligible. This means, physically, that the effects of compressibility on the density produce greater variations in the inertia of a unit volume of air than the stratification does. In practical cases α may be important and must therefore be retained in general, but for the same reasons as we neglect β without g, we can usually neglect most of the terms in g/c^2.

If we put $\left(1+2\frac{g}{c^2}\zeta\right)^{-1} = 1-2\frac{g}{c^2}\zeta$ and neglect terms in g^2/c^4 compared with $g\beta/u_0^2$ we obtain from (5.6.5)

$$\nabla^2\zeta-\left(\alpha-\frac{1}{2}\beta-\frac{g}{c^2}\right)(\nabla\zeta)^2+\left(2\alpha-\beta-\frac{g}{c^2}\right)\frac{\partial\zeta}{\partial z}+\frac{g\beta}{u_0^2}\zeta\left(1-2\frac{g}{c^2}\zeta\right)=0. \tag{5.6.6}$$

Since the middle terms are usually negligible anyway except when they contain α, we see that the main effect of compressibility is in the last term. The term in ζ^2 is comparable with that in ζ when ζ is of the order of 5 km in the atmosphere. Thus for application to the atmosphere we write the equation in the form

$$\nabla^2\zeta+\alpha\left[2\frac{\partial\zeta}{\partial z}-(\nabla\zeta)^2\right]+\frac{g\beta}{u_0^2}\zeta\left(1-2\frac{g}{c^2}\zeta\right)=0. \tag{5.6.7}$$

5.7 THE EQUATION OF CONTINUITY IN A COMPRESSIBLE ATMOSPHERE

The following discussion in equally valid if the earth's rotation is taken into account. This is represented by the vector **f**, and the equation of motion is (see section 4.3)

$$\frac{\mathrm{D}\mathbf{v}}{\mathrm{D}t}=-\frac{1}{\rho}\,\mathbf{grad}\,p+\mathbf{g}+\mathbf{v}\times\mathbf{f} \tag{5.7.1}$$

so that

$$\begin{aligned}\mathbf{v}\,.\,\mathbf{grad}\,p &= -\rho\mathbf{v}\,.\,\frac{\mathrm{D}\mathbf{v}}{\mathrm{D}t}+\rho\mathbf{v}\,.\,\mathbf{g}\\ &= -\rho\frac{\mathrm{D}}{\mathrm{D}t}\frac{1}{2}q^2-g\rho w,\end{aligned} \tag{5.7.2}$$

w being the vertical velocity. For adiabatic motion (see (1.6.9))

$$\frac{\mathrm{D}p}{\mathrm{D}t}=c^2\frac{\mathrm{D}\rho}{\mathrm{D}t} \tag{5.7.3}$$

and for continuity

$$\frac{\mathrm{D}\rho}{\mathrm{D}t}=-\rho\,\mathrm{div}\,\mathbf{v}, \tag{5.7.4}$$

and from these three equations we obtain

$$\mathrm{div}\,\mathbf{v}=-\frac{1}{\rho c^2}\frac{\partial p}{\partial t}+\frac{1}{c^2}\left[\frac{\mathrm{D}}{\mathrm{D}t}\frac{1}{2}q^2+gw\right]. \tag{5.7.5}$$

The term in $\partial p/\partial t$ is negligible comparred with that in g, because if the local rate of change of pressure is 1 mb min^{-1}, which is only likely to be exceeded at the passage of a sharp front or tornado, the two terms are comparable for a vertical velocity of w of only 15 cm s^{-1}. This is very small compared with what would actually occur if the pressure changes were that large. Thus the vertical displacement of the air in the hydrostatic pressure field produces much greater divergence of velocity in practice than local pressure changes. The term in the material rate of change of $\frac{1}{2}q^2$ represents the pressure changes which must occur when the air is accelerated. These again are of the order of one or two millibars within any flow field under consideration, and are negligible because that corresponds to vertical displacements of 10 to 20 m, which are greatly exceeded in cases of interesting dynamics. Consequently the equation of continuity takes the form

$$\text{div } \mathbf{v} = \frac{g}{c^2} w, \tag{5.7.6}$$

which states that the only important cause of expansion of air parcels is their vertical displacement in the gravitationally produced pressure field.

In sections 5.3 to 5.7 we have investigated possible simplifications of the wave equation without restricting it to small amplitude displacements. Actually the most useful form of the equation will be seen to be very quickly obtained by assuming the displacements in the vertical direction to be small; but that would be to miss the usefulness because it is valid in some important cases when the amplitude is large and the flow contains regions of reversed velocity.

5.8 THE EFFECTS OF THE EARTH'S ROTATION, AND A DISCUSSION OF THE NEED TO ASSUME SMALL AMPLITUDE

We now use a notation appropriate to a small perturbation (u, v, w), τ, ϖ from an undisturbed stream (U, V, 0), τ_0 and ϖ_0 in which the undisturbed quantities are all functions of z. We neglect products of small quantities. Because waves of large amplitude are dominated by other effects, the earth's rotation is then negligible, and so the perturbation technique is appropriate when the rotation is included. The problem of large amplitude waves with rotation has so far eluded any comprehensible analysis. It is important to appreciate that we seek solutions more to understand the mechanism than to calculate wavelengths or wave shape in particular cases.

The adiabatic equation is

$$\frac{\mathrm{D}\tau}{\mathrm{D}t} = 0 \tag{5.8.1}$$

and the equation of motion is (see (4.1.9) and (1.6.6))

$$\frac{\mathrm{D}\mathbf{v}}{\mathrm{D}t} = -\frac{1}{\tau}\,\mathbf{grad}\,\varpi + \mathbf{g} + \mathbf{v} \times \mathbf{f}. \tag{5.8.2}$$

A wave motion proportional to e^{ikx} is assumed to be taking place so that when applied to the disturbance

$$\frac{\partial}{\partial x} = ik \tag{5.8.3}$$

but is otherwise zero. The motion is assumed to be two dimensional: this is a limitation in the sense that solutions proportional to $e^{\mu y}$ can be obtained, and represent Kelvin waves travelling along a wall which exerts a pressure on the fluid to counteract the deviating force. These waves can travel in the sea along a coastline or depth discontinuity, but are such a special phenomenon that we shall not discuss them here. Such waves could probably occur in the atmosphere along a mountain range, but propagated stable waves which are not stationary relative to the ground are not a very significant phenomenon in the atmosphere. Some cases unaffected by rotation are mentioned later in this chapter, but they have never been reported to be of the Kelvin type in the atmosphere. Of course, in order to have the undisturbed stream in geostrophic balance, ϖ_0 does vary in the y-direction. Thus the equations for the undisturbed motion are

$$0 = -\frac{1}{\tau_0}\frac{\partial \varpi_0}{\partial x} + Vf, \tag{5.8.4}$$

$$0 = -\frac{1}{\tau_0}\frac{\partial \varpi_0}{\partial y} - Uf, \tag{5.8.5}$$

$$0 = -\frac{1}{\tau_0}\frac{\partial \varpi_0}{\partial z} - g. \tag{5.8.6}$$

We have neglected the horizontal components of the earth's rotation because they produce no observable effect: f is the vertical component of **f**, which we assume not to vary horizontally by a significant amount, so that our conclusions are not valid in detail for an area extending over a wide range of latitude.

The full vertical component of (5.8.2) is

$$(\tau_0 + \tau)\left[(U+u)\frac{\partial}{\partial x} + (V+v)\frac{\partial}{\partial y} + w\frac{\partial}{\partial z}\right]w = -\frac{\partial}{\partial z}(\varpi_0 + \varpi) - g(\tau_0 + \tau), \tag{5.8.7}$$

and subtracting out (5.8.6) while neglecting second-degree terms in the perturbation, and writing a prime for $\partial/\partial z$, we get

$$\tau_0 ikUw + \varpi' + g\tau = 0. \tag{5.8.8}$$

The x- and y-components obtained in a similar way are

$$\tau_0 ikUu + \tau_0 U'w + ik\varpi - \tau_0 fv = 0, \tag{5.8.9}$$

and

$$ikUv + V'w + fu = 0. \tag{5.8.10}$$

The equations for adiabatic change (5.8.1), and continuity (5.7.6), are

$$ikU\tau + \tau_0' w = 0, \tag{5.8.11}$$

$$iku + w' - \frac{g}{c^2} w = 0. \tag{5.8.12}$$

From the last three equations v, τ and u may be expressed in terms of w and so (5.8.9) gives ϖ in terms of w thus:

$$ik\varpi = \tau_0 U\left[\frac{g}{c^2} w - w'\right] + \tau_0 U' w + \tau_0 f\left[\left(\frac{V'}{ikU} - \frac{fg}{k^2 U c^2}\right) w + \frac{f}{k^2 U} w'\right]. \tag{5.8.13}$$

This may be differentiated with respect to z and compared with (5.8.8) to eliminate ϖ. The result is expressed below as (5.8.16) with the following notation:

$$\alpha = \frac{U'}{U}, \quad \beta = -\frac{\tau_0'}{\tau_0}, \quad \alpha_1 = \frac{V'}{U}, \quad \beta_1 = \frac{T'}{T}, \quad F^2 = \frac{f^2}{k^2 U^2} \tag{5.8.14}$$

$$c^2 = \gamma RT, \quad \text{which gives} \quad \frac{\partial}{\partial z}\frac{g}{c^2} = -\beta_1 \frac{g}{c^2}. \tag{5.8.15}$$

These quantities represent the basic features of the undisturbed physical situation. α and α_1 may be of the same order of magnitude, and lie between zero and 10^{-5} cm^{-1} in the atmosphere, although they might rise to as large as 10^{-2} cm^{-1} in shallow layers, or even, in billow-producing situations, to 10^{-1} for short periods. β_1 is of the same order of magnitude as β, which in large depths averages around 10^{-7} cm^{-1} or up to 4 times this value in the stratosphere or in shallow layers. g/c^2 is about 10^{-6} cm^{-1}, and in mid-latitudes f is about 4×10^{-5} s^{-1}.

The elimination of all variables but w gives the equation

$$\begin{aligned}(1 - F^2)w'' &- \left[\frac{g}{c^2} + \beta - F^2\left(\alpha + \beta + \frac{g}{c^2}\right) - \mathrm{i}F\alpha\right] w' \\ &+ \left[\frac{g\beta}{U^2} - \frac{U''}{U} - k^2 + \alpha\left(\beta - \frac{g}{c^2}\right) + \frac{g}{c^2}(\beta + \beta_1)\right. \\ &\left. - \mathrm{i}F\left(\alpha_1(\beta + \alpha) - \frac{V''}{U}\right) - F^2 \frac{g}{c^2}(\alpha + \beta - \beta_1)\right] w = 0.\end{aligned} \tag{5.8.16}$$

The terms in F are negligible unless F is of the order of unity, which means that the time taken to travel one wavelength is of the order of a day. When that time is a small fraction of a day the effect of the earth's rotation is negligible. This is an important result and applies to a wide range of phenomena besides stable waves, and to convection phenomena in particular. We have noted elsewhere that geostrophic forces can intensify temperature gradients (see sections 4.7.(3), 4.8 and 8.5) but billows, which are on a smaller timescale, are set off by smaller timescale aspects of the medium-scale motion. Cloud streets are influenced by the geostrophic wind, but the unstable aspects of their motion are not generated by the deviating force, as some people suppose.

When F, which is a kind of Rossby number defined in (5.8.14), is of the order of unity the important terms in (5.8.16) are

$$(1-F^2)w'' + (F^2\alpha + \mathrm{i}F\alpha_1)w' + \left[\frac{g\beta}{U^2} - \frac{U''}{U} - k^2 - \mathrm{i}F\left(\alpha\alpha_1 - \frac{V''}{U}\right)\right]w = 0, \tag{5.8.17}$$

while when $F \ll 1$ they are

$$w'' + \left(\frac{g\beta}{U^2} - \frac{U''}{U} - k^2\right)w = 0. \tag{5.8.18}$$

When the earth's rotation is important, the scale of the motion is likely to be such that the terms in α, α_1, U'' and V'' are not important. Then the equation is

$$(k^2U^2 - f^2)w'' + g\beta k^2 w = 0, \tag{5.8.19}$$

and we see that $\partial/\partial z$ is of the order of $\left(\dfrac{g\beta k^2}{k^2U^2 - f^2}\right)^{1/2}$, which has magnitude of the order of $(g\beta/U^2)^{1/2}$, i.e. more or less the same as when the earth's rotation is not important. The waves may aptly be called **Queney waves** because they were first extensively studied by Queney in 1947 [8]. They have a vertical wavelength (within a multiple of 2 or 3) of 2π km and a streamwise wavelength of the order of $2\pi(U/f)$, $= 2\pi \times 250$ km when U is of the order of 10 m s^{-1} in mid-latitudes.

In this case u and v are of the same order of magnitude according to (5.8.10), while u/w is about the ratio of the horizontal to vertical wavelength, according to (5.8.12). Thus the lateral and longitudinal displacements are about 250 times the vertical and the slope of the streamlines is small.

In Queney waves the dominant forces are those due to static stability and the earth's rotation, and the displacements are along surfaces inclined to the horizontal with a gradient of the order of 1:250.

To express (5.8.18), which is the short wavelength case in which f is negligible, in terms of the vertical displacement ζ, we note that

$$w = \frac{\mathrm{D}\zeta}{\mathrm{D}t}, \quad = \mathrm{i}kU\zeta \text{ when linearised,} \tag{5.8.20}$$

so that

$$w' = \mathrm{i}k(U'\zeta + U\zeta'), \tag{5.8.21}$$

and

$$w'' = \mathrm{i}k(U''\zeta + 2U'\zeta' + U\zeta''). \tag{5.8.22}$$

From this we obtain the equation for the displacement in the form

$$\zeta'' + 2\alpha\zeta' + \left(\frac{g\beta}{U^2} - k^2\right)\zeta = 0, \tag{5.8.23}$$

which is the linearised version of (5.6.7) assuming ζ to be proportional to cos kx.

In the case where $\alpha = 0$ and $c^2 = \infty$ (no shear, incompressible, or ζ is small enough for $g\zeta/c^2 \ll 1$) equation (5.6.7) is linear and can be written

$$\zeta'' + \left(\frac{g\beta}{U^2} - k^2\right)\zeta = 0, \tag{5.8.24}$$

which is identical with the form taken by (5.8.23). The assumptions of perturbation theory are therefore not necessary to obtain this equation. In particular it is not necessary that $u \ll U$ etc. and (5.8.23) is a good approximation to the full equation for fairly large displacements and is exact for arbitrarily large displacements when $\alpha = 0$ and $g\beta/U^2$ is independent of z. The classical derivation of the wave equation by the usual small perturbation methods is therefore both misleading about the conditions in which it is valid, and unduly restrictive as to its application.

Equation (5.6.7) is obviously non-linear in ζ because of the terms explicitly containing ζ^2. But if α, β and u_0 are functions of z_0 they become functions of ζ when there is a displacement, so that the most important restrictions that have to be imposed in practice to obtain a linear equation are on the stream profile rather than on the amplitude of the wave. It is important that variations in $g\beta/U^2$ (sometimes called the Scorer parameter because its variations with height are very important in determining the character of lee waves) generate non-linear terms in ζ when the amplitude is large.

5.9 LINEAR WAVES OF LARGE AMPLITUDE IN AN INCOMPRESSIBLE FLUID

Equation (5.4.3) can be written

$$\begin{aligned} \nabla^2 z_0 + S(\nabla z_0)^2 + \ell^2 z &= \text{constant along a streamline} \\ &= S + \ell^2 z_0, \end{aligned} \tag{5.9.1}$$

if $z = z_0$ far upstream, where

$$S = \alpha - \frac{1}{2}\beta = \frac{1}{\rho^{1/2}u_0}\frac{d}{dz_0}\left(\rho^{1/2}u_0\right), \quad \ell^2 = g\beta/u_0^2. \tag{5.9.2}$$

We now investigate the possibility that there may be a variable $\psi = \psi(z_0)$ in terms of which this equation is linear. Thus

$$\nabla z_0 = \frac{dz_0}{d\psi}\nabla\psi$$

$$\nabla^2 z_0 = \left(\nabla\psi\frac{d}{d\psi}\right)^2 z_0 = \frac{dz_0}{d\psi}\nabla^2\psi + \frac{d^2 z_0}{d\psi^2}(\nabla\psi)^2$$

$$(\nabla z_0)^2 = \left(\frac{dz_0}{d\psi}\right)^2 (\nabla\psi)^2, \tag{5.9.3}$$

and the Wave Equation (5.9.1) becomes

$$\nabla^2\psi + (\nabla\psi)^2 \frac{d\psi}{dz_0}\left[\frac{d^2 z_0}{d\psi^2} + \left(\frac{dz_0}{d\psi}\right)^2 S\right] = \frac{d\psi}{dz_0}\left[S + \ell^2 (z - z_0)\right]. \tag{5.9.4}$$

The term in $(\nabla\psi)^2$ disappears if

$$\frac{d^2 z_0}{d\psi^2} + \left(\frac{dz_0}{d\psi}\right)^2 \frac{d}{dz_0}\ln\left(\rho^{1/2}u_0\right) = \frac{1}{\rho^{1/2}u_0}\frac{\partial}{\partial\psi}\left(\rho^{1/2}u_0\frac{\partial z_0}{\partial\psi}\right) = 0. \tag{5.9.5}$$

This is the case if

$$\rho^{1/2}u_0\frac{dz_0}{d\psi} = \text{constant} = 1, \tag{5.9.6}$$

say, which may be written

$$\frac{d\psi}{dz_0} = \rho^{1/2}u_0, \quad \text{or} \quad \psi = \int \rho^{1/2}u_0\, dz_0. \tag{5.9.7}$$

This result is due to Long (1953), and the equation becomes

$$\nabla^2\psi - S\rho^{1/2}u_0 - \ell^2 (z - z_0)\rho^{1/2}u_0 = 0, \tag{5.9.8}$$

which has the form

$$\nabla^2\psi - F(\psi) - G(\psi)z = 0. \tag{5.9.9}$$

If this equation is linear in ψ, F and G must be linear, and so

$$F(\psi) = (S - \ell^2 z_0)\rho^{1/2}u_0 = A\psi + B$$

$$G(\psi) = \ell^2 \rho^{1/2}u_0 = C\psi + D \tag{5.9.10}$$

where A, B, C and D are constants, independent of z_0, and the equations (5.9.10) restrict the profiles of ρ and u_0 to certain functional forms of z_0 in the undisturbed stream. The wave equation then becomes, with (5.8.3),

$$\frac{d^2\psi}{dz^2} = (A + k^2)\psi + B + (C\psi + D)z; \tag{5.9.11}$$

and eliminating $\ell^2 \rho^{1/2} u_0$ and S from the equations (5.9.10), (5.9.7) and (5.9.2)

$$\frac{\mathrm{d}^2\psi(z_0)}{\mathrm{d}z_0^2} = A\psi(z_0) + B + \big(C\psi(z_0) + D\big)z_0, \tag{5.9.12}$$

which has solutions of the same functional form as (5.9.11). If A, B, C and D are chosen, (5.9.12) could be solved for $\psi(z_0)$ so that if (5.9.11) is then solved, the streamlines ψ = const, i.e. z_0 = constant, can be drawn. The stream profiles of ρ and u_0 can be determined from (5.9.10) but clearly profiles which are of interest can only be found by intelligent integration of (5.9.10) with trial values of the constants A, B, C and D.

It is fairly easy to find particular cases which are of interest, the most obvious being $S = 0$, $B = C = 0$, $\psi \propto z_0$. This requires that both ℓ^2 and $\rho^{1/2}u_0$ be constant and independent of z_0, which is the case discussed in the following pages. Several cases with more variable values of u_0 and ρ as functions of z_0 can be contrived, but they all require the medium to be bounded by rigid walls above and below because no solution which does not become exponentially large for large z has yet been found. The case in which the disturbance remains bounded for large z, mentioned at the end of section 5.5., is that in which S is a constant and u_0 is approximately proportional to z_0; this is not of practical significance.

The solutions of (5.9.12) must be monotonic functions of z_0 if they are to be meaningful because ψ is a stream function defined by (5.9.7). Moreover u_0 must not become zero, for if it did ℓ^2 would be infinite and the solution would have an infinity of zeros in the neighbourhood. What happens at such a level is discussed in section 5.17. All the solutions of (5.9.11) and (5.9.12) are like the Airy functions, which are either oscillatory or exponential and cannot be algebraic for large z.

The alternative approach of representing a real atmosphere by several superposed layers in each of which the solution is simple is useful only when the displacement is small at the interfaces, because the boundary conditions are non-linear. Normally they are applied at the undisturbed position of the interface but they become intractably dependent on ζ if applied at the displaced position.

Yih has employed linear solutions of the kind implied by this discussion in cases where a liquid is bounded on top by a free surface which, because of the very large density decrease, behaves as a rigid surface. The chief merit of this kind of solution as far as the atmosphere is concerned is in elucidating some of the mechanisms that are important in the atmosphere, as Long has demonstrated. It is to this aspect of the subject that we now turn.

5.10 STANDING WAVES AND ROTORS

The phenomena will be best exemplified by the simplest cases: the possibility of more complicated cases is obvious but no essentially new mechanisms are revealed by them. We use the equation for the vertical displacement ζ because it is linear in the cases treated. The equation for w, the vertical velocity, is linear only for small amplitudes, in the form (5.8.18), because the relationship between ζ and w is non-linear, namely

$$w = \frac{D\zeta}{Dt} = u\frac{\partial \zeta}{\partial x} + w\frac{\partial \zeta}{\partial z} \tag{5.10.1}$$

in the two-dimensional case. We start with equation (5.6.7) and treat the particular case of a uniform stratified incompressible stream between horizontal rigid boundaries.

The wave equation in this case is, with $\zeta \propto \cos kx$,

$$\frac{\partial^2 \zeta}{\partial z^2} + (\ell^2 - k^2)\zeta = 0 \tag{5.10.2}$$

and if we write

$$\frac{g\beta}{u_0^2} = \ell^2, \quad \nu^2 = \ell^2 - k^2, \tag{5.10.3}$$

the solution satisfying the boundary conditions $\zeta = 0$ at $z = 0, h$ is

$$\zeta = A \sin \nu z \cos kx, \tag{5.10.4}$$

with

$$\sin \nu h = 0, \quad \text{or} \quad \nu h = n\pi. \tag{5.10.5}$$

This last equation determine the wavelength $2\pi/k$ for each value of n. In Fig. 5.10.i streamlines for $n = 1$ are shown. The profile of ζ has one node within the fluid when $n = 2$, and the flow in each half of the fluid is then like that shown but with opposite phase in the two halves.

There is only one possible value of k if $\pi < \ell h < 2\pi$, none if $\ell h < \pi$, two if $2\pi < \ell h < 3\pi$, and so on. Although there could be a large number of possible values of k, there are usually only one or two in practical cases, and when there are more, not all the modes are excited to a significant amplitude.

There can be regions of reversed flow, usually with closed circulations, if the horizontal velocity is somewhere negative, which means that

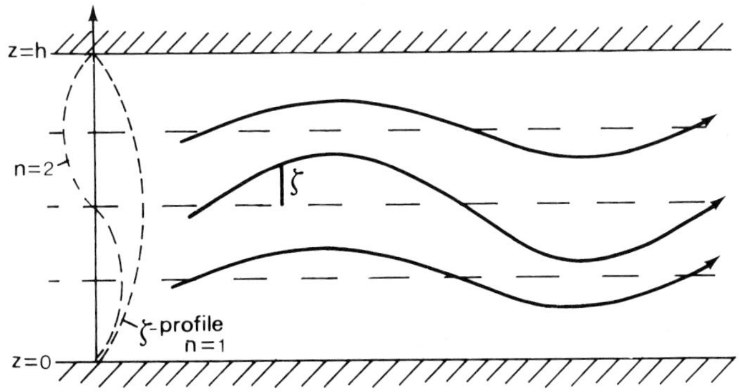

Fig. 5.10.i Coordinates and boundaries for a stratified stream between rigid boundaries. The streamlines relate to the profile of ζ for $n = 1$.

$$1-\frac{\partial \zeta}{\partial z}=\frac{\partial z_0}{\partial z}<0, \quad \text{by (5.1.4)},$$

or

$$\frac{\partial \zeta}{\partial z}>1. \tag{5.10.6}$$

In this case the condition is

$$A\nu \cos \nu z \cos kx > 1 \quad \text{somewhere,} \tag{5.10.7}$$

which means that

$$A\nu > 1 \quad \text{somewhere.} \tag{5.10.8}$$

In the case of the first mode $\nu h=\pi$ and so the condition becomes

$$A > h/\pi, \tag{5.10.9}$$

and the reversed flow occurs in the neighbourhood of where $\cos \nu z = \cos kx = 1$ or $\cos \nu z = \cos kx = -1$. In this case there is a closed circulation around these points, each of which is on the boundary, and the circulations are called rotors.

The stagnation points in the middle of the circulations are given by

$$\cos kx = \pm 1, \quad A\nu \cos \nu z = \pm 1 \tag{5.10.10}$$

and those on the nodal surfaces by

$$\cos \nu z = \pm 1, \quad A\nu \cos kx = \pm 1. \tag{5.10.11}$$

The line on which the velocity is vertical, which is the boundary of the region in which u is negative and there is backward flow, is given by

$$A\nu \cos \nu z \cos kx = 1. \tag{5.10.12}$$

Fig. 5.10.ii shows the flow pattern when $n=2$. The rotors on the nodal surface $z=h/2$ are double-yolked, and the boundary of the reversed flow is shown by a dashed line.

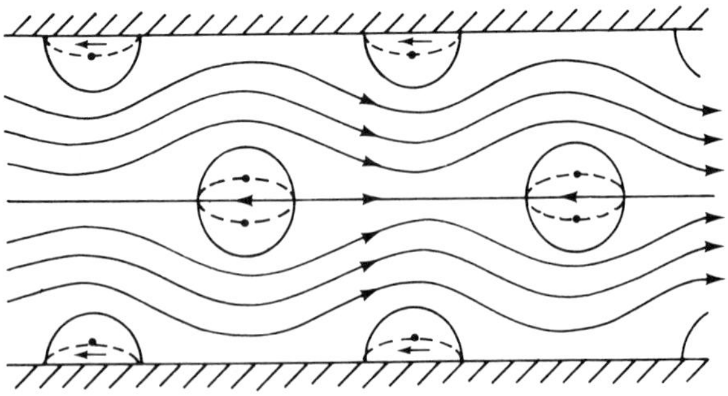

Fig. 5.10.ii The pattern of streamlines for the case $n=2$ when the amplitude is large enough for the waves to have rotors. The dotted lines are the boundaries of the regions of reversed flow. The stagnation points at the centres of the rotors are marked and the flow circulates around them: the rotors in the middle of the stream are therefore double-yolked.

Where the flow is reversed, the fluid is statically unstable if it is stable elsewhere, so that it is to be expected that buoyant convection would occur within the rotor until the density had been made more or less uniform inside. This in turn would produce a small density discontinuity on the boundary of the rotor, which might produce billows by the mechanism described in Chapter 6.

The fluid inside the rotors is described by the mathematics as having a density equal to some of the fluid on the opposite side of the nodal surface. Thus the fluid in the rotors on the bottom is described as having a density, and a value of z_0, from below the bottom. Since this is impossible it must contain fluid which is not statically unstable, and the theory is strictly incorrect in this detail. In the double-yolked rotors on the central nodal surface the fluid is described as having the same density as fluid in some part of the system, but it is unlikely that if the waves were set up with a large enough amplitude to have rotors the density would be thus distributed.

An important feature of rotor flow is that the flow is very swift in between the stagnation points on the nodal surfaces, and that the crests and troughs of the streamlines are not similarly shaped, the lines close to the nodal surfaces being flattened, while those over or below the rotors having greater curvature.

Experimentally the motion is most readily demonstrated by moving an obstacle along the top of a tank of static stratified liquid. Fig. 5.10.iii shows the disturbance of a moving obstacle by means of a time exposure.

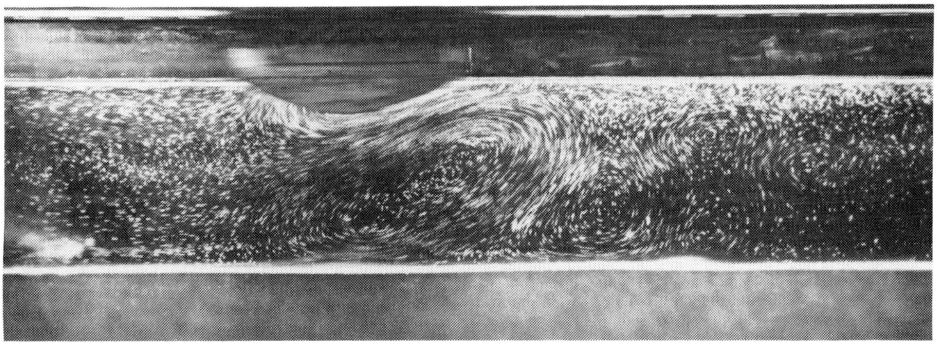

Fig. 5.10.iii A time exposure reveals the vector velocity of particles floating in a water tank. Stratified by the addition of salt. The 'mountain' is moving along to the left, so that the velocity field for a stream 'over' a fixed obstacle is obtained by subtracting the velocity of the mountain from the whole pattern. The disturbance is small on the right because it is near the end of the tank from which the obstacle was started from rest.

5.11 WAVE TRAPPING

(1) An infinite atmosphere with a single discontinuity of static stability

It is assumed that the displacement is small enough for the compressibility of the air to be neglected, in which case we can use equation (5.10.2) throughout. We require the amplitude to tend to zero as z tends to infinity and to be zero at $z = -h$. In the upper layer, therefore, the solution is of the form (omitting cos kx)

$$\zeta = A\,\mathrm{e}^{-\mu z} \tag{5.11.1}$$

where

$$\mu = +\left(k^2 - \ell_s^2\right)^{1/2}, \tag{5.11.2}$$

the alternative solution being proportional to $\mathrm{e}^{\mu z}$ and unacceptable for large z. The value of ℓ^2 in the upper (superior) layer is ℓ_s^2.

In the lower layer we need to have a sinusoidal solution as in section 5.10 in order to have a zero at $z = -h$ and to fulfil the conditions at the boundary between the layers at $z = 0$, namely that the displacement ζ be continuous and that the pressure be continuous. The pressure is given by

$$\frac{\varpi}{\tau} + \frac{1}{2}q^2 + gz = \text{constant}. \tag{5.11.3}$$

We write

$$\begin{aligned} q^2 &= u^2 + w^2 \\ &= \left(\frac{\rho_0}{\rho}u_0\right)^2\left[\left(\frac{\partial z_0}{\partial z}\right)^2 + \left(\frac{\partial z_0}{\partial x}\right)^2\right] \\ &= \left(\frac{\rho_0}{\rho}u_0\right)^2\left[\left(1 - \frac{\partial \zeta}{\partial z}\right)^2 - k^2\zeta^2\right] \end{aligned} \tag{5.11.4}$$

and consider the simplest kind of discontinuity possible at $z = 0$, namely one in which ρ_0 and u_0 are continuous so that ρ is also continuous. Because ζ is continuous, the continuity of ϖ and τ across the interface requires, in view of the equations (5.11.3) and (5.11.4), that $\partial\zeta/\partial z$ also be continuous. This result is strictly correct regardless of the amplitude of the displacement of the interface which, being $z_0 = 0$ in this case, lies at $z = \zeta$.

In the lower layer, in which $\ell^2 = \ell_i^2$,

$$\zeta = B\sin\nu(z + h), \quad \text{which } = 0 \text{ at } z = -h, \tag{5.11.5}$$

where

$$\nu = +\,(\ell_i^2 - k^2)^{1/2} \tag{5.11.6}$$

and so, putting ζ and $\partial\zeta/\partial z$ equal to the same value according to equations (5.11.1) and (5.11.2), we obtain at $z = 0$,

$$A = B\sin\nu h, \qquad -A\mu = B\nu\cos\nu h, \tag{5.11.7}$$

whence

$$\boxed{\nu\cot\nu h = -\mu.} \tag{5.11.8}$$

The details of the solution are illustrated in Fig. 5.11.i.

There will be a rotor on the bottom boundary ($z = -h$) if

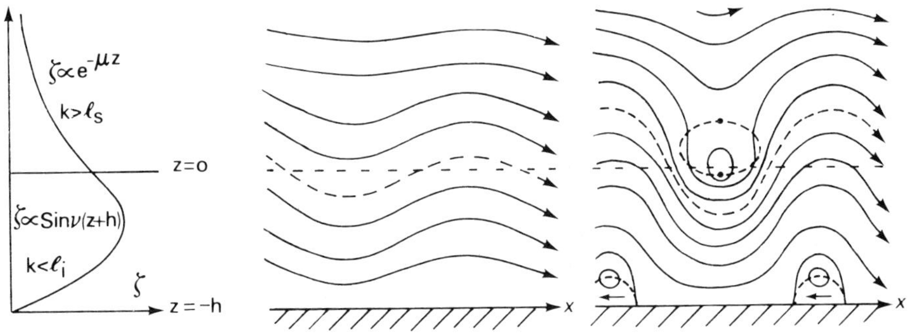

Fig. 5.11.i On the left is the profile of ζ for the two-layer case, the upper layer extending to infinity. In the middle, the shape of the streamlines is shown for small amplitude, and on the right where there are rotors. The dashed streamlines shows the displaced position of the interface. The boundary of the region of reversed flow (also shown by a dashed contour) passes through the two stagnation points and inside this contour the fluid is statically unstable.

$$\frac{\partial \zeta}{\partial z} = B\nu \cos kx > 1 \text{ somewhere,}$$

or

$$B\nu > 1 \tag{5.11.9}$$

and there will be one at the interface ($z = 0$), where the value of $\partial\zeta/\partial z$ has a maximum, if

$$B\nu \cos \nu h > 1. \tag{5.11.10}$$

In Fig. 5.11.i the case of no rotors is illustrated together with the case of rotors at both surfaces. The rotor on the interface is smaller, and if (5.11.9) is satisfied but not (5.11.10) there is a rotor on the bottom boundary only.

The equation (5.11.8) determines k from the given parameters of the problem, namely ℓ_i, ℓ_s and h. In Fig. 5.11.ii the continuous curve represents the function $\nu \cot \nu h$ as a function of νh, while the dashed lines represent the function μ as a function of νh, these being circles because

$$\mu^2 = \ell_i^2 - \ell_s^2 - \nu^2. \tag{5.11.11}$$

If the radius of the circle lies between $\frac{1}{2}\pi$ and $\frac{3}{2}\pi$, there is one value of νh at which $\nu \cot \nu h = -\mu$. If it lies between $\frac{3}{2}\pi$ and $\frac{5}{2}\pi$ there are two values. In general there are n values if

$$\frac{(2n-1)\pi^2}{2h} < \ell_i^2 - \ell_s^2 < \frac{(2n+1)\pi^2}{2h}. \tag{5.11.12}$$

If $\ell_s^2 > \ell_i^2$, no solutions of this kind, with a zero on the bottom and an exponentially decreasing value for large z, can be obtained; and in all cases where there is a solution

$$\boxed{\ell_i > k > \ell_s.} \tag{5.11.13}$$

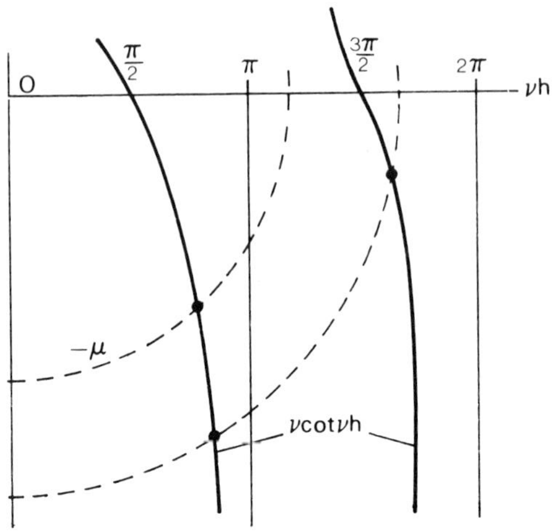

Fig. 5.11.ii Solutions obtained by plotting the left (continuous curve) and right (dashed) sides of (5.11.8) as functions of νh.

(2) The general requirement for trapped waves

The previous case is a particular example of waves trapped in the part of the airstream close to the ground. The requirement for at least one wave to be possible is (5.10.24) with $n = 1$, namely

$$\boxed{\ell_{\mathrm{i}}^2 - \ell_{\mathrm{s}}^2 > \pi^2/4h^2\,.} \tag{5.11.14}$$

In the more general case the possibility of the existence of waves can be investigated by considering the case of small amplitude. This is because it is unlikely that a case would be complicated in a way which would make it different from the simple cases where if a periodic wave is possible for a large amplitude it is possible for small amplitude, although the converse is not always true. For small amplitude we may reverse the transformation from w to ζ set out in equations (5.8.20)–(5.8.23) and obtain the equation for w, subject only to the approximation required in (5.8.20), namely u, $w \ll U$. Then, from (5.8.18)

$$\frac{\partial^2 w}{\partial z^2} + (\ell^2 - k^2)w = 0 \tag{5.11.15}$$

where now the parameter is

$$\ell^2 = \frac{g\beta}{u_0^2} - \frac{1}{u_0}\frac{\mathrm{d}^2 u_0}{\mathrm{d}z_0^2}\,. \tag{5.11.16}$$

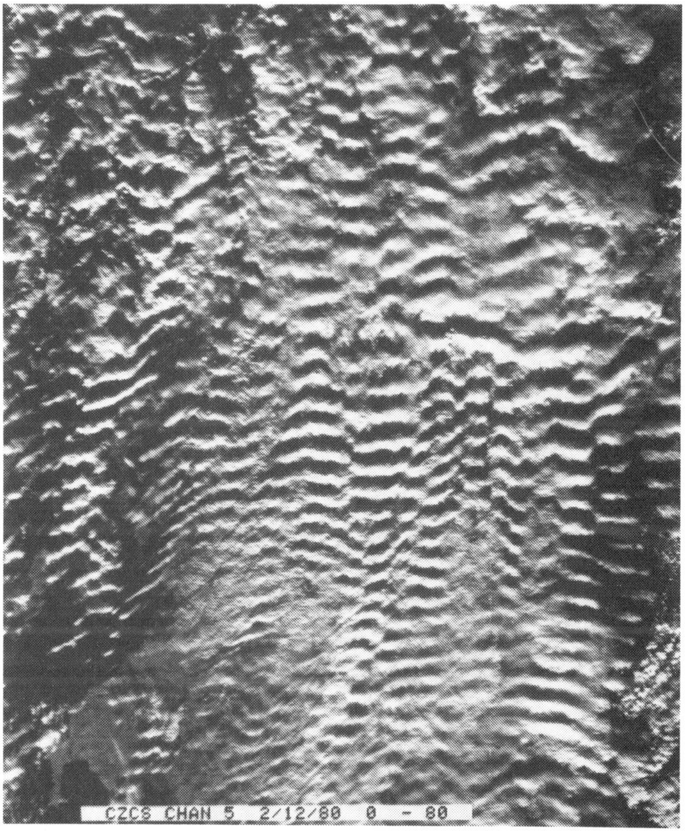

Fig. 5.11.iii Satellite CZCS (Nimbus 7) Channel 5 at 1057, 2.12.80. Waves in a cloud layer in a wind from the north over Scotland, showing a 'ladder' structure in which wave trains side by side do not exactly fit together. No mathematical boundary conditions for this condition have been proposed other than a narrow angle of spread of ship-wave patterns (see section 6.6): the wavelength appears to be slightly different on the two sides of the dividing line.

The profile of w, like that of ζ, must asymptote to zero as z tends to infinity for a wave to be trapped, so that $\partial^2 w/\partial z^2$ must be a positive multiple of w, and so $k^2 > \ell^2$. On the other hand, in order to curve back to zero at the ground like the profile of ζ in Fig. 5.11.i, the curvature of the profile must be reversed in the lower layers of the stream. So in some lower layers we must have either $\ell^2 > k^2$ in a fairly deep layer, or else ℓ^2 must exceed k^2 by a large amount in a less deep layer. Thus in all cases the value of k is equal to a value between the extreme values of ℓ in the profile of ℓ, and there must be enough difference between these extremes for a condition like (5.11.14) to be satisfied.

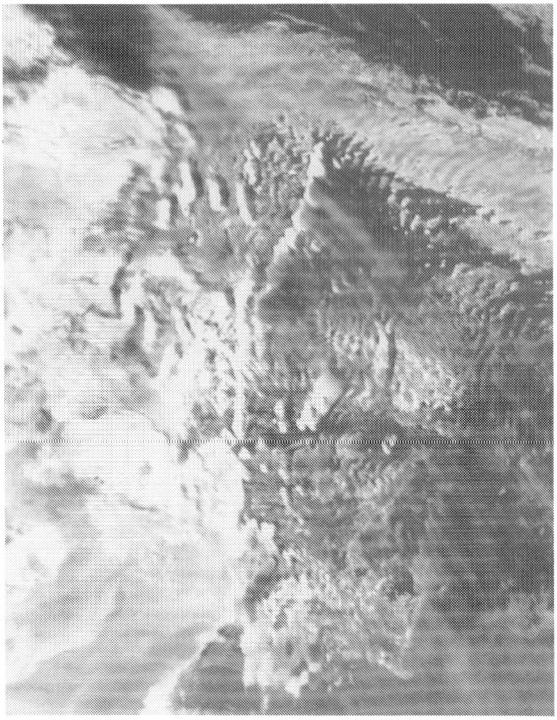

Fig. 5.11.iv This satellite picture received at 1352 on 11.5.82 shows an airstream with two characteristic wavelengths (and two slightly different orientations in this case) over Iceland. The higher-level waves of longer length produce orographic cirrus trails leading off to the SE.

5.12 TRAPPING BY A DENSITY DISCONTINUITY

We shall see in Chapter 6 that a discontinuity of velocity cannot be maintained, but that discontinuities of density may exist. Methods by which they may be created in the atmosphere are described in sections 6.4, 11.1(2) and 11.5(2).

In steady motion in two dimensions on each streamline we have

$$\frac{p}{\rho}+\frac{1}{2}q^2+gz=\text{constant} \tag{5.12.1}$$

and in the case of adiabatic motion we may replace p/ρ by ϖ/τ and obtain an analogous result. At an interface the pressure p is continuous and since ϖ is a function of p only for a gas of uniform composition [see (1.6.4)], ϖ must also be continuous at a density discontinuity where ρ and τ are discontinuous. The variable part of p obtained from (5.12.1) is the same on the two sides of the discontinuity. With q and z varying along the streamline, and with z put equal to ζ by choosing the zero of z to be at the undisturbed position of the discontinuity, we thus obtain, with i and s denoting the inferior and superior layers,

$$-\rho_s u_0^2 \frac{\partial \zeta_s}{\partial z} + g\rho_s \zeta = \text{constant} - p_s$$
$$= \text{constant} - p_i$$
$$= -\rho_i u_0^2 \frac{\partial \zeta_i}{\partial z} + g\rho_i \zeta. \tag{5.12.2}$$

Here we have put $\zeta_i = \zeta_s = \zeta$, and in each layer by the variable part of q^2, to first order in ζ, is $-2u_0^2 \partial\zeta/\partial z$. Thus

$$\rho_i u_0^2 \frac{\partial \zeta_i}{\partial z} = g\zeta(\rho_i - \rho_s) + \rho_s u_0^2 \frac{\partial \zeta_s}{\partial z} \tag{5.12.3}$$

expresses the continuity of pressure across the discontinuity of density. There are two cases of special interest:

(i) when the upper layer has a very small density and is a gas of approximately constant pressure above a liquid, we may put $\rho_s = 0$ or p = constant on the surface. In this case

$$\frac{\partial \zeta_i}{\partial z} = \frac{g\zeta_i}{u_0^2} \quad \text{at the surface.} \tag{5.12.4}$$

This particular result implies that we are discussing external waves, that is waves on the free surface, and for them the internal stratification is of negligible importance. If we are to discuss internal waves we have to impose the boundary condition that the free surface is displaced negligibly compared with the internal displacements and the boundary condition becomes $\zeta = 0$ there. The pressure variations along the surface are maintained by a displacement of it which is small and is calculated from (5.12.4) after $\partial\zeta/\partial z$ has been obtained on the assumption that ζ is zero there.

(ii) when the density discontinuity is a small fraction of the density we may put

$$\rho_s = (1 - \epsilon)\rho_i \tag{5.12.5}$$

and neglect squares of ϵ. Then (5.12.3) may be written

$$\frac{\partial \zeta_i}{\partial z} = \frac{\epsilon g \zeta}{u_0^2} + \frac{\partial \zeta_s}{\partial z}. \tag{5.12.6}$$

A discontinuity of density behaves like a very thin layer in which β is very large, and so the slope of ζ may be reversed on passing through it. We may therefore have, for example, two layers with static stability small in the upper layer and zero in the lower layer and a stable density discontinuity at the interface. This is a common situation in the atmosphere when sunshine warms the ground and produces thermal convection in the lower layer. In that case,

$$\zeta_s \propto e^{-\mu z}, \qquad \zeta_i \propto \sinh k(z+h) \tag{5.12.7}$$

so that the waves vanish at great height and are of zero amplitude at the ground $z = -h$. The interface condition (5.12.6) is then, at $z = 0$,

$$k \coth kh = \frac{\epsilon g}{u_0^2} - \mu \tag{5.12.8}$$

if u_0 is constant throughout, and μ is defined by (5.12.6). There is a value of h satisfying this condition of $\coth kh > 1$, i.e. if

$$\epsilon g > u_0^2(k+\mu). \tag{5.12.9}$$

Theoretical treatments tend to regard the determination of k as the objective of the analysis, but in practice the length of waves can be measured if they are observed and u_0 and μ can then be obtained from standard soundings of the air, approximately. The theory thus determines the value which ϵ must have exceeded, the actual value being obtainable from (5.12.8) if h is known roughly.

6

Lee waves, mountain waves, dispersed waves

6.1 COMPARISON WITH OBSERVATION

We have so far derived the flow equation for various cases of steady motion, which means for airflow through a standing (stationary) wave pattern. In this chapter we shall be concerned more with the origin of these waves. Most usually the standing waves are lee waves, and result from flow over a hill somewhere upstream. Clearly, if we add a linear velocity to the coordinate system the waves could be seen as travelling over level ground (or sea), but in that case we would require an impulse to start them which may be a downdraught of air from a higher-level storm moving with a different velocity, or an explosion of a volcano or the sudden arrival of a meteorite.

Whatever the cause it is likely that some sort of integral will be used to represent it, which means that we have to use a linearised flow equation in order to be able to add the component parts of the integral. As is usual in all fluid flow studies, observations of the phenomenon are very useful in making clear the conditions under which our approximations are valid, and this is a necessary step in establishing that our explanations are correct.

One of the most important observed phenomena is separation of the flow from the surface, and in what way our obstacle is not truly two-dimensional. There are also regions where even the linearised theory is not applicable, and where some speculation has to be made in conformity with such tracers of the airflow as may be available.

6.2 THE CLASSICAL CASE

The use of a Fourier Integral representation of a disturbance of flow which is governed by a linear equation has become almost conventional. The most popular is illustrated in Figs 6.2.i and 6.2.ii, which represent the displacement of the lower boundary of the flow by the quantity ζ_0 given by

$$\begin{aligned}\zeta_0 &= ab\int_0^\infty \mathrm{e}^{-kb+\mathrm{i}kx}\,\mathrm{d}k \\ &= \frac{ab}{-b+\mathrm{i}x}\left[\mathrm{e}^{-kb+\mathrm{i}kx}\right]_{k=0}^{\infty} \qquad (6.2.1)\end{aligned}$$

$$= \frac{ab(b + \mathrm{i}x)}{b^2 + x^2}$$

$$= \frac{ab^2}{b^2 + x^2} + \mathrm{i}\,\frac{abx}{b^2 + x^2}. \tag{6.2.2}$$

The parameters a and b represent the lengths indicated in the figures.

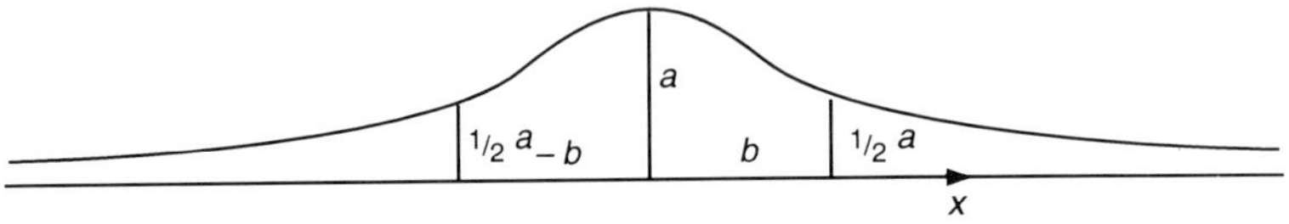

Fig. 6.2.i The function ζ defined by (6.2.2) real part.

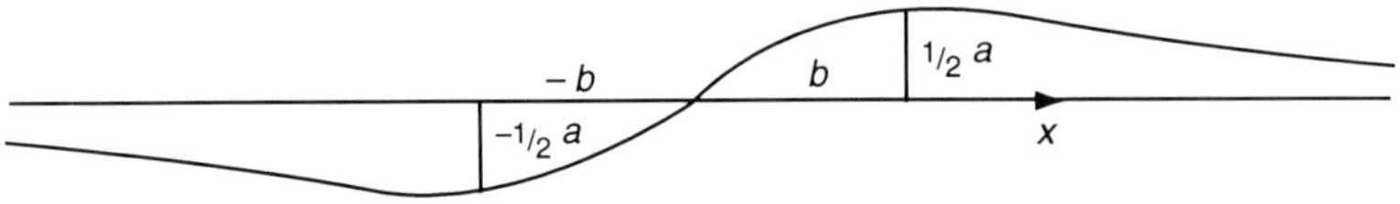

Fig. 6.2.ii The function ζ defined by (6.2.2) imaginary part.

The equation to which this is to be applied as a bottom boundary condition is (5.6.7), and the most generally applicable is that in which ζ appears linearly. Thus $(\nabla\zeta)^2$ and $2\frac{g\beta}{\mu_0^2}\cdot\frac{g}{c^2}\zeta^2$ are neglected and this requires the assumption that the vertical displacement ζ is a small quantity. We therefore use the equation (5.8.23), which is

$$\frac{\partial^2\zeta}{\partial z^2} + S\frac{\partial\zeta}{\partial z} + (\ell^2 - k^2)\zeta = 0 \tag{6.2.3}$$

in which

$$S = 2\alpha - \beta - \frac{g}{c^2}, \qquad \ell^2 = \frac{g\beta}{u_0^2}. \tag{6.2.4}$$

The use of this is best illustrated by the simplest case where S is zero and ℓ^2 is a constant. The airstream can then be crudely represented by superposed layers in each of which these assumptions are valid. We choose an upper layer of the stream extending up to infinite height, in which $\ell^2 = \ell_s^2$, and a lower layer in which $\ell^2 = \ell_i^2$, and $\ell_i > \ell_s$ so that as in the cases illustrated in Chapter 5, there exists at least one standing wave number k^*

$$\ell_i > k^* > \ell_s. \tag{6.2.5}$$

The solution of the equation in which the variables x and z are separated, so that we may write for the general value of k required in obtaining the lower boundary condition, the form of ζ in the two layers is

$$\zeta_s = e^{ikx} C\, e^{-\mu_s z} \tag{6.2.6}$$

$$\zeta_i = e^{ikx}(A\, e^{i\nu z} + B\, e^{-i\nu z}) \tag{6.2.7}$$

where an upper boundary condition requiring that $\zeta_s \to 0$ as $z \to \infty$ has been imposed and the real part is taken as the appropriate solution, when the obstacle mountain is described by the real part of (6.2.2) as illustrated in Fig. 6.2.i.

6.3 EARLIER FORMULATIONS

The Fourier Integral formulations of disturbances producing waves were made late in the 19th century; but the applications to the atmosphere made in response to the needs of aviation are more recent. The two earliest were by Lyra [10] and Queney [8], and there have been many others which are the same in principle since Scorer [11] pointed to the importance of the wind profile, showing that it did not depend, even primarily, on discontinuities or great variations in the temperature (i.e. density) profile of an airstream.

Lyra was well aware of the observations by glider pilots of standing waves caused by mountains, and their use in soaring. He chose the simplest possible model, namely a stably stratified airstream with uniform wind velocity. In effect he used a simplified version of Fig. 6.2.i, namely a delta function or a brick-shaped profile requiring a step up followed by a step down.

The delta function is represented by the case $b = 0$, and is in effect a thin wall of finite height. A sloped step up is given by

$$\zeta_0 = \int^z \mathrm{d}x \int_0^\infty a\, e^{-kb+ikx}\, \mathrm{d}k \qquad \text{real part} \tag{6.3.1}$$

$$= \int^z \frac{a(b+ix)}{b^2+x^2}\, \mathrm{d}x \qquad \text{real part}$$

$$= \int^z \frac{ab}{b^2+x^2}\, \mathrm{d}x$$

$$= a \tan^{-1}\frac{x}{b} \tag{6.3.2}$$

where the constant of integration is chosen to make the displacement zero at $x = 0$ (Fig. 6.3.i) and the boundary is at $\mp a$ at $x = \mp\infty$, and becomes a more abrupt step up at $x = 0$ when $b \to 0$.

The corresponding imaginary part (derived from Fig. 6.2.ii) is

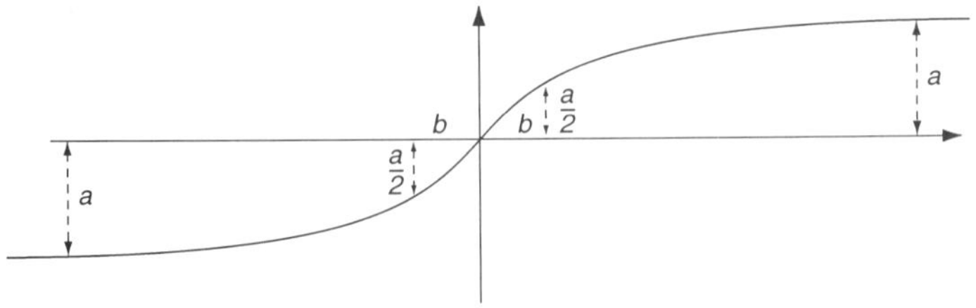

Fig. 6.3.i The ground contour represented by $\zeta_0 = a \tan^{-1} \frac{x}{b}$.

$$\int^x \frac{ax}{b^2 + x^2} \, dx = a \ln(b^2 + x^2), \tag{6.3.3}$$

The wall is represented by placing a step up at $x = -b$ and a step down at $x = b$ so that a step up and a step down take place from and to $x = \pm 0$ when b is infinitesimal. When b is of negligible magnitude the Fourier formula for a 'narrow wall' is

$$\zeta_0 = \int_0^\infty a \, e^{ikx} \, dk.$$

Although the wavelength of stationary waves dominates the pattern, which is to say that it is a property of the airstream, not of the obstacle, the amplitude and horizontal extent on a given streamline increase upwards, and energy is continuously being transferred upwards and lost from the lower layers of the current as it passes far downstream. Thus a steady state is never reached in any practical sense. In reality the flow would separate from the rectangular or sharp corners of the wall and is of too small amplitude near the ground to represent experience. An example of Lyra's calculated flow pattern is shown in Fig. 6.3.ii.

A significant treatment of this subject was written by Queney [8], in which the flow over a two-dimensional hump was shown in several papers without any lee waves, there being no trapping mechanism in the airstreams which were of uniform stratification and wind speed. The pattern shown in Fig. 6.3.iii, page 170, is repeated upwards as z increases and the energy lost in travelling through the pattern travels upwards only—although the formulation contains no energy source because it depends on the square of the velocity and is therefore neglected in the linearisation! In the absence of dissipation the phenomenon of an infinite train of lee waves can only be supported if the waves are **trapped** in the lowest layer. It was first thought that this would require a discontinuity of density in the airstream which would make it analogous to the well known waves on the surface of water.

But it was demonstrated (Scorer [11]) that a train of lee waves which would be virtually undamped would extend a long way downstream from the obstacle and that this result could be achieved without any discontinuity of temperature or windspeed by a non-uniformity of the coefficient ℓ^2 described in equation (5.9.2). Thus it became obvious that trapped lee waves would be set up in airstreams which were stratified in a way which is

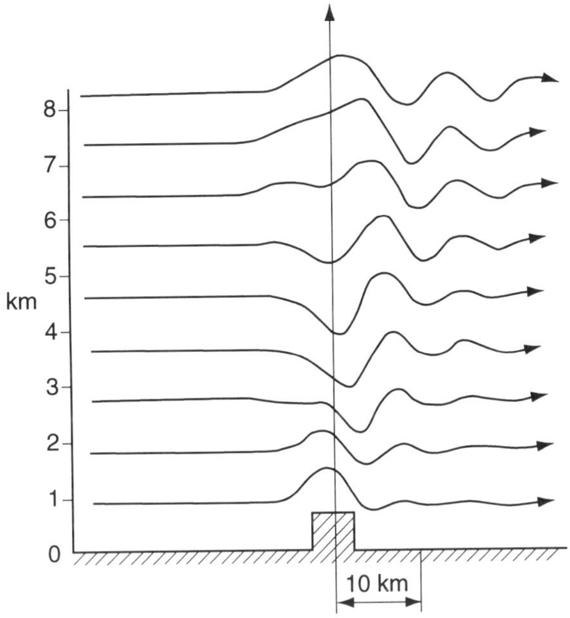

Fig. 6.3.ii Apparent lee waves over a sharp-cornered obstacle: Lyra's calculation.

very common in the westerlies of temperate latitudes, and in many other geographical areas, and of course in the many cases where a discontinuity of temperature had been observed.

6.4 PRACTICAL EXAMPLES OF CALCULATED WAVE PROFILES

There are considerable difficulties in handling this exercise with confidence if the disturbance from horizontal streamlines is not small, and it is intended to use the Fourier Integral method. Nevertheless the indications of what a larger amplitude may be like can be useful. It is necessary in such cases to draw the flow lines correctly. It has been common practice in drawing flow patterns simply to displace each horizontal streamline by a distance ζ which was computed at the undisplaced position. But the displacement should be **computed at the displaced position**. This is particularly important, because levels at which $\zeta = 0$ must not be displaced at all, and streamlines at other levels must not be displaced through these levels. (See section 6.15 on blocking.)

The first **nodal surface** above the ground in lee waves is where the flow remains horizontal, and as the amplitude is increased, the waves become more box-shaped with rapid flow round the edges and closed rotors in the middle. This is a rather far-fetched situation because in order to obey the formula, the fluid in the interior of the rotors has to be fetched from the other side of a nodal surface: for the label on the streamline enclosing a rotor is the same as that on the nodal surface. Also, about half of the fluid in a rotor is (theoretically) statically unstable, and this region is outlined in Fig. 6.4.i.

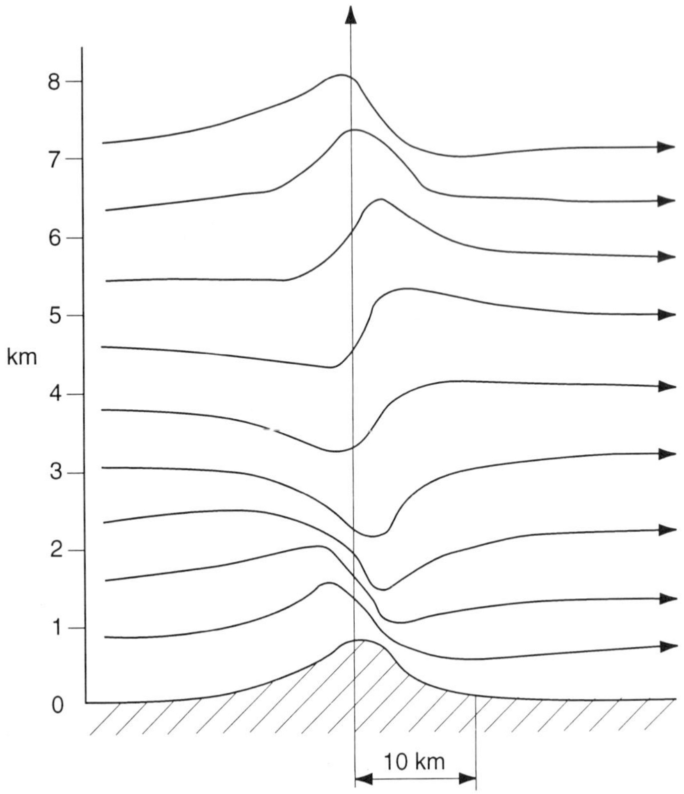

Fig. 6.3.iii Queney's calculation.

It has to be assumed that the interior of the rotors would be composed of air from the same layer as the immediate surroundings which would be gravitationally neutral and possibly rather turbulent and also not having such close streamlines and certainly not having the calculated density gradients. Thus the story is told of the glider pilot Larry Edgar whose aircraft was broken by passing through severe turbulence on the upwind side of the rotor, and who descended slowly by parachute along the track indicated in Fig. 6.5.iv (page 181) and landed close to the wrecked glider.

The airstream profiles of wind and temperature are usually measured by radiosonde balloon with a roughly constant rate of rise through the air, which in calm air would be measured by the observed rate of decrease of air pressure. In the computed flow pattern of Fig. 6.4.ii (page 171) for a three-layer airstream (Scorer [12]) the tracks of two rising balloons and the temperature profiles which they would report are shown against the profile of the undisturbed airstream.

This airstream has an adiabatic layer at the bottom, for this is normal in daytime below cloudbase. It is assumed that above the condensation level the air is only partially clouded and therefore statically stable. It is quite probable that in such a case the airflow over a ridge would not be as simple as indicated and possibly not even steady (Fig. 6.4.ii).

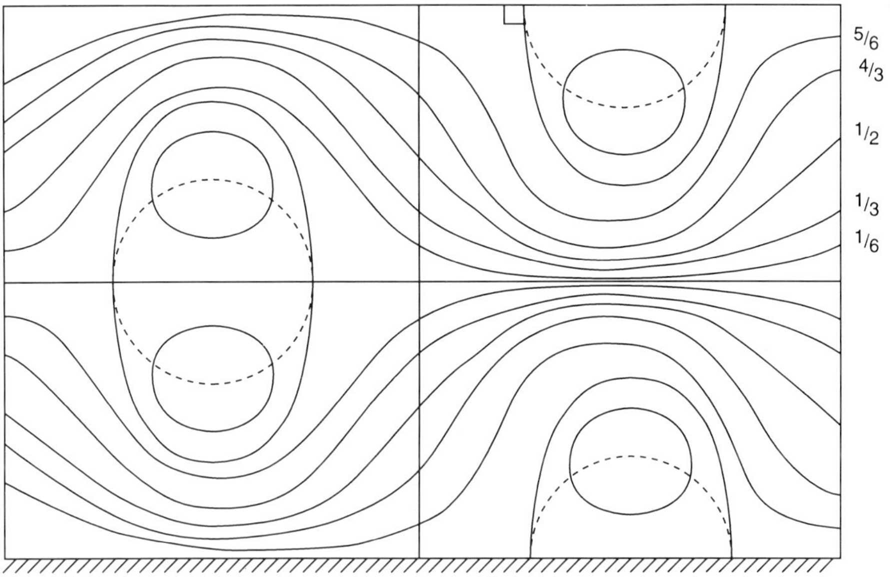

Fig. 6.4.i Wave-train flow with theoretical rotors having gravitationally unstable flow inside the dashed boundary. (See also Fig. 5.10.ii.) In large amplitude theory the streamlines cannot cross the level at which the displacement is zero, i.e. a nodal surface.

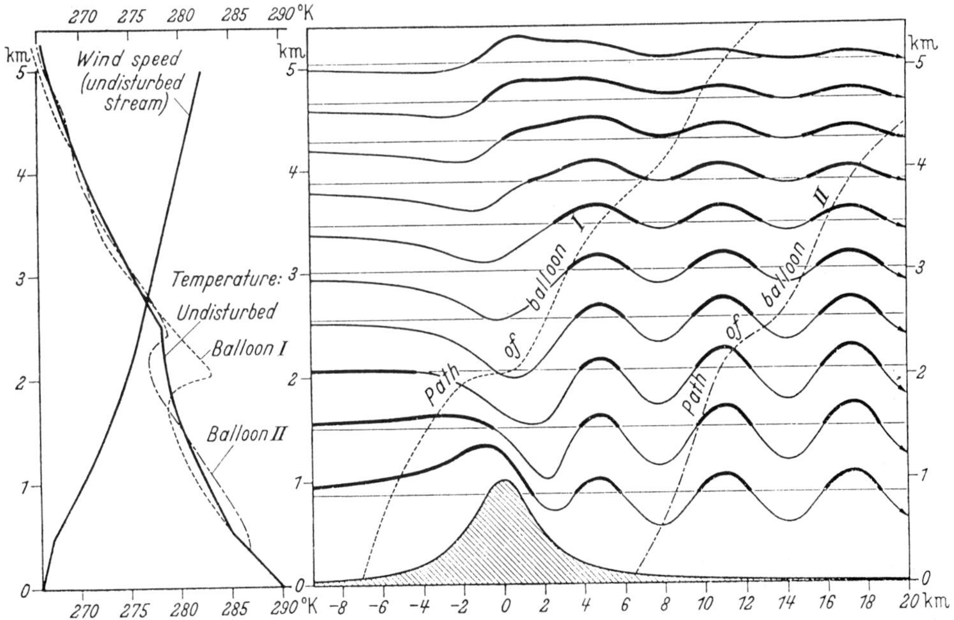

Fig. 6.4.ii Airflow over a ridge under conditions favourable for the occurrence of a train of lee waves. Velocity and temperature profiles of the undisturbed airstream are shown on the left. The lapse rate is adiabatic in the lowest 500 m. Possible tracks of radiosonde balloons with a rate of ascent of 5 m s^{-1} are shown together with the temperature profiles they would measure.

In order to avoid algebraic complexity the case will be illustrated by assuming that there are only two separate layers indicated by the subscripts, 'i' for the lower (inferior) layer and 's' for the upper, the governing equations being those given in section 6.2.

Equations (6.2.6) and (6.2.7) become

$$\zeta_i = e^{ikx}(A\,e^{i\nu z} + B\,e^{-i\nu x}) \tag{6.4.1}$$

$$\zeta_s = e^{ihx}C\,e^{-\mu z} \tag{6.4.2}$$

where

$$\nu = \pm\left(\ell_i^2 - k^2\right)^{1/2} \qquad \mu = -\left(k^2 - \ell_s^2\right)^{1/2}.$$

When $k > \ell_s$, this ensures that the disturbance decreases to zero at great height, but when $k < \ell_s$ we would have to use $+i\nu_s$ in the place of $-\mu$ so that waves in this latter range are propagated upwards. This results from consideration of the case of standing waves 'open to the sky', which means not trapped in the lower layers as in the examples in sections 5.11 and 5.12, for the use of the Fourier Integral requires us to consider the whole range of wavelengths.

In Fig. 6.4.iii the waves, with fluid moving from left to right, are governed by

$$\zeta = e^{i(kx+\nu z)} \tag{6.4.3}$$

so that on the upslope part of the shaded streamline the speed is greater than on the downslope part and there is a horizontal force to the left on the air above and to the right on the surface below, the pattern being steady. This means that the air above is retarded and the effect is carried upwards, theoretically retarding the whole airstream. Energy is thus communicated upwards, and when the waves were being set up energy would have been propagated upwards from the ground.

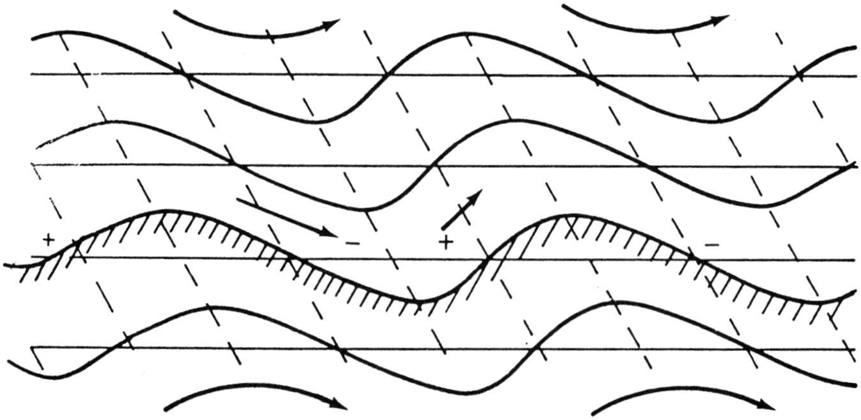

Fig. 6.4.iii Waves with the crests in surfaces inclined upstream have a negative pressure anomaly on the downslope where the velocity is larger. The fluid below any given streamline therefore exerts a force in the upstream direction on the fluid above it.

However the wave fronts appear to be slanted backwards like the dashed lines and by analogy with acoustic waves or surface waves on water viewed from above (see sections 2.1 and 6.6) the waves would be expected to be in the position shown if the flow had been from right to left, and the diagram would indicate a flow of energy upstream and downwards. The difference in mechanism in this case is that the pressure variations are determined predominantly by the gravitational weight of the air above. The pressure is increased where the air not far above has been raised and cooled to increase the density. Further up, the air directly above has actually been lowered and warmed but the process is operating at a lower density and therefore has a smaller counter-effect. An error was made in Scorer's 1949 paper, where the main attention was on the lee waves, for which (6.4.1) was the appropriate formula. The effect on the calculation of the mountain wave is evident in (6.4.21), where $\mathrm{i}\nu_s$ would have been replaced by $-\mathrm{i}\mu_s$.

The interface is placed at $z = 0$, and the ground at $z = -h$. At the interface both ζ and $\partial\zeta/\partial z$ are continuous. This leads to

$$C = A + B; \qquad \nu(A - B) = \nu_S C \tag{6.4.4}$$

and for the shorter wavelengths

$$\nu(A - B) = \mathrm{i}\mu C \tag{6.4.5}$$

and so with ν_s being interchanged with $-\mathrm{i}\mu$ for the appropriate wavelengths

$$2\nu A = (\nu + \nu_s)C; \qquad 2\nu B = (\nu - \nu_s)C. \tag{6.4.6}$$

At $z = -h$ the disturbance is

$$ab\int_0^\infty \mathrm{e}^{-kb+\mathrm{i}hx}\,\mathrm{d}k \tag{6.4.7}$$

and for the following calculation of the streamlines we retain the real and imaginary parts and separate them only in the final answer where the real part only is of interest.

The ground shape is given by

$$\begin{aligned}\zeta(-h) &= \frac{ab}{-b + \mathrm{i}x}\left[\mathrm{e}^{-kb+\mathrm{i}kx}\right]_0^\infty \\ &= \frac{ab(b + \mathrm{i}x)}{b^2 + x^2}\end{aligned} \tag{6.4.8}$$

of which the real part is

$$\frac{ab^2}{b^2 + x^2} \tag{6.4.9}$$

In order that ζ_i shall have this value at $z = -h$ we find

$$\zeta_i(z) = \int_0^\infty ab\,e^{-kb+ikx}\,\frac{(v+v_s)e^{ivz}+(v-v_s)e^{-ivz}}{(v+v_s)e^{-ivh}+(v-v_s)e^{ivh}}\,dk$$

$$= ab\int_0^\infty e^{-kb+ihx}\,\frac{v\cos vz + iv_s\sin vz}{v\cos vh - iv_s\sin vh}\,dk \tag{6.4.10}$$

in which v and v_s are functions of k.

The special interest of this integral is that the integrand has a pole for the value of k, for which we write k^*, and which satisfies

$$\boxed{v\cos vh - iv_s\sin vh = 0,} \tag{6.4.11}$$

which is seen to be the same as (5.11.8) when v_s is replaced by $i\mu$. The residue at this pole therefore represents the train of lee waves set up by the mountain (6.4.9). These waves are exponentially small at great height where they satisfy (6.4.1). The residue which is provided by this pole is to be multiplied by $2\pi i$ as will be explained below, and the waves are

$$\zeta = 2\pi ab\,e^{-k^*b}\sin k^*x\,\frac{v^*\cos v^*z - \mu^*\sin v^*z}{\left[\dfrac{\partial}{\partial k}(v\cos vh + \mu\sin vh)\right]_{k=k^*}} \tag{6.4.12}$$

for we can replace $-iv_s$ by μ and take the real part of the multiplier $2\pi i\,e^{ikx}$ to give a value whose dependence upon b, which determines the width of the mountain, is expressed by the factor $b\,e^{-k^*b}$, which has a maximum for a variable b when $k^*b = 1$. This gives the mountain width which raises the greatest amplitude lee waves. This is when $2\pi b$ is the natural standing wavelength of the airstream, as illustrated in Fig. 6.4.iv. The details of the boundary conditions leading to the result (6.4.12) are given at the end of the analysis.

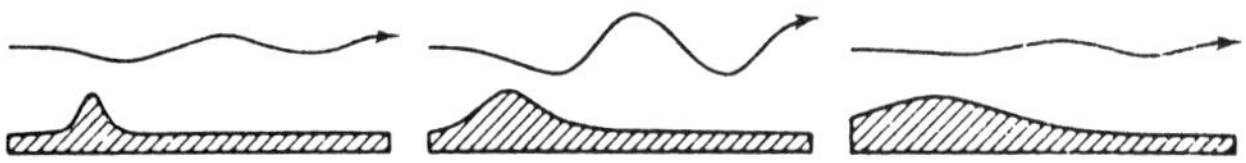

Fig. 6.4.iv The amplitude of the lee waves generated by an obstacle of given height is a maximum when the half width is about the wavelength/2π, for the shape given in Fig. 6.2.i.

To evaluate (6.4.12) we note that

$$v^{*2} = \ell_i^2 - k^{*2} \qquad \mu^{*2} = k^{*2} - \ell_s^2$$

and so

$$\frac{\partial v^*}{\partial h} = -\frac{k^*}{v} \qquad \frac{\partial \mu^*}{\partial h} = \frac{k^*}{\mu} \tag{6.4.13}$$

and

$$v^*\cot vh = -\mu^*$$

i.e. (6.4.11) so that the lee wave becomes, in the lower layers,

$$-2\pi \frac{ab\nu^{*2}\mathrm{e}^{-k^*b}}{k^*(1+\mu^*h)} \cot \nu^* h \sin \nu^*(h+z) \sin k^* x \tag{6.4.14}$$

and this is equal to the value in the upper layer at $z = 0$, and this is

$$-2\pi \frac{ab\nu^{*2}\mathrm{e}^{-k^*b}}{k^*(1+\mu^*h)} \cos \nu^* h \, \mathrm{e}^{-\mu^* z} \sin k^* x. \tag{6.4.15}$$

The integral for the rest of the mountain wave is

$$\zeta_\mathrm{i} = ab \int_0^\infty \mathrm{e}^{-kb+\mathrm{i}kx} \frac{\nu \cos \nu z + \mathrm{i}\nu \sin \nu z}{\nu \cos \nu h - \mathrm{i}\nu \sin \nu h} \, \mathrm{d}k \quad \text{real part} \tag{6.4.16}$$

and we now turn to a simple approximation to evaluate this. There is a negative exponential, e^{-hb}, and an oscillating factor $\mathrm{e}^{\mathrm{i}kx}$, and a factor which is slowly varying from its value at $k = 0$, at which its derivative w.r.t. k is zero, being composed of functions of k^2. So a good approximation is obtained by putting $k = 0$ in all but the oscillatory and negative exponential parts. It will be a good approximation except where x is small. We replace μ by $-\mathrm{i}\nu$ as in (6.4.10) and note that by (6.4.2)

$$\nu = \ell_\mathrm{i}, \; \nu_\mathrm{s} = \ell_\mathrm{s} \tag{6.4.17}$$

and after some detailed algebra we get

$$\zeta_\mathrm{i} = \frac{ab^2 p}{b^2 + x^2} \left(\ell_\mathrm{i}^2 \cos \ell_\mathrm{i} h \cos \ell_\mathrm{i} z - \ell_\mathrm{s}^2 \sin \ell_\mathrm{i} h \sin \ell_\mathrm{i} z \right)$$
$$- \frac{abxp}{b^2 + x^2} \ell_\mathrm{i} \ell_\mathrm{s} \sin \ell_\mathrm{i} (z+h) \tag{6.4.18}$$

where

$$p^{-1} = \ell_\mathrm{i}^2 \cos^2 \ell_\mathrm{i} h + \ell_\mathrm{s}^2 \sin^2 \ell_\mathrm{i} h \tag{6.4.19}$$

and in the upper layer

$$\zeta_\mathrm{s} = ab \int_0^\infty \mathrm{e}^{-kb+\mathrm{i}hx+\mathrm{i}\nu_\mathrm{s} z} \frac{\nu}{\nu \cos \nu h - \mathrm{i}\nu_\mathrm{s} \sin \nu h} \, \mathrm{d}k \quad \text{real part,} \tag{6.4.20}$$

which gives

$$\zeta_\mathrm{s} = \frac{ab^2 p \ell_\mathrm{i}}{b^2 + x^2} \left(\ell_\mathrm{i} \cos \ell_\mathrm{i} h \cos \ell_\mathrm{s} z - \ell_\mathrm{s} \sin \ell_\mathrm{i} h \sin \ell_\mathrm{s} z \right)$$
$$+ \frac{abxp\ell_\mathrm{i}}{b^2 x^2} \left(\ell_\mathrm{i} \cos \ell_\mathrm{i} h \cos \ell_\mathrm{s} z + \ell_\mathrm{s} \sin \ell_\mathrm{i} h \sin \ell_\mathrm{s} z \right) \quad \text{real part.} \tag{6.4.21}$$

Boundary conditions

The integral (6.4.10) is represented in the complex k-plane (Fig. 6.4.v) by the path from the origin to ∞ on the real axis. If this path is deformed into a radius, at angle θ, to ∞ and

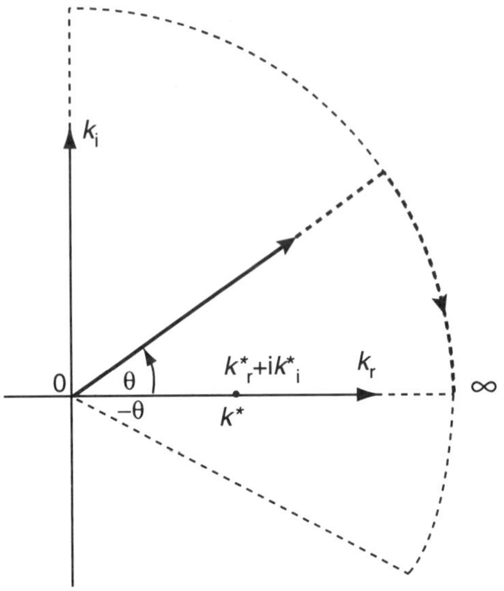

Fig. 6.4.v The integral to be evaluated for the streamlines is taken from the origin to +∞ along the real axis. The integrand has a pole at $k = k^*$ on the real axis and so the contour may be deformed along a radius at angle θ and along the circular arc at ∞. If there is damping the amplitude decreases exponentially in the downstream ($+x$) direction and this is represented by a factor $e^{ik_i^* x}$, so that the residue is from a pole at $k = (k_r^* + ik_i^*)$ which is within the sector above the axis.

then along the circle at ∞ to the real axis; then back to the origin completes the circuit, which encloses half the pole at $k = k^*$, which is on the real axis. The residue at the pole represents a complementary function which may be added to the integral without altering the condition that $z = -h$ is a rigid boundary because it is zero there. As it stands, when x is positive, the integral along the deformed path contains zero contribution from the arc at ∞ because of the negative exponential $e^{-kb-k_i x}$ when the variable k has a positive imaginary part k_i: the lee wave term would be half the residue representing a train of lee waves extending downstream to $x = +\infty$. In the upstream direction where x is negative we would have to draw the deformed contour in the negative half of the k-plane to obtain the negative exponential at ∞ (for large negative x) and that would require us to evaluate the negative half of the residue at the pole on the real axis, and this would represent a train of waves extending upstream to $x = -\infty$. To obtain a realistic solution we need to add a train of waves from $x = -\infty$ to $x = +\infty$, which satisfies the boundary condition at $z = 0$ and at very large z. This cancels the upstream waves and doubles that downstream.

This device may be thought of as representing the result we would have obtained if the pole had been entirely within the positive imaginary half-plane giving the result in (6.4.14). It is as if the pole were at $k^* + im$, where m represents the rate of dissipation of the lee (standing) wave in the downstream direction, which would also have dissipated any waves in the oncoming stream in a long distance upwind over level ground. This is a classical device explained by Lamb in section 227 of the 1895 edition of *Hydrodynamics* (page 398), which has been used widely since then in problems in which a disturbance composed of a range of wavelengths excites a resonant wavelength of the system. In the

(a)

(b)

Fig. 6.4.vi (a) Four parallel wave clouds in the lee of a hill near the central coast of Wales. (b) Lee waves showing striations along the wind direction through the waves; east of Detroit.

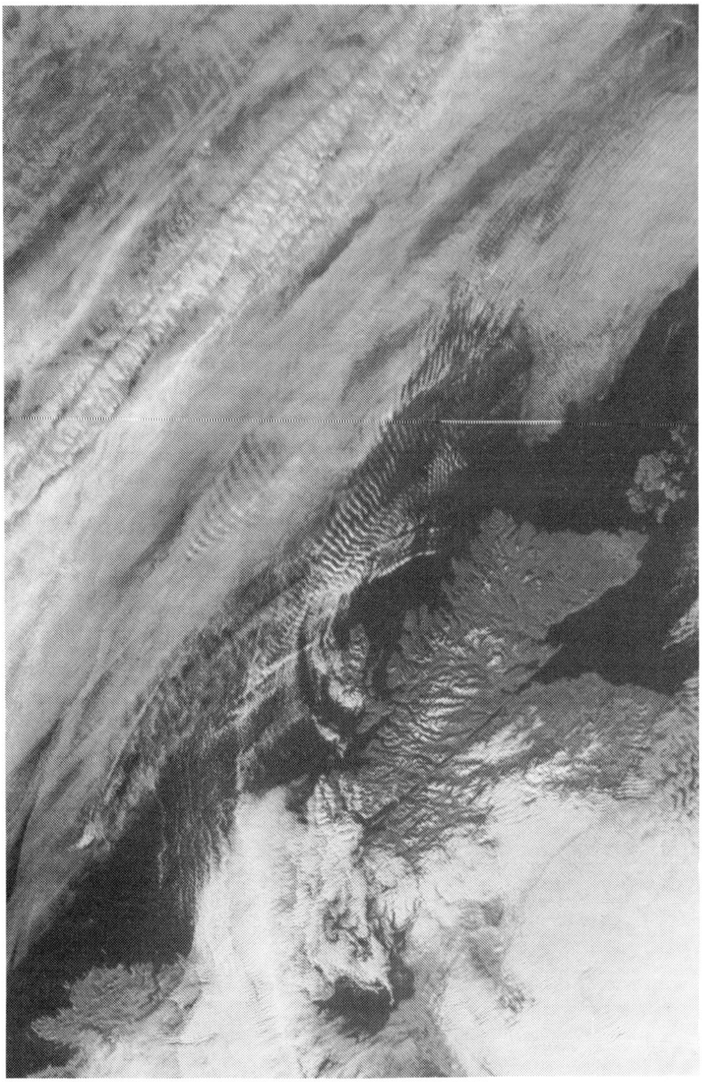

Fig. 6.4.vii NOAA 6, 0923,28.1.81.Vis. Waves in a cloud layer increasing in wavelength from southern Scotland, across the Outer Hebrides, to St Kilda and beyond to the front where there are spike clouds (see 8.6. ix–xiv). (Photo courtesy of University of Dundee.)

present case it is appropriate because viscosity was neglected in setting out the flow equation, on the grounds that it is negligible, for the waves are very slowly dissipated (see Fig. 5.11.ii). The adding of a factor which represents a gradual decrease in the amplitude is a crude way of avoiding the inclusion of viscosity in the equation of motion, and it is more important to represent the oncoming airstream as having no waves, for this is easy to observe, and it is mathematically correct to add the wave train so as to produce this result.

In Fig. 6.4.ii the streamlines were calculated for a model airstream which includes an adiabatic layer in the lowest 500 m using an analysis similar to that described in this section, and the lines are thickened where the displacement is upwards to indicate where wave clouds would be possible, and it is important to note that the latent heat of condensation has not been represented in the model. The tracks of two possible balloons show that both would have 'measured' an inversion, but at different heights close to 2 km. This will be seen to suggest different amplitude profiles for the lee waves in a manner discussed in section 6.7.

The profiles of wind speed and temperature were chosen to be quite ordinary, but a much stronger wind at higher levels would reduce the wave amplitude there significantly. The wind there may be so strong that the whole disturbance would be very small, but much argument has been expended on what boundary conditions ought to be applied there. In a personal discussion, Professor Queney gave his opinion that the details of these calculations are all unreliable in the sense that it may falsify the result to assume that the flow is steady, and that only in some circumstances are wave clouds observed to be steady.

This criticism is especially valid in cases where the wind decreases to near zero, or at least to a near-zero component across the ridge. Researches by Gerbier and colleagues in France (Gerbier and Berenger [13]), in which rubber balloons enclosed in tight silk covers which gave them a constant volume and made them reach a constant pressure level and travel through waves while being tracked by double theodolite or radar, produced the results shown in section 6.10, where the unsteadiness is depicted by eddies which were found prevalent in some layers of low wind speed over Mont Ventoux.

The details in Fig. 6.4.ii for small values of x close to the ridge were obtained not by accurate numerical evaluation of the formulae but by ensuring that the flow above the ridge was smooth. A realistic and accurate evaluation would have been impossible because the lee wave term has a discontinuity of slope at $x = 0$ from zero to $-\sin k^* x$, which could have been offset by a similar discontinuity in the values of the formulae (6.4.18) and (6.4.21) which are not good approximations near $x = 0$. As in Lyra's case (Fig. 6.3.ii) there should be waves in the uppermost layer increasing upwards, but the existence of these waves in that case depended mainly on the contribution of the larger values of k (shorter wavelengths) in the steep-sided obstacle which were neglected in Queney's case of a smooth-sided obstacle (Fig. 6.3.iii).

6.5 LARGE AMPLITUDE WAVES

An attempt to model a case of large amplitude was prompted by a glider flight by pilot Lary Edgar and meteorologist Harold Klieforth, who soared to the then record height of 44 000 ft in the Sierra wave over the Owens Valley in the lee of the Sierra Nevada in California.

The main obstacle was the large amplitude of the interface between two layers of the numerical model where so far the boundary equations had been applied at the undisturbed position of the interface.

At the main interface where the amplitude would be large, not only were ζ and $\partial\zeta/\partial z$ made continuous, but $\partial^2\zeta/\partial z^2$ was also made continuous, and then the first two only were made continuous at another interface at a higher altitude where the amplitude of the vertical displacements was much smaller. Thus the model was an improvement on the conventional linearised theory, and it made it possible to use the large amplitude theory within the lowest layer which contained a large rotor (Scorer and Klieforth [14]).

First of all, the shapes of the lee waves were calculated for three different amplitudes and then, for the largest amplitude, the flow over a ridge was calculated with lee waves, and the pattern was compared with a pattern derived from good observations (see Figs 6.5.i, 6.5.ii, 6.5.iii and 6.5.iv). Of special interest is the fact that with waves of increased amplitude for which the formulae are sinusoidal they become steeper and sharper in the troughs and smoothed out in the crests where rotors appear.

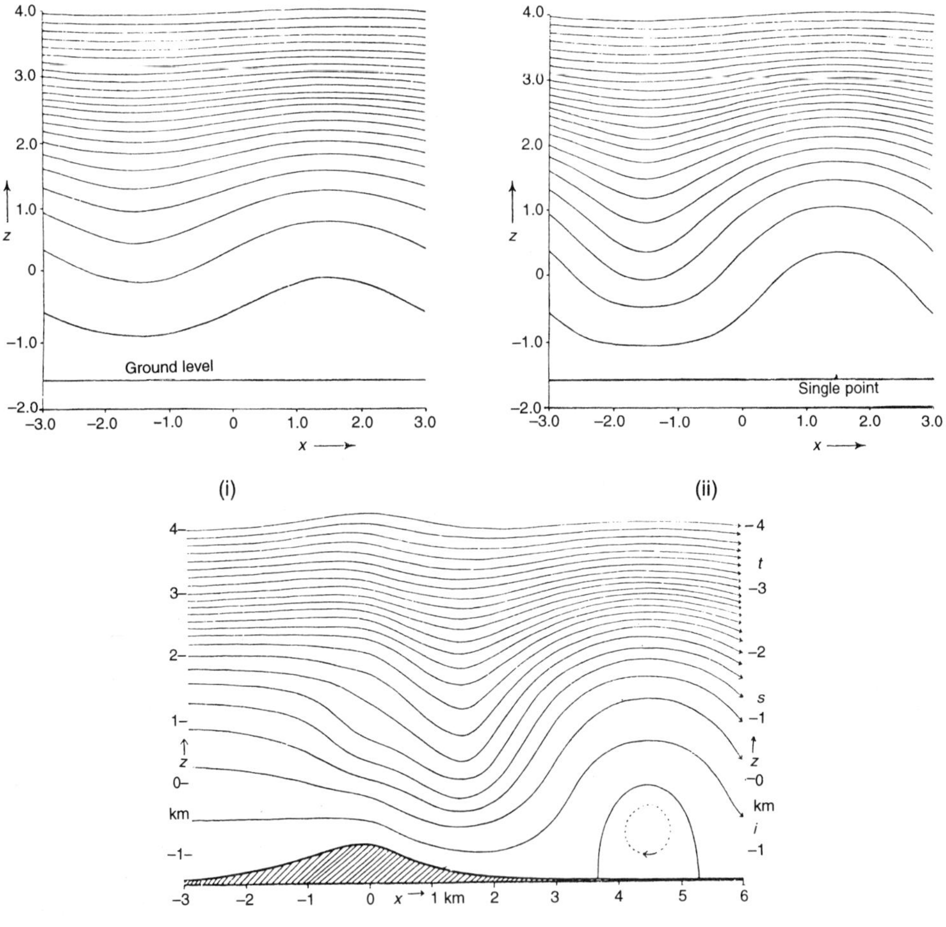

(iii)

Fig. 6.5.i, ii, & iii These show the streamlines for waves (i) of small amplitude and (ii) of amplitude large enough to produce a point under the wave crest where the wind is zero at the ground. (iii) When the amplitude of the waves is larger than in (ii) a rotor is produced, which appears under the first lee wave behind a ridge. The rotor in the sensitive case (illustrated) is higher than the mountain.

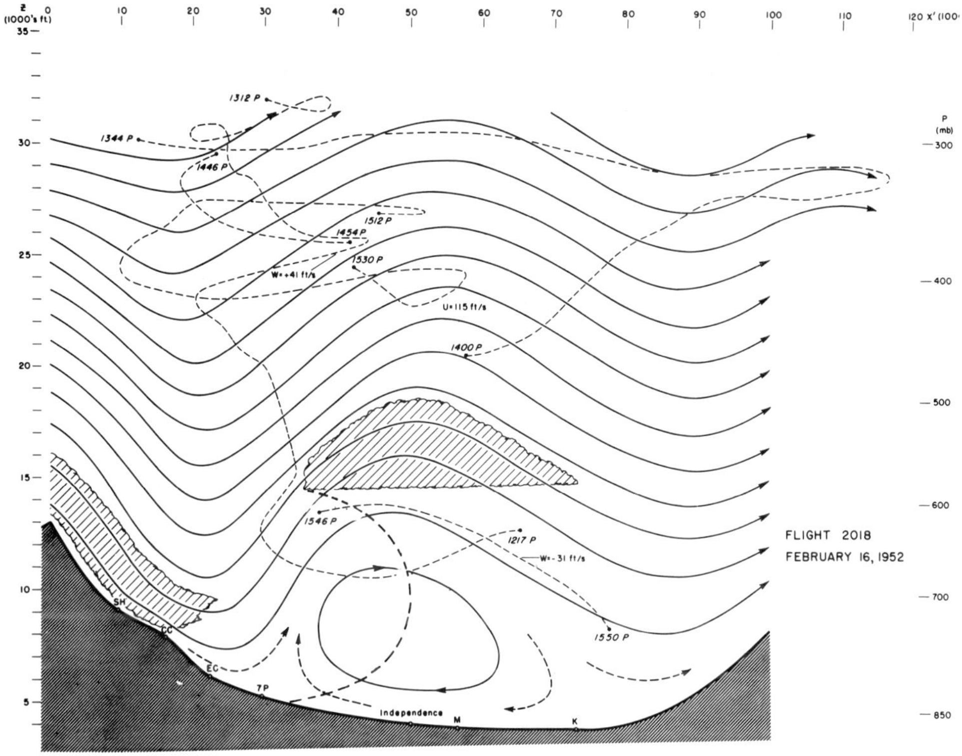

Fig. 6.5.iv An illustration of the flow pattern deduced from several flights in the 'Sierra Wave Project' on 16.2.52 in the Owens Valley in the lee of the Sierra Nevada, California. The dashed line from the upwind edge of the wave cloud indicates the track of Edgar parachuting down after turbulence wrecked his glider. The scene of these events is shown in Plate 1 in the colour section between pages 352 and 353 and in Fig. 6.5.v which is a photograph by Klieforth.

The flow over the Owens Valley was photographed by Robert Symons from a fighter plane in which he soared with feathered propeller, and that black-and-white picture has been coloured by an artist to match the tones of a colour picture of the same area taken by Harold Klieforth. This is shown in Plate 1 in the colour section between pages 352 and 353.

Many occasions with a large Sierra wave have been explored, and on one occasion (25 April 1955) Larry Edgar was soaring close to the upwind edge of the lee wave cloud when the turbulence caused his glider to break apart and he descended more slowly by parachute than did the wreckage, which fell rapidly. First he was carried rapidly eastwards in the west wind below the lee wave cloud, and then at about 2500 m (1300 m above the ground) he encountered a layer of calm air below which he drifted westwards in a wind estimated to be about 25 kt. He finally landed below the leading edge of the roll cloud on the west side of the valley. See track in Fig. 6.5.iv.

Although the flow near the ground may be different in some respects from that in the theoretically large amplitude flow, particularly in and close to the cloud, the flow at greater heights is often found to be very smooth.

Fig. 6.5.v A large amplitude lee wave in the Owens Valley. (Photo by H. Klieforth.)

6.6 LEE WAVES SPREADING OUT FROM AN ISOLATED OBSTACLE

The pattern of waves behind a ship often represents a large fraction of the wave drag as the area occupied by the wake widens. (See Plate 1 in the colour section.) An isolated obstacle can be synthesised by a system of ridges at all angles and passing through the same centre where their added amplitude constitutes an isolated obstruction. The height of the obstacle may be determined by the functional form of the individual ridges so as to decrease quite rapidly with distance from the centre and to have an amplitude which varies with the angle of inclination to the stream. The simplest case to illustrate this principle is the calculation of the basic pattern of lee waves due to an obstacle concentrated at an isolated point on a water surface. The dispersion is represented by the relationship between wave speed, U, on deep water and wavelength $(2\pi/k)$, namely

$$\boxed{U^2 = g/k} \tag{6.6.1}$$

in which surface tension is ignored and the water depth is large enough to make no difference (which in practice means larger than the wavelength).

If a ridge-like disturbance is inclined at angle ϕ to the normal to the stream (or ship velocity on stationary water), the component $U\cos\phi$ produces lee waves to which the component $U\sin\phi$ is added parallel to the ridge to complete the stream without altering the pattern. If the lee waves are proportional to $-\sin kx$, x being distance downstream in a direction perpendicular to the ridge (see Fig. 6.6.i), then the following two equations are derived, for a point P whose polar coordinates are (r, θ), from the definition of x and (6.6.1):

$$X = r\cos(\phi+\theta); \qquad k = g/U^2\cos^2\phi \tag{6.6.2}$$

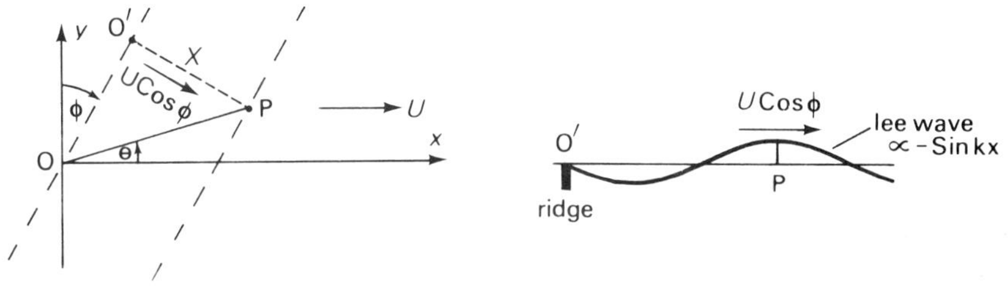

Fig. 6.6.i The left diagram shows the coordinates for the lee wave at P parallel to the ridge OO′ inclined at angle ϕ to the normal to the stream. A set of ridges at different angles ϕ together makes an isolated obstacle at O: the lee waves are calculated by summing the individual lee waves. In the right diagram is the elemental lee wave in vertical section through O′P.

We now envisage the disturbance centred on O to be made up of a system of ridges at angle ϕ, $-\pi/2 < \phi < \pi/2$, each of which produces its own lee waves. The total wave pattern behind the disturbance is the sum of all the lee waves. If P is on the crest of a wave in the total wave pattern, the locus of P is the envelope of the individual lee wave crest lines, on which

$$kX = N \tag{6.6.3}$$

where N is the phase, in this case a crest. Equally the nodes and troughs of the total pattern are the envelopes of the individual node and trough lines, ϕ being the parameter determining the individual lee waves. The envelope is discovered as the locus of the intersection of lines (6.6.3) with the lines for the neighbouring value of ϕ, and this is obtained by solving (6.6.3) together with its derivative

$$\frac{\partial}{\partial\phi}(kX) = 0, \tag{6.6.4}$$

These two equations are readily found, from (6.6.2), to be

$$r\sec^2\phi\cos(\phi+\theta) = a = NU^2/g \tag{6.6.5}$$

$$2\tan\phi = \tan(\phi+\theta). \tag{6.6.6}$$

If we now put $x = r\cos\theta$, $y = r\sin\theta$, the locus of the phase lines in the total lee wave pattern is found to be

$$x = a\cos\phi(2 - \cos^2\phi)$$

$$y = a\sin\phi(1 - \sin^2\phi), \tag{6.6.7}$$

the parameter ϕ ranging from $-\pi/2$ to $\pi/2$.

The most interesting feature of this wave pattern, which is shown in Fig. 6.6.ii, is that it is confined between two values of θ. Since from (6.6.6)

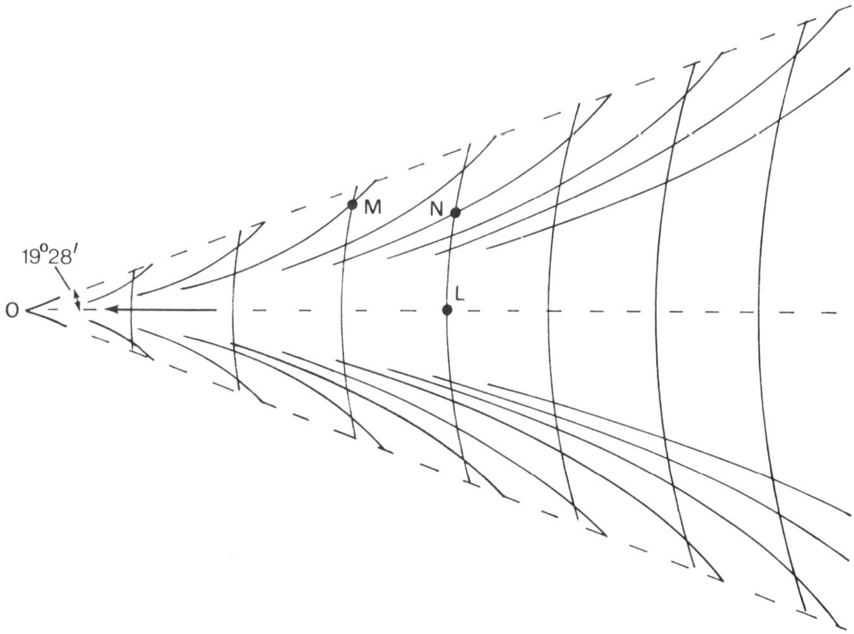

Fig. 6.6.ii The crest lines of the total lee wave pattern of a disturbance centred at O travelling over the surface of deep water. The lines of constant phase for other phases are similar. The points where the diverging phase lines cross the transverse ones, such as M, are where the maximum amplitude of the wave occurs. A second maximum is sometimes found at N, but this must not be confused in the case of a ship with the wave crests of type M from the stern of the ship. The transverse waves at points like L are large, but in the case of a well-designed ship travelling at its design speed the total disturbance primarily due to the bow and the stern is minimised there to reduce the wave resistance. See Figs 6.6.iii (a) and (b) for the dominance of the transverse or diverging waves.

$$\frac{y}{x} = \tan\theta = \frac{\tan\phi}{1 + 2\tan^2\phi} \tag{6.6.8}$$

it is evident that θ is zero when $\phi = 0$ or $\pm\pi/2$, and its maximum occurs when

$$\tan^2\phi = \tfrac{1}{2} \tag{6.6.9}$$

and

$$\tan \theta_{\max} = 2^{-3/2}$$

or

$$|\theta_{\max}| = 19°28'. \tag{6.6.10}$$

All the phase lines in the wave pattern have the same shape and are obtained from successive values of N, which is increased by 2π in proceeding from one crest to the next.

These formulae are simple because of the simple nature of equation (6.6.1), which expresses the relationship between wave length and wave speed on deep water. In a stratified fluid the simplest relationship is (5.10.9) for the case of flow between horizontal rigid boundaries. (6.6.2) is then replaced by

$$X = r\cos(\phi + \theta); \quad k^2 = \frac{g\beta}{U^2 \cos^2 \phi} - \frac{n^2\pi^2}{h^2} \tag{6.6.11}$$

and (6.6.3) and (6.6.4) become correspondingly much more complicated, while as ϕ approaches $\pm\pi/2$ we may put n successively equal to 2, 3, ... and obtain higher modes of lee waves contributing to the pattern. The contribution of these higher modes is usually negligible because they are restricted to a small range of ϕ and their amplitude is usually smaller in the two-dimensional case from which they are derived. They will therefore be neglected in further discussion and we shall write $n = 1$ in (6.6.11) and corresponding equations for other cases.

Some details of the wave pattern are revealed by a different approach as follows. An arbitrarily shaped disturbance at O may be represented by

$$\zeta(t, \theta) = \int_{-\pi/2}^{\pi/2} \Phi(\phi) f(X)\, \mathrm{d}\phi \tag{6.6.12}$$

where $f(X)$ is the profile of the individual component ridge and $\Phi(\phi)$ as its amplitude, varying with ϕ. Thus if $\Phi = 1$, so that all the ridges are the same, and have the cross-section of a thin 'wall' of unit cross-section area, at distance r from the origin, the height of the obstacle given by (6.6.12) is proportional to r^{-1}. If the 'wall' has width b then the obstacle is flat topped for $r < b$ and for $r > b$ its height decreases and becomes proportional to r^{-1} at large r.

If the lee waves of the ridges are proportional to $-\sin kX$ then the total wave pattern is

$$\int_{-\pi/2}^{\pi/2} \Phi(\phi) \mathrm{i}\, \mathrm{e}^{\mathrm{i}kX} \zeta(z, \phi)\, \mathrm{d}\phi \tag{6.6.13}$$

where $\zeta(z, \phi)$ represents the variation of the component waves with height, which may vary with ϕ.

To evaluate this integral we make use of the principle of stationary phase. We write

$$kX = k^* X^* + \tfrac{1}{2}(\phi - \phi^*)^2 (k^* X^*)'' + 0(\phi - \phi^*)^3 \tag{6.6.14}$$

where the asterisk denotes the value of the functions obtained by putting $\phi = \phi^*$, and ϕ^* is the value of ϕ for which

$$(kX)' = 0, \quad \text{i.e. } (k^* X^*)' = 0 \tag{6.6.15}$$

where the prime denotes $\partial/\partial\phi$. Since e^{ikX} is an oscillatory function of ϕ, the major contribution to the integral will be made in the neighbourhood of the value of ϕ satisfying (6.6.15) because, since the phase kX is stationary only at that point in the range of integration, e^{ikX} is very slowly changing. A good approximation to the integral is therefore obtained by replacing the variables $\Phi(\phi)$ and $\zeta(z, \phi)$ by the values they have at $\phi = \phi^*$. We also make use of the formula

$$\int_{-\infty}^{\infty} e^{im^2\xi^2}\, d\xi = \frac{\sqrt{\pi}}{m}\frac{1+i}{\sqrt{2}} = \frac{\sqrt{\pi}}{m}\, e^{i\frac{\pi}{4}} \tag{6.6.16}$$

where i may be replaced by –i throughout. Thus if m^2 is negative, the right-hand side is replaced by $\dfrac{\sqrt{\pi}}{|m|}\, e^{i\frac{\pi}{4}}$. In our case, therefore, the approximate value of (6.6.13) (real part) is

$$\Phi(\phi^*)\zeta(z,\phi^*)\left(\frac{2\pi}{\left|(k^*X^*)''\right|}\right)^{1/2} \cos\left(k^*X^* \pm \frac{\pi}{4}\right), \tag{6.6.17}$$

the upper or lower sign being taken according to whether $(k^*X^*)''$ is positive or negative. ϕ^* is now our parameter, and we see that for the locus of constant phase $k^*X^* = \text{constant} = N$, which is (6.6.3). Also (6.6.15) is equivalent to (6.6.4). The maximum value of θ in (6.6.8), which is (6.6.4) written for the particular case of the wave pattern on deep water, was obtained by differentiation with respect to ϕ. This means that at the outer boundary of the pattern, $(k^*X^*)''$ is zero and changes sign as ϕ passes through the value giving the maximum of θ. Consequently the sign in (6.6.17) is changed there and $\pi/2$ is added to the phase N. For this reason the waves in Fig. 5.15.ii are drawn so that the transverse and diverging waves are out of phase there by $\pi/2$.

In arriving at (6.6.17) we have replaced the limits of integration in (6.6.13) by $\pm\infty$, on the grounds that the oscillations of the integrand well away from the point of stationary phase contribute negligibly.

The fact that the amplitude is large near the boundary is made plain in (6.6.17), although the approximation fails at the boundary where $(k^*X^*)''$ is zero. Therefore we would expect to see the waves most pronounced in this neighbourhood. A more accurate calculation of the value of the integral in that region using the term in $(k^*X^*)'''$ and $(\phi-\phi^*)^3$ by a method described by Scorer [15] has been given by Warren [16]. The formula (6.6.17) also indicates how the amplitude of the waves is related to the function Φ. The transverse waves, which are generated for $0 < |\phi| < \phi^*$, are greatest when the obstacle has a larger extent across the stream, whereas an obstacle elongated in the downstream direction has larger values of Φ when $|\phi| > \phi^*$. Thus a long narrow boat, such as a rowing eight, produces only the diverging waves, while a squat flat-sterned boat generates only the transverse waves. These features are illustrated in Figs 6.6.iii and 6.6.iv.

The same kind of treatment may be applied to the lee waves behind a hill in a stratified stream capable of supporting standing waves. An example was given by Scorer and

Fig. 6.6.iii(a) The wake of a wide tugboat in Cherbourg Harbour as seen from the deck of the liner *Queen Mary*. The transverse waves are dominant. Acceleration of the ship alters the angle speed.

Fig. 6.6.iii(b) The diverging wave pattern set up by a narrow boat on the river Seine, photographed from the top of the Eiffel Tower in the reflected sunshine.

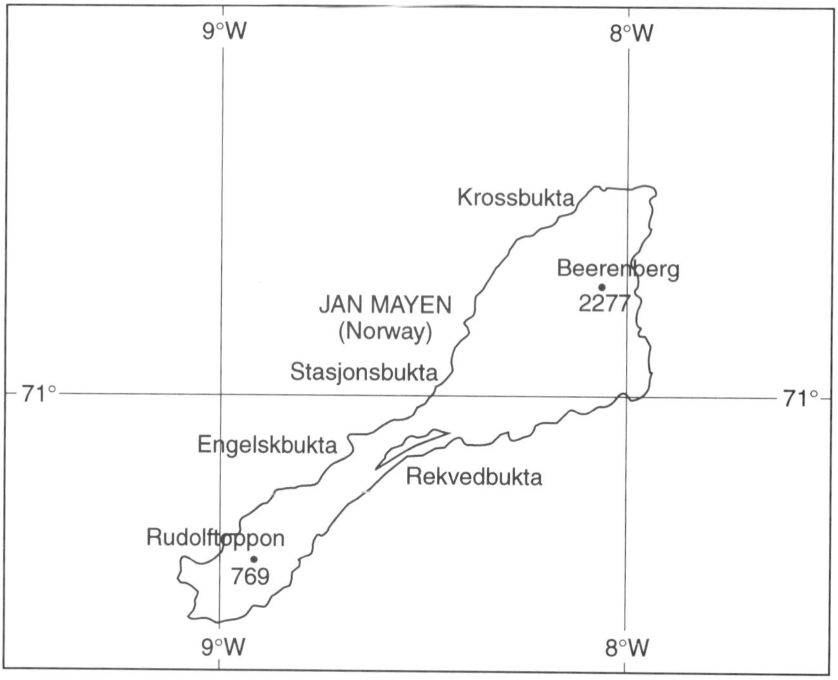

Fig. 6.6.iv(a) The coastline and peaks of the island Jan Mayen which produces frequent vortex streets and/or lee waves in layer cloud, depending on cloud height.

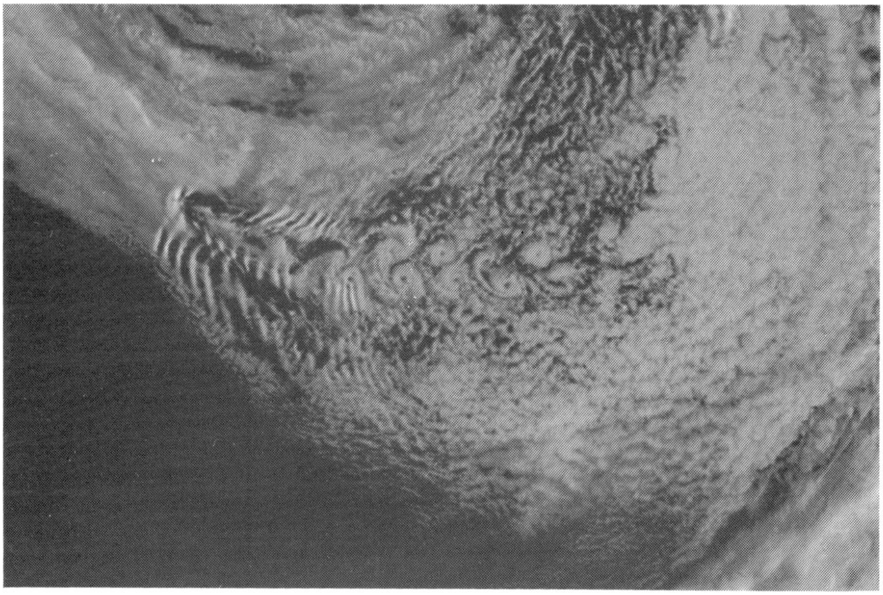

Fig. 6.6.iv(b) A striking example of the flow around Jan Mayen changing to flow over the island as the cloud layer deepens and begins to form waves, in place of a vortex sheet. The flow is curved round a low pressure area to the NE of the island. (Courtesy Czech Met. Service.)

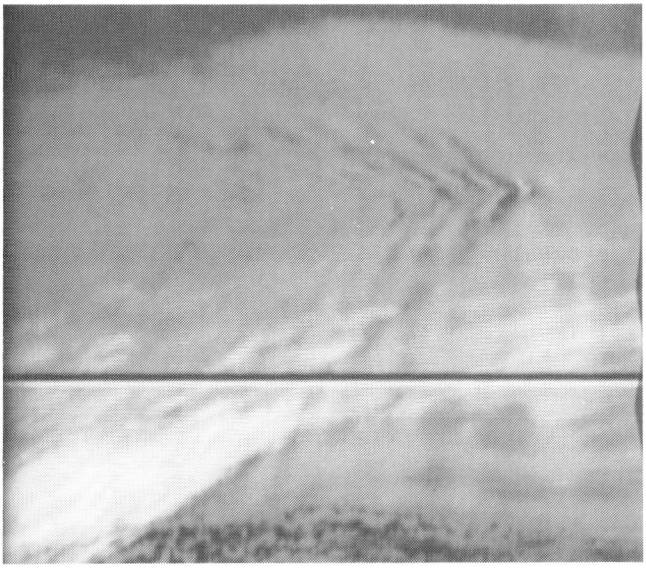

Fig. 6.6.iv(c) The waves in the lee of Jan Mayen with a dominant set of diverging waves and some very wide transverse waves in a thin, higher, cloud layer spreading almost directly across the wind direction (from NE). NOAA5 0927, 1.9.76. Infra Red. (Courtesy University of Dundee.)

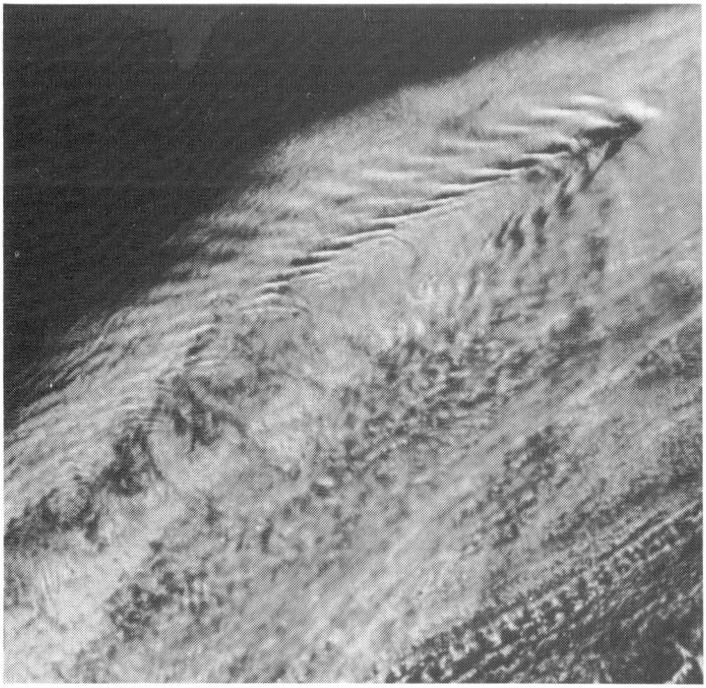

Fig. 6.6.iv(d) A mixture of different wave systems in a N wind from Jan Mayen, with some residual vortices fard down-wind. NOAA5, 1220, 1.9.76, Visible. Note the dark beam from the NW almost along the line at the cross-wind waves of picture (c). (Courtesy of University of Dundee.)

Wilkinson (1956). The maximum value of θ beyond which there is no disturbance varies according to the circumstances and in the example cited it was shown that the wave pattern could spread as much as $\pm45°$ or only to $\pm1°$ if the parameters were suitably chosen. Pictures of lee wave cloud and patterns are shown in Figs 5.11.iii, 6.6.iii, 6.6.iv, and 6.6.v.

The pattern is not symmetrical when $\Phi(\phi)$ is not symmetrical about $\phi = 0$. The form of disturbance due to an oval obstacle inclined at an angle to the stream is given by Scorer [18]. It is clear that there exist innumerable possible different cases, the main features of which may be elucidated by these methods. It should also be remembered that methods of this kind are not worth pursuing in much greater detail because it is impossible in any practical case to specify the boundary conditions and the airstream profile with any precision. Berkshire [17] has applied similar methods to the study of three-dimensional wave patterns in the stratosphere, in order to obtain qualitative results, but there can be little prospect of calculations of this kind being of any use except in the examination of causes in a well documented case long after the event.

Fig. 6.6.v NOAA9; 1337, 21.2.87, Chan2. A typical scene in Western or Central Europe, with lee waves very readily created over the mountains of any of British Isles, France, Iberia, Italy, Scandinavia etc. In this case, Sardinia, Sicily, South Italy, Albania and Greece have well developed wave trains. (Courtesy University of Dundee.)

6.7 PROFILE OF LEE WAVE AMPLITUDE

It was shown in section 5.11 that a condition for a train of lee waves to occur is that they are trapped close to the ground; and this requires an exponential decrease upwards in an upper layer. The profile of such a layer has a positive curvature because $\zeta \propto e^{-\mu z}$, and ζ decreases at a decreasing rate upwards. But the sign of this curvature must change near the ground in order that the slope be positive and ζ decrease to zero at the ground. ζ is, of course, a continuous function of height.

In Fig. 6.7.i (pages 192–3) we show the amplitude profiles of the waves of seven different two- or three-layer airstreams which all have the same wavelength. The amplitude maximum occurs either at a discontinuity of the quantity ℓ^2 or in a layer in which it is larger than above and below. All streams are assumed to have crossed the same two-dimensional ridge, and we find that the amplitude varies in a way that is neither simple nor obvious: it certainly does not depend on the parameters of the ridge or the wavelength which are included in the factor $ab\,\mathrm{e}^{-k^*b}\sin k^*x$ in the equation (6.4.12). In all seven cases the upper layer had the same value of ℓ_s^2. The values for the lower layers are given in Table [6.7.I].

Table 6.7.I

Case	1	2	3	4	5	6	7
Discontinuity of density at 2 km interface	3%	2%	1%	—	—	—	—
Depth and	—	—	—	—	0.25	0.37	0.65 km
ℓ^2 in middle layer	—	—	—	—	5.0	5.0	3.25
Depth and	2.0	2.0	2.0	2.0	1.75	1.63	1.35 km
ℓ^2 in lowest layer	0.0	1.0	1.79	2.04	0.0	1.08	0.0
Relative excitation $\left[\frac{\partial \zeta}{\partial h}\right]^{-1}$	0.197	0.255	0.098	0.125	0.156	0.129	0.151

6.8 AIRFLOW BEHIND AN ISOLATED HILL

The pattern of waves produced in a stably stratified stream by passage over an isolated hill may be deduced in the same way as the ship-wave pattern, except that the more realistic the model, the more complicated the mathematics, although the method is the same.

From the preceding sections we may state that for the two-dimensional ridge

$$\zeta_1 = \frac{a^2 b}{b^2 + x^2};$$

the lee waves are given by

$$L = -2\pi a^2 \frac{\zeta_{z,k}}{\partial \zeta_{1,h}/\partial k}\,\mathrm{e}^{-kb}\sin kx \tag{6.8.1}$$

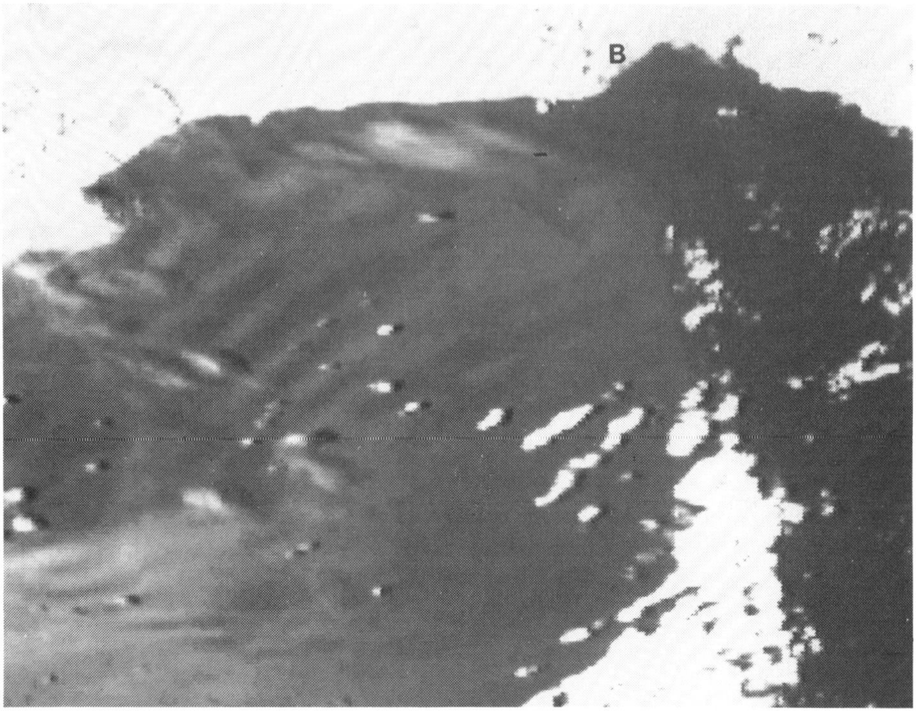

Fig. 6.7.i(a)

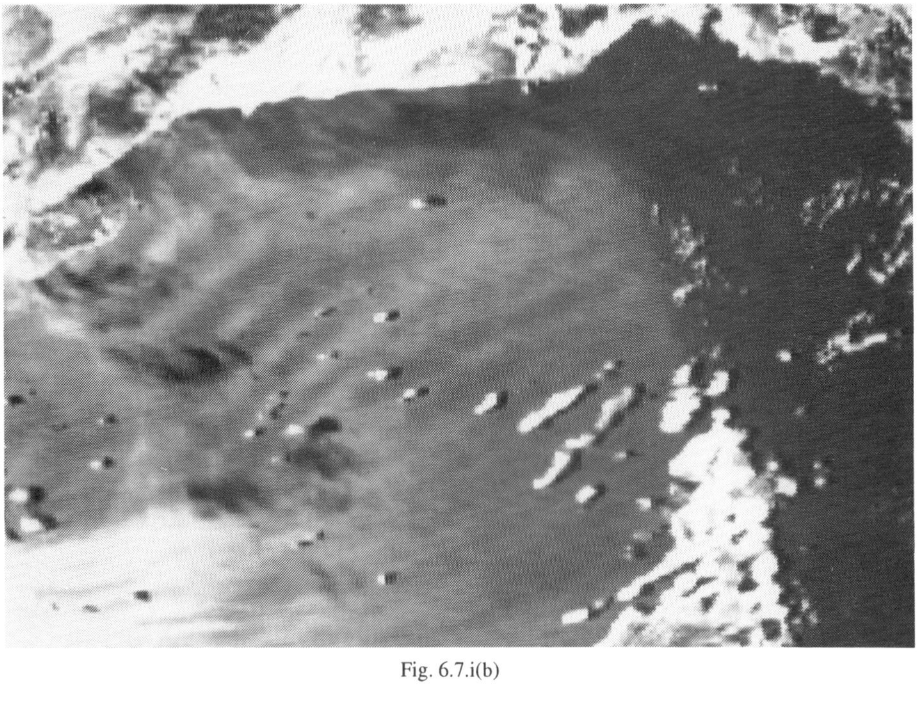

Fig. 6.7.i(b)

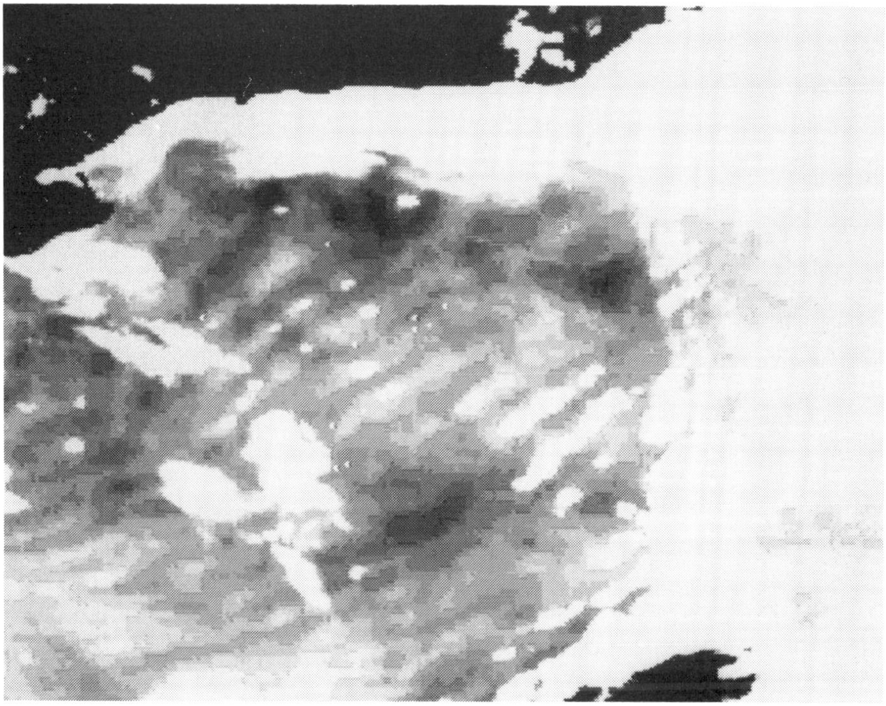

Fig. 6.7.i(c)

Legend for pictures 6.7.1(a), (b), (c). Waves seen in the sea surface NOAA7, 1402, 13.6.82. (a) Chan2 (vis). The land is shown brightly in order to emphasise the glint from the sea which is brightest under the lee waves between the coast of Spain and the Balearic island of Parma. The wind is lighter under the wave crests and so the glint is concentrated into a narrower angle than where the sea is roughened.
(b) Chan3 (near IF). This wavelength is more sensitive to glint than Chan2. At the same time some of the highest clouds contain ice crystals which are larger than the droplets of low clouds and absorb this wavelength. Note also the darkness of the cloud shadows. The glint is brighter in relation to the land than in Chan 2.
(c) This Chan4 (thermal IR) shows brightness in the wave crests and a slight enlarging is apparent in the lowest cloud. This channel does not record glint (which is the visible wavelengths) but it does record IR emissions from water vapour. The wave amplitude is a maximum at the inversion below which the air is much moister than above—which is a typical situation in the Mediterranean Sea. The water vapour is colder than the sea below it, so that it appears brighter in the wave crests where the depth of the moist layer is greater, and the wave amplitude is a maximum at the inversion as shown in the profiles calculated for Fig. 6.7.ii.
(Photographs courtesy of the University of Dundee.)

where $2\pi/k$ is the lee wavelength. The subscript '1' denotes the value at the ground, so that the differentiation w.r.t. k must be performed before the lee wavelength is inserted in the formula.

Using ϕ as variable angle about a vertical axis, we may define an isolated hill by

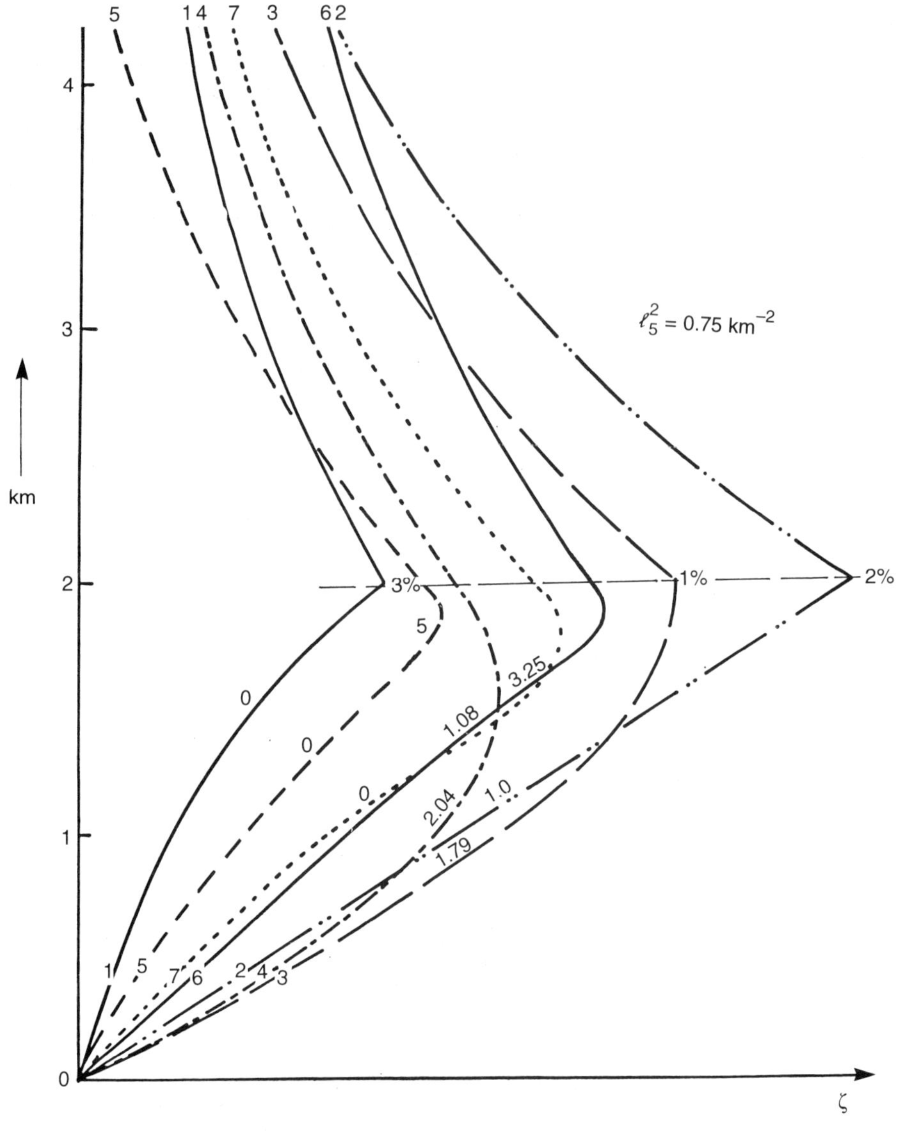

Fig. 6.7.ii The amplitude of the lee waves of the same mountain barrier is shown to vary in a complicated manner from one airstream profile to another. The same wavelength of the lee waves emerges for all of these different airstreams: clearly the drag force on the airstream due to the mountain is not dependent particularly on the wavelength. The waves in cases 1, 5, 7, 6, 2 have a fairly sharp maximum at or just below 2 km.

$$\zeta_1 = \int_{-\frac{\pi}{2}}^{\frac{\pi}{2}} \Phi(\phi)F(x)\,d\phi \tag{6.8.2}$$

where $X = r\cos(\theta + \phi)$, and Φ represents the magnitude of the component ridge inclined at angle ϕ, and $\phi = 0$ when it is perpendicular to the stream flow. (r, θ) are polar coordinates in a horizontal plane and we could use any suitable function $F(X)$ in the place of the already familiar $a^2b/(b^2+x^2)$ to represent the cross-section shape of the component ridge. In performing these calculations for the first time, familiar functions were used to save computing time before high-speed electronic computers were available. It was shown that the field could be explored that way, but it soon became more profitable to examine regularly the satellite pictures of the patterns because they were soon seen to be very common. It serves little purpose to compute ordinary examples of a commonplace phenomenon because the models are not discriminating enough.

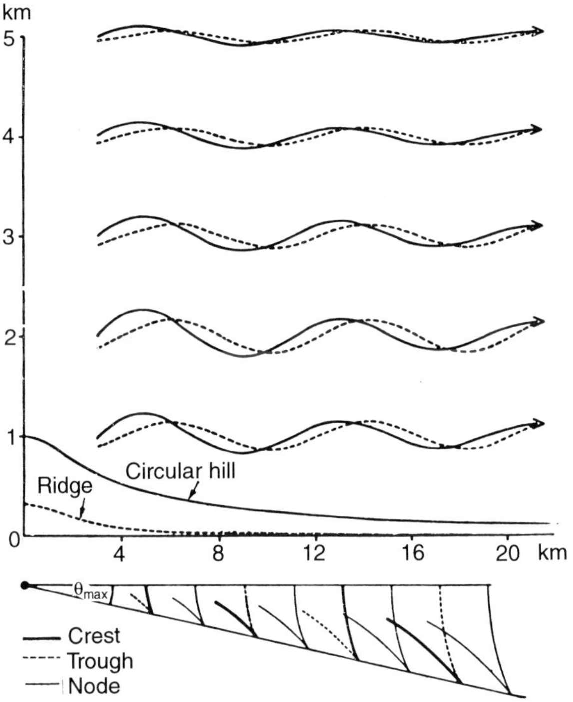

Fig. 6.8.i The wave pattern of an isolated hill: The lee waves of an isolated circular hill in the central vertical plane are shown compared with the waves produced in the same airstream by a symmetrical two-dimensional ridge which has the same amplitude as the second wave due to the hill. The lower diagram shows half of the wave pattern in plan. It is confined within a wedge-shaped region. The amplitude of the waves is proportional to the mountain height and they are both arbitrary. The height scale therefore only applies to the streamlines; the airstream has $\ell_s^2 = 0.5\ \mathrm{km}^{-2}$ above 2 km, and $\ell_r^2 = 1.5\ \mathrm{km}^{-2}$ below 2 km, and is such that $\theta_{max} = 12.3°$. [R. S. Scorer and M. Wilkinson, **82**, 354 (1956).]

The lee wave system was readily shown to be susceptible to the principle of stationary phase (as used in section 6.6), and it was found that the angle within which the wake of an isolated hill would be found might be almost any angle between 0 and $\pi/2$. In the case of ship waves on deep water the angle is well known to be 19°28′. Most interesting is that

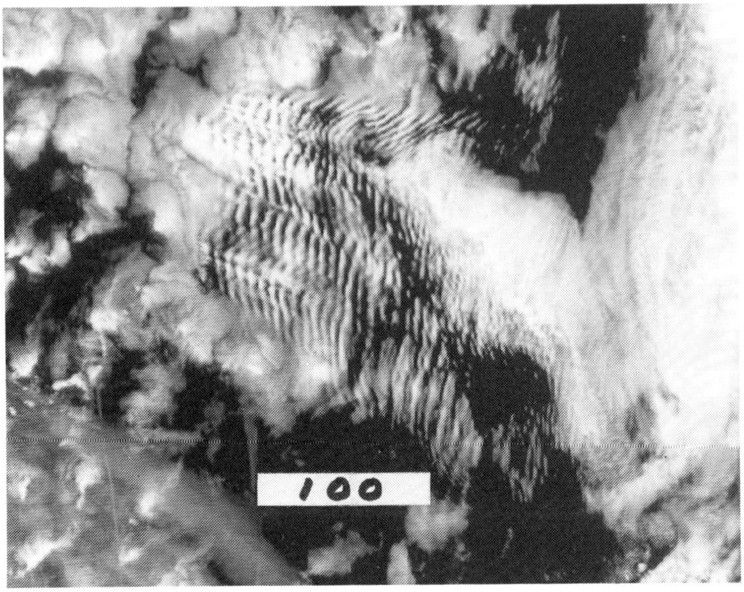

Fig. 6.8.ii Lee waves of the Faroe Islands showing a multiple ship-wave structure with a dominance of narrowly spreading transverse waves NOAA7; 1359, 17.7.82, Chan.2. (Courtesy University of Dundee.)

this angle may appear to be almost as large as 90° in some cases, and this is because when the waves are trapped in a variously layered airstream the variety of possible regions of stationary phase is greatly increased when the upper layers are made to include the stratification of the stratosphere.

A particular case is taken from an original paper (Scorer and Wilkinson [19]) showing the streamlines in the central plane downwind from an isolated circular hill, or an oval hill. The wake of the circular hill in this case was confined within a wedge whose half-angle was about 12.3°.

Below the vertical cross section the flow pattern shows in plan the layout of the crests and troughs and it is seen that at the maximum angle enclosing the wake the phases of the diverging wave and the transverse wave are of opposite sign. Observations seem to indicate that in most cases either the diverging or the transverse waves are dominant, depending on the shape of the obstacle. (See Figs 6.6.iii and 6.6.iv.)

6.9 CRITICISMS OF THE THEORY: CASES OF LARGE AMPLITUDE

The waves over and near to the obstacle depend to a large extent on the shape of the obstacle. The real difficulty about applying the kind of theory described in the preceding pages lies in the nature of the assumptions. Thus we have assumed that the profile of the stream far upstream of the obstacle can be specified and can be assumed to be undisturbed. If an obstacle is placed suddenly in a stream it is observed in **all** cases to send a wave, or bore, travelling upstream which alters the height of the Bernoulli surfaces in the stream arriving at the obstacle. It cannot be assumed that a given stream and a given obstacle are compatible and that no disturbance of that kind will occur before a steady state is reached.

It has also been assumed that a steady-state can be reached; yet no argument has ever been given to justify this. This is particularly important in the case of separation of the flow from the surface, because periodic flow may then occur (see section 6.11.viii), rather like the shedding of vortices behind a cylinder.

Nevertheless waves over mountains and lee waves do occur, although waves upstream have never been recorded without doubt. The formulae obtained for the lee waves, for example in sections 5.9 and 6.5, show that the lee wave amplitude can be as large as, or even larger than, the obstacle height. In the case of waves trapped in a two-layer model, the formulae, such as (6.4.14) and (6.4.21), show that when the stream profile is such that waves are trapped, the disturbance over the mountain may rise to a larger maximum than in a stream in which no waves are trapped.

In exceptional cases (such as Fig. 6.5.iv) the lee waves may be considerably larger than the height of the obstacle, and the parameters in the formulae can be juggled to achieve this result in a theoretical example. In practice the largest lee waves occur when the airstream flows down from a plateau without separation from the surface. This is because the waves represent energy added to the stream. In our solutions this energy appears to be added without any source. While this can be accounted for in the case of flow between two rigid boundaries by a pressure drop in the downstream direction, in the case of a stream of infinite depth in which no disturbance is imposed at the top, this explanation is not acceptable. In the atmosphere there could be an acceleration of the flow by a deviation towards low pressure across the isobars, but such an acceleration cannot be effective in a distance which is very small compared with that travelled by the fluid in a day. On the other hand, the theory is a small amplitude one for the case of no upper boundary, so that energy considerations are not applicable because energy is represented by second-degree terms which are neglected.

The most effective cause of large lee waves other than descent from a plateau is to have several parallel ridges spaced at a whole number of lee wave lengths apart. Thus the wave amplitude may be doubled by a second obstacle one wavelength downstream of the first (Fig. 6.9.i).

A second obstacle placed at an odd number of half wavelengths downstream could exactly cancel the lee waves of the first obstacle, and leave none. In general a second obstacle merely alters the phase of lee waves. In Fig. 6.9.i two mountains 1 and 2 have two airstreams depicted, the one with dashed lines having half the lee wavelength of the other. When the mountain 2 is 1 or 3 wavelengths downstream, the lee waves are added, but at $\frac{1}{2}$ or $1\frac{1}{2}$ wavelengths downstream, the lee waves cancel out.

Since the component of velocity along a two-dimensional ridge is assumed unaffected by the ridge and retains its undisturbed value in each Bernoulli surface, a given airstream profile may produce lee waves when it crosses two parallel ridges at one angle but not when it crosses the same ridges at a different angle. A rare combination of circumstances may cause very large lee waves to occur, and then a very strong wind is produced in the wave trough with almost no wind, or even reversed flow in a rotor, at the surface under

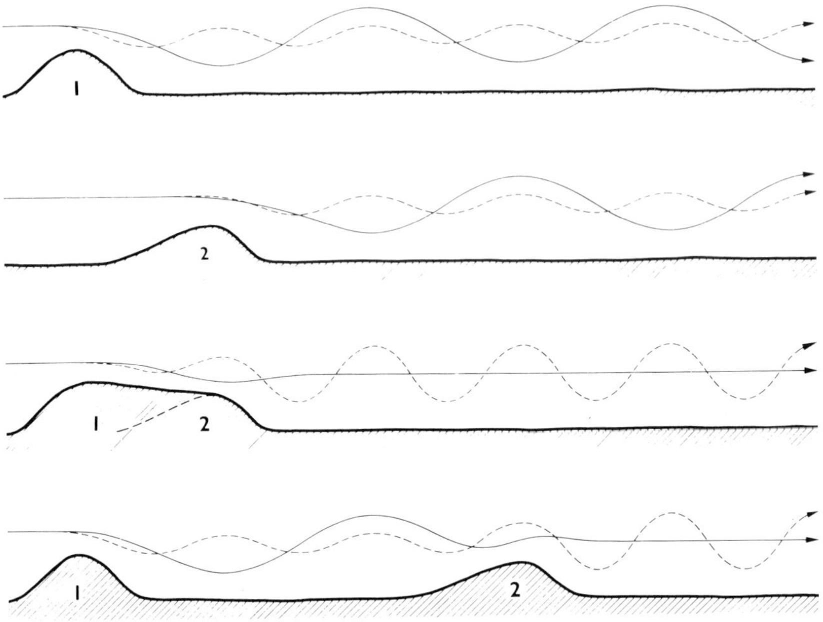

Fig. 6.9.i The mountain 1 produces the lee waves shown for two airstreams, one having twice the wavelength of the other. When mountain 2 is added the lee waves may be increased or decreased. In the case shown they are doubled for the shorter and cancelled out for the stream having the longer wavelength.

the wave crests. Such a lee wave removed the roofs from houses in a suburb of Sheffield although some had stood undisturbed for over 50 years. The gale which achieved this was considered the more remarkable because of the areas of calm on each side of it not far away. The rare circumstance in this case was actually the coincidence of the lee wavelength in the airstream with the spacing of three successive ridges.

There is also a local tradition in the Isle of Man that if the roof is blown off a barn or house this is because the fairies are displeased at its location in their territory. The theory is 'proved' when the roof is blown off again, after repair, by a very localised wind which does not occur elsewhere, even on other occasions.

During a scare about the presence of Colorado beetles in Central Europe, a few were found in Bohemia, and enthusiastic officials asserted that they had been placed there by ill-intentioned visitors from a neighbouring country, because otherwise 'would they have been found in such neat rows?' A meteorologist explained that the beetles would have been more likely to settle in the calmer air under a lee wave than in the strong winds which would be found under the wave troughs. The area on the upwind side of the mountain ridge was known to have been infected by the beetles recently. Pests have no respect for political (in this case ideological) boundaries.

One very common feature of lee waves is that they are more pronounced at night. The sky is typically filled with wave clouds at dawn, and they disappear and are replaced by convection clouds during the day. As the sun sets, the wave clouds often return. Billows associated with wave clouds show the same diurnal variation. The most pronounced gales in lee waves occur at night, and it is at night that fairies remove the roofs from barns.

One of the most striking examples of the onset of waves in the evening is experienced at the gliding site at Great Hucklow, near Sheffield (see Roper [20]), close to the area where the roofs were blown off. The site is situated on the top of the second ridge and experiences lee waves in the evening. These used to be called evening thermals because they were thought to be associated with warm air rising up the west-facing slope which was the last piece of ground to be exposed to the rays of the setting sun. Actually the site often finds itself in the almost calm air under the wave crest of the first lee wave from the ridge upwind (see Fig. 6.9.ii). While it is difficult to get a glider aloft by the simple device of catapulting it from the top of the ridge into an upslope wind, it is often very easy for those already airborne at a few hundred metres above the site to find the upslope side of the wave and soar to greater heights with it.

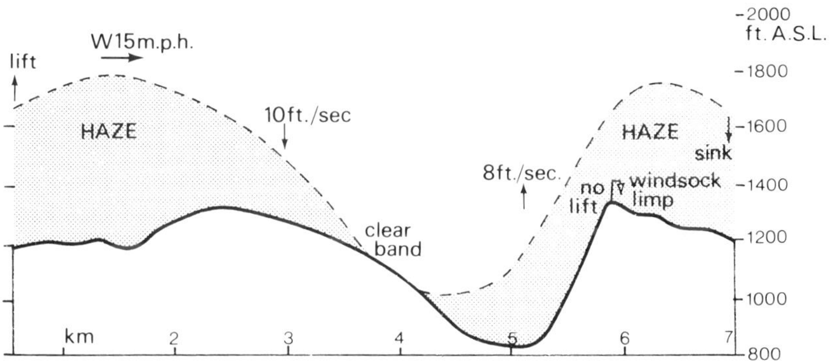

Fig. 6.9.ii The waves in an airstream are often visible at a haze top. This example, from Roper (1952), shows the observed behaviour on an evening when waves were re-established and the wave occurred so as to produce a calm over the second ridge top. Because lift was found by glider pilots, the phenomenon was given the same 'evening thermal' at a time when waves were not well understood. This 'clear band' could have been partly an optical illusion, the haze not being visible in the shadow on the east side of the hill in the evening.

The early explanation of this phenomenon was based on the assumption that destruction of the stable stratification in the lowest layers by sunshine on the ground during the morning was reversed, thus decreasing the value of ℓ^2 near the ground and making trapping impossible. It now seems more likely that it is the result of the ending of separation of the flow from the lee side of the obstacle. This is discussed in section 6.10. The reason why the first explanation in terms of waves is unlikely to be correct is that the stable stratification produced by cooling the ground affects only a very shallow layer of air close to the ground during the first few hours. On the other hand, the warming of the ground in the morning can alter the stratification over a greater depth in a short time.

In the Eden Valley in the north of England, where lee waves have long been known, the cloud in the first lee wave moves away from the mountain ridge during the morning and finally disappears, local lore asserting that it never crosses the river (see Manley [21]). Thus the lee wave length increases (k^* decreases) as would be expected, because the depth of the stable layer at the bottom of the airstream is reduced from below by daytime convection, and the time required for an oscillation is increased. Alternatively we may argue, according to the ideas set out in section 5.11(2), that since the layer of large β is reduced in thickness, a smaller value of k is required in order to achieve a zero at the ground. This is because that increases the curvature of the ζ-profile in the layer in which $\ell > k$ and reduces the curvature in the opposite direction in the layers in which $k > \ell$. As the layer is thinned, it becomes impossible to achieve a zero of ζ at the ground, and the lee waves disappear.

A common experience of glider pilots, noticed very early in the days of exploiting lee waves for soaring, is that the lee wave amplitude has a maximum at about the height of the top of a layer of cloud if one is present. Very often the cloud top is the upper limit of a layer of air which has been stirred by convection from the ground, and the density (temperature) discontinuity which exists at its top is the cause of the presence of waves (see section 6.7). As is explained in section 13.1(2), cloud top is usually a level from which heat is lost by radiation into space and so the layer below it is continuously cooled, and a stable temperature discontinuity is established. Since this is, in effect, a shallow layer of very great static stability, it is one in which the curvature of the ζ-profile is very large and negative, and so it is almost inevitably a level at which the profile has a maximum (see section 6.7).

Since a layer of cloud may be continuous, that is to say it is thick enough not to be evaporated in the wave troughs, the fact that waves are present is sometimes not visible from the ground. If the layer of air in which the cloud is formed is well mixed, all stream surfaces have the same condensation level and the cloud base is horizontal. On rare occasions a moist layer lies above a drier layer, and then waves are visible on the cloud base.

A common observation in mountainous areas is that when the air streams down off a plateau, sometimes a layer of cloud existing over the plateau disappears because the air at the level of the cloud also descends. But equally often a layer of cloud is seen over the low ground and not over the plateau, or even occasionally there is a hole in a layer of cloud over a mountain ridge and not elsewhere. This is called the Föhn window (Lücke) (Fig. 6.9.iii) because the blue sky above can be seen only through two or more cloud layers in the neighbourhood of the downslope region. The formula (6.4.12) for the displacement of the streamlines near to the ridge shows that the phase of the displacement is reversed with a period of $2/\ell_s$ in z, and it is a detail of the case whether the displacement is upwards or downwards over the edge of a plateau at the level at which the cloud layer happens to be.

6.10 LAYERS OF ZERO WIND VELOCITY

If at some altitude the wind is zero relative to the ground and lower down the air is streaming over an obstacle, we cannot find an acceptable solution of the wave equation in such a case because ℓ^2 becomes infinite when u_0 is zero. Also, unless $\partial u_0/\partial z_0$ is zero at that altitude, α also becomes infinite.

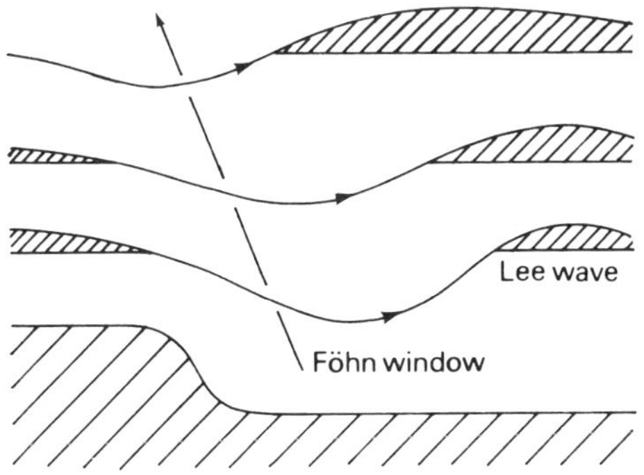

Fig. 6.9.iii Whether there is a lee wave visible in cloud at a low level or not there is sometimes a higher layer at which the air ascends above its condensation level when the air at ground level descends. A 'window' to the sky then appears over the lee slope, and is often referred to in the literature as a 'fohn lucke'.

The mathematical solutions of the equation (5.6.7) or (5.8.13) have nodal surfaces packed ever closer together as the altitude of zero u_0 is approached from below (Scorer 1951), and the magnitude of ζ exceeds the distance between nodal surfaces, close to the zero of u_0, which means that rotors are implied. The only alternative is to have a cat's-eyes pattern (see Chapter 8) with a row of vortices at the level of zero u_0 with waves above and below (see Fig. 8.2.i bottom diagram). See also Queney [22] and Gerbier and Berenger [13].

Various suggestions have been made as to what probably happens. Several authors, e.g. Booker and Bretherton (1967), have given the level of zero u_0 the rather unhelpful and unspecific name 'critical level'. They guess that any wave energy propagated from below is absorbed at this level, which implies that momentum is communicated to the air there and that the waves are not reflected. The theory used to back up this idea is very speculative, is valid only for very small wavelengths, and cannot be applied to typical lee wavelengths in the atmosphere. Since momentum would be communicated to the air in which the kinetic energy is absorbed, the presence of layers of zero u_0 is thought to imply that the air at that level is decelerated by this absorption and that much of the drag of a mountain is applied to the air at that level. This conclusion is based on incomplete mathematics not valid for the wavelengths in question.

Perhaps the most important possibility is that if the wind gradient is large below the zero of u_0 it might be confined to a sufficiently shallow layer for it to behave like a discontinuity of velocity with zero velocity above, in the case of waves whose length is large compared with the depth of the shear layer. Although this would mean that it would be likely to break down into billow motion (see Chapter 8) it could still, for longer wavelengths, behave like a statically stable discontinuity of density and velocity with stagnant air above. In that case it could support standing waves, and the air could be in smaller scale unsteady motion near the top of the layer. This possibility was envisaged in some early calculations (Scorer [11(1)]), and waves having such characteristics have been observed from time to time by glider pilots (e.g. at the Long Mynd[23]).

Although the theory is incapable of a proper explanation, several authors have described the phenomena in diagrams which summarise their cumulative experience. The diagrams in Figs 6.10.i(a)–(d) are from a paper by Gerbier and Berenger (1961), and Forchtgott (1949) gave diagrams based on observations of moving rotors.

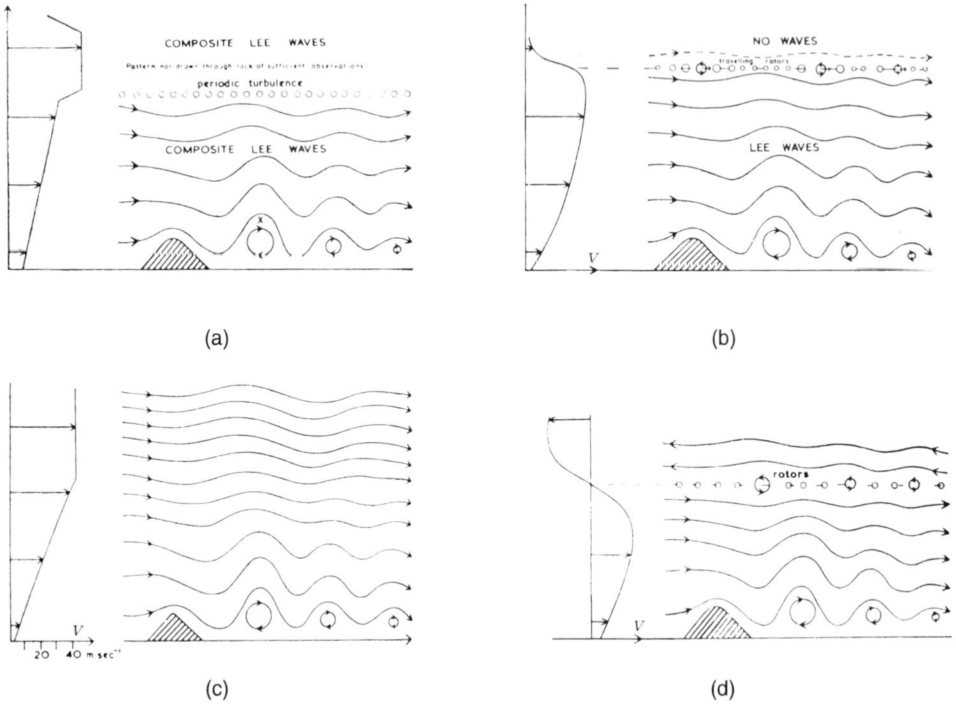

Fig. 6.10.i(a, b, c, d) Schematic diagrams indicating the flow pattern observed with four different velocity profiles in studies by Gerbier and Berenger in the French Alps. It is a matter of argument whether the rotors observed at the level of zero velocity are billows or rotors of the kind envisaged by Queney (1955) as 'cat's eye' rotors. Rotors decrease in size downwind in lee waves, and in the first diagram there is a longer wavelength evident at higher altitude.

6.11 SEPARATION OF THE FLOW FROM THE SURFACE

The theory of waves has so far made no useful progress with unsteady wave phenomena, and none is in prospect because of separation. The phenomena are so complicated and varied that they are best described by simple diagrams.

(1) Salient edges

In this case the topography has a sharp edge which induces separation. This can sometimes be due to a wall. It may take place on the lower or upper side of a cliff top or in a street with tall buildings on both sides (Fig. 6.11.iv).

(2) Corner eddies

These are eddies formed in a boundary layer at the foot of a steep hill or a building (Figs 6.11.i and 6.11.iv). They are sometimes referred to as 'bolster' eddies. It can be demonstrated that they are due to the velocity profile of a boundary layer, and that the separation is like that produced in a boundary layer by placing a solid surface upstream of

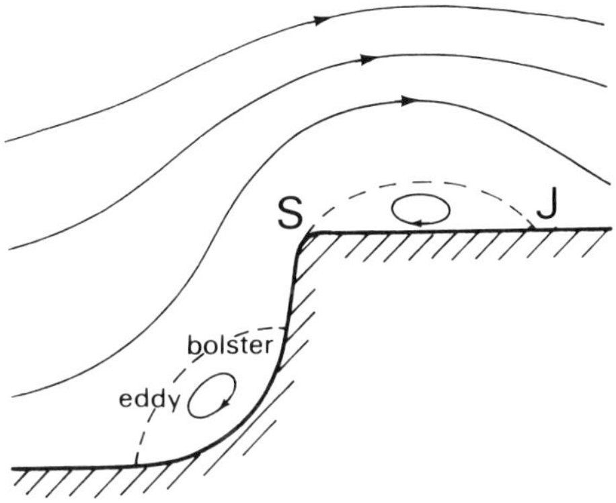

Fig. 6.11.i Separation of flow at a cliff top S and jointing at J. Where the flow diverges on the upwind side of a steep slope or cliff a 'bolster' eddy may occur because of the separation.

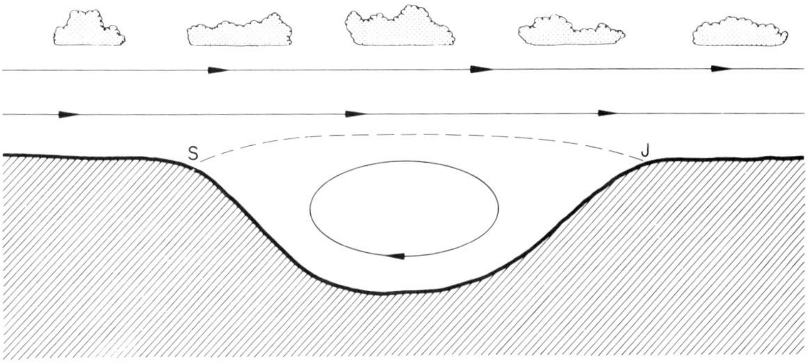

Fig. 6.11.ii Flow separating on a lee slope, the eddy filling a valley. There is no effect on the flow above.

a wedge. Bolsters are then induced at the apex of the wedge: the effect is illustrated by Figs 6.11.v and 6.11.vi.

(3) Fixed and moving lee eddies

When the flow separates on the lee side of an obstacle because the main stream is widening out and moving towards higher pressure, the separating streamline is usually unsteady because it is in a region of maximum vorticity. The eddies are sometimes steady, particularly if there is a good salient edge to fix the point of separation, but with a fluctuating outer boundary. Others are shed, often periodically. Some parts of a coastline have a sharp cliff top (Fig. 6.11.vii(a)) where an eddy is fixed, while on other well

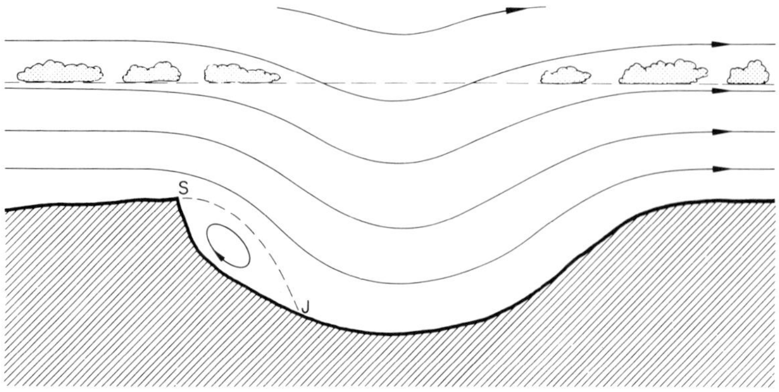

Fig. 6.11.iii A small eddy may be produced by a salient edge but the large valley may nevertheless produce downmotion which evaporates the cloud over it.

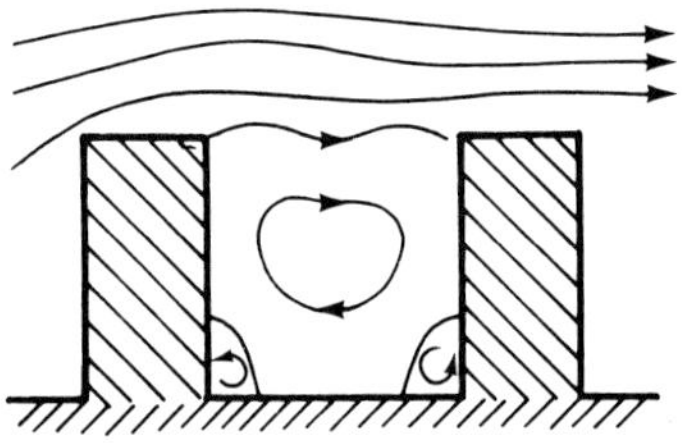

Fig. 6.11.iv In a street, standing eddies may occur, including corner eddies driven by the main one. But the exterior flow is not usually steady, and contains eddies shed by other obstacles upwind; furthermore it contains a wind component along the street, which influences the pattern of separation. Since the architecture is also variable, this diagram is only a first indication of the kind of motion to be looked for.

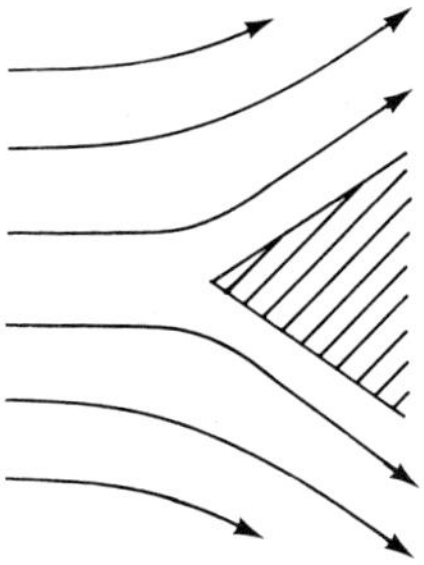

Fig. 6.11.v This shows the flow produced by a sharp corner facing the wind.

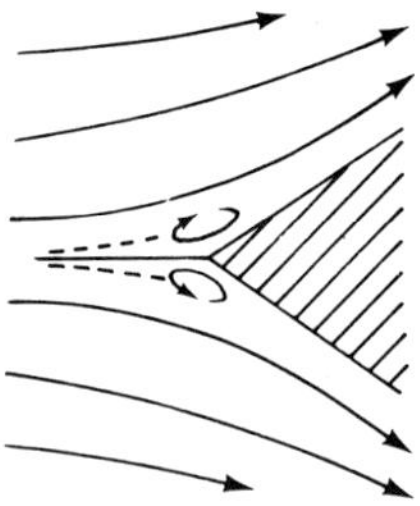

Fig. 6.11.vi If a solid surface is placed upstream of the apex, the drag of its surface may lead to the formation of bolster eddies.

rounded parts (Fig. 6.11.vii(b), (c)) sometimes there are eddies and sometimes not. The presence of eddies can have a significant effect on lee waves, for the position at which rejoining takes place may be determined by minimising the wave energy.

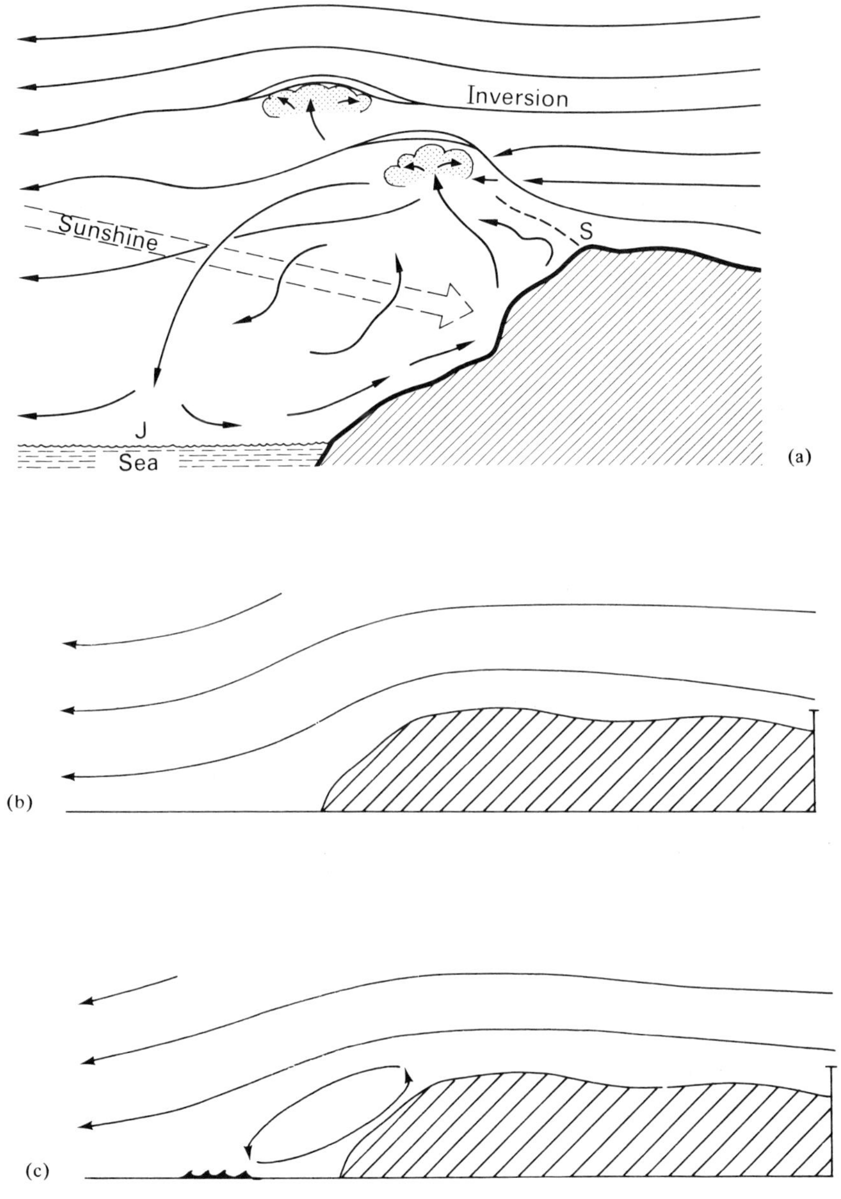

Fig. 6.11.vii Diagram (a) depicts the complicated situation where a sun-facing slope may have separation at the top caused by a salient edge or by an anabatic wind, or both. The anabatic wind produces thermals which are intermittent, and sometimes produce cumulus clouds. The eddy is very unsteady, and at the point of joining the wind may be gusty and creates cat's paws on the sea surface, or on the opposite side of a valley in bracken, grass, or a standing crop. (b) on the other hand shows the simplest situation of fairly smooth flow out to sea. (c) shows the simpler case of an eddy, and the flow often oscillates between (b) and (c) in a way that can be detected by putting wind socks or smoke generators on the slope.

Fig. 6.11.vii(d) This picture was taken from the position S in diagram (a) with a rolling cloud obscuring the sun. The eddy on the sunny face of the cliff was continuously shifting so that the flow had a large amount of unsteadiness: an east wind blew across the Isle of Man.

On the other hand, wave clouds have often been observed to come and go periodically, and periodic upsurges of wind followed by slackening are experienced on the lee side of hills. This could be due to the periodic shedding of eddies from a hill, and in this case, the first lee wave cloud is observed to drift slowly down wind and disappear, to be replaced at a point closer to the hill by a new cloud. This sequence is depicted in Fig. 6.11.viii, which is taken from Scorer (1955), which is based on an account by Wills of his observations while on night work at the Radar Establishment.

6.12 CURL OVER

'Curl over' refers to the behaviour on top of a cliff up which the wind is blowing. Such a cliff top is a favourite site for glider launching and so pilots wish to land downstream of the cliff edge but close to it. Whether there is simply a widening of the stream tubes (Fig. 6.12.i) and a consequent reduction in wind strength, or whether there is an actual reversal of wind because of an eddy, does not make much difference to the pilot, who in either case experiences a decrease in airspeed as he lands. The phenomenon is therefore called a '*clutching hand*', because it causes a rapid increase in rate of descent (see Scorer 1951(b)).

The eddy formation is more common in stable air, for example when the ground is cold, and less common when it is sunny and the air is turbulent near the surface. But in any case it is a rather unsteady phenomenon, and even if the eddy is more or less permanent, its boundary is likely to be subject to billow instability (Chapter 8).

Fig. 6.11.viii Wave clouds have often been observed to come and go periodically, and periodic upsurges of wind followed by slackening are experienced on the lee side of hills. This could be due to the periodic shedding of eddies from a hill, and in this case, the first lee wave cloud is observed to drift slowly down wind and disappear, to be replaced at a point closer to the hill by a new cloud. This sequence is based on an account by Wills of his observations while on night work at the Radar Establishment.

6.13 CAT'S-PAWS

The point at which the flow rejoins the surface is often seen on water where it looks like a giant paw quietly sliding forward. The point moves downstream behind an eddy that has separated and creates a sudden increase in wind strength which ruffles the water surface. Cat's-paws indicate that the release of eddies is not simply a downstream drift of the flow pattern due to stationary eddies, but contains fronts (see section 9.9 and Fig. 4.8.ii for comparison) behind which air with greater velocity descends to the surface (see Fig. 6.13.i).

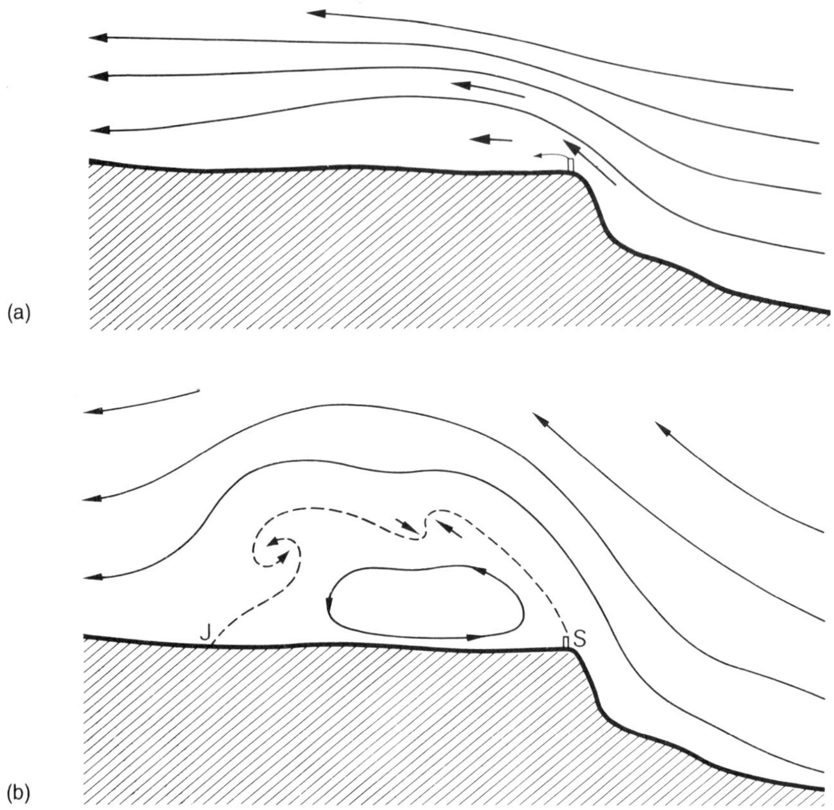

Fig. 6.12.i (a) Flow at a hill top widening out but not separating. The motion aloft in which there may be waves influences this behaviour and (b) depicts an eddy with billow-like vortices travelling down the separation streamline.

6.14 DIURNAL SEPARATION CONTROL

Anabatic and katabatic winds have a very marked effect on separation phenomena. A katabatic wind, which begins when the ground cools as the sun goes down, is very shallow, but is nevertheless sufficient to prevent separation from the lee slope of a hill. The katabatic wind controls the separation like boundary layer control depicted in Figs 2.9.v and 2.9.ix. The most interesting aspect of this is that as a result lee waves may begin to occur, and the large wind speed which occurs in the first lee wave trough at the foot of the hill has often been thought to be the katabatic wind itself. A well developed example was described by Lee and Neumark (1952) at Dunstable, where a glider soared to 4000 ft in the lee wave behind a 200 ft downslope.

Anabatic winds very readily induce separation, particularly if the lee slope is facing the sun (see Fig. 5.18.vii(a)). This is a major cause of the disappearance of wave clouds by day. We may seek an explanation of the 'evening thermal' (see Fig. 6.9.ii) as a consequence of separation quelled by a katabatic effect.

Fig. 6.12.ii At the top of the ridge at Camphill gliding site. The wind socks indicate the separation of the flow coming up the slope, which is the other side of the wall on the right in this view.

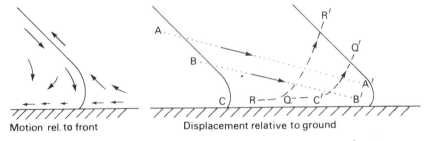

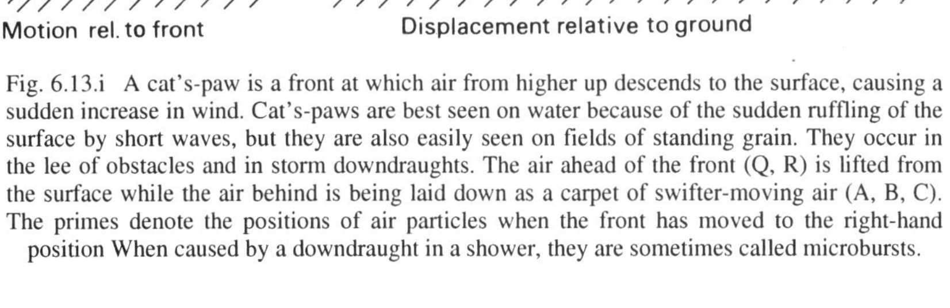

Fig. 6.13.i A cat's-paw is a front at which air from higher up descends to the surface, causing a sudden increase in wind. Cat's-paws are best seen on water because of the sudden ruffling of the surface by short waves, but they are also easily seen on fields of standing grain. They occur in the lee of obstacles and in storm downdraughts. The air ahead of the front (Q, R) is lifted from the surface while the air behind is being laid down as a carpet of swifter-moving air (A, B, C). The primes denote the positions of air particles when the front has moved to the right-hand position When caused by a downdraught in a shower, they are sometimes called microbursts.

6.15 BLOCKING

According to the theory outlined earlier in this chapter, the disturbance in the neighbourhood of a ridge has a phase variation in the z-direction with a period $2\pi/\ell$. This means that there must be a nodal surface at some height not greater than $2\pi/\ell$ above the undisturbed bottom of the stream, and if a mountain ridge extends above this height there is no flow of the stream which can be described by the theory, even in its arbitrary amplitude form. Long suggests that probably blocking occurs in that case, for there is no possible systematic displacement of the air above the ridge which can produce a low enough pressure for the surface air to reach there (the Froude number is much larger than unity).

Blocking is the main cause of Föhn winds, in which the air descends from an altitude not far from that of the ridge top on the upwind side to the surface on the lee side (Fig. 6.15.i). The warmth of Föhn winds is ascribed in many textbooks to the release of the latent heat of rain falling on the ridge. It is shown in section 1.6(2) that this explanation is wrong in most cases even when there is rain, but it is obviously wrong in the many cases where there is not rain and more so when there is not even any cloud above the mountain ridge. Typical Föhn and Chinook winds occur when a cold air mass is blocked by a mountain range.

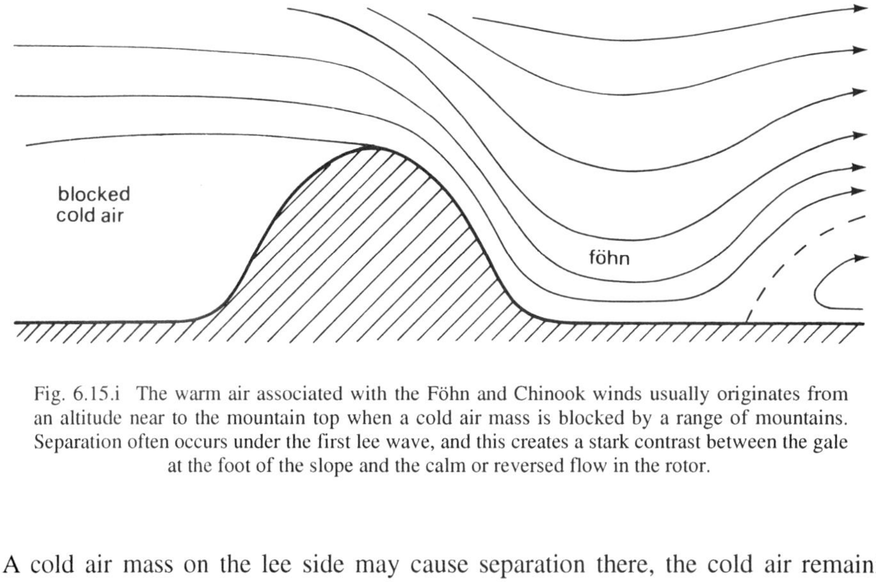

Fig. 6.15.i The warm air associated with the Föhn and Chinook winds usually originates from an altitude near to the mountain top when a cold air mass is blocked by a range of mountains. Separation often occurs under the first lee wave, and this creates a stark contrast between the gale at the foot of the slope and the calm or reversed flow in the rotor.

A cold air mass on the lee side may cause separation there, the cold air remaining stagnant in a valley on the lee side of the mountain. The top surface in such a case is usually a cloud top, radiation from which produces a stable density discontinuity (see section 1.1(2)), and it is tilted by lee wave motions in the airstream above it. Sometimes, even if it is not significantly tilted but in the strong winds of a wave trough, billows may be produced at the stable layer. The presence or otherwise of anabatic winds on the part of the lee slope above the stable layer may cause a ruffling of the top of the layer, particularly if the cloud there is very thin and the anabatic wind is being generated by the sunshine on the slope below the layer (Fig. 6.15.ii). Sometimes shallow pools of very cold air may remain in hollows or under the light surface winds in lee wave crests, with the boundary between them and the warm Föhn air moving around and causing very large changes in temperature, both up and down. This is most common in the first warm Föhn of spring when cold air is first carried away but the ground is still very cold.

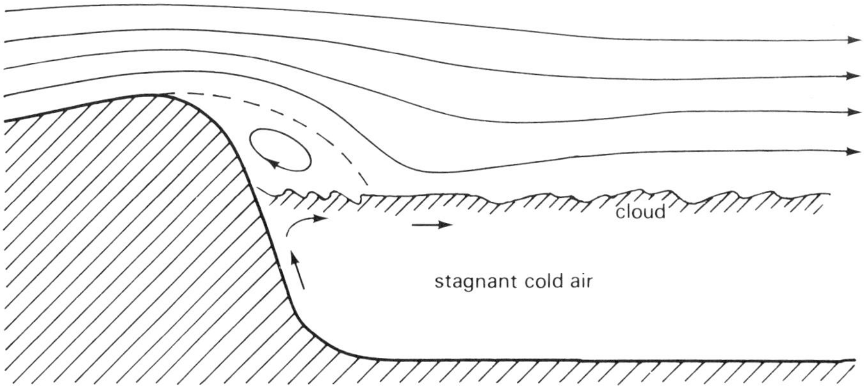

Fig. 6.15.ii Often a cold air mass occupies the valley or plain behind a mountain range. It is often capped by a cloud layer, and the position of billows or other disturbances on the cloud top is determined by the behaviour of eddies behind the mountain crest.

6.16 THREE-DIMENSIONAL EFFECTS

There is almost no systematic way of discussing the flow behind a three-dimensional obstacle of arbitrary shape when separation occurs. Each case has to be observed to be discussed. On buildings with courtyards or towers, on ships with a variety of salient edges, on the ends of mountain ridges, the flow depends not only on the shape but also on the level of turbulence in the air and its stratification. The best advice is to learn to use any tracer that is going to elucidate the flow: this is not evasive advice because the effectiveness of a tracer depends very much on the user, and there is lots of fun to be had with smoke, soap bubbles, balloons, snow, flags, etc. A flow pattern suggested by Forchtgott (1950) for a circular hill is shown in Fig. 6.16.i, based on observations.

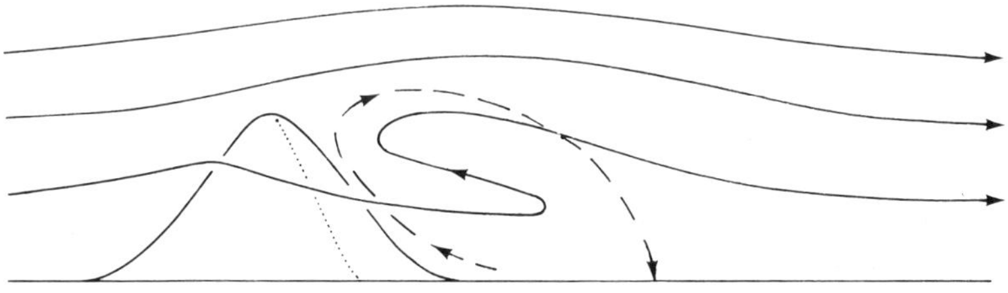

Fig. 6.16.i The flow up to the lee slope of an isolated hill may not be in the form of a lee eddy but a circulation through the lee slope region in which some particles are recirculated perhaps a few times before escape.

6.17 SEPARATION VORTICES

A line of separation may be the base of a vortex sheet, if the airstreams approaching from the two sides have different components of velocity along the line. The upward acceleration from the line, which is a line of stagnation points, stretches the vortex lines which are vertical and end on the line. The intensity of the sheet is thus increased and individual vortices may become quite violent. Dust whirls and whirls of blown snow often occur on the lee slopes, and sometimes bits of loose turf or tufts of vegetation are carried into the air by a vortex. These vortices are usually shortlived. The motion is depicted in Fig. 6.17.i, and an example is shown in 6.17.ii.

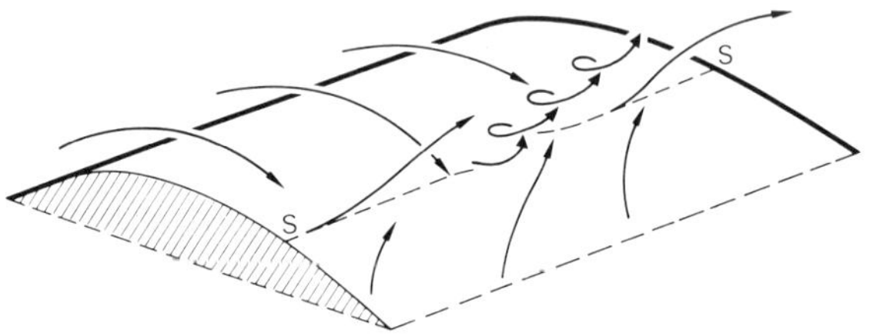

Fig. 6.17.i When separation occurs on the lee slope, vortices are sometimes intensified by the stretching, and dust or snow devils may be formed. Similar vortices have been observed on the ascending side of a rotor.

Fig. 6.17.ii A snow devil on the lee side of a hill, in the Colorado Rockies, where the flow has just separated in the manner of Fig. 6.17.i.

7

Wave dispersion and global effects

7.1 DISPERSION OF A PRESSURE PULSE

When a large explosion takes place, depending on the mechanism by which it is produced, a body of hot air may be created at the same time. We are not studying here the convection resulting from the creation of a buoyant air mass but the pressure pulse which travels outwards when the initial expansion makes room for the increased volume. The entry of nuclear bombs on the military scene in the 1940s and continuing into the early 1960s prompted renewed interest in the two natural explosions which had recently taken place in 1883 and 1908 since the setting up of instruments sensitive enough to record the pressure fluctuations they produced. While information about bomb blasts remained secret, the records of the blasts of the volcano Krakatoa and the Great Siberian Meteorite were available for comparison with any new theoretical results. Information about the total energy of those two cases could make possible comparisons with A-bomb and H-bomb blasts.

The theoretical plan was straightforward: the initial explosion at ground level would send out a hemispherical pressure (probably a shock) wave, but as the distance travelled increased, it would acquire the characteristics of pressure surge shaped like a circular vertical wall displaced outwards, and would lose all details which might indicate the altitude of the original explosion. Thus it would appear as if it had been started by an explosion on a vertical line producing a cylindrical shock with a vertical axis of symmetry. The passage of the pulse would occupy only a small fraction of the distance from the point of origin. It would cause an outward displacement as it passed, and it would appear from observation at a point passed as if it were part of a plane wave in a vertical plane. It had started as an introduction of volume and would raise the height of the atmosphere locally and spread out as a gravity wave (external, like a surface water wave).

Thus there were good reasons to expect the equations describing the progress of the pressure pulse to be separable into one describing the dispersion of the pulse in the direction of propagation and another describing the variation of its behaviour in the vertical direction. The variation in the third (lateral) direction would be trivial because the whole system would have axial symmetry and would be independent of the azimuthal coordinate.

Being an acoustic wave system its velocity would be dependent on the temperature, and would therefore advance more rapidly in the lower troposphere and in the ozonosphere (in the region of 50 km above the surface) than in the intermediate layers of the atmosphere, particularly the layers at and immediately above the tropopause. The wave fronts would therefore be refracted so as to become inclined upwards from the troposphere and downwards from above towards the lower stratosphere. The atmosphere as a whole would thus act as a waveguide along which the pulse would be transmitted with the speed of sound at some intermediate level, corresponding to some sort of average or intermediate temperature.

The whole pulse would change so as to consist of an increasing number of oscillations which would take an increasing time to pass a fixed point. It would travel to the antipodes of the source and then back again. G. I. Taylor had already shown how the wave front returning to the source would be distorted from a circle into a three-leafed clover by its passage through the trade winds. By travelling the full 40 000 km round the world, the pulse would be extended to a series of oscillations lasting at most an hour or two, and this indicates that the earth's rotation would have little effect on the mechanics. This is quite different from the mechanics of the semi-diurnal tide in the atmosphere which takes half a day to travel one wavelength

Dispersion occurs because the speed of propagation varies significantly with wavelength. This is inevitable because the wave involves the whole depth of the atmosphere. The pressure results from the weight of the whole depth of air above. It is in that sense that an external wave, like a surface wave on water, is unlike a lee wave, which is a true internal wave which is trapped by layers far below the region where the conditions at the top of the atmosphere can be said to apply. Pure sound waves, which are not dispersive, are propagated by movements along the direction of propagation, so that gravity is not involved.

We shall assume that the pressure fluctuations we are studying are of small amplitude. A concentrated explosion of 1 km^3 will produce an intense shock wave to begin with, but after travelling 10 km while the wave is still hemispherical the outward displacement is only 5 m. It is a mere 2.5 cm at a distance of 100 km. Beyond that, it is valid to linearise the equations.

The 1908 event has been called the impact of the Great Siberian Meteorite, on the assumption that it was a fragment of comet or asteroid because it was accompanied by a cloud of dust particles which were high enough to be seen in bright sunshine at midnight in northern latitudes. No traces of solid material have been found in any exploration of the site of impact. It is now described as the Tunguska explosion after the name of the nearby river, and it is thought that there was a powerful down burst of very hot air which impacted on the forest, felling all the trees radially outwards.

The pulse is represented by the Fourier spectrum of waves which originated as the spectrum of the explosion, and is dispersed according to an equation which depends on the atmospheric model employed. The trouble arises because such an integral is customarily evaluated by the principle of stationary phase and by evaluating the second derivative of the dispersion equation. That relationship expressing the dependence of the wave velocity on the wavelength turns out to describe a slow variation with an almost zero second derivative. The differentiation of a variable whose value has been determined

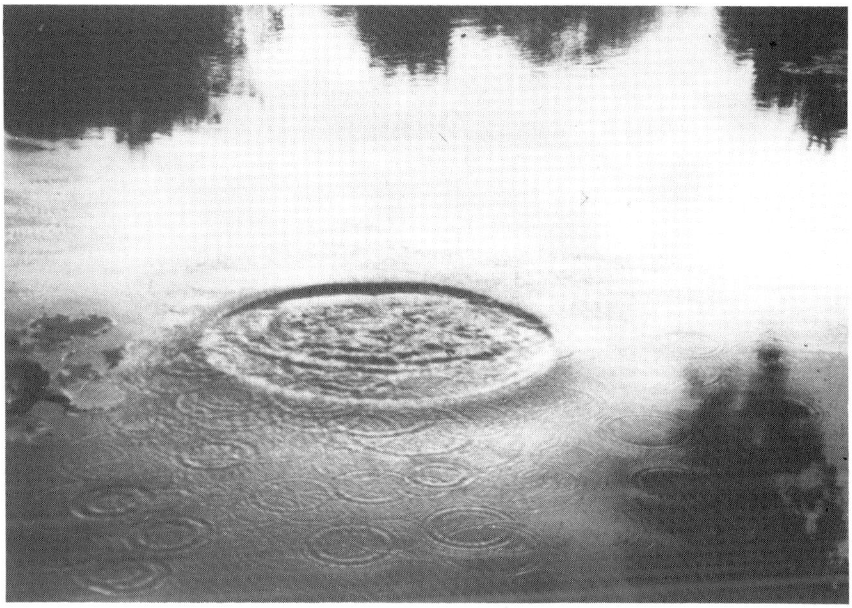

(a)

(b)

Fig. 7.1.i Dispersion of the splash made by two stones in a pond seen when the long waves have travelled faster than the shorter, after a few seconds.

only as a computed table of values is a questionable procedure in the first place, and until it has been explored it is not known whether it is better to calculate the answer at very small intervals of the coordinate or with a very great accuracy but at larger intervals. The second course was adopted, and the equation expressed in terms which made it suitable for integration by a new method recently devised by Hartree which was extremely suitable for a second-order equation with no term in the first derivative, although the choice would seem scarcely relevant in the presence of modern high-speed electronic machines. The problem was to choose a wavelength and by successive approximations determine the speed which caused the boundary conditions to be satisfied.

Today the problem, which arose in an attempt to deduce the whereabouts of an explosion from the pressure records, from which the power of the explosion might be determined even though most of the pulses were mixed in with all the natural pressure fluctuations in the weather. It would not be of particular value to repeat the calculations with a better model because we no longer live in a world with similar unknown secrets and unlike much earlier days we do not depend on such indirect methods to discover the basic structure of the atmosphere.

7.2 THE DISPERSION MODEL

The simple model of the atmosphere contained no variations with latitude or longitude, and the stratosphere was assumed to have a constant temperature from the tropopause upwards. Thus the acoustic waves including those of audible wavelength which are reflected back to the ground by the warm layers of the ozonosphere, which generally have an increasing temperature from about 30 km up to about 50 km, were not taken into account so that there is a **cut-off frequency** such that the shorter component wavelengths become lost to the calculation. When the Tunguska explossion occurred, five new sensitive barograms had been installed in various parts of the British Isles as an investigation to provide more information about higher frequencies than were being recorded by the standard weekly barographs in service at the time. The records of the pulse which showed some variations from place to place, and another from Leningrad, are shown in Figs 7.2.i and 7.2.ii.

The final figure in this set of records (Fig. 7.2.iii) is the pressure trace which Dr Whipple [16] made as the best possible composite of the five microbarograms made on

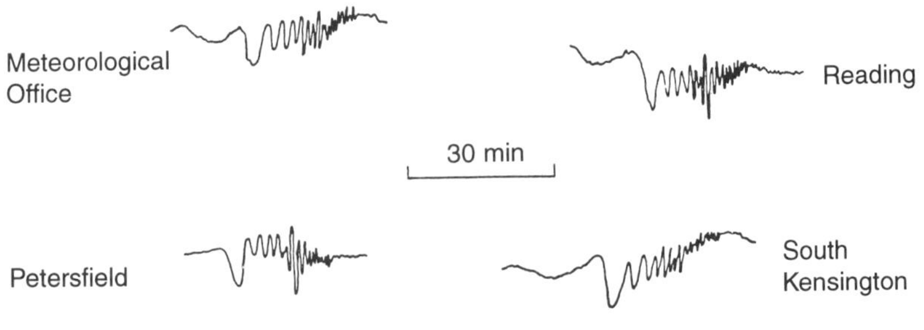

Fig. 7.2.i The records of the pulse from the meteorite which landed in Siberia on 30 June 1908, as it reached the newly installed microbarographs in England.

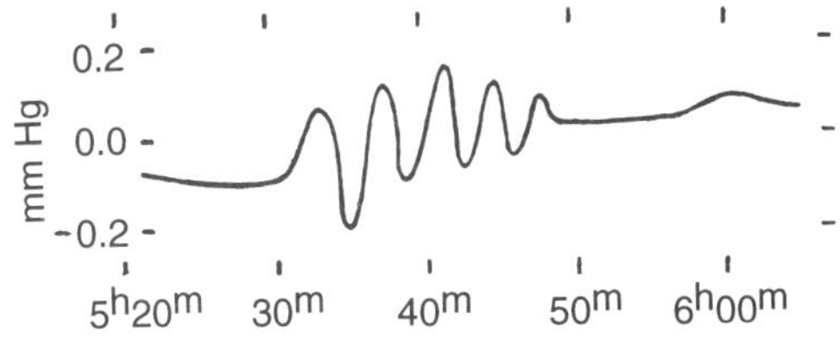

Fig. 7.2.ii Diagram of record of air waves caused by the explosion on the occasion of the Tunguska explosion on 30 June 1908 according to the records of the Sprung microbarograph at the Slutsk Geophysical Observatory near Leningrad.

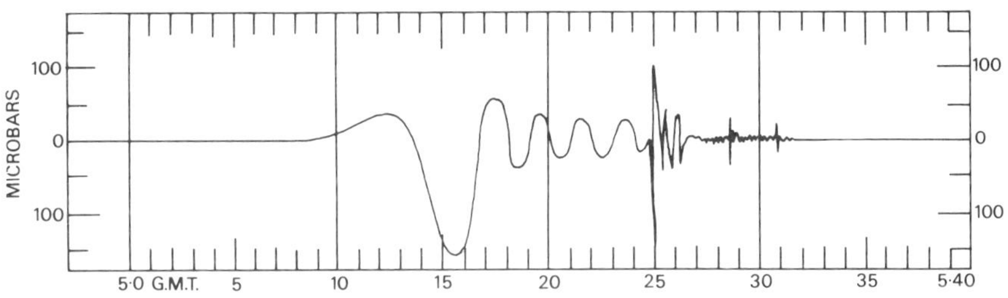

Fig. 7.2.iii A composite diagram derived from the figures 7.2.i to represent the essentials of the pressure pulse as it passed over Britain.

similar instruments in the British neighbourhood. In this it is made clear that the higher frequencies represent the reflection of the pulse from the warmer layers which are not represented in our model, and which may be described as beyond the cut-off frequency of it.

7.3 DERIVATION OF THE DISPERSION RELATIONSHIP

Spherical polar coordinates (θ, ϕ, z) are placed so that the explosion occurs at the pole $(0, 0, a)$. z then becomes height in the spherical atmosphere, with (u, v, w) being the air velocity components (Fig. 7.3.i). z is assumed to be small compared with a, and the dynamics of the pulse are confined at all stages within a small range of θ, so that they are the same as if it was being transmitted over a flat earth, through otherwise calm air.

A periodic motion proportional to $e^{i\sigma t}$ was considered as a component of the Fourier analysis to be used later, and then

$$\mathbf{grad} = \frac{\partial}{a\,\mathrm{d}\theta}, \quad \frac{\partial}{a\sin\theta\,\mathrm{d}\phi}, \quad \frac{\partial}{\partial z}, \tag{7.3.1}$$

$$\mathrm{div}\ (u, v, w) = \frac{\partial(u\sin\theta)}{a\sin\theta\,\partial\theta} + \frac{\partial v}{a\sin\theta\,\partial\phi} + \frac{\partial w}{\partial z} \tag{7.3.2}$$

$$\frac{\partial}{\partial t} = i\sigma, \tag{7.3.3}$$

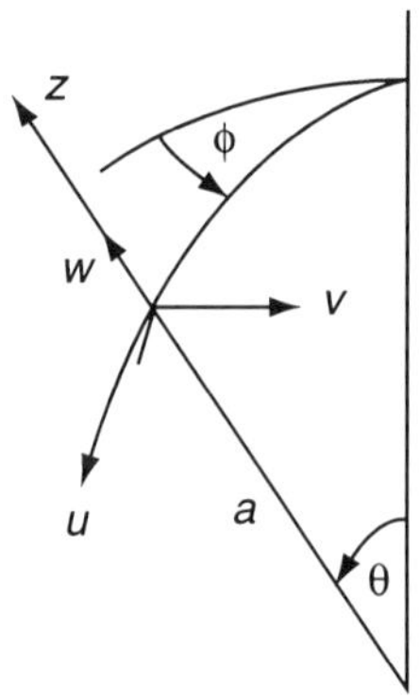

Fig. 7.3.i System of coordinates.

this last equation being appropriate when applied to the quantities ρ, p, T, τ, ϖ, u, v, w, which have the same definition as in Chapter 1 and represent the disturbance due to the pulse. The products of the small quantities are neglected. The subscript zero indicates the undisturbed value and subscript 1 denotes the value at the ground .

The equation of state for the air is

$$p_0 + p = (\rho_0 + \rho)R(T_0 + T), \tag{7.3.4}$$

and by writing differentials of the undisturbed state as representing the disturbance

$$\varpi = \mathrm{d}\varpi_0 = \frac{R}{p_{01}}\left(\frac{p_0}{p_{01}}\right)^{-1/\gamma} \mathrm{d}p_0 \tag{7.3.5}$$

and since

$$\tau_0 = \frac{1}{T_0}\frac{p_0}{p_{01}}\left(\frac{p_0}{p_{01}}\right)^{-\frac{1}{\gamma}} \tag{7.3.6}$$

we have

$$p = \mathrm{d}p_0 = \frac{p_0}{\tau_0 T_0 R}\,\mathrm{d}\varpi_0 = \frac{\rho_0}{\tau_0}\varpi. \tag{7.3.7}$$

Because the potential temperature is invariant in adiabatic changes

$$\frac{\mathrm{D}}{\mathrm{D}t}(\tau_0 + \tau) = 0 \tag{7.3.8}$$

and because the undisturbed state is assumed to have zero horizontal gradient we must have

$$\mathrm{i}\sigma\tau + \tau_0' w = 0 \tag{7.3.9}$$

where we have used a prime for $\partial/\partial z_0$.

If the earth's rotation is ignored on the grounds that the whole dispersed pulse passes a parcel of air within a small fraction of a day the equation of motion is

$$\frac{\mathrm{D}}{\mathrm{D}t}(u,v,w)=-\frac{1}{\rho_0+\rho}\ \mathbf{grad}(p_0+p)+(0,0-g)$$

$$=-gw+\frac{\mathrm{i}\sigma}{\tau_0}\varpi. \tag{7.3.10}$$

The adiabatic equation may be written

$$\frac{\mathrm{D}}{\mathrm{D}t}(p_0+p)=c^2\,\frac{\mathrm{D}}{\mathrm{D}t}(\rho_0+\rho)$$

where $c^2=\gamma R\tau_0$, and the equation of continuity is

$$\frac{1}{\rho_0+\rho}\frac{\mathrm{D}}{\mathrm{D}t}(\rho_0+\rho)+\mathrm{div}\,(u,v,w)=0 \tag{7.3.11}$$

and from these $\rho_0+\rho$, then p_0+p, may be eliminated to give

$$\frac{1}{c^2}\left(-gw+\frac{\mathrm{i}\sigma}{\tau_0}\varpi\right)+\mathrm{div}\,(u,v,w)=0$$

or

$$\left(\frac{\cot\theta}{\alpha}+\frac{\partial}{u\partial\theta}\right)u+\frac{\partial}{a\sin\theta\partial\phi}v+\left(\frac{\partial}{\partial z}-\frac{g}{c^2}\right)w+\frac{\mathrm{i}\sigma}{c^2\tau_0}\varpi=0. \tag{7.3.12}$$

From the definition of ϖ and τ in (1.6.10) and (1.6.11) the components of the equation of motion are found to be

$$\mathrm{i}\sigma u+\frac{1}{\tau_0 a}\frac{\partial\varpi}{\partial\theta}=0 \tag{7.3.13}$$

$$\mathrm{i}\sigma v+\frac{1}{\tau_0 a\sin\theta}\frac{\partial\varpi}{\partial\phi}=0 \tag{7.3.14}$$

$$\left(\mathrm{i}\sigma-\frac{g\tau_0'}{\mathrm{i}\sigma\tau_0}\right)w+\frac{1}{\tau_0}\frac{\partial\varpi}{\partial z}=0 \tag{7.3.15}$$

where we have subtracted the undisturbed hydrostatic equation from the third equation. We can now eliminate the three components of velocity from the equation (7.3.12) and this leads to

$$\left(\frac{\cot\theta}{a}+\frac{\partial}{a\partial\theta}\right)\left(\frac{\mathrm{i}}{\sigma a\tau_0}\frac{\partial\varpi}{\partial\theta}\right)+\frac{\partial}{a\sin\theta\partial\phi}\left(\frac{\mathrm{i}}{\sigma a\tau_0\sin\theta}\frac{\partial\varpi}{\partial\phi}\right)$$

$$+\left(\frac{\partial}{\partial z}-\frac{g}{c^2}\right)\left(\frac{\mathrm{i}\sigma}{\sigma^2+g\tau_0'/\tau_0}\frac{1}{\tau_0}\frac{\partial\varpi}{\partial z}\right)+\frac{\mathrm{i}\sigma}{c^2\tau_0}\varpi=0. \tag{7.3.16}$$

It has next to be assumed that the horizontal variations of the undisturbed state of stratification of the air can be ignored while we study the passage of the pulse—which is equivalent to saying that the propagation of the pulse is not significantly affected by such variations because they only produce convective motions which have a negligible effect on the pulse travelling through the air with the speed of sound. There are, of course, distortions of the pulse due to such variations, and to wind systems which extend over considerable distances and distort the wave fronts whose passage we are studying. But that is a different mechanism from that which disperses each bit of the wave front in the same way.

Thus we can multiply the whole equation (7.3.16) by $\sigma\tau_0/\mathrm{i}$, which assumes that we can deal with each value of σ separately and add them together later on the grounds that the equations are linear and so solutions may be added together and still give a solution. Thus

$$\left(\frac{\cot\theta}{a^2}+\frac{\partial}{a^2\partial\theta}\right)\frac{\partial\varpi}{\partial\theta}+\frac{1}{a^2\sin^2\theta}\frac{\partial^2\varpi}{\partial\phi^2}$$
$$=-\tau_0\left(\frac{\partial}{\partial z}-\frac{g}{c^2}\right)\left(\frac{\sigma^2}{\sigma^2\tau_0+g\tau_0'}\frac{\partial\varpi}{\partial z}\right)-\frac{\sigma^2}{c^2}\varpi \qquad (7.3.17)$$

and the dependence on z (right-hand side) is separated from the horizontal variation. The second term may be put equal to zero for the case in which the disturbance is centred upon the axis $\theta=0$, and is independent of ϕ, and $v=0$. The variations in the direction of θ increasing are separated from the upward variations (z increasingly), and each side of the equation (7.3.17) must be equal to the same constant (independent of z and of θ). Thus we may write

$$\varpi=\overline{\varpi}(\theta)\varpi(z)$$

and if the constant is put equal to k^2 we have, by putting $r=a\theta=$ the great circle distance from the pole,

$$\left(\cot\frac{r}{a}+\frac{\partial}{\partial r}\right)\frac{\partial\overline{\varpi}}{\partial r}=k^2\varpi \qquad (7.3.18)$$

and

$$\varpi''-\left(\frac{g\tau_0''+\sigma^2\tau_0'}{g\tau_0'+\sigma^2\tau_0}+\frac{g}{c^2}\right)\varpi'+\left(\frac{g\tau_0'}{\tau_0}+\sigma^2\right)\left(\frac{1}{c^2}-\frac{k^2}{\sigma^2}\right)=0. \qquad (7.3.19)$$

With these two equations the boundary conditions determine a solution which defines the disturbance throughout the whole atmosphere. In earlier treatments, which were incomplete, Taylor [24] and Pekeris [25] used the divergence rather than the modified pressure as the dependent variable. The choice in this later work was based on the special interest in the pressure at ground level, and is more related to other work in this book.

The pulse is imposed by determining the boundary conditions on the velocity, for

$$u = \frac{\mathrm{i}}{\sigma\tau_0}\frac{\partial\varpi}{\partial r}, \qquad w = \frac{\mathrm{i}\sigma}{\sigma^2\tau_0 + g\tau_0'}\varpi'. \tag{7.3.20}$$

To obtain numerical results it is necessary to choose a particular model of the atmosphere, and this was chosen (Scorer [16]) to have a constant lapse rate of temperature up to the tropopause above which the atmosphere was considered to be isothermal. The variables were transformed so that the equations could subsequently be solved to get a relationship between k and σ by numerical methods using a hand desk calculator. There was necessarily an upper boundary condition that the disturbance should decrease exponentially to zero at great height.

If we had studied the propagation of a disturbance in the form of a plane wave (in a vertical plane) across a plane surface with the same representation of gravity, the components would have separated in the same way as in this case. It turns out that the equation (7.3.19) requires a condition at great height, and this must require that the disturbance be exponentially small. When this is applied and the particular profiles described in the last paragraph are used (tropopause at about 10 km, isothermal atmosphere above and a tropospheric lapse rate of 5.6°C per km, which roughly represents the atmosphere in middle latitudes) we find that the frequency must be less than σ_c, which is defined by

$$2\pi/\sigma_c = 111 \text{ s}. \tag{7.3.21}$$

The exact figures of the calculation can be obtained from the original paper, but it must be remembered that the tropopause height can normally vary from about 6 km in polar regions to about 17 km in equatorial latitudes, so that the value represents only temperate areas in this case.

For a given frequency (or wavelength) a wavelength (or frequency) was calculated to fit the boundary conditions, and the result represented the dispersive property of the atmosphere. The value given in (7.3.21) gives the **cut-off frequency**, σ_c, and shorter wavelengths than about 35 km are not trapped in this model. This wavelength may seem rather large, and the oscillations computed and displayed in the figures seem to have an upper limit of about 2 minutes for the periods of oscillation. Indeed it appears that Whipple contrived, in Fig. 7.2.iii, to fit the records of the microbarographs to that period, with all the more rapid oscillations placed in the later part which he recognised as being reflected from much higher altitudes. The waves caused by dispersion of a pulse, which was originally a single outward displacement at a very powerful shock wave, gradually lengthen with distance travelled.

In the case chosen for study in 1949 the wave velocity varied almost linearly with frequency and the change as the pulse travelled in the horizontal direction was represented by Bessel's equation for radiating outwards, according to equation (7.3.18), in circles. The relative intensity of the pulse decreased with distance from the source, taken as the pole, to the 'equator'.

The initial explosion was represented by a vertical velocity at the ground close to the source which could introduce a chosen volume instantaneously or gradually into the atmosphere. Thus the upward velocity was given by

$$w_1 = B\int_0^\infty \mathrm{d}\sigma \int_0^\infty \mathrm{d}k\, \Sigma(\sigma,t)K(k,r)\mathrm{e}^{ikr}\left(\mathrm{e}^{\mathrm{i}\sigma t} + \mathrm{e}^{-\mathrm{i}\sigma t}\right) \quad \text{real part only.} \tag{7.3.22}$$

The functions Σ and K gave to w_1 a finite value near to $r = 0$, $t = 0$ and a negligible value elsewhere. If

$$w = W(z); \qquad \varpi = \Psi(z) \tag{7.3.23}$$

are solutions of equations (7.3.19) and (7.3.20) which satisfy the upper boundary condition, then the solution is

$$w = B\int_0^\infty \mathrm{d}\sigma \int_0^\infty \mathrm{d}k\, \Sigma K\, \mathrm{e}^{ikr}\left(\mathrm{e}^{\mathrm{i}\sigma t} + \mathrm{e}^{-\mathrm{i}\sigma t}\right)\frac{w}{W_1}$$

and

$$\varpi = B\int_0^\infty \mathrm{d}\sigma \int_0^\infty \mathrm{d}k\, \Sigma K\, \mathrm{e}^{ikr}\left(\mathrm{e}^{\mathrm{i}\sigma t} + \mathrm{e}^{-\mathrm{i}\sigma t}\right)\frac{\Psi}{W_1} \tag{7.3.24}$$

and the pressure fluctuation is obtained from the equation (7.3.7).

The calculation first of all involves a relationship between σ and k, which has to be found numerically. For small r the equation is Bessel's equation of order zero and an instantaneous point source at $r = 0$, $t = 0$ is represented by the well-known integral (used by Pekeris)

$$\varpi_1 = 2B\int_0^\infty \cos\sigma t\, \mathrm{d}\sigma \int_0^\infty J_0(kr)k\, \mathrm{d}k \tag{7.3.25}$$

and the volume introduced is

$$\int_0^\infty \mathrm{d}t \int_0^\infty 2\pi r w_1\, \mathrm{d}k = 8\pi^2 B \tag{7.3.26}$$

and so

$$p_1 = \frac{1}{4}\pi\rho_{01}B\int_0^{\sigma_c} F(\sigma)J_0(kr)k\left(\mathrm{e}^{\mathrm{i}\sigma t} + \mathrm{e}^{-\mathrm{i}\sigma t}\right)\mathrm{d}\sigma. \tag{7.3.27}$$

Adjustments have to be made when r is large, and $r^{-1/2}$ has to be replaced by $(a\sin\theta)^{-1/2}$ because the wave is expanding over a sphere and not cylindrically over a flat plane.

To evaluate the integral (7.3.27) using the principle of stationary phase* which had been suggested by Pekeris for this problem, it would have been necessary to evaluate k'', by which is signified $\partial^2 k/\partial\sigma^2$. But we are interested in the long wavelengths because the short wavelengths are not trapped by our model atmosphere, and anyway the longest wavelengths travel fastest and so they contribute the initial, and most easily detectable, part of the pulse arriving at large distance.

* As set out by Lamb (1932 edn art 241).

The principle of stationary phase is valid, as usually described, only if

$$|k'''| \ll |k''|, \tag{7.3.28}$$

but this is not satisfied because $k'' \to 0$ as $\sigma \to \infty$. However, $k^{\text{iv}} \to 0$ also, and so the method is adequate if $kx - \sigma t$ is expressed as a cubic in σ instead of the usual quadratic. But this required the relationship between σ and k to be evaluated numerically to at least 7 significant figures—a trivial exercise using modern high-speed computers—so that the third derivative could be obtained. The other details of the method are similarly dated items of numerical analysis such as calculating k^2/σ^2 at equal intervals of σ^2 to obtain a smooth tabular function differentiable by formulae involving numerical differences.

The results are summarised in Figs 7.3.v–ix.

One result that emerged was that the meteorite communicated to the atmosphere as much energy as would be represented by an explosion of about 1 km^3. It is noticeable that at distances of more than 10 000 km, one quarter of the earth's circumference, the amplitudes of the second and subsequent oscillations are increasing with distance because the belt round the earth occupied by the pulse is contracting towards the antipodes of the explosion. But because of the irregularity in the profile of the atmosphere the pulse must become much more ragged and incoherent than what the calculation represents.

No dissipation has been included in the calculated pulse forms. It is probable that the total energy generated by the impact was a few times greater than that which went into the pulse. The original paper (Scorer [16]) and those referenced therein contain many more details of interest.

7.4 BAROTROPIC MOTION AND ROSSBY WAVES

Air motion is described as **barotropic** when it is the same at all heights above the ground. Thus there is no thermal wind in such flow (see section 4.5). Nor is there any shearing in horizontal surfaces. The model imagined as being barotropic is quite impossible to realise as persisting for a significant time or to find exemplified in the real atmosphere. The idea is based on the possibility that the flow in the atmosphere might persist, or that the closest possible similarity to actual flow might persist for a useful length of time. Thus a 'barotropic model' was constructed, and assuming that each separate column of air retained its vorticity in the vertical direction, it was possible by numerical means to calculate the distribution of vorticity at a later moment. It is arguable that this model, by introducing a specific calculation, would give an improvement over the contemporary methods of extrapolation of the current weather situation and lead to better forecasts. But by comparison with the sophisticated methods of Sutcliffe described in Chapter 4, there was little to entice forecasters into the barotropic trap.

One particular idea taken from the barotropic model is of great interest, namely the idea of wavelike oscillations due to variation with latitude of the Coriolis force, since called **Rossby waves**. The theory depends on the fiction that an airstream remains barotropic when it flows to a different latitude. Thus the total vorticity component in the vertical direction is assumed to remain constant, and this is approximate because the

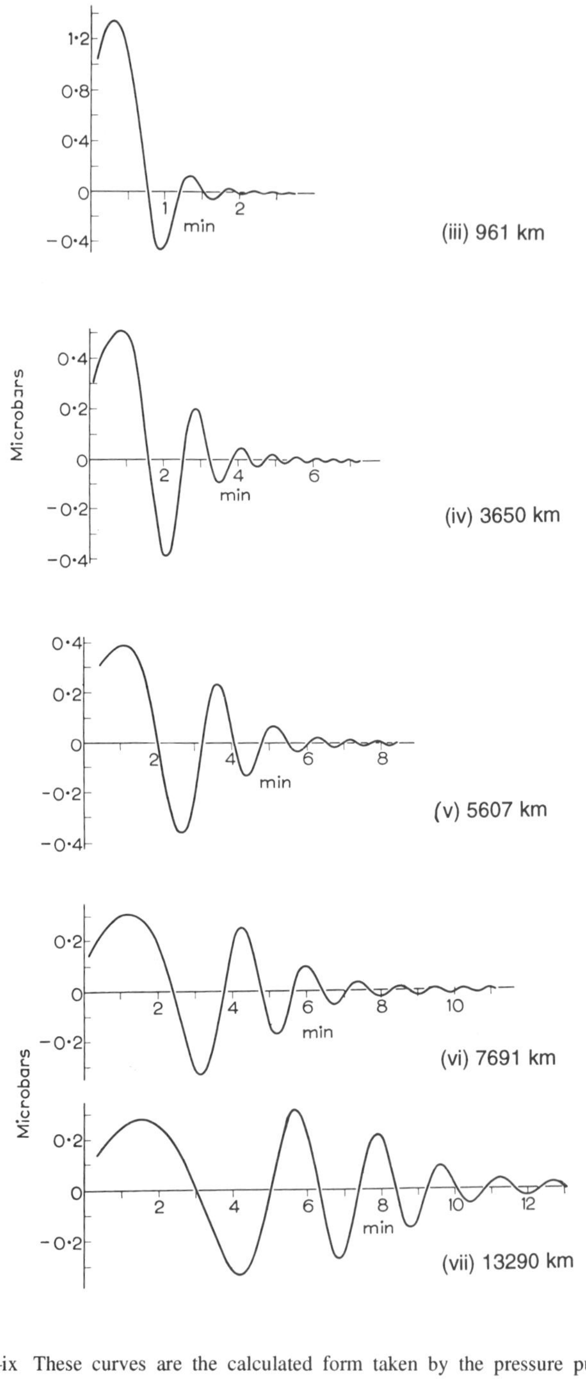

Figs 7.3.v–ix These curves are the calculated form taken by the pressure pulse due to the Tunguska explosion at various distances over the earth's surface. In the first case the pulse lasted about 2 minutes and extended over about 40 km of air and travelling through it at about 320 m s^{-1}.

horizontal component of the earth's rotation is conventionally ignored because it produces only a vertical force which is very small compared with the gravitational force.

The absolute vorticity is composed of two parts: the vorticity, ζ, relative to the axes fixed to the earth, and the vertical component of the earth's rotation in space, f. And so we can say

$$\zeta + f = \text{constant along a particle path.} \qquad (7.4.1)$$

Every parcel of air thus carries its absolute vorticity with it.

The two-dimensional flow in lee waves follows the 'same' equation, which merely states that because there is no component of motion in the y-direction, that component is changed only by the horizontal stratification of density. But we also make the Boussinesq approximation and neglect the effect of that stratification except where it is multiplied by gravity (which is equivalent in this case to saying that gravity contributes significantly more to the pressure variations along a streamline where the height changes, than is represented by the change in stagnation pressure) and so equation (5.2.6) becomes

$$\eta - \frac{g\beta}{u_0^2} = \text{constant along a streamline.} \qquad (7.4.2)$$

We have seen that rotors may occur and be correctly represented as part of large amplitude waves. Furthermore, no assumptions have been made in deriving equation (7.2.1) and so large amplitude waves may be said to occur in this kind of barotropic motion, and may be expected to contain rotors. These are observed as **blocking anticyclones**.

In equation (7.4.1) the coordinates may be chosen with x measured eastwards and y measured northwards, and so locally we may have

$$f = 2\Omega \cos\theta = \beta y \qquad (7.4.3)$$

where Ω is the earth's angular velocity, and the coordinate y is equivalent to $-z$ in equation (7.4.2), and θ is the colatitude. There is therefore a close analogy between these two wave situations. The Rossby waves over the Atlantic Ocean may be thought of as waves in the lee of the Rocky Mountains, or of the disturbance caused in the westerly airflow across the American East Coast and/or the Gulf Stream. β is the conventional symbol for the latitudinal variation of the Coriolis parameter. In some treatments the theory of Rossby waves is treated as if there were a restriction to small amplitude, which is not the case.

The wavelength of Rossby waves is $2\pi\left(\frac{U}{\beta}\right)^{\frac{1}{2}}$ where U is the wind speed and $\beta = \partial f/\partial y$. Because Rossby waves are usually set up by large geographical features they are most likely to be roughly stationary; for this reason the anticyclones appear to divert the flow around them, and so are called blocking anticyclones. In so far as blocking cyclones on the equatorial side of an airstream have not been described, this is because they are much less common and also because they have not been looked for as such. Also β and its variations is of smaller magnitude.

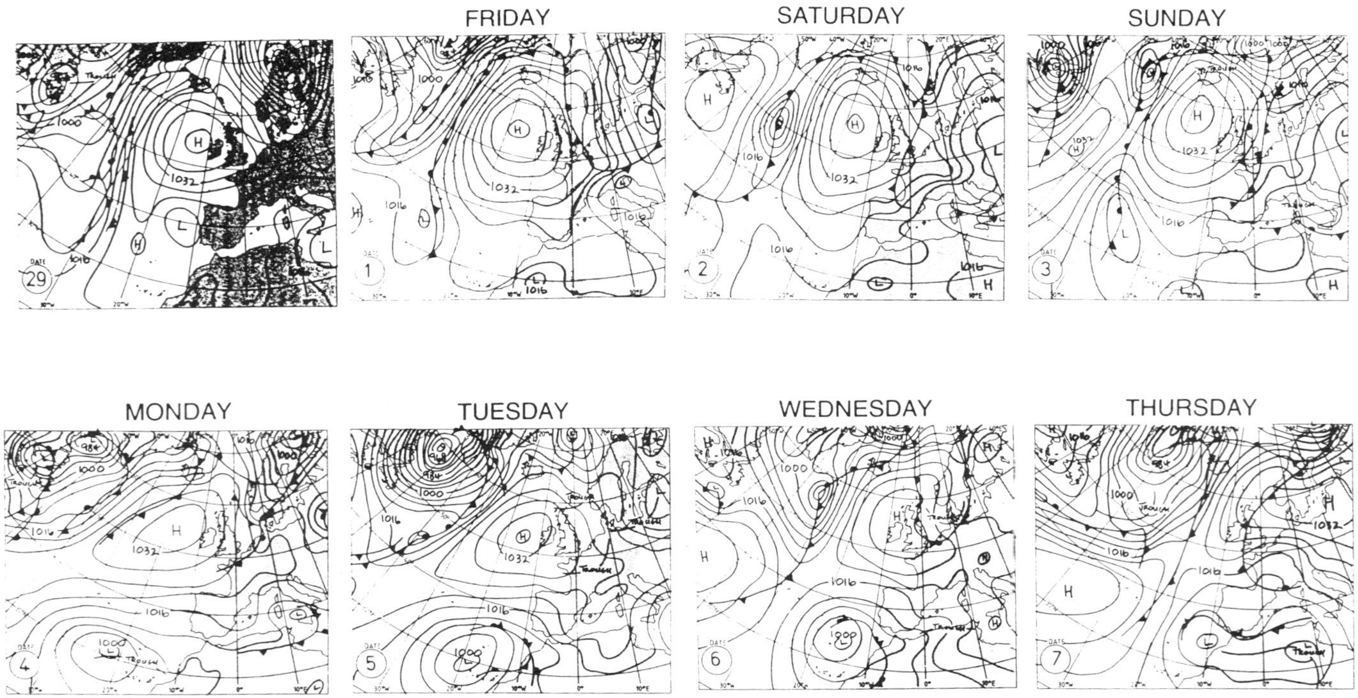

Fig. 7.4.i (Blocking weather features) Midday charts, 29 Feb–7 March 1996. Blocking anticyclone, further north than the climatological Azores anticyclone seen on the chart of 7 March. This blocking high pressure lasted, typically, about a week in the same place, with cyclones travelling in the wave around the north side.

7.5 STORM PULSES

When a strong downdraught is produced by a large thunderstorm, and it descends from a layer moving horizontally with a different velocity from the surface air, it delivers a pressure pulse to the lower layer and this travels as a gravity wave. Such a pulse will only be observed at a large distance if it is trapped in the lower layers in the manner described in section 5.11(2) for lee waves. These conditions are commonly satisfied by the temperature profile, but a factor most likely to determine the direction for trapped propagation is the wind profile. Thus they would be likely to travel against a strong wind at higher levels.

An unusually good example is illustrated in Fig. 7.5.i, which shows the variations of surface wind speed and direction which passed over Abingdon (south of Oxford) early on 5 July 1952 as a result of a pulse generated in a storm in the English Channel during the previous night. This phenomenon is not to be confused with a cold outflow which is the front of an outflowing cold air mass, and more like a sea breeze front (see section 11.9).

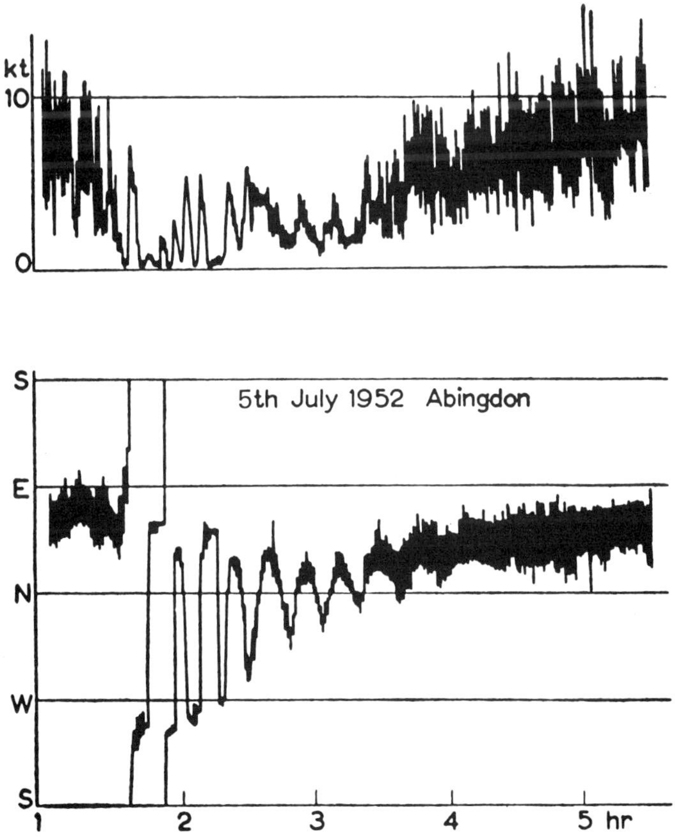

Fig. 7.5.i The wind speed and direction recorded at Abingdon at the passage of gravity waves emanating from nocturnal thunderstorms over the English Channel. [Reproduced by courtesy of the late I. J. W. Pothecary, Meteorological Office.]

Storm pulses are considerably dispersed because of the large range of the variation of wave speed with wavelength. Reference may be made to the 'morning glory' in this connection, because it travels rather like a bore with the appearance of dispersion of the energy behind the leading wave (see *Spacious Skies*, p. 144) but does not show normal dispersal effects.

7.6 THE SEMI-DIURNAL PRESSURE WAVE

The planet **Mercury** always has the same face towards the sun, and the moon always has the same face towards the earth, because the gravitational field of the larger body exerts gravitational forces which distort the shape of the smaller, which cause changing forces which dissipate the energy of rotation until the shape remains almost constant. Although the kinetic energy of rotation is dissipated, the angular momentum of the earth–moon system remains constant: thus although the moon has lost angular momentum of rotation about its own axis, the moment of its momentum about the earth is correspondingly increased and it is now further away from the earth than it once was.

Venus, which has no moons, is rotating so slowly that it takes about 243 Earth days to complete one rotation about its own axis, but completes one rotation around the sun in 224.7 earth days. Further it is rotating in the retrograde direction about its own axis, unlike all the other planets except Uranus, whose axis is approximately in the plane of the ecliptic. Venus has no liquid ocean but may have tidal effects in its atmosphere because the winds have been recorded as causing clouds in the upper atmosphere to encircle the planet in about 4 earth days. Nor did we know whether or not its rotation is in the last stages of slowing down, or has been zero during a recent epoch. The mass of the atmosphere gives it a surface pressure about 90 times that on earth. Until we know how the winds are caused we have little hope of finding any waves that there may be in the atmosphere.

Mars has one rotation in just over 24 earth hours, but it is much further away from the sun and has neither ocean nor massive atmosphere. Although there are occasional clouds, particularly wave clouds, it is possible that the white winter polar caps are made largely of solid CO_2. The atmospheric mass is about a tenth of the earth's and there is no indication that the rotation is being appreciably slowed by tidal friction. A very small planet at a large distance from the sun is less subject to the effects of the gradient of the sun's gravity. The two moons of Mars are small and complete their orbits in just over 7.5 h at a distance of 23 500 km. Because of the low gravity, the hydrogen has long escaped from the atmosphere, which is now composed mostly of CO_2. There is no liquid water on the surface and all that is visible is occasional clouds or as part of the polar ice cap. There may be water underground; but if any did appear on the surface it would be quickly frozen at night or evaporated by day.

There are tides in the **earth's** ocean due to the combined gravitational filed of the moon and the sun. The moon's component is greater than the sun's owing to the greater variation in the gravitational field; and this is quite evident in the ocean tides. The atmospheric tide due to the moon's gravity is only a very small fraction of a millibar. The solar semi-diurnal tide has an amplitude of the order of 2 mb in equatorial areas. This unexpected outcome was at first explained by suggesting that the atmosphere was selectively resonant to the solar frequency which was therefore greatly amplified (Wilkes and Weekes [26]). This seemed to be unlikely because of the actually very small lunar component, and Kelvin suggested that the solar semi-diurnal tidal wave was caused by solar heating, which seems very reasonable. The resonance suggestion is relevant because it was discovered that simple atmospheric models possessed a resonant period of 12 h. Even so there seems to be such great damping of the wave by wind systems and very great variation of the wave outside the tropics where the winds are most variable (Fig. 7.6.i).

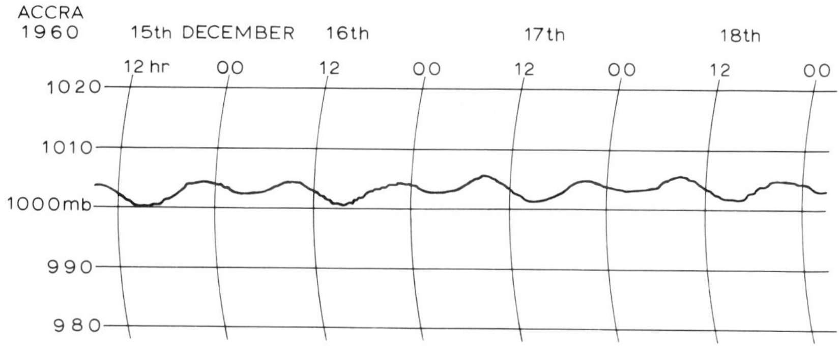

Fig. 7.6.i A typical tropical barogram showing the pressure variations due to the solar semi-diurnal tide in the atmosphere.

Many purists who have tried to find a resonant period which fits the solar semi-diurnal tide very accurately, but does not amplify the lunar tide at all, have lost interest because the phenomenon has not been fully understood. What remains obvious is that the couple of the solar gravitational field on the semi-diurnal solar atmospheric tide is such as to accelerate the earth's rotation. The tide does not alter position with the phase of the moon and so its influence on the earth's rotation averages out to zero. The height of the ocean tide is not well enough known globally to reveal the magnitude of the couple, but it is certain that the blockage of the tide by the land has a decelerating effect on the earth's rotation through the combined gravity field of the sun and moon. However, the recent development of a satellite from which the height of the ocean surface can be measured to an accuracy of about 15 cm may make it possible soon to measure the gravitational couple of the sun and moon and estimate their **decelerating** effect.

It is possible that the **accelerating** effect of the sun on the atmospheric tide may have caused the slowing down of the earth's rotation to cease in recent epochs. The length of the day varies according to the rotational momentum of the atmosphere, because the total angular momentum of the earth (solid, oceanic and atmospheric) is assumed to be constant from day to day. At the astronomical level of accuracy the variations are attributable to the changes in atmospheric angular momentum, and daily forecasts of the length of the day based on contemporary global numerical models are available to astronomers.

Thus several research problems are seen to be in the in-tray. The tidal waves are significantly affected by the Coriolis forces, but the equations of motion separate in the same way as for the problem of the dispersed pulse described earlier in this chapter, and while modern computers make the solution of such problems much easier, the geographical variations of the static condition of the atmosphere presents thc same difficulty as before.

7.7 EFFECTS OF TIDAL FRICTION

The tidal period in the oceans is determined by the interaction of the solar and lunar periods. The spring and neap tides occur when the gravitational fields are acting together or in opposition respectively. The period between the spring and neap tides is determined by the length of the lunar month, and this is determined by the radius of the lunar orbit. Because the gradient of the moon's gravitational field exceeds that of the sun in the earth's oceans, an examination of the tidal deposits in deltas variations in the spring tidal period can be deduced from the thickness of the deposits. In order to assign a date to the laminations in what is now a sedimentary rock we need to place the rock in the geological timetable and compare it with other similar deposits elsewhere on the earth. Thus the laminations provide detail about what took place in that era, which is hundreds of millions of years before the present, just as the cores obtained from ice in Antarctica and Greenland and from mineral and biological deposits on ocean beds are used to discover climatic events of the past hundreds of thousand years.

From such deposits, Sonett *et al.* [27] deduced that the length of the year was about 400 days during 345 to 395 Ma BP. If the earth's orbit round the sun has remained unchanged, which is likely because the sun's gravitational field is not synchronised with that of the moon, and the tides are synchronised with the latter, only the moon exerts a consistent couple on the earth's tides, which reduces the rotation speed of the earth and transfers the angular momentum to the moon, which in turn increases the radius of the moon's orbit.

During about 900 Ma the distance of the moon has increased by about 4000 km (a mean retreat rate of about 3.25 cm a year) and the earth day has increased from about 18 to 24 h. This calculation allowed for some frictional drag due to the sun's gravitational field, but is bound to be rough because of the movement of the continents, which must have affected the drag due to the moon through possible resonance of the tidal wave in the oceans of the time. By comparison with the known factors which have been only roughly accounted for, there is also the possibility that the moon's rotation speed about its own axis was not exactly one lunar month (with the same face towards the earth). The original paper describes in considerable detail the approximations which had to be made to be in accord with the different observation sites which were in different positions with respect to the tidal waves surging around the continents of the time.

8

Instability of vortex flows

'*Vot is billows?*'—G. F. Handel.

8.1 INSTABILITY OF A VORTEX SHEET—'K–H INSTABILITY'

The classical discussion of the instability of a horizontal surface at which the density and velocity are discontinuous is as follows:

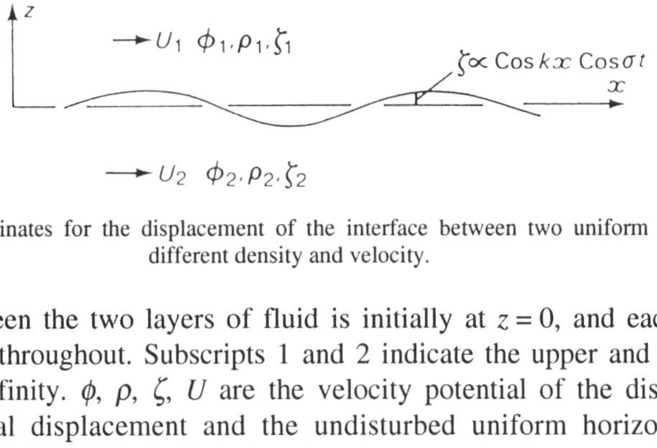

Fig. 8.1.i Coordinates for the displacement of the interface between two uniform layers of different density and velocity.

The interface between the two layers of fluid is initially at $z = 0$, and each layer is in irrotational motion throughout. Subscripts 1 and 2 indicate the upper and lower layers, which extend to infinity. ϕ, ρ, ζ, U are the velocity potential of the disturbance, the density, the vertical displacement and the undisturbed uniform horizontal velocity respectively, and the displacement of the interface is assumed to be proportional to cos kx and to cos σt, represented by $e^{i(kx-\sigma t)}$, the real part of the equation being taken (Fig. 8.1.i)

The fluids are supposed incompressible so that the equation of continuity is

$$\text{div } \mathbf{grad}\, \phi = \nabla^2 \phi = 0. \tag{8.1.1}$$

Since the disturbance is assumed to be small, so that all powers of the disturbance higher than the first can be ignored, the Lagrangian operator is, for two-dimensional motion, with velocity $(U + u, w)$

$$\frac{D}{Dt} = \frac{\partial}{\partial t} + (U+u)\frac{\partial}{\partial x} + w\frac{\partial}{\partial z} = i(-\sigma + kU) \tag{8.1.2}$$

when applied to the disturbance, which is taken to be proportional to $e^{i(kx-\sigma t)}$. To satisfy (8.1.1) the disturbance must be proportional to $e^{\pm kz}$, the positive sign holding in the lower layer and the negative sign in the upper layer; otherwise the disturbance would be infinitely large for large positive or negative z. From this it follows that

$$w = \frac{\partial \phi}{\partial z} = \frac{D\zeta}{Dt} = -i(\sigma - kU)\zeta \tag{8.1.3}$$

$$\begin{aligned} &= -k\phi \qquad \text{for } z > 0 \\ &= \ \ k\phi \qquad \text{for } z < 0. \end{aligned} \tag{8.1.4}$$

and so

$$\phi_1 = i\frac{\sigma - kU_1}{k}\,\zeta_1; \quad \phi_2 = -i\frac{\sigma - kU_2}{k}\,\zeta_2. \tag{8.1.5}$$

By Bernoulli's equation (1.7.9) with $\rho =$ constant and

$$q^2 = \left(U + \frac{\partial \phi}{\partial x}\right)^2 + \left(\frac{\partial \phi}{\partial z}\right)^2$$

we have

$$\frac{p_1}{\rho_1} + U_1\frac{\partial \phi_1}{\partial x} + \frac{\partial \phi_1}{\partial t} + g\zeta_1 = \text{constant in the upper fluid,} \tag{8.1.6}$$

with a similar equation with subscript 2 for the lower layer. The constants U_1^2 and gz for each streamline (at height $z + \zeta_1$) are absorbed into the right-hand side, and the second-order terms in ϕ neglected for small disturbances. The variable part of the pressure is given by

$$-p_1 = \rho_1\left[i(kU_1 - \sigma)\phi_1 + g\zeta_1\right] = \rho_1\left(\frac{(\sigma - kU_1)^2}{k} + g\right)\zeta_1 \tag{8.1.7}$$

and

$$-p_2 = \rho_2\left[i(kU_2 - \sigma)\phi_2 + g\zeta_2\right] = \rho_2\left(\frac{(\sigma - kU_2)^2}{k} + g\right)\zeta_2. \tag{8.1.8}$$

We have already imposed sufficient boundary conditions at large positive and negative z, which would be slightly more complicated mathematically if they had been replaced by rigid boundaries at finite distance. In order to obtain the dispersion equation, which relates σ to k, or the wavelength to the wave velocity, we apply the physical constraints at the interface, namely

$$p_1 = p_2, \quad \zeta_1 = \zeta_2, \quad \text{at } z = 0.$$

Consequently,

$$\rho_1\left(\frac{(\sigma - kU_1)^2}{k} + g\right) = \rho_2\left(-\frac{(\sigma - kU_2)^2}{k} + g\right), \tag{8.1.9}$$

which is a quadratic equation for σ of which the solutions are

$$\sigma = \frac{k}{\rho_1 + \rho_2}\left(\rho_1 U_1 + \rho_2 U_2 \left[\pm\ (\rho_1 U_1 + \rho_2 U_2)^2 - (\rho_1 + \rho_2)\left[\rho_1 U_1^2 + \rho_2 U_2^2 - \frac{g}{k}(\rho_2 - \rho_1)\right]\right]^{1/2}\right). \tag{8.1.10}$$

If σ turns out to be complex then the disturbance, which is proportional to $e^{i\sigma t}$, will contain a real positive exponential which indicates instability because the disturbance will grow without limit. When a finite size is achieved, the higher-order terms become important and the present mathematics is no longer applicable, and we shall study subsequent development by other means. Instability occurs when the quantity inside the larger square bracket in (8.1.10) is negative, and this is the case if

$$(\rho_1 U_1 + \rho_2 U_2)^2 < (\rho_1 + \rho_2)\left[\rho_1 U_1^2 + \rho_2 U_2^2 - \frac{g}{k}(\rho_2 - \rho_1)\right],$$

i.e.

$$\frac{(\rho_2 - \rho_1)(\rho_2 + \rho_1)}{\rho_1 \rho_2}\frac{g}{k} < (U_1 - U_2)^2. \tag{8.1.11}$$

This condition shows that there is always a range of wave number, k, giving unstable waves if $U_1 \neq U_2$. The condition (8.1.11) is fulfilled for all wave numbers if $\rho_2 < \rho_1$, which is obvious because it is the case of static instability. Instability in the statically stable case is called Helmholtz instability (often, but unnecessarily, coupled with the name of Kelvin and abbreviated to K–H).

When the discontinuity of density is of small magnitude and statically stable we may write

$$\rho_2 - \rho_1 = \Delta\rho, \quad \rho_2 + \rho_1 = 2\rho, \quad \rho_1\rho_2 = \rho^2 \tag{8.1.12}$$

and so if $U_1 - U_2 = \Delta U$, the condition (8.1.11) is that instability occurs when

$$k > k_c = 2g\frac{\Delta\rho}{\rho}\Big/(\Delta U)^2. \tag{8.1.13}$$

which introduces a form of Richardson number (see (2.14.4)).

Waves of length shorter than $2\pi/k_c$ are unstable, i.e. if

$$\mathrm{Ri} = \frac{2g\Delta\rho}{k\rho(\Delta V)^2} = \frac{k_c}{k} < 1.$$

This treatment has many shortcomings in practice. If the critical wavelength is very small the gradients of velocity are increased and viscous effects become important.

Furthermore, a discontinuity of velocity becomes smudged into a thin layer with a large gradient of velocity, or vortex layer. The treatment is not valid when the wavelength is small and comparable with the thickness of the layer, if only because we have assumed each layer to be in irrotational motion, of uniform speed, in the undisturbed state. The theory gives no indication of the nature of the development when the waves become large; and in the case of the surface of water ruffled by the wind the effects of surface tension can be overriding.

More sophisticated treatments by Drazin [28], Miles, Howard and Warren [29] and others have considered the case in which there is a transition zone from one layer to the other, and have arrived at criteria corresponding to (8.1.13) but with a lower limit for the wavelengths that are unstable. They thus find the maximum value for a quantity similar to that on the right of (8.1.13) for which some instability is possible. They are also concerned with criteria for the onset of instability. Our objective is different. We are concerned not with the exact criterion for instability but with its mathematical form so that we may study what has to happen in nature to bring the instability about. We are also concerned to know what the subsequent motion is like. This approach waas originated in 1971.

8.2 ANALYSIS OF VORTICITY DISTRIBUTION

We represent the vortex layer by a layer of thickness Δz in which there is a velocity gradient η, which is the vorticity component normal to the planes of the motion. There is also a density gradient represented by β, so that (see Fig. 8.2.i(a) and (b)).

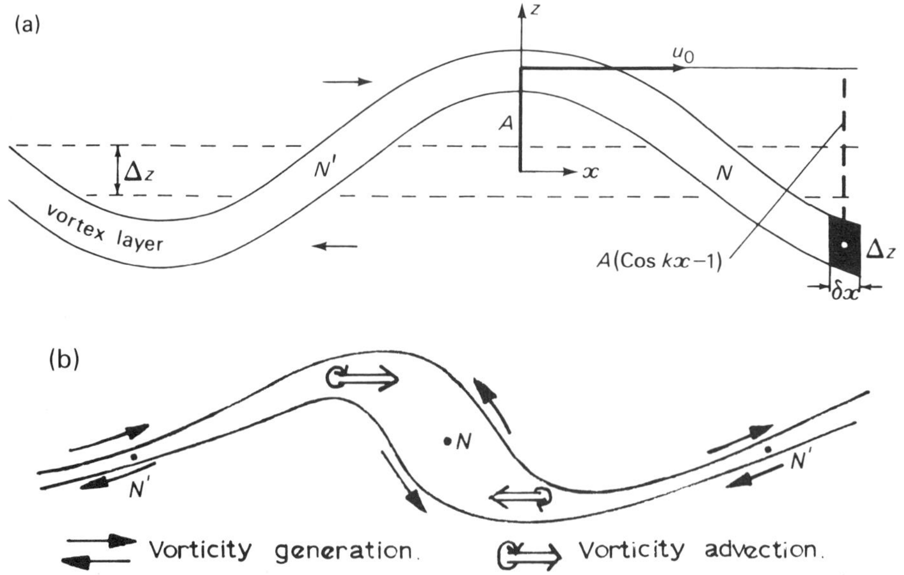

Fig. 8.2.i (a) & (b) Replacement of the discontinuity by a vortex layer with a sinusoidal displacement. The velocity of the crest of a wave is calculated from the vorticity. Vorticity accumulates at the downward nodes such as N, and moves away from N', as indicated in (b).

$$\Delta\rho = \beta\rho\Delta z, \qquad \Delta U = \eta\Delta z. \tag{8.2.1}$$

The layer is now subjected to a vertical displacement $\zeta = A\cos kx$, and to a first approximation a particle in the centre of the layer will be subject to a horizontal velocity $u_0 \cos kx$, having been initially at rest with the fluid moving in opposite directions above and below. This motion is due entirely to the vorticity contained in the layer, and the material of the layer is subject to its own velocity field.

The coordinates of an element of the layer, of length δx, relative to the crest of the wave which is at displaced height A, are $[x, A(\cos kx - 1)]$. It possesses vorticity $\eta\Delta z\delta x$, and imposes on the particle at the crest of the wave a horizontal component of velocity equal to

$$\frac{\eta\Delta z\delta z}{2\pi x}\,\frac{A(1-\cos kx)}{x}. \tag{8.2.2}$$

To obtain this we have assumed that $A(1 - \cos kx)$ is small compared with x. The velocity of the particle at the wave crest, denoted by u_0, is therefore

$$\begin{aligned} u_0 &= \int_{-\infty}^{\infty} \frac{\eta\Delta zA(1-\cos kx)}{2\pi x^2}\,\mathrm{d}x \\ &= \frac{\eta\Delta zA}{\pi}\int_{-\infty}^{\infty}\frac{\sin\frac{1}{2}kx}{x^2}\,\mathrm{d}x \\ &= -\frac{\eta\Delta zA}{\pi}\left[\frac{1}{x}\sin^2\frac{1}{2}kx\right]_{-\infty}^{\infty} + \frac{\eta\Delta zA}{2\pi}\int_{-\infty}^{\infty} k\frac{\sin kx}{x}\,\mathrm{d}x. \end{aligned} \tag{8.2.3}$$

The first term in this last expression vanishes because it represents the integral of an odd function of x which vanishes at infinity. The integral in the second term is a well known form and is equal to $k\pi$. The horizontal component of the velocity of the fluid in the vortex layer is therefore given by

$$u = u_0\cos kx = \tfrac{1}{2}\eta\Delta zAk\cos kx. \tag{8.2.4}$$

The vertical velocity at the crest is zero because the vorticity producing the motion is mirrored at the z-axis. This is an interesting disagreement with the classical treatment described above, which appears to allege that the amplitude of the waves grows but the shape (sinusoidal) remains unaltered. It may be noted that the wave velocity is zero if the two layers have undisturbed velocity $\pm U$, for if ρ_1 is roughly equal to ρ_2 the real part of σ is zero in (8.1.10). Also w is seen in (8.1.3) to be out of phase with ζ so that the vertical velocity of the crest is zero although it is growing exponentially. The growth mechanism is therefore not treated in the classical theory.

Since u in (8.2.4) varies in the horizontal direction, the vorticity carried by this velocity tends to accumulate at the downward-sloping nodes such as N in Fig. 8.2.i and is drawn away from the upward-sloping nodes N'. The rate of accumulation of vorticity per unit length of the layer is equal to $-\eta\Delta z\dfrac{\partial u}{\partial x}$, which is equal to

$$\tfrac{1}{2}(\eta\Delta z)^2 Ak^2 \sin kx. \tag{8.2.5}$$

At the same time the vorticity in the layer is being altered by the action of gravity on the sloping density gradient, at a rate represented by the term $\mathbf{R}\times\mathbf{g}$ in equation (1.4.1). This rate of creation of new vorticity is therefore $\beta g\dfrac{\partial\zeta}{\partial x}$ per unit are of the cross-section of the layer, which is a rate per unit length of layer equal to

$$g\beta\Delta z\frac{\partial}{\partial x}A\cos kx = -g\beta\Delta z Ak\sin kx. \tag{8.2.6}$$

If the sum of (8.2.5) and (8.2.6) is positive when $\sin kx$ is positive there will be an accumulation of vorticity at the downward-sloping nodes, towards which it is being advected. There will be a corresponding flight of vorticity from the upslope nodes. Since such a redistribution of vorticity will tend to make the crests move upwards and the troughs downwards, the disturbance will grow. Equally, if the sum of (8.2.5) and (8.2.6) is negative when $\sin kx$ is positive there will be an increase of vorticity at the upslope nodes and the crests will be moved downwards. Hence the disturbance will not grow, but will travel as a stable wave or oscillate about the equilibrium undisturbed position. The condition for instability is therefore

$$\tfrac{1}{2}(\eta\Delta z)^2 k > g\beta\Delta z$$

or

$$k > k_c = 2g\beta/\eta^2\Delta z. \tag{8.2.7}$$

It is necessary to add that a thick vortex layer is not unstable for very short wavelengths. If we imagine that a mechanism is operating which gradually increases η, the vorticity of the layer, the wavelength which will become unstable first will be at the long wavelength end of the unstable range represented in (8.2.7). This is, of course, the same condition as (8.1.13) because of (8.2.1). If we now call the thickness of the vortex layer h, (8.2.7) implies that

$$\frac{1}{2}k_c h = \frac{g\beta}{\eta^2} = \text{Ri} \tag{8.2.8}$$

where Ri is the Richardson number of the layer (see section 9.5).

The classical studies of Miles, Howard and Warren [29] have shown that instability cannot set in unless the Richardson number is less than 0.25 somewhere in a fluid, and so if it is roughly equal to this value we have that, roughly,

$$\frac{2\pi}{k_c} = 4\pi h. \tag{8.2.9}$$

This gives a crude estimate of the wavelength most likely to appear in a vortex layer of thickness h. The mechanism whereby the Richardson number is decreased so as to cause instability is discussed later in this chapter.

Any calculation leading to a conclusion like (8.2.9) must of necessity be very rough because actual layers do not have a uniform Richardson number throughout but one that varies through the depth of the layer. Criteria about instability must depend not only on the value of the Richardson number but on its distribution. This is because there is no necessary instability in a stream in which $Ri < 0.25$ everywhere, the obvious example being the case of uniform density and uniform shear in which $Ri = 0$ everywhere. The instability arises, according to this analysis, because there is a concentration of vorticity into a layer. The vorticity is collected into the nodes by the instability and the layer is rolled up into **billows**, which is the name given to this phenomenon to distinguish it from gravity waves caused by an outside disturbance such as a hump in the boundary, or violent convection. **Billows** always occur several at a time and, as we shall see, are the principal cause of mixing in a stably stratified medium.

The observed phenomenon, very common in cloud forms, is striking because of the sudden appearance of a disturbance of a clearly defined single wavelength. There is not an existing unstable situation subject to a fortuitous perturbation leading to the sudden appearance of the instability. Rather, a gradual destabilisation takes place, and the disturbance grows as soon as a very small range of wavelengths becomes unstable. Viscosity plays no significant role in the determination of the smallest wavelength that is unstable in the atmosphere or ocean where the phenomenon is very common (see Woods and Wiley [31]). Even if the destabilisation continues after the first unstable waves have appeared, the growth rate will simply be increased, other wavelengths will not appear because the initial unstable wave has a start on them and will be completely dominant because the growth is exponential.

8.3 THE GROWTH OF THE CAT'S-EYES PATTERN

We shall now make a rough study of the development of the unstable waves with a view to getting an estimate of the final outcome. According to the classical theory the waves grow in amplitude only. We have seen that the growth is caused by, and itself causes, the accumulation of vorticity at the nodes in which the upper fluid is moving down the slope. A series of experiments by Thorpe [30] showed many of the important features of billows which are not easy to observe in nature. Woods [32] demonstrated their existence under water, and Scorer [33] has illustrated the various cloud forms that are created by billow motion.

As a result of the stretching of the vortex layer at the upslope nodes, the rotation is intensified at the downslope nodes towards which the vorticity is carried. The subsequent evolution is rapid, and is indicated in Fig. 8.3.i.

At the third stage shown the fluid at A is denser than that at B and so the fluid in the rolled-up vortices becomes statically unstable and becomes mixed up. In the final stage the cat's-eyes pattern is achieved. The mixed fluid is gathered into discrete rolls with a wave motion above and below, in which the phase of the waves has shifted a quarter of a wavelength downstream on each side so that the wave is of opposite phase above and below.

In the third stage it is possible to observe a considerable winding up of the vortex into a sort of 'swiss roll', especially when the density difference is small and the shear large.

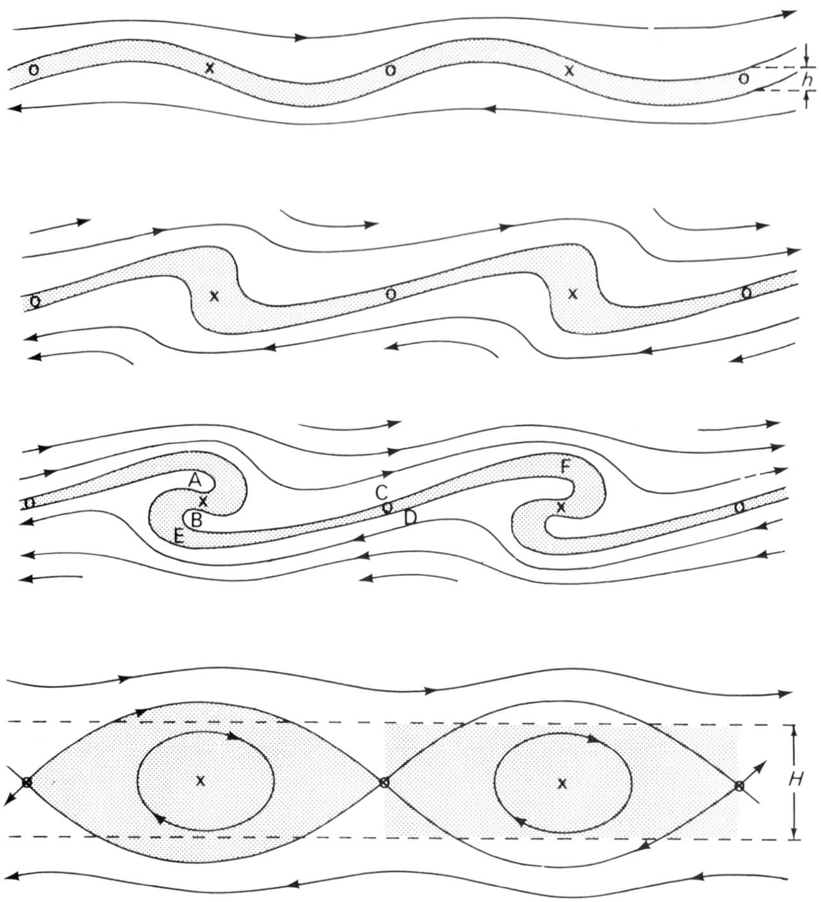

Fig. 8.3.i The evolution of billows from a sinusoidal perturbation of the vortex layer, which rolls up at the downslope nodes. The fluid in the rolls is gravitationally unstable and soon begins to mix up to form discrete lumps of mixed fluid in the cat's-eyes pattern shown in the last diagram: eventually this fluid flattens out under gravity into a layer of depth H indicated on the right of that diagram.

When the density difference is large, mixing begins soon after stage 2. In the theory the formulae would be modified by the introduction of a factor tanh kd, d being the depth of one of the fluid layers if it were of finite depth. When $kd > 1.5$, tanh $kd = 1$ with an error of less than 10 per cent, and so if the depth of the layer exceeds about a quarter of a wavelength the presence of a rigid boundary has little effect on the wavelength. In Thorpe's experiments this effect was present in shallow layers, but in deeper layers the wavelength was determined by the thickness of the vortex layer, as would be expected from the theory.

In the final stage the vorticity is obviously distributed throughout the mixed fluid in a manner that cannot be determined. The flow outside is not much affected because the amount of the vorticity is the same, whatever its distribution. Therefore it is not a bad approximation to assume that it is all collected at the centres of the cells, and on that basis the classical formula for the motion of a row of vortices can be used. The complex potential is given by Lamb ([35], Art. 156) to be

$$\phi + i\psi = -2iA \log \sin \tfrac{1}{2}k(x + iz) \tag{8.3.1}$$

and the horizontal and vertical velocities are

$$U + u = Ak \frac{\sin\ h\,kz}{\cosh kz - \cos kx}$$
$$\rightarrow Ak \operatorname{sgn} z\left[1 + \exp\left(-k|z|\right)\cos kx\right] \tag{8.3.2.}$$

for large z, and

$$w = Ak \frac{\sin kx}{\cosh kz - \cos kx} \rightarrow Ak \exp\left(-k|z|\right)\sin kx \tag{8.3.3}$$

for large z, U being the velocity which transports the fluid horizontally and u the part dependent upon x. The value of U at a distance of a few wavelengths above and below the pattern does not affect the motion as tanh $kz \rightarrow 1$. Above and below the pattern the motion is sinusoidal except very close to the cell boundary. Lamb discussed the stability of the pattern and showed that any disturbance which takes one or more of the vortices out of line will grow. This is because they would be subject to a velocity transporting them further out of line in exactly the same way as we saw that the distortions of the vortex layer grow. When there is no stabilising density gradient, as in Lamb's case, disturbances of all wavelengths are unstable; but in our case, where there is a stable density gradient, and in the atmosphere and ocean where the layers are usually stably stratified throughout, the unstable wavelengths are excluded. The point of noting Lamb's analysis is to obtain a rough estimate of the volume within the closed circulations. It turns out (Scorer [34]) that the streamlines in the final stage have equation

$$z = \frac{1}{k}\cosh^{-1}\left[\exp(\psi/A) + \cos kx\right] \tag{8.3.4}$$

and those marking the boundary of the circulations are given by

$$\psi = A \log 2. \tag{8.3.5}$$

The centres of the circulations have $\psi = -\infty$, but of course in practice they rotate solidly near the centre with the same total vorticity. If the fluid becomes mixed within the circulations and then subsequently flattens out as a horizontal layer of thickness H, equating the cross-section areas before and after

$$\pi H = \int_0^{\pi/k} 2\cosh^{-1}[2 + \cos kx]\,dx \tag{8.3.6}$$

from which

$$\frac{2\pi}{k} \approx 2.7\,H, \tag{8.3.7}$$

and comparing this with (8.2.9) we have

$$H = 4.6\,h. \tag{8.3.8}$$

The effect of the whole evolution, therefore, is to multiply the thickness of the vortex layer by the factor 4.6, and to mix it up into a uniform fluid, with a fairly large density step at the top and the bottom of the new layer.

The largest density gradient produced in the sequence of events is first around the upslope node, the fluid between C and D in stage 3 being stretched and thinned. Later a large density discontinuity is produced all the way from E to F on each wave, and it is that shape that is clearly seen by radar when billows occur in the atmosphere (see Fig. 8.3.ii). Still later the vortices become visible when the mixing first begins, and large density gradients are produced in the early stages of mixing by the juxtaposition of masses of different density. Finally the fluid within the cat's-eye circulations becomes more uniform, and the radar sees the outer boundary of the cells where the density increment becomes concentrated.

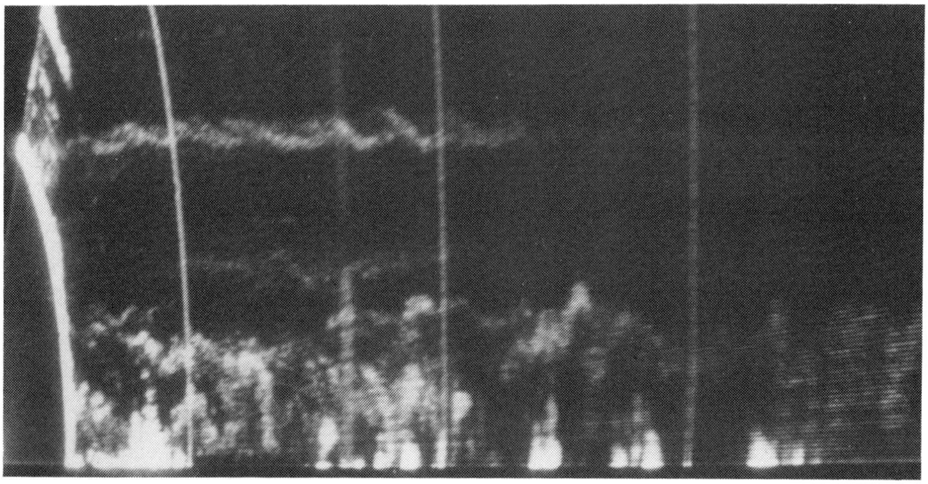

Fig. 8.3.ii A radar picture of billows at 3.5 km with convection below in clear air. The range lines are at 5 n mi intervals, and the near area is blanked off to avoid glare. [Photograph by courtesy of J. J. Hicks, Johns Hopkins University.]

8.4 THE GENERATION OF BILLOW INSTABILITY ON SLOPING SURFACES

The generation of vorticity in the atmosphere is dominated by the effect of gravity. In equation (1.4.1) the effects of viscosity and divergence of velocity are quite negligible, and in two-dimensional motion there is no gradient of $\mathbf{v}$ in the direction of $\boldsymbol{\omega}$, so the vortex lines are not stretched but remain perpendicular to the planes of the motion. The acceleration of the fluid is small compared with gravity because it is produced by buoyancy forces (Boussinesq approximation) and the vorticity equation becomes

$$\frac{D\boldsymbol{\omega}}{Dt} = \mathbf{R} \times \mathbf{g} \tag{8.4.1}$$

or

$$\frac{D\eta}{Dt} = q\frac{\partial \eta}{\partial s} = g\beta \sin\psi \tag{8.4.2}$$

where η is the magnitude of the vorticity, s is measured along a particle trajectory, q is the fluid speed, β measures the magnitude of the density (or potential temperature) gradient and ψ is the inclination of the isentropic surfaces (the stream surfaces in adiabatic motion) to the horizontal (Fig. 8.4.1).

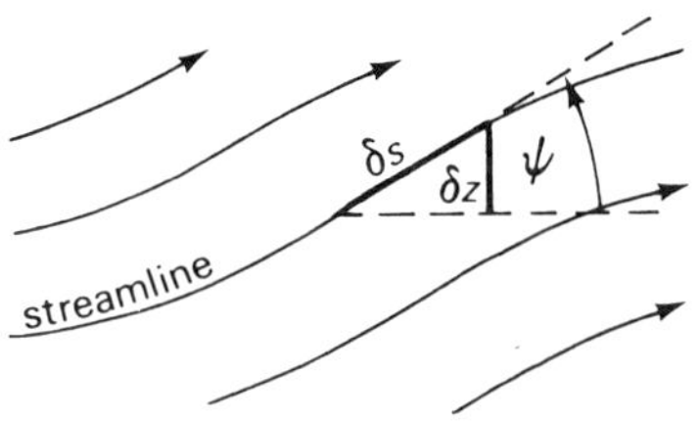

Fig. 8.4.i Coordinates for the derivation of the equation (8.4.3) for the growth of vorticity in inclined flow.

If the motion is steady

$$\frac{\partial \eta}{\partial s} = \frac{g\beta}{q}\sin\psi \tag{8.4.3}$$

and for a small element of arc, $\delta s = \delta z/\sin\psi$; therefore

$$\delta\eta = \frac{g\beta}{q}\delta z. \tag{8.4.4}$$

In steady motion in two dimensions the speed and the density (potential temperature) gradient are both proportional to the closeness of the streamlines and so we can write

$$\frac{q}{\beta} = \text{const} = \frac{q_0}{\beta_0} \tag{8.4.5}$$

the subscript 0 referring to some position far back along the streamline.

The situation is of special interest when there is a thin layer in which β_0 is large and in which there is, to begin with, very little shear, so that it is stable. If it is tilted, in a wave motion produced by flow over a mountain for example, the additional vorticity generated is directly related by (8.4.4), to the vertical displacement of the streamline from its original level, and the Richardson number is given by

$$\text{Ri} = \frac{g\beta}{(\eta_0 + \delta\eta)^2} = g\beta \Big/ \left(\eta_0 + \frac{g\beta_0}{q_0}\delta z\right)^2. \tag{8.4.6}$$

When η_0 is small so as to be of little importance (and Ri is initially very large and the flow stable), we have, in the displaced state,

$$\mathrm{Ri} = \frac{q^2}{g\beta(\delta z)^2} = \frac{qq_0}{g\beta_0(\delta z)^2}. \tag{8.4.7}$$

In motion through a given wave pattern, with given δz, the smallest values of Ri will be found in a layer where β takes its largest value, and also, of course, where δz is largest and q smallest. But δz and q are circumstances determined by factors on a much larger scale than the depth of the layer and β in it.

It was known from the very early days (around 1950), in which clear air turbulence was studied by flight of aircraft through it, that it seemed to occur most often in layers of air where static stability was greatest. This was a source of puzzlement until it was appreciated (Scorer [33]) that, although β was a stabilising influence, the vorticity η which it produced appeared as η^2 in the Richardson number. Thus as soon as the stable layers are tilted, the large static stability becomes a destabilising influence. (See Figs 8.4.ii and iii.)

In unsteady motion the equation corresponding to (8.4.4) is

$$\delta\eta = g\beta\delta t \sin\psi \tag{8.4.8}$$

where δt is the time for which the tilt ψ is maintained. This explains why the factor q^{-1} appears in (8.4.4), for a slow motion through a given wave pattern will maintain the tilt of a parcel of air for a longer time.

It has often been thought that a horizontal acceleration of the flow, which decreases the spacing of the streamlines, has a significant effect on the Richardson number. If the vorticity is perpendicular to the flow it remains unaltered in the accelerated flow because the vortex lines remain the same length. Meanwhile the value of β is increased as the isentropic surfaces are brought closer together. Consequently the Richardson number is increased and the layer is stabilised. The motion is therefore destabilised by a deceleration.

On the other hand, if there is only a streamwise component of vorticity, ξ, in a flow not varying in the transverse (y-) direction, it is increased by an acceleration, and stretching, in the direction of the stream (x). Thus using (1.4.6) for $\boldsymbol{\omega}$ we have

$$q\frac{\partial \xi}{\partial s} = \xi\frac{\partial q}{\partial s}, \quad \text{or} \quad \frac{\partial \xi}{\xi} = \frac{\partial q}{q} \tag{8.4.9}$$

and

$$\mathrm{Ri} = \frac{g\beta}{\xi^2} = \frac{g\beta_0 q/q_0}{\xi_0^2\left(1+\delta\xi/\xi_0\right)^2} = \frac{g\beta_0}{\xi_0^2}\frac{q}{q_0}\left(1+\frac{\delta q}{q_0}\right)^{-2} \tag{8.4.10}$$

$$= \mathrm{Ri}_0\left(1-2\frac{\delta q}{q}\right) \tag{8.4.11}$$

roughly. In this case of transverse shear, an increase in speed therefore decreases the Richardson number even though the static stability is increased. Since this is a contrary effect to that on the component η, it is not likely that acceleration can be taken with confidence to be the cause of a change in Ri: in any case the effect is likely to be small compared with any effect of tilting when β is large.

In the most general kind of motion likely in the free atmosphere or ocean, accelerations are likely to be small. However, tilting can produce large effects, especially in long wavelengths or in the special case in which eddies are shed from the lee of mountains and the waves travel downstream, and are therefore slow moving relative to the air.

In the equation for lee waves the most significant terms are (e.g. (5.8.24))

$$\nabla^2\zeta + \left(\frac{g\beta}{u_0^2} - k^2\right)\zeta = 0 \qquad (8.4.12)$$

where ζ is the vertical displacement in two-dimensional motion in the xz-plane. At layers where β is much larger than at other levels, the curvature of the profile of ζ as a function of z has a maximum curvature, unless ζ happens fortuitously to be small at that level. These levels are therefore likely to be those at which the gradient of the ζ-profile is reversed and ζ has a maximum magnitude (see section 6.7). Thus in addition to being the level at which Ri is most increased when tilting occurs, the layers of large β are usually those at which the wave amplitude, and therefore δz, is a maximum. This again has a maximum effect on Ri. Mountain waves are therefore a very likely site for the occurrence of billows. Their evolution is depicted in Fig. 8.4.ii.

Indeed glider pilots have known for a long time that the top of a layer of stratocumulus is usually the level at which lee waves have their maximum amplitude. This is because the radiation from the cloud top cools all the air below (by downward convection) and creates a layer of large static stability at cloud top, where billows are often seen; in fact, cloud layer tops are their most common site in waves. (See Scorer *Clouds of the World*, Chapter 6).

Fig. 8.4.ii The diagram shows the typical stage of evolution of billows in the crest of a wave, and below in the trough of a wave where they are not usually visible in cloud.

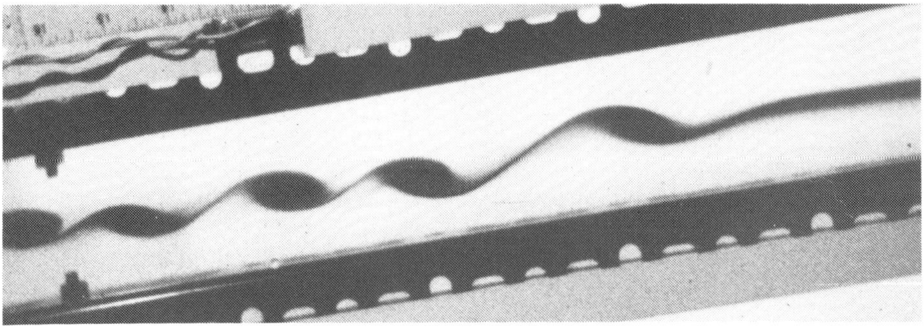

(a)

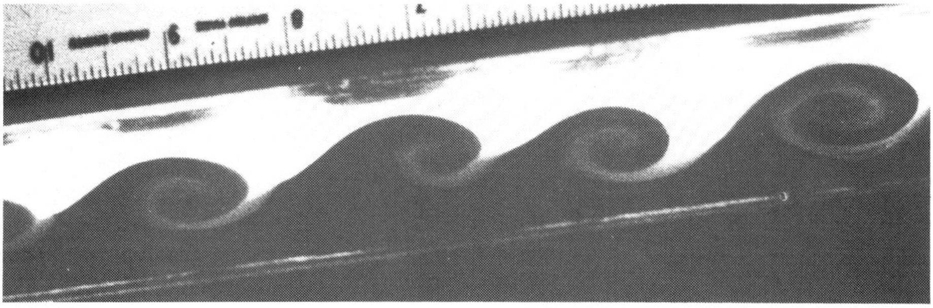

(b)

(c)

Fig. 8.4.iii Tilted tank experiments by S. A. Thorpe with a density discontinuity in the middle. When the tank is moved to the tilted position the lighter fluid flows up to the right. (a) There is a layer of dye at the discontinuity which illustrates the movement of vorticity towards the downward-sloping nodes as in 8.2.i. (b) The accumulated vorticity causes a rolling up of the layer into the cat's eyes pattern. (c) A second experiment with a larger discontinuity of density at middle interface; the whole layer below the interface is dyed.

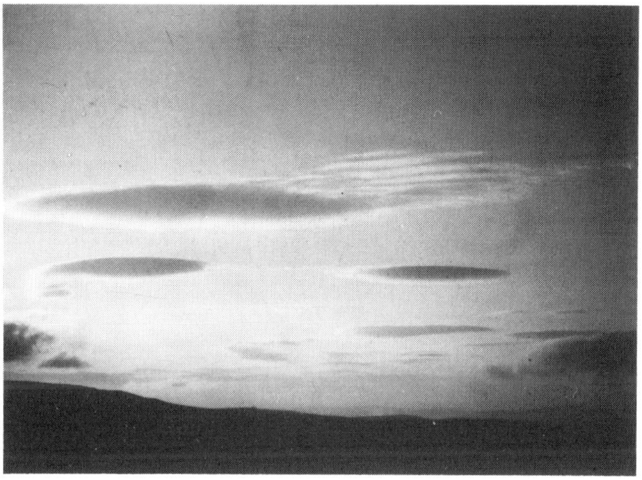

Fig. 8.4.iv Wave clouds in Yorkshire over the hills generating billows in a wave crest as in 8.4.ii.

Fig. 8.4.v Longer wavelength billows developed in a wave crest where cloud increased the instability. (Photo E. Jennings.)

8.5 THE DYNAMICAL GENERATION OF INSTABILITY IN LAYERS OF LARGE STATIC STABILITY

In this section we illustrate a very economical type of argument which can be used in many stability problems where there is no viscosity or boundary present to introduce particular dimensions into the dispersion equation for waves in a medium. The results about to be obtained can, of course, be obtained by classical perturbation theory. They are

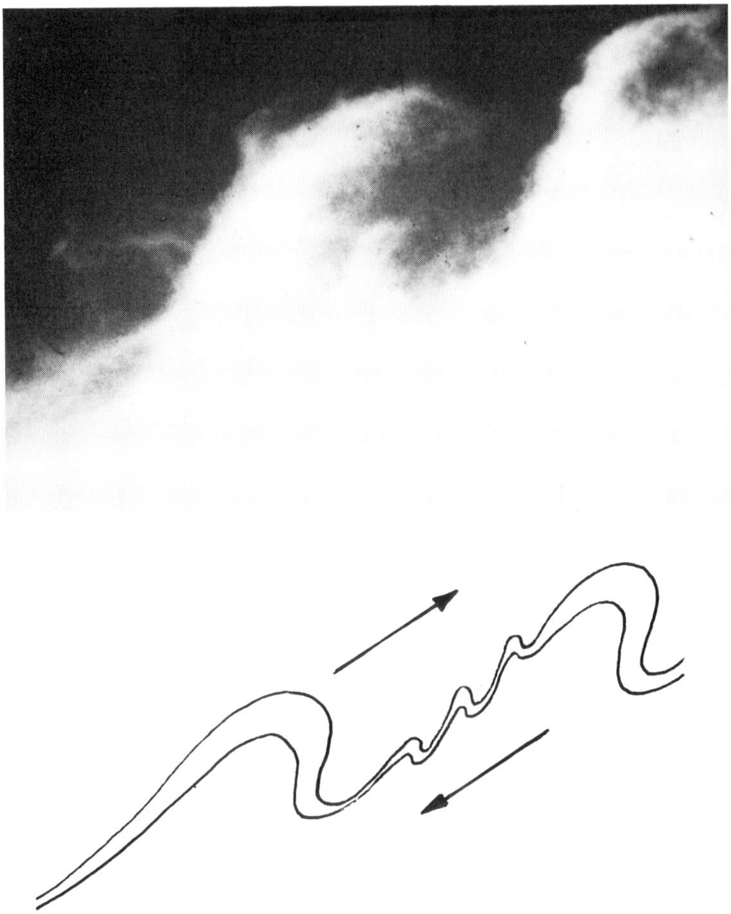

Fig. 8.4.vi A detail from a cine-film of billows forming on the upslope side of a wave over the Pie-du-Midi (Pyrenées), and where the density discontinuity is strong. Smaller wavelength billows form on the crest of the larger ones, as illustrated in the sketch below.

mainly the result of the curved flow in space, and so the Coriolis force plays a major role in the argument.

We consider the case of an airstream moving horizontally in geostrophic balance, but with static stability β and vertical and horizontal gradients of wind $\partial u/\partial z$ and $\partial u/\partial y$, the flow being in the x-direction. We then consider the work done against the various forces which arise when a displacement is made. If this work turns out to be negative the displacement will support itself and represents a form of instability. As we shall see, some of these types of displacement produce large gradients of potential temperature which, as we have just seen, are likely causes of billows and therefore of clear air turbulence, which is billows in almost all cases.

A parcel displayed a vertical distance c would experience a buoyancy force $g\beta c$, and so in being displaced this force would absorb an amount of work equal to

$$\int_0^c g\beta c = \frac{1}{2} g\beta c^2. \tag{8.5.1}$$

As a result of this displacement it would have a velocity anomaly $c\frac{\partial u}{\partial z}$, and would therefore experience a deviating force due to the earth's rotation, which differs from that on its surroundings by amount $fc\frac{\partial u}{\partial z}$, where f is the vertical component of the earth's rotation. From this force it would receive work if it travelled a distance b in the direction of the force equal to

$$fbc\frac{\partial u}{\partial z}. \tag{8.5.2}$$

But in being displaced a distance b in the y-direction it would acquire a velocity anomaly $b\left(\frac{\partial u}{\partial y} - f\right)$ in the x-direction ($b \times$ absolute vorticity), and if f is assumed to be constant over the distance of the displacement, the work done in the displacement b in the y-direction by the corresponding anomaly of the deviating force is

$$\int_0^b fb\left(\frac{\partial u}{\partial y} - f\right) \mathrm{d}b = \frac{1}{2} f\left(\frac{\partial u}{\partial y} - f\right) b^2. \tag{8.5.3}$$

If the total work done in the displacement (b, c) in the y- and z-directions is positive, such displacements are unstable, and circulations around the streamlines can develop because a positive amount of work would also be done in the displacements $(-b, -c)$ of other parcels, necessary to complete the circulations. Adding the three components (8.5.1), (8.5.2) and (8.5.3) we obtain the condition for instability, namely

$$\frac{1}{2} f\left(\frac{\partial u}{\partial y} - f\right) b^2 - fbc\frac{\partial u}{\partial z} - \frac{1}{2} g\beta c^2 > 0. \tag{8.5.4}$$

The flow is unstable when $c = 0$ if $\frac{\partial u}{\partial y} > f$. This is simply a statement that in space the flow, which is rotating with the earth, has circulation decreasing along the outward normal to the curved flow (see section 8.6). Putting this in local terms, the direction of rotation of the tangent moving with the particle is in the opposite direction, in space, to the vorticity of the fluid, the component normal to the earth's surface only being considered. We have also ignored the curvature of the streamlines relative to the earth (and therefore the cyclostrophic force). More typically, $\partial u/\partial y$ is of the order of $\frac{1}{2} f$.

But c/b can be negative, and so also can $\partial u/\partial z$. A real value of c/b exists to satisfy (8.5.4) if

$$\left(\frac{\partial u}{\partial z}\right)^2 \geq g\beta\left(1-\frac{\partial u}{\partial y}\bigg/f\right). \tag{8.5.5}$$

The case of just no instability is determined by writing equality in (8.5.4) and (8.5.5), and then

$$\frac{c}{b}=\left(\frac{f\left(f-\dfrac{\partial u}{\partial y}\right)}{2g\beta}\right)^{1/2}. \tag{8.5.6}$$

This quantity is the aspect ratio of the secondary cells which would develop in this geostrophically moving airstream. To get an idea of the shape we put in typical values, as follows:

$$f = 5\times10^{-5}\,\mathrm{s}^{-1} \qquad \text{(at latitude 45° roughly)}$$

$$g = 10^3\ \mathrm{cm\ s}^{-2} \qquad \text{roughly}$$

$$\beta = 10^{-7}\ \mathrm{cm}^{-1}\ \ (\text{when } \partial T/\partial z \text{ is about } \tfrac{1}{3}\Gamma)$$

$$\frac{\partial u}{\partial y}=\frac{1}{2}f = 2.5\ \mathrm{m\ s}^{-1}\ \text{per 100 km}$$

and then

$$\frac{c}{b}=2.5\times10^{-3}. \tag{8.5.7}$$

The range of values of $\partial u/\partial y$ and β which are common make this quantity vary by a factor of about 3. The corresponding value of the vertical gradient of wind speed given by equality in (8.5.5) is $(\frac{1}{2}g\beta)^{1/2}$, and so

$$\frac{\partial u}{\partial z}=0.7\times10^{-2}\ \mathrm{s}^{-1}, \tag{8.5.8}$$

which represents a gradient of 7 m s^{-1} per km. This is quite common, and gradients larger than this would increase the degree of instability. An increase in $\partial u/\partial y$ to more nearly equal to f reduces c according to (8.5.6), and so the cells are flattened, but even when $\partial u/\partial y = 0$ the value in (8.5.7) is not altered significantly for the purposes of this discussion. In the case when $\partial u/\partial y = 0$, the criterion (8.5.5) is sometimes expressed in the form of a Richardson number, thus

$$\mathrm{Ri}=\frac{g\beta}{(\partial u/\partial z)^2}\leqslant 1, \tag{8.5.9}$$

and (8.5.4) can be written

$$Q^2 + \frac{2}{\mathrm{Ri}^{1/2}} Q + 1 < 0 \tag{8.5.10}$$

where

$$Q = \frac{c}{b} \frac{(g\beta)^{1/2}}{f}, \tag{8.5.11}$$

and if Ri < 1, (8.5.10) is satisfied for the range of values of c/b satisfying

$$1 - (1 - \mathrm{Ri})^{1/2} < \mathrm{Ri}^{1/2} Q < 1 + (1 - \mathrm{Ri})^{1/2}. \tag{8.5.12}$$

The sign of $\partial u/\partial z$ is immaterial, and the places where a large magnitude is likely to occur are where a jet stream is being decelerated as it catches up slower-moving air. In the northern hemisphere, where f is positive, and where $\partial u/\partial y$ is positive on the left of the jet stream, the most likely place for this instability to occur is on the left of a jet stream, particularly at the exit, because there the velocity anomaly is in excess of the value assumed in calculating (8.5.3) where geostrophic flow was assumed. Indeed ageostrophic flow will certainly enhance these effects.

The consequence of this secondary motion in flat cells is to distort the temperature field so as to produce layers of increased and decreased vertical gradient where it was previously uniform. Since layers where β is large are more stable statically they will be the boundaries of cells and so the value of β will be further increased there. According to (8.5.7) the cells are very flat, and if they are 100 km in horizontal extent they will have a depth of the order of $250 \overset{\times}{\div} 3$ metres, the more stable layers being thinner.

Since they are produced by the deviating force of the earth's rotation they require a time of the order of a day or two at least to appear and it is unlikely that they would be directly responsible for the low Richardson numbers which could cause billows. We need to look to mountain waves or fairly vigorous motion near fronts, where the isentropic surfaces become more inclined, for the causes of moderate or intense clear air turbulence.

Nevertheless very light clear air turbulence is often observed far from mountains and fronts, and is not uncommon in the stratosphere, with a wavelength of the order of 100–200 m. Two comments need to be made in this connection. First of all, if the tilting of the stable layers is so slight that the Richardson number is only very slowly reduced, the degree of instability (growth rate) will be very small because the disturbances will have plenty of time to grow without the instability becoming greater. In the case of mountain waves the tilting can be so large that a very large growth rate is produced before disturbances have had time to grow. When one of these slight disturbances has taken place it leaves a very stable layer at its top and bottom, and if the original tilting is still present, slight turbulence could well be produced there. Consequently very slight turbulence could be a feature of a layered air mass for many hours.

Secondly, it must be recognised that billows are usually quickly evolving in time so that an observation of clear air turbulence may reveal very definite periodicity if the billows are traversed at an early stage. However, if the flight through them is along the length and at a stage when the cat's-eyes pattern is degenerating into a horizontal layer of depth H, the turbulence will not appear to be organised. It is probably not necessary to invoke the possibility of any other kind of turbulence in clear, stably stratified, air.

8.6 INSTABILITY OF THREE-DIMENSIONAL VORTICITY

(1)

The aim of this investigation is to provide guidance in framing expectations for the evolution of a varied flow pattern of an inviscid incompressible fluid of non-uniform density with non-uniform vorticity. This can be generalised to the atmospheric case of a fluid compressed by its own weight on a rotating earth. We start with the following equation for steady flow (the relevant case of equation (1.4.1):

$$\mathbf{v}\,.\,\mathbf{grad}\;\boldsymbol{\omega} = (\boldsymbol{\omega}\,.\,\mathbf{grad})\mathbf{v} + \mathbf{R}\times(\mathbf{g}-\mathbf{f}), \qquad (8.6.1)$$

which indicates that the vortex lines are carried by the fluid motion, and that new vorticity arises when the density gradient is not parallel with $\mathbf{g}-\mathbf{f}$. The special case of this was illustrated by G. I. Taylor's experiment in which initially a layer of oil was at rest above a layer of denser water. The system was then accelerated downwards with an acceleration in excess of gravity, $\mathbf{f} > \mathbf{g}$, so that $\mathbf{g}-\mathbf{f}$ changed sign and then any small disturbances made previously on the interface became intensified. The small disturbances may be regarded either as small vertical displacements of parcels which were then subject to acceleration in the same direction, or as small rotational displacements of bits of the interface (which would have been necessary to make the linear vertical displacements) and they became subject to rotational acceleration in the same direction and gained increased vorticity.

The rotational displacement may be measured by an angular displacement, ϕ, about an axis in the interface, and if q is the magnitude of $\mathbf{v}$ and we represent the secondary vorticity by $2q\dfrac{\partial\phi}{\partial s}$, then the relevant terms in (8.6.1) are

$$q\frac{\partial}{\partial s}\left(2q\frac{\partial\phi}{\partial s}\right) = \frac{1}{\rho}\frac{\partial\rho}{\partial n}(g-f)\sin\phi \;\approx\; \mathbf{R}\,.\,\mathbf{n}(\mathbf{g}-\mathbf{f})\phi \qquad (8.6.2)$$

and we are concerned with the acceleration around the axis of the original displacement.

We now return to the notation of Chapter 3 because we are concerned with a fluid in motion, and not initially at rest as in Taylor's experiment, but in steady motion in which the vorticity lies in surfaces of constant density (Bernoulli surfaces, see (3.2.7)) (Fig. 8.6.i).

The situation about to be studied is not completely general. Every steady inviscid flow has Bernoulli surfaces which contain all the vortex lines through a streamline (or all the streamlines through a vortex line). In the atmosphere, every air mass is modified towards having the isopycnics and the Bernoulli surfaces horizontal, and this means that in many cases the isopycnics are the Bernoulli surfaces, because that is how they will remain for some time (see section 1.5). Here we investigate instabilities of this case.

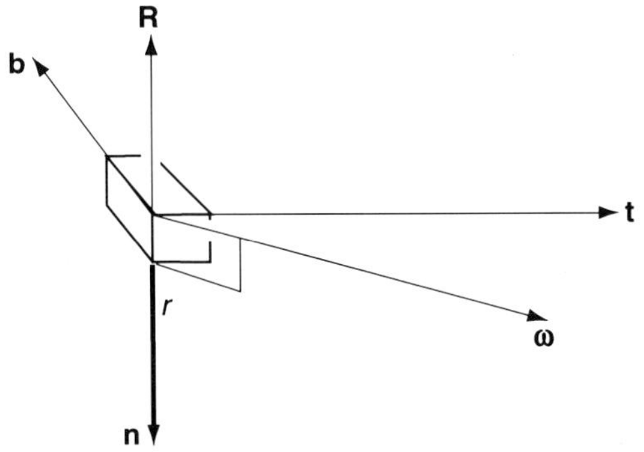

Fig. 8.6.i The vector triad **t**, **n**, **b** (tangent, normal, binormal) in relation to the density gradient vector $\mathbf{R}\left(=\frac{1}{\rho}\,\mathbf{grad}\,\rho\right)$ and the vorticity vector $\boldsymbol{\omega}$, which together with **t** defines the Bernoulli surface (see (3.2.7)).

The vector **R** is initially in the plane normal to **t**, and is subject to a small rotational displacement ϕ around the streamline (tangent **t**). In this case equation (8.6.1) takes the form (3.2.3), namely

$$\mathbf{v}\,.\,\mathbf{grad}\,\frac{\omega_{\mathrm{s}}}{q} = 2\kappa\boldsymbol{\omega}\,.\,\mathbf{n} + \kappa q\mathbf{R}\,.\,\mathbf{b} + \frac{1}{q}\mathbf{R}\times\mathbf{g}\,.\,\mathbf{t}, \tag{8.6.3}$$

and we neglect the last term on the assumption that we are dealing with a case of steady flow, so that there already exists a pressure field which cancels out the effect of gravity. Any instability will be indicated if the secondary vorticity $\boldsymbol{\omega}_{\mathrm{s}}$ is subject to an acceleration as a result of the small rotational displacement, i.e. an increase in ϕ in the downstream direction which will be equal to $q\frac{\partial}{\partial s}2q\frac{\partial\phi}{\partial s}$.

It is presumed that the existing flow in the neighbourhood is defined by the local vorticity, of which the vector lies in the tangent plane to the flow (containing **t** and normal to **n**) and the Bernoulli surface (containing **t** and $\boldsymbol{\omega}$) is normal to **R**. For the purpose of investigating the local stability we assume that the local flow is the same as a part of the flow of coaxial cylinders rotating around the axis with circumferential velocity v and velocity w along the axis. In effect we are regarding the locality of our test piece of fluid as behaving like a bit of a helical vortex system (Fig. 8.6.ii). To study that, we consider various coordinate systems for the position vector **x**, and the path length s along the particle path (Fig. 8.6.iii).

In Fig. 8.6.iii the particle of fluid under consideration has the vector triad, **t**, **n**, **b** carried with it along its helical path. We also have fixed ones

$$\mathbf{x} = (x, y, z) = (r\cos\theta, r\sin\theta, z), \tag{8.6.4}$$

and a cylindrical polar system (r, θ, z); in both the z-axis is the axis of the local helix. For the particle r is a cosntant radius of the circular cross-section of the helix, and the

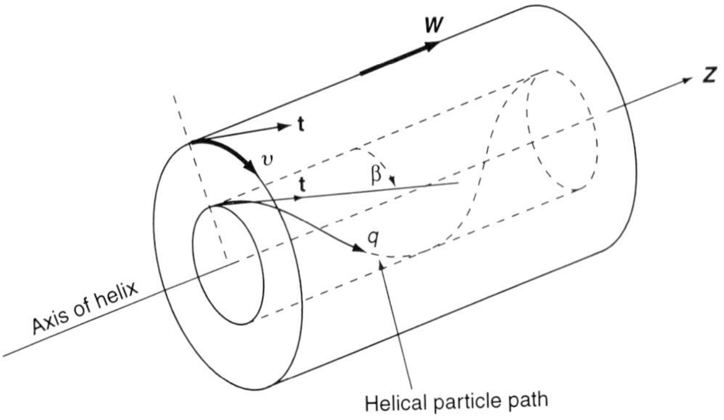

Fig. 8.6.ii The helical vortex system.

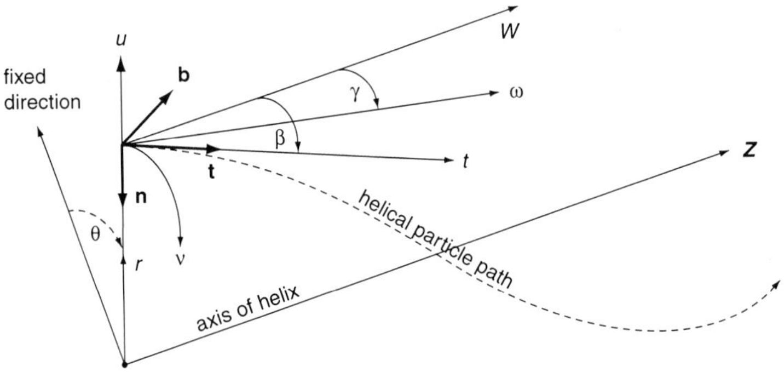

Fig. 8.6.iii Details of the helical system.

Bernoulli surface, containing the vectors $\mathbf{t}$ and $\boldsymbol{\omega}$. The velocity components along the Cartesian directions are $(0, v, w)$, and the triad is given by

$$\mathbf{t} = \frac{\partial \mathbf{x}}{\partial s} = \left(-r\sin\theta\frac{\partial\theta}{\partial s}, r\cos\theta\frac{\partial\theta}{\partial s}, \frac{\partial z}{\partial s}\right) \tag{8.6.5}$$

$$\mathbf{v} = q\mathbf{t} = q(0, \sin\beta, \cos\beta)$$

in cylindrical coordinates where β is the angle between the tangent and the direction of the axis.

In fixed Cartesians

$$= q(-\sin\beta\sin\theta, \sin\beta\cos\theta, \cos\beta)$$

$$= (0, v, w) \tag{8.6.6}$$

and so

$$r\frac{\partial\theta}{\partial s} - \sin\beta = \frac{v}{q}$$

$$\frac{\partial z}{\partial s} = \cos\beta \tag{8.6.7}$$

Then by (3.1.3)

$$\frac{\partial t}{\partial s} = \kappa\mathbf{n} = \left(-\sin\beta\cos\theta\frac{\partial\theta}{\partial s}, -\sin\beta\sin\theta\frac{\partial\theta}{\partial s}, 0\right)$$

$$= \frac{1}{r}\sin^2\beta(-\cos\theta, -\sin\theta, 0) \quad \text{by (8.6.7)}$$

$$= \frac{1}{r}\sin^2\beta(-1, 0, 0) \quad \text{in cylindrical coordinates} \tag{8.6.8}$$

and since $\mathbf{n}$ is a unit vector,

$$\kappa = \frac{1}{r}\sin^2\beta = \frac{v^2}{q^2 r} \tag{8.6.9}$$

Likewise the torsion may be calculated by (3.1.7) and (3.1.8):

$$\mathbf{b} = \mathbf{t}\times\mathbf{n}$$

$$= (0, -\cos\beta, \sin\beta) \quad \text{in cylindrical coordinates}$$

$$= (\cos\beta\sin\theta, -\cos\beta\cos\theta, \sin\beta) \quad \text{In Cartesians} \tag{8.6.10}$$

$$\frac{\partial\mathbf{b}}{\partial s} = \left(\cos\beta\cos\theta\frac{\partial\theta}{\partial s}, \cos\beta\sin\theta\frac{\partial\theta}{\partial s}, 0\right)$$

$$= \frac{1}{r}\sin\beta\cos\beta(\cos\theta, \sin\theta, 0) = -\tau\mathbf{n} \tag{8.6.11}$$

and so

$$\tau = \frac{1}{r}\sin\theta\cos\beta = \kappa\cot\beta = \frac{vw}{q^2 r} \tag{8.6.12}$$

Having established the relationship of the local vorticity and the particle trajectory to the helical vortex, composed of the same vorticity and particle path, we may consider the stability by applying a small rotational displacement and observing its evolution downstream (Fig. 8.6.iv).

The vectors $\boldsymbol{\omega}$ and $\mathbf{R}$ are carried to positions $\boldsymbol{\omega}_1$ and $\mathbf{R}_1$. In order to discover the consequence of this we use (8.6.3) and omit the last term, which concerns only the effects of the component of the gravity field in the direction of $\mathbf{t}$, which is the rotation axis. Thus from Fig. 8.6.iv we obtain $\mathbf{R}_1.\mathbf{b}$ and $\boldsymbol{\omega}_1.\mathbf{n}$ in terms of ϕ for the displaced position.

$$\mathbf{v}.\mathbf{grad}\frac{\boldsymbol{\omega}_s}{q} = q\frac{\partial}{\partial s}\frac{2}{q}\frac{\partial\phi}{\partial s} = 2\kappa\boldsymbol{\omega}_1\mathbf{n} + \kappa q\mathbf{R}_1.\mathbf{b}$$

$$= 2\kappa(-\boldsymbol{\omega}.\mathbf{b})\phi + \kappa q\mathbf{R}.(-\mathbf{n})\phi$$

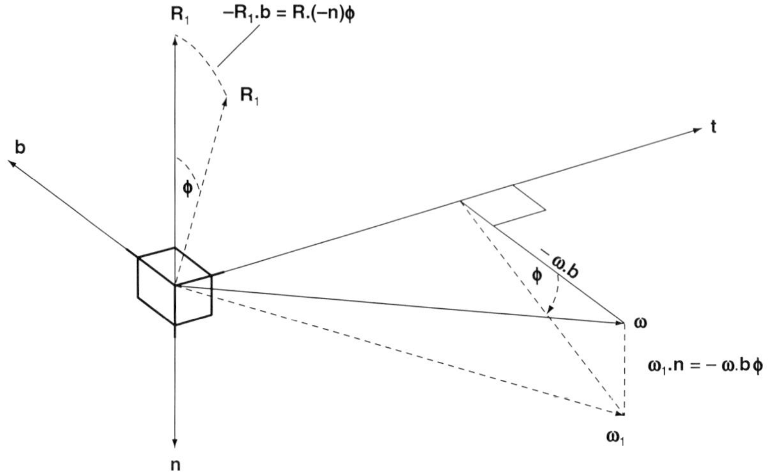

Fig. 8.6.iv The rotational displacement, ϕ, of the plane containing the vorticity, $\boldsymbol{\omega}$, about the line containing the tangent, **t**, and carrying with it the density gradient vector, **R**, and the vorticity vector, places these two vectors in the positions $\boldsymbol{\omega}_1$ and $\mathbf{R}_1$. Right angles are indicated.

or

$$\frac{\partial^2\phi}{\partial s^2} = \left(-\frac{\kappa}{q}\boldsymbol{\omega}.\mathbf{b} + \frac{\kappa}{2}\mathbf{R}.\mathbf{n}\right)\phi = \frac{\kappa}{q}\mu^2\phi \tag{8.6.13}$$

and this represents an unstable situation if

$$\mu^2 = -\boldsymbol{\omega}.\mathbf{b} + \tfrac{1}{2}q\mathbf{R}.\mathbf{n} > 0 \tag{8.6.14}$$

because by definition the curvature κ and the speed q are positive. This requires that **n** should be directed towards the centre of curvature and **b** completes the moving triad (**t**, **n**, **b**). **R** is directed outwards, on the grounds that this is much more stable than if **R** is directed inwards towards greater density. Thus when there is no vorticity ($\boldsymbol{\omega} = \mathbf{0}$) the term in **R** means that the centrifugal force would cause instability if **R** is directed inwards.

We may note that, as in the more conventional displaced parcel theory, the pressure field in which the displaced material finds itself is not disturbed by the initial displacement.

In this case of axial symmetry in which the quantities are functions of r only, the **curl** of any vector **a** is

$$\textbf{curl a} = \left(0, -\frac{\partial a_z}{\partial r}, \frac{1}{r}\frac{\partial}{\partial r}(ra_\theta)\right) \text{ in cylindrical coordinates} \tag{8.6.15}$$

and so

$$\boldsymbol{\omega} = \textbf{curl v} = \left(0, -\frac{\partial w}{\partial r}, \frac{1}{r}\frac{\partial}{\partial r}(vr)\right) \text{ in cylindrical coordinates} \tag{8.6.16}$$

and from (8.6.10), noting that $v = q \sin \beta$, $w = q \cos \beta$, we get

$$\begin{aligned}\boldsymbol{\omega}\,.\,\mathbf{b} &= \frac{\partial w}{\partial r}\cos\beta + \frac{1}{r}\sin\beta\frac{\partial}{\partial r}(vr) \\ &= \frac{\partial w}{\partial r}\cos\beta + \frac{\partial v}{\partial r}\sin\beta + \frac{v}{r}\sin\beta \\ &= \frac{\partial q}{\partial r} + \frac{v^2}{qr}.\end{aligned} \tag{8.6.17}$$

The coefficient μ^2 in (8.6.13), whose sign determines the stability, is

$$\mu^2 = -\frac{\partial q}{\partial r} - \frac{v^2}{qr} - \frac{1}{2}\frac{q}{\rho}\frac{\partial \rho}{\partial r} \tag{8.6.18}$$

for an incompressible fluid of density ρ which is initially a function of r only. But in the equilibrium flow the equation of motion is

$$\frac{v^2}{r} = \frac{1}{\rho}\frac{\partial p}{\partial r}, \tag{8.6.19}$$

whence

$$\mu^2 = -\frac{1}{\rho q}\frac{\partial}{\partial r}\left(p + \tfrac{1}{2}\rho q^2\right) = -\frac{1}{\rho q}\frac{\partial p_0}{\partial r} \tag{8.6.20}$$

where p_0 is the stagnation pressure measured by a pitot tube fixed to the Cartesian coordinate system.

The implication of this last result is that **flow in which the stagnation pressure decreases outwards is unstable for small rotational displacements of fluid parcels around the streamline**. Therefore a stable configuration is one in which the lowest stagnation pressure fluid is closest to the axis. We have already noted in section 3.5 that Hawthorne observed the tendency of the fluid to move into this configuration in a straight pipe after a bend (see Fig. 3.5.vii).

The consequences of this result are much further reaching than appears at first sight because there are cases in which the sign of μ^2 **depends on the choice of the coordinate** system, although this obviously has nothing to do with the mechanics. Thus

$$p_0 = p + \tfrac{1}{2}\rho q^2 = p + \tfrac{1}{2}\rho(v^2 + w^2) \tag{8.6.21}$$

and

$$\frac{\partial p_0}{\partial r} = \frac{\partial p}{\partial r} + \frac{1}{2}\frac{\partial \rho}{\partial r}(v^2 + w^2) + \frac{1}{2}\rho\left(2v\frac{\partial v}{\partial r} + 2w\frac{\partial w}{\partial r}\right). \tag{8.6.22}$$

In the helical motion there is an absolute value for v, the circumferential velocity, which must be measured relative to a non-rotating coordinate system, and angular velocities have absolute values. But there is no absolute value for w, to which we may add a constant w_0 by moving the coordinate system with speed w_0 in the negative direction of z. In the case in which $\partial\rho/\partial r$ is zero, this alters only the term $\rho w \partial w/\partial r$ into $\rho(w+w_0)\partial w/\partial r$, and its magnitude and sign may be chosen freely provided $\partial w/\partial r$ is not zero. In that case the sign of $\partial p_0/\partial r$ depends on the zero chosen for w, and so does the sign of μ^2, which means that **there are always some disturbances which are unstable when $\partial w/\partial r$ is non-zero.**

The effect of adding w_0 to w is to alter β, the pitch of the helix. The dynamics must, of course, remain unaltered by the change in the velocity of the coordinate system, but by altering the direction of **t**, which is determined by the ratio v/w, we are merely studying a different perturbation of the flow, which was a rotation about **t**.

Clearly the motion may be stabilised or destabilised by changing the sign and magnitude of $\partial\rho/\partial r$, instability being made more likely by having the density decrease outwards, for then the term $\dfrac{1}{2}\dfrac{\partial\rho}{\partial r}(v^2+w^2)$ in (8.6.22) is negative. If we now ignore the effect of density gradients we may revert to the equation (8.6.13) in the form

$$\frac{\partial^2\phi}{\partial s^2} = -\frac{\kappa}{q}\boldsymbol{\omega}\,.\,\mathbf{b}\phi. \tag{8.6.23}$$

The sign of $\boldsymbol{\omega}\,.\,\mathbf{b}$ is determined by the position of **t** in relation to $\boldsymbol{\omega}$ in the plane which is tangent to the cylinder of the flow. If $\boldsymbol{\omega}$ lies between **t** and **b** the scalar product $\boldsymbol{\omega}\,.\,\mathbf{b}$ is positive and the motion is stable. If $\beta < \gamma$ the motion is unstable (see Fig. 8.6.iii).

The motion is thus unstable for rotations around an axis in the tangent plane (normal to **n**) which lies between the vorticity vector and the direction of the axis of the helix appropriate to the local motion (see Fig. 8.6.v). In principle this helical axis may be found by noting the values of κ and τ from which β may be obtained from (8.6.12), and the axial direction is at angle β to **t** in the plane normal to **n**. The quantities, κ, τ, **t** and **n** can be found at any point in the neighbourhood of which the trajectory is defined using only first derivatives of the velocity field, the fact that the local motion is regarded, for convenience, as part of a helix does not require the rest of the motion be like the rest of the helix.

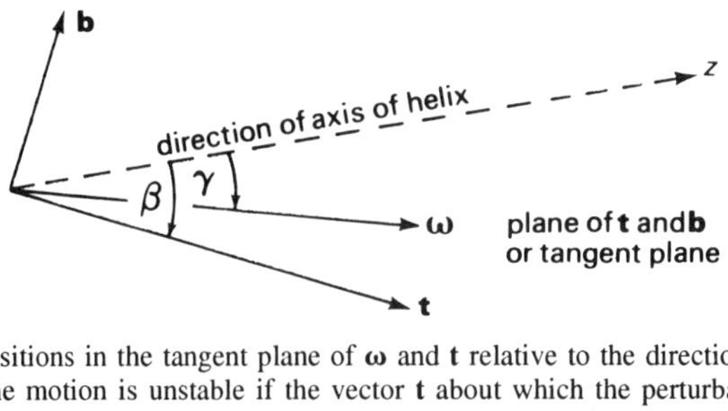

Fig. 8.6.v The positions in the tangent plane of $\boldsymbol{\omega}$ and **t** relative to the direction of the axis of the local helix. The motion is unstable if the vector **t** about which the perturbation rotates the fluid lies between the vorticity and the direction of the axis.

We have pursued this argument assuming the helix to be right-handed, and β positive, as in Fig. 8.6.vi. The diagrams corresponding to those given in Fig. 8.6.vi when β or γ is negative are simply mirror images of those shown which indicate the directions about

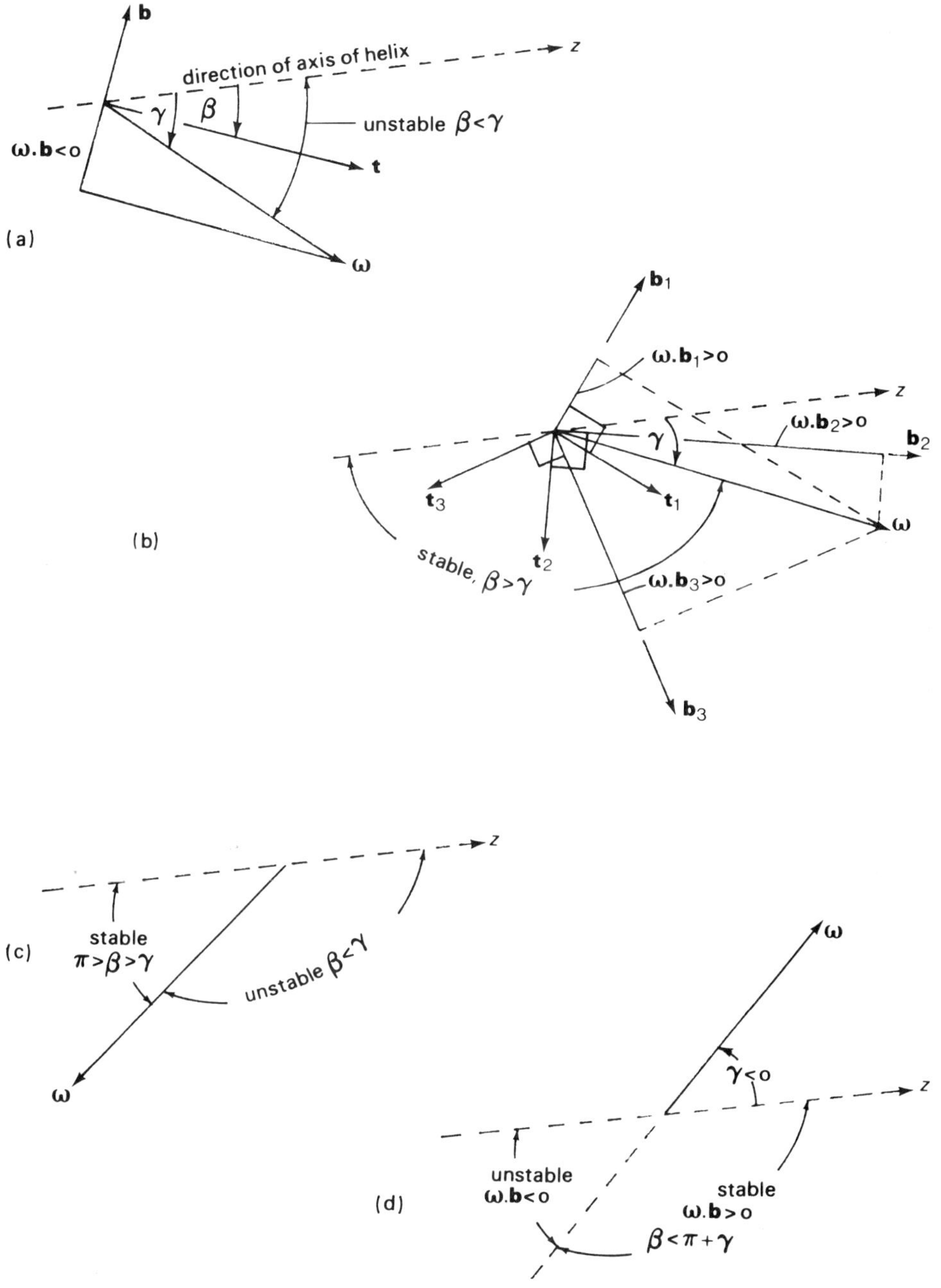

Fig. 8.6.vi The various directions about which rotations are unstable for a given flow (given helical axis and given vorticity vector). In (a) we have instability. For the three direction $\mathbf{t}_1$, $\mathbf{t}_2$, $\mathbf{t}_3$ in (b) we have stability. In (c) and (d) the results are shown for backward and left-handed vorticity. The directions indicated in this diagram are in the tangent plane to the Bernoulli surfaces of the flow.

which rotations are unstable and stable for various positions of **ω** relative to the axis of the helix. When **t** is pointing backwards along the axis it simply means that w is negative.

To interpret these results in terms of a general disturbance we imagine helical flow on coaxial cylinders disturbed by the formation of a bump, or blister, on one of them. Fig. 8.6.vii shows the contours of such a blister. The blister is formed as a hump on one of the Bernoulli surfaces by rotation of the fluid around the direction of the tangent to the contours so that this direction is unstable where it lies between the direction of the axis and the direction of the vorticity vector. The part of the blister marked unstable will steepen, while the stable part of the disturbance will disperse as stable waves. It is to be expected that the blister would lengthen in the direction of greatest instability.

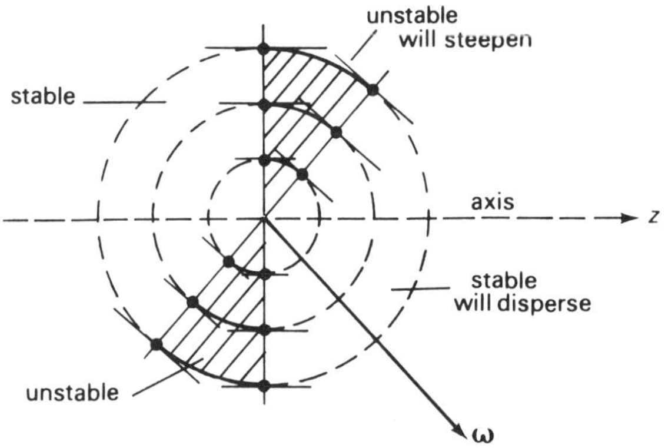

Fig. 8.6.vii The parts of the circular blister on a Bernoulli surface in the unstable region will steepen. The blister is formed by a twist at every point around the tangent to the contour of the blister as indicated by the short tangent lines. At the points where the tangents are drawn longer the disturbance is neutral: it grows in the shaded area marked 'unstable'.

When there is no density gradient

$$\frac{\kappa}{q}\mu^2 = -\frac{\kappa}{q}\boldsymbol{\omega}.\mathbf{b} = \frac{\sin^3\beta}{vr}\omega\sin(\gamma-\beta), \tag{8.6.24}$$

and so because $\mathrm{d}z = \mathrm{d}s\cos\beta$, and $q\mathrm{d}t = \mathrm{d}s$, for constant q, equation (8.6.25) may be represented in the following three forms:

$$\frac{\mathrm{D}^2\phi}{\mathrm{D}s^2} = \frac{\omega}{vr}\sin^3\beta\sin(\gamma-\beta)\phi = \mu_s^2\phi, \tag{8.6.25}$$

$$\frac{\mathrm{D}^2\phi}{\mathrm{D}t^2} = \frac{v\omega}{r}\sin\beta\sin(\gamma-\beta)\phi = \mu_t^2\phi, \tag{8.6.26}$$

$$\frac{\mathrm{D}^2\phi}{\mathrm{D}z^2} = \frac{\omega}{vr}\frac{\sin^3\beta}{\cos^2\beta}\sin(\gamma-\beta)\phi = \mu_z^2\phi. \tag{8.6.27}$$

These represent the growth of a disturbance following a parcel of fluid along its trajectory s, in time t, or along the axis z. If we have identified an unstable disturbance by adding a velocity w_0 to the coordinate system, it is possible that in any practical case this would mean that the growth as z or s increases might appear to be large. However, because the coordinate system has a large velocity relative to the apparatus, the parcel could well be a long way from the region of interest before any significant growth had occurred. This is most likely to be the case when γ, the angle between the vorticity vector and the axis, is small, and a large velocity w_0 has to be imposed in order to make **t** lie between them.

In practice, therefore, it is most sensible to fix the coordinate system to any apparatus we may have and use (8.6.26) to find out which circumstances create instability. Perturbations caused by the presence of the apparatus will cause secondary rotations around the tangent to the flow relative to the apparatus. Thus, if there is swirling flow inside a straight circular pipe, an imperfection on the pipe surface is likely to set off instability if the tangent lies at a smaller angle to the axis than the vorticity vector.

The vortex lines are, in the situation being considered, helices, and the angle γ is increased for a given distribution of circumferential velocity (or swirl) $v(r)$ by increasing the magnitude of $-\partial w/\partial r$, the circumferential component of vorticity (see 8.6.16). In the case of a swirling flow passing down a circular pipe, with no boundary layer on the outer wall at the entrance, the flow rapidly becomes unstable because the retardation of the fluid at the outer wall produces low stagnation pressure fluid there, and this will tend to move towards the centre of the pipe as mentioned in section 3.5 and earlier in this section in remarks following (8.6.20). The vorticity vector may be directed along the axis if the swirl is a solid rotation (as it must be in the centre of the flow), or there may be no vorticity, if $v \propto r^{-1}$ and w is independent of r (see Fig. 8.6.viii). Further down the pipe, w decreases towards the boundary which gives $\boldsymbol{\omega}$ a right-handed circumferential component, and v decreases towards the boundary giving $\boldsymbol{\omega}$ a backward-pointing component. Thus γ exceeds 90°, and whatever the direction of **t**, it must lie between 0° and 90° in forward-moving right-handed swirling flow. Consequently the boundary fluid moves inwards, and in the case where the original swirling flow is irrotational except for a small

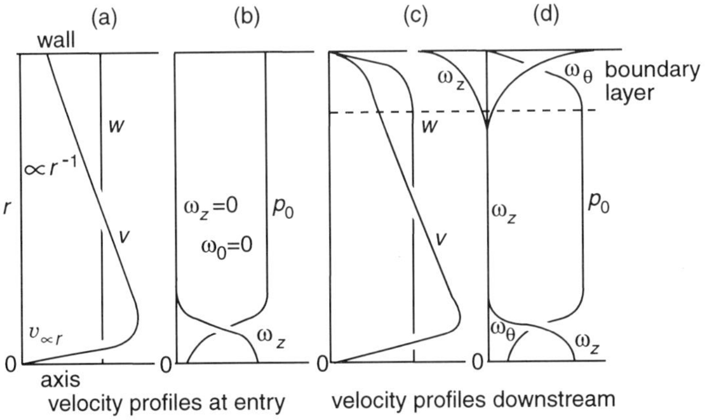

Fig. 8.6.viii Profiles of velocity components (a), stagnation and pressure vorticity (b), at entry and downstream (c), (d) when irrotational swirling flow enters a pipe.

solidly rotating core, this fluid will very rapidly penetrate to the core. The vortex will then appear to break down or explode according to the visualisation employed.

This kind of breakdown will not occur if the vorticity vector has large magnitude and is directed forwards along the axis initially because the instability will then be confined to the boundary layer. In the case of the main body of the fluid being in the neutral, irrotational state to begin with (except for a small rotational core), the onset of breakdown will be delayed until the boundary layer is thick enough for the instability to overcome the stabilising effect of viscosity which damps small disturbances, and the global effect of the rigid outer boundary which damps large ones. When the instability sets in, therefore, there is quite a large mass of fluid ready to peneetrate to the core as soon as a disturbance begins to grow, and the breakdown appears like an explosive global phenomenon involving the whole mass rather than the boundary layer alone.

To find the most unstable disturbance we find the maximum of μ_t by differentiating (8.6.26) with respect to β and setting $\partial\mu_t/\partial\beta = 0$. This gives $\beta = \frac{1}{2}\gamma$, and from the equation itself it is obvious that $\mu_t = 0$ when $\beta = 0$, or γ. Thus **the most unstable disturbances are those which are a rotation of the fluid around lines bisecting the angle between the axis and the vorticity vector**. This means that the flow is most unstable if γ is contrived to be 2β where β is the angle made by the flow relative to a coordinate system at rest relative to the disturbing influence.

To make the flow as stable as possible it can be seen from the instability diagrams (Fig. 8.6.vi) that in right-handed swirl it is best to have vortex lines which have a left-handed screw, i.e. $\gamma < 0$. This is achieved by having w increasing in magnitude with radius, which ensures that the fluid on the outside has the largest stagnation pressure. However this flow would be upset by a disturbing body such as a bullet or a fish or an obstacle drawn along by a wire which was overtaking the flow, for relative to it w would be increasing in magnitude towards the core of the flow.

An interesting application of equation (8.6.27) can be made to the case of swirling flow with initially uniform axial velocity w, and a large vorticity directed along the axis. This is not explosively unstable as in the case just discussed by small initial vorticity in the main bulk of the fluid, and we may imagine that γ increases from zero to a small positive value as the boundary layer is formed. The most rapidly growing disturbance has been shown to be with $\beta = \frac{1}{2}\gamma$, the one which achieves the greatest amplitude at the point of exit from a given apparatus is the one having the largest value of μ_z.

$$\frac{\partial}{\partial\beta}\mu_z^2 = \frac{\omega}{2vr}\frac{\sin^2\beta}{\cos^3\beta}[5\sin(\beta-\gamma)+\sin(3\beta-\gamma)] \tag{8.6.28}$$

which is zero where $\beta = 0$ (minimum) and where

$$5(\gamma-\beta) \approx 3\beta-\gamma$$

$$\beta = \tfrac{3}{4}\gamma \tag{8.6.29}$$

(2) Spike clouds, illustrating the essential instability of 3-D vorticity

The clouds shown in this group of satellite pictures (Figs 8.6.ix–xiv) have been variously called 'warm front instability' and 'spike clouds' because of the tendency to form in particular places and to have a particular appearance. The times are given to the nearest minute in G.M.T. with the date in each case, followed by the spectral channel.

8.7 INSTABILITY OF PLANE ROTATING FLOW

Two particular cases of swirling flow are of special interest, namely those in which there is no motion along the z-axis initially, where the motion is in circles around a common axis. In the first case (Fig. 8.7.i, p. 270) the vorticity is directed in the same direction as the rotation, in which case the circulation increases with radius, and v decreases more slowly than r^{-1}, or increases with r, and $\omega_z = \frac{1}{r}\frac{\partial}{\partial r}(vr) > 0$. Here $\gamma = 0$ and there are no unstable disturbances (Fig. 8.7.i). If an axial component w which varies with r is introduced, immediately some unstable disturbances exist. If $\omega_z = 0$, then $v \propto r^{-1}$, and if an axial component of velocity which varies with r is introduced, ω has only one component, namely ω_0, and the most unstable disturbances appear to be rotations around helices with $\beta = 45°$; but this is not a correct conclusion when $v = 0$ because $\kappa = 0$ and the growth rate is zero.

The second case is that in which ω is directed in the negative direction of z (assuming v to have a right-handed rotation). In this case $\gamma = 180°$ and it so happens that if $w = 0$ (in the most obvious coordinate system) then $\beta = 90°$, and the most unstable disturbances are toroidal, namely rotations around the circular streamlines (Fig. 8.7.ii). The circulation and stagnation pressure decrease outwards in this case, and v decreases more rapidly than r^{-1}.

This instability is well known, and has been explained by linearly displaced parcel theory. According to that argument, in the unstable situation the moment of momentum about the axis decreases outwards, so that if a parcel is given a radial displacement by means of a radial force its moment of momentum is conserved and it finds itself with a greater circumferential velocity than its new neighbours. In that case the pressure field which exerts a radial force on the fluid initially exactly balancing the centrifugal force is too small to move the displaced parcel in the same circular path as its new surroundings. It will therefore be accelerated outwards and the disturbance is unstable. An inward displacement puts the parcel in a pressure field which exerts a force towards the axis in excess of the centrifugal force.

This argument is a good explanation of the phenomenon, but it does not enable us to deduce that the most unstable disturbance is the toroidal one, nor does it tell us the effects of variations of w with r.

An interesting corollary of this result is that if a stream with foward shear (velocity increasing away from the boundary) is set into wave motion by the passage of the stream over an uneven bottom, the motion is unstable to toroidal disturbances in the troughs of the waves where the centre of curvature of the flow lies in the direction of the increasing stream velocity (Fig. 8.7.iii). The growing disturbance takes the form of corrugations of the Bernoulli surfaces along the direction of the flow. The instability lasts while the fluid passes through the wave trough, and is stabilised in the following crest. However the dispersion of the stable waves in the crest provides an ideal disturbance for the further growth of the waves in the following trough.

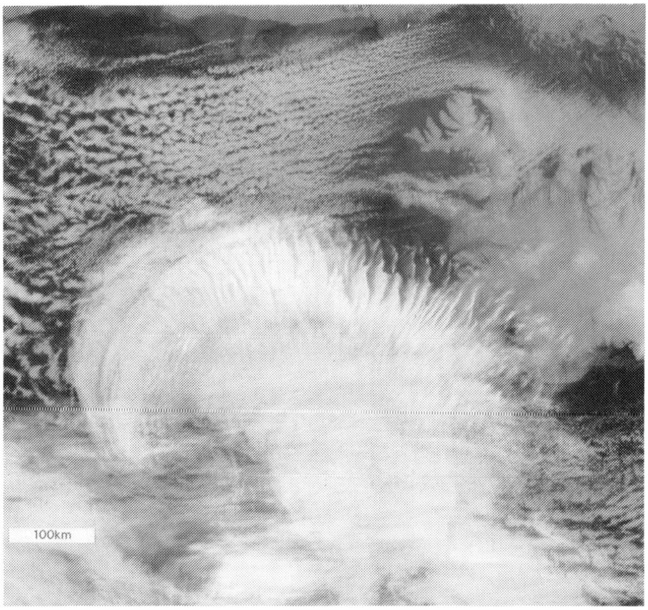

(a)

(b)

Fig. 8.6.ix (*cont.*)

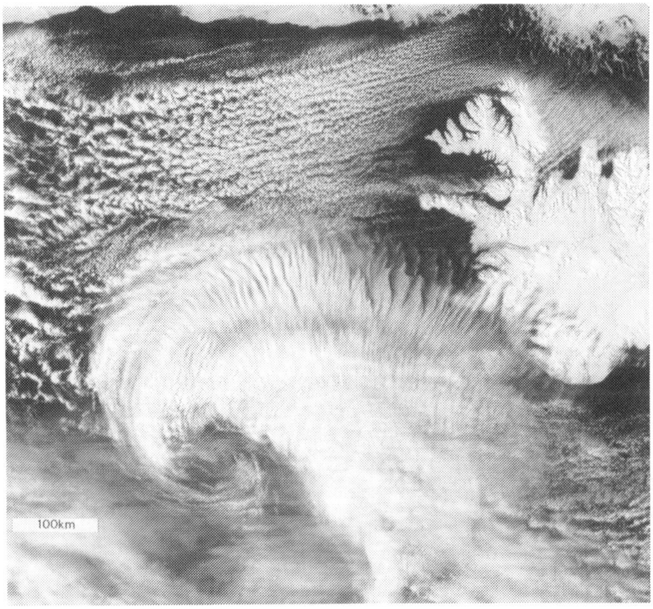

(c)

Fig. 8.6.ix 1602,11.4.88;(a)4, (b)3–, (c)2. The frontal cloud of a cyclone southwest of Iceland. There are billows in the centre left of the main cloud mass. In (b) the wavelength discriminates well between the larger cloud particles (black) and the smaller, which are bright, and in the extreme NE the fragments are melting ice. The mountains of NW Iceland protrude above the inversion at the top of the low clouds, except for the smallest, which produce lee waves. (a) shows colder water (melted ice) which floats on warmer salty water. (Courtesy University of Dundee.)

In the troughs, **b** is directed into the paper so that $\boldsymbol{\omega}.\mathbf{b} > 0$, but in the wave crests the directions of **n** and **b** are reversed and **b** is outwards from the paper. Since $\boldsymbol{\omega}$ is unchanged, $\boldsymbol{\omega}.\mathbf{b} < 0$ in the crests.

This type of instability is sometimes called **Görtler instability** because Görtler vividly discribed how it occurs in the boundary layer of a fluid travelling over a concave boundary, e.g. Görtler [34]. Scorer and Wilson [35] showed that it might occur in standing waves in the atmosphere where a shear layer is contained in the undisturbed stream. In this connection it should be noted that the shear (vorticity) produced in a wave trough is so as to stabilised the flow and prevent the occurrence of Görtler instability; it is therefore necessary that the flow should contain the shear where it is moving horizontally if the instability is to occur further downstream. Billow instability is, in fact, much more likely to occur in standing waves, and is indeed very common.

It may be noted that equation (8.6.35) shows that μ_z becomes infinite in the case $\beta = 90°$, which means simply that because there is no motion in the z-direction, an unstable disturbance will grow indefinitely within the apparatus.

(a)

(b)

Fig. 8.6.x (*cont.*)

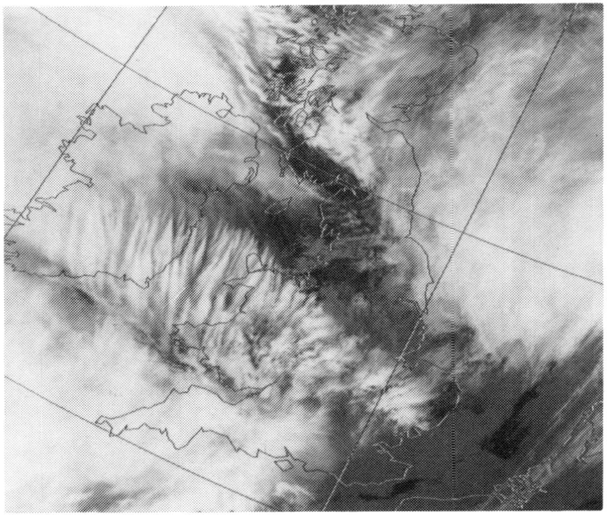

(c)

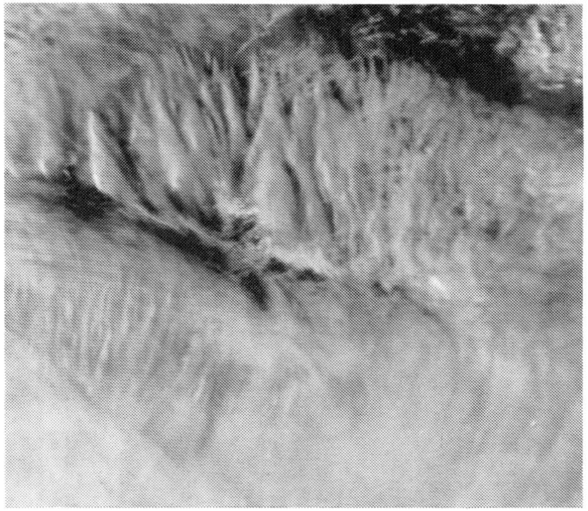

(d)

Fig. 8.6.x (a) 0840, (b) 1605/1424,17.10.84;4. The spike clouds have grown up between these two pictures showing the more general scene. (b) is a composite made from two successive passes about 100 minutes apart. (c) is at 1424;4 enlarged × 3 and (d) is at 1605;1 enlarged × 4.5. (c) has coastline exposed and (d) shows new growth to the south.

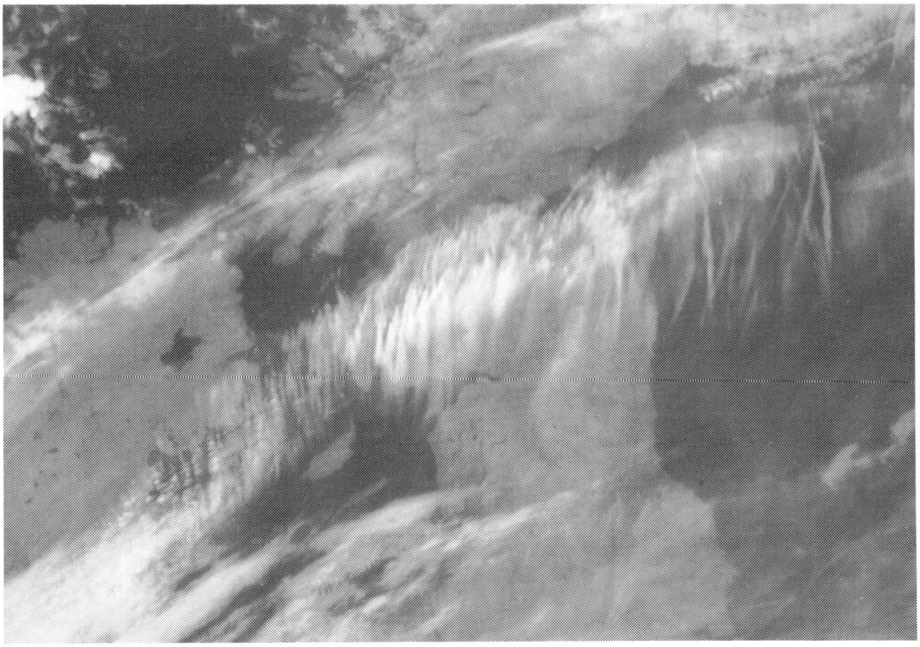

Fig. 8.6.xi 1840,4.10.84; 4. To give an idea of the typical size of these clouds this one is taken over the British Isles. The spike systems are of a size which prevents them being recognised from an aeroplane close by and they are usually obscured by lower cloud from being seen from below. On the right, over the North Sea, are some inverted V-shaped structures which are quite common. The NE coast of England and the NW coast and the Isle of Man can be seen. (Courtesy University of Dundee.)

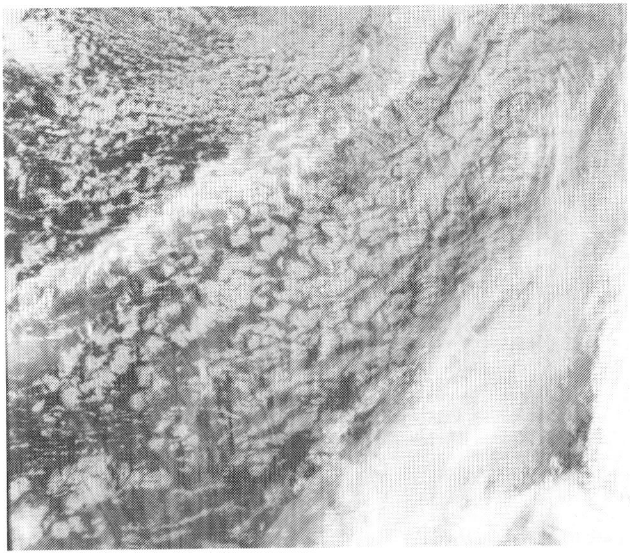

(a)

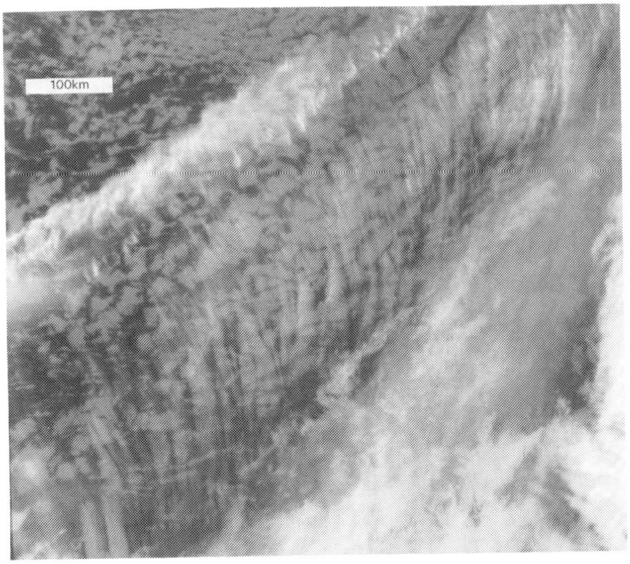

(b)

Fig. 8.6.xii 0419,27.3.87; (a)3+, & (b)4. Two views of spikes at a very early stage when they are transparent to IR (3+ and 4) at 45°N 12°W with no daylight to interfere with the use of channel 3+ as a temperature measure. (Courtesy University of Dundee.)

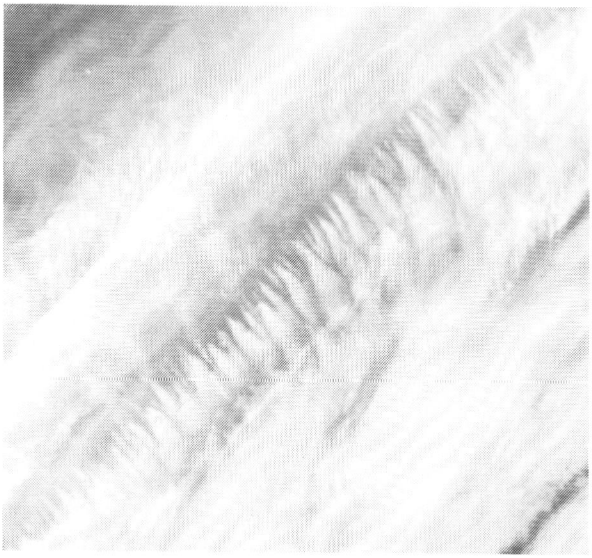

(a)

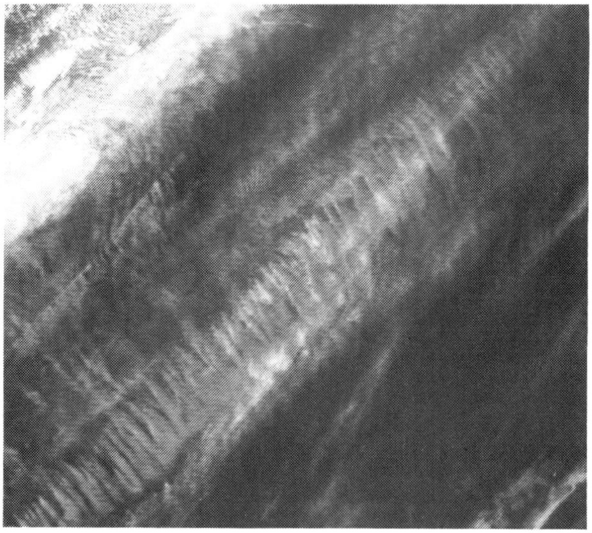

(b)

Fig. 8.6.xiii (*cont.*)

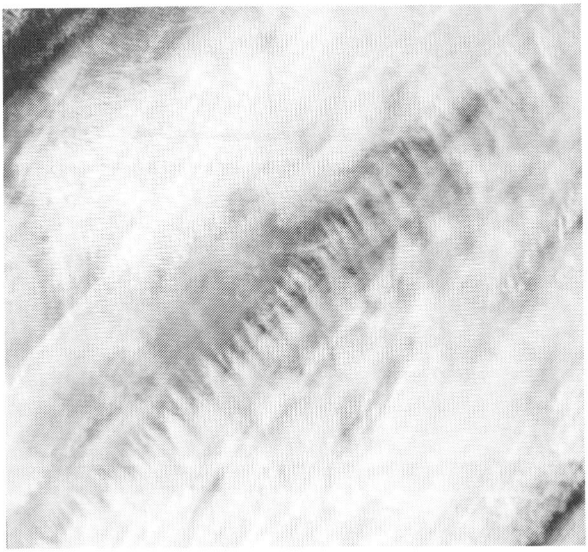

(c)

Fig. 8.6.xiii 1731,26.2.88; (a)4, (b)3–, (c)2. A front lying SW–NE across Newfoundland, centred at about 53°N 47°W. This is typical of many well defined developing fronts. (Courtesy University of Dundee.)

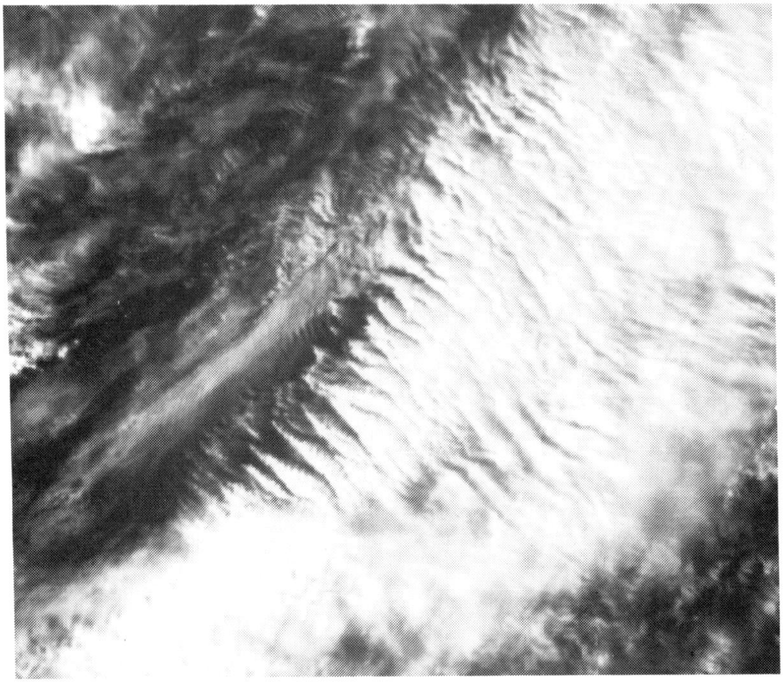

Fig. 8.6.xiv 1431,17.9.81; 2. In mid-Atlantic, spikes often have billows indicating sloping motion, typical of the North Atlantic conveyor belt.

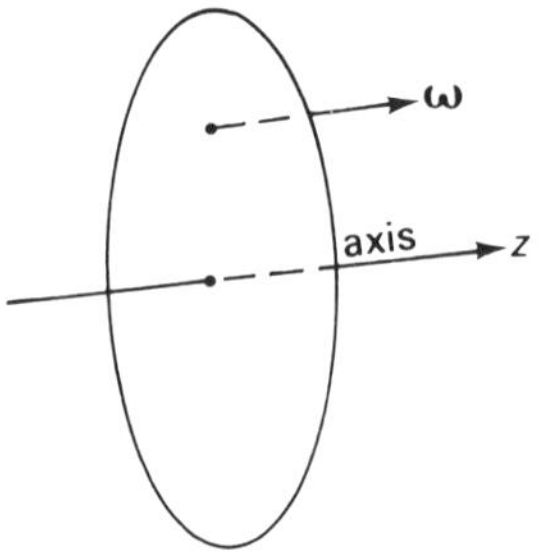

Fig. 8.7.i Plane flow in circles with circulation increasing outwards is stable to toroidal disturbances.

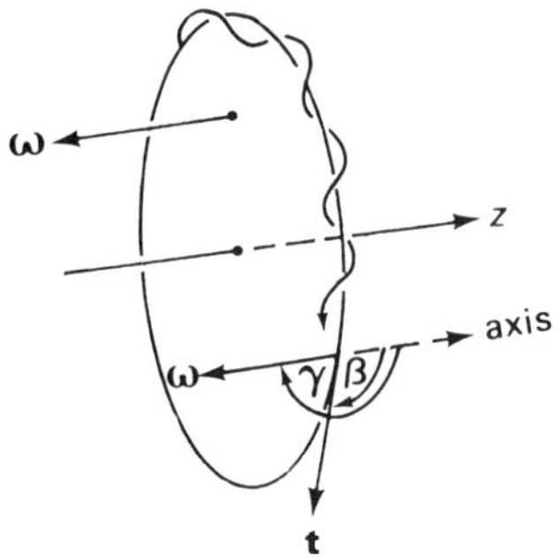

Fig. 8.7.ii Toroidal disturbances are the most unstable when the circulation of plane flow decreases outwards.

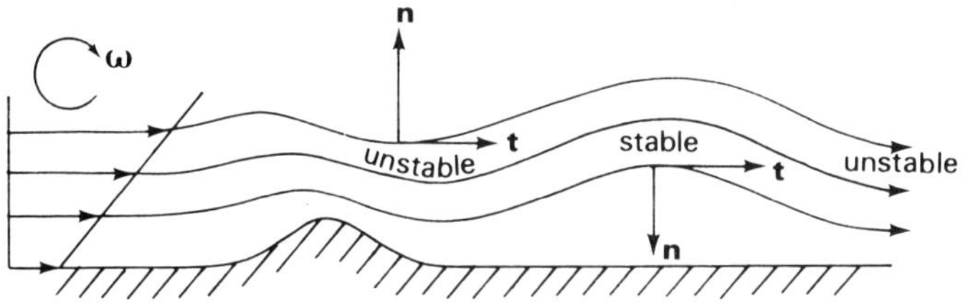

Fig. 8.7.iii Regions of instability in waves in a stream with shear.

8.8 ROTATIONS AROUND THE PRINCIPAL NORMAL AND BINORMAL

Rotations around the binormal are the same as rotations around the tangent with an appropriate axial velocity w_0 added. The equation for the growth of the binormal component of vorticity is (see Scorer 1967)

$$\frac{\mathrm{D}}{\mathrm{D}t}\frac{\boldsymbol{\omega}\,.\,\mathbf{b}}{q} = -2\tau\boldsymbol{\omega}\,.\,\mathbf{n} - \kappa q\mathbf{R}\,.\,\mathbf{t} \qquad (8.8.1)$$

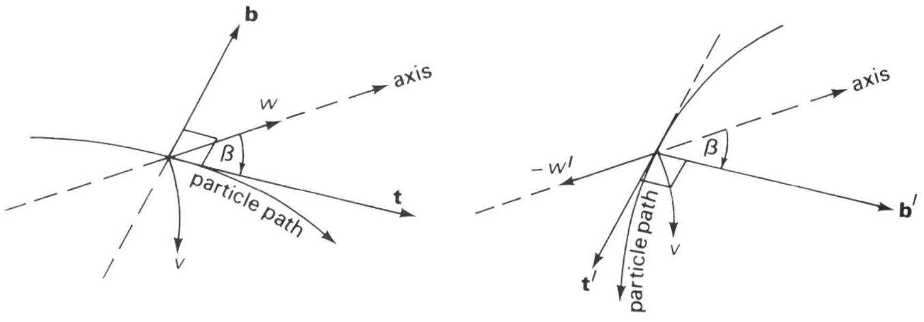

Fig. 8.8.i Coordinates changed so that rotation around **b** becomes a rotation arount **t′**, by a suitable addition of axial velocity.

and this can be transformed directly into (8.6.1) for steady flow by imagining that an appropriate w_0 is added and the new coordinates are denoted by a prime (Fig. 8.8.i). Then

$$\mathbf{t}' = -\mathbf{b}, \quad \mathbf{b}' = \mathbf{t}, \quad \beta' = \beta + 90°,$$

$$\mathbf{v}' = \mathbf{v}, \quad \frac{v}{w} = \tan\beta, \quad \frac{v'}{w'} = \tan\beta' = -\cot\beta,$$

$$w' = -v\tan\beta, \quad q' = \frac{v}{\cos\beta} = q\tan\beta,$$

$$\kappa' = \frac{1}{r'}\sin^2\beta' = \frac{1}{r}\cos^2\beta = \kappa\cot^2\beta = \tau\cot\beta,$$

$$\omega = \omega', \quad \mathbf{R} = \mathbf{R}', \quad \mathbf{n} = \mathbf{n}',$$

$$\tau' = \kappa'\cot\beta' = -\kappa\cot^2\beta\tan\beta = -\kappa\cot\beta. \tag{8.8.2}$$

Equation (8.8.1) applied to the new coordinates is

$$\begin{aligned}\left(\frac{\mathrm{D}}{\mathrm{D}t}\right)'\frac{\boldsymbol{\omega}'.\mathbf{b}'}{q'} &= -2\tau'\boldsymbol{\omega}'.\mathbf{n}' - \kappa' q'\mathbf{R}'.\mathbf{t}' \\ &= 2\kappa\cot\beta\,\boldsymbol{\omega}.\mathbf{n} - \tau\cot\beta q\tan\beta\mathbf{R}.(-\mathbf{b}) \\ &= \cot\beta(2\kappa\boldsymbol{\omega}.\mathbf{n} + \kappa q\mathbf{R}.\mathbf{b}) \\ &= \frac{q}{q'}\frac{\mathrm{D}}{\mathrm{D}t}\frac{\boldsymbol{\omega}.\mathbf{t}}{q}.\end{aligned} \tag{8.8.3}$$

In steady flow when there is no tangential acceleration q is a constant and we can say that

$$\left(\frac{\mathrm{D}}{\mathrm{D}t}\right)'\boldsymbol{\omega}'.\mathbf{b}' = \frac{\mathrm{D}}{\mathrm{D}t}\boldsymbol{\omega}.\mathbf{t} \tag{8.8.4}$$

and (6.7.1) gives the same information as (8.6.1), which must be the case because it is concerned with the same component of vorticity seen in a different coordinate system.

The equation for the normal component of vorticity is readily shown (Scorer 1967) to be

$$\frac{\mathrm{D}}{\mathrm{D}t}\frac{\boldsymbol{\omega}.\mathbf{n}}{q} = 0 \tag{8.8.5}$$

to the first degree in ϕ, and so the helical **swirling flow is neutral to rotations around the principal normal**, i.e. rotations in which the position of the Bernoulli surfaces remains unchanged.

We have thus demonstrated that the discussion of the local instability in curved flow is completely covered by the equation for rotational displacements around the tangent, namely (8.6.1).

8.9 EPILOGUE

All the phenomena of instability we have analysed exemplified the principle that an orderly pattern of vorticity breaks down into an apparently disorganised one. This accords with the second law of thermodynamics, which makes clear the reduction of orderly patterns of kinetic energy, through viscosity, to heat. The spike clouds are like billows, streets and cellular correction.

In a fluid medium like the air and water we have mainly had in mind this reduction will be seen to take place by the transfer of kinetic energy of larger scales of vorticity to eddies of smaller scale, so that the vorticity becomes chaotic before viscosity becomes important and converts the energy into heat.

The next few chapters will be concerned with the creation of orderly vorticity and its conversion into **chaotic vorticity**, which we shall take as our definition of what is rather unspecifically called turbulence. The chief consequence of this is that any quantity carried by the air or water is diffused, and, as we used to hope, also becomes dispersed.

Part 2

Turbulent phenomena, clouds and dispersion

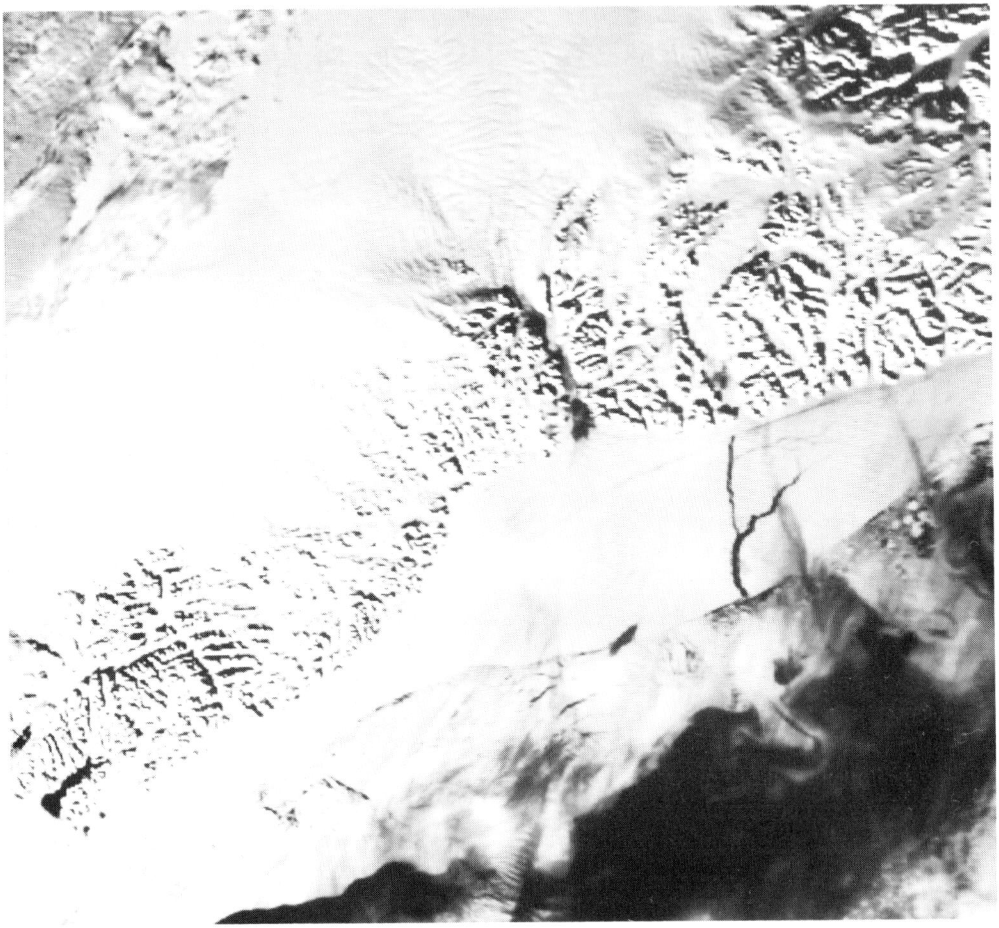

This is the same scene as the frontispiece of Part 1 but made in Channel 2 which is very close to eye vision. In addition to the clear water in the main katabatic outflow (which is not at all visible) there is an unfrozen estuary in the far SW corner. Here also, as in the case of the water under the big crack in the fast ice, the warmer salt water has penetrated to the surface. The shadows in the snow-covered mountain ridges are shown as relatively dark; more also, on the flat fast ice there are shadows of cirrus clouds.

9

Turbulence and diffusion

9.1 THE OUTSTANDING DIFFICULTY

In the first (1895) edition of *Hydrodynamics*, Lamb wrote (Art. 311), 'It remains to call aggention to the chief outstanding difficulty of our subject.' The same sentence heads the completely rewritten treatment of the last (1932) edition (Art. 365). Interesting though it is, Lamb's discussion is not particularly illuminating to the student, whereas Brunt [7], by contrast, has managed an orderly presentation which is a mixture of good physical concepts and simple mathematical treatment. The simplicity is illusory: it is a tribute to Brunt's unsurpassed skill as a teacher that a sense of certainty seems to be imparted and that the analysis conveys the impression of achievement. By making no pretence that all motions which someone has categorised as turbulence are susceptible to a single treatment, and by considering certain selected situations where light can be shed on the mechanics, one can both advance and avoid the error of thinking that **turbulence** must be one thing and that any property it is discovered to possess in one context must be possessed by it in another. Turbulence has almost acquired the characteristics of a 'Loch Ness monster' (see section 15.1).

Writing in 1962, Sutton also said that '**turbulence** ... constitutes the outstanding difficulty of hydrodynamics' (*The Challenge of the Atmosphere*, p. 129). Since there are many turbulent motions for which a full analysis can never be obtained, we have to employ a new approach. Batchelor's *The Theory of Homogeneous Turbulence* is a good example of a treatment of a selected topic in the vast realm of turbulent motions, and it has often been unfairly criticised because its techniques have very limited usefulness in practical cases where the turbulence is generally not homogeneous. It is however very enlightening on the process of decay, which we are scarcely concerned with in this book.

Prandtl's mixing length theory gives a simple mathematical description of turbulent motion near boundaries and the transfer of momentum to the fluid; this, together with Reynolds's eddy transfer theory, stimulated the solution of problems by *K*-theory (well exemplified in Brunt's treatment). The feeling of warm familiarity which *K*-theory engenders in engineers and others who are well versed in the theory of heat and electrical conduction and of viscous fluids made inevitable the extravagant flirtations with that theory in which some authors have engaged (e.g. Hinze [36]), and which remains widely used through lack of familiar alternative.

A completely different approach to turbulence, started by Reynolds's study of the onset of turbulence in pipes, is concerned with hydrodynamic instability. Much of this work is very sophisticated and erudite, and discusses the conditions in which small disturbances will grow and cause a transition from laminar to turbulent flow. We need to know something about this approach because the transition is very common in the atmosphere, but it must be emphasised that, although engineers may seek to establish fairly exact criteria for the transition, we are more concerned with the mechanisms which change the state from stable to unstable than with the mechanics of transition itself. Nature is not concerned with the critical state in which there may or may not be eddies, but engages in a destabilisation or stabilisation over many minutes or hours. When the motion becomes unstable it is so violent that it bears no resemblance to the form of the growing disturbances at the initial moment of breakdown of laminar flow.

In Chapter 10 we treat details of cases in which there are some useful simple certainties. Although we shall make some appropriate remarks about turbulence in the neighbourhood of solid boundaries, there are many texts on this part of the subject which we shall make no attempt to summarise. Turbulence away from or in the absence of boundaries is alamost completely neglected in the literature. Indeed many discussions of such cases (for example, of the motions which cause inhomogeneities of refractive index in the atmosphere and give rise to the scattering of radio waves) are bedevilled by concepts developed in connection with boundary layer turbulence or wind tunnel turbulence created by flow-through grids. The free use of the word turbulence to categorise such motion is one cause of the mischief. Some people always assume that the theory of turbulence taught and used in one context must be applicable to complicated fluctuating motions in a quite different context, especially when the same instruments (e.g. hot wire anemometers) are used in both cases.

It is regrettably necessary to warn the reader against two errors of logic which are common. The first is that a statement about one particular kind of turbulent motion may be true (i.e. is not necessarily untrue) even though it may be shown to be obviously untrue in a different kind of turbulent motion. Thus it has been suggested that since flying insects have no navigational ability, turbulence must disperse them just as it disperses smoke and aphids in the atmosphere. But some insects are concentrated by motion which in other contexts may validly be called turbulence (see section 15.2). Equally some complicated wave motions may have many of the appearances of turbulent motion but may yet fail to disperse or diffuse anything carried by the fluid. The best example of 'turbulent' motion causing a concentrtion of a substance carried by the air motion is the fallout of rain and the deposition of pollution carried with the incipient rain.

It has been shown that decaying homogeneous turbulence has, under certain circumstances, a range of eddy sizes in which their energy, E, is related to their wave number, k, according to

$$E \propto k^{-5/3}. \tag{9.1.1}$$

If a specimen of turbulence is then measured and found to obey this relationship it is sometimes argued that it must be homogeneous and decaying, and that the eddies observed lay in the range for which (9.1.1) holds. Furthermore, these statements are sometimes said to be roughly true when (9.1.1) is found to be roughly true. The absurdity of such a conclusion is clear when we note that a single sharp change in velocity may be analysed into Fourier components. Then

$$E \propto k^{-6/3}, \tag{9.1.2}$$

which, in practical cases where fluctuations of velocity are subjected to Fourier analysis, is scarcely distinguishable from (9.1.1). The logical incompleteness of the conclusion should be apparent to all scientists; yet the temptation to assert that an explanation which appears to predict the observed fact must be correct is often found irresistible.

9.2 DEFINITIONS OF TURBULENCE

It is not my purpose to arrive at a definition of turbulence; too much breadth at conferences has already been expended on that. My aim is simply to clarify some of the ideas so that two things will be plain. First, that there is no single definition which ought to be universally adopted, and secondly that there is enough incompatibility among the definitions to make us wary of applying to other cases conclusions arrived at in the study of some example of motion called turbulent.

The idea of turbulence arises because some motions cannot be precisely described. Furthermore, it would not be useful to have a precise description of them because it would be too complicated. What we desire is to talk about the effects of such motions in practical terms, which means arriving at overall or statistical descriptions or conclusions without discussing the details. To some people this has apparently meant that we may not or cannot discuss the details of the mechanics and must therefore represent the effects of turbulence by parameters. However, there is no guarantee that a representation by parameters has any value, for in some cases we can only evaluate the parameters afterwards, and predictions are not facilitated. Lamb's description of it as the 'outstanding difficulty' carries the implication, or at least the hope, that one day this difficulty will be resolved. In fact the resolution seems to be taking the form of learning to live with impossibilities rather than inventing a tractable theory.

This coming to terms without limitations is actually much more satisfactory than it may seem because it amounts to making our objectives sensible. It is obvious that the deeper we probe into something, the more complex it seems to be, and in the case of indescribably complicated motion the only road to simplification is to decide that there shall be limits to the complexity we are prepared to study. Life is too short to spend it sorting out the infinite details of the flow on a single occasion, and anyone who did analyse them could have no assurance that the next occasion would be the same, nor that anyone else would be interested anyway.

The degree of complexity beyond which we decide not to go in our studies is a matter of choice. Up to that point we think of the flow as being laminar and susceptible to description in all detail, but beyond that point we call it turbulence, possessing complexities whose overall effect only will interest us.

Thus we arrive at the idea that there is in existence a mean motion with some fluctuations added. The mean motion is fully described, but of the fluctuations we only have statistical information. Consequently there is an element of choice: we decide how to make the division between the mean motion an the fluctuations. There is no reason why two people should not make different divisions, either because they set up different criteria for doing so, or because they had different purposes for their theory. As a result the turbulence (the fluctuating part) is not a clear-cut objective property of the motion but a matter of arbitrary definition for our convenience. We may be driven by the logic of the situation to make the definition in a certain way, and then it may appear to be a recognition of a real feature of the flow. However the logic is not general; it is peculiar to the situation and its relationship to us, to our skills, and our wants.

A quite different definition to the effect that 'turbulence is a complex motion which causes diffusion' is a very practical one. It recognises that when we stir the milk into the tea or the dye into the paint it is the result of a very complex pattern of motion that uniformity of composition is arrived at, and that the mixing is much more effective than molecular diffusion. Diffusion is an irreversible event. Furthermore, when we wave a spoon about in the tea we produce spoon- or cup-sized motions which quickly degenerate into much more complicated motions with smaller-size scales associated with them. The motion which spreads tobacco smoke would be described as turbulence: it is not long before the smell can be detected in every corner of a room. This is also an arbitrary definition in the sense that we have already made up a definition of diffusion, which we think of as a complicated motion which transports fluid particles around and mixes them up irreversibly. Sometimes turbulence is described as 'the random part of the motion', and again there is implied a previous choice as to what is random, for random merely means 'containing details we shall not enquire into'.

The turbulence diffuses the fluid itself, and with it any characteristic attached to it, such as colour, chemical constitution, together with airborne pollution such as smoke, smell, water vapour and so on. Fluid parcels also possess heat, density, linear and angular momentum, and kinetic energy of rotation and of linear motion relative to their surroundings. There are features not attached permanently to the parcels: some may be diffused by molecular transfer mechanisms, or by the action of pressure gradients at a much greater rate than would occur through the molecular diffusion of the material. To cite extreme cases, fluctuations of density may be communicated through a fluid at the speed of sound by pressure waves, whereas fluctuations of concentration of a pollutant can only be moved by the movement of the material.

The velocity of the fluid at a point might be fluctuating because of **eddies**, which transport the fluid elements or because of **waves** passing in all directions and causing all the fluid elements to oscillate about their mean positions and return to them when the fluctuations have finished. In the first case, any property carried by the fluid parcels woould be diffused; in the second case, the distribution would be as it was before the waves began to pass through the fluid.

There is a very obvious difference between these two types of motion. Waves can be thought of as being produced by the movement of the boundary of the fluid, and if for the moment we consider the case in which the viscosity produces negligible effects, any surface within the fluid may be thought of as the boundary (of the region of interest, say). If the fluid is at rest initially, or is in motion but possesses no vorticity, and is of uniform density (or potential temperature in the case of adiabatic motion) no vorticity will be produced. The motion will therefore be irrotational throughout and will also be reversible. Thus if there were a boundary of very complicated shape moving past the fluid, or if the boundary were flexible and made to move in a very complicated way, the fluid would be subjected to very complicated fluctuations of velocity and the parcels subjected to very complicated changes of shape. However, the complexity of the motion would correspond with the complexity of the motion of the boundary, and would not be more complicated at any stage. Also, the fluid could be brought back exactly to its initial position by simply reversing the movement of the boundary back to the beginning if the fluid were also incompressible.

By contrast, a body of fluid with a fixed boundary enclosing it, set into eddying motion by some means such as a spoon moved through it, would diffuse throughout its whole volume any concentrated material which initially polluted only a small part of it. In the absence of viscosity the motion which would produce this diffusion could take place in the reverse direction, but the prescription for setting up that motion which would reverse diffusion that had already taken place becomes more and more complicated as time proceeds. It is easy to distinguish between earlier and later states through which the fluid passes when diffusion occurs, but in the case of the irrotational incompressible fluctuating motion the later state is essentially no more complicated than an earlier state. Likewise, if waves were travelling back and forth either as sound or gravity waves without any dissipation or loss of energy in a box containing fluid, it would be impossible to tell by inspection whether a cine film of it was being shown forwards or backwards. A casual observer might describe a wavy sea surface or a wobbling jelly as being in turbulent motion because of the random nature of the velocity fluctuations of a particle or at a point, but the motion is essentially different from that in which eddies are causing diffusion. Our discussion will be restricted to cases where vorticity is present and where the direction in time is obvious.

This leads us to what is probably the simplest definition of turbulence, namely **chaotic vorticity**. This still retains the arbitrary human element in the word chaotic, for often the experienced eye sees order in a scene which to the untrained observer is chaos. However, there is a degree of complexity beyond which no one wishes to probe when analysing the details of the motion of a fluid. When there is a fairly complicated pattern of vorticity present, and in some cases when there is a very simple pattern (such as a plane vortex sheet), the motion will become progressively more complicated. The element of arbitrariness in our definition then is considerably reduced because the motion will be turbulence before very long by any definition, unless viscosity stops it becoming complex by destroying it. Thus if we waft a spoon once through a bucket with a blob of ink in it we set up turbulence and the blob becomes a diffusing cloud which obviously qualifies for the title 'turbulent'. If we waft the spoon through a bucket of goldent syrup in the same way a blob of black treacle in its midst would at worst be slightly distorted, but would certainly not assume a shape of indescribable complexity. An artist could reproduce its shape in all essentials with ease and a technician could measure it. This is because the viscosity is so large that eddies of the size of the bucket, or smaller, cannot turn through more than a small fraction of one rotation before they are brought to a standstill.

A very simple way of determining whether a distribution of vorticity is chaotic or not is as follows. We determine a grid of points such that we are not interested in investigating motion on a smaller scale than the grid spacing. If the vorticity vector is highly correlated over several neighbouring grid points it means that an orderly pattern of motion is present on a scale larger than the grid spacing. If we make a Fourier series such that it takes the value of the velocity at all the grid points along one coordinate line, the motion must be on a much larger scale than the grid length if only the first few terms in the series are required to give a good fit. If, on the other hand, the final term used corresponding to wavelengths of twice the grid length has an amplitude as large as the others, the motion has not been approximately described by giving the values at the grid points. It means that if the number of grid points were doubled, twice as many terms might be required. Therefore much of the motion was being missed by using the original grid size, this being the part of the motion we didn't want to know about except in so far as it affected the motion on a larger scale, for example by diffusing momentum and behaving like viscosity.

The special feature of motion possessed by vorticity is that it contains energy of motion in the fluid which cannot be removed from it by suitable motion at the boundaries, and which equally cannot be set up simply by motion of the boundaries. The motion of the spoon through the cup of tea or bucket of water is special in that it has salient sharp edges or lines along which the flow separates from the boundary, and vorticity can stream into the fluid from the lines of separation.

Before studying the mechanics of the turbulent motion composed of chaotic vorticity it is necessary to point out that progress can be made only if the motion can be divided effectively into a mean motion, which is intelligible, and a random motion, which is chaotic. We may wish to set the dividing length, or time, scale between the two parts of the motion for our own convenience, but it is possible that there may be limits set by the nature of the motion itself. Thus there may be a very orderly motion on a large scale such as the flow down a straight pipe of circular section. Superimposed on this there may be fluctuations which clearly cause the diffusion of a filament of ink released into the pipe. There are no eddies much larger than the radius a of the pipe, and no fluctuations of period longer than na/U, where n is not much larger than unity, and U is the average speed derived from the flux $\pi a^2 U$. The natural division into a mean flow, $u(r)$, and fluctuations is achieved by defining $u(r)$ as the average speed at radius r either at a fixed section over a time much longer than na/U or over a length of pipe much greater than r.

Such a natural division is not necessarily possible because eddies of all sizes may be present. This would mean that if a small number of smaller eddies or a large fraction of one larger eddy was sampled in an attempt to measure the mean by taking an average over a certain length or time, the value obtained would depend very much on the starting and ending points of the sample. If a very large number of eddies is included in the sample, another half eddy will make a negligible difference to the mean value obtained: equally, if only a very small fraction of an eddy is sampled, a slightly larger fraction will obtain very nearly the same value. Then samples obtained from different parts of this large eddy would merely serve to reveal the shape of the eddy, which would be seen as part of the mean motion. Thus eddies whose scale is very different from the sampling scale sort themselves naturally into random fluctuations, which we do not study in detail, and aspects of the mean motion. The technique fails or is very limited in usefulness if there are eddies present of scale near to the scale of the sampling procedure, which in our previous example was the grid spacing.

9.3 REYNOLDS STRESSES; MIXING LENGTH; DEGENERATION OF EDDIES

The mean value of a quantity q which varies with s, which may be a space or time coordinate, is defined at $s = s_0$ by

$$\overline{q}(s_0) = \frac{1}{2\sigma}\int_{s_0-\sigma}^{s_0+\sigma} q(s)\,\mathrm{d}s. \tag{9.3.1}$$

This is the average value of q over the interval of length 2σ surrounding the point or instant s_0. To be meaningful this definition must make $\overline{q}$ independent of σ over a fairly wide range of values of σ. If σ is too large we shall be unable to detect variations of $\overline{q}$ from one value of s_0 to a nearby value; if σ is too small, $\overline{q}$ will be found to fluctuate rapidly as s_0 varies, and the whole purpose of our definition is to define a mean value quite distinct from the fluctuations. The success of the following analysis depends on the existence of a significant range of values of σ which lead to the same value of $\overline{q}$. Thus σ must lie in a gap in the spectrum of eddy size present in the motion.

We may now write

$$q = \overline{q} + q' \tag{9.3.2}$$

where q' is the fluctuation. By definition

$$\overline{q'} = 0 \quad \text{and} \quad \overline{\overline{q}} = \overline{q}. \tag{9.3.3}$$

For two components of velocity v_i and v_j we have

$$\begin{aligned}\overline{v_i v_j} &= \overline{\left(\overline{v_i} + v_i'\right)\left(\overline{v_j} + v_j'\right)} \\ &= \overline{\overline{v}_i\overline{v}_j} + \overline{\overline{v}_i v_j'} + \overline{\overline{v}_j v_i'} + \overline{v_i' v_j'} \\ &= \overline{v}_i\overline{v}_j + \overline{v_i' v_j'}\end{aligned} \tag{9.3.4}$$

for by (9.3.3)

$$\overline{\overline{v}_i v_j'} = v_i\overline{v_j'} = 0.$$

The equations of motion (1.3.26) and continuity $(\partial v_i/\partial x_i = 0)$ for an incompressible fluid may be combined to give the following:

$$\begin{aligned}\rho\frac{\mathrm{D}v_i}{\mathrm{D}t} &= \rho\left(\frac{\partial v_i}{\partial t} + v_j\frac{\partial v_i}{\partial x_j}\right) \\ &= \rho\left(\frac{\partial v_i}{\partial t} + \frac{\partial}{\partial x_j}(v_i v_j)\right) \\ &= \rho F_i + \frac{\partial}{\partial x_j}p_{ij},\end{aligned} \tag{9.3.5}$$

and if we now take the mean value of each term, the fluctuations only appear in the product term and we have, for the case in which ρ is a constant,

$$\rho\left(\frac{\partial v_i}{\partial t}+\frac{\partial}{\partial x_j}(\bar{v}_i\bar{v}_j)\right)=\rho\bar{F}_i+\frac{\partial}{\partial x_j}\left(\overline{p_{ij}}-\rho\overline{v_i'v_j'}\right). \tag{9.3.6}$$

This equation is of the same form as (9.3.5), but is an equation for the acceleration of the mean motion in terms of the mean body force $\bar{F}_i$, and the gradients of the mean stresses $\overline{p_{ij}}$ with an additional term $-\rho\overline{v_i'v_j'}$ added. This additional term is called the Reynolds stress, and (9.3.6) is **Reynolds's equation** for the mean motion.

The quantity p_{ij} is the force in the i-direction on unit area normal to the j-direction, and represents the rate of transfer of i-momentum in the j-direction by the viscous stresses. The mean value of the viscous stress is defined so that fluctuations in the stress have zero mean value. This is the same as the stress that would be produced by viscosity if the velocity were everywhere equal to the mean velocity if there are no fluctuations in viscosity. The quantity $\rho\overline{v_i'v_j'}$ is the average rate at which i-momentum represented by $\rho v_i'$ is transported by the velocity v_j' in the j-direction. It is therefore called the **eddy stress**, and when there are eddies it is usually very much larger than the mean viscous stress $\overline{p_{ij}}$, except when $i=j$, for then the mean normal pressure is much larger than the eddy normal pressure represented by $\rho\overline{v_i'^2}$.

The mechanism of the eddy shear stress is simple. If there is a gradient of the mean velocity v_i in the j-direction, the fluctuation v_i' is correlated with the fluctuation v_j' by the gradient. Thus if v_x increases in the y-direction (see Fig. 9.3.i), a parcel of fluid moving in the y-direction possesses, on the average, a smaller v_x than the surroundings into which it is moving, and it therefore carries a negative anomaly of x-momentum.

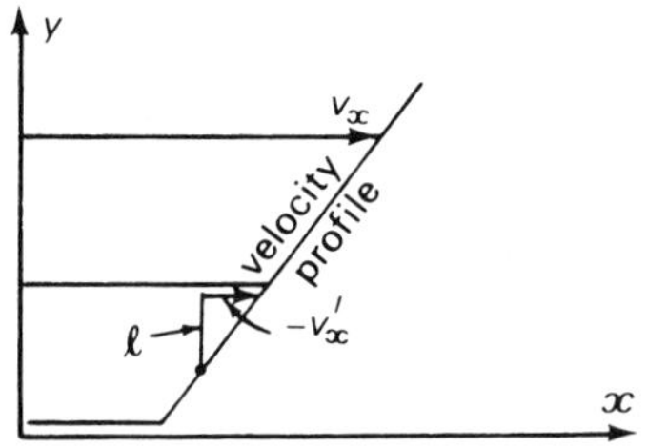

Fig. 9.3.i If the gradient of mean velocity is positive, a parcel displaced a distance ℓ in the positive direction acquires a negative velocity anomaly $v_x' = -\ell\,\partial v_x/\partial y$.

If we imagine the parcel to travel a distance ℓ, called the **mixing length**, before it mixes with its new surroundings, its velocity anomaly, v_x', is $-\ell\,\partial v_x/\partial y$. The mixing length is not the same for all parcels but only represents a statistical average; furthermore it could well be that the length is different for the transfer of different properties. Thus the parcel may share its momentum with its surroundings more readily through pressure gradients than it shares its heat, or its content of some pollutant that has ultimately to be shared by molecular or small-scale mixing.

It has sometimes been argued, as it was originally by Prandtl, that the velocity fluctuations have no preferred direction, and therefore v'_y is the same as v'_x. This is based on observations, backed up by some theory, that the velocity fluctuations quickly tend to this state, but the argument which now follows is valid provided only that the various components are in the same proportion at all distances from the boundary. The shear stress τ, which is the force in the direction along the boundary on unit area parallel to the boundary, is equal to the rate of transfer across the boundary of momentum along it. Thus if we accept Prandtl's assumption that the only quantity which determines the eddy size and the mixing length is y, the distance from the boundary

$$\tau = \rho \overline{v'_x v'_y} = \kappa \rho y^2 \left(\frac{\partial v_x}{\partial y} \right)^2, \tag{9.3.7}$$

which is sometimes written as

$$\kappa \rho y^2 \frac{\partial v_x}{\partial y} \left| \frac{\partial v_z}{\partial y} \right| \tag{9.3.8}$$

in order to ensure the correct sign. Physical considerations make it obvious what the sign is. κ is a non-dimensional number which relates ℓ to y, and if the turbulence is generated by the work done against the eddy shearing stress and is dissipated into heat via smaller eddies, it may be supposed that equilibrium is reached and that κ always takes the same value. The exercise is simply one of dimensional analysis in that case, and there are many alternative derivations. Thus von Karman proposed that the eddy viscosity was determined by an appropriate combination of the features of the velocity profile which generates the turbulence, and suggested that

$$\tau = \kappa \rho \left(\frac{\partial v_x / \partial y}{\partial^2 v_x / \partial y^2} \right)^2 \left(\frac{\partial v_x}{\partial y} \right)^2. \tag{9.3.9}$$

This formulation, however, amounts to the same thing, for in the case of both (9.3.7) and (9.3.9) we find that if τ is supposed independent of y, so that there is no pressure gradient in the x-direction, the velocity profile is given by

$$v_x = \frac{1}{\kappa} \left(\frac{\tau}{\rho} \right)^{1/2} \log y + \text{const.} \tag{9.3.10}$$

This is the logarithmic profile of velocity near a boundary and is to be expected when no other boundary is near, and at a sufficient distance from the boundary for the flow to be turbulent. That means that the eddies must be large enough for them not to be dominated by viscosity. It also means that the velocity gradient at larger distances from the wall is smaller because the flow is subject to a greater eddy viscosity: this is clearly not the case in many natural flow fields, and there are other causes for the decrease of velocity gradient, such as are described in Chapter 4. κ is called **von Karman's constant** and is equal to about 0.4.

We are thus led to consider what happens to turbulence, and what is the real mechanism of the Reynolds stress, for the fluid is a continuum and the mixing length theory is no more than a picture for the dimensional analysis. The results can be obtained directly from the assumptions as to what factors determine the motion, and in this case it is assumed that the eddies and the motion are on a sufficiently large scale for the effects of viscosity to be negligible. They are valid for a large enough Reynolds number, which in this case is

$$\mathrm{Re} = \frac{(\tau/\rho)^{1/2} y}{\nu}, \tag{9.3.11}$$

there being no other velocity specifiable for the system than $(\tau/\rho)^{1/2}$, which is called the **friction velocity** and is often denoted by u_*. Clearly viscosity becomes important for small y. The kinematic viscosity ν was defined for equation (1.3.28).

A turbulent field of velocity is composed of eddies, or fluctuations, of a wide range of sizes. The scale of the turbulence can be measured by its largest and its smallest eddies. A large eddy will cause correlations between the velocities of particles at two different points in it, and so the macroscale of turbulence is the distance at which there is no correlation of the velocity fluctuations. In the atmosphere this scale is larger than the size of the earth beacuse wind systems such as the trade winds and westerlies extend right round it.

Small eddies are quickly dissipated by viscosity; indeed it is possible to create smaller eddies in water or air than in treacle because of this. The greater the energy input, the smaller the eddies that are created, and if the turbulence is left to itself the size of the smallest eddy increases as the motion is turned into heat by viscosity. If ε is the rate at which energy is being fed into the eddies which are dissipating it as heat, then the quantity

$$\left(\nu^3/\varepsilon\right)^{1/4} \tag{9.3.12}$$

is a length, which may be regarded as a measure of the smallest eddy size.

If the eddies are regarded as a spectrum of scales of motion which can be represented by a Fourier analysis, it would be expected that the interaction of two eddy sizes would be to create both larger and smaller eddies. Furthermore, at large enough scales for viscosity to be negligible, the flow is theoretically reversible because there is no dissipation to heat, and so the motion which generates small eddies out of large ones can be reversed and generate large ones out of small ones. This cannot happen in practice because the situation in which only the larger ones exist is special, and time cannot be reversed so as to make the flow tend towards it. For example the complicated flow generated in billows, or any other unstable flow pattern, cannot be reversed because the complex pattern which would lead to it cannot be set up. We now discuss how this comes about, for it is part of the mechanics of increasing disorder and degeneration of motion into heat.

A characteristic of vortex motion is that the line of particles joining two given particles is ultimately stretched. Thus, in the case of plane shearing motion in which the velocity is (Fig. 9.3.ii)

$$v_x = \alpha y, \quad v_y = v_z = 0, \tag{9.3.13}$$

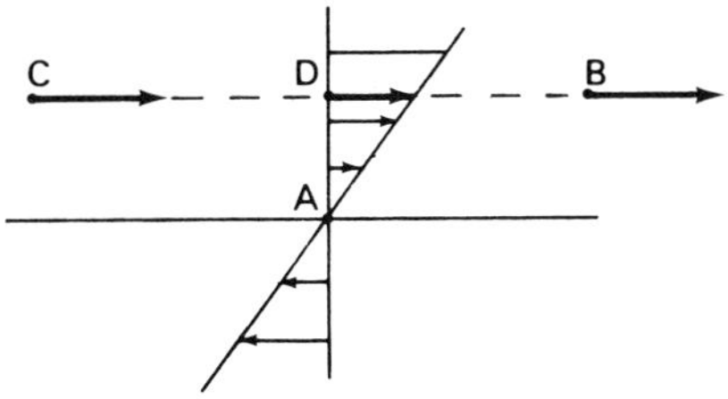

Fig. 9.3.ii Even the line AC, which is compressed by the motion at first, is ultimately subjected to stretching from the position AD.

the particles at A and B are moved apart, whereas the particles at A and C are moved nearer together until the upper particle arrives at D, and from then onwards they are moved apart, for ever. If the particle at B were allowed to move away for a century and the flow then put into reverse, after a century of moving towards A it would move away from A, for ever!

Turbulence contains vorticity because it is a motion in which kinetic energy has been added *in situ* without increasing the orderly transport of the fluid (section 1.1). It is **chaotic vorticity**, because if it were orderly it would be possible to describe it in every detail. In calling it chaotic we decide not to describe every detail because it serves no purpose to do so, although theoretically it could be done. Thus, if we were to describe the distribution of vorticity by giving the vorticity vector at every point of a grid, there would be an orderly pattern which we could subtract out, together with a departure from it which we would call the fluctuations. The fluctuations at neighbouring points of the grid would not then be correlated. There would be eddies present which would be on a scale smaller than the grid length and could not be described by specification of velocity or vorticity at the grid points. The intensity and some other properties of the small eddies would, however, be described statistically by the information at the grid points.

An eddy stress is a mechanism whereby energy of the mean motion, or of a large eddy (which is the same thing as part of the mean motion if we choose to probe into it in detail), is transferred to the eddies. The vortex lines are carried with the motion, except on the small scale, where each component of the vorticity is conducted by the viscosity as well. The eddies are motions around the vortex lines. After a finite time, which in practice means at all times except in specially contrived situations which cannot persist indefinitely, the length of every vortex line is being increased by the motion due to all the others and the density of vorticity is continuously increasing. This is made obvious by consideration of the kinetic energy of an individual eddy represented by a single vortex line, for when a cyclinder of fluid surrounding a segment of the line is stretched, its angular momentum is conserved. Fluid originally at distance r_0 with angular velocity ω_0 around the line (see Fig. 9.3.iii) has angular velocity ω_1 when it is at distance r_1, where

$$\omega_1 = \frac{r_0^2}{r_1^2}\omega_0. \tag{9.3.14}$$

The initial and final kinetic energies of unit mass are therefore

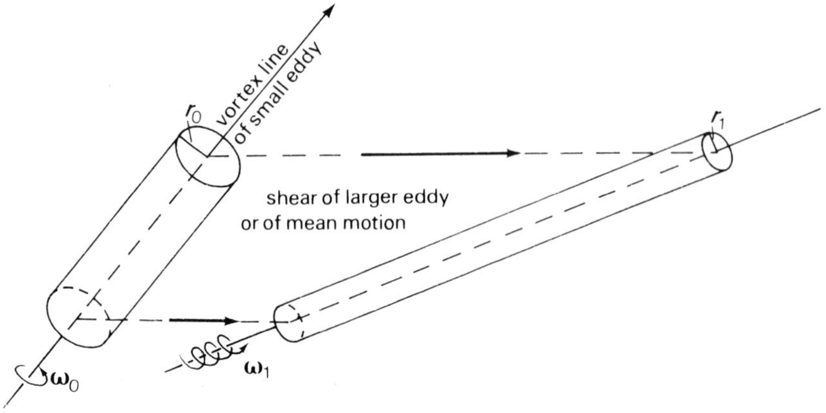

Fig. 9.3.iii Stretching of vortex lines and increase of kinetic energy.

$$\frac{1}{2}(r_0\omega_0)^2 \quad \text{and} \quad \frac{1}{2}(r_1\omega_1)^2 = \frac{r_0^2}{r_1^2}\frac{1}{2}(r_0\omega_0)^2, \qquad (9.3.15)$$

and so the kinetic energy is increased in the ratio r_0^2/r_1^2.

As a consequence of the increase in density of vorticity the viscous dissipation of the motion is increased. Therefore there is a continuous reduction in the sizes of the eddies and a transfer to them of kinetic energy from the larger eddies or from the mean motion at the same time.

It is now evident that **turbulence** which produces a Reynolds stress **cannot be isotropic**: if it were, for every eddy that was being stretched there would be another that was being compressed along its vortex lines. Such a situation cannot persist because after a finite time the eddies could no longer be paired in this way and the majority would be in a state of being stretched. The fluctuations of velocity produced by an eddy consisting of a single vortex line can be considered for the case of plane shear flow, and the argument is the same for any flow containing vorticity. On the near side of the vortex line AB (Fig. 9.3.iv) a positive value of v_x' is produced and is associated with a negative value of v_y'. On the paired vortex line A′B′ a positive v_x' is associated with a positive v_y' and so the momentum transfer $\rho\overline{v_x'v_y'}$ is of the opposite sign. But after a finite time, A′B′ takes up the former position of AB, which has moved by then into the position A_1B_1, and both are giving rise to a shear stress of the same sign. Vortex lines parallel to either of the axes produce no eddy motion along that axis and no shear stress; a vortex line along Ox produces no v_x' and one along Oy no v_y'. Also a vortex line in any direction in the planes of the motion (y = const) remains unstretched in the case of this simple motion.

Unit length of vortex line is stretched at the maximum rate if it is inclined at 45° to the planes of the motion, which is important in the production of sound in a compressible fluid (see section 9.11). The shearing motion itself is giving such eddies energy at the maximum rate, and they may therefore be thought of as shear-induced. On the other hand, vortex lines produced by flow around obstacles are created parallel to the lines along which separation occurs. This is shown in Fig. 9.3.v. In general this is most likely

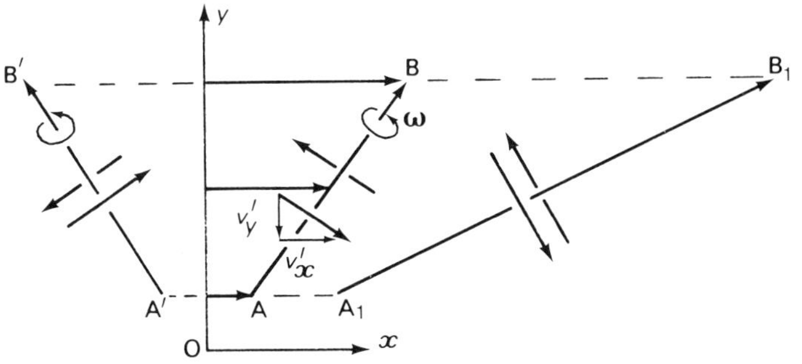

Fig. 9.3.iv Velocity fluctuations due to the vortex lines of a small eddy.

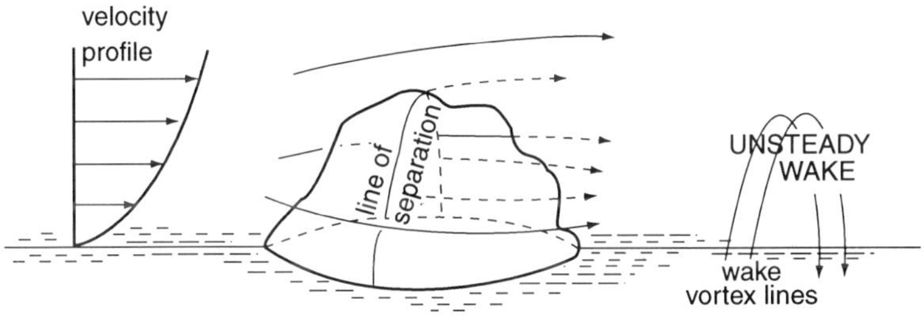

Fig. 9.3.v The vortex lines shed in the wake of an obstacle from the line of separation have a vertical component and are therefore stretched by the shear due to boundary drag in the main stream, and produce an eddy shear stress as soon as the vortex sheet in which they lie becomes billowed.

to be at right angles to the flow. Therefore obstacles in the flow produce vorticity with a significant component at right angles to the shear vector produced by the boundary on which they stand. Mechanically produced turbulence is thus produced by roughnesses on the boundary, for their wakes extract energy from the mean shear flow produced by the boundary. The mean shear flow itself makes no contribution to the chaotic vorticity unless it is unstable, but it is likely to become unstable to smaller, newly generated, disturbances if it becomes curved (see section 8.10).

By contrast, in shear flow over the ground with horizontal streamlines the eddies produced by gravity necessarily have vortex lines which are horizontal, and are therefore like the vortex lines produced by viscosity in creating no Reynolds stress. However, when a thermal (see Chapter 10) penetrates to a higher level through shear, the vortex lines are tilted and do then produce a stress (see section 11.6).

9.4 PROPERTIES OF TURBULENCE: DIFFUSION AND DECAY

The most important feature of turbulence is that it causes diffusion of any physical property carried by the fluid, such as heat, momentum or pollution. This is due to the stretching of all lines of particles so that any compact mass of fluid is drawn out in a complicated way. Thereafter, when the thickness of the sheet is reduced to less than the size of the smallest eddy, further diffusion is effected by molecular motion. The turbulence achieves diffusion by a mixing motion, but the concentration gradient of any property originally occupying a limited compact volume of the fluid is always increased by the turbulence. This is because surfaces of different constant concentration are brought closer together as the sheets of polluted fluid are stretched and made thinner by both large- and small-scale eddies, and by the mean motion.

Because all concentration gradients are increased, so also are all velocity gradients, and the rate of dissipation of energy is increased in a given eddy by the reduction in its size. The Reynolds number of an eddy remains constant because when its size is reduced in accordance with (9.3.14), its characteristic velocity $r\omega$ is increased, and their product is the circulation. The Reynolds number, $r^2\omega/\nu$, thus remains constant. It may be thought of as a measure of the number of rotations it has to make while its velocity is halved, and so its lifetime is reduced when it is made smaller and its angular velocity increased. The dimension along the vortex line is increased, but that is irrelevant, being large compared with the diameter anyway.

If a section of a turbulent fluid were taken and contours of constant vorticity component normal to the section drawn, the viscosity would be seen to diffuse and cancel out areas of negative and positive vorticity. As time passes, the scale of the pattern is reduced and it becomes more complicated, the areas of positive and negative vorticity becoming smaller and more intermingled, and therefore more rapidly reduced by viscosity. The rate of dissipation by viscosity is determined by the rate at which small intense eddies are produced and intensified by stretching. The size of the smallest eddies is determined by the rate at which energy is fed down the size scale, which determines how small an eddy becomes before it performs those rotations which will completely exhaust its kinetic energy. The rate of cascading of energy down the size scale is determined not at all by how it is dissipated, but by the rate at which it produces smaller eddies from big ones.

In between the shear of the mean motion (or of the large eddies into which energy is fed by some sort of instability), and the smallest eddies (whose size is proportional to $(\nu^3/\varepsilon)^{1/4}$), there may be a large range of eddy sizes having an energy spectrum equal to $\overline{u'^2}E(k)$, where $\overline{u'^2}$ is the mean square of the velocity fluctuations and k is the wave number. If this is a function of k and ε (which has the dimensions of veloc$^2 \times$ time^{-1}), which seems reasonable because ε is the manifestation of its effect, we find that the only combination of k and ε having the dimensions of acceleration (which is required because $\int \overline{u'^2}E(k)\,\mathrm{d}k$ has dimensions of veloc^2 being the kinetic energy, the density being taken as unity in this discussion) is

$$\overline{u'^2}E(k) \propto \varepsilon^{2/3}k^{-5/3} \tag{9.4.1}$$

This is **Kolmogorov's –5/3 law**. It has been applied with incorrect logic in a multitude of situations. It is applicable only to decaying turbulence in the **inertial range**, usually called subrange, of eddy sizes remote from both the largest and the smallest eddies. It is **valid only for homogeneous turbulence**, which means that it is not usually valid for shear flows, particularly when near to a boundary.

Since there is a considerable body of theory valid for homogeneous isotropic turbulence, there is a temptation to apply it in situations where turbulence seems to be present. Many people have sought to demonstrate that the turbulence is homogeneous and isotropic and in the inertial range by measuring the spectrum and finding it to be roughly proportional to $k^{-5/3}$. But their logic is faulty here because many quite different fluctuating systems have very similar spectra, if only because their scale is roughly the same. The long wavelength end of any spectrum in nature is determined by the length of the sample used to determine it, and the form of the fluctuations is determined by the degree of phase correlation of the different wavelengths. Thus a single discontinuity of velocity has a spectrum proportional to $k^{-6/3}$, which is scarcely distinguishable in any practical case from $k^{-5/3}$. The process of averaging and of expressing all the features of a turbulent flow by statistics only hides many important characteristics. In nature there are usually boundaries nearby, or else the medium is stratified, or it has a significant mean velocity gradient, or there are no gaps in the spectrum which make the taking of averages and the definition of a mean easy. All these factors make it very unlikely that the turbulence is either homogeneous or isotropic. It is usually undergoing either rapid development from a state of no turbulence to one of intense turbulence, or destruction by viscosity or by radiation of waves.

Any medium capable of transmitting waves can carry away the energy of turbulence, and thereby destroy it. A compressible fluid radiates sound waves from the pressure fluctuations caused by the stretching of the vortex lines: the pressure fluctuations also produce displacements which radiate gravity waves in a medium with a density stratification. In the former case, the sound waves represent an almost completely negligible fraction of the energy of the flow; in the latter case, quite a large fraction of the energy can be radiated as gravity waves.

9.5 RICHARDSON'S CRITERION FOR THE NON-PERSISTENCE OF TURBULENCE

In a stratified fluid, turbulent motion displaces parcels of fluid so as to create potential energy. Thus if a layer of air which is stratified with warm air on top is stirred up by turbulence, the potential temperature becomes more nearly uniform and so the upper layers are cooled and the lower layers warmed. The gravitational potential energy of the whole mass is increased by the lifting of its centre of gravity. At the same time, if there exists a horizontal shear flow, this mean flow will do work against the eddy stresses, and energy will be fed into the turbulence. Richardson argued that if the energy supplied in this way were not as much as the work being done against gravity, the turbulence would die out. Actually it would die out with a lower degree of stratification because a considerable fraction of the turbulent energy is used up in the cascade to smaller eddies and its subsequent dissipation by viscosity.

If a parcel of unit mass is displaced a vertical distance ζ in a stratified fluid of static stability $g\beta$, the work done against gravity is

$$\int_0^{\zeta} g\beta z \, dz = \frac{1}{2} g\beta\zeta^2 \tag{9.5.1}$$

because the restoring force due to a displacement z is $g\beta z$. The kinetic energy of the relative motion of the parcel displaced a distance ζ, in a shear of magnitude $\mathrm{d}U/\mathrm{d}z = \eta$, is

$$\tfrac{1}{2}(\eta\zeta)^2. \tag{9.5.2}$$

If the expression (9.5.1) exceeds (9.5.2), the turbulence cannot be maintained, and so there will be no turbulence if the Richardson number, which is the ratio of these quantities, denoted by Ri, exceeds unity, i.e. if

$$\mathrm{Ri} = \frac{g\beta}{\eta^2} > 1. \tag{9.5.3}$$

The converse is obviously not true, and it is incorrect to say that laminar flow in which $\mathrm{Ri} < 1$ will be unstable and become turbulent, for when $\beta = 0$ and η is the same at all heights the motion is stable and has a uniform vorticity, η, which cannot be redistributed to set up eddies. On the other hand, we saw in Chapter 8 that if the vorticity is concentrated into layers, these layers may be rolled up into vortices which are themselves gravitationally unstable. We also saw that such instability may be induced in layers in which Ri is initially infinite if they are tilted. In nature the interesting criterion is not whether a flow is unstable, but whether there are mechanisms operating which will change it so as to make it so.

Many authors have shown that the flow must have a Richardson number less than $\frac{1}{4}$ somewhere if it is to be unstable, an idea conjectured by G. I. Taylor; again, it must be remarked that many flows with smaller Richardson numbers are stable. The most concise demonstration of this result is given by Howard (1961; the short paragraph 2 of the paper).

The Richardson number is a local number which may vary across a flow field, and it has no global significance, such as the Reynolds number has, for in the case of viscous steady flow the motion is coupled throughout. In stratified fluids the flow of each layer is less dependent on the others.

Attempts have been made from time to time to find a similar criterion for the stability of cylindrical flow in which centrifugal forces play a role rather similar to gravity. These have been unsuccessful because the instability is essentially more complex. In horizontal shear flow the most unstable disturbance is a two-dimensional one in vertical planes containing the shear vector. In cylindrical flow unaffected by gravity the acceleration $\mathbf{f}$ changes direction along the path of the fluid. Consequently axisymmetric (toroidal) disturbances take on a different nature from two-dimensional ones in which the disturbance is in planes perpendicular to the axis of the undisturbed motion. Thus the criterion developed in section 8.6 cannot be expressed in terms of a single Richardson number.

9.6 TURBULENCE GENERATED BY STANDING WAVES

Momentum can be transferred across the mean flow by steady gravity waves provided that the surfaces containing the crests are inclined to the vertical (see section 6.4). In the theory it is supposed that the momentum flux is upwards, away from the obstacle on the ground which experiences a force in the direction of the flow, called wave drag. This is the **radiation condition**, which is appropriate in a medium which extends to infinity. In the atmosphere two mathematical difficulties arise. The first is that the decrease of density upwards means that the waves slowly tend towards very large amplitude, making the problem very non-linear. Thus it is usually supposed that the waves overturn (break) or are absorbed by the very large viscosity at heights in the atmosphere above 150 km. The second difficulty arises when the velocity of the waves relative to the air is zero at some altitude. This is very common simply because it means that at that height the wind is zero relative to the obstacle setting up the waves. It is not as great an issue as the two-dimensional theory suggests because in three dimensions a rotation of the wind with height produces a zero wind relative only to an infinitesimal two-dimensional component of the obstacle.

Nevertheless, in the two-dimensional case energy is being transmitted upwards, and cannot pass through the level of zero velocity (see section 6.10). This energy must therefore be added to the air at this level of zero velocity and must manifest itself as energy resident in the motion of that air, namely vorticity, which is turbulent. There is no acceleration of the flow between this level and the ground, but the air here must be subject to an upstream acceleration due to the momentum it absorbed, and the mean velocity profile modified as a result.

9.7 APPLICATION OF TURBULENCE THEORY TO THE ATMOSPHERE

Reynold's equation is only valid if there is a range of eddy sizes in the atmosphere which contain almost no energy, otherwise without such a gap in the spectrum the definition of the mean value is not applicable. This difficulty appears to have been circumvented by the general assumption first that whatever the turbulence is like, its effects can be represented by an eddy transfer coefficient (*K*-theory), and secondly that the motion obeys the same equations as for a molecular transfer coefficient, but with a coefficient 10^3 or 10^4 times as great. However, the problem of defining the mean is not circumvented by not mentioning it, and the coefficient has no meaning without that definition. Nor does the representation by the coefficient give any information about how the coefficient varies in space, with height in particular, nor about the rate of transfer in different directions. The rate of transfer in the atmosphere often varies by several orders of magnitude within a very few hours, and is very different for different transferred quantities. Thus momentum can be transferred by wave motions in a stratified fluid while pollution is not transferred at all. On the other hand, gravitationally produced turbulence may transfer material much more effectively than momentum (see section 11.6). The application of eddy transfer coefficients to the atmosphere is a convenient way of describing mechanisms that are not understood, but it has little real usefulness. Any situation in which it appears to be useful is likely to be one in which a prediction is made that a future occasion will be rather like a past one when the effects were measured.

An exception to this general condemnation of *K*-theory as being of little use is where new formal theoretical results are obtained. An example is Saffman's (1962) treatment of the effect of transverse shear aloft on the distribution of pollution at the ground downwind of a source. He calculated the path taken by most or all of the pollution reaching the ground at a point displaced sideways from the line downwind of the source. This could be via the upper layers of air with turbulent transfer upwards and transverse displacement with the mean upper wind, followed by downward turbulent transfer to the ground, rather than by horizontal transfer by eddies across the mean wind direction near the ground. The formula has no practical use in the field as a method of prediction: but the theory demonstrated the possibility of this kind of dispersion away from the control line downwind.

The variety of forms of turbulence is legion, and their effects so different, that there can be no single theory for 'turbulence'. We study various mechanisms which are common in the atmosphere, and are often called turbulence, in some later chapters of this book.

9.8 TURBULENCE IN ROTATING FLUIDS

The question arises whether turbulence can produce effects in rotating fluids which are fundamentally different from effects of viscosity. The outstanding problem of how hurricanes are started is a most important practical question. A hurricane is a centre of low pressure caused by rotation, and the air moving into the vortex near the sea surface is cooled by the adiabatic expansion which increases the flux of heat and water vapour from the sea beneath which is warmer than this cooled air. The low-pressure centre is therefore a source of heat to the air, most of which appears in the release of latent heat in the production of copious rain. The heat source contributes to the maintenance of the low pressure, which is produced by the sucking down of the tropopause and the consequent presence of a core of warm stratospheric air (see section 14.10).

Once the hurricane is started it is therefore easy to find mechanisms which will keep it going, but these mechanisms depend on the existence of a low-pressure centre. Therefore we look for mechanisms which will cause the concentration of vorticity and thereby produce small vortices with low-pressure centres.

The predominant feature of areas in which hurricanes begin is the presence of large cumulonimbus clouds. In addition to their vertical air currents which produce rain, the air from which the rain is released spreads out horizontally at the tropopause with speeds of the order of 5–10 m s^{-1}. At the same time, downdraughts are produced by rain and these spread out at the sea surface and maintain the eye. Large parcels of air are therefore transferred horizontally by these anvils (the spreading out cloud tops), and these may be studied in a crude way by mixing-length theory.

A mass of fluid rotating initially as a solid annulus between radii a and b with angular velocity Ω, subject to the radial displacement of parcels of fluid which are subsequently mixed with their new surroundings, would achieve a final distribution of velocity in which all parcels had the same moment of momentum about the axis. This happens in a tea cup, where the radial displacements are caused by the secondary flow in the boundary layer at the bottom (see section 3.8), but in that case there is a decrease in the total kinetic energy and angular momentum of the fluid.

In the case we are envisaging there is an energy source in the buoyancy forces produced by the release of latent heat in the clouds. Work is done by gravity, and when the cloud anvils spread out at the tropopause (or at some lower altitude) they possess kinetic energy of horizontal motion which is used up in moving across the isobars against the stable stratification due to the rotation. As the radially displaced parcels become mixed into their new surroundings, the final state of zero vorticity is achieved, in which the circumferential velocity v is equal to k/r. Since all the forces due to buoyancy are vertical and due to mixing are internal, the total moment of momentum remains constant and consequently

$$\tfrac{1}{2}\pi\Omega(b^4 - a^4) = \pi k(b^2 - a^2) \tag{9.8.1}$$

or

$$k = \tfrac{1}{2}\Omega(b^2 + a^2). \tag{9.8.2}$$

Initially $v = r\Omega$, finally $v = k/r$, and $\partial p/\partial r = v^2/r$; from which we find that, if Δp_1 and Δp_0 are the final and initial pressure differences between radii a and b, with unit density, then

$$\Delta p_1 = k^2 \frac{b^2 - a^2}{2b^2a^2}, \qquad \Delta p_0 = \frac{1}{2}\Omega^2(b^2 - a^2); \tag{9.8.3}$$

or

$$\Delta p_1 = \frac{1}{4}\frac{(b^2 + a^2)^2}{b^2a^2}\Delta p_0 \simeq \frac{1}{4}n^2\Delta p_0 \tag{9.8.4}$$

if $b = na \gg a$.

The vorticity was originally distributed uniformly with magnitude $\frac{1}{2}\Omega$; finally it takes the form of two vortex sheets at $r = a, b$, each sheet containing half the original vorticity, assuming that the fluid inside $r = a$ and outside $r = b$ has not been altered.

There thus exists the possibility that convection clouds confined to a finite region might generate a centre of very low pressure in the middle by concentrating half the vorticity there. An interesting experiment to investigate this matter was devised by McEwan (1976).

The difficulty of testing this possibility arises mainly because there is no information to indicate whether the displacements in anvils are indeed primarily in a radial direction in the region of convection. In the next section we show that without some preferential orientation there is no effect.

9.9 MIXING-LENGTH THEORY APPLIED TO A ROTATING FLUID

We assume that the flow is circular (Fig. 9.9.i) with circumferential velocity $v(r)$. A particle is now given additional initial radial and tangential velocities u_0', v_0', and moves in the environmental pressure field for time τ, after which it mixes with its new environment. The velocity anomaly which it possesses after time τ determines the effect of the mixing on the mean motion. This is a momentum mixing theory, and follows Prandtl's model. The stress is represented by $\overline{u'v'}$ and it is assumed that the velocity anomalies are small. The pressure gradient $\partial p/\partial r$ is equal to v^2/r.

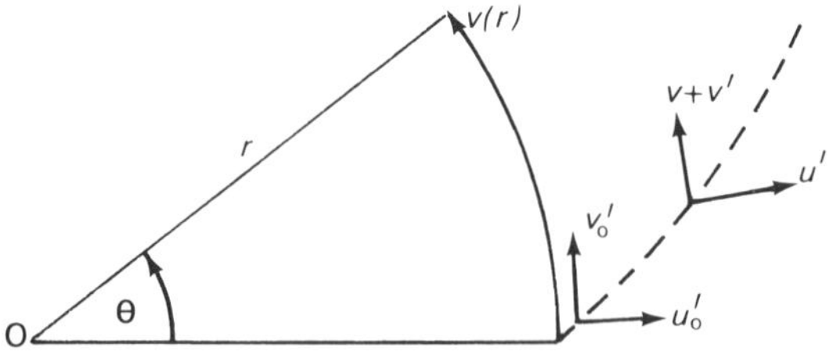

Fig. 9.9.i The dashed line shows the track of a particle displaced by being given an initial perturbation of velocity (u_0', v_0').

The equations governing the motion of the particle are Bernoulli's equation and the conservation of angular momentum, namely

$$\frac{\partial}{\partial r}\frac{1}{2}(\mathbf{v}+\mathbf{v}')^2+\frac{v^2}{r}=0 \tag{9.9.1}$$

and

$$v+v'=k'/r \tag{9.9.2}$$

where k' is a constant for the particle in its free trajectory in the pressure field of the mean flow. We differentiate (9.9.2) and then eliminate k'. Since

$$\mathbf{v}+\mathbf{v}'=(u', v+v') \tag{9.9.3}$$

we deduce that

$$(v+v')\frac{\partial}{\partial r}(v+v')+u'\frac{\partial u'}{\partial r}+\frac{v^2}{r}=0$$

and

$$\frac{\partial}{\partial r}(v+v')=-\frac{k'}{r^2}=-\frac{v+v'}{r},$$

so that

$$-\frac{(v+v')^2}{r}+\frac{v^2}{r}+u'\frac{\partial u'}{\partial r}=0.$$

Therefore

$$u'\frac{\partial u'}{\partial r}=v'\left(\frac{2v}{r}+\frac{v'}{r}\right) \tag{9.9.4}$$

and

$$\frac{\partial v'}{\partial r}=-\left(\frac{v+v'}{r}+\frac{\partial v}{\partial r}\right)=-\left(\zeta+\frac{v'}{r}\right), \tag{9.9.5}$$

where

$$\zeta = \frac{\partial v}{\partial r} + \frac{v}{r} = \text{ vorticity of the mean flow.} \tag{9.9.6}$$

The stress may be written as the time mean value of

$$u'v' = u_0'v_0' + \Delta r\left[\frac{\partial}{\partial r}(u'v')\right]_0 + \frac{1}{2}(\Delta r)^2\left[\frac{\partial^2}{\partial r^2}(u'v')\right]_0 + O(\Delta r)^3 \tag{9.9.7}$$

and

$$\Delta r = u_0'\tau + \frac{1}{2}\left[u'\frac{\partial u'}{\partial r}\right]_0 \tau^2 + O(\tau^3) \text{ after time } \tau \tag{9.9.8}$$

because $\partial u'/\partial t = u'\partial u'/\partial r$. Thus we obtain

$$\begin{aligned} u'v' = u_0'v_0' &+ \tau\left[u'\left(u'\frac{\partial v'}{\partial r} + v'\frac{\partial u'}{\partial r}\right)\right]_0 \\ &+ \frac{1}{2}\tilde{\tau}^2\left[v'\left(\frac{2v}{r} + \frac{v'}{r}\right)\left(u'\frac{\partial v'}{\partial r} + v'\frac{\partial u'}{\partial r}\right)\right. \\ &\left. + \left(u'\frac{\partial^2 v'}{\partial r^2} + 2\frac{\partial u'}{\partial r}\frac{\partial v'}{\partial r} + v'\frac{\partial^2 u'}{\partial r^2}\right)\right]_0 + O(\tau^3). \end{aligned} \tag{9.9.9}$$

We assume that the eddies are symmetrical about $r = \text{const.}$ and $\theta = \text{const}$, which means that

$$\overline{u_0'v_0'} = 0; \tag{9.9.10}$$

furthermore, on taking the mean at $t = 0$, $\overline{u_0'} = \overline{v_0'} = 0$, and so $v'^2u', v'^3, \frac{v'^3}{u'}, \frac{v'^4}{u'}, \frac{v'^5}{u'}$ all vanish in the mean at $t = 0$. Therefore

$$\overline{u'u'\frac{\partial v'}{\partial r}} = \overline{-u'^2\left(\zeta + \frac{v'}{r}\right)} = \overline{-u'^2}\zeta, \tag{9.9.11}$$

$$\overline{u'v'\frac{\partial u'}{\partial r}} = \overline{v'^2\left(\frac{2v}{r} + \frac{v'}{r}\right)} = +\overline{v'^2}\frac{2v}{r}, \tag{9.9.12}$$

and

$$v'\left(\frac{2v}{r}+\frac{v'}{r}\right)\left(u'\frac{\partial v'}{\partial r}+v'\frac{\partial u'}{\partial r}\right)$$

$$=\left(\frac{2v}{r}v'+\frac{v'^2}{r}\right)\left[-u'\left(\zeta+\frac{v'}{r}\right)+\frac{v'^2}{u^2}\left(\frac{2v}{r}+\frac{v'-}{r}\right)\right]$$

$$= 0 \text{ in the mean at } t = 0. \qquad (9.9.13)$$

Also we find that

$$u'^2u'\frac{\partial^2 v'}{\partial r^2}=-u'^3\left(\frac{\partial\zeta}{\partial r}-\frac{\zeta}{r}-\frac{2v'}{r^2}\right),$$

$$u'^2\frac{\partial u'}{\partial r}\frac{\partial v'}{\partial r}=-2u'v'\left(\frac{2v}{r}+\frac{v'}{r}\right)\left(\zeta+\frac{v'}{r}\right), \qquad (9.9.14)$$

and

$$u'^2v'\frac{\partial^2 u'}{\partial r^2}=u'v'\left[-\frac{v'^2}{u'^2}\left(\frac{2v}{r}+\frac{v'}{r}\right)^2-\left(\zeta+\frac{v}{r}\right)\left(\frac{2v}{r}+\frac{v'}{r}\right)\right.$$

$$\left.+v'\left(\frac{\partial}{\partial r}\frac{2v}{r}-\frac{\zeta}{r}-\frac{2v'}{r^2}\right)\right],$$

all of which vanish in the mean at $t = 0$ because they are composed of odd powers of u'. Consequently,

$$\overline{u'v'}=\left[-\overline{u_0'^2}\zeta+\overline{v_0'^2}\frac{2v}{r}\right]\tau+O(\tau^3). \qquad (9.9.15)$$

Calculation also shows that the lowest powers of u' and v' in the term in τ^3 are

$$\frac{4v\zeta}{3r}\left[\overline{u_0'^2}\zeta-\overline{v_0'^2}\frac{2v}{r}\right]\tau^3, \qquad (9.9.16)$$

the higher terms being of the order of v'/v in comparison with those retained. Since this has the same factor as the term in τ, no new phenomena are introduced by retaining it because it will vanish too if the term in τ vanishes. We may now consider various cases.

(1) Isotropic eddies: $\overline{u_0'^2}=\overline{v_0'^2}=q'^2$.

$$\overline{u'v'}=-q'^2\tau\, e_{r\theta} \qquad (9.9.17)$$

where $e_{r\theta}$ = fluid deformation rate $=\dfrac{\partial v}{\partial r}-\dfrac{v}{r}=\zeta-2\dfrac{v}{r}$, and this is zero only in rigid rotation. Thus the eddies will produce a stress as long as the rotation is not the same as for a rigid body, and so they have the same effect as viscosity.

(2) Eddies composed of radial displacements only: $\overline{v_0'^2} = 0$.

$$\overline{u'v'} = -q'^2\tau\zeta, \tag{9.9.18}$$

and the stress vanishes only when $\zeta = 0$, which is when $v = k/r$. In order to achieve the result envisaged in section 9.8, therefore, the eddies must produce predominantly radial rather than tangential displacements of fluid parcels.

(3) Eddies composed of tangential displacements only: $\overline{u_0'^2} = 0$.

$$\overline{u'v'} = \frac{2v}{r}q'^2\tau = (\zeta - e_{r\theta})q'^2, \tag{9.9.19}$$

and the stress only vanishes if r is very large. This may be taken to imply that eddies of this kind, in which parcels are pushed in both directions along the tangents to the flow, tend to straighten out the flow, presumably because all particles are displaced radially outwards as a result of displacements along the tangents. Certainly no effect on the profile $v(r)$ is indicated. ζ can only be equal to $e_{r\theta}$ when the flow is straight.

(4) Rectilinear flow: $r = \infty$.

In this case ζ is equal to the shear and we obtain the classical result of mixing-length theory, namely

$$u'v' = -q'^2\tau\frac{\partial v}{\partial r}, \tag{9.9.20}$$

and the mixing length being $q'\tau$.

9.10 GENERATION OF VORTICES BY EDDIES

We see from the cases described in section 9.9, that with isotropic eddies, namely anvils which spread out equally in all directions, the region in which they are generated tends to be made to rotate like a rigid body. If its total angular momentum is not changed and it is not initially rotating like a solid, the consequence will be to alter the velocity at the boundary of the region and generate a vortex sheet there. Since air in the trade winds does not originate recently all from the same latitude and since the intensity of convection is not uniform, it is to be expected that vortex sheets, and individual vortices as a consequence, will be generated on the boundaries of regions of intense convection.

The eddies might be made non-isotropic by the alignment of the convection clouds into streets so that they are more nearly two dimensional, the main horizontal displacements being perpendicular to the lines of cloud. Any deformation of the air tends to draw out a compact region of air into a long line so that it is not surprising that clouds in the tropics are usually observed to be arranged in lines. Streets also tend to form along the wind shear, usually not far from the wind direction. Even a block of air in rectilinear flow which is uniform relative to the earth possesses the earth's rotation in addition, and so the state of zero vorticity may be approached. This means that in the northern hemisphere a uniform stream with lines of cumulonimbus along it could generate zero absolute vorticity in itself by accelerating the fluid on the left and decelerating that on the right. This could generate vortex sheets, and subsequently individual vortices along the boundaries of the region of convection.

Attempts have been made in the laboratory to produce results of this kind, but it should be appreciated that the outcome must depend either very much on the amount of anisotropy in the horizontal velocities generated in the eddies, or upon large spatial variations in the intensity of the eddies (either isotropic or anisotropic).

A criticism of this crude and simple analysis is that it does not appear to conserve the vorticity of the parcels displaced by the eddies. As in the case of classical mixing-length theory however, the ultimate state which will not be change is the same whether the parcels conserve either momentum (or moment of momentum in this case) or their vorticity, because the final state has uniform or zero vorticity.

9.11 ELASTIC PROPERTIES OF VORTEX TUBES IN TURBULENCE

A vortex is capable of transmitting waves along its length. To study this property we take the simplest possible case. With cylindrical polar coordinates (r, θ, z) we suppose a fluid velocity $(0, r\Omega, 0)$ in the undisturbed state where $\Omega = \Omega(r)$, and add a small perturbation velocity (u, v, w). The equations of motion in the three directions for uniform density are:

$$\frac{\mathrm{D}u}{\mathrm{D}t} - \frac{(v + r\Omega)^2}{r} = -\frac{1}{\rho}\frac{\partial p'}{\partial r} = -\frac{1}{\rho}\left(\frac{\partial p_0}{\partial r} - \frac{\partial p}{\partial r}\right), \tag{9.11.1}$$

$$\frac{\mathrm{D}}{\mathrm{D}t}(v + r\Omega) + \frac{u(v + r\Omega)}{r} = 0, \tag{9.11.2}$$

$$\frac{\mathrm{D}w}{\mathrm{D}t} = -\frac{1}{\rho}\frac{\partial p'}{\partial z} = -\frac{\partial p}{\partial z}, \tag{9.11.3}$$

where for convenience we have represented the perturbation pressure by ρp, thus the pressure is

$$p' = p_0 + \rho p. \tag{9.11.4}$$

If there is a perturbation proportional to $\exp \mathrm{i}(kz - \sigma t)$, corresponding to waves travelling along the axis of rotation with velocity σ/k, then with axial symmetry,

$$\frac{\mathrm{D}}{\mathrm{D}t} = \frac{\partial}{\partial t} + u\frac{\partial}{\partial r} + w\frac{\partial}{\partial z}$$

$$= -\mathrm{i}\sigma \text{ for the perturbation linearised,} \tag{9.11.5}$$

$$= u\frac{\partial}{\partial r} \text{ for the undisturbed motion.} \tag{9.11.6}$$

The linearised equations are

$$\mathrm{i}\sigma u + 2\Omega v = \frac{\partial p}{\partial r} \tag{9.11.7}$$

$$-\mathrm{i}\sigma v + u\frac{\partial}{\partial r}(r\Omega) + u\Omega = 0 \tag{9.11.8}$$

$$\mathrm{i}\sigma w = \frac{\partial p}{\partial z}, \tag{9.11.9}$$

and the equation of continuity is

$$\frac{\partial}{\partial r}(ru) + \frac{\partial}{\partial z}(rw) = 0. \tag{9.11.10}$$

Eliminating p, v and w, we find that u satisfies

$$\frac{\partial^2 u}{\partial r^2} + \frac{1}{r}\frac{\partial u}{\partial r}\left[\frac{1}{r^2} + k^2\left(1 - \frac{4\Omega^2 + 2r\Omega\Omega'}{\sigma^2}\right)\right]u = 0 \tag{9.11.11}$$

where

$$\Omega' = \partial\Omega/\partial r. \tag{9.11.12}$$

If waves can be trapped in the tube and propagated along it, u must vanish on the axis, must also have an oscillatory form for small values of r so that it may rise to a maximum, and must have an exponential form for large r so that the perturbation may die away at infinity. The analysis is simplified if we write

$$\phi = r^{1/2}u, \qquad x = kr, \qquad S = 2\Omega/\sigma \tag{9.11.13}$$

and then in a region in which Ω is constant (9.11.11) takes the form

$$\frac{\partial^2 \phi}{\partial x^2} = \left(\frac{3}{4x^2} + 1 - S^2\right)\phi. \tag{9.11.14}$$

In a region of zero vorticity in which

$$\Omega = \frac{k}{r^2}, \qquad \Omega' = -\frac{2\Omega}{r} \tag{9.11.15}$$

we have

$$\frac{\partial^2 \phi}{\partial x} = \left(\frac{3}{4x^2} + 1\right)\phi. \tag{9.11.16}$$

The simplest case in which waves are trapped is one in which a region of solid rotation (constant Ω) is surrounded by irrotational motion extending to infinity. If the coefficient of ϕ in (9.11.14) for some values of r is negative and in (9.11.16) for large r is positive, there will exist solutions of the kind desired. This will happen if the region in which Ω is constant extends to large enough r. The larger the value of S (and Ω), the smaller the radius in which waves are trapped. It is required at the very least that $S^2 > 1$, i.e.

$$2\Omega > \sigma, \tag{9.11.17}$$

and so tubes of this kind cannot trap oscillations of frequency greater than the vorticity of their core (2Ω). Much smaller frequencies are trapped at a small radius which satisfies

$$S^2 > 1 + \frac{3}{4x^2}. \tag{9.11.18}$$

9.12 TURBULENCE OF A HOT SUBSONIC JET OF AIR

If we now imagine a turbulent shear flow, such as that shown in Figs 9.12.i–iv, which are shadowgraphs of subsonic jets emerging into still air, the shear represents a very small frequency disturbance, and so distortions of the ends of tubes which are visible as tangled spaghetti will be propagated along them. If the end of a tube is pulled away at one end, a stretching wave will be propagated along it as if it were elastic. The pressure on its axis is reduced when the stretching occurs because of the increased rotation, and as the tension of the tube, represented by the force on a cross-section of it, is increased. If such a stretching were relaxed, the recently stretched tubes would provide a tension equivalent to a system of forces tending to restore the distortion just imposed on the spaghetti. Hence we would expect this turbulence to have viscoelastic properties since the tension would die away in being propagated along the tubes in an uncoordinated way because of the tangled structure of the eddies.

In Fig. 9.12.i the jet emerges surrounded by a vortex sheet which rapidly breaks down into a tangled mass of eddies. The most intense eddies are those wnich are most rapidly stretched in the shear zone, and these are inclined at 45° to the axis. The rapid stretching increases the rotation, which decreases the pressure on the axis. There is a corresponding increase in volume of the eddy which is almost explosive in character, and which emits a sound pulse. In this way **jet noise** is generated. The theory of jet noise simply calculates the pressure fluctuations which are generated as if the fluid were incompressible, and then applies them to a compressible fluid. With the greatest fluctuations being in eddies aligned at 45° to the axis, the sound wave fronts will tend to be oriented in the same direction and propagated in the direction at right angles, and will travel predominantly out along the 45° cone from the orifice. This explanation assumes that there will be larger eddies of intermediate size which stretch the small eddies in bundles, and this can be seen to be happening in the figure.

The shadowgraph picture shows there to be a smallest scale of turbulence in which the large temperature gradient of the original vortex sheet is rolled up, initially in billows, and the core made uniform by molecular conduction. The most intense vortices have the lowest axial density because of the low pressure. This turbulence is clearly anisotropic because of the large shear.

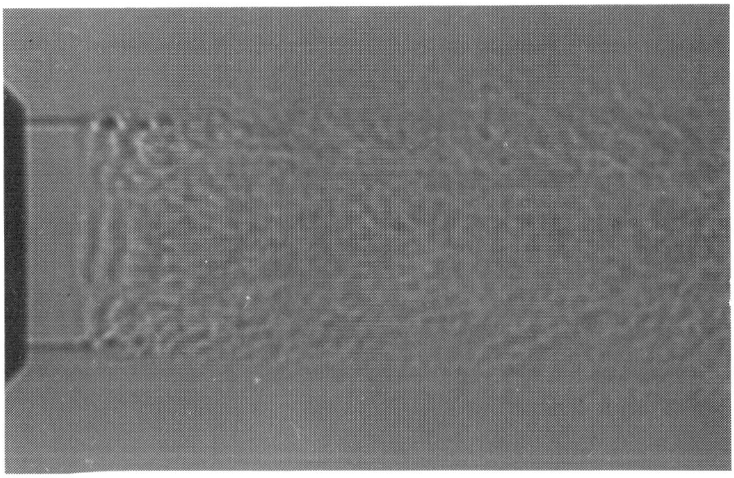

Fig. 9.12.i The jet emerges from its nozzle as a circular cylindrical vortex sheet which quickly develops circular billows around it: these become chaotic and the turbulence is most intensely developed where vortices are inclined at about 45° from the jet axis.

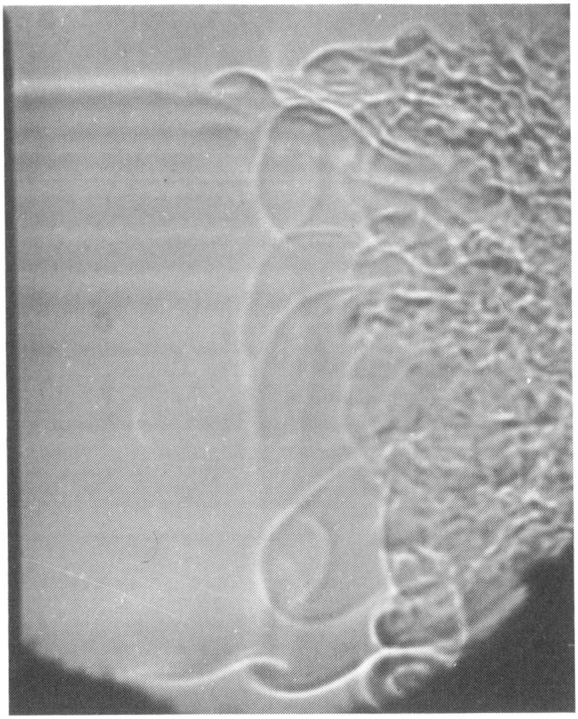

Fig. 9.12.ii A close-up of part of the billows becoming like tangled spaghetti, seen through a circular aperture vary close to the jet.

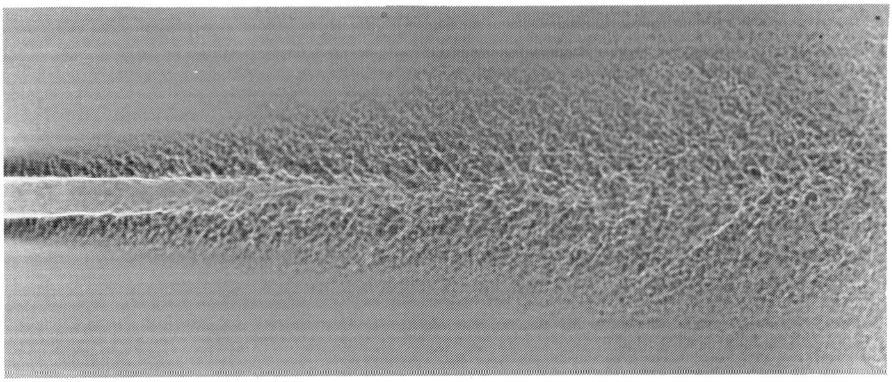

Fig. 9.12.iii A more distant view so that we can see the stretching of the vortex tubes dominating the pattern at 45°, which is the angle at which the stretching and intensification of the vortices is a maximum.

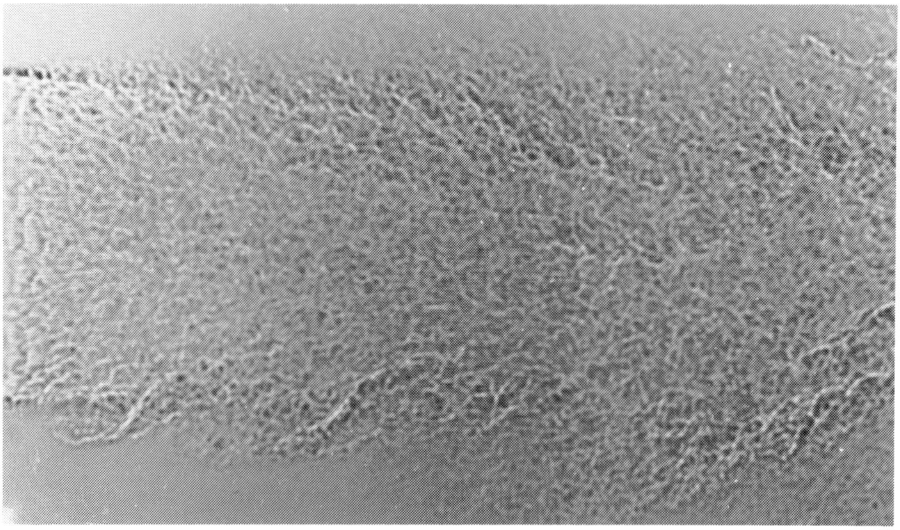

Fig. 9.12.iv In this example occasional 45° vortices achieve greater intensity. The emission of noise from the jet is perpendicular to the line of maximum stretching, which is out on conical surfaces with the centre of the jet as axis. The noise can be reduced by increasing the depth of the mixing layer so as to increase the size of the vortex tubes and prevent the isolation of the most stretched vortices from the rest of the flow, so that the expansion of the tube is lessened. The mixing layer is increased by not having a precisely circular nozzle but one like the outline of a flower with several petals, to increase the region of mixing of the jet into the air.

Figs. 9.12 Shadowgraph pictures by courtesy of J. Mollo-Christensen.

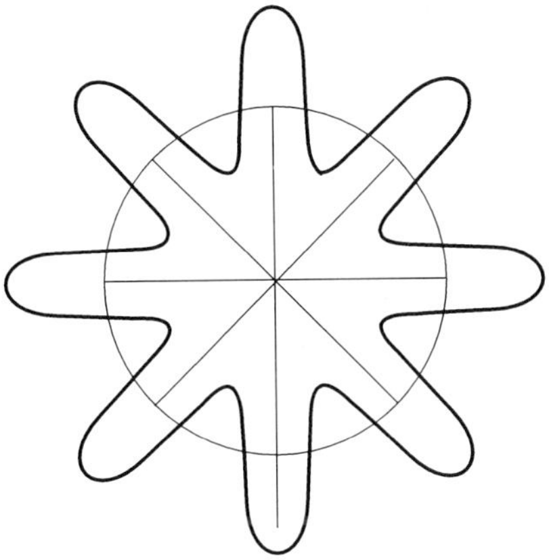

Fig. 9.12.v The outline of the jet nozzle section designed to reduce the noise, particularly in the high frequencies, produced by an exhaust gas jet into air.

10

Partly turbulent flow

10.1 A RISING BUBBLE OF AIR IN WATER

The flow is **partly turbulent** when fluid is injected with momentum, or with buoyancy which creates momentum, into an otherwise undisturbed fluid environment.

Our first example is that of a void introduced into a uniform medium otherwise at rest. The word 'void' comes from the study of fluidised beds consisting of heavy particles suspended by their aerodynamic drag in an upward stream of fluid. As a whole, the medium behaves very like a fluid of larger density. Thus a medium of sand particles suspended in an upward stream of air can have a free surface above which there are no sand particles and upon which larger bodies of smaller density than the sand and air combined can float as if on a liquid. In order to maintain the fluidised bed, air has to be pumped through small holes or through a porous surface at the bottom, and it is easy, particularly if the minimum flux of air needed to fluidise the bed is much exceeded, to cause bubbles of air to rise through the bed like a bubble of air in water. Such a bubble is called a void because it contains no particles of the bed. We now discuss a large air bubble in water.

The air in the bubble has a density about one thousandth of the density of water, and so the motion of the air inside the bubble produces negligible variations of the pressure inside. This pressure may therefore be regarded as a constant. If a steady bubble shape is maintained, the flow of the water around it must be such as to produce a constant pressure on the surface. We have seen (section 2.13) that on the front of a body moving through a fluid the flow is approximately irrotational. We have also seen that the pressure on a sphere moving vertically through water of density ρ with velocity w is (e.g. (2.3.1) but including the gravity term)

$$p = p_0 - g\rho z - \frac{9}{8}\rho w^2 \sin^2\theta, \qquad (10.1.1)$$

where p_0 is the pressure far from the sphere, at the level where $z = 0$, and is a constant (see Fig. 10.1.i(a)).

Large bubbles are observed to be approximately spherical on top and to have a more or less flat bottom on which there is an irregular motion. They are lens-shaped, and the rim is in fluctuating motion although the upper surface is smooth (see Fig. 10.1.iii).

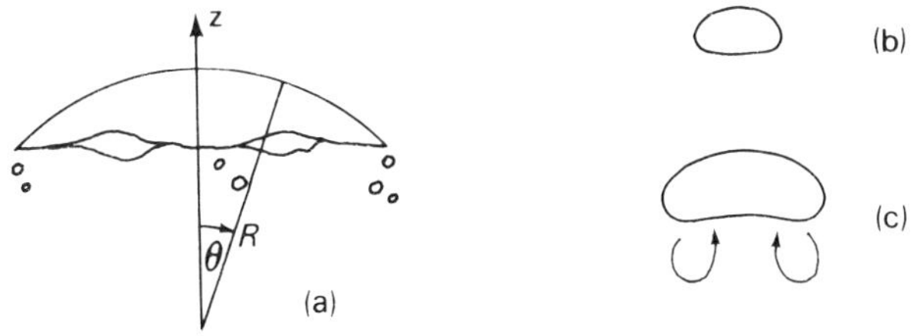

Fig. 10.1.i(a) The coordinates used for a large (10 cm, say) spherical cap bubble The base of the bubble is irregular and unsteady, with small bubbles circulating in the wake. (b) The shape of a small bubble (< 2 cm) for which surface tension is important and makes the periphery rounded. (c) When the fluid viscosity is increased (Reynolds number decreased) a moderately large bubble has a rounded periphery and may have a concave lower surface.

The effect of surface tension is not a significant factor in determining their overall shape when the bubbles are a few centimetres in diameter, but the rim of those less than about 2 cm in diameter is noticeably rounded (Fig. 10.1.i(b)). If a more viscous liquid than water is used, the turbulence in the wake is reduced and the wake is made more nearly steady. It then appears to contain a ring distribution of vorticity beneath the bubble, and the base of the bubble is humped up in the middle (Fig. 10.1.i(c)).

The zero of z is put at the top of the bubble. In order for p to be equal to a constant, since

$$z = -R(1 - \cos\theta) \tag{10.1.2}$$

and

$$\sin^2\theta = (1 - \cos\theta)(1 + \cos\theta) \approx 2(1\cos\theta), \tag{10.1.3}$$

roughly for θ less than about 30°, and very accurately for much smaller values of θ, it is seen from (10.1.1) that the pressure is a constant provided that

$$w = \frac{2}{3}(gR)^{1/2}, \tag{10.1.4}$$

where R is the radius of curvature of the upper surface of the bubble.

This formula for the velocity of rise of the bubble, first given by R. M. Davies and G. I. Taylor [37], gives it in terms of the size of the bubble provided the angular aperture of the bubble (the value of θ at the rim) is known. Bubbles can be produced with apertures ranging from around 25° up to over 40°, and so the volume is not determined by knowledge of R alone.

Large bubbles with diameters up to around 20 cm can be made fairly easily by filling a dome-shaped saucer with air under water and allowing it to upturn so as to pour the air upwards.

These bubbles have many fascinating properties. They can be studied by causing them to rise through patches of dye in a water tank so that the motion in the wake is made visible. The wake is roughly parabolic in shape as illustrated in Fig. 10.1.ii, with a strong upcurrent up the middle where some particles move upwards at about twice the speed of

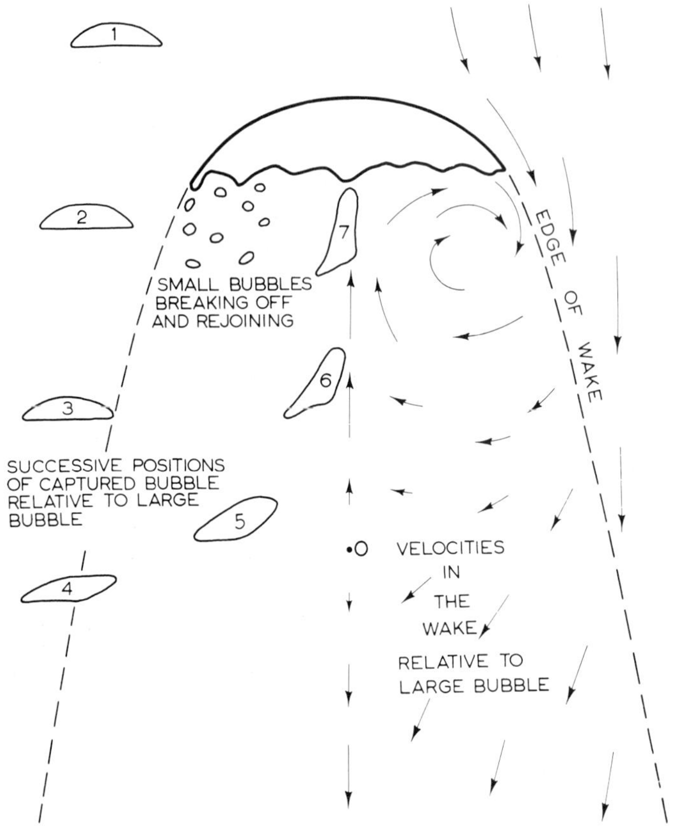

Fig. 10.1.ii On the left are shown 7 successive positions of a smaller bubble overtaken by an ascending larger one and a group of smaller bubbles which circulate close to the edge, sometimes breaking away from and sometimes rejoining the main bubble. On the right is shown the mean velocity around which there are fluctuations associated with the unsteady rim and base of the main bubble. The velocities are relative to the bubble and 0 is a stagnation point.

the bubble. If a small bubble is released ahead of a large one it can be seen first to be pushed aside as the large one overtakes it, but later its edge will impinge on the edge of the wake. Soon after that it is carried rapidly up the centre of the wake and is distorted in shape by the deformation of the water around it. When the velocity of the small bubble becomes large the importance of virtual mass is apparent, for a small bubble rising up the wake with a sufficient speed relative to the water can be observed to pass through the large bubble and emerge from its upper surface! This is in contrast to the behaviour of a bubble with small velocity relative to the water, which behaves as if it possessed negative mass (see section 2.11).

If we wish to extend this analysis to a bubble of liquid of different density ρ' which is immiscible with water, we might expect a formula for the velocity such as

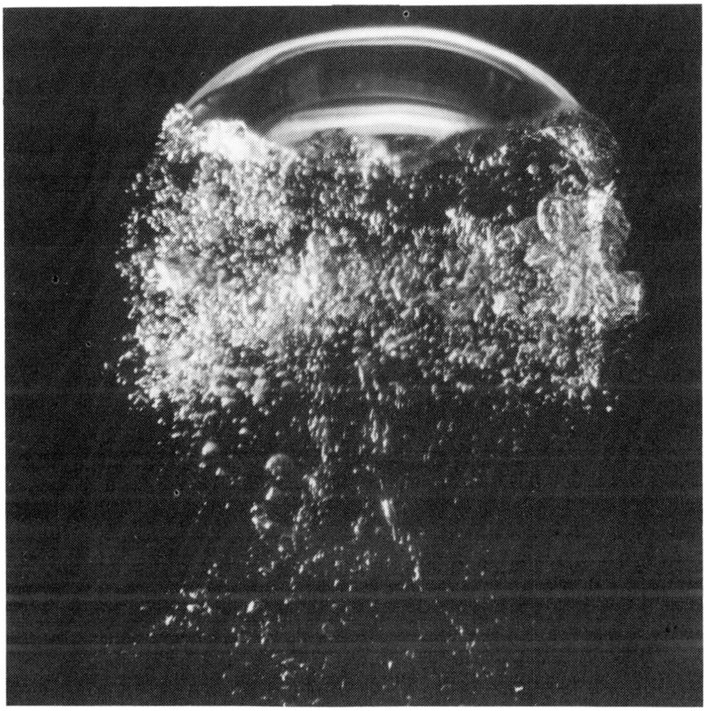

Fig. 10.1.iii A bubble of air of about 15 cm radius rising through water. The position of the small bubbles in the wake indicates that, although it is very turbulent, it is dominated by a vortex ring circulation.

$$w = \frac{2}{3}\left(g\frac{\rho - \rho'}{\rho}R\right)^{1/2}, \tag{10.1.5}$$

where we have replaced g by the buoyancy force due to the density difference. This result may be invalidated by the inertia forces in the interior liquid, which produce a variable pressure on the bubble surface, by skin effects, or by mixing across the surface. In this book we are only concerned with fluids which are miscible, such as hot air in cool air or salt water in fresh water, and with cases in which the density difference is small.

10.2 AXISYMMETRIC THERMALS

A thermal is a body of buoyant fluid in a fluid with which it **can** mix. The name was given by glider pilots to those masses of warm air rising from hot ground warmed by sunshine in which they can soar. In the atmosphere buoyant convection produces a very complicated motion (Chapter 11), but here we are concerned with the behaviour of one isolated thermal in an otherwise undisturbed environment. Fig. 10.2.i shows successive pictures of two laboratory thermals.

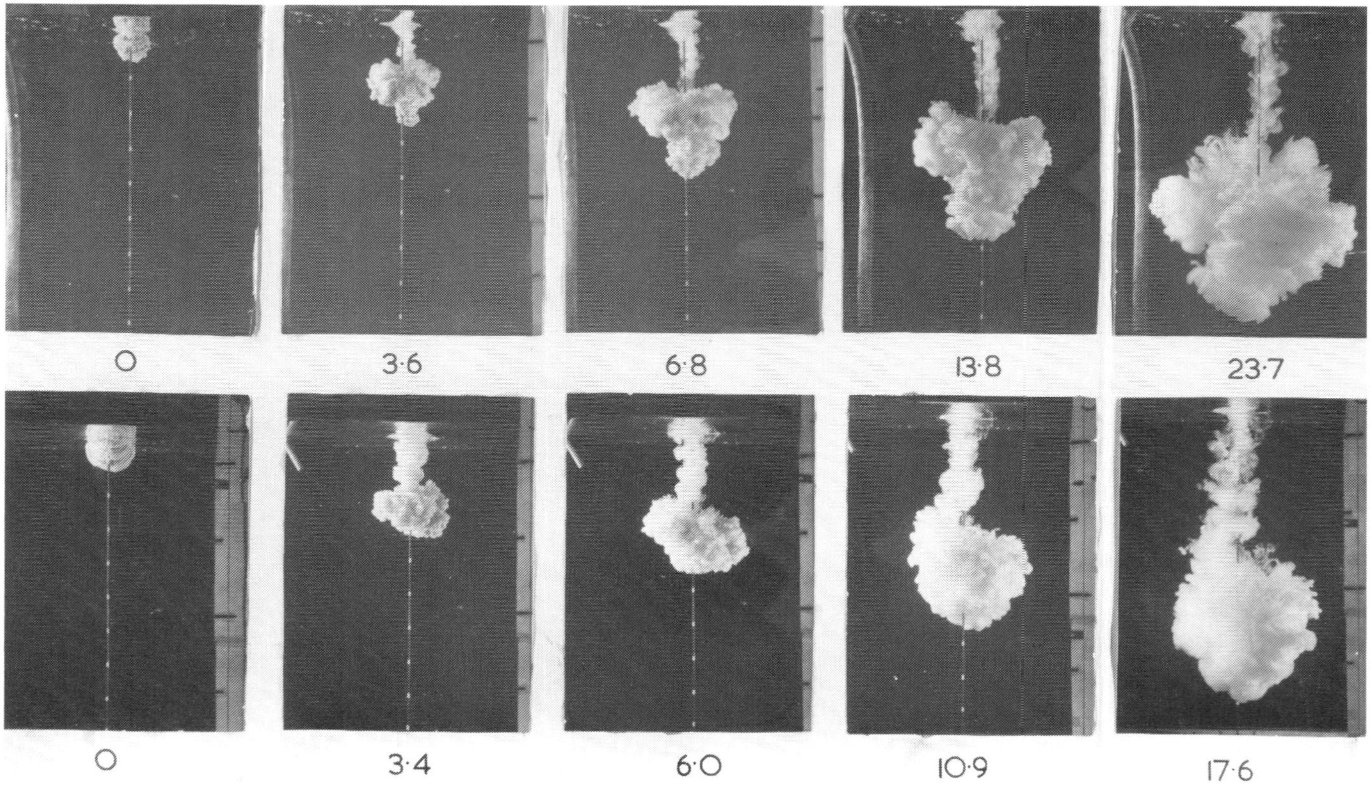

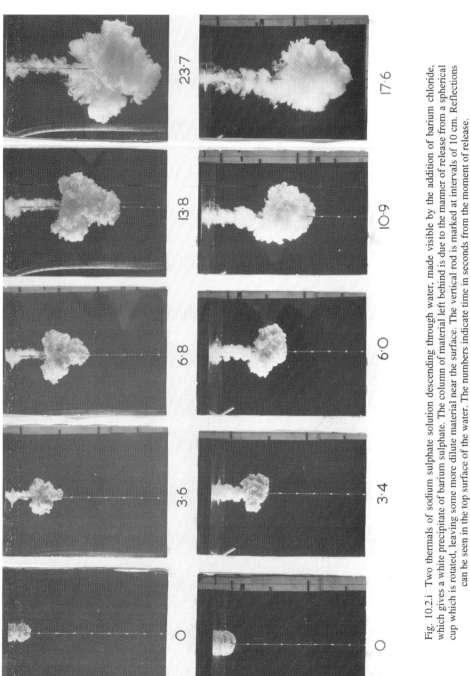

Fig. 10.2.i Two thermals of sodium sulphate solution descending through water, made visible by the addition of barium chloride, which gives a white precipitate of barium sulphate. The column of material left behind is due to the manner of release from a spherical cup which is rotated, leaving some more dilute material near the surface. The vertical rod is marked at intervals of 10 cm. Reflections can be seen in the top surface of the water. The numbers indicate time in seconds from the moment of release.

A very simple theoretical treatment is sufficiently useful where more complicated theories have led to no more useful results, largely because thermals differ from one another quite considerably in a way that cannot be predicted.

The assumptions of the theory are justified by its success. We observe that a thermal in water made visible by a white precipitate or dye is apparently in turbulent motion and resembles a cumulus cloud which has a cauliflower-like appearance on its upper surface. We therefore assume that the molecular viscosity is of no importance. We also observe that the thermal grows in size as it moves vertically, but retains roughly the same shape. Because of the turbulence we suppose that the initial distribution of density in the thermal when it was small has a negligible influence on the distribution later on when it is much larger, and when some of the individual eddies can be seen to be larger than the thermal was at its creation or moment of release. It is reasonable to suppose therefore, that the original configuration is forgotten (provided that it was compact) and that **all thermals tend to take up the same shape, with the same mean velocity and density distributions**. All the velocities, including typical eddy velocities, will then be determined by the total buoyancy and overall size of the thermal: thus to have the right dimensions the velocity must be of the form

$$\boxed{w = \mathrm{C}(gBR)^{1/2},} \tag{10.2.1}$$

where B is the average buoyancy such that if the volume is mR^3, the total buoyancy (or weight deficiency) is $gBmR^3$. C is a constant which depends on how w and R are defined. Thus if R is defined as the radius of the largest horizontal cross-section and w is w_c, the vertical velocity of the cap or uppermost point of the thermal, we find from experiments and C is about 1.2.

Because the rate of growth of the radius, $\mathrm{d}R/\mathrm{d}t$, is a velocity, it must have similar mathematical form to w. Since the total buoyancy is constant,

$$gB = gB_0\, R_0^3/R^3, \tag{10.2.2}$$

where subscript 0 refers to the values at a particular moment and so we have

$$R\frac{\mathrm{d}R}{\mathrm{d}t} = \mathrm{C}'\left(gB_0 R_0^3\right)^{1/2}, \tag{10.2.3}$$

or

$$R^2 \propto t, \tag{10.2.4}$$

the constant of proportionality being $2\mathrm{C}'(gB_0R_0^3)^{1/2}$. C′ is a different constant, namely that appropriate to the rate of increase of radius.

If z is the height of the uppermost point (the cap) of the thermal, by the same argument we have that

$$z^2 \propto t \tag{10.2.5}$$

and so we can write

$$z = nR. \tag{10.2.6}$$

This conclusion is a direct consequence of (10.2.1), which was obtained by a dimensional argument. If other factors such as viscosity, ν, influenced w, we could not arrive at such a simple result. This is because wR/ν, which is the Reynolds number, is dimensionless and so C, instead of being a universal constant, could be a function of the Reynolds number.

In (10.2.4) and (10.2.5) we have assumed that the constant of integration is chosen so that R and z are zero at $t = 0$. Consequently the thermal is behaving as if it once possessed an infinitesimal size and was growing up from a virtual point origin. This is clearly impossible, principally because the density cannot be negative or infinite, and so B defined by

$$B - \frac{\rho - \rho'}{\rho} \tag{10.2.7}$$

cannot exceed unity. Therefore the motion cannot possibly be extrapolated backwards beyond the size at which the total buoyancy is achieved by zero interior density.

On the other hand, we can imagine a negative thermal, namely a body of heavier fluid, to the density of which there is no theoretical upper limit. But in that case we would be no longer justified in making the Boussinesq approximation and introducing B only through g, because the motions of the fluid inside the thermal produce pressure forces proportional to the density. Since B is itself a non-dimensional number, in that case C could be a function of it, just as it might be of the Reynolds number. **The analysis is therefore valid only if the Boussinesq approximation is valid** and the accelerations of the fluid are small compared with g (see section 2.12). Since the fluid accelerations are produced by the buoyancy forces in the first place all we need to ensure is that

$$B \ll 1 \tag{10.2.8}$$

In practice the temperature excesses in an atmospheric thermal are of the order of 1°C and so B is around 0.003. In a model experiment using saturated salt water to produce a negative buoyancy in fresh water, B is around 0.15 at first; but it soon becomes much less when mixing occurs. If alcohol, or some other fluid lighter than water and miscible with it, is used to produce positive buoyancy, again B is comparable with unity only at the very early stages. Thus it does not matter whether experimental measurements are made on thermals of positive or negative buoyancy. In the early stages the thermals are gradually assuming the final form which we have supposed they will take up, and measurements at that stage will be strictly relevant only to the particular case and the kind of apparatus used to produce it.

To conduct an experiment we have to release the thermal without disturbing the surrounding fluid. This is readily done with a hemispherical cup made of thin metal and partly filled with buoyant fluid, and which is pivoted about a horizontal diameter. The cup is then inverted rapidly so as to release a spherical-cap-shaped body of buoyant fluid. If it is released from the surface, the level of fluid inside and outside the cup must be made equal, otherwise a vertical impulse is given when the buoyant fluid is 'dropped' in. The cup must be overturned rapidly for otherwise a sideways impulse is given to the thermal as the fluid is poured from the cup.

Successive positions of the thermal can be recorded by cine photography and then traced successively on to graph paper, as in Fig. 10.2.ii. The value of z at which $R = 0$ can then be determined. If we plot z^2 against t we find that the points lie on or very close to a straight line. z can be defined as the most forward visible point of the thermal, and it is observed that as successive knobbles on the surface of the thermal energy from its surface the extreme point advances faster than the average. At the same time it is seen that the rate of advance decreases as the foremost knobble is displaced by the circulation of the motion towards the side. Thus the fluctuations from the straight line represent the turbulence in the thermal itself, though otherwise they are found to follow a straight line well. Using (10.2.6), equation (10.2.5) can be written

$$z^2 = \frac{2nC}{m^{1/2}}\left(gB_0mR_0^3\right)^{1/2} t = t/k \tag{10.2.9}$$

in which $gB_0mR_0^3$ is the total buoyancy, or weight deficiency. To check the validity of this theory experimentally, thermals of several different total buoyancies were measured and from each a value of k was determined. This was plotted against the total buoyancy. In the first set of experiments done (Scorer [38]) the total buoyancy varied from 1 to 45 gm wt so that k varied by a factor of about 7, and the quantity $2n\mathrm{C}/m^{1/2}$ varied by up to 16% on either side of the mean value (Fig. 10.2.iii).

In order to see whether C is a universal constant we need to examine the variations of n and m. According to the rather simple idea that all thermals will tend to a universal shape, it is to be supposed that n and m are universal constants giving the distance from the virtual origin to the cap and the volume of the thermal is a multiple of a power of the horizontal radius. The shape factor m is difficult to measure. It is roughly equal to 3 because a thermal is flattened by comparison with a sphere for which the value is $4\pi/3$,

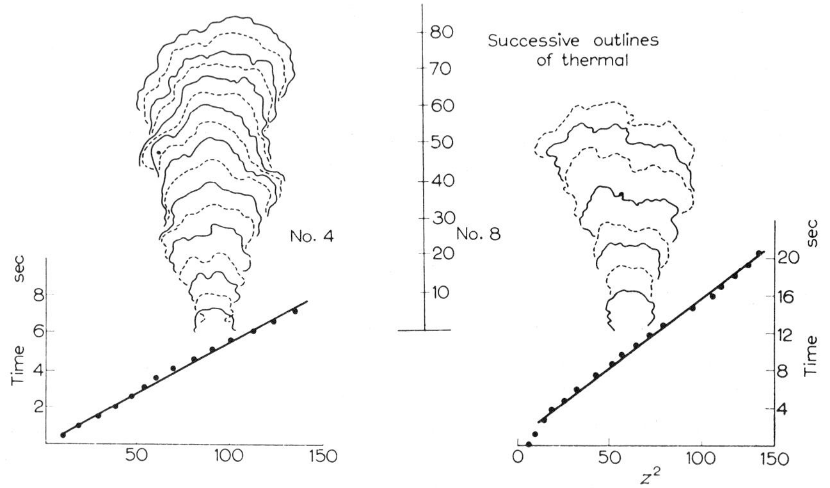

Fig. 10.2.ii Outlines of successive positions of the two thermals drawn from cine film, below which are plotted graphs of z^2 against time.

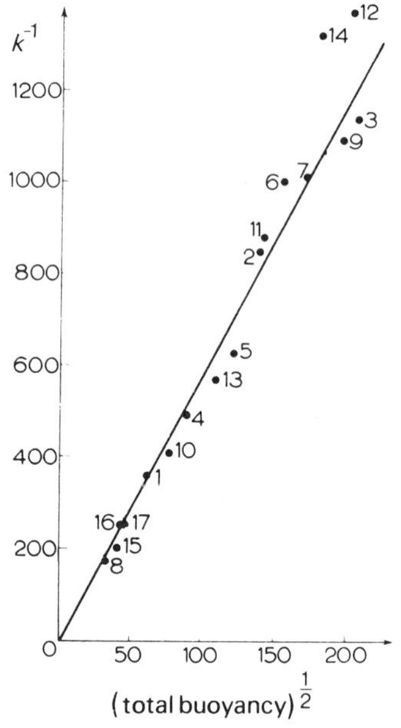

Fig. 10.2.iii A plot of total buoyancy against the quantity $1/k$ measured from the rate of advance of the thermal. The quantity C in equation (10.2.1) is determined from equation (10.2.9) and the slope of the line in this diagram. The number is the serial number of the experiment.

and any variations which might be used will have a small effect on the calculations since it appears as $m^{1/2}$. On the other hand n is easy to measure and shows considerable variation from thermal to thermal, even though it usually appears to remain constant for the measurable life of a given thermal. In the early experiments it was found to vary between 2.9 and 5.0. Since then Richards has obtained values up to 7.3 and down to 2. There does appear to be a correlation between the value of $2n\mathrm{C}/m^{1/2}$ and $n^{1/2}\,\mathrm{C}$ is roughly constant (Fig. 10.2.iv). A theoretical rationalisation of this is that instead of putting the volume equal to mR^3 as in (10.2.2) we should put it equal to mR^2z/n, in which case equation (10.2.1) becomes

$$w = \mathrm{C}\left(gB_0R_0^2z_0n\right)^{1/2}\Big/z = \frac{\mathrm{C}n^{1/2}}{m^{1/2}}\ (\text{total buoyancy})^{1/2}t, \tag{10.2.10}$$

or

$$z^2 = 2\frac{\mathrm{C}n^{1/2}}{m^{1/2}}\ (\text{total buoyancy})^{1/2}t. \tag{10.2.11}$$

Fig. 10.2.iii shows the manner in which k^{-1} was found to vary with total buoyancy $(gB_0mR_0^3)^{1/2}$. Fig. 10.2.iv shows how the ratio of these quantities for each experiment was related to n, and the curve represents a function proportional to $n^{1/2}$. Some of the variation is thereby accounted for, but no smooth line representing a simple function can be drawn to pass closer to all the points.

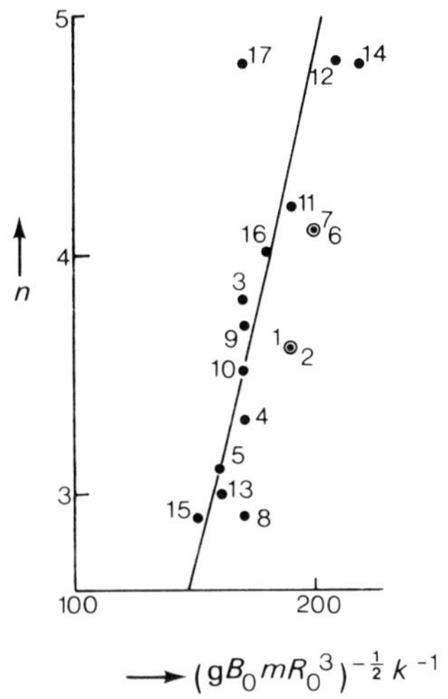

Fig. 10.2.iv The variation of the slope of the line from the origin to points in Fig. 10.2.iii plotted as a function of n. The curve represents a function proportional to $n^{1/2}$.

On the other hand, Richards [39] added many more experiments to those included in Fig. 10.2.iv and found that the quantity $k^{-1}(B_0 m R_0^3)^{-1/2}$ varied like $n^{3/2}$ over a greater range of values without increased scatter. His results imply that the constant C (in equation 10.2.1 and 10.2.10, for example) is not strictly constant and equal to about 1.2, but is more accurately given by

$$C = 0.36 m^{1/2} n^{1/2} = 0.63 n^{1/2}, \tag{10.2.12}$$

if m is taken to be about 3.

It is not very useful to take the variation in n into account anyway, because we do not at present have any very certain idea of what determines n. The points in Fig. 10.2.iii lie close to the line which gives

$$2Cn/m^{1/2} = 180 g^{-1/2} \quad \text{CGS} = 5.7, \tag{10.2.13}$$

and with $n = 4$, $m = 3$, we thus obtain, approximately,

$$C = 1.2, \tag{10.2.14}$$

or

$$Cn^{1/2} = 2.4. \tag{10.2.15}$$

One significant conclusion from the results shown in Fig. 10.2.iii is that there is no tendency to deviate from a straight line through the origin in any consistent way. This may be cited as evidence that no viscous effects were important in the experiments, and since such measurements as have been made on clouds lead to the conclusion that the value of C is about 1 although their total buoyancy is about 10^{10} times as great and the medium was air, it appears that in all cases the turbulent effects dominated the viscous effects (see, for example, Malkus and Scorer [40]).

A Reynolds number for a thermal is wR/ν, which is a constant according to (10.2.10), since $z \propto R$. Thus the ratio of the inertia forces to the viscous forces is constant for a given thermal, and if the viscous forces are negligible at one stage in its life they are negligible throughout its life. It could be argued that they may be important in determining the internal velocity pattern without significantly affecting the parameters such as C, which refer to its overall behaviour. If, for example, a small drop of ink in a glass of water turned out to have the same value of C, this would not prove that it was turbulent. The belief that the laboratory thermal with total buoyancies of a few grams weight in water is similar to atmospheric thermals is justified because we can see the turbulent eddies in action.

10.3 PROPERTIES OF THERMALS

The value of n sometimes changes rather abruptly in the life of a thermal. A rather larger than usual protuberance may appear on one side and become detached by not being completely entrained into the circulation as it moves round to the rear. The value of n then decreases. The value of n sometimes increases when the thermal becomes widened by one or two large protuberances which do become completely entrained in the rear. There therefore appears to be no certainty that a thermal will retain its characteristic value of n even though it seems to have preserved it up to a certain moment.

The circulation of a thermal may be defined as the circulation around a circuit as shown in Fig. 10.3.i. This is composed of a section of its axis of symmetry through it together with a loop completing the circuit in the irrotational flow outside where its position does not affect the magnitude of the circulation. The magnitude of this circulation which is given by **the integral,**

$$\oint_{\mathrm{OAPO}} \mathbf{v}\,.\,\mathbf{ds} \propto wR \tag{10.3.1}$$

is constant throughout the life of the thermal because $w \propto R^{-1}$ for any characteristic velocity. The circulation is only changed by the action of forces which produce accelerations around the circuit, as indicated by the first term on the right-hand side of equation (1.1.15), and the only forces which can operate in this case are the buoyancy forces. Consequently this result implies that **there are no buoyancy forces on the fluid up the axis of symmetry**.

At first sight this seems to contradict the obvious fact that generally some buoyant fluid can be found on the axis somewhere. It can only be explained by the assumption that the eddies produce Reynolds stresses which act like viscosity, reducing the circulation so as to offset exactly the increase caused by buoyancy on the axis. An alternative way of expressing the same fact is to say that in a thermal a path must exist leading up through the middle of the thermal on which there is no buoyancy (i.e. no variation from the external density). **The circulation** measured using such a path, wherever it may be at the moment of measurement, **is then a constant for the thermal**.

Although a compact body of buoyant fluid is released, after a short time a hole through the middle appears, which can easily be seen when the buoyant material is made cloudy (Fig. 10.3.ii). A simple way to do the experiment is to use either milk or a fine

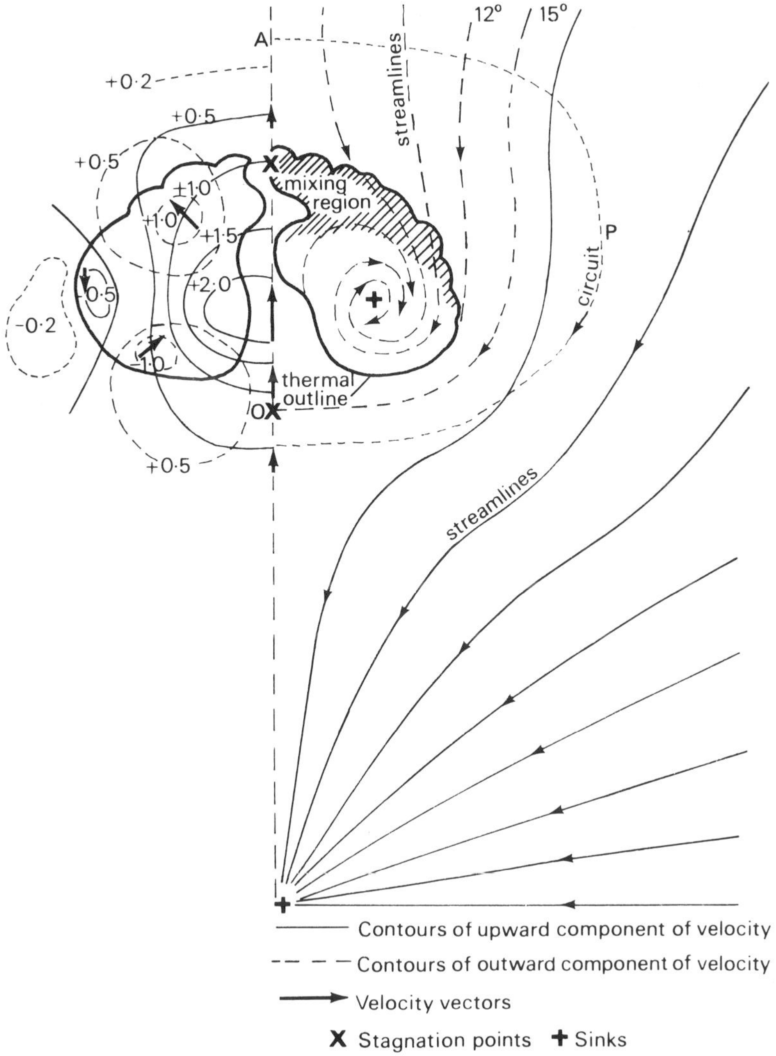

Fig. 10.3.i The right half of the diagram shows the flow of the fluid relative to the outline of the thermal which is kept fixed in size, but with turbulent protuberances moving around it and the flow up its axis wobbling from side to side as it decelerates towards the cap to form protuberances. Fluid inside a cone of roughly 12° angle is entrained into the front, and between that and a cone of roughly 15° it is entrained into the rear. Fluid outside this is represented as moving into a sink at the origin. The fluid inside goes into a sink at the point where the fluid is instantaneously at rest relative to the thermal outline. The circuit marked is completed up the axis and is used for defining the circulation of the thermal, which is constant. On the left, contours of upward and outward velocity components are shown expressed as multiples of the rate of the rise of the cap, and vector velocities are indicated at significant points.

white precipitate, easily obtained by adding a little barium chloride to a thermal prepared from sodium sulphate (used instead of the usual sodium chloride to make the water

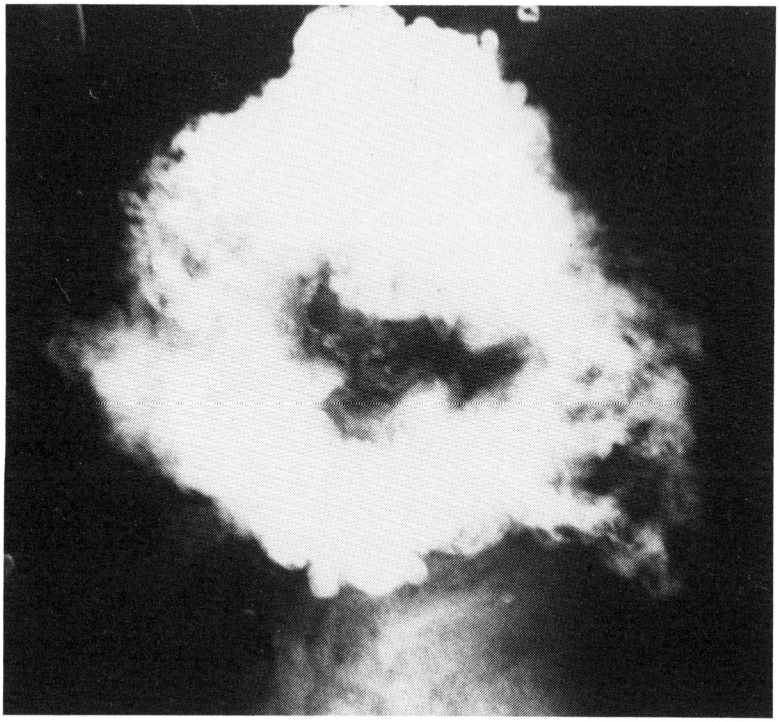

Fig. 10.3.ii A view of a thermal from above showing the doughnut shape with a clear view through the middle.

denser) so as to form an optically dense precipitate of barium sulphate. The adding of silver nitrate to a sodium chloride solution to form a precipitate of silver chloride is not recommended. Silver salts are expensive and silver chloride tends to flocculate and settle in the spherical cup if there is any delay in releasing the thermal while internal motions are allowed to die away.

It used to be thought that the turbulent protuberances on a thermal were produced primarily by the unstable density gradient there as seen in the first picture of Fig. 10.2.i, but this is in fact a very minor cause. In the first place, the most vigorous protuberances appear near the centre of the cap of the thermal where the unstable density gradient is least; and secondly, identical protuberances appear on the front of a puff, where there are no density gradients (see section 10.6).

Initially a thermal is a compact body of warm air or other buoyant fluid. When it is released in a model experiment in water it quickly develops the cauliflower appearance on its surface, and this must be due to static instability. But the vorticity generated in the form of rings around the buoyant fluid, which cause it to penetrate through the surroundings, also cause fluid from the surroundings to penetrate from the rear through to the front so that the buoyant fluid forms a torus-like mass (see Fig. 10.3.ii). It differs from both a laminar vortex ring and a spherical vortex, as well as from anything in between.

This is because the fluid penetrating through the middle becomes turbulent in the region of deceleration and diffluent curved flow, and mixes with the buoyant fluid. The fast-moving stream through the middle meanders about in the diffluent region and draws buoyant fluid into the eddies. This stream therefore distributes itself all around the front of the thermal with protuberances of the mixture passing round the circulation first on one meridian, then on another. (See Figs 10.3.iii, p. 318 and 10.4.ix, p. 327.)

Thus there is a turbulent region all over the front of the thermal, and marked particles of fluid in its path can be seen to be mixed into the front. There is a cone of roughly 12° semi angle from the virtual origin of the thermal such that all particles inside it are mixed into the front of the thermal as it advances. Another cone of angle roughly 15° is such that all the particles between it and the smaller cone flow into the rear of the thermal as it advances. All particles outside this larger one are not engulfed by the thermal. The angles of these cones must not be regarded as having unique values because they vary with the angle of spread of the thermal itself. This may pass up a cone of angle as small as 8° or as large as 26°, but is most commonly around 16°. About 60% of the fluid entrained enters the front; but this is not to be taken as an exact figure, and individual cases could easily range from 50 to 70%.

All thermals are thought of as being geometrically identical, the only differences being in size, R, and in the magnitude of their buoyancy, B (which is expressed as a fraction of the weight of the environmental fluid it displaces). A cine film of one thermal could be projected to a size and with a speed to represent any thermal, and we may think of the thermal as being fixed in size with the world around it shrinking into the virtual origin or being entrained into the thermal. This representation is given in the right half of Fig. 10.3.i and in Fig. 10.4.vi (p. 324).

10.4 FURTHER THERMAL EXPERIMENTS

(1) Tracks within thermals

A thermal turns itself inside out as it rises about $1\frac{1}{2}$ diameters, which means that every parcel of it passes through the mixing region and becomes diluted when it travels that distance. Inside, the core shaped like a torus is continually displaced backward through the thermal because fluid is removed by mixing at the front and transferred to the rear.

If a particle inside a thermal has a very small fall speed relative to the surrounding fluid, it will remain inside because some exterior air below the thermal is always entrained into it. With a larger fall speed its position in the thermal would determine whether or not it would fall out the next time round. Woodward [41] studied this question and her conclusions are summarised in Fig. 10.4.ix and Fig. 10.4.x (on p. 327).

A question of considerable interest in connection with the upward transport of pollution when the gases containing it are buoyant is how much will be carried through an 'inversion', as a density discontinuity is usually called in this context (see Fig. 10.4.iv). Richards [39] found the following results experimentally in water tanks. Y, called the yield, is the fraction of the material of the thermal which continues to ascend after passing through the inversion; S is the amount of stratification. It is expressed as the ratio of the density difference across the inversion to the average density difference which the thermal would have had from the surroundings at the level of the inversion if

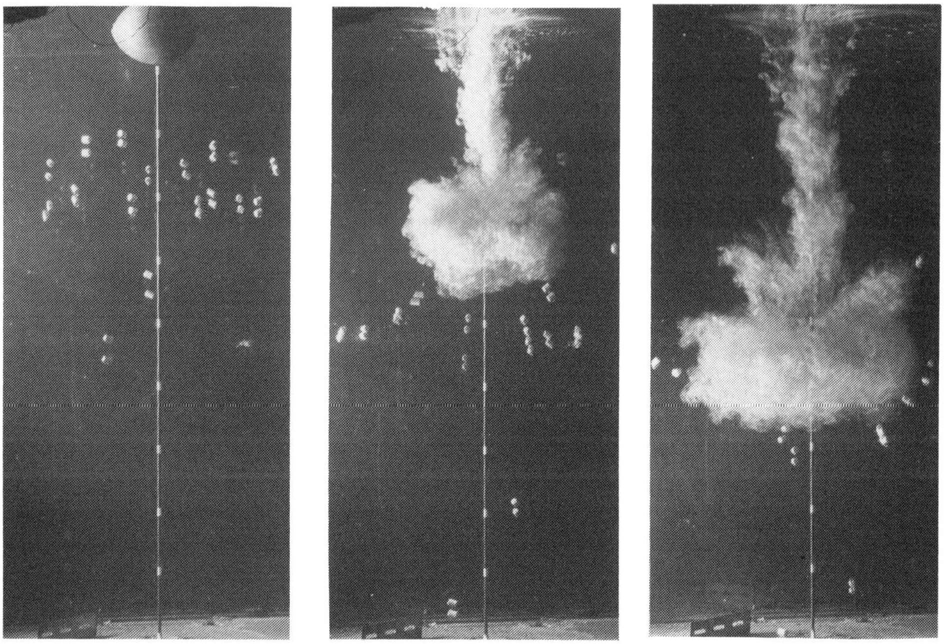

(a) (b) (c)

Fig. 10.3.iii In each picture there are two exposures about 0.5 s apart. (a) shows the free rate of sink of some slowly sinking particles. In (b) and (c) an idea of the velocity field outside the thermal can be gained. The particles immediately in front of the thermal are displaced downwards as much as the faster-sinking ones lower down, while those at the side are in an upcurrent. Those that have disappeared have entered the advancing 'front' of the thermal, not the rear. The markings on the vertical rod are at 10 cm intervals.

the inversion had not existed but instead the layer in which the thermal started had continued. For axisymmetric thermals it was found that (see Fig. 10.4.ii, p. 320)

$$Y = 0.95 - \tfrac{1}{2}S \quad \text{when} \quad 0.1 < S < 1.9$$

$$Y = 0 \quad \text{when} \quad S > 1.9$$

$$\mathrm{Y} = 1 \quad \text{when} \quad S = 0 \qquad (10.4.1)$$

and for 2D (i.e. cylindrical) thermals (see section 10.8 for the 2D cases)

$$Y = (0.93S + 1)\exp(-93S), \qquad (10.4.2)$$

Y being zero for practical purposes when S exceeds about 7.5.

The distribution of density in a thermal is not uniform, and clearly a thermal will not penetrate a sufficiently strong inversion, but it will completely or almost completely pass through one sufficiently weak. These equations express what happens for inversions of intermediate strength. Axisymmetric thermals are less effective at penetrating,

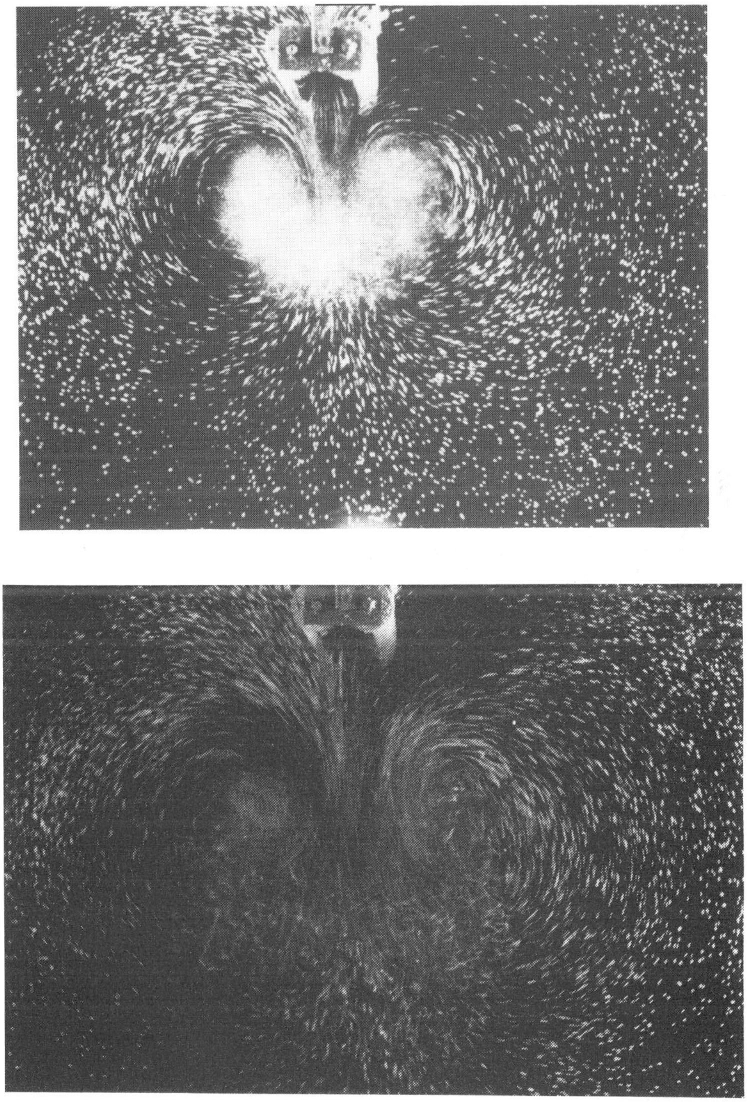

Fig. 10.4.i The motion due to a two-dimensional thermal revealed by a time exposure of particles illuminated in a vertical section. The region of turbulence is marked by more erratic vectors, the region of buoyancy is marked by slight cloudiness, which is more evident in the upper photograph closer to the release trough.

particularly when S exceeds unity. This is probably because in the 2D case the fast-moving fluid in the centre of the thermal has about the average density of the thermal (see Fig. 10.4.iii) whereas in the axisymmetric case it has about the same density as the surroundings (see Fig. 10.3.ii). The forward momentum of fluid parcels is more closely correlated with their buoyancy in the 2D case. Furthermore, after penetrating, many of the 2D terminals became a row of separate fragments which would, of course, again

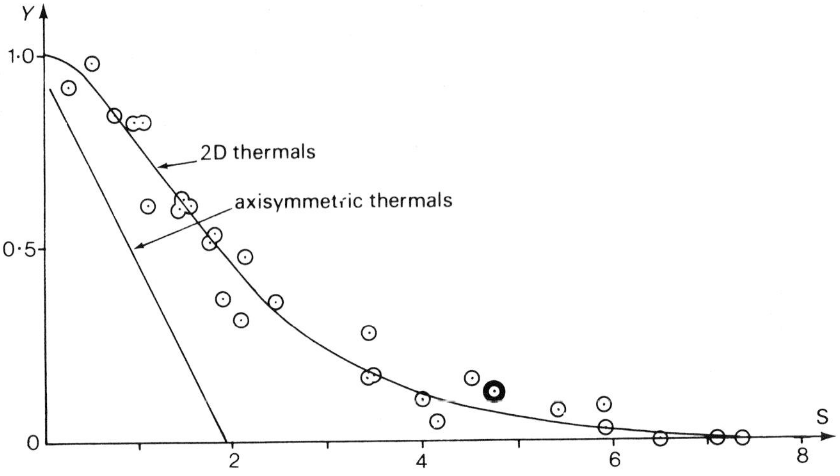

Fig. 10.4.ii Graphical representation of the experimental results of Richards [39] represented in equations (10.4.1) and (10.4.2).

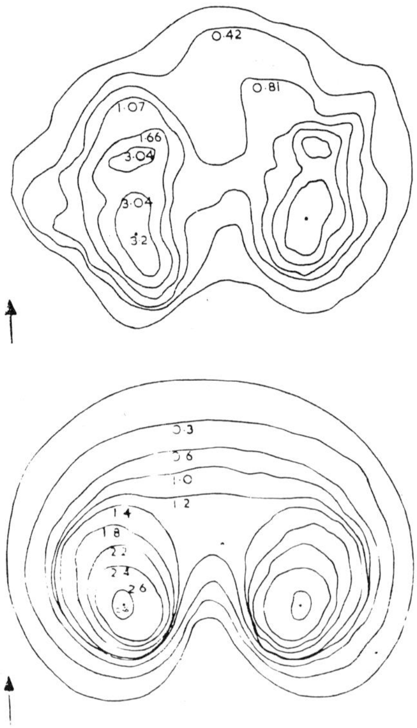

Fig. 10.4.iii Two typical mean cross-sections of the density of a two-dimensional thermal expressed as multiples of the mean buoyancy, measured by Richards [39].

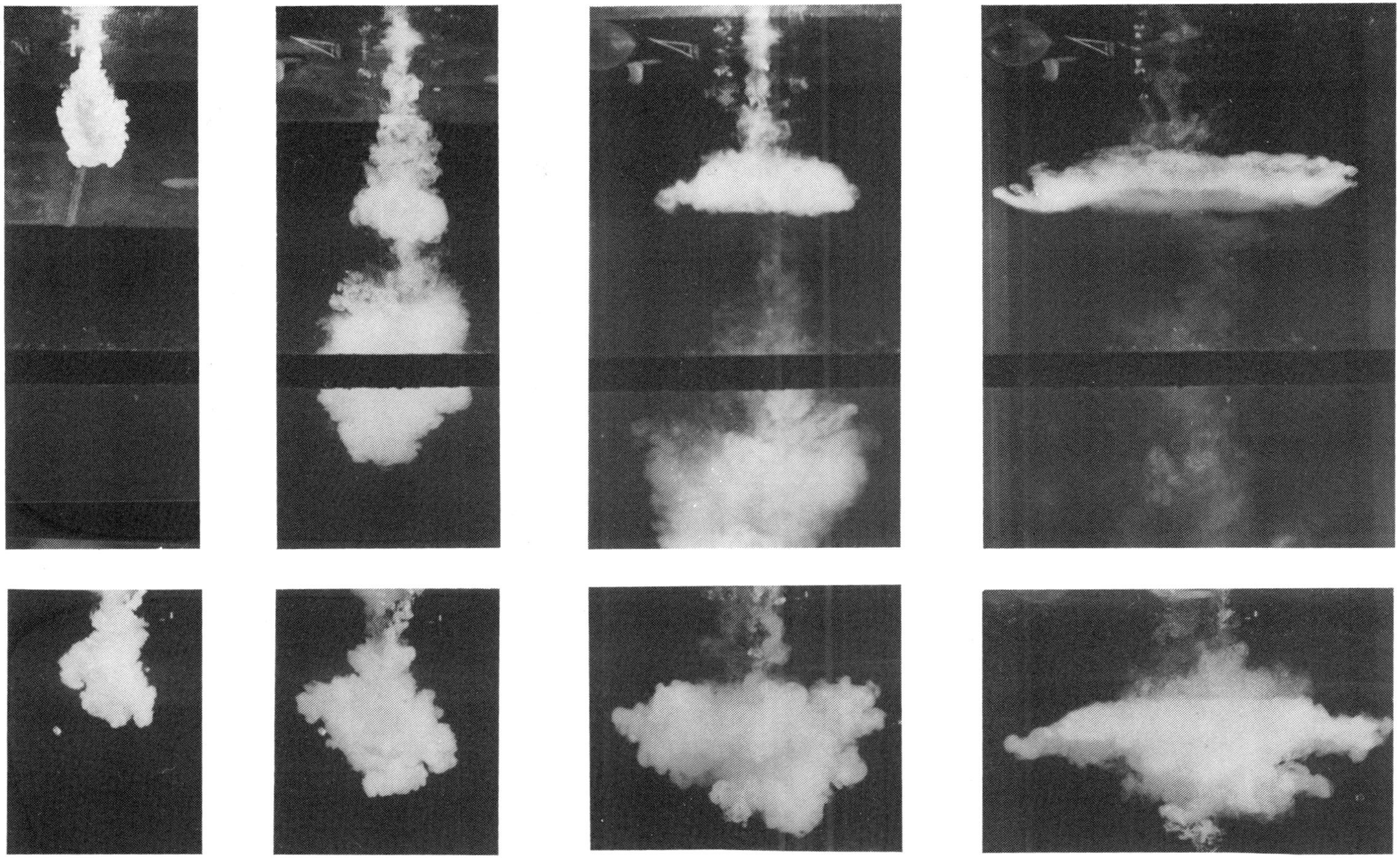

Fig. 10.4.iv Successive photographs of two laboratory thermals partly penetrating (upper) and not penetrating (lower) an inversion.

amalgamate into a 2D thermal as they grew larger by entrainment. Thus the variations in density along the length of the cylindrical thermal act in favour of the most buoyant parts penetrating although the material would be stopped if it were uniformly distributed. This explains how some material can penetrate even when S is much larger than 2.

The flow pattern can in any case only be represented rather roughly because thermals differ from one another and have large fluctuations in their own motion. The chief cause of fluctuations is where the exterior fluid advancing up the middle from the rear meanders first one way and then another, producing large knobbles, or protuberances, on the front cap. These are then displaced round the sides as their successors follow them. Each large knobble seems to have smaller knobbles on it with yet smaller knobbles on them, and so on, each knobble behaving rather like the larger one on which it sits. This appears to be a characteristic of the region of decelerating motion because turbulent puffs (see section 10.6) behave in a similar way without buoyancy.

Surrounding the thermal, which can be defined easily by making the fluid visible, is a region of irrotational motion produced as if by a body moving through it. There is no wake such as is found behind a rigid body because there is no drag on its surface. The velocity field outside can be crudely represented as due to a doublet of strength μ with potential $\mu\cos\theta/r^2$. This is the same as the irrotational flow (see section 2.2) due to a sphere of radius a moving with velocity U, where $2\mu = Ua^3$. In the context of a thermal, the doublet strength is proportional to wR^3. Since wR is constant, the flow is becoming stronger outside at a given distance r from the centre of the thermal in proportion to R^2, which increases like t. This is because momentum is being added to the motion at a constant rate equal to the total buoyancy. Thus although the thermal itself is decelerating ($w \propto t^{-1/2}$) it is getting larger and the exterior velocity at a fixed distance from its centre is increasing.

The velocity field as measured experimentally by Woodward [41] is incorporated in Fig. 10.4.i. It can now be appreciated that because the motion, or any other feature of a thermal, can be displayed in relation to the outline of the buoyant fluid, this outline may as well remain fixed in a diagram. Fig. 10.3.i (right half) is drawn to show the motion of particles relative to this outline. Particles are either entering the thermal and becoming incorporated into it or are passing round it and entering a sink at the virtual origin of the thermal. It is not possible to define precisely in such a diagram a streamline which would separate particles entering the front from those entering the rear because the knobbliness of the thermal makes this boundary diffuse.

In using these diagrams it must be appreciated not only that the angles at which the thermals expand differ but so also does the shape. Thus a very flat thermal expanding up a cone of the same angle as a less oblate one would have a smaller value of n, because that relates the distance of the cap from the origin to the maximum horizontal radius.

In the upper part of the thermal near to the top boundary the motion is very unsteady. At the same time a horizontal surface of fluid through which the thermal penetrates is stretched around it, by the steady component of the motion, as shown in Fig. 10.4.ix (p. 327). Thus the mixing is achieved by the simultaneous mechanisms of penetrative eddy motion and stretching out of compact bodies of fluid. During the ascent of 2 or 3 diameters, a single blob of marked fluid may be mixed into the whole body of the thermal.

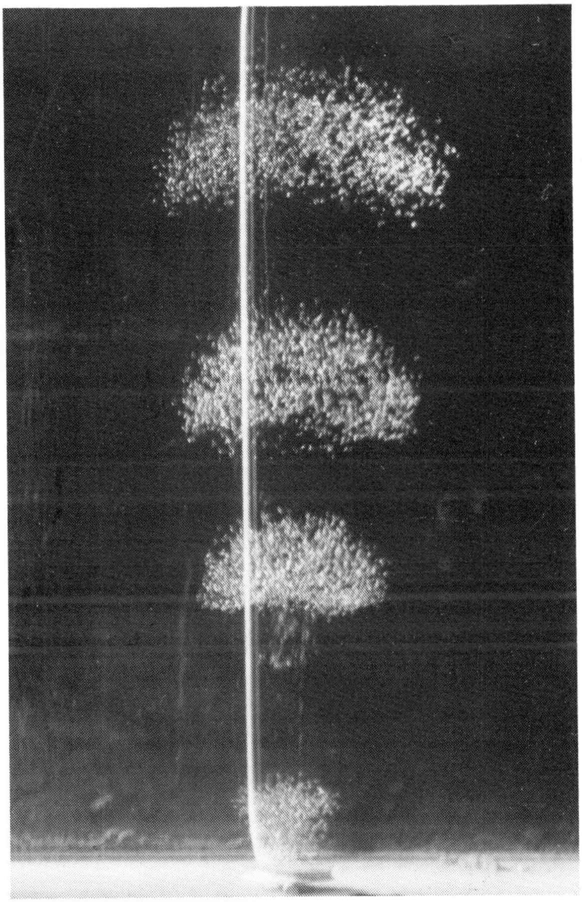

Fig. 10.4.v Four shots of a cloud of air bubbles as it rises through water. In the second, the column of bubbles from the later moments of the release are catching up with the cloud which has the familiar thermal shape. The growth is along the familiar cone.

Particles with a fall speed relative to the fluid in which they are immersed may be carried around the core of the thermal several times before they fall out of it. Woodward used her measurements to calculate Fig. 10.4.x, which shows where a particle would have to be so as not to fall out of the bottom of the thermal. An interesting consequence of the inflow into the base of a thermal concerns those particles which are carried up by a thermal and will eventually fall out as the rate of its rise decreases. In descending through the inflow they are carried towards the axis. Thus they all fall back to the starting level close to the virtual origin, even though they may have been spread out over a much larger horizontal distance while they circulated within the thermal.

(2) Clouds of bubbles

Raindrops or hail falling from a cloud, which was built by thermal convection, produce

Fig. 10.4.vi The cloud of bubbles is released so as to pass up through a region made cloudy by a fine precipitate. Because the individual bubbles have a significant terminal rate of rise through water the cloud surges ahead of the water dragged up in its wake.

downdraughts which spread out on the ground. We shall see in section 11.6 that when this fallout is descending below cloud in unsaturated air, a much greater downward force than the drag of the particles may be produced by the cooling produced as the fallout evaporates. But in this section we are concerned only with alternative methods of modelling a thermal by forming a cloud of bubbles of air in a water tank. (See Fig. 10.4.v.)

Compressed air was forced into a flat circular metal container with a large number of holes made in the side facing upwards. When the air pressure was raised, air was forced out of the holes to form a cloud of air bubbles. The experiment is illustrated in Fig. 10.4.v, where the vertical pipe leading to the container at the bottom of the tank is seen with four positions of the cloud shown in one picture. The growth up a vertical cone is obvious. At position 1 the cloud is still emerging and at position 2 the bottom of the cloud is seen just before entering the base of the thermal. Positions 3 and 4 are confined within the same cone. In the picture Fig. 10.4.vi a bubble cloud is seen after rising through a layer of whitened water, which is seen rising upwards behind the thermal; but unlike the thermals of buoyant fluid (in Fig. 10.2.i or the large air bubble in (Fig. 10.1.iii) it is leaving clear fluid in its wake which has passed through the bubble cloud because the small air bubbles have a rate of rise which is significant when compared with the rise of the thermal itself. No success was achieved in making the bubbles in the cloud smaller and more numerous without at the same time making the thermal occupy so much of the tank that its rise through undisturbed water could not be studied.

If a second bubble cloud was released close behind the first it was not left behind; but in the rising water in the wake of the first it rose into the rear of the first, and emerged from the top and then widened out to become the first, through which the other rose in its

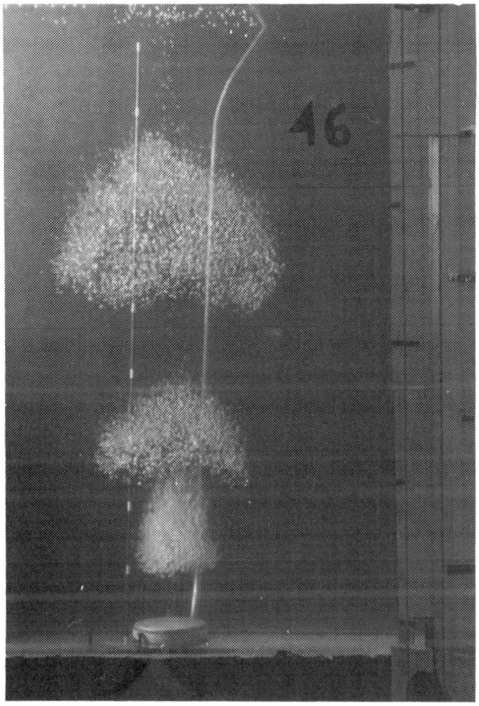

Fig. 10.4.vii In this case we see that a column of bubbles which has formed the thermal shape at its top now has a second cloud with no tail of bubbles at the bottom of its tail. The consequence is that by the time of the second exposure the originally lower thermal has risen through and is spread out above the originally upper thermal which is in its turn now rising up through the originally lower bubble cloud.

turn. This provided a method of determining how far a thermal would rise while all its buoyant material made a complete circuit of the vortex ring. When viewed from above, the bubble cloud had a clear hole through its middle, like the other thermals. The buoyancy of the bubble cloud was measured by capture in an inverted funnel.

10.5 FLYING IN THERMALS

In the development of soaring, glider pilots have had a variety of pictures in their minds as they have exploited thermals in the atmosphere to gain height. At first thermals were thought of as more or less spherical masses of air called bubbles, which rose through the environment. If a glider pilot was lucky enough to enter near the top he would then be able to sink down through it until he emerged from the bottom. Thereafter he would no longer be able to gain height with it.

Since solid bodies passing through air have turbulent wakes, it was thought that thermals also had wakes, and as a pilot emerging from the bottom would feel turbulent air this accorded with the pattern of experience. It fitted with the idea that a thermal was rather like an air bubble in water as well. This analogy saw all the buoyancy as being

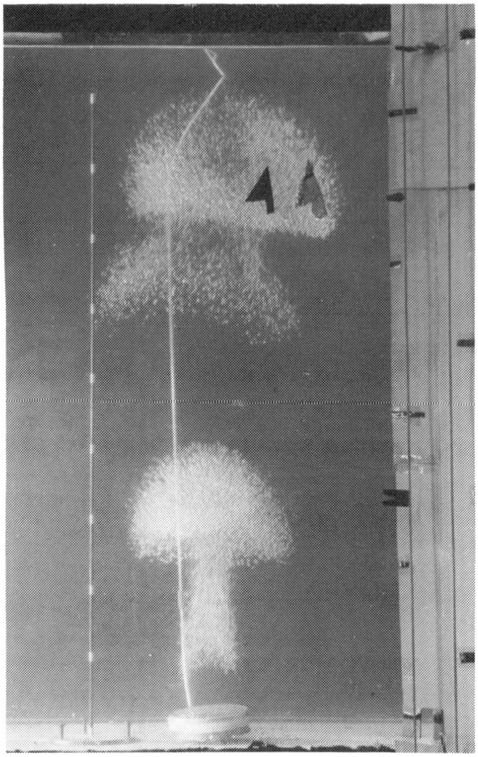

Fig. 10.4.viii At the bottom we see that a column of bubbles (which imitates the 'sausage-shaped' thermal imagined at one time by some glider pilots) forms a spherical-topped thermal at its top (like a new plume), and begins to form a second concentration of bubbles at the bottom of the column. But at the time of the second flash the whole system has turned inside out and the lower thermal now gathering together to rise up through the middle of the upper one is, in fact, the upper one of the positions shown by the first flash. Thus it may be argued that a thermal which at one moment has the shape of a vertical sausage could not retain that shape.

at the top with a ring vortex behind this lens-shaped buoyant mass in which smooth updraughts could sometimes be found. Below was a turbulent wake with the upward velocity gradually decreasing at greater distance below the thermal itself. The analogy with the air bubble (Fig. 10.1.iii) was modified by saying that, since the warm air can be mixed with the environment, mixing (clearly seen to take place on the top of a cumulus cloud) would occur and would erode the bubble. This would cause it to decrease in size and eventually waste away. This theme is developed in the chapter on clouds (Chapter 13). Such an eroded thermal would have a vortex ring-like wake with turbulence below it.

It was a common experience that on first entering a thermal, finding its centre by circling and manoeuvring to obtain the maximum updraught appeared to be easy. Turbulence was experienced when it became difficult to centre on the thermal and find

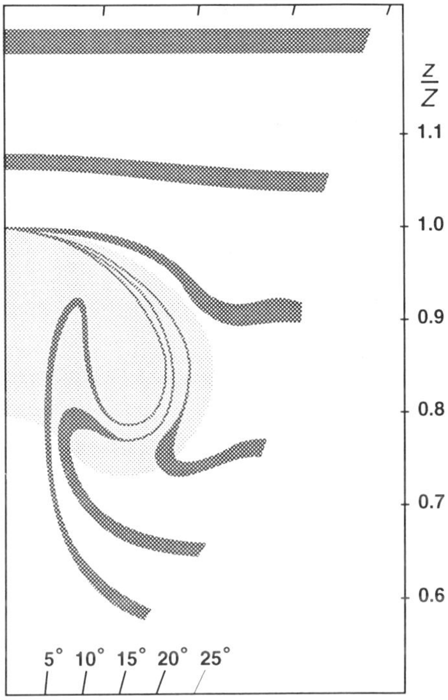

Fig. 10.4.ix Successive positions of an initially horizontal layer of fluid above a thermal shown relative to the growing thermal (Woodward [41]).

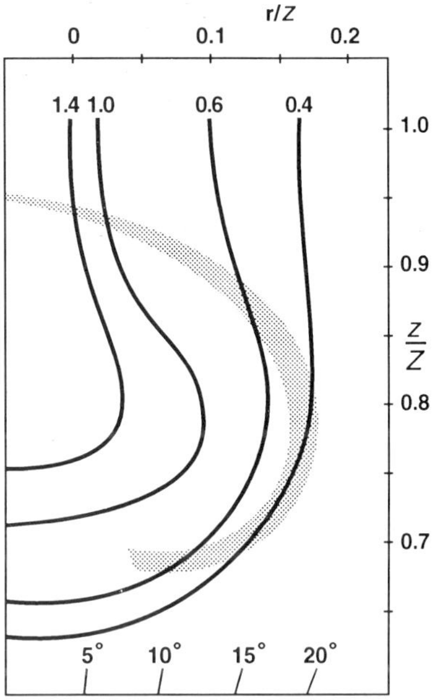

Fig. 10.4.x The boundaries of the region outside which a particle with a given fall speed will fall out of the bottom of the thermal. The thermal outline is indicated by shading. The fall speeds are expressed as multiples of the velocity of rise of the cap of the thermal (Woodward [41]).

the best updraught. Some pilots believed thermal upcurrents to be elongated like a sausage in the vertical direction, but that particular idea has not had much support. This is because no one has ever produced a buoyant mass of fluid which could retain such a shape: the lower part always caught up with the upper part and made a more spherical shape. The question whether thermals can be plumes of warm air (see section 10.10) is discussed in Chapter 11.

When laboratory experiments on thermals were performed it became obvious that the turbulence was on the top (the front) and that there was no wake below (in the rear) at all. At about this time the conventional picture, with its strong upcurrents near the top and a turbulent wake with weak upcurrents below, was being questioned because the less experienced pilots at gliding championships had acquired the habit of watching the champions and, when they circled, of circling under them. Instead of having the advantage over novices the champions soon found themselves surrounded by them, unable to get away. This was remarkable because when the champions came into the turbulent region it appeared that the others had a stronger upcurrent lower down, although they should have been down beyond the thermal altogether.

The argument was then advanced that thermals tended to travel in groups, in series up the same path for some reason, and that the novices below who seemed to be doing so well were in the next thermal of the series. At the same time, since thermals are due to rising hot air, it was supposed that temperature measurements would reveal vertical velocity to be well correlated with temperature excess. Some pilots fitted thermometers on their wingtips and tried to locate updraughts by turning towards higher temperatures indicated by a sensitive dial showing the temperature difference between the wingtips. They met with no success that could not be ascribed to a greater awareness of vertical velocity differences between the wingtips, and there were many frustrating failures: frustrating because pilots continued to believe that the method must be successful since excess temperature and strong updraught should go together.

In fact the most rapidly rising air is the coolest of all, because the environment of thermals is slightly stably stratified (see section 11.4). Hence that air which has been displaced upwards the greatest distance either ahead of the thermal or by entrainment up its centre is coldest.

All the experiences of pilots except those gained near to the ground were resolved and put in place by the thermal model which emerged from the laboratory experiments. It is rather funny that pilots used to speak of having reached the top of a thermal when it became turbulent, meaning that they had gone as high as they could in that one and imagining that they were sinking out of the bottom of it. They had in fact reached its top.

Captain N. Goodhart [42] pointed out the very important fact that in the region of confluence beneath the buoyant air, at the point O in Fig. 10.3.i, where the horizontal indraught in a thermal is as strong as the updraught, a pilot circling with his wings tilted at 45° to the horizontal will be able to use the indraught component as an updraught. Equally, in the diffluent region, which is the turbulent part of the thermal, he will regard the outflow as if it were a downdraught when circling. He will also find it very easy to centre on the lower part of the updraught because of the inflow. However, above the middle of the thermal it will be more difficult to do so, even though the vertical component of the velocity of the air is the same as below.

In using a thermal the pilot is most likely to discover it by entering near the bottom. When he banks to circle in the upcurrent he will find that it appears to increase on account of the confluence. By contrast it will tend to disappear if he circles around it at the top, in the diffluence. He exploits the thermal by rising in the upcurrent, whose maximum strength is equal to about 2.2 × the rate of rise of the thermal. When the air becomes turbulent it is time to look for another thermal because he cannot continue to ascend at more than about half the former rate: at the same time those circling below in the same thermal will be able to use it to great advantage.

In making use of the diagram showing the velocity distribution in a thermal it must be appreciated that in the upper part, where the velocity is fluctuating, it may be possible to soar momentarily near the cap in an upcurrent twice as strong as the rate of rise of the cap, and that this upcurrent may quickly fade as it emerges from the top. The magnitudes given in Fig. 10.3.i represent averages for the meandering upcurrent.

If all thermals were like the model they would be very intense and very small near to the ground, and would have a velocity which decreases like z^{-1} upwards. In fact there must be modifications both near the ground where the thermals originate and higher up where they must overlap as they increase in size. These are discussed in section 11.3.

The exploitation of thermals by birds and insects is discussed in sections 14.2. 14.3, 14.4 and 14.6.

10.6 PUFFS

A puff is a turbulent body of fluid ejected into an otherwise undisturbed environment. A vortex ring is not a puff because it is not turbulent. Assuming that there is a unique configuration and that the total momentum remains constant, we must have the following relationships between its forward velocity w, its size R and the distance travelled z:

$$wmR^3 = \text{constant} = w_0 m R_0^3 \tag{10.6.1}$$

$$\frac{dR}{dt} \propto w = \frac{dz}{dt}. \tag{10.6.2}$$

From this it follows that

$$z = nR \tag{10.6.3}$$

$$\frac{w}{w_0} = \frac{R_0^3}{R^3} \tag{10.6.4}$$

and

$$z^4 = 4 w_0 R_0^3 n^3 t. \tag{10.6.5}$$

The subscript 0 denotes values at a particular time, mR^3 is the volume and n is a constant (at least for each puff) defining the angle of spread. The above argument can be presented in this abbreviated form because we have knowledge of the consequences of the relevant suppositions gained from our study of thermals. There is a dimensional argument hidden away in (10.6.1), which includes the supposition that viscosity is of no importance.

The circulation is proportional to wR, which is proportional to z^{-2}. Consequently the circulation, and therefore **the Reynolds number, is decreasing** like z^{-2} or $t^{-1/2}$. This means that before long the viscous forces will take over **and the flow will become predominantly laminar**.

There are two other important features of puffs which must be mentioned. Richards [39] measured the velocity distribution in and around puffs made in water in the laboratory and found that it was roughly the same as in a thermal, to within the accuracy of measurement possible and meaningful in such a situation. The overall appearance is also very similar, there being a mixing region at the front as in a thermal. Typically $m = 3$ and $n = 4$, but n varied between 1.7 and 6.3, which is about as much as in the case of thermals. The pattern of motion, therefore, is not significantly dependent on how it is produced, for the thermal looks the same even though it is decelerating less because the buoyancy is feeding in new momentum.

The second feature is that they are difficult to produce in the laboratory because it is easier to produce a vortex ring from a circular orifice. At the edge of an orifice there is separation of the flow and a vortex sheet streams away from it. If a short collar-shaped piece of vortex sheet is emitted it quickly rolls up into a vortex ring, especially if the length emitted is small compared with radius of the ring. Richards emitted a slug of fluid longer than the width of the orifice from a rectangular orifice, and in this way ensured that the flow was turbulent. Other configurations are possible, but two things must be remembered. Some distance is needed to form the model puff, and the motion will soon become laminar because of the rapid decrease of Reynolds number. Good puffs are made by the firing of guns.

10.7 BUOYANT PUFFS

If we assume that the configurations of a thermal and a puff are precisely the same, it may be expected that a puff possessing buoyancy will in due course become a thermal if the original momentum is vertical because no change in form is required. In that case the momentum is of the form

$$kwmR^3 = kw_0 mR_0^3 + gB_0 mR_0^3 t \tag{10.7.1}$$

where k is a number which gives the momentum as a multiple of the velocity multiplied by the volume. Using (10.6.2) we therefore obtain

$$z^4 = 4n^3 R_0^3 \left(w_0 t + \frac{1}{2k} gB_0 t^2 \right), \tag{10.7.2}$$

which for small t is like (10.6.5) but for large t is like (10.2.11). The circulation and the Reynolds number are proportional to wR, which in terms of time is given by

$$wR = \frac{n^{1/2} R_0^{3/2}}{2} \frac{w_0 + \frac{gB_0}{k} t}{\left(w_0 t + \frac{1}{2} \frac{gB_0}{k} t^2 \right)^{1/2}} . \tag{10.7.3}$$

This decreases when t is small like $t^{-1/2}$, but tends to the limit $n^{1/2} R_0^{3/2} g^{1/2} B_0^{1/2} / (2k)^{1/2}$ for large t.

For the study of puffs emitted with buoyancy at an angle to the vertical the reader is referred to the independent work of Richards [39].

10.8 TWO-DIMENSIONAL (FLAT) THERMALS AND PUFFS

An exactly analogous treatment may be used to study two-dimensional cases having a line source. A two-dimensional thermal can be created in a laboratory by having a cylindrical trough mounted on a horizontal axis which is quickly inverted after being loaded with buoyant (heavy) liquid. It is found that it is better to have either a divided trough or two parallel troughs which tip their fluid towards each other in order to avoid imparting a net sideways momentum to the fluid. As for the axisymmetric thermal, considering unit length, we now have

$$z = nR \tag{10.8.1}$$

and

$$w = \mathrm{C}(gBR)^{1/2}, \tag{10.8.2}$$

with

$$BR^2 = B_0 R_0^2 \tag{10.8.3}$$

for the conservation of total buoyancy. This leads to

$$\frac{\mathrm{d}z}{\mathrm{d}t} = \mathrm{C}\left(ngB_0 R_0^2 / z\right)^{1/2}$$

or

$$z^{3/2} = \frac{3}{2}\mathrm{C}(ngB_0 R_0^2)^{1/2} t; \tag{10.8.4}$$

and since

$$w \propto z^{-1/2} \tag{10.8.5}$$

the circulation and **the Reynolds number** are proportional to

$$wz \propto z^{1/2} \propto t^{1/3}, \tag{10.8.6}$$

which **increases with distance and time. Two-dimensional thermals therefore remain turbulent**, and if not initially fully turbulent will become so.

The general configuration of 2D thermals is very similar in cross-section to axisymmetric thermals and puffs, except that the angle of spread is greater, a typical value of n being 2. At the same time there is no complete penetration to the front of exterior fluid entrained at the rear; otherwise there could be no buoyancy forces operating along the axis of symmetry to increase the circulation according to (10.8.6). Richards has found that for 2D thermals n may range between 1.3 and 3.2 while for non-buoyant 2D puffs it ranges between 2.6 and 5.7, with typical values of about 2 and 4 respectively.

Fig. 10.4.i (p. 319) shows a 2D thermal with its motion pattern made apparent by taking a time exposure of bright particles in an illuminated cross-section.

The **2D puff** satisfies the following:

$$z = nR \tag{10.8.7}$$

and

$$wR^2 = w_0 R_0^2, \tag{10.8.8}$$

according to which

$$z^3 = 3n^2 w_0 R_0^2 t \tag{10.8.9}$$

and

$$wR = n w_0 R_0^2 z^{-1} \propto t^{-1/3}, \tag{10.8.10}$$

showing that the Reynolds number decreases, although more slowly than in the case of the axisymmetric puff.

For a **2D buoyant puff** the equation analogous to (10.7.1) is

$$kwmR^2 = kw_0 mR_0^2 + gB_0 mR_0^2 t \tag{10.8.11}$$

from which

$$z^3 = 3nR_0^2 wt + \frac{1}{2}\frac{gB_0}{k}t^2 \tag{10.8.12}$$

and a calculation for the Reynolds number shows a decrease followed by an indefinite increase. It slows down like a puff but becomes a thermal.

Without a specific experiment we have no assurance that a thermal which is very slow because it originated as a puff which had been slowed down and which became laminar at low Reynolds number will necessarily become turbulent again. However, the implications of this argument are that it will do so when it becomes large enough, and the Reyholds number of the eddies becomes large on this account.

10.9 JETS

In this section it will be assumed that the density is uniform throughout and buoyancy effects will be considered later (section 10.10.–13).

(1) Steady axisymmetric jets

The style of analysis is now established and we may proceed quickly with the cases of different kinds of continuous source of momentum or buoyancy.

The axisymmetric jet of fluid entering an otherwise motionless fluid of the same density is easily observed to be conical. This is inevitable if the only factors determining the motion are the momentum flux and the distance from the orifice. We may be at distances so far from the orifice that all trace of its shape and of the velocity profile across it have been lost because of the eddies, and the eddies may be intense enough for the viscous effects to be negligible. If these two provisions are met we must find that the width R is proportional to distance from the virtual origin, and that the eddy size characteristic of the motion at any point must also be proportional to it. Therefore

$$z = nR. \tag{10.9.1}$$

Likewise all velocities at a given distance bear the same relationship to one another as at all other distances. If fluid emerging from the orifice is marked (see Fig. 10.9.i), a cone, whose radius we take to define R, will be seen clearly setting the boundary beyond

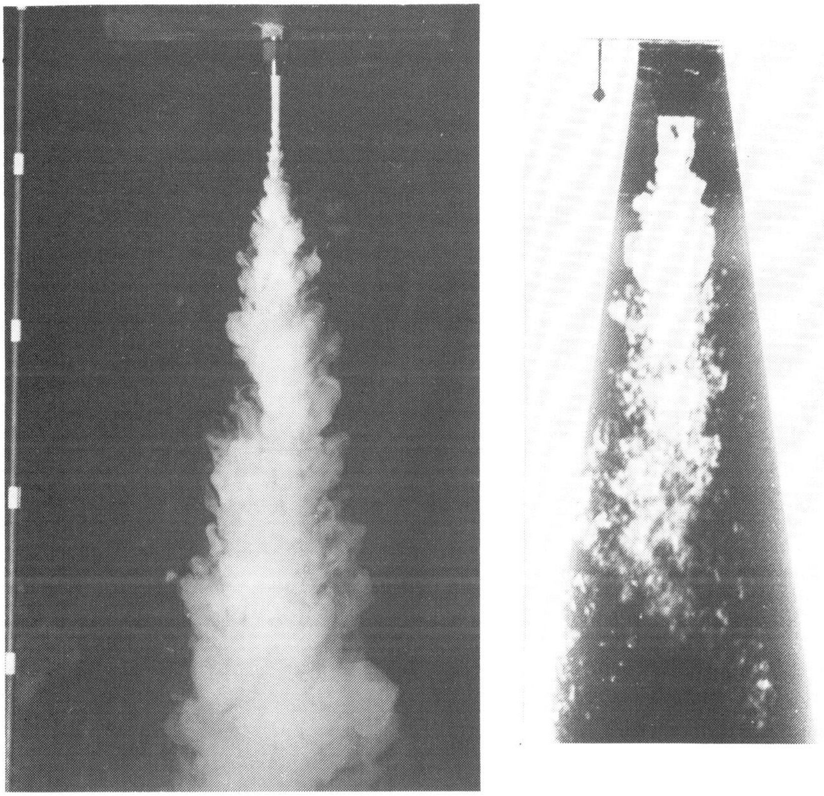

Fig. 10.9.i Buoyant fluid emerging from a nozzle into an otherwise undisturbed tank of water. Notable features which are referred to in 10.9–13 are

(1) the slight convergence of the fluid as it moves away from the nozzle according to (10.13.7);

(2) the beginning of instability when the velocity reaches the value given by (10.10.1), the constant being given in Table 10.15.1;

(3) the growth along a cone with eddy size increasing in proportion to the cone width;

(4) the shape of the eddies indicating a more rapid flow up the axis of symmetry.

A jet possessing no buoyancy but emitted with larger momentum from the orifice has a very similar appearance except that the turbulence begins close to the nozzle.

In this case the fluid was salt water containing $BaSO_4$ precipitate, having a larger density than the fresh water in the tank, but the appearance would be the same with upward motion and buoyancy. The scale of this experiment is indicated by the 10 cm intervals on the left of the picture.

The photo on the right is of a plume of colourless salt water. Its background is a dark edge, but the illumination from each side is refracted so as to reveal the plume. This shadowgraph shows the converging stage of a wider plume emerging with a velocity lower than that appropriate to its buoyancy, as described in the case of Fig. 10.10.ii.

which marked fluid does not penetrate. Outside it the motion is irrotational, even if it can be observed to be unsteady near to the cone. Inside the cone the average motion and pattern of fluctuations is similar at all distances: this is a consequence of the simplicity of the situation which excludes any influence of compressibility. Thus the analysis may not be valid for supersonic jets.

Since the momentum flux is proportional to the square of the velocity and the cross-section area

$$w^2 z^2 = w_0^2 z_0^2 = \text{constant}, \tag{10.9.2}$$

or

$$w \propto z^{-1}. \tag{10.9.3}$$

The Reynolds number is equal to

$$\frac{wR}{\nu} = \text{constant}, \tag{10.9.4}$$

but the circulation (measured up the axis, out along a radius and back to the origin through the irrotational fluid) is not a useful concept because it is logarithmically large at $z = 0$. The constancy of the Reynolds number means that **if the jet is turbulent near the orifice it will remain turbulent up to infinity** (unlike a puff).

The inflow velocity, denoted by U at the boundary at distance R, may be assumed to be horizontal. For dimensional reasons it is proportional to w and therefore to z^{-1} and R^{-1}. This accords with the continuity of momentum flux across a section normal to the axis, for the increment is zero for a displacement dz in the direction of the axis. It also accords with the equation of continuity, for the increment in upward flux of volume must equal the horizontal flux inwards through depth dz. Thus (see Fig. 10.9.ii)

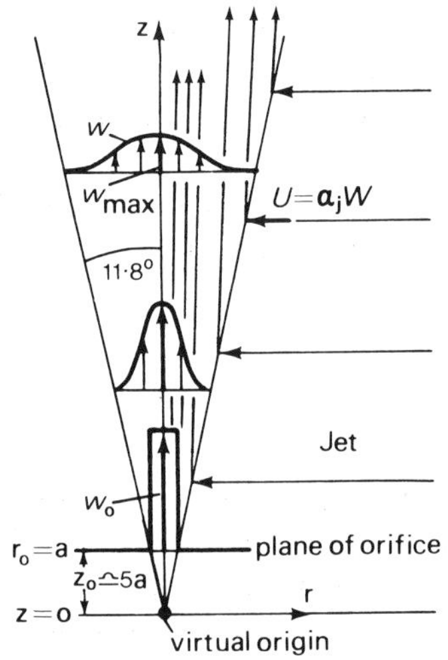

Fig. 10.9.ii Coordinates used to describe a jet. At the orifice the velocity has the 'top-hat' profile, but this quickly becomes bell-shaped and is fairly well represented by a gaussian or cosine curve. On the right are shown the mean streamlines. The virtual origin, at the vertex of the enveloping cone, is about 5 orifice radii from the orifice.

$$\mathrm{d}(wR^2) \propto 2\pi RU\,\mathrm{d}z. \tag{10.9.5}$$

We use the proportionality sign because we have not defined precisely what velocity is represented by w.

The flow is fluctuating from time to time and from place to place so that we may write

$$w = \overline{w}(r,z) + w'(r,z,t) \tag{10.9.6}$$

where $\overline{w}$ represents the time-average value at a point. w has a characteristic profile at each value of z defined by the function $w_1(r)$, so that

$$\overline{w} = w_1(r)z^{-1} \tag{10.9.7}$$

and

$$w' = w_1'(r)z^{-1}f(t). \tag{10.9.8}$$

Observation shows that the vertical velocity falls to around zero at any point from time to time; therefore w' is of the same order of magnitude as $\overline{w}$. The consequence of this is that the time-averaged momentum flux, which is

$$\begin{aligned}\overline{\int_0^R \int_{\theta=0}^{2\pi} w^2 r\,\mathrm{d}\theta\,\mathrm{d}r} &= \overline{2\pi\int_0^r \left(\overline{w}^2 + w'^2 + 2\overline{w}w'\right)\mathrm{d}r},\\ &= 2\pi\int_0^R \left(\overline{w}^2 + \overline{w'^2}\right)\mathrm{d}r,\\ &= \dot{M}_\mathrm{m} + \dot{M}_\mathrm{f}\end{aligned} \tag{10.9.9}$$

where θ is the azimuth in the plane z = constant. $\dot{M}_\mathrm{m}$ and $\dot{M}_\mathrm{f}$ are the momentum fluxes attributable to the mean and fluctuating motions respectively. It is assumed that the time averaging is done over a long enough period for all averages to be independent of θ, so that $\overline{w'} = 0$. Since w' is of the same order of magnitude as $\overline{w}$ the expression (10.9.9) is very dependent on the eddies.

If it be supposed that the jet contains some sort of passive pollution of concentration σ, but with no buoyancy, whose flux is constant up the jet, then

$$w\sigma R^2 = \text{constant}, \tag{10.9.10}$$

and so

$$\sigma \propto w \propto R^{-1} \propto z^{-1}. \tag{10.9.11}$$

It appears therefore that pollution is diluted in exactly the same manner as momentum, but without making measurements there is really no justification for any assumption that the profiles of w and σ are the same. On the contrary, it is clear that in the case of a puff the correlation is very poor because the fastest-moving fluid on the axis is the least polluted. The assumption that momentum is diluted in the same way as any quantity such as pollution attached to the fluid is known as **Reynolds's analogy**. It is only justified as far as dependence upon z of the quantities is concerned, and it implies nothing about correlation between σ' and w'. If we write similar equations to (10.9.7) and (10.9.8) for σ, the pollution flux is seen to be

$$2\pi\int_0^R\left(\overline{w}\overline{\sigma}+\overline{w'\sigma'}\right)\mathrm{d}r. \tag{10.9.12}$$

In this expression the second term contributes a smaller fraction of the whole than the second term in (10.9.9) because w' and σ' may not be well correlated. On the other hand, $\overline{\sigma}$ and $\overline{w}$ are observed to have very similar profiles (i.e. are proportional to roughly the same function of r).

The motion outside the jet may be assumed to be horizontal if the orifice is in a plane horizontal wall so that there is a constraint present not contributing any dimensional factors. In that case the radial inward velocity takes the value

$$u = UR/r \tag{10.9.13}$$

at all values of z and $r > R$. This ensures that the pressure field can exist to produce this motion, the isobaric surfaces being cylinders. Inside the jet the time average streamlines are approximately as shown in Fig. 10.9.ii, and have been measured by several workers since Schmidt [44] first embarked on this kind of analysis.

The form the streamlines take where the orifice is not in a rigid plane boundary capable of supporting this pressure distribution has not been analysed. This analysis is not valid for the case of an orifice at the end of a long tube of negligible diameter because nothing has been specified about the inflow velocity around the tube. Any other configuration such as an orifice at the tip of a cone, or at the end of a pipe at a finite distance from a plane boundary, would introduce additional non-dimensional numbers which would destroy the simplicity of the analysis. However it cannot be correct in practice for a nozzle on the end of a long tube because the pressure field behind the nozzle is not the same as in front due to entrainment, and so the outside flow cannot be plane.

The value of n in (10.9.1) is, for practical purposes, equal to 5. It cannot vary from case to case, as it does for isolated thermals and puffs, because it is a phenomenon averaged over a sufficiently long time to obtain a unique value. In theory, therefore, all jets are similar and a cine film of one can be used to represent them all by varying the speed and magnifying it as required. The exterior motion in the case where there is a plane wall surrounding the orifice is the same as for a uniform distribution of sinks along the axis, i.e. a line sink. Hence it is reasonable to suppose that the motion outside the jet when there is no wall is roughly that due to a semi-infinite line sink. At points close to the jet boundary this is not very different from the case of an infinite line sink because of the small angle of spread.

(2) Two-dimensional (flat) jets

A **two-dimensional jet** is produced by a line source of momentum. The constancy of momentum flux requires that

$$w^2 z = \text{constant}, \tag{10.9.14}$$

and so

$$w \propto z^{-1/2} \propto R^{-1/2}. \tag{10.9.15}$$

The Reynolds number is proportional to wz, which increases up the jet like $z^{1/2}$, and so **it is always turbulent beyond a certain point**, although it may not be nearer the source. We are, however, in some difficulty because the inflow velocity decreases upwards like $z^{-1/2}$, which means that the same horizontal pressure gradient is not required to set up the motion at all heights. Since the motion is two dimensional there need be no horizontal acceleration towards the jet if the exterior motion is strictly horizontal. Thus horizontal exterior motion is a theoretical possibility although it is rotational and therefore could not be set up by the jet in otherwise motionless surroundings. In practice there is likely to be some descending motion outside the jet because its finite existence prevents it setting up the theoretical state which has no pressure gradients.

(3) New axisymmetric (conical) jets

A **new jet**, sometimes called a starting jet, (Fig. 10.9.iii) is susceptible to simple analysis. It consists of an upper portion which looks like the front of a puff and a lower conical portion which looks like a jet. It retains its similarity so that its total volume is proportional to z^3. The total upward momentum is proportional to the time since it began, and since all velocities are proportional to one another the upward velocity of its cap can be taken as representative. Therefore

$$wz^3 \propto t \tag{10.9.16}$$

and

$$w = \frac{dz}{dt}, \tag{10.9.17}$$

so that

$$z^2 \propto t. \tag{10.9.18}$$

This relationship, and (10.10.10) below, was accurately confirmed by Spackman [42]. The Reynolds number, and the circulation measured around a circuit such as MNPQM in Fig. 10.9.iv, is a constant because

$$w \propto z^{-1} \tag{10.9.19}$$

as in the jet. The behaviour of the top dome-shaped part is more like a thermal than a puff because momentum is fed into its base to retain (10.9.18) and (10.9.19); otherwise it would be retarded according to (10.6.5).

If the source continues, the behaviour at fixed distances from it becomes like that of a jet as the top moves further away; if the source stops, the fluid already set in motion becomes a puff.

Spackman found the following features:

For a jet, $\theta = 11.8° \pm 0.3°$, giving a mean value of n equal to 4.76. Using measurements of Ricou and Spalding [43], Spackman found that $\dot{M}_f / \dot{M}_m = 0.74$, which is the ratio of the momentum fluxes due to the fluctuations and the mean flow. For a new jet, Spackman obtained a value for the constant of proportionality in (10.9.18), namely

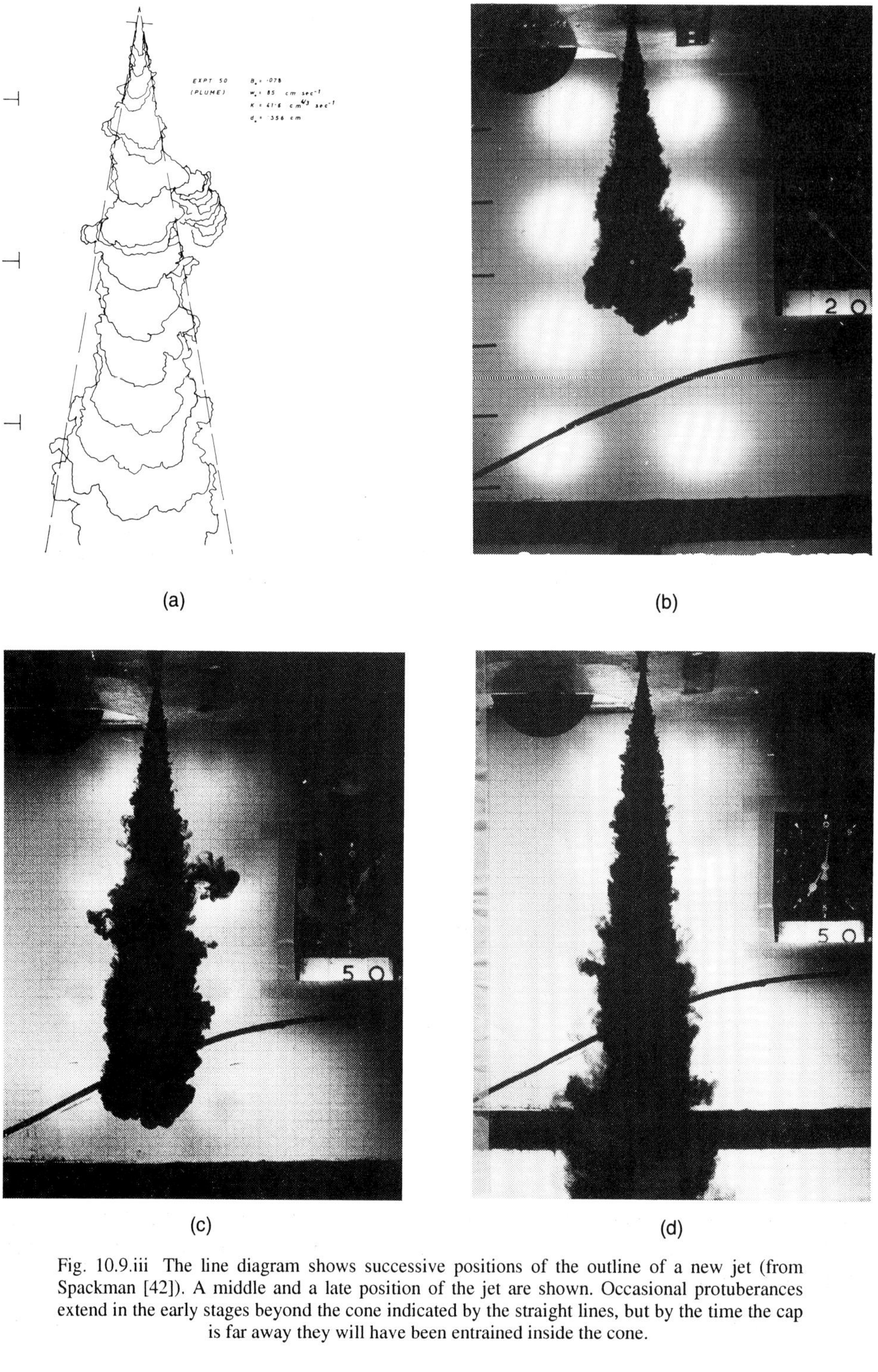

Fig. 10.9.iii The line diagram shows successive positions of the outline of a new jet (from Spackman [42]). A middle and a late position of the jet are shown. Occasional protuberances extend in the early stages beyond the cone indicated by the straight lines, but by the time the cap is far away they will have been entrained inside the cone.

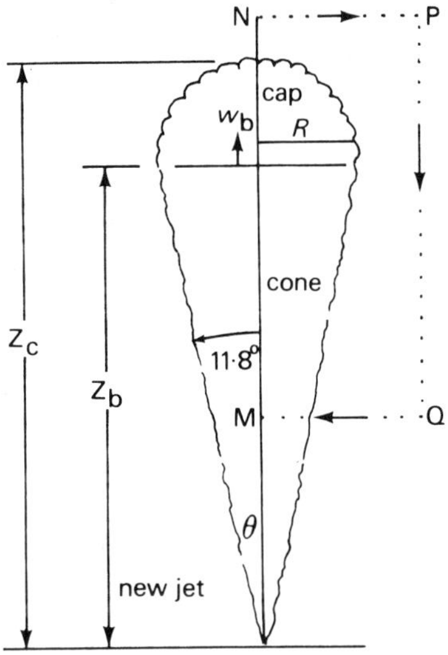

Fig. 10.9.iv Definitions and dimensions for a new (conical) jet. The circuit MNPQM may be used to define the circulation of the system, M being the mid-point between the virtual origin and the base of the cap, or some similarly defined point.

$$z_c^4 = 180\frac{\dot{M}}{\rho}t^2 \qquad (10.9.20)$$

where $\dot{M}$ is the momentum flux through the orifice. Shape factors were as follows:

$$z_c/z_b = 1.3 \qquad \text{vol}_{\text{cap}} = 0.012z_c^3,$$

$$dz_c/dt = 2W_b \qquad \text{vol}_{\text{cone}} = 0.021z_c^3 \qquad (10.9.21)$$

Spackman also made estimates of the entrainment velocity on the cap, and found it to vary from a maximum on the axis to the value 0.11 W_b at the junction with the cone, with a mean value on the cap of 0.6 W_b. W_b is the mean vertical velocity at $z = z_b$, the mean being defined as for (10.11.3), below. Rather more than a quarter of the entrained fluid enters over the cone. The fluctuations of velocity in the irrotational flow outside are illustrated in the time exposure picture in Fig. 10.9.v.

10.10 PLUMES

For convenience the motion produced by a steady source of buoyancy is called a **plume**, the word jet being used for the case of no buoyancy. An axisymmetric **conical plume** has a constant flux of buoyancy (or pollution), denoted by B, in the direction of its axis, and the increase in momentum flux over a height dz is equal to the buoyancy force on the slab of thickness dz. Thus (see Fig. 10.10.i)

$$wBR^2 = \text{constant}, \qquad (10.10.1)$$

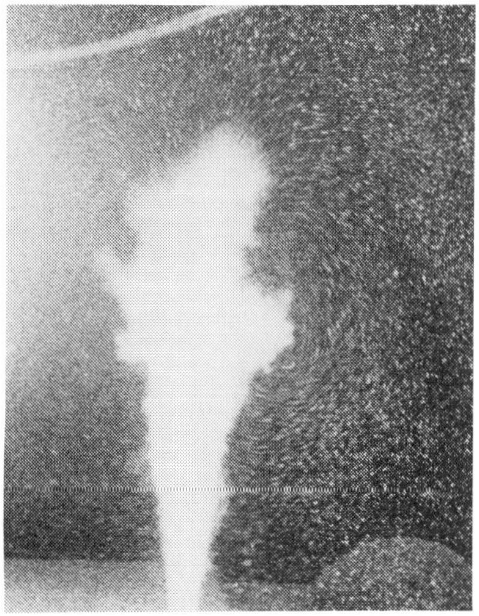

(a)

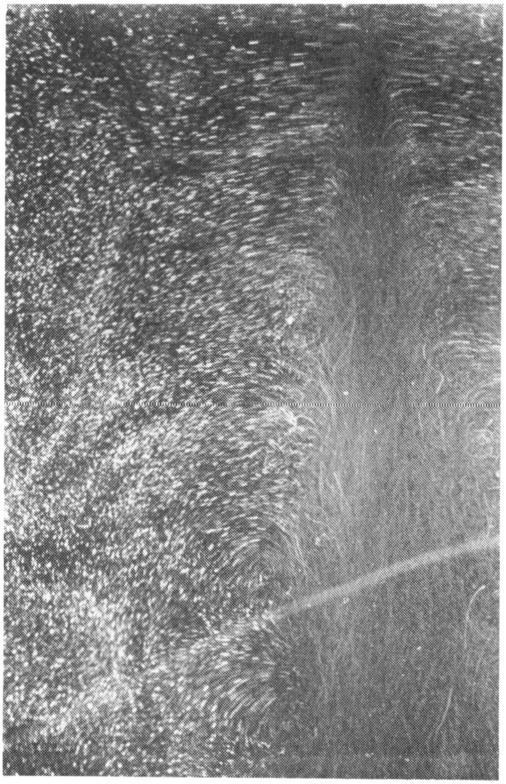

(b)

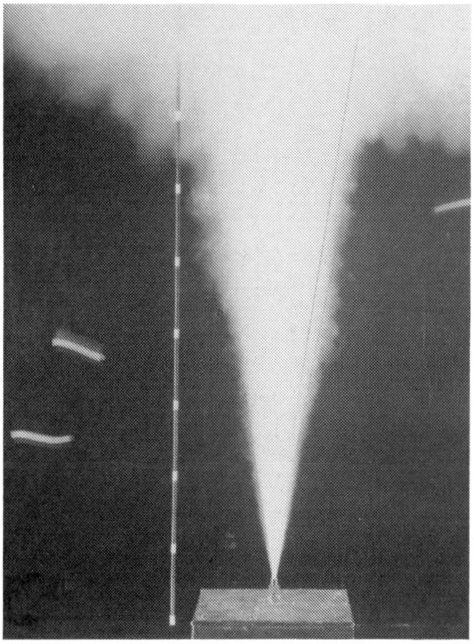

(c)

Fig. 10.9.v (a) A time exposure of a new jet, the displacement of particles floating in the water indicating the velocity vector. Note the variations around the large protuberances and the outflow ahead of the cap.

(b) A time exposure of a plume of unmarked fluid showing the displacements of particles illuminated in a plane of the motion. Note particularly the greatly increased velocity inside the plume and the disturbance of the exterior fluid due to the larger eddies. (From Spackman 1964.)

(c) In this time exposure the displacement towards the plume can be seen. Strictly horizontal flow is not to be expected with a buoyant plume and the finite extent of the tank.

and

$$d(w^2R^2) \propto BR^2\, dz. \tag{10.10.2}$$

Since, for the same reasons as before, $R \propto z$, these equations require that

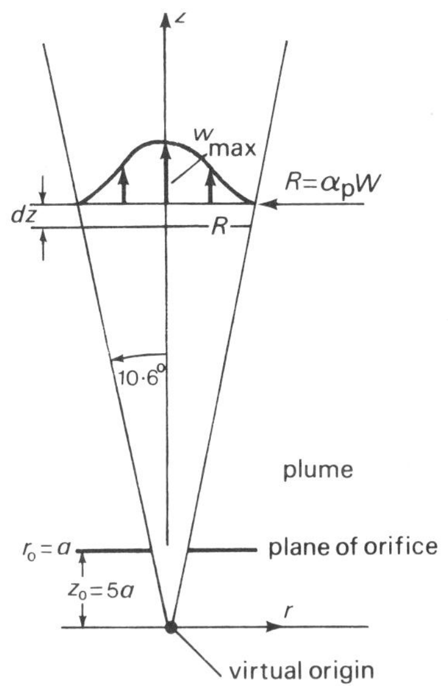

Fig. 10.10.i Coordinates and definitions for a plume.

$$w \propto z^{-1/3} \tag{10.10.3}$$

$$B \propto z^{-5/3}. \tag{10.10.4}$$

Thus it is seen by comparison with (10.9.11) that a plume slows down less rapidly than a jet, because of the continuous application of the buoyancy force, but it is diluted more rapidly because of the additional mixing induced by density gradients. Since the velocity is not proportional to R^{-1} the inflow is different at different heights at a given radius outside the plume, and it is therefore not strictly horizontal.

The most extensive measurements on gaseous plumes, which confirmed the relationships (10.10.3) and (10.10.4), were made by Rouse, Yih and Humphreys [45]. It is impossible to set up a plume with laminar flow because it is essentially unstable, the Reynolds number being proportional to wz, i.e. to $z^{2/3}$, which increases rapidly up the plume. That line of argument is not complete, however, for it only implies that there could be a laminar region lower down in a plume which is turbulent higher up, because the Reynolds number decreases downwards. This question is discussed further in section 10.12. Certainly, experience confirms the conclusion that plumes cannot be prevented from becoming turbulent.

A **two-dimensional (flat) plume** is produced by a horizontal line source of buoyancy. The equations corresponding to (10.10.1) and (10.10.2) are

$$wBR = \text{constant} \tag{10.10.5}$$

and

$$d(w^2 R) \propto BR\, dz, \tag{10.10.6}$$

which, with $R \propto z$, imply

$$w = \text{constant} \tag{10.10.7}$$

i.e.

$$z \propto t$$

and

$$B \propto z^{-1}. \tag{10.10.8}$$

Equation (10.10.7) means that the maximum, mean and typical eddy velocities are the same at all heights although the profile, which is of the same mathematical form, is spread over a width proportional to height. The inflow velocity is the same at all heights.

If the inflow velocity is u, and the plume is set up in a cross-wind of strength u_0, it may be argued that the plume is simply translated sideways as it rises, that its centre rises up a plane inclined at angle $\tan^{-1} u_0/w$ to the vertical and that the inflow is $u + u_0$ and $u - u_0$ on the two sides. This idea is naive in the sense that there is no single upward velocity, w, which will determine the angle at which it will be inclined. Moreover the plume is inclined to gravity so that the pressure field at each height is altered. When u_0 exceeds u no inflow can occur on the downwind side, and so the plume is attached to the ground there. A very practical application of this ideas was made by G. I. Taylor in FIDO, the system for clearing fog from an airfield runway in no wind. A line source of heat in the form of paraffin burners was installed on each side of the runway, and each produced a wind inwards at first equal to u. In effect they entrained all the air in between, quickly coalescing to form a single source located half way between them at approximately the same height (see Fig. 10.10.ii). There was as a consequence, a region of heated air in which the fog was evaporated over the runway and in which an aircraft could be landed visually. The flow pattern is similar in all such cases; all that was required was enough heat to evaporate the fog in the entrained air. When the buoyancy flux ($\propto wBR$) is doubled, since $w^2 \propto gBR$, w^3 is doubled. With an inflow of only $2^{1/3}u$ the plumes do not lie down on the ground.

A two-dimensional plume is very turbulent because the Reynolds number is proportional to z and increases rapidly upwards as the width increases.

A **new axisymmetric plume** is similar to a new jet in appearance but has a total upward momentum increasing at a rate proportional to the total buoyancy already emitted. Thus with characteristic buoyancy B and a constant emission rate beginning at $t = 0$

$$\frac{\mathrm{d}}{\mathrm{d}t}(wR^3) \propto t \tag{10.10.9}$$

or

$$wz^3 \propto t^2$$

and

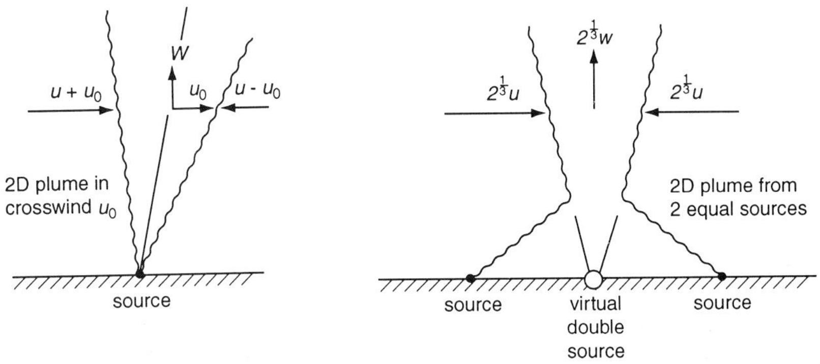

Fig. 10.10.ii A two-dimensional plume in a cross-wind u_0 shown adding this velocity to the inflow velocity u, which is the same at all heights. Thus two parallel line sources of heat (buoyancy) form a single plume by entraining each other: they cover the ground between them with a region of heated air, above which the plume rises as if from a double-strength source at the middle. The behaviour of two parallel line jets is uncertain. The momentum flux is unknown because it depends on the pressure at the ground between them.

$$\frac{\mathrm{d}}{\mathrm{d}t}(BmR^3) = Ft$$

or

$$Bz^3 \propto t$$

where mR^3 is the volume and F is the buoyancy flux at the orifice.

Since $w = \dfrac{\mathrm{d}z}{\mathrm{d}t}$ and $R \propto z$ we must have

$$z \propto t^{3/4}, \tag{10.10.10}$$

$$w \propto z^{-3/4} \propto t^{-1/4}, \tag{10.10.11}$$

and

$$B \propto z^{-5}. \tag{10.10.12}$$

The laws of dilution of buoyancy (and pollution) and momentum appear to be the same as for an axisymmetric steady plume (see (10.10.3) and (10.10.4)), but z_c is the overall height, and w and B are characteristic values of the new plume. The thermal-like upper part is fed with buoyancy and momentum from below and is slowed down less rapidly than a thermal in which $w \propto z^{-1} \propto t^{-1/2}$, $B \propto z^{-3}$. The Reynolds number increases like $z^{2/3}$ and so the phenomenon must become turbulent.

A **two-dimensional (flat) new plume** follows similar relationships, which now take the form

$$\frac{\mathrm{d}}{\mathrm{d}t}(wR^2) \propto gBR^2 \propto t, \tag{10.10.13}$$

which leads to

$$z_c \propto t \tag{10.10.14}$$

$$w = \text{constant} \tag{10.10.15}$$

and

$$B \propto z^{-1}. \tag{10.10.16}$$

These resemble (10.10.7) and (10.10.8) as expected.

It may be noted that for any of the buoyancy-driven systems we may write

$$w \propto (gBR)^{1/2} \tag{10.10.17}$$

as an alternative to one of the equations used above. This is based on simple dimensional analysis.

Spackman found that for a new axisymmetric plume, in the notation defined at the end of section 10.9,

$$\theta = 10.4° \pm 0.6°;$$

$z_c^4 = 39Ft^3$, where F = buoyancy flux at the orifice;

$$\left.\begin{aligned} \text{vol}_{\text{cap}} &= 0.0084 z_c^3, \\ \text{vol}_{\text{cone}} &= 0.017 z_c^3 \end{aligned}\right\} \quad \begin{aligned} &m' = 0.0254 \\ &\text{where } m' z_c^3 \text{ is the volume} \end{aligned}$$

$$z_c/z_b = 1.3$$

$$dz_c/dt = 1.3 W_b.$$

Spackman also estimated the entrainment and found that the contribution over the cone is more important than over the cap. This is because it has a mean value of 0.26 W_b on the cap, which is about the value at the top of the cone. About 60 per cent of the fluid entrained enters over the cone.

10.11 BUOYANT JETS

In an axisymmetric plume there is a momentum flux appropriate to the buoyancy flux which causes it. If the momentum flux exceeds the appropriate value the system will behave like a jet in that the momentum will decrease more rapidly and the dilution proceed less rapidly than in a plume. When the appropriate value is reached it will behave like a plume.

We desire an analysis for the region in which the transition takes place. Because the dimensional analysis produces single-power relationships between the quantities, it is often assumed that this type of relationship is a natural one, and some researchers assume a power relationship and use experiments to measure what the power is. It should be emphasised that when the situation is as simple as the cases we have been discussing, that is simple enough to produce single-power proportionalities, the analysis will always reveal what the law is. We must perform experiments to find out all the other relationships, in particular those relating to the geometrical shape of the phenomena, the profiles of velocity and buoyancy across horizontal sections, and the magnitudes of, and correlations between, fluctuations of velocity and buoyancy (or pollution).

If w and b are the instantaneous values of upward velocity and buoyancy, the time-averaged fluxes of momentum and buoyancy are

$$mR^2W^2 = \int_0^\infty 2\pi r\overline{w^2}\,\mathrm{d}r = \int_0^\infty 2\pi r\left(\overline{w}^2 + \overline{w'^2}\right)\mathrm{d}r \tag{10.11.1}$$

and

$$\frac{1}{g}\Phi = kR^2WB = \int_0^\infty 2\pi r\overline{wb}\,\mathrm{d}r = \int_0^\infty 2\pi r\left(\overline{w}\overline{b} + \overline{w'b'}\right)\mathrm{d}r \tag{10.11.2}$$

where the bar denotes time averaging and the prime the departure from the time average. Φ is defined by (10.11.2); m and k are constants of proportionality which have to be measured. R is defined, as at the beginning of section 10.9, to be the radius beyond which the material of the buoyant jet does not penetrate at a given height. W and B are a mean velocity and buoyancy defined so that the flux of velocity (momentum at unit density) is πR^2W and the total buoyancy in a section of thickness dz is gBR^2 dz. The mean horizontal inflow velocity at height z and radius R is defined as U, and **the entrainment ratio** α is defined by

$$U = \alpha W. \tag{10.11.3}$$

The equation of continuity may be expressed by stating that the change in upward volume flux is equal to the inward volume flux; thus

$$\frac{\mathrm{d}}{\mathrm{d}z}(\pi R^2W) = 2\pi RU = 2\pi\alpha RW. \tag{10.11.4}$$

If we put in the formulae for the vertical variation of R, W and U with z, which for a jet are (10.9.1) and (10.9.2) we find that, using (10.11.3), for an axisymmetric jet

$$\alpha = \frac{1}{2n}. \tag{10.11.5}$$

On the other hand if we use (10.10.3) and (10.10.4) we find that for an axisymmetric plume

$$\alpha = \frac{5}{6n}. \tag{10.11.6}$$

The numerical quantities α, n, m, k are constants for each separate case but there is no means of telling what their values are and whether they differ from one case to another except by observing the phenomena.

The profiles of velocity appear to be more or less gaussian. This cannot be strictly true of the gaussian profile at large values of r where, in the case of the axisymmetric jet at any rate, $\overline{w}$ is identically zero when $r > R$. Perhaps a single wavelength of a cosine-shaped curve (which is very similar to a gaussian curve except at the ends) is a more accurate representation, but people prefer the gaussian formula because it is familiar and easier to handle mathematically. In any case the profiles measured for a plume and a jet are sufficiently alike for it to be assumed that in equation (10.11.1) m takes the same value for both. It is very much easier to observe n (because it is equal to z/R) and it does appear to take more or less the same value for plumes and jets. Therefore (10.11.5) and (10.11.6) imply that the entrainment ratio for a plume is equal to $\frac{5}{3}$ times its value for a jet, and it appears that buoyancy is a direct cause of increased entrainment through the creation of vorticity. This conclusion depends (through (10.11.3)) on the definition of W (in (10.11.1)) and that is why the equality of values of m in the two cases is important. Some treatments have been based on the assumption that α is a kind of universal constant on the grounds that mixing is caused by the velocity difference between the jet and its environment, and it has not been noted that this requires a different value of either m or n for jets and plumes.

If, for the moment, (10.11.2) is regarded as being applicable to flux of pollution of concentration b in the case of a jet with no buoyancy (it being noted that buoyancy and pollution are both carried with the fluid and are diluted identically so that (10.11.2) may refer to either quantity in a plume), then to make any progress with the theory of a buoyant jet we need to assume that (10.11.2) applies throughout with the same value of k. At present there is no good evidence as to whether this is a correct assumption, but its consequences are of some interest.

The vertical rate of change of momentum flux is equal to the buoyancy, thus

$$\frac{d}{dz}(mW^2R^2) = gBR^2, \tag{10.11.7}$$

and we can write equality because we have defined the quantities for this purpose. The buoyancy flux is constant so that by (10.11.2)

$$WgBR^2 = \frac{1}{k}\Phi = \text{constant}. \tag{10.11.8}$$

Therefore (10.11.7) becomes

$$\frac{d}{dz}\left(mW^2\frac{z^2}{n^2}\right) = \frac{\Phi}{kW} = \frac{\Phi z}{k(W^2z^2)^{1/2}}. \tag{10.11.9}$$

This has solution

$$\frac{3}{2}(W^2z^2)^{3/2} = \frac{n^2\Phi}{km}\frac{1}{2}(z^2 + c^2) \tag{10.11.10}$$

where c is a constant of integration. It can be reshaped as

$$W^3 = C^3\left(\frac{1}{z} + \frac{c^2}{z^3}\right), \tag{10.11.11}$$

where

$$C^3 = \frac{3n^2}{4km}\Phi. \tag{10.11.12}$$

Equations (10.11.3) and (10.11.4) give α in the form

$$\alpha = \frac{1}{n}\left(1 - \frac{1}{6}\frac{z^2 + 3c^2}{z^2 + c^2}\right). \tag{10.11.13}$$

For the jet stage of the phenomenon where $z \ll c$ we find

$$W = Cc^{2/3}z^{-1}, \qquad \alpha = \frac{1}{2n}; \tag{10.11.14}$$

while for the plume stage where $z \gg c$

$$W = Cz^{-1/3}, \qquad \alpha = \frac{5}{6n}. \tag{10.11.15}$$

The buoyancy may be derived from (10.11.7) and is

$$gB = \frac{n^2\Phi}{kz(z^2 + c^2)^{1/3}}, \tag{10.11.16}$$

which is proportional to z^{-1} and $z^{-5/3}$ in the jet and plume stages respectively.

These formulae can be applied straightforwardly in the case where the momentum is in excess of that of a plume of the given buoyancy, the transition from a jet to a plume taking place at heights near $z = c$.

If a jet has negative buoyancy, (10.11.12) shows that C^3 is negative, which means that for W to be positive in (10.11.11) c^2 must be negative, and the range of z for which the plume rises is restricted. If we replace c^2 by $-c'^2$, we see that the velocity vanishes at $z = c'$, after which it must fall back towards the source and the analysis is of doubtful meaning. The entrainment ratio is

$$\alpha = \frac{1}{n}\left(1 + \frac{1}{6}\frac{3c'^2 - z^2}{c'^2 - z^2}\right), \tag{10.11.17}$$

which is the same as for a jet when $z \ll c'$, but which decreases with z and becomes zero when $z^2 = \frac{3}{5}c'^2$. This expresses literally the idea that, when positive, buoyancy increases mixing, and therefore when negtive reduces it to zero before the upward velocity falls to zero. The model is unrealistic between this level and $z = c'$ because 'detrainment' is required to keep the shape conical while it is being decelerated.

Above $z = c'$ the implication of the model is that, where the buoyancy is greater than in a plume of the given velocity, entrainment is increased above the value for a plume. The entrainment is infinite at $z = c'$ where $W = 0$ because it is implied that mixing occurs on account of the buoyancy even when there is no velocity relative to the surroundings. But this conclusion is questioned in section 10.13 on grounds that it is not correct to assume that $R \propto z$.

10.12 SIMILARITY IN STRATIFIED SURROUNDINGS

In the cases so far considered it appears that the rates of change of all quantities depend only on local conditions, that is on the values of the various quantities involved at the particular value of z. This has sometimes been taken to imply that, because the theory accords so well with observations, the rates of change do actually depend only on local values. All that has happened is that the similarity ensures they can be expressed that way, although the mechanics are such that the motion depends very much on what is going on above and below as well as on what is happening at the same level. Thus we cannot assume that the similarity and the power laws hold for a more general stratification.

When the surroundings are stratified according to

$$\beta = \frac{1}{\rho}\frac{\partial \rho}{\partial z} \tag{10.12.1}$$

in the incompressible case, the equations for the momentum and buoyancy fluxes in the more general case than that covered by (10.11.7) and (10.11.8) are

$$m\frac{\mathrm{d}}{\mathrm{d}z}(W^2R^2) = gBR^2 \tag{10.12.2}$$

and

$$k\frac{\mathrm{d}}{\mathrm{d}z}(WR^2B) = -WR^2\beta. \tag{10.12.3}$$

The last of these equations expresses the fact that the buoyancy is reduced by an upward displacement in a stratified environment. These become

$$\frac{\mathrm{d}R}{\mathrm{d}z} + \frac{R}{W}\frac{\mathrm{d}W}{\mathrm{d}z} = \frac{RgB}{2mW^2} = \frac{\mathrm{F}}{2m}, \tag{10.12.4}$$

and

$$\begin{aligned}\frac{\mathrm{d}R}{\mathrm{d}z} + \frac{3R}{W}\frac{\mathrm{d}W}{\mathrm{d}z} &= -\frac{R\beta}{kB} - \frac{R}{\mathrm{F}}\frac{\mathrm{dF}}{\mathrm{d}z} \\ &= -\frac{\mathrm{S}}{k\mathrm{F}} - \frac{R}{\mathrm{F}}\frac{\mathrm{dF}}{\mathrm{d}z},\end{aligned} \tag{10.12.5}$$

where

$$\mathrm{F} = gBR/W^2 \tag{10.12.6}$$

is a Froude number relating the buoyancy and inertia forces, and

$$S = g\beta R^2 / W^2 \tag{10.12.7}$$

is a stability number relating the static stability to inertia forces. We then obtain

$$\frac{\mathrm{d}R}{\mathrm{d}z} = \frac{1}{n} = \frac{3\mathrm{F}}{4m} + \frac{R}{2\mathrm{F}}\frac{\mathrm{dF}}{\mathrm{d}z} + \frac{\mathrm{S}}{2k\mathrm{F}}, \tag{10.12.8}$$

and from the continuity equation in the form (10.11.4) we get

$$\begin{aligned}\alpha &= \frac{\mathrm{d}R}{\mathrm{d}z} + \frac{R}{2W}\frac{\mathrm{dW}}{\mathrm{d}z} \\ &= \frac{5\mathrm{F}}{8m} + \frac{R}{4\mathrm{F}}\frac{\mathrm{dF}}{\mathrm{d}z} + \frac{\mathrm{S}}{4k\mathrm{F}} \\ &= \frac{1}{2n} + \frac{\mathrm{F}}{4m}.\end{aligned} \tag{10.12.9}$$

The particular case of no stratification is recovered by putting β and S equal to zero. If F is then constant, and if it is not zero (10.12.6) corresponds to (10.10.17); so $\frac{1}{n} = \frac{3\mathrm{F}}{4m}$, and $\alpha = \frac{1}{2n} + \frac{1}{3n}$ for a plume. If F is zero we simply get $\alpha = \frac{1}{2n}$ for a jet. If n is constant when β is zero, (10.12.8) can be solved to give

$$\mathrm{F} = \frac{4m}{3n}\frac{z^2}{z^2 + c^2}, \tag{10.12.10}$$

which gives the variation of F in the case of section 10.11.

In practice, plumes depart very little from the conical shape in stratified surroundings (except close to the level at which they cease to rise), and it is probably a realistic assumption to put n constant. We have to assume that m and k are also constant in order to make any progress with the analysis. This implies constancy of the profiles of mean and fluctuating velocity and buoyancy.

If the additional assumption is made that S is constant then (10.12.8) may be written as

$$-2n\,\mathrm{d}z/z = \mathrm{dF}\bigg/\left(\frac{3\mathrm{F}^2}{4m} - \frac{\mathrm{F}}{n} + \frac{\mathrm{S}}{2k}\right) \tag{10.12.11}$$

and integrated to give

$$\mathrm{F} = \frac{2m}{3n}\left(\sigma + 1 - \frac{2\sigma c^{2\sigma}}{z^{2s} + c^{2\sigma}}\right) \tag{10.12.12}$$

where

$$\sigma^2 = 1 - \frac{3n^2\mathrm{S}}{2mk}. \tag{10.12.13}$$

The supposition that S is constant implies a relationship between W and β through (10.12.7), and is very artificial. (10.12.12) then gives B as a function of z through (10.12.6); also it reduces to (10.12.10) when β, and consequently S, is zero and $\sigma = 1$.

W is then given by integration of (10.12.4) in the form

$$W^3 = \mathrm{C}^3 z^{-\sigma-2}(z^{2\sigma} + c^{2\sigma}), \tag{10.12.14}$$

which is (10.11.7) when $\sigma = 1$.

The constant of integration c is given by putting in some known set of values denoted by subscript 1; thus

$$\frac{gB_1R_1}{W_1^2} = \mathrm{F}_1 = \frac{2m}{3n}\left(\sigma + 1 - \frac{2\sigma c^{2\sigma}}{z_1^{2\sigma} + c^{2\sigma}}\right), \tag{10.12.15}$$

and C is then determined by (10.12.14) in the form

$$\mathrm{C}^3 = \frac{W_1^3 z_1^{\sigma+2}}{2\sigma c^{2\sigma}}\left(\frac{2m}{3n}(\sigma+1) - \mathrm{F}_1\right). \tag{10.12.16}$$

This special case has been given because it is susceptible to complete solution; but it is utterly artificial in requiring S to be a constant, although this is not impossible in reality of course, and might be set up in a laboratory. It also possesses similarity.

A special case of this last result is that in which F is also a constant, and represents a plume possessing similarity in stratified surroundings. Then, with $\sigma = 0$, we have

$$\frac{gBR}{W^2} = \mathrm{F} = \frac{2m}{3n},$$

$$\frac{g\beta R^2}{W^2} = \mathrm{S} = \frac{2mk}{3n^2}, \tag{10.12.17}$$

and

$$W^3 = \mathrm{C}^3 z^{-2}$$

as a solution, which is possible only if the initial conditions make c equal to zero.

These similarity solutions do not actually make the assumption that everything is determined at the particular value of z, but mould the surroundings and initial conditions so that this is so. On the other hand, if we make the assumption (which is not justified by any observations and is clearly wrong in cases where there is any sort of overshoot) that everything is determined at the local level only so that the profiles across the plume are fixed and m and k are constants, equations (10.12.2) and (10.12.3) may be combined to give

$$\frac{\mathrm{d}}{\mathrm{d}z}\left(W\frac{\mathrm{d}}{\mathrm{d}z}(W^2z^2)\right) = -\frac{1}{mk}g\beta W z^2 \tag{10.12.18}$$

which can be integrated numerically to give W as a function of z if β is given as a function of z. B is then given by (10.12.2), and α by (10.12.9).

The assumptions that are required to validate such a numerical solution are such that the much simpler approach of Chapter 12 probably gives more enlightenment. In any case, it is difficult to imagine a case other than a sinister military one in which this kind of theory would be useful before the event. Afterwards events speak more eloquently than such predictions, as when unintended fires occur.

More recently, however, a method of mining the metalliferous deposits on the bed of the Red Sea requires a slurry of unwanted material to be returned to the sea; the slurry is denser than the sea water, which is stratified slightly except close to the bed where the salinity is much greater. These formulae are applicable in the region of slight stratification.

Special solutions of (10.12.18) can easily be found. For example if

$$\beta = \frac{mk}{W}\sum_{\lambda}\frac{a_\lambda}{z^\lambda} \tag{10.12.19}$$

the solution is

$$W^3 = \frac{3}{2}z^{-3}\left(k_0 + \frac{1}{2}k_2 z^2 + \sum_{\lambda}\frac{a_\lambda}{(3-\lambda)(5-\lambda)}z^{5-\lambda}\right) \tag{10.12.20}$$

where k_0 and k_2 are constants of integration related to the conditions imposed. This is more general than the solutions in which W^3 is related to only one or two powers of z, but it is relevant to problems only in that it can be used to exemplify conical plumes in stratified surroundings, making all the necessary assumptions described above, because β has not been allowed to be specified independently of W.

10.13 LAMINAR PLUMES

In section 10.11 we discussed at length the case of a plume starting with an excess of momentum. The analysis was unrealistic for the case when it starts with an excess of buoyancy, and required 'detrainment'. An approach more in accord with observation in this second case is to assume that when possessed of excessive buoyancy a plume accelerates and becomes narrower. The flow then remains more or less uniform across the whole cross-section and is bounded by a free stream surface (see Fig. 10.13.i) on which the pressure is the same as the outside hydrostatic value, so that the exterior fluid is not disturbed.

If the interior and exterior values are denoted respectively by subscripts i and a, and p_0 is the pressure at some starting level,

$$p_a = p_0 - g\rho_a z. \tag{10.13.1}$$

Since the flux of volume up the plume, with uniform velocity W over a cross-section having radius R, is constant

$$\pi R^2 W = \text{const} = Q, \tag{10.13.2}$$

and the pressure inside is given by

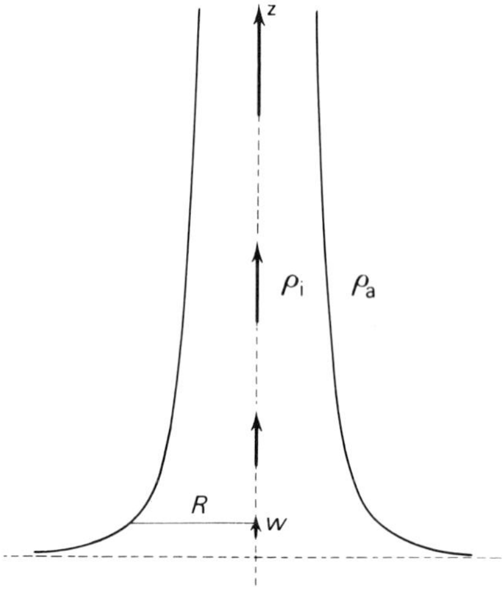

Fig. 10.13.i Coordinates for an accelerating laminar plume.

$$p_i = p_0 - g\rho_i z - \tfrac{1}{2}\rho_i W^2, \tag{10.13.3}$$

assuming that z is measured from where $W = 0$. On the boundary $p_a = p_i$ and so

$$W^2 = 2gz\frac{\rho_a - \rho_i}{\rho_i} = 2gBz, \tag{10.13.4}$$

where

$$B = \frac{\rho_a - \rho_i}{\rho_i}. \tag{10.3.5}$$

Consequently, according to (10.13.2)

$$R^4 = \frac{Q^2}{\pi^2 W^2} = \frac{Q^2}{2\pi^2 gBz} \tag{10.13.6}$$

and

$$W \propto R^{-2}, \quad R \propto z^{-1/4}. \tag{10.13.7}$$

Thus the buoyant fluid accelerates away from its source as a contracting stream. The buoyancy flux is constant and the momentum flux increases until it reaches the value appropriate to a turbulent axisymmetric plume of its own size, after which it becomes a plume. Fig. 10.9.i is a picture of the beginning of the turbulence in the form of meanderings of the plume in a case where viscosity probably prevents the occurrence of much smaller eddies. As in the case of all our other motion patterns, compact bodies of

Plate 1. The Sierra Wave in the lee of the Sierra Nevada (and Mt Whitney) looking at the dust being blown off the Owens Valley into the first lee wave and rotor, which are stationary in front of the Inyo Mountains. This picture was taken by Robert Symons, who had soared in a fighter plane with feathered propellers in the lee wave upcurrent. The colours in this picture were added by an artist to match the tones of a photograph of the same area by Harold Klieforth.

Plates 2, 3. Mexico City on a day when the wind has replaced the polluted air by fresh clean air. But too often the air moves very little in the 'hollow' in which the city is situated among the mountains. (Courtesy of H. A. Bravo.)

Plates 4, 5. We look down on a shallow medium-level cloud which has a low albedo because it is too old to contain a large number of the smaller droplets and most of the scattered sunshine is scattered forwards and is not seen from above. The illumination we see originated from the ground and is coloured accordingly. But the droplets are of a size where single scatter produces a 'cloud bow' (Plate 4), and a 'glory' surrounding the shadow of the 'plane' (Plate 5).

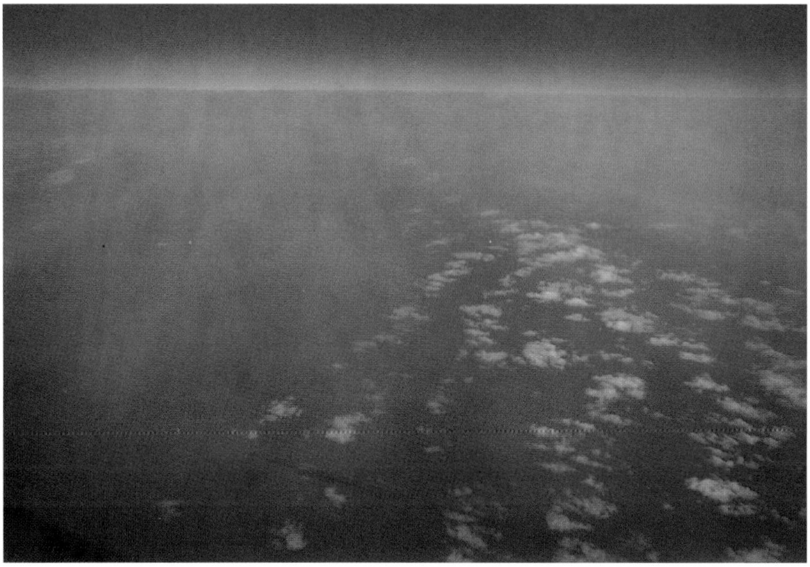

Plate 6. Thin cirrus cloud seen from just above the tropopause over the eastern Mediterranean. The colour contrasts sharply with the very white cumulus below which is scattering the sunshine back upwards, while the cirrus allows it to pass downwards and is seen illuminated by the brownish light from the earth below.

Plate 7. Autumn late evening mist forming as the ground cools and cools the calm air by proximity and radiative exchange. (Courtesy of A. & J. Verkaik.)

Plate 8. Mother-of-pearl cloud over Norway. This is a typical northern hemisphere stratospheric cloud similar to the PSCs of the Antarctic continent. This cloud is found typically at heights between 17 and 30 km. (Courtesy of E. Hesstvedt.)

Plate 9. The tip of a propeller, like the tip of a wing exerting an aerodynamic force, sheds a vortex; and this is made visible by condensation of cloud in the low-pressure centre of the vortex if the air is near to saturation with water vapour. Here, cloud condensed on the low pressure of the vortex shed from the propeller tip at a moment of maximum thrust on take-off from London airport. Because vortex lines cannot begin or end in the fluid these and every subsequent rotation of the propeller will drape its vortices half above and half below on to the leading edge of the wing all the way to the destination.

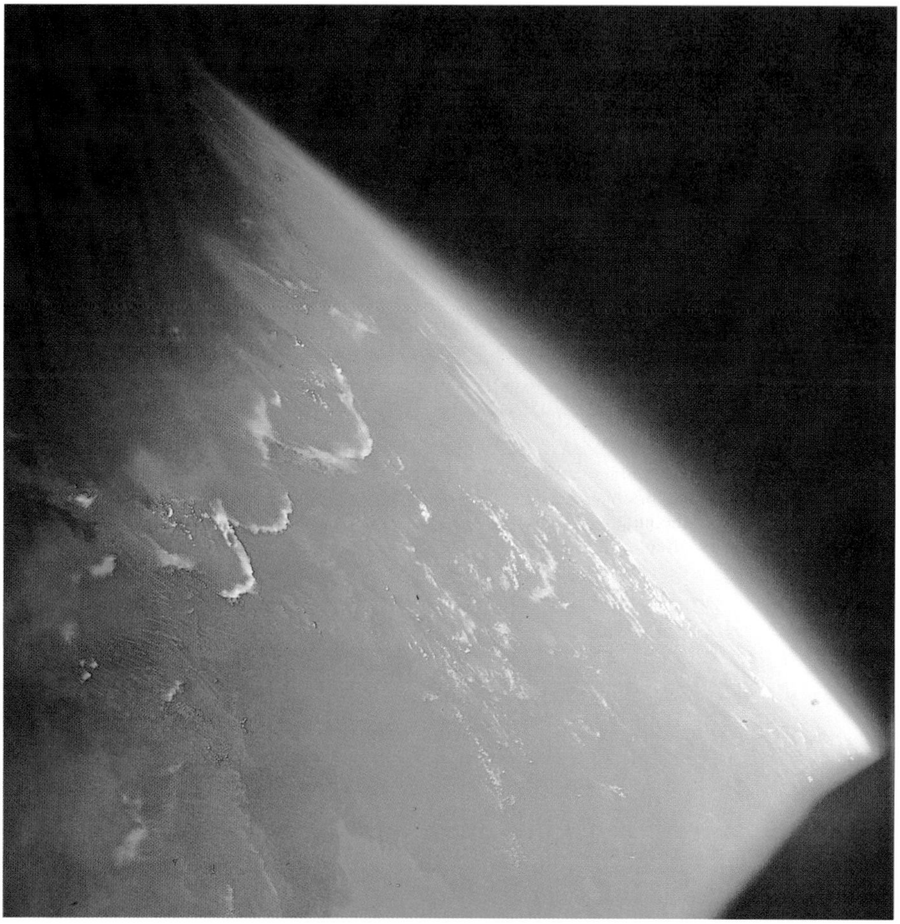

Plate 10. The tops of the wave clouds over the Andes of Bolivia as seen by Gemini VII on 12.12.65 from above Lake Poopo, the clouds on the left being wave clouds, like those in the next picture (Plate 11), with sunset illumination. (Courtesy of NASA.)

Plate 11. Wave clouds seen at the edge of the steep Rockies, near Boulder, Colorado. They are a winter phenomenon often with strong winds down the eastern face of the Rockies. They illustrate well how the humidity is distributed into layers by the strong wind shear. This type of wave cloud is often called 'a pile of plates', particularly when the cloud is isolated. (Courtesy of J. P. Lodge.)

Plate 12. Fallstreaks from a supercooled water cloud which has been seeded by the passage of an aircraft from Washington airport. The seeding with fragments of ice as the droplets freeze creates a hole in the cloud. In this case a mock sun is produced by refraction in the ice crystals at 22° to the left of the sun (horizontally).

Plate 13. Hadhramaut as seen from Gemini spacecraft. This desert area of southern Yemen receives torrential rain occasionally but only a small annual rainfall. Cape Gardafui (the Horn of Africa) can be seen across the sea. A cyclone in the Arabian Sea (see Fig. 14.11.iv) could deposit enough moisture to cause a swarm of desert locusts to hatch out and another could transport it a few weeks later to a destination in Africa or Asia to start a plague. The cloud seen here is typical coastal stratus but promises no rain. (Courtesy of NASA.)

Plate 14. The glacier of Mt Denali in Alaska presents a fine picture to the air traveller who lands at Anchorage. The lower end is gradually melting in the warm air at sea level, and it becomes 'dirty' as the material carried in the ice begins to appear at the surface.

the buoyant fluid are stretched out and carried first to one side then another and the eddies grow in size further along the plume.

Motion of this kind could well take place in calm conditions inside a cooling tower. The phenomenon is not to be confused with the vena contracta (see section 2.4), which is a phenomenon due to the curvature of the streamlines as a fluid emerges from an orifice into a region of constant pressure. This case is gravity dominated, and the curvature of the streamlines is of no importance.

The narrowing of a plume can often be observed, but only when the wind is calm and the narrowing causes the visible smoke, or whatever it is, to rise vertically. On such occasions the plume is obviously narrower than usual and does not occupy a cone. The same can be seen over a bonfire, where the smoke plume narrows before supplying a series of thermals with the characteristic cauliflower shape on top.

In Chapter 12 we shall revert to this topic when we study the motion above cloud base where the condensation of cloud droplets releases a very significant amount of heat. But the same phenomenon may occur when water, which has been at a considerable depth where the pressure may be at several atmospheres, may become saturated with dissolved gas. The most obvious for such a case is CO_2 because the threat of global warming has provoked ideas on how to dispose of a surplus of it. CO_2 can be compressed to form a liquid at the temperature and pressure combinations that commonly occur deep in the sea.

'On 21 August 1986 a massive release of carbon dioxide from Lake Nyos in Cameroon killed about 1700 people. A similar event occurred on 15 August 1984, at Lake Monoum, also in Cameroon.' So wrote Youxue Zhang [46]. Evidently the water at the bottom of the lake became saturated with CO_2 which had come out of the ground and became unstable after a small upward displacement of some of it and became filled with bubbles of CO_2, thereby forming a thermal. Further lifting caused more bubbles to be formed so that this bottom water became a very powerful source of buoyancy. Using an analysis having much in common with this chapter, the writer argued that on reaching the water surface at atmospheric pressure the water had become a froth of CO_2, and because the potential energy available as the gas rose to the surface, the mixture could well have reached an upward velocity of about 50 m s^{-1} in Lake Monoum and as much as 90 m s^{-1} at Lake Nyos. Very little difference is made by assuming a lower atmospheric pressure, for the pressure decrease in the two lakes from the bottom to the top is about 10 and 20 atmospheres respectively.

On emerging as a fountain of water and CO_2, the mixture rose to a height of over 200 m and the water droplets now became a cause of negative buoyancy so that a mass of almost pure CO_2 spread out on the lake and drained over the enclosing hill and caused a considerable loss of human and animal life through suffocation by CO_2.

The phenomenon is so unusual that a volcanic disturbance in the bottom of the lake was first proposed as explanation, but the paper referred to shows convincingly that the supersaturation of the bottom water provided such potential instability that the enormous amount of CO_2 could be released, and the cloud of water droplets which caused it to spread out on the surface after being emitted at such high velocity was responsible for the concentration remaining long enough to cause the disaster.

10.14 RECIRCULATING JETS

The phenomenon of recirculation may be illustrated by considering a jet confined in a parallel-sided tube. If a jet of cross-section area A_2 and mean speed U_2 enters a tube of identical fluid moving with speed U_1 across the area A_1 surrounding the jet, and some distance downstream the flow across area $A_1 + A_2$ has average speed U_3, since the flux of material is constant we must have

$$A_1U_1 + A_2U_2 = (A_1 + A_2)U_3. \tag{10.14.1}$$

There is a rise in pressure Δp from the point of emergence of the jet to a point downstream where the two streams are thoroughly mixed equal to the decrease in momentum flux, so that

$$A_1U_1^2 + A_2U_2^2 = (A_1 + A_2)\left(\frac{A_1U_1 + A_2U_2}{A_1 + A_2}\right)^2 + \frac{\Delta p}{\rho}(A_1 + A_2). \tag{10.14.2}$$

Thus

$$\Delta p = \rho A_1A_2\left(\frac{U_1 - U_2}{A_1 + A_2}\right)^2 > 0 \tag{10.14.3}$$

if $U_1 \neq U_2$.

This rise in pressure will produce stagnation of the surrounding fluid if

$$\Delta p > \frac{1}{2}\rho U_1^2, \tag{10.14.4}$$

i.e. if

$$\frac{U_2 - U_1}{U_1} > \frac{A_1 + A_2}{2A_1A_2}, \quad \text{or} \quad \frac{U_2}{U_1} > \frac{A_1^2 + A_2^2}{2A_1A_2}, \tag{10.14.5}$$

which is obviously possible if U_2 is large enough. The right-hand side of (10.14.5) is a minimum when $A_1 = A_2$, and then stagnation in the outer stream occurs if $U_2 > U_1$. The flow separates from the wall at the stagnation point and a region of recirculation surrounds the jet from that point on until it has filled the tube. The conditions for recirculation are very easy to establish in practice.

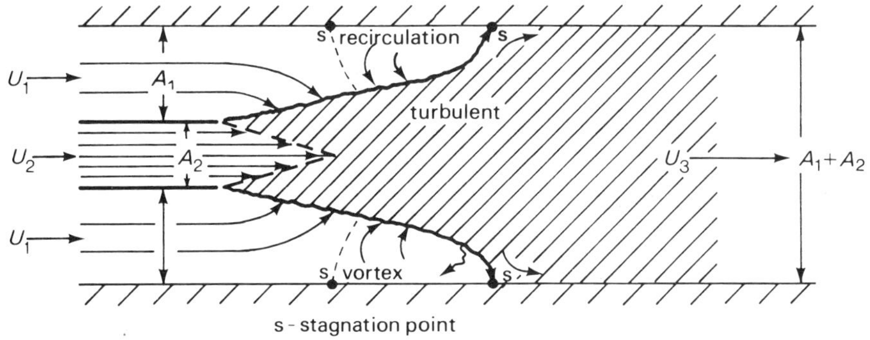

Fig. 10.14.i Enclosed jet with recirculation: a laminar jet entering an annular laminar stream becomes turbulent with a potential core diminishing to nothing at about 5 nozzle diameters downstream. The surrounding fluid stagnates at a point S before the turbulent cone reaches the wall, and a recirculation vortex is established.

Experimentally a steady symmetrical flow like that illustrated in Fig. 10.14.i is difficult to establish in two dimensions because transverse uniformity, i.e. strictly symmetrical motion, is difficult to set up and the jet tends to attach itself to one wall. In the axisymmetric case the flow fluctuates but the recirculation is easy to establish.

10.15 SUMMARY OF EXPERIMENTAL VALUES

See Table 10.15.i on page 356.

$$z = nR, \qquad w = \mathrm{C}(gBR)^{1/2} = \mathrm{C}'(gBz_c)^{1/2},$$

$$\mathrm{Vol} = mR^3 \text{ (axisymmetric)} = m'z_c^3$$

$$= mR^2 \text{(cylindrical)}$$

where z is either the height of the cap (z_c) or the distance up the axis, w is the velocity of the cap or a representative velocity, B is the overall mean buoyancy or buoyancy at height z, R is the maximum radius or half width. F is the buoyancy flux and $\dot{M}$ the momentum flux at the source. Also $w \propto z^{\alpha}$, $B \propto z^{\beta}$, $\sigma \propto z^{\beta}$, $z \propto t^{\lambda}$, and $\mathrm{Re} \propto z^{\delta}$. **When $\delta > 0$ the motion becomes turbulent, when $\delta < 0$ it becomes viscous-dominated with increasing z or t.** σ is the concentration of pollution and is proportional to buoyancy. In a plume, B_{max} and W_{max} are the maximum values for a given z. The values in brackets are extreme measured values. Subscripts m and f refer to mean and fluctuating components respectively.

Table 10.15.i

	$z = nR$		$w = C(gBR)^{1/2} = C'(gBz_c)^{1/2}$		$\text{Vol} = mR^3 = m'z_c^3$		$w \propto z^{\alpha}$	$\frac{B}{\sigma} \propto z^{\beta}$	$z \propto t^{\gamma}$	$\text{Re} \propto z^{\delta} \propto wz$	
Point source Axisymmetric	n	θ	C	C′	m	m'	α	β	γ	δ	
Puff	(1.7) 4 (6.3)				3		–3	–3	$\frac{1}{4}$	–2	
Thermal	(2) 4 (7.3)		1.2		3		–1	–3	$\frac{1}{2}$	0	
Jet	4.76	11.8° ± 0.3					–1	–1		0	$\dot{M}_f = 0.74\dot{M}_m$
Plume	5.51	10.4° ± 0.6	$w_{max} = 1.4(gB_{max}z)^{1/2}$		$gB_{max} = 11z^{-5/3}F^{2/3}$		$-\frac{1}{3}$	$-\frac{5}{3}$		$\frac{2}{3}$	$F_f = 0.4F_m$
			$= 4.7^{-1/3}F^{1/3}$		$\cong 3gB$						
			$\cong 3\overline{w}$								
New jet	4.76					0.033	–1	–1	$\frac{1}{2}$	0	
New plume	5.51		0.04	0.02		0.025	$-\frac{1}{3}$	$-\frac{5}{3}$	$\frac{3}{4}$	$\frac{2}{3}$	
Cylindrical (flat line source)											
Puff	(2.6) 4 (5.7)			$\text{Vol} = mR^3$			–2	–2	$\frac{1}{3}$	–1	
Thermal	(1.3) 2 (3.2)		1.0	(1.8) 2.4 (3.0)			$-\frac{1}{2}$	–2	$\frac{2}{3}$	$\frac{1}{2}$	
Jet							$-\frac{1}{2}$	$-\frac{1}{2}$		$\frac{1}{2}$	
Plume							0	–1		1	
New jet							$-\frac{1}{2}$	$-\frac{1}{2}$	$\frac{2}{3}$	0	
New Plume							0	–1	1	1	

11

Buoyant convection in the dry atmosphere

11.1 ATMOSPHERIC CONVECTION: REGIMES AND OBSERVATIONS

While convection is occurring there are great variations in temperature horizontally. Because thermals entrain outside air into their centres, the air with the greatest updraught does not have any significant buoyancy, but is surrounded by a ring of air that has the maximum buoyancy. Outside that there may be air that is temporarily descending at the maximum rate in the neighbourhood, but this may be on the outer reaches of a thermal and have a positive temperature anomaly. In between thermals the air may be slightly stably stratified and very slowly subsiding, being warmed as a result even though the warm air rising from below is not actually being mixed with it.

A sample taken by a rising balloon will differ according to whether the balloon rises up in a thermal or instead rises through the slowly subsiding surroundings.

Because of great variations in the nature of the surface, the temperature close to the ground varies accordingly. Thus the temperature over a dry street may be 5°C or more warmer than that over a grassy field nearby, and a quantity such as the maximum daytime shade temperature is very much characteristic of the surface.

The situations we shall be discussing will therefore have features extremely dependent on whether we are concerned mainly with the different temperature of different parcels of air of different origin, or with the vertical profile of the horizontally averaged conditions. Even when we are discussing the latter, the properties will depend very much on the existence of inhomogeneities.

Near the ground the lapse rate may become very high, and in the bottom inch or two, where mirages occur on a hot day, convection occurs on a small scale, and is inhibited by viscosity and radiation. Rather higher up we may enter a similarity regime where these physical properties of the fluid have no influence. As we approach the condensation level, the influence of clouds becomes dominant, and their effects will be discussed in Chapter 14.

The rate at which the bottom temperature rises makes a big difference to the geometry of convection. It is very different over grassland, desert or sea, and in the morning, afternoon or evening. The reader must be warned not to assume that the conclusions reached by any of the treatments now following are applicable in particular cases. It is for each observer, practitioner, consultant or researcher to determine for himself what regime he is actually concerned with. Many discussions have been frustrated by assertion and counter assertion concerning the geometry of convection, whether it is composed of plumes or thermals. Clearly it could look like either if the scope of the observation is restricted, but the fitting of a model to observations is no guarantee that the model is correct.

11.2 SLOW CONVECTION: THE RAYLEIGH NUMBER

Near the rigid boundary at the bottom of the atmosphere the motion is inhibited by viscosity. Equally it is inhibited by thermal conductivity, which conducts the heat upwards without motion when the boundary is hot, and also reduces the horizontal gradients of temperature without which vertical motion would not occur. When vertical motion does occur it is resisted by viscosity away from the boundary because it produces stresses which resist up and down motion side by side. If the up and down motions are moved further apart the horizontal temperature gradients are reduced and the horizontal motion is increased, and this is resisted by viscosity near the boundary. Above the up and down motion there must be other horizontal motion in the opposite direction, and so a cellular structure is completed. The cells are very roughly square in cross-section for otherwise the resistance to motion just mentioned would be greater. The precise shape depends on whether they are square or hexagonal in plan, and on what conditions occur at the upper and lower boundaries. There is a considerable literature about the properties of cellular convection between horizontal boundaries or fluid surfaces but it is not relevant to the atmosphere except in the bottom inch or so. Even there most theories are irrelevant because the radiative heat flux far exceeds the convective heat flux in that layer when the surface is hot. Also the conduction conditions beneath the surface have a big effect on the geometry of the convection above.

The driving force of this laminar convection is gravity operating on the relative density anomaly $\Delta\rho/\rho$. If we imagine that there is a particular density gradient at which motion begins, then a length h, representing the depth of the layer, must be involved. If $\Delta\rho$ is associated with the buoyancy anomaly resulting from a vertical displacement of a particle at distance h, then the critical gradient is increased by the conductivity, κ, and the viscosity, ν, which must enter the equations in an identical manner. To obtain the dimensions of a gradient of buoyancy (accel/length) the relationship must take the form

$$\frac{g\Delta\rho}{h\rho} = \mathrm{Ra}\,\frac{\kappa\nu}{h^4} \tag{11.2.1}$$

where Ra is non-dimensional and is called the Rayleigh number. Its value is determined by what phenomenon is under consideration. For the onset of motion with rigid perfectly conducting boundaries the classical analysis gives $\mathrm{Ra} = 27\pi^4/4$, which is of the order of 10^3. Because of its dependence on h^3 we deduce that very large gradients may be set up in shallow layers close to the heated boundary. In practice the maximum temperature exceeds the critical value and heat is also lost upwards by radiation. Even in relatively dry climates this radiation is mainly absorbed by the water vapour in the lowest 2 m of air. (See Plate 7 and Chapter 13.) For this reason the molecular coefficients κ and ν are not relevant to the geometry of convection in the atmosphere, and it is no use pretending that an eddy conductivity and viscosity will do instead. This is because it is the eddies themselves that are under investigation which cause the transfers of heat and momentum: there is no uniformity of such coefficients over the space occupied by the individual eddies, and the eddy transfer coefficient concept is useless in this context.

The theory of laminar cellular convection is a good example of inbreeding of ideas. All the interesting problems have been generated as a result of laboratory attempts to demonstrate the validity of a theory which was designed to be applied to the laboratory experiments: it has little direct application elsewhere. (But see Ship Trails in Chapter 14.) In particular the direction of the cellular motion, in so far as it has meaning in a cellular pattern (which in some cases it does not), has been explained in terms of the variation of viscosity and conductivity with temperature. In the case of the atmosphere the direction is always determined by the fact that parcels of air break out of the very hot layer near the hot ground and travel upwards. They entrain environmental fluid by mixing as they go, and rise penetratively through the environment. Various suggestions have been put forward that these parcels are plumes near the ground (within a very few metres or less) and become more like thermals higher, up, probably passing through a new though temporary plume stage before again rising as a thermal. As we shall see, the actual configuration depends on the rate of heating and the nature of the surface.

11.3 SIMILAR FULLY TURBULENT CONVECTION

A theoretical state which has attracted much interest is that in which the only dimension determining the parcel size, or geometrical scale of the buoyant convection, is distance above the ground. The analysis proceeds without reference to the geometry as follows: The vertical velocity, w, is given by

$$w = C(gBz)^{1/2} \tag{11.3.1}$$

while

$$B \propto \frac{z}{\theta}\frac{\partial \theta}{\partial z} \tag{11.3.2}$$

because $\Delta\theta = z\partial\theta/\partial z$ for adiabatic vertical displacements and $B = \Delta\theta/\theta$. θ is the potential temperature. If we now assume geometrical similarity the same fraction of any horizontal surface is occupied by upcurrents and if the flux of buoyancy is the same at all heights so that the heat fed in at the bottom passes out at the top, we must have that

$$wB = \text{const} \tag{11.3.3}$$

or

$$\frac{1}{\theta}\frac{\partial \theta}{\partial z} \propto z^{-4/3}. \tag{11.3.4}$$

If the whole layer is being warmed up slowly then the heat flux decreases linearly and we have an expression of the form $C_1 - C_2 z$ in the place of the constant in (11.3.3). Then

$$\frac{1}{\theta}\frac{\partial\theta}{\partial z} \propto \left(\frac{C_1 - C_2 z}{z^4}\right)^{1/3}. \tag{11.3.5}$$

The similarity has now been lost because an additional height has been introduced, namely that at which the buoyancy flux goes to zero, and the convection does not exist above it. But this makes very little difference to the result (11.3.4) near the bottom boundary. Indeed, with the kind of measurements that are available in the bottom half of the layer it would be very difficult to detect any departure from (11.3.4) unless there was a large rate of change of the situation in time. The well-known result is obtained by noting that $1/\theta$ varies much less than $\partial\theta/\partial z$ and so, roughly, (11.3.4) is

$$\partial\theta/\partial z \propto z^{-4/3}, \quad B \propto z^{-1/3}, \quad w \propto z^{1/3}. \tag{11.3.6}$$

The significant point about this result is that it has been obtained without reference to the geometry of the convection, and solely on the basis of a simple dimensional argument which generates (11.3.1) and (11.3.2). The argument has been presented in a variety of forms dependent on a supposed geometry, but even if (11.3.4) is observed to be true in any particular case, nothing at all is proved about the geometry.

If the convection consists of thermals, plumes or new plumes then we must suppose that the number of them in unit volume decreases so that the same fraction of a horizontal surface at any height may be occupied by upcurrents. This can only be achieved by their amalgamation when the cones along which they rise overlap. In consequence of this amalgamation the vertical velocity increases like $z^{1/3}$ instead of decreasing as z increases, as it must in any isolated thermal, plume or new plume. This is because the buoyancy decreases more slowly when some of the entrained fluid contains buoyancy.

11.4 SIMILAR CONVECTION WITH A SUBSIDING UNSTRATIFIED ENVIRONMENT

The following treatment is taken from Scorer (1969).

If σ is the fraction of a horizontal area occupied by upward motion, and w and w_e are the vertical velocities of the upcurrents and their environment respectively, if also there is no upward transfer of volume (i.e. no total horizontal convergence, neglecting the density anomalies in this context),

$$\sigma w + (1-\sigma)w_e = 0. \tag{11.4.1}$$

If $\bar{\tau}$ is the mean value of τ, $= \theta^{-1}$, at any height, and τ' and τ'_e are the anomalies of θ^{-1} in the upcurrents and the environment, then

$$\bar{\tau} = \sigma(\bar{\tau} - \tau') + (1-\sigma)\ (\bar{\tau} - \tau'_e),$$

or

$$\sigma\tau' + (1-\sigma)\tau'_e = 0, \tag{11.4.2}$$

where by definition τ' is positive, and so τ'_e is negative. For a liquid we would replace τ by ρ.

The buoyancy flux, omitting the factor g, is

$$F = \sigma w \tau' + (1-\sigma) w_e \tau'_e = \frac{\sigma}{1-\sigma} w \tau'. \qquad (11.4.3)$$

Also

$$B = \tau'/\bar{\tau} \qquad (11.4.4)$$

and by the formula (11.3.1), which holds for all configurations of the convective parcels,

$$w\tau' = \mathrm{C}\left(g \frac{\tau'}{\tau} r\right)^{1/2} = \mathrm{C}\left(g\tau'^3 z / n\bar{\tau}\right)^{1/2} \qquad (11.4.5)$$

where C and n are the values appropriate to whatever the parcel configuration may be, as described in Chapter 10. Since

$$\bar{\tau} = \tau_e - \tau'_e = \tau_e + \frac{\sigma}{1-\sigma} \tau', \qquad (11.4.6)$$

where τ_e is the constant value of τ in the adiabatically subsiding unstratified environment, ignoring the region close to the ground in which τ' is comparable with τ, we may put $\bar{\tau} \simeq \tau_e$, so that (11.4.3) and (11.4.5) give

$$F = \frac{\sigma \mathrm{C}}{1-\sigma} \left(\frac{g}{n\tau_e} \tau'^3 z \right)^{1/2}. \qquad (11.4.7)$$

Thus if F is independent of z, we must have

$$\tau' \propto z^{-1/3} \qquad (11.4.8)$$

as in (11.3.6). If we differentiate (11.4.6) we get

$$\frac{\partial \bar{\tau}}{\partial z} = \frac{\sigma}{1-\sigma} \frac{\partial \tau'}{\partial z} \qquad (11.4.9)$$

or

$$-\frac{\sigma \tau'}{1-\sigma} \Big/ \frac{\partial \bar{\tau}}{\partial z} = -\tau' \Big/ \frac{\partial \tau'}{\partial z} = 3z. \qquad (11.4.10)$$

This result is required in section 11.6. It leads to

$$\frac{\partial \tau'}{\partial z} \propto \frac{\partial \bar{\tau}}{\partial z} \propto z^{-4/3} \qquad (11.4.11)$$

which is equivalent to (11.3.4), and (11.3.6) follows.

It is important to note that the total environment penetrated by the thermals is slightly unstably stratified. Thermals, with a decreasing excess temperature, undergo a more 'unstable' lapse rate. The subsiding surroundings have a natural lapse rate: sampling must be made representative to obtain the lapse rate of the total environment. If in an actual case the surroundings were slightly stably stratified, the sinking motion would cause an apparent warming. Indeed this warming is very real in practice, but this similarity analysis is no longer strictly applicable, for an extra linear dimension is introduced defining the stratification.

11.5 RISE OF THERMALS IN WIND SHEAR

A thermal rising through wind shear $\partial U/\partial z$ experiences a horizontal acceleration relative to its environment in the negative direction of U of magnitude $w\partial U/\partial z$. It also experiences an Archimedean acceleration of its environment downwards of magnitude gB. It is therefore subject to a total acceleration which is the vector sum of those two, of magnitude

$$g^*B = gB/\sin\theta = w\frac{\partial U}{\partial z}\Big/\cos\theta \tag{11.5.1}$$

where θ is the inclination of the axis of the thermal to the vertical (Fig. 11.5.i).

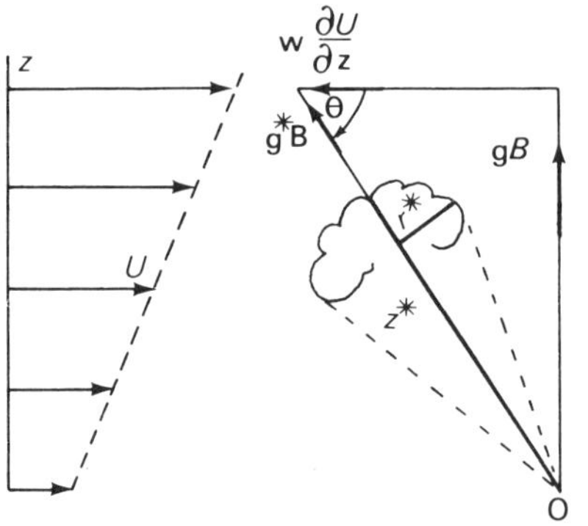

Fig. 11.5.i A thermal rising in wind shear represented by the velocity profile on the left experiences a horizontal acceleration of its environment equal to $w\partial U/\partial z$, and a downward acceleration of its environment of magnitude gB. These combine tc produce a total acceleration g^*B at an angle θ to the horizontal, and its axis of symmetry is inclined accordingly.

Assuming the mechanics of the inclined thermal to be the same as that of a vertically accelerated one, its velocity w^* relative to the surrounding air is related to its vertical and relative horizontal velocities, w and u, by

$$w^* = \frac{w}{\sin\theta} = \frac{u}{\cos\theta} = C(g^*Br^*)^{1/2} = C\left(\frac{g}{\sin\theta}B\frac{r}{\sin\theta}\right)^{1/2} \tag{11.5.2}$$

where r^* is its radius measured perpendicular to its axis of symmetry. From this

$$w = C(gBr)^{1/2} \tag{11.5.3}$$

and

$$\frac{u}{\partial U/\partial z} = \frac{w\cot\theta}{w^{-1}gB\cot\theta} = \mathrm{C}^2 r = \mathrm{C}^2 z/n. \tag{11.5.4}$$

The argument of this section is based on the properties of a thermal, and could not be repeated for a plume or a new plume except by means of much more extravagant assumptions. This is because, although the motion of the top of a new plume rising through wind shear would be much the same as for a thermal, the top would be carried downwind faster than the bottom and there would remain no axis of symmetry. We have to assume in this treatment that the variation in U across the depth of the thermal is not significant and is small compared with w, which requires that

$$r\frac{\partial U}{\partial z} \ll w. \tag{11.5.5}$$

11.6 THE RATIO OF THE EDDY TRANSFER COEFFICIENTS FOR MOMENTUM AND HEAT

The previous sections provide the means to calculate the magnitudes of the rates of vertical transfer of horizontal momentum and heat or material. The transfer coefficient of buoyancy K_H is related to the flux F given by (11.4.3), and the mean gradient (according to Fickian diffusion theory, or K-theory) by the equation which defines it, namely

$$F = \frac{\sigma}{1-\sigma} w\tau' = -K_H \frac{\partial\bar{\tau}}{\partial z}, \tag{11.6.1}$$

and so by (11.4.10)

$$K_H = -\frac{\sigma}{1-\sigma} w\tau' \frac{\partial\bar{\tau}}{\partial z} = 3wz. \tag{11.6.2}$$

By the definition of U as the mean wind, if u and u_e are the horizontal velocity anomalies of the thermals and of the environment, the equation analogous to (11.4.2) is

$$U = \sigma(U-u) + (1-\sigma)\ (U-u_e) \tag{11.6.3}$$

and so

$$\sigma u + (1-\sigma)u_e = 0. \tag{11.6.4}$$

By the definition of K_M, the vertical transfer coefficient of horizontal momentum

$$-K_M \frac{\partial U}{\partial z} = -\sigma uw - (1-\sigma)u_e w_e, \tag{11.6.5}$$

which is a measure of the Reynolds stress described in section 7.3.

By means of (11.5.4)

$$K_M = \frac{\sigma}{1-\sigma}\frac{uw}{\partial U/\partial z} = \frac{\sigma}{1-\sigma}\frac{\mathrm{C}^2}{n} wz. \tag{11.6.6}$$

Therefore, by (11.6.2)

$$\frac{K_M}{K_H} = \frac{\sigma C^2}{(1-\sigma)3n}. \tag{11.6.7}$$

This argument does not require that there be a very deep layer in which similarity exists; it is merely only applicable in such a layer. Even if the flux were not constant with height it would only lead to a small change in the factor 3 in (11.6.7), which comes from (11.4.10) and depends on τ_e being independent of z in (11.4.6). The numbers C^2 and n are characteristic of the nature of the convective parcels: if they are like thermals we may use (10.2.13) so that with $n = 4$ and $m = 3$ we have that

$$\frac{C^2}{n} \approx 0.4. \tag{11.6.8}$$

The value of σ is extremely uncertain, and probably varies with circumstances. Some radar plan pictures of thermals (e.g. Fig. 11.6.i) obtained by Konrad and Kropfli [47] indicate values of σ between 0.1 and 0.6, which is probably excessive because the method collects thermals which exist in a finite layer and not at an infinitesimally thin surface. In a layer of cumulus cloud σ could be very occasionally as large as 0.5 or as small as 0.01: but the large values probably arise in cases where there is horizontal convergence and a non-zero value of the mean vertical velocity. In most circumstances we can probably be confident that $\sigma < 0.3$. If $\sigma = 0.1$, which seems a likely value immediately under cumulus,

$$\frac{K_M}{K_H} = 0.014. \tag{11.6.9}$$

If $\sigma = 0.5$,

$$\frac{K_M}{K_H} = 0.13, \tag{11.6.10}$$

which seems to be about the upper limit. Any larger value of the ratio, of the order of unity for example, must be associated with a different kind of eddy.

The cause of such a low value of the ratio was explained in section 9.3. It is, simply, that buoyancy produces vorticity which is horizontal, and eddies composed of horizontal vorticity are not stretched by shear of a horizontal wind. However, closed horizontal vortex lines transfer volume vertically and therefore very effectively transfer mass or buoyancy without producing a shear stress because they extract no energy from the mean shear. In section 9.5 we saw that the tilting of the thermal axis by wind shear produces inclined vortex lines, and these are then stretched by the mean flow and extract energy from it. The small eddies in thermals continuously destroy the energy of the thermal so that a large proportion of this energy is of recent production. This means that a large proportion of the vorticity present is perpendicular to gravity which produced it.

At the other extreme, namely in a stably stratified fluid, as soon as a horizontal vortex line tilts the isopycnic surfaces out of horizontal, vorticity of opposite sign begins to be

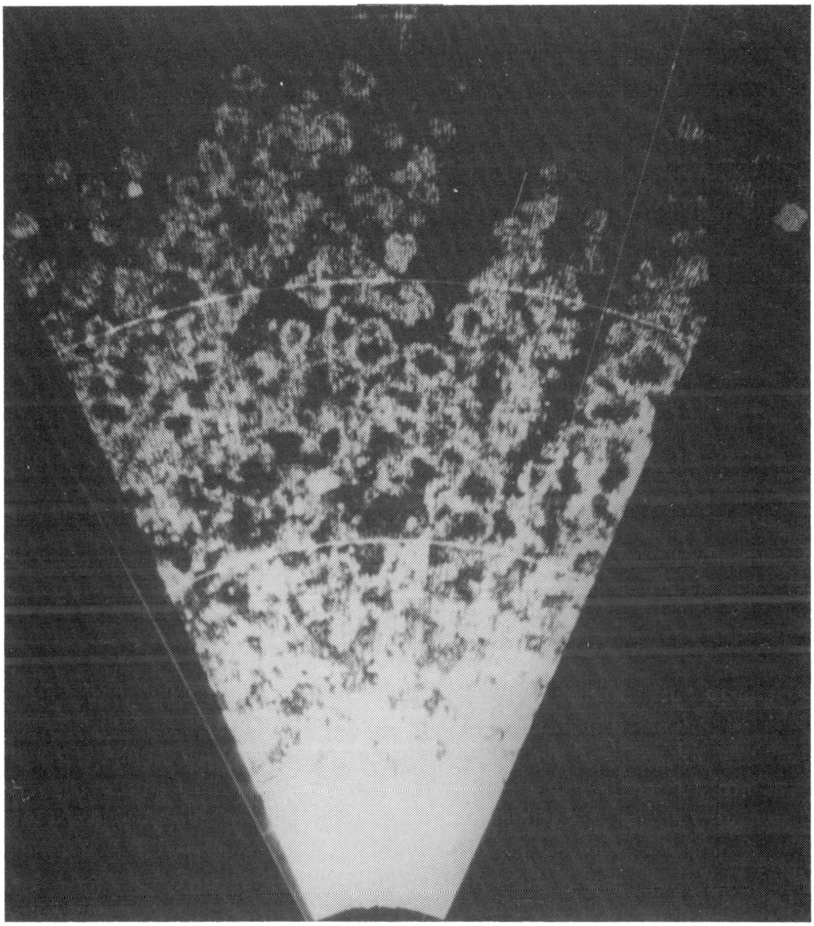

Fig. 11.6.i A radar photo of thermals by Konrad and Kropfli taken at 2° elevation with range markers at 5 n. mi intervals. The unsteadiness in time, variability with height, and dependence upon intensity make it difficult to find a representative value of σ even for a single occasion. This occasion was one of great convective intensity; although the fraction of the area occupied by active thermals is difficult to determine, it does appear to decrease with height. See also Fig. 8.3.ii. [Photograph courtesy of Johns Hopkins University.]

generated. This is simply to say that a stable stratification opposes penetrative vertical motion and puts it into reverse. Therefore if eddies are generated by some means, the horizontal component of vorticity is destroyed, but the other component remains and can produce eddy stresses by becoming tilted in shear flow. Turbulence which produces vertical mass and heat transfer cannot persist with stable stratification: either the stratification or the penetrative motions are destroyed. The extreme case is when the only eddies are stable waves. Then horizontal momentum can be transferred (see section 6.4) without any heat or mass transfer, and then

$$K_M/K_H = \infty. \tag{11.6.11}$$

In numerical weather forecasting the finite difference approximations inevitably introduce artificial smoothing equivalent to a diffusion mechanism. This is evidently sufficient to represent any vertical transfer of momentum, but not of heat and water vapour. In practical cases therefore K_M is put equal to zero while K_H is made infinite in situations in which convection is known to be active, namely when the lapse rate exceeds the wet adiabatic. K_H then refers to wet bulb potential temperature and humidity mixing ratio.

11.7 THE DOUBTFUL MEANING OF AN EDDY TRANSFER COEFFICIENT

The use of an eddy transfer coefficient is valid only if an appropriate gap in the spectrum of eddy sizes exists over which the sampling may be done (see section 9.3). Moreover, it is not correct to suppose that a long enough sampling time will necessarily provide a meaningful parameter, even though the value does not depend on the sampling time. Atmospheric convection does not take place when the air is stably stratified, for example at night. The average lapse rate measured over a period of several days is usually stable: it never becomes very unstable because thermals start up and transfer heat upwards even when it is slightly unstable. Since there is no convective heat transfer downwards when it is stable, there is a net transfer of heat upwards by convection over a long time. The average heat transfer is up the average gradient and so the transfer coefficient is negative. This nonsense was remarked upon by Gold two generations ago, but his comment still remains unheeded by those devoted to finding out what the coefficient is. Its value is very large when convection is active, but zero when it is not: its average value is meaningless.

11.8 FORMS OF BUOYANT CONVECTION

It is so difficult to obtain slow convection, even in the laboratory, that it almost never occurs in the atmosphere. A slight increase in the heat supply causes the motion even in a viscous liquid to become unsteady with parcels rising from the heated boundary through the surroundings and flattening out at the top. The motion is therefore penetrative: in a shallow viscous layer in the laboratory the turbulent nature of atmospheric convection is not simulated because viscosity prevents the eddies which mix a thermal into its surroundings from operating.

(1) Streets and cells

Cloud streets may consist of very shallow clouds, as if they were merely the top of a motion that goes on below and is not significantly affected by the buoyancy effects of condensation. Although some cloud lines do contain large, even raining, clouds, here we are concerned with the small variety, exemplified in Fig. 11.8.i. Cloud streets occur over land temporarily in the morning, and occasionally in the evening, when the vigour of the convection is not great and is confined to a layer of more or less uniform depth. This influences the spacing of the clouds: they cannot be supported closer together because the upcurrents supporting them would overlap, and would leave no space for downcurrents

Fig. 11.8.i(a) An aerial photograph of cloud streets over Oxfordshire.

Fig. 11.8.i(b) An aerial view of streets (on the right) over Dorset which are aligned along the wind from near-right to far-left. And beyond are lee waves of the cliffs of the Dorset coast, and these are aligned across the wind. (Photo J. C. Nielan.)

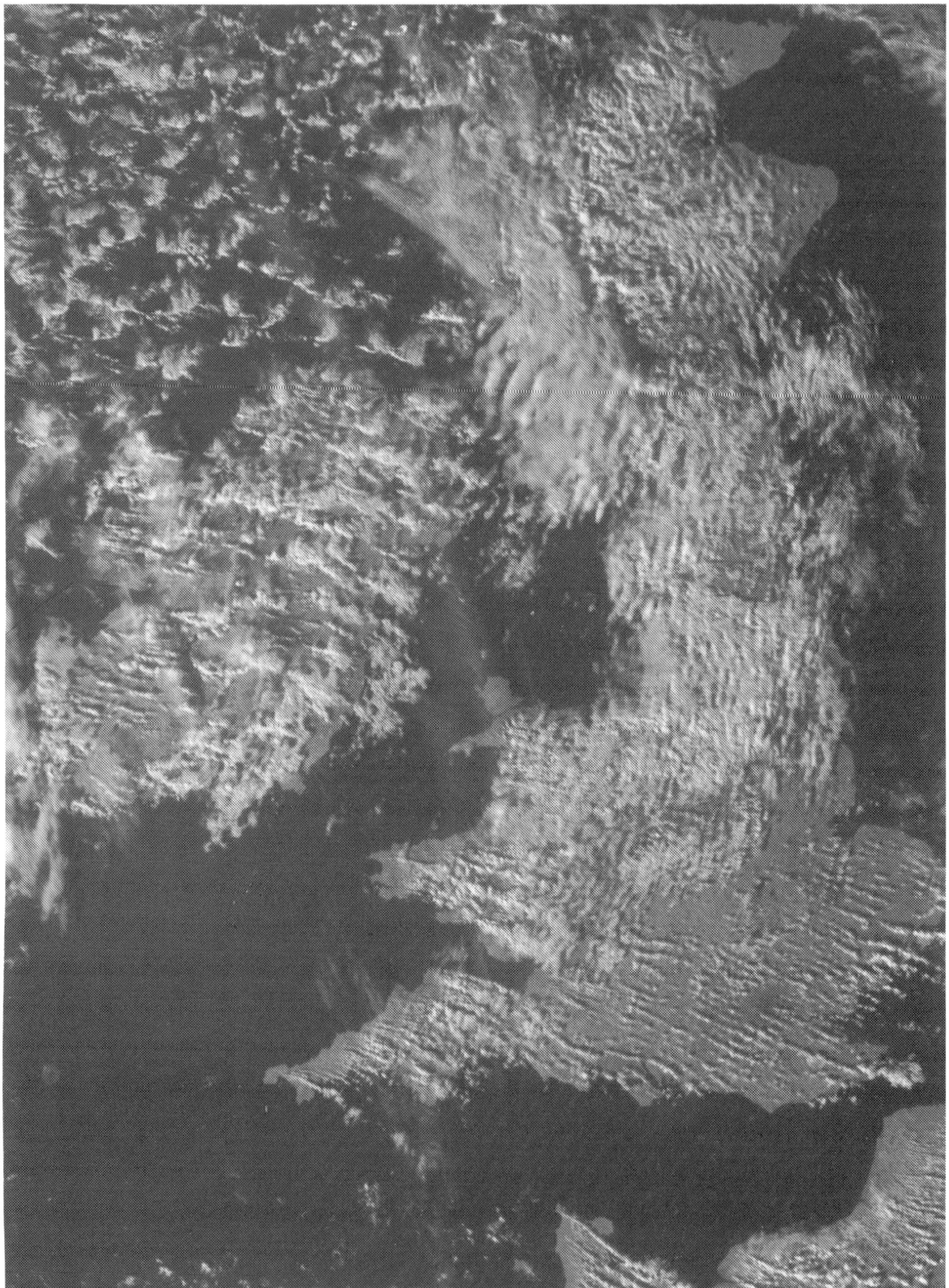

Fig. 11.8.i(c) The same simultaneous streets in a wind from the WNW and lee waves aligned at right-angles over the British Isles. (Courtesy of the University of Dundee: 1451, 10.2.80, Vis.)

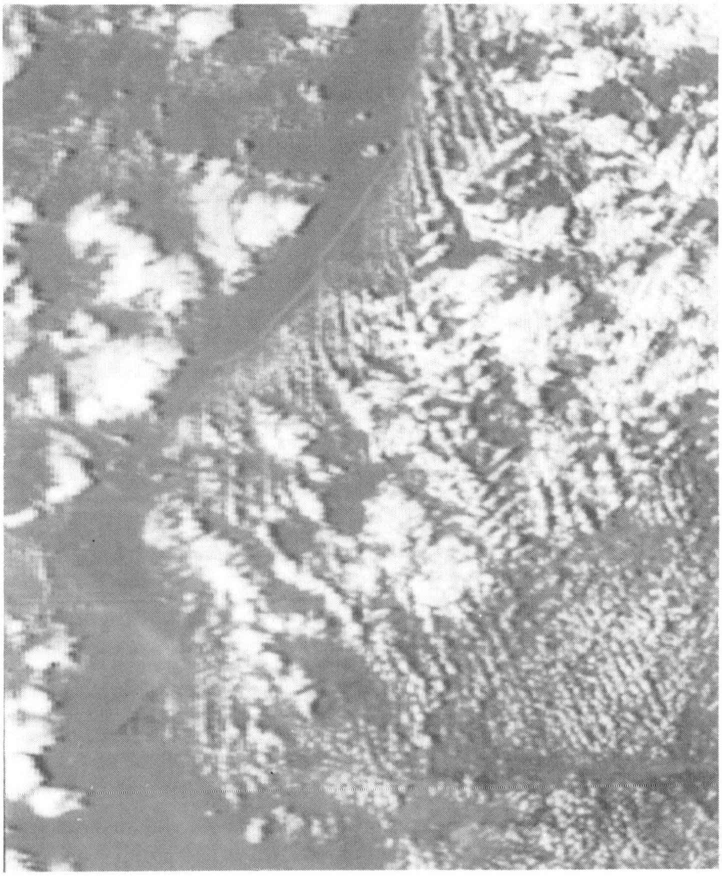

Fig. 11.8.i(d) The same over Iberia in a wind from NW. (Courtesy the University of Dundee: 1512, 5.10.84, 2.)

even if they were initiated as tiny thermals at the ground, while if they were further apart fresh thermals would pop up in between. It does not require much sophistication to appreciate this argument, and if we allow about as much width at the top for downcurrents as upcurrents and assume that in rising from near the ground to the top of the layer the width of a thermal is about doubled, we find that with $n = 4$, as for thermals, this doubling of width occurs in an ascent equal to the width at the top. Thus the cell width is about twice its depth. Cloud streets usually are spaced about twice the cloud base height apart, and the inevitability of cells being about this wide in no way validates the argument sometimes advanced that because convection cells in a liquid of uniform viscosity also have roughly this shape, the mechanics in the atmosphere are analogous. Certainly the eddy mixing in the atmosphere varies greatly across the system. There have also been arguments put forward concerning the relevance of the Prandtl number to cell shape (the eddy Prandtl number is K_M/K_H for a uniformly turbulent layer). However we have no measurements which would give any idea of the horizontal transport of momentum and heat by eddies such as would be required if an analogy with the case of a viscous conducting fluid in the laboratory were to be made. Some people, however, prefer a wrong theory to an admission that we have no mathematical formulation of the detailed mechanics. This is silly because occasions differ so greatly in the atmosphere that a precise theory would be of no use, especially as an eddy transfer theory reveals nothing, being only a representation of a kinematical idea and not of basic dynamics.

Streets are common over the sea, particularly where the airstream is being slowly warmed, and the layer is of uniform depth over a large area. The warming is slow so that interactions between thermals can achieve a more or less steady state. More vigorous convection does not arrange itself in streets but clouds grow to a great variety of sizes. Vigorous convection in a layer capped by an inversion through which clouds do not penetrate is rare, but it usually generates a complete layer of anvil cloud anyway.

Streets are formed in a matter of 10–30 minutes and are arranged in rows when the clouds first appear so that their pattern already exists in the air motion below. This is seen when air moves from cool land to warm sea where the situation is suitable, or when the streets first form in the morning. The time is comparable with the ascent time for parcels through the layer. Thus the mechanics of streets cannot be connected with the earth's rotation, which requires a much larger fraction of a day to produce effects. One theory of streets supposes them to arise in the 'Ekman layer', which is a layer with a theoretical wind profile corresponding to a steady horizontal wind whose speed and direction are constant at each height over a large area and are determined by a balanceof the pressure gradient force, the Coriolis force, and the shear stresses (see section 4.10). This wind profile is unstable to longitudinal perturbations, which are said to correspond to the streets. The profile depends on a constant eddy momentum transfer coefficient, which does not exist when streets are presented. Furthermore there is a large vertical exchange of fluid so that the supposed steady wind profile would not be present. Finally, of course, the air particles pass round the cells in a very small fraction of a day. Angell and Pack (1963; 1967) made balloon measurements showing circulation times of the order of 30 min, which agrees with deductions from observed air velocities in clouds or upcurrents in streets of the order of 1–10 m s^{-1} found by glider pilots. This means that the Coriolis force is not a significant factor in the dynamics.

Streets appear to lie nearly along the wind: the motion of particles is helical (Fig. 11.8.ii), and it is generally thought that the arrangement in longitudinal rolls is caused by the drag of the ground. The difficulty with transverse rolls is that adjacent rolls would place the upcurrent of one beside the downcurrent of the next. This is because they would have to rotate in the direction of the vorticity in the shear produced by ground drag, whereas adjacent longitudinal rolls have opposite directions of rotation. In the free air, on the other hand, a shear layer can generate billows all with the same direction of rotation because the upper and lower surfaces of the overturning layer can be distorted to separate the rolls (Fig. 8.3.ii right of diag.). In layers of altocumulus arranged in long cells the motion is either billows of the kind described in Chapter 8, where the shear is large, or else the overturning is very slow, and is more usually arranged in squares or not arranged

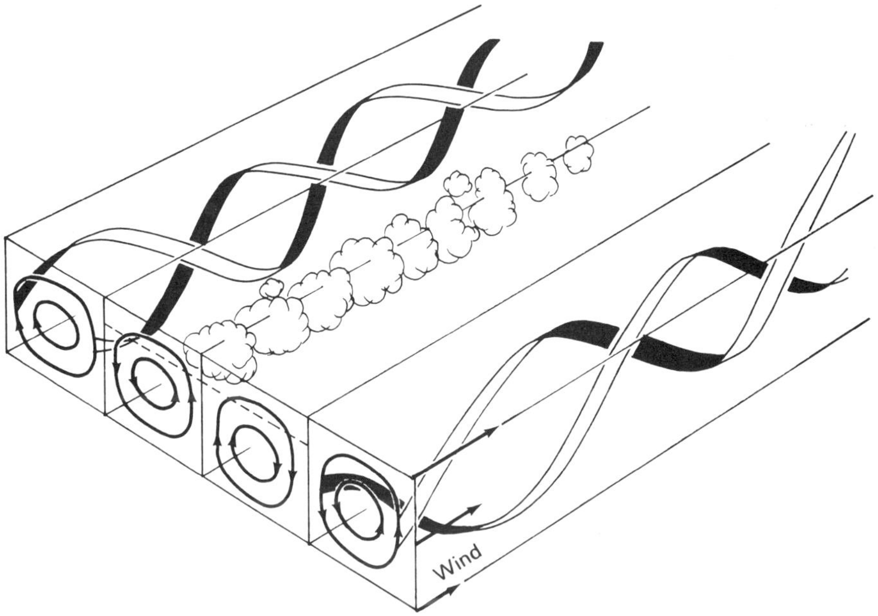

Fig. 11.8.ii Illustration of the helical average motion of particles in convection streets. Clouds may be formed at the top of the upcurrents either in the cells or penetrating a little above them: they are only represented in the middle upcurrent in this diagram. If the condensation level were at the dashed line and the flow were laminar a cloud layer would occupy the whole area above that level up to the top of the cells.

at all. This contradicts the old idea, drawn from laboratory experiments in smoke between sheets of glass in relative motion, that the cells are longitudinal when the shear is large, but transverse when it is small. There is no evidence that longitudinal rolls occur in thin layers of altocumulus at all, and a rigid boundary is probably a requirement for them to be produced.

It has often been suggested that the shadows of the cloud streets over land cause the heating of the ground to be arranged in lines, with the new upcurrents coming from the sunny areas. This could cause a migration of the lines across their length, as is often observed. This is less likely over the sea, where sunshine has little effect on the small time scale involved.

When there is no wind, and therefore no wind shear, small convection clouds are isolated and arranged with equal spacing unless the terrain is variegated.

(2) Plumes and lines

Solitary lines of cloud are often formed downwind of some obstacle or heat source. Fig. 11.8.iii shows the kind of convective motion which is like a thermal taking off from a superadiabatic layer at the ground, part of which is pushed upwards by flow over a hump (see section 3.6). Equally a heat source such as a large steel works where a hot slag heap

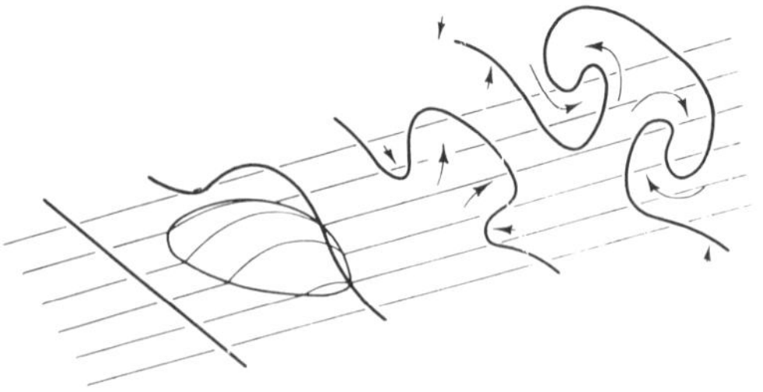

Fig. 11.8.iii The development of a thermal taking off from an unstable layer, initiated by a hump on the ground.

is created, or a lake in a burst of cold air from polar regions in the autumn, may produce a line of convection clouds downwind. Oceanic islands may do the same, but more commonly they produce streamers of cirrus cloud: the low-level air is slow moving so that the island in the sunshine is a continuous isolated heat source, and the tops of large anvil cumulus are carried away by a strong wind at say 10 km or higher. These streamers can be observed clearly by fixed satellite photographs and estimates of the wind made accordingly.

Sometimes it seems that a small cumulus cloud, which appears at the top of the ascent of a thermal and which owes its position to what goes on below, consists of a succession of thermals coming up from a source moving along more slowly than the cloud. Fig. 11.8.iv shows how in wind shear the downcurrent coming in from above is predominantly from one side when warm air takes off from the superadiabatic layer at the ground. The shear causes the colder air descending to the ground to travel along scooping up the slower-moving warmer air at the ground, and so sending up thermals from a source point which moves faster than the surface wind but slower than the speed of travel at the cloud level.

Clouds of this kind show the motion of a thermal rising in wind shear depicted in Fig. 11.5.i and Fig. 13.5.iv.

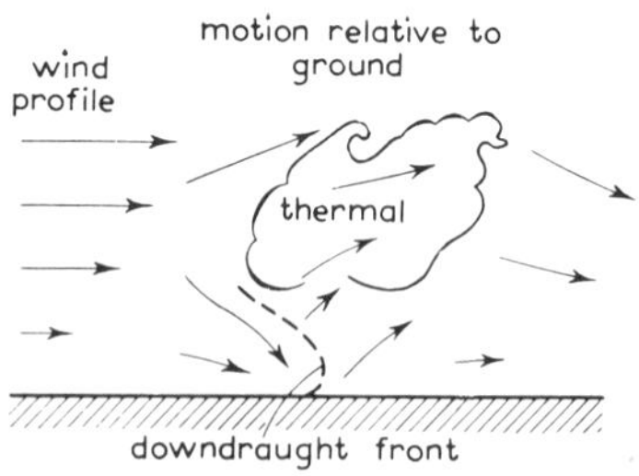

Fig. 11.8.iv When the wind increases with height the downdraught close to an upcurrent taking off from the ground is moving faster than the warm air at the ground, and scoops it up, feeding the cumulus cloud with fresh thermals.

(3) Thermal sources

There are several well known sources of thermals, some already described in (2) above. Others are produced by variations in the nature of the ground surface. Standing ripe corn can collect a reservoir of warm air which is carried away by the arrival of a cool gust which bends the corn down. Green vegetation, by contrast, remains cool by transpiration, so that it is a source of latent rather than sensible heat. Bare dry earth, rock, concrete or sand beside it may therefore initiate thermals. A favourite among glider pilots is the wind-shadow source: it is supposed that in the lee of a wood or cliff there is a body of slow-moving air which gets hotter than the surrounding air because it does not have its heat diffused upwards by turbulence as much as in the case of fast-moving air. Therefore from time to time it generates an accumulation of buoyancy which causes it to rise as a thermal.

(4) Rotating convection

The desert provides a very hot surface but one in which very little heat is stored. Infrared radiation from the surface is the agent of a large fraction of the heat loss, and this is absorbed and reemitted over a depth of a few metres mainly by the water vapour present. Consequently an unstable layer is built up, often to a depth of a few tens or even hundreds of metres, without there being any well marked thermal source at the ground. As soon as a small thermal rises, cooler air descends to the surface from higher up and the surface is cooled significantly as only the uppermost very few millimetres are hot. The air spaces between the sand particles make the sand layer a very effective insulator. Thus at a depth of a very few centimetres there is no perceptible diurnal variation of temperature in spite of variations of the order of 50°C at the surface. Consequently one thermal is not followed by others which feed it, and large thermals are inhibited.

In this situation any system able to sweep up heat from the ground as it travels over it can persist and perhaps grow. Such a system is the dust devil. The rotation must be initiated by a variation in the drag of the surface offered to the wind so that a horizontal gradient of wind is produced. Convergence taking place in this air produces vertical stretching of the vortex lines and an intensification of the rotation. The result is the generation of secondary inflow as described in section 3.8, so that heat from the surface over which the system travels is continuously fed into it, predominantly on the side where the circumferential velocity is in the direction of the wind. Thus a cyclonically rotating dust devil moves to the right of the wind (Fig. 11.8.v). A dust devil may remain stationary in a suitable position: one account described how a dust devil sat on the end of a causeway across a depression in the ground being built by lorry loads of sand brought to the point reached—only to be whisked away by the whirlwind. Swinn (see Wills [48], p. 78) gives a picturesque description of dust devils in Egypt, including their hollow appearance. This could be due to the centrifuging of particles from the centre, but it also appears that the air close to the axis has often been drawn into the vortex from the top, and is colder and free of dust. On the other hand, the dust is sometimes said to absorb sufficient sunshine to increase the temperature of the air containing it. This could be true of a dust devil a few hundred feet high, for then the dust would be airborne for several minutes, and the dust-free air in the vortex core would be cooler.

Dust devils are understood in general terms but many details remain elusive and not susceptible to even the most sophisticated theory. There are great variations in the importance of friction, conduction and the transfer of heat from the hot dust to air into which it is carried and from the shallow layer of hot air a few millimetres thick which is

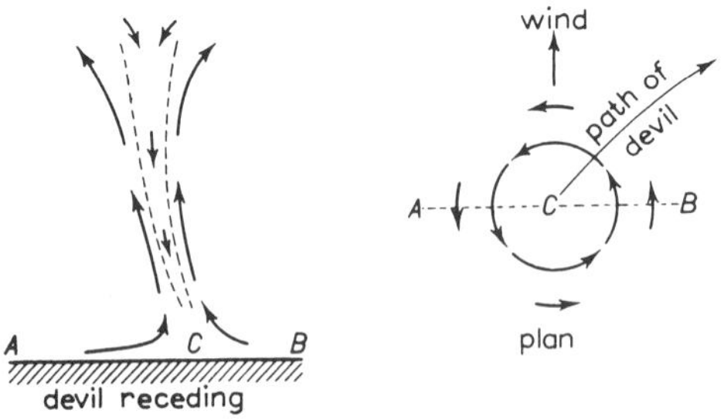

Fig. 11.8.v The motion in a dust devil in vertical section and plan. The base moves with the wind but also towards the forward-moving side. There is sometimes a downcurrent in the middle, especially when the system is new and not very tall.

almost stagnant before the arrival of the winds of the vortex. Devils vary from a foot or two to several hundred metres in height, sometimes extending up to the condensation level and being capped by cloud (Ives 1947). Of course they are great fun, readily associated with spirits by primitive people, and equally liable to facile explanation in a technological age.

(5) Anabatic winds

This name is given to the winds which may be produced on a slope heated by sunshine. They are best developed when the air mass against the slope is stably stratified, because the warmed air is thereby prevented from rising far from the sloping ground and is warmed as it continues up the slope. Often the anabatic flow is stopped at the snow line, above which the surface reflects the sunshine and remains colder than the adjacent air, perhaps with a katabatic flow on it. Haze generated in a valley travels up the slopes and spreads out into the stratified air mass at various levels according to the stratification (Fig. 11.8.vi). A well defined haze top is generated at the upper limit of the anabatic wind, which is often the general level of the mountain tops.

On occasions when there is no strong stratification above the mountain tops the anabatic flow feeds cumulus clouds above them (Fig. 11.8.vii and Fig. 11.8.viii). For

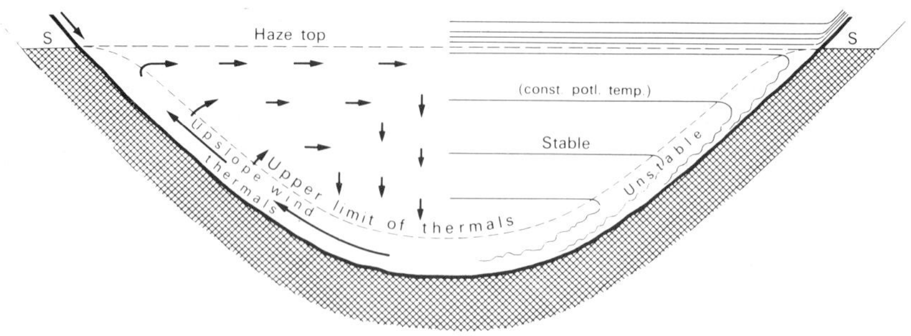

Fig. 11.8.vi The air motion (left) and stratification (right) in a valley with anabatic winds blowing up the sides. S denotes the snow line above which there is a shallow cold layer of air and possibly a katabatic wind.

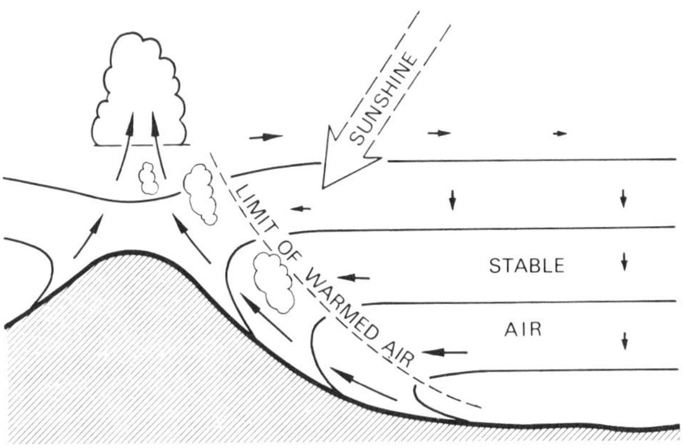

Fig. 11.8.vii Upslope winds feeding cumulus clouds over mountains when the air mass is slightly stable.

Fig. 11.8.viii A valley in Bohemia where the effluent from a fairly small industry escapes into the clean air and is not trapped in a smog because of its buoyancy. (Photo M. Koldavsky.)

more detailed treatment reference is made to *Clouds of the World* and *Air Pollution*. The more intrepid glider pilots who are prepared to fly close to a mountain slope have developed the art of soaring in anabatic winds.

(6) Effects of mechanical stirring: Föhn

When air flows over rough terrain and is mechanically stirred, the stirred layer acquires a uniform potential temperature. When the layer initially had a stable stratification, the effect is to cool the upper part of the layer and warm the lower part. This may sometimes explain why there is a higher air temperature at the ground on the lee side of hills than on the upwind side (Föhn effect), particularly if the terrain is rough enough to stir it over a depth of a few hundred metres. The most significant effect is to produce an inversion at the top of the stirred layer, which may then be the upper limit of convection on the lee side. The inversion may also be intensified by the presence of cloud produced at the level of maximum cooling: this cloud radiates into space in the infrared and is cooled thereby, giving rise to downward convection in the layer and a progressive cooling of it. This sometimes leads to a situation in which smog occurs. Again the reader is referred to *Air Pollution*.

11.9 SEA BREEZE FRONTS

Sea breezes are a commonplace of elementary geography and of the experience of coast dwellers. They have also been referred to for countless years, at inland places to which they sometimes penetrate, as the cause of a change of air mass during the afternoon or evening. That they arrive with a sudden drop in temperature has also been long known, but because knowledge of their upper air structure is lacking, this has not been thought of as being due to fronts.

When air over the land is warmed by morning convection, the penetration of sea air inland generates a temperature discontinuity as the cool air undercuts the warm. This discontinuity creates difficulty in any computer-based study using a grid of points of finite spacing, and the history of the subject contains several attempts at analytical representation of the steady state achieved after a long time. Because of mixing on the interface (which is rather like that on top of a thermal, with lobes of the advancing fluid entrapping the fluid into which it advances), the bulk parameters of the phenomena are the useful ones as in the case of thermals.

The speed, V, of advance of the front relative to the fluid it undercuts is related to the buoyancy, $B\ (= \Delta\rho/\rho)$, and the height, d, reached by the head of the advancing fluid before it sinks to a lower level further back (see Fig. 11.9.i) by

$$V = \mathrm{C}(gBd)^{1/2} \tag{11.9.1}$$

where C is found to be roughly about 2/3. V is defined as $U_\mathrm{w} + U_\mathrm{c}$, which is the sum of the speeds with which the warm and the cold air at some distance approach the front. The result (11.9.1) is true for laboratory experiments and sea breeze fronts in the atmosphere, mainly observed by Simpson [49] near Lasham using a glider. The cold air, having a higher humidity, was often clearly defined by its greater haziness.

The region of maximum upcurrent above the nose of magnitude about $0.7V$ is surprisingly narrow in the atmosphere, on many occasions giving lift to a glider only over a strip 100 m wide. This apparent effect may be due to the presence of irregularities on the advancing front in which mushroom-shaped lobes of cold air entrap the warm air. The mixed zone is buoyant relative to the cold air. In model experiments the mixing looks

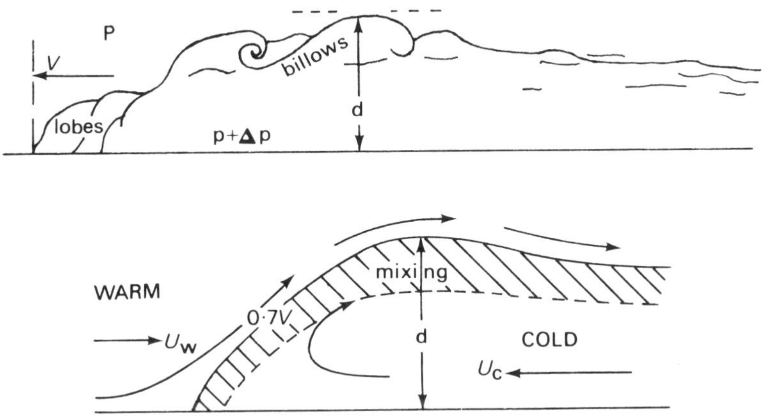

Fig. 11.9.i Schematic representation of a sea breeze front with density difference $\Delta\rho$, maximum height d, and speed V relative to the warm air into which it is moving. Mixing occurs at the front where warm air is entrapped in the lobes, and on the top where billow-like motions occur. The maximum upcurrent is of the order of $0.7V$.

very like billows, with a wavelength fractionally in excess of d, when seen in a vertical plane of the motion illuminated by light from a slit. The billows travel backwards on the head of the advancing current. Fig. 11.9.ii shows a model experiment of the phenomenon.

The sea breezes observed by Simpson typically had values of d equal to 700 m, $\Delta\rho$ corresponding to about 1°C temperature difference and speeds of about 3 m s^{-1}. Land breezes caused by the cooling of the land at night are much shallower, like katabatic winds, because the nocturnal cooling affects only a few metres when there is not much wind otherwise. When there is a significant wind the land breeze produces only a small effect. However land breezes can have significant consequences, notably by driving banks of fog out across the coast. These are observed early in the day and are sucked inland and evaporated during the morning. They are common on the coasts of the Western Mediterranean, near the Straits of Gibraltar particularly.

Similar density currents occur when an avalanche causes snow to be lifted from the surface and cool the air. Outflows from rain storms also have well defined fronts, and perhaps the most spectacular of these are the Haboobs, or dust storms, of the Sudan. The rain storms which generate these may be 100 miles away and may not be observed by those who experience the Haboobs.

The air particles flowing through the sea breeze front system pass through it in a small fraction of a day, the upcurrents frequently being around 3 m s^{-1} or larger. Consequently, Coriolis forces play a negligible part in their detailed mechanics, although the direction of the air motion at the coast is seen to veer after a few hours of sea breeze. Thus the formula (4.8.3) for the slope of a front given in Chapter 4 is irrelevant. Because of the motion inside the cold air the slope is not 60° as would be suggested by Fig. 2.4.ii(a). An overhanging nose is observed to varying extents and in model experiments is clearly associated with the viscous drag of the bottom boundary.

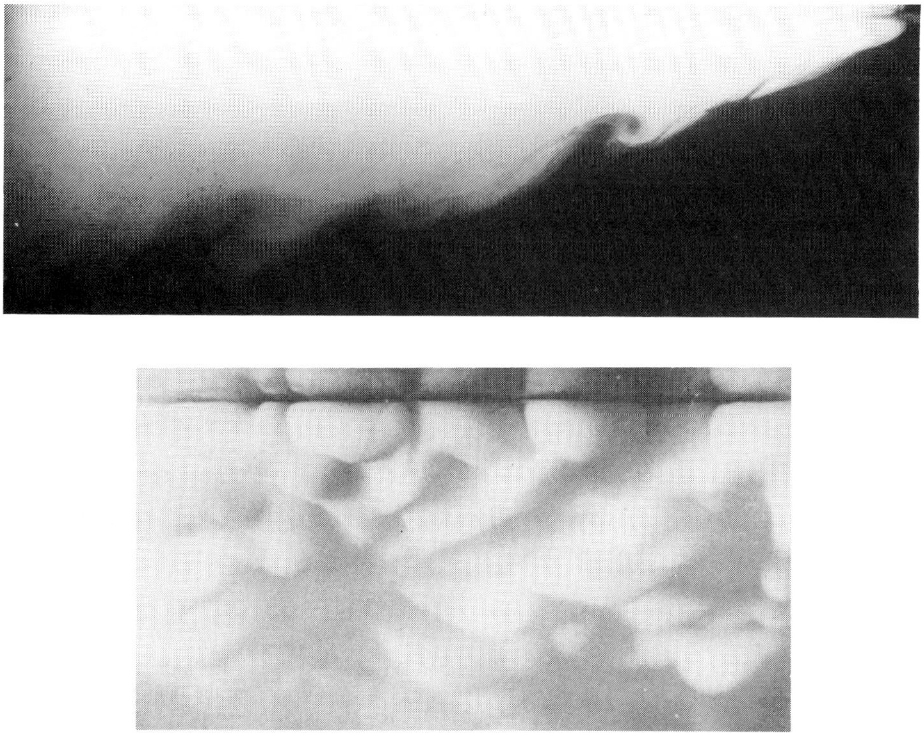

Fig. 11.9.ii A model experiment by H. O. Anwar. It differs from the experiments of Simpson (1969) in having lighter, whitened, fluid advancing along the free surface of a water tank. The views are from the side and front. Because of the absence of friction at the boundary there was less overhang than when the boundary is rigid (see Simpson 1972). In both views, reflections can be seen in the surface. [Photographs by courtesy of H. O. Anwar, Hydraulics Research Laboratory.]

11.10 AIRCRAFT DOWNWASH AND CONTRAILS

In order to support its weight an aircraft leaves behind it a downwash which for practical purposes consists of a pair of equal and opposite vortices and a motion pattern as shown by the streamlines in the left half of Fig. 11.10.i. Relative to the vortices the motion is as shown in the right half of Fig. 11.10.i, where it is seen that a body of **accompanying fluid** (AF) exists which circulates around the vortices, and travels downwards with them. The surrounding fluid passes round as if the AF were a solid body.

In a neutrally stratified atmosphere these vortices may descend to the ground, where they come under the influence of their image vortices and this causes them to move apart close to the ground. But usually before that happens they are disrupted by ambient turbulence and by instabilities inherent in them. Trailing vortices usually contain an axial component of velocity which varies with distance from the axis, and this destabilising according to section 8.6, and as a result the vorticity is spread away from the axis. Furthermore, a vortex pair is unstable because if a sinusoidal disturbance is created it will

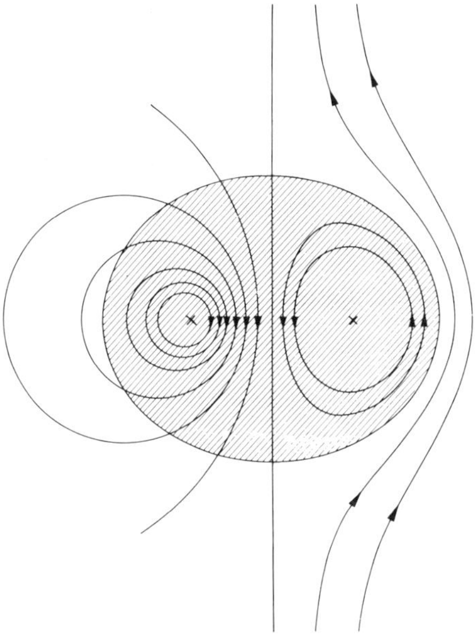

Fig. 11.10.i The streamlines of the motion produced by two parallel line vortices. On the right is the motion relative to the vortices, and on the left is the motion relative to the fluid at a great distance. The shaded area shows the fluid which travels with the vortices, called the 'accompanying fluid' (AF).

grow. The most unstable such disturbance occurs in planes at 45° to the plane containing the vortices, so that the parts of the vortices which are brought closer together and displaced downwards are subject to an acceleration in that direction. This result is demonstrated by classical small perturbation methods, which usually make various simplifying assumptions about how to calculate the motion field due to the sinuous vortices. Because of the difficulties it is impossible to pursue the calculation when the disturbance becomes large.

The atmosphere above cloud base is stably stratified, and this means that the accompanying fluid acquires buoyancy when it has descended any distance through the air. As a consequence a density discontinuity is produced at the outer surface of the accompanying fluid and this generates a vortex sheet there with the direction of the vorticity as indicated in the left half of Fig. 11.10.iii. The magnitude of this vorticity increases from zero at the bottom to a maximum near the top as the fluid flows around this boundary having vorticity generated in it all the way from the lower stagnation point.

After some time the amount of vorticity on the boundary has grown large enough to alter the flow pattern. If we apply equation (11.3.4), using (11.3.5), which are obtained on the assumption that the fluid acceleration is small compared with gravity, the vorticity is obtained by integration. Actually the assumption just mentioned is not all that restrictive

Fig. 11.10.ii The visible effect of the aircraft downwash on a cloud top along which the aircraft (B17—one of the few aircraft with a good backward-facing view) was flying. [Photo courtesy V. G. Plank.]

because the greater part of the fluid acceleration is due to the curvature of the flow and is nearly normal to the isopycnics. Therefore it does not produce much vorticity. Thus we find that at height z measured from the level of the vortices the vorticity is

$$\eta = \frac{g\beta_0}{q_0}\left(z + \sqrt{3}R\right). \qquad (11.10.1)$$

This is because it is zero at the bottom stagnation point, S, in Fig. 11.10.iii, where $z = -\sqrt{3}R$, and the downward velocity of the fluid is the same there as that of the vortices. The speed q is the speed along the boundary of the AF relative to the vortices, each of which produces a circumferential velocity field of magnitude $K/2\pi r$ at distance r from itself.

We assume that approximately the value of $\Delta\rho$ across the boundary remains constant for a particle as it passes round: this is not strictly correct, partly because of the up and down motions induced in the outside fluid which is stratified. The amount of vorticity in the transition layer at the boundary is obtained by multiplying (11.10.1) by the thickness δ, so long as it is small. Putting

$$\eta\delta = H \qquad (11.10.2)$$

and

$$\beta_0\delta = \frac{\Delta\rho}{\rho} \qquad (11.10.3)$$

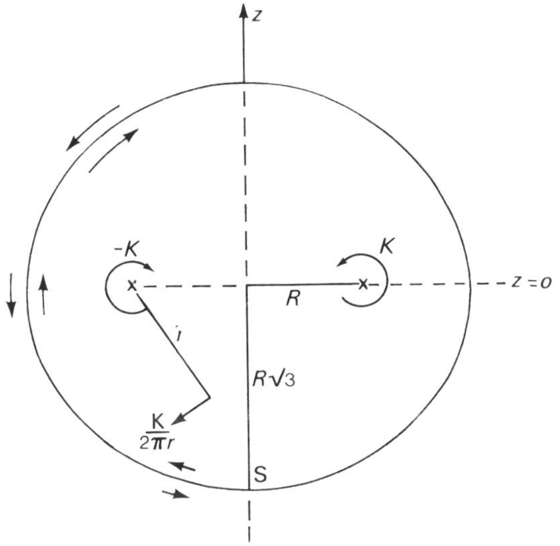

Fig. 11.10.iii Coordinates for a vortex pair having circulations $\pm K$, a distance $2R$ apart. S is the lower stagnation point on the boundary of the accompanying fluid: it travels downwards with the same speed as the vortices. The vorticity generated on the boundary by the density discontinuity increases as the air flows round from the lower to the upper stagnation point.

H is now the vorticity in unit length of the boundary and is

$$H = \frac{g\Delta\rho}{q_0\rho}\left(z + \sqrt{3}R\right). \tag{11.10.4}$$

The velocity produced by this new vorticity is obtained by integrating its effect around the boundary. On the boundary z is proportional to R multiplied by a function of position which represents the geometrical shape of the system. The length of the boundary is likewise proportional to R. Before calculating it we consider its effect. The vorticity on the left half of the boundary produces a horizontal velocity at the left vortex which is greater in magnitude than and opposite in direction to that produced by the right half. The upper left quadrant has larger vorticity than corresponding points in the lower left quadrant, and since it has the effect of moving the left vortex to the right the net result will be to move the two vortices towards one another. As a consequence each will cause the other to descend more rapidly. The velocity due to the vorticity H is proportional to H/R, and when integrated round the boundary the total effect is proportional to RH/R. Consequently

$$\frac{\mathrm{d}R}{\mathrm{d}t} \propto -H \propto -g\frac{\Delta\rho}{\rho}\frac{R}{q_0} \propto -g\frac{\Delta\rho}{\rho}\frac{R^2}{K} \tag{11.10.5}$$

because the velocity q_0 on the boundary, being substantially that due to the vortices, is proportional to K/R.

If we now suppose there to be a stable stratification of the environment β, as the vortex system descends with velocity proportional to K/R, the density anomaly is given by

$$\frac{\mathrm{d}}{\mathrm{d}t}\left(g\frac{\Delta\rho}{\rho}\right) \propto g\beta\frac{K}{R}. \tag{11.10.6}$$

Then from (11.10.5) we have

$$g\beta\frac{K}{R} \propto \frac{\mathrm{d}}{\mathrm{d}t}\left(\frac{K}{R^2}\frac{\mathrm{d}R}{\mathrm{d}t}\right), \tag{11.10.7}$$

i.e.

$$g\beta\frac{1}{R} \propto \frac{\mathrm{d}^2}{\mathrm{d}t^2}\frac{1}{R},$$

so that

$$R \propto R_1 \mathrm{e}^{-\alpha\sqrt{g\beta}t}. \tag{11.10.8}$$

R_1 is the initial value of R and α is a constant determined by the geometry of the system, and we shall not be concerned to evaluate it here.

Clearly the system cannot last. As R is decreased, the rate of descent increases until $\Delta\rho$ becomes so large that the boundary vorticity H begins to produce important effects to be discussed in a moment.

An alternative approach was made by Scorer and Davenport [50]. They followed a theme initiated by J. S. Turner [51] in his study of buoyant vortex rings, and the reader is referred to his book *Buoyancy effects in fluids*. According to this approach the downward momentum of the vortex system is (see Lamb, 1932, Art. 152, eq. 6) equal to $2\rho KR$. Since an upward buoyancy force is operating on the accompanying fluid this momentum is decreased. Since K is constant, R must decrease. By ignoring the effect of the vorticity in the boundary on the momentum, this led to the conclusion that

$$R = R_1 \operatorname{sech} \frac{\sqrt{mg\beta}}{2\sqrt{2\pi}}t, \tag{11.10.9}$$

which is similar to (11.10.8) except close to the start. The number m is defined so that mR^2 is the cross-section area of the AF, and it is equal to 11.5. We may therefore set α equal to $(11.5/8\pi)^{1/2}$, i.e. 0.68.

The effect of the vorticity in the boundary is to alter the position of the upper stagnation point and bring it inside the AF. Consequently the fluid containing the new vorticity is removed and left behind, and the volume of the AF decreases so that the vortices come closer together. The streamlines are shown in Fig. 11.10.iv.

As the vorticity is fed towards S, the original upper stagnation point, it modifies the velocity field and carries itself away upwards as a thin double vortex layer. Here the total vorticity is zero and is soon cancelled out by the proximity of equal vorticity of opposite sign.

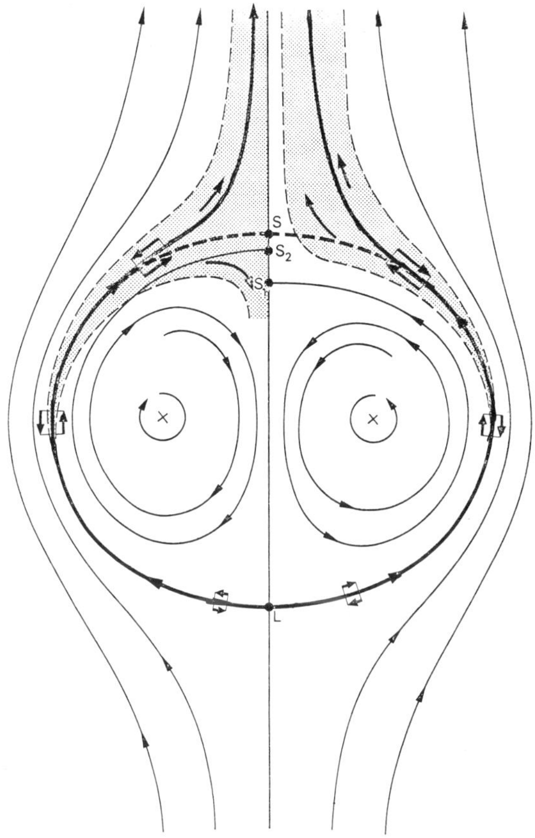

Fig. 11.10.iv The boundary on which the vorticity is generated is shown by the heavy line. This is detrained at the top of the AF when the upper stagnation point is lowered from its starting position, S, to S_1. The shaded area represents what may be imagined to be the mixing region produced by instability on the vortex sheet at the boundary. Most of this is detrained, e.g. in the right half of the diagram. But if the stagnation point rises to S_2, part of this mixed fluid may circulate down the centre of the AF. This story is continued in sections 14.3 and 14.4.

However, the vortex layer is unstable and as it becomes more intense it will break into billows and become thicker. Scorer and Davenport showed that although initially the stagnation point moved down to a position S_1 so that the mixed layer would all be fed way upwards, the stagnation point would move upwards again after a time which, with typical atmospheric stratification, would be about two minutes. At the same time the vortex layer became thicker. In consequence some of the mixed fluid would then begin to be fed into the middle of the accompanying fluid, as indicated in the left half of Fig. 11.10.iv, with the stagnation point at S_2. As soon as this happens the motion in the accompanying fluid becomes unstable in the sense described in section 8.7 and illustrated in Fig. 8.7.ii.

We now recall that the original vortex pair was unstable for sinusoidal disturbances even before buoyancy became important. Consequently, small perturbations would grow.

However, because of the decrease of R given in (11.10.8) and the consequent exponential increase in the vertical velocity of the vortices, $K/2\pi R$, this sort of disturbance will occur on account of the buoyancy forces alone, and it is impossible to distinguish the two effects simply by observation. The theory just outlined can be used to interpret the visible appearance of condensation trails behind aircraft. Because it involves initially a discussion of the physics of condensation, that part of the treatment is left until section 14.3 and section 14.4.

12

Air pollution problems

12.1 GENERAL PRINCIPLES

Most of the time the level of air pollution is acceptable, and although the efforts to achieve clean air since the 1950s have been largely successful the margin of acceptability continually recedes as the population and all its traffic and industry inexorably grow. The regulations made for today soon become inadequate to preserve the level of cleanliness we hope to achieve. Because the trends are not reliably predicted it is assumed that the mood of today will remain in control. Yet the history of the second half of the 20th century has been neglect of the realities of human evolution, however much they are cried from the housetops: the generation born since 1945 has carried its freedom to the point of irresponsibility, not believing that it can be as foolish as its grandparents, that it is doing better, that it is bound to be doing better, and that their good intentions do not lead to hell—not for us anyway!

The first job is to understand the mechanisms by which the atmosphere treats pollution, and then to study the possibility of forecasting the consequences of our proposed actions.

Before governments thought of legislating to control pollution, mankind exploited the benefits of using fire and treated the obnoxious effects as something to stay away from as much as possible: thus there were strong objections to the burning of coal because it contains much more sulphur than wood. The higher temperatures attainable gave much greater production of wealth and military power and was therefore irresistible so that by the turn of the 19th century all the leading countries were absolutely dependent on a supply of coal. No answer has been given to the question 'what shall we depend on when the coal and oil run out?' and idealists shy away by being convinced that there will be some alternative source of energy but that as far as they are concerned it will not be nuclear power. Religions usually end up saying that the real trouble has been human selfishness, while the theological imperatives appear quite unconvincing to the rest. In that argument the time scale has been so effectively cut down to human size that fear is only aroused by immediate and evident danger, and even philosophers fail to teach the transitory nature of all human endeavour and its dependence upon the fact that our extravagance is historically very very recent, if only because as little as two centuries ago there was only one of us where there are six of us today. This view of the human predicament is nevertheless well known but is not acted upon by any governments. A growing few try their best to gain experience in living in a sustainable way and gain a little ground when society at large is forced to admit its unsustainability. But the essential intellectual blockage which denies to us the right to think further ahead than the stock exchange is dismissed as a failure of the feel-good factor which ought to be restored by better public information. This book seeks to improve public understanding by giving a correct view of one little bit of the great scientific endeavour.

12.2 MAXIMUM GROUND LEVEL CONCENTRATION

The purpose of tall stacks is to get rid of the products of the combustion which is useful to us. We first seck to predict the **maximum ground level concentration** which will be produced by this single polluter. For this purpose we assume that it stands on a flat plain in a steady uniform wind U, and emits at height h above the ground. It is assumed that some mechanism causes the plume to be progressively diluted and will be made acceptable at a point beyond which we need to have no further concern. The wind is assumed to be turbulent because it is blowing over rough ground and so it is reasonable to assume that the pollutant will be steadily diluted according to some diffusion coefficient K.

At this stage it is necessary to specify precisely how the concentration of the pollutant is to be measured. If a sample of air is analysed chemically the sample size has to be specified, probably as a sample of air drawn through the analysing system during, say, during 10 minutes at a height of 10 m above a representative piece of ground. It is important to take into account that the ground or the vegetation growing on it may absorb an unknown amount of a reactive gas such as SO_2. Indeed it was discovered that when the hawthorn was growing, the concentration of SO_2 was measured to be twice as great on the upwind side of a hedge. The sampling must be done so as to record the spatial variations over a large area downwind of a stack, and some sampling must be done on the upwind side in order to discover what background value the chimney is adding to.

The most obvious feature of samples of effluent from a stack is the large range of variation observed as the weather changes from day to day. Furthermore, there are very obvious variations with time of day. In sampling it is desirable to make the sample representative of the doses which may cause harmful effects. Thus there may be **whiffs** during which the concentration rises to high levels which may be enough to hospitalise, or even kill, a person suffering serious bronchial disadvantage, but which would be ignored by a healthy athlete. Thus some picture of the frequency and duration of strong whiffs must be an important objective of a sampling experiment. The complement of whiffs is the **long-term average**; for example, 24 h samples should be taken by automatic samplers which may be tended daily or weekly, or even longer—such as a chemically prepared surface which can record monthly or annual averages which would be relevant to the damage done to leather, paint or textile surfaces.

Samples should be taken **indoors**, and immediately outside any building, such as a hospital, where they may be concern about the effects. Obviously every case is a matter for appropriate design of the array of sampling apparatus to be used. This text is not designed for that purpose, but we shall describe the various mechanisms which it is necessary to be aware of when planning a survey or assessing the usefulness of a formula derived from particular observations.

The real atmosphere is found to have a wide range of eddy motions, varying in size from a minimum of a centimetre or so, smaller ones being destroyed by viscosity, up to individual cyclones and anticyclones of the order of 1000 km across and persisting sometimes for several days. This latter group provides what we call **the wind** while we are thinking about the diffusion around a particular source by much smaller eddies. This provides the valuable idea due to Sutton according to which we note that we proceed to larger distances and longer times it is the larger eddies only which make a significant contribution to the dispersion: at that stage we can neglect the diffusion by small eddies which contribute by spreading the pollution rather widely at first and more slowly later when distances from the source are large compared with the eddies. To cover as wide a range of cases as possible therefore we assume that the **spread is down a cone**—so that the eddy size which is important is a size proportional to the distance from the source. The effect is illustrated in Fig. 12.2.i, the paraboloid being the shape a plume would have if it were spread according to diffusion theory (see, for example, the dimensional analysis in section 2.9).

Although we may see cones spreading out at a variety of angles according to the smoothness and strength of the wind, we often see sinuous plumes, as if the long thin cone was subject to large eddies, which nevertheless remained within an **enveloping cone** as in Fig. 12.2.ii, which may be seen as representing a top or side view. The concentration of pollutant is then an average of several whiffs and other variations. The concentration profile is shown in Fig. 12.2.iii for the section of the plume above the point A where the plume first impinges on the ground. From that point the ground level concentration rises to a maximum at M at which point it is composed of the expected profile for width b, which is composed of the sum of the expected value at the distance of M together with the addition of that due to an imagined plume at the image position in the ground. This is added on the assumption that there is **no absorption of the effluent**

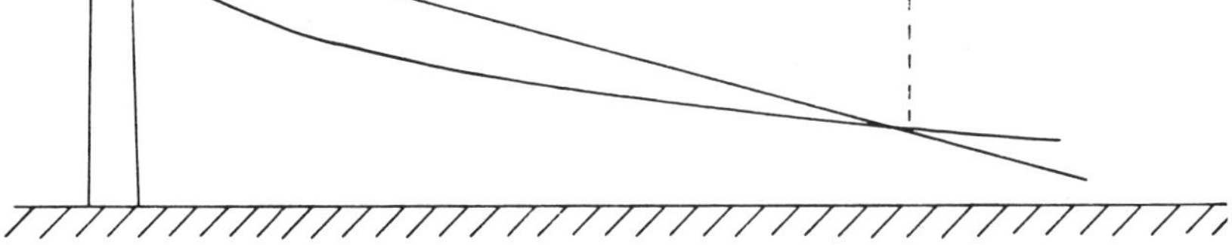

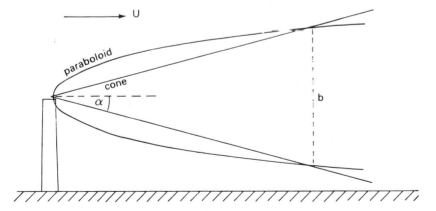

Fig. 12.2.i The average pollution across a plume is the same when its cross-section area is b^2 whether it has spread out along a cone or along a paraboloid. If the shape of the cross-section is the same in case of different spreading rates a single linear dimension, b, is enough to specifity the area.

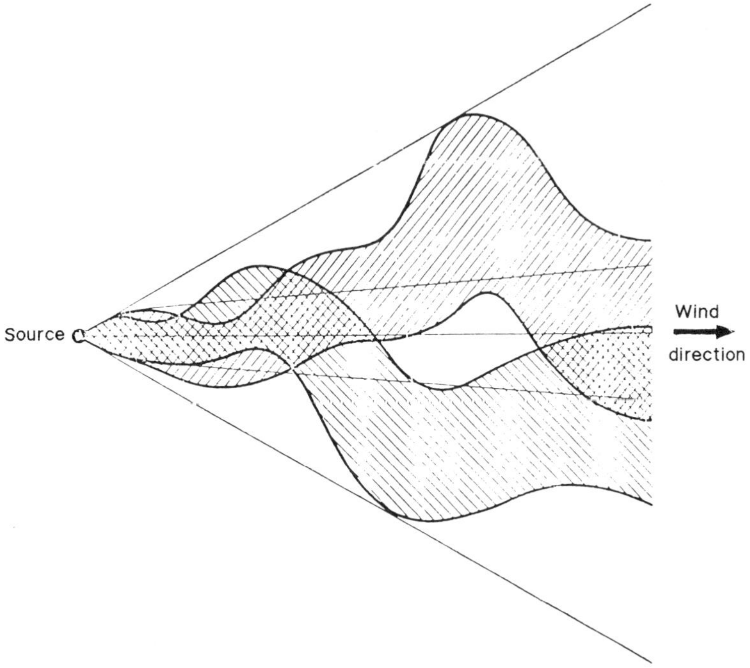

Fig. 12.2.ii The widening plume. If the plume were widened by eddies of only one size it would widen along a parabola which might represent the spreading well at first but would soon become much too narrow. As the weather situation changed, the wind direction at the source would vary and produce wider sinuosities than were apparent in the instantaneous plume at the beginning. The appropriate sampling time therefore increases with distance, so as to include several sinuosities, which would be a representative sample.

on the ground—which may, of course, not be the case. In the case of a reflected plume the strength at the ground is taken to be twice what it would have been if there were no ground. The most important result from this analysis is that the concentration at A is the same whatever the distance from the chimney, because the dilution has taken place over the area of the cross-section above A which is determined by the height of the chimney (and the assumption that the section is circular) regardless of the value of the angle. The pictures of Fig. 12.2.iv support the cone idea because the smoke does not follow a sinuos path but each piece of the plume seems rather to follow a straight line roughly as it circulates round the eddy which is transported bodily forward. According to this model the iso-lines of ground level concentration would make a map like Fig. 12.2.v.

If some additional height h' is given to the plume by the addition either of greater efflux velocity or of buoyancy due to a higher temperature, we would draw the cone enveloping the plume as having an apex at height $h + h'$, and slightly upwind at the virtual source as shown in Fig. 12.2.vi. The plume is also diluted by a factor $1/U$, because the length of plume receiving an amount of pollution Q in unit time is Q/U. The cross-sectional area above A is proportional to $1/(h + h')$, and so the maximum ground level concentration is given by the formula

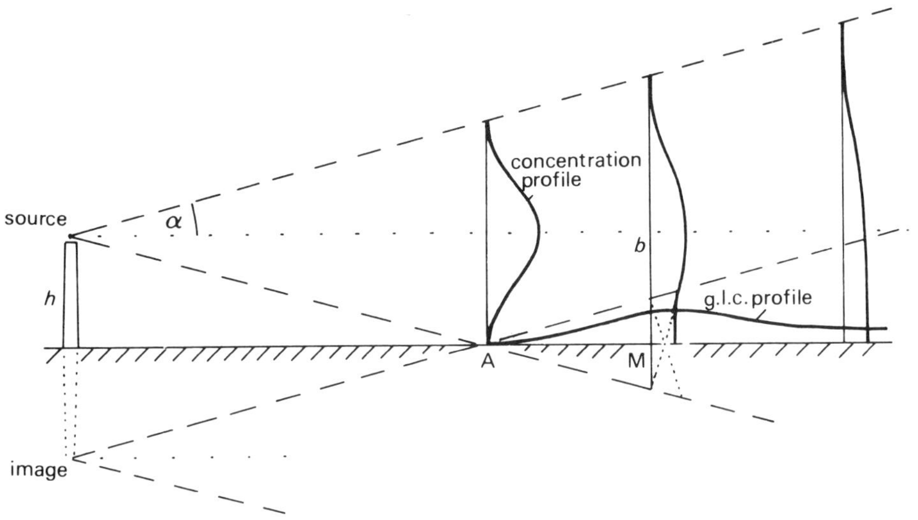

Fig. 12.2.iii If there is no absorption of pollution at the ground the concentration is as if there were an image source below ground. The profile across the plume is shown by the heavy line at three positions for conical spreading. At M the ground level concentration (g.l.c.) reaches a maximum, and is the sum of the two dotted profiles. The downwind g.l.c. profile is shown using a vertical coordinate for its magnitude.

$$P_{\max} = N \frac{Q}{U(h+h')^2} \tag{12.2.1}$$

in which N is a number to be found by experiment/observation.

The value of N lies between 0.05 and 0.5, and it is not misleading to assume a value of about 0.15, and we cannot be assertive about the value except that it is probably within a factor of about 3.

Such an uncertain formula is not useless, for it emphasises the value of increasing the height, and it states that the only other feature over which we have some control is a reduction of Q, the rate of emission of what we wish to avoid.

12.3 MODELS OF THE DIFFUSION OF A CHIMNEY PLUME

It was demonstrated in section 12.2 that if we consider the effect of several eddies each of which distributes part of the plume in a slightly different way from the others, the result of considering a large number of eddies will be the same as if there had been a constant diffusion of the pollution analogous to the molecular diffusion of gases which occurs when the partition between two chambers containing different gases is removed: each gas becomes uniformly distributed throughout the double chamber.

In smoothing (averaging out all the whiffs and other variations in the concentration of pollution) we may be representing the real phenomenon as remaining below a threshold of concentration which we think has to be achieved before some particular form of

Fig. 12.2.iv An example of black smoke making the way a sinuous plume is carried by the wind visible. The plume from Podsbrook Colliery came from a furnace designed to burn cheap coal—revealing the path of its pollution by smoke. These pictures were taken a few seconds apart so that the progress of the sinuosities can be followed. The enveloping cone was wider than average, and the point of first impact on the ground was only 4 or 5 chimney heights downwind.

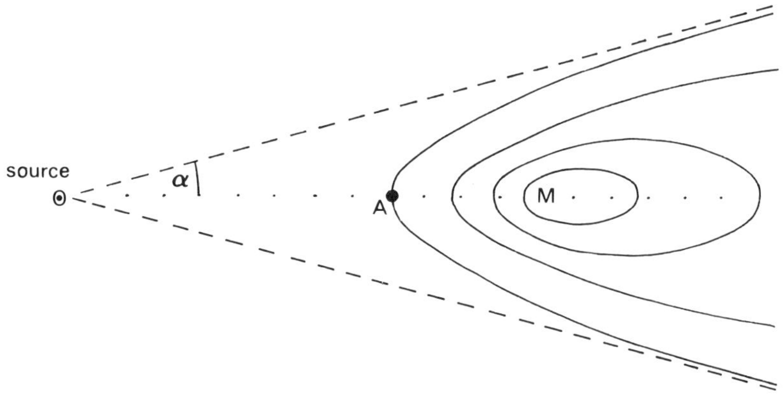

Fig. 12.2.v Ground level concentration. The projection of the plume cone on to the ground is along the dashed lines, A is the point of first impact and M is the approximate position of the occurrence of P_{max}.

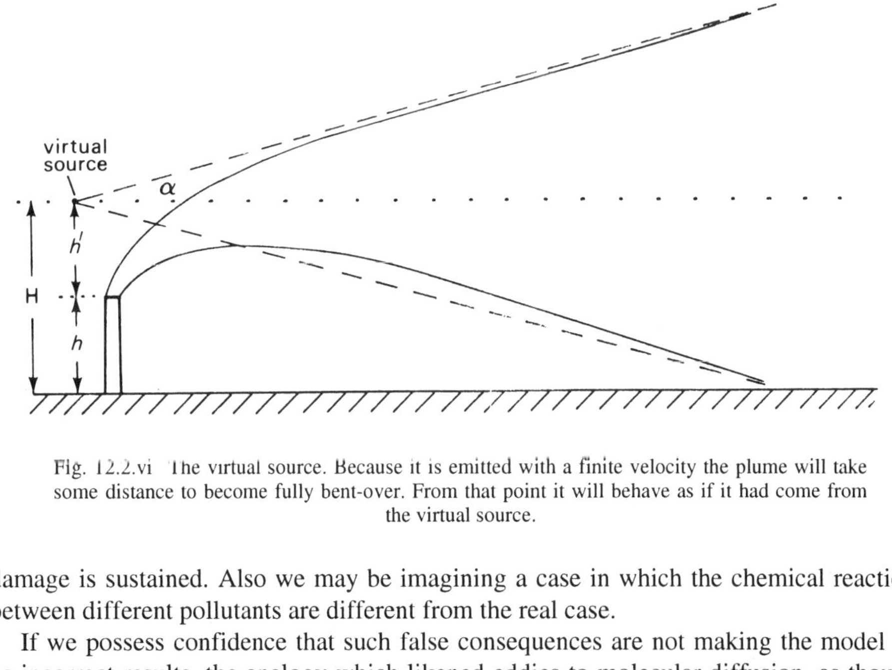

Fig. 12.2.vi The virtual source. Because it is emitted with a finite velocity the plume will take some distance to become fully bent-over. From that point it will behave as if it had come from the virtual source.

damage is sustained. Also we may be imagining a case in which the chemical reactions between different pollutants are different from the real case.

If we possess confidence that such false consequences are not making the model tell us incorrect results, the analogy which likened eddies to molecular diffusion, as thought of in the kinetic theory of gases, has the danger that the molecular phenomenon is repeatable in the laboratory where the diffusion coefficient was originally measured. But two occasions of the diffusion of a chimney plume are almost never the same in detail. The laboratory event depends on the temperature; the outdoor event depends on many facets of the weather.

The most important elements of the weather determining how a chimney plume is carried by the wind are the wind direction and velocity, both of which may vary with height, and the stratification of temperature. The details of these are scarcely ever the same on two occasions. That means that if we are to have a formula which describes the shape and concentration of a plume it must be far more complicated than anything so far contemplated. A single plume is the simplest form of air pollution for which we can make a useful model. Indeed it is remarkable that O. G. Sutton wrote in his book *The Challenge of the Atmosphere* that when Reynolds had explained the analogy between molecular viscosity and the transport down the gradient of pollution (or whatever the air contains) many scientists thought that all that was required to solve the dispersion problem was to measure the eddy diffusion coefficient of the atmosphere; for then we could use all theories of conduction of electricity, heat and anything that flows down the gradient and apply it to the dispersion in the atmosphere.

The interesting aspect of this story is that people started to measure the 'eddy diffusion coefficient' and fairly quickly found that it varied by a factor of up to 1000 from one occasion to another. It was not long before this was understood as being due to variations of the meteorological factors mentioned in the previous paragraph, but the concept has not died. It has been defined; therefore it exists. It has been defined as the ratio of the rate of transport to the gradient. A measurement that is peripheral to any discussion has been treated as fundamental, and its value is conventionally denoted by the symbol K (usually chosen to represent a constant—which this ratio definitely is not). Sutton assumed that the value, once found for the transport of one quantity, could be treated as valid for another—in this case he advocated the assumption that it would have the same value for the vertical transport of heat (defined as potential temperature) as for the vertical transport of horizontal momentum. This assumption made possible a theory for the vertical transport of pollution.

From this beginning was developed the famous mixing-length theory, which postulated that the eddies consisted of 'small' parcels of fluid which moved across the lines of constant concentration a distance which was defined as the mixing length (and therefore existed) usually denoted by ℓ. The theory assumed that after moving the defined distance it was completely mixed into its new surroundings, and therefore any quantity carried by the parcel was carried the same distance and all such quantities had the same mixing length, and the same 'turbulent transport coefficient' K.

Doubts about momentum having the same coefficient as other quantities such as water vapour or any kind of pollution because the momentum could be transferred by the pressure field around the parcel without material mixing. A famous controversy arose between the two most eminent theoreticians, Prandtl and Taylor, in the later 1930s, because Taylor argued that the parcel would carry its vorticity, which it would conserve until it mixed whereas it could share its momentum before it was physically mixed with its new surroundings. If vorticity and momentum had the same mixing length, Taylor's assumption gave a value of K for momentum which was twice the value according to Prandtl's assumption. This was an important argument because it applied to the simplest and most important case—plane rectilinear shear flow. It produced a discussion about what was a 'transferable quantity' in the sense of having the same transfer coefficient or mixing length as all the others; thus momentum was described as not being a transferable quantity, which many of us saw as being a game with words which did not help in understanding reality; for momentum certainly is transferred.

It is misleading to refer to K, or ℓ (mixing length), as a coefficient because they are not evaluated in a laboratory as a property of the air in which the diffusion is occurring, but the value is an order of magnitude guess which is to be used when we have no other method of knowing the value until we have made the observations where we have tried to use it to predict the result.

The guess was necessarily based on measurements of a previous case or cases which may or may not have been close to the case to which it was applied.

Throughout all these cases the only reliable assumption has been that the total amount of the pollutant remains constant and all that we are doing is guessing where it goes to. The use of K as a coefficient is not really analogous to the use of the coefficient of molecular viscosity, ν. That is the statistical result of the behaviour of billions of identical molecules and we have a reliable kinetic theory to back up our belief that we have measured a physical property of the substance itself. The K-theory depends on the assumption that our measurement of the value of K is based on the behaviour of a relatively small number of eddies which are not necessarily identical to each other, so that our statistical result is no more than a guide to representation of the process by a parameter which may have to be adjusted when we have studied a few cases.

12.4 JET AND THERMAL RISE

The following discussion is based on the assumption that the effect of adding upward momentum or buoyancy to a plume can be correctly represented by adding a quantity h' to the stack height. As we shall see, this idea is stimulating but has very strict logical limits. Even so, the outcome has a useful message.

A **bent-over plume** which has significant upward momentum or buoyancy is assumed to rise like a two-dimensional (i.e. line source) puff or thermal described in sections 10.6 and 10.7. Thus each cross-section is assumed to grow as if it were part of a system rising from an instantaneous line source as illustrated in Fig. 12.4.i, and has not yet reached the stage where the base has begun to descend, following the outline of the cone as suggested in Fig. 12.2.vi. The upward velocity is assumed to be appropriate for the line source puff or thermal until it has reached a height where its upward velocity has decreased to the value of the turbulent eddies in the wind stream, which is assumed to be $\lambda \times$ the wind speed U. After reaching that height it is assumed to become a **passive plume** so that it is from there, at height $h + h'$, expanding along a cone. This enables us to assign a value to h'. In order to give a value to the point at which the plume may be said to be bent-over we assume that this is where the upward velocity is equal to λU, where λ is a numerical factor to be determined by observation and is expected to be approximately 0.1. While it may typically be equal to 0.1 where the ground is not particularly rough, it could be around 0.4 in a built-up area or even as high as 1.0 when there is strong convection in a light wind. But it could be as small as 0.001 when the chimney stands well above very smooth ground or when the lower layers are stably stratified.

To illustrate the method we consider first the case of zero buoyancy. Subscript zero denotes values at the orifice whose radius is a, σ is the pollution concentration. Subscript 1 denotes values at the height where the jet becomes bent-over and $w = \mu U$. In the bent-over part, the height is denoted by y measured from a virtual origin at the height where

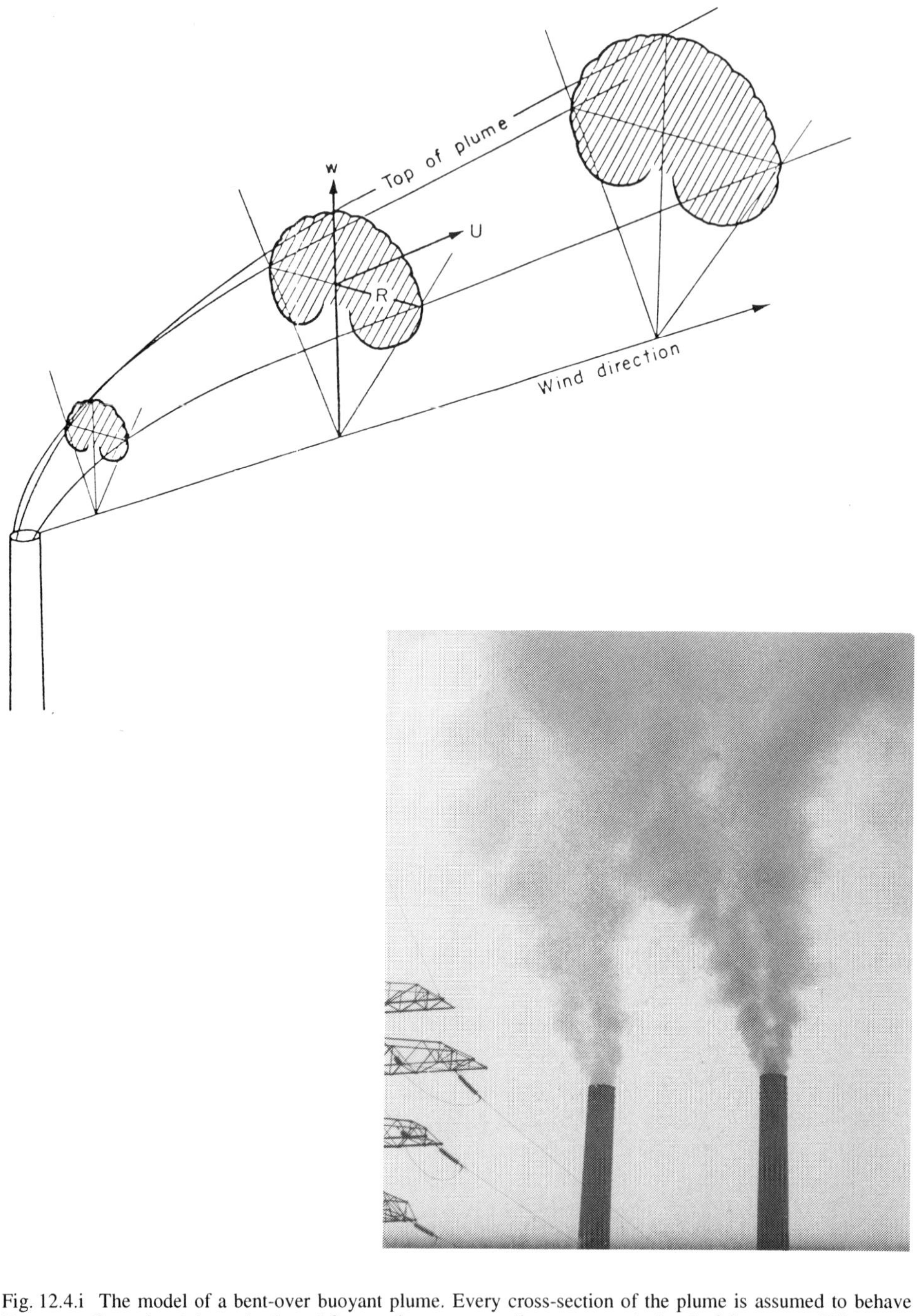

Fig. 12.4.i The model of a bent-over buoyant plume. Every cross-section of the plume is assumed to behave like a two-dimensional thermal released from an instantaneous line source at the level of the source. This means that it widens in proportion to its height above the source line and displays the bifurcated form with clear air entrained from below into the centre, and with mixing by small eddies all over the upper surface. (The theory is discussed in full in Chapter 10.)

the cylindrical puff which is equivalent to the cross-section of the bent-over plume would have originated. It is seen from section 10.15 that typically n is equal to about 4 and so the width, r, of the bent-over part is given by

$$y = 4r. \tag{12.4.1}$$

In the vertical part, by

$$\begin{aligned} w &= w_0 z_0/z, \quad w_1 = \mu U = w_0 z_0/z_1, \\ \sigma &= \sigma_0 z_0/z, \quad \sigma_1 = \sigma_0 z_0/z_1 = \sigma_0 \mu U/w_0, \\ z_1 &= z_0 w_0/\mu U. \end{aligned} \tag{12.4.2}$$

The flux of pollution is

$$F = \sigma_0 w_0 \pi a^2. \tag{12.4.3}$$

In the bent-over part the cross-section area (see section 10.15) is taken to be $2.4r^2$ and so the pollution flux, when the effluent is travelling with the wind U, is

$$F = 2.4r^2 \sigma U. \tag{12.4.4}$$

Using (12.4.1) and (12.4.3) we therefore deduce that

$$2.4\left(\frac{y_1}{4}\right)^2 \frac{\sigma_0 \mu U}{w_0} U = \sigma_0 w_0 \pi a^2,$$

or

$$y_1 = 4\left(\frac{\pi}{2.4\mu}\right)^{1/2} \frac{w_0 a}{U}. \tag{12.4.5}$$

Since σ and w are proportional throughout, the above argument is the same if applied to the conservation of upward momentum.

In the bent-over part, according to (10.8.8)

$$w = w_1 y_1^2/y^2, \tag{12.4.6}$$

so that the height y' at which the bent-over jet becomes passive, where $w = \lambda U$, is

$$\begin{aligned} y' &= y_1 (w_1/\lambda U)^{1/2} \\ &= 4\left(\frac{\pi}{2.4\lambda}\right)^{1/2} \frac{w_0 a}{U} \cong 4.5\ w_0 a/\lambda^{1/2} U. \end{aligned} \tag{12.4.7}$$

Taking $z_0 = 5a$, i.e. $n = 5$ for a jet, the rise from the orifice before the plume becomes passive is equal to

$$h' = y' - y_1 + z_1 - z_0$$

$$= a\left(\frac{w_0}{U}\left(\frac{4.5}{\lambda^{1/2}} - \frac{4.5}{\mu^{1/2}} + \frac{5}{\mu}\right) - 5\right). \qquad (12.4.8)$$

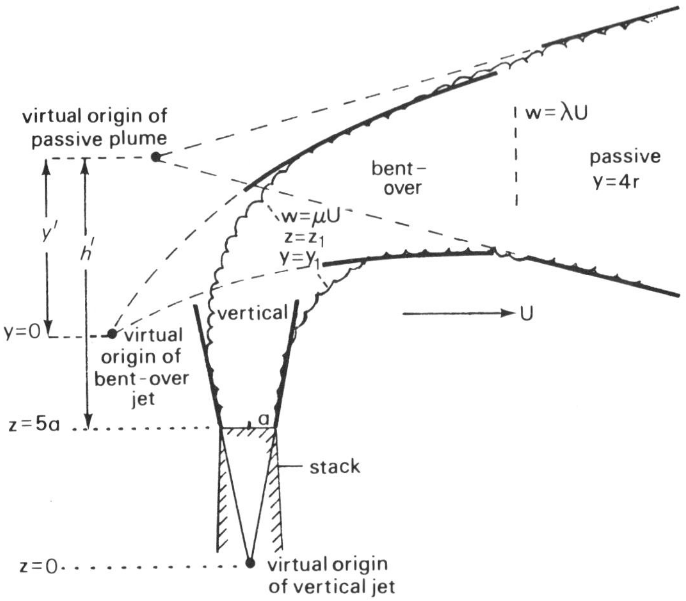

Fig. 12.4.ii Stages in the dispersion of a bent-over jet. In the jet stage the virtual origin is within the stack. It becomes bent-over after the point where $w = \mu U$, at which $z = z_1$ and $y = y_1$. It behaves like a 2D puff until the ambient eddies take over when w has decreased to λU, and has virtual origin at height $y = 0$. When $w < \lambda U$ it behaves like a passive plume, diffusing along a cone originating at height h' above the stack top, where $y = y'$.

This formula is irrelevant if $w_0 < U$, for the plume is bent over from the orifice. In discussing it therefore it is to be assumed that h' is positive, presumably because $w_0 > U$, μ is about unity and λ is much less. If the orifice is narrowed while the volume flux is unaltered we have that

$$w_0 a^2 = \text{constant} \qquad (12.4.9)$$

so that a decrease in a is equivalent to an increase in w_0. It is most effective when λ is much smaller than μ, which is in conditions of low ambient turbulence. In this most effective case the **jet rise** is

$$h_j' \cong 4.5\frac{w_0 a}{U\lambda^{1/2}} \qquad (12.4.10)$$

and so the effect of a large efflux velocity is to lift the effluent by jet rise a few orifice diameters above the stack top. This is an effective method of avoiding the capture of effluent by the eddies in the lee of the stack. When such capture occurs it is called **'flagging'** because the plume looks like a flag on the stack (Fig. 12.4.iii). It is sometimes also called 'downwash'.

Fig. 12.4.iii A 'flagging' plume. The efflux velocity is not large enough compared with the wind to prevent smoke entering the wake of the chimney. When this happens frequently the chimney becomes blackened on the outside and the effective height of the stack is reduced.

At greater distances than the chimney height from the chimney, assuming there is no flagging, the most important effect is the rise due to buoyancy after bending over has occurred. Cases in which the other smaller factors are included were discussed by Scorer [52] but we now calculate the dominant component of h', namely y', for the case of

Fig. 12.4.iv The power station at Kingston, Tennessee, where a test of a narrowed nozzle on no. 4 was made before the station size was doubled (to supply military needs). It proved more effective to extend the chimney height because less fan energy was used.

significant buoyancy. This is the **thermal rise**. In this case, instead of (12.4.1) we have roughly, according to Richards's measurements for a 2D thermal referred to in Chapter 10, that in the bent-over plume

$$y \cong 2r, \tag{12.4.11}$$

$$w = 1.0(gBr)^{1/2}, \tag{12.4.12}$$

and the buoyancy flux is

$$F = 2.4r^2 gBU. \tag{12.4.13}$$

If we put $w = \lambda U$, $y = y'$ and eliminate B and r we obtain the thermal rise as

$$h_b' \cong y' = \frac{2F}{2.4\lambda^2 U^3} \cong 0.8F/\lambda^2 U^3. \tag{12.4.14}$$

In view of the roughness of most of the approximations it is scarcely worth including the values of y_1, z_1 and z_0 in this formula. Since at the orifice

$$F = \pi a^2 w_0 g B_0 \tag{12.4.15}$$

the value of h' given by (12.4.14) exceeds that given by (12.4.10) in the ratio

$$\frac{h'_{\mathrm{b}}}{h'_{\mathrm{j}}} = \frac{0.8\pi a g B_0}{4.5\lambda^{3/2}U^2}, \tag{12.4.16}$$

and the thermal rise is much more significant than any jet rise if

$$gB_0 \gg 3\lambda^{3/2}\, U^2/a. \tag{12.4.17}$$

The most significant feature of these formulae is their extreme dependence on λ, which is probably the least well known factor. This is not a criticism of the formulae: it merely indicates that the plume rise h' is very dependent on the level of ambient turbulence. This fact cannot be altered by any other theory except one which seeks to define h' without reference to λ.

In (12.4.14), h' is seen to be proportional to F, the buoyancy flux. Measurements made by Lucas and others [53] indicate that for large chimneys the dependence is more nearly like $F^{1/4}$. But that could be due to their assumption that h' depends on a power of U different from U^3 and also to the fact that taller chimneys have larger values of F and may have a stronger wind to contend with. Furthermore, the definition of h' may interfere: for example, if it is defined as the height reached by a buoyant plume at a given distance downwind or given time from efflux, then according to (10.8.5) we have for a two-dimensional thermal

$$z \propto t^{2/3}, \tag{12.4.18}$$

while distance downwind is given by

$$x = Ut. \tag{12.4.19}$$

Therefore the height h' attained at $t = T$ and $X = UT$ is

$$h' \propto T^{2/3} \propto X^{2/3}U^{-2/3}, \tag{12.4.20}$$

which is a very different power of U from that in (12.4.14). Thus there have never been adequate measurements to check any proposed formula for h', and certainly more which have been able to eliminate the dependence of the results on the unknown power of U and on the way in which h' is defined. In this treatment the dependence upon λ is very great, which is to be expected. In practice λ is likely to decrease with height, and this again emphasises the importance of tall stacks. Somewhat different powers of U can be obtained by other models of a buoyant plume.

The formula (12.4.14) indicates that if $B_0 = 0.3$, $g = 10^3$, $w_0 = 2 \times 10^3$, and $a = 300$ CGS (which is typical for a power station) when $\lambda = 10^{-2}$ and $U = 3 \times 10^2$ (reasonable for a clear evening when the ground has begun to cool) we find that $h' = 170$ km. On the other hand, when $\lambda = 10^{-1}$, $h' = 1.7$ km.

In either case the plume rise is very large, but if $U = 10^3$ CGS (10 m s^{-1}) the results are about 4.6 km and 46 m respectively. These answers are quite realistic except that a neutrally stratified atmosphere of sufficient depth for it to happen does not exist in all cases.

In practice, however, the plume is not diffused downwards by small eddies, but by large ones. Hence all theories which given an important place to a definite value of h' must be viewed with the greatest scepticism.

Bosanquet pointed out that if $h' \propto U^{-1}$, which it was according to one of his suppositions, then when the wind was such that the maximum ground level concentration according to (12.2.1) took its greatest value, the value of h' was h. Thus a chimney emitting buoyant effluent could always be safely thought of as being twice as high as it actually was. This ludicrous result is not so much wrong as irrelevant to reality, because it is eddies large compared with plume width which effectively bring the effluent to the ground. At very large distances the chimney height makes no difference to the pollution concentration, especially if its upward diffusion is limited by a stable layer, which to a large extent it usually is at cloud base. Furthermore, if $h' \propto U^{-3}$, then Bosanquet's result is that when the wind is such as to maximise $P_{max,}$ $h' = h/5$, which means that we can safely always add on $h/5$ to a stack height in calculations of ground level pollution. This is a very different result from Bosanquet's original one and is more realistic, but still silly because both results are clearly false for infinitesimal buoyancy. This shows that the functional form of (12.2.1) is inadequate for representing thermal rise.

The upward momentum of the effluent at efflux is w_0 per unit mass. The upward momentum acquired by buoyancy gB_0 during time t after efflux is gB_0t. The buoyancy effects are therefore more important than any effects due to efflux velocity from time

$$t = w_0/gB_0 \text{ onwards.} \tag{12.4.21}$$

For an excess temperature of around 100°C, B_0 is of the order of $1/3$. For an efflux velocity of the order of 10 m s^{-1}, the buoyancy effects are dominant from about 3 s after efflux, or from 30 m downwind from the stack in a 10 m s^{-1} wind. For large stacks, buoyancy is clearly the only significant contributor to plume rise.

12.5 NEGATIVE PLUME RISE: EFFECT OF LIQUID WATER CONTENT

If a plume contains liquid water at efflux and it is evaporated subsequently as a result of mixing into the surroundings, the latent heat of evaporation is lost, and a negative buoyancy may be created. The plume then sinks to the ground, and behaves as if it were travelling with negative buoyancy (see Fig. 12.5.i) from some imaginary source higher than the actual chimney top. Fig. 12.5.ii gives the cooling effect. The isopleths give the temperature deficit which the effluent would have to have at this imaginary emission point in order to have the same temperature deficit as the actual plume after the water has evaporated. More practically it may be regarded as the factor by which the plume will have been diluted into the surroundings when it has a temperature deficit of 1°C.

Usually the liquid water is present because of a gas washing operation. Here, water is evaporated into the gases, but further cooling occurs by the loss of some sensible heat to the water. This emerges from the bottom of the washing chamber warmer than it went in and there is a further loss through the walls of the stack. The result is the condensation of a cloud of droplets. In the diagram, T_1 is the temperature at which the gases are saturated and T_2 is the temperature to which they are cooled before efflux. The side scale shows the

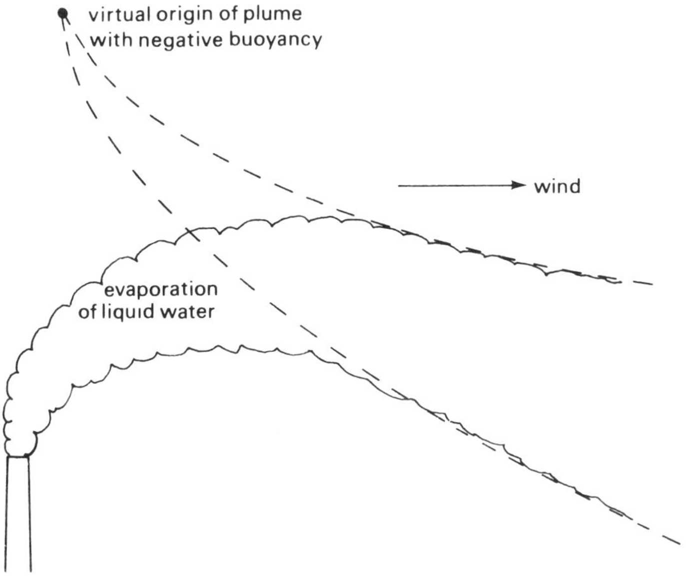

Fig. 12.5.i When a plume acquires negative buoyancy as a result of the evaporation of liquid water contained in it at efflux it behaves as if it were descending from a virtual origin some distance above the chimney top. This behaviour is common for plumes from which pollution has been removed by wet gas washing.

mass of water present as vapour in 1 kg of air when the gases are saturated at T_1 from which the mass condensed on being cooled to T_2 may be deduced. Typically, T_2 is about 25°C and T_1 about 40°C, and the cooling has the same effect as if the effluent had emerged about 80°C colder than the environment, implying that $B_0 \approx -1/3$. No surprise is warranted if such a plume descends to the ground within 5–10 stack heights of the stack, or even nearer in the light winds. An example is shown in Fig. 12.5.iii.

12.6 NARROWED STACK TOP; MULTIPLE FLUE STACK: COLD INFLOW

This is a cone-shaped addition which narrows the orifice diameter at a chimney top and thereby increases the efflux velocity and decreases the likelihood of flagging. The question arises whether this also increases the effective chimney height by giving additional upward momentum to the effluent. The virtual origin of the emerging jet is raised to about $5a$ below the top if the radius a is reduced. This is likely to be a very small advantage at the cost of the power required to eject the same amount of gas at greater velocity. If the effluent is already warmer than the surrounding air according to section 10.11 it will be come a buoyant plume travelling up the same cone very soon, and the main consideration should be to get the effluent cleanly from the stack top.

Many stacks of older design (Fig. 12.4.iii) can be seen with gaudy architectural features at the top, creating a more turbulent wake than a cleanly finished one. Sometimes

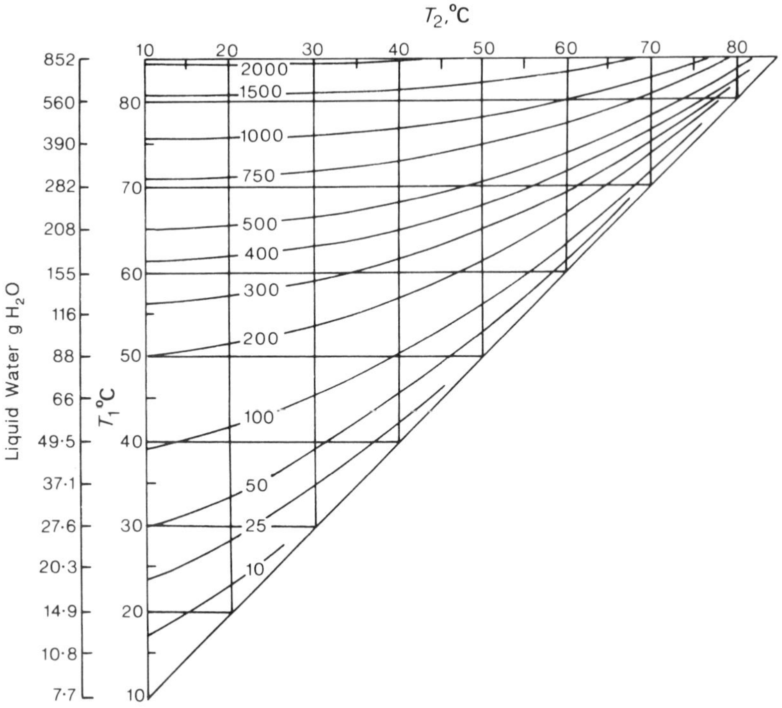

Fig. 12.5.ii The isopleths show the cooling °C which results from the evaporation of liquid water in a plume as airborne droplets. T_1 is the temperature at which the effluent gases are saturated, with no liquid water, and T_2 is the temperature to which they are cooled before efflux, the liquid water being condensed as cloud. Alternatively, the isopleths may be thought of as the dilution factor at which the plume would be 1°C colder than it would have been if it had contained no liquid droplets at efflux. For example, if the saturation temperature of flue gases is 60°C and they are cooled to 33°C before efflux then after the liquid water has been evaporated the plume will behave as if it had been emitted 300°C colder: or, after being diluted 300 times by ambient air, it will be 1°C colder than if it had been emitted without liquid droplets. The side scale shows the mass of water vapour present in grams with 1 kg of air when the air is saturated at T_1.

a flat disk is built round the orifice, and this prevents the effluent entering the immediate wake (see Fig. 12.4.iii).

The benefits of buoyancy in the effluent and a large stack height can be combined in a multiple flue stack. When all flues are in use the buoyancy is added to give an enhanced thermal rise when compared with that of each on its own, and retains the height advantage when only one flue is in use. If the multiple stack is not built higher than each built separately there might be an increased ground level concentration in a very strong wind which, being correspondingly turbulent, would cause downward diffusion at an earlier stage than in lighter winds, thereby preventing the occurrence of useful thermal rise. But it should be noted that the dilution by increased wind strength would reduce this effect.

If a bottle of water is held upside down it empties and air flows in, but if a downward-pointing tap is turned on, the efflux velocity of the water is usually too large to allow air

Fig. 12.5.iii The washed plume of Battersea power station being brought to the ground as a result of negative buoyancy.

Fig. 12.5.iv When only two sets (of four) were operating and only one was operating washing—the unwashed plume exploits its buoyancy to rise through an early morning inversion; while the washed plume is bifurcated also, but is trapped at the inversion as at a ceiling, and the vorticity carries the two halves apart. (This phenomenon became an attraction to visitors.)

Fig. 12.5.v This appears to be a plume from a gas furnace with a large component of water vapour, the condensation of which creates buoyancy. The bifurcation is very prominent. [Photo by Ryuji Kimura, Tokyo.]

Fig. 12.6.i Eggborough power station with its early multiple flue stack (206 m). With the four flues, its corners were rounded to give better aerodynamics. Each flue has a nozzle at the top to give an exit velocity which avoids flagging.

to penetrate into the pipe. Therefore at some value of the efflux velocity as it is decreased the air must begin to flow in. The same phenomenon occurs at a chimney mouth: when gas of large buoyancy is emerging with a rather small velocity, cold air from outside can penetrate down the stack. The cold inflowing air is mixed with the rising warm gas as it penetrates down, and eventually it reaches its maximum penetration beyond which its much reduced negative buoyancy is too small to carry it against the upcurrent.

The phenomenon was studied by Jorg and Scorer [54], who found that the critical mean efflux velocity below which inflow occurred depended not only on the buoyancy of the effluent but also on its velocity profile. If the velocity in the boundary layer was increased by flattening the profile for a given volume flux, the buoyancy difference needed for inflow was increased. No non-dimensional relationships were obtainable for this reason, but it was clear that in turbulent flow the diameter of the orifice was not generally a determining factor, although the thickness of the boundary layer was.

The distance d penetrated into a tube by exterior fluid was found roughly to fit the formula

$$\frac{Bg\nu}{\left(d^2/D^2+8\right)^3 V^3} = 10^{-6} \qquad (12.6.1)$$

where D is the tube diameter, V the mean velocity far inside, B the buoyancy and ν the kinematic viscosity. Although this provides an order of magnitude for the critical velocity at which d is zero for a given B, and provides also an idea of the depth of penetration, it has considerable limitations in the spread of values upon which it is based. This is because there are many extraneous influences such as tube interior roughness to be reckoned with. In practice there is nearly always a crosswind, and then the experiments showed that the inflow usually took place at the sides rather than at the up- or downwind points of the orifice.

The problem has been a real one. Oil-fired furnaces used to produce black acid smuts, particularly when starting up from cold and with metal stacks. On low load the efflux velocity was reduced in greater proportion than the buoyancy, and so (12.6.1) shows that penetration is more likely. Also, when penetration occurs in cold weather it can be enhanced by cooling through the stack wall. The result was that the chimney wall was cooled below the acid dewpoint (H_2SO_4) and black carbon deposited on the wall soaked up the acid. When the efflux velocity was then increased, particles a few millimetres in size were carried out by the airstream. The concentrated sulphuric acid burnt holes in clothing, in the paint of car roofs, and in other objects. The remedy has been to maintain sufficient efflux velocity and to insulate the chimney so that the temperature is kept above the acid dewpoint. Various bevelled tops have also been found to reduce the likelihood of cold inflow.

Although cold inflow does sometimes occur in cooling towers, the more likely phenomenon is separation of the outflow from the wall at the level of smallest diameter in a hyperboloidal tower. Actually, very little is known about inflow and possible penetration to the bottom in cooling towers. However it is probably not a serious problem anyway because of the fairly uniform distribution of the heat source horizontally across the bottom in a conventional wet cooling tower. Thus the heat source distribution prevents a large fraction of the volume becoming cold.

The picture in Fig. 12.6.ii is of the experiment from which (12.6.1) was derived.

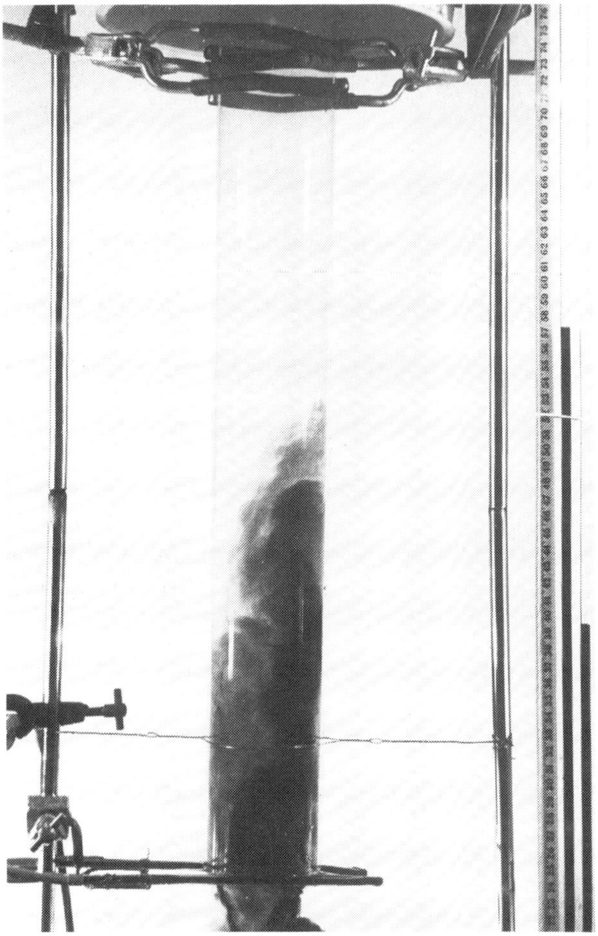

Fig. 12.6.ii A model experiment of cold inflow by Jorg (Jorg and Scorer [54]). Salt water is descending the glass tube (8.5 cm outside diameter) which is immersed in a tank of fresh water. Around the open bottom of the tube are six jets of dye. Any inflow of the buoyant fresh water up the tube carries dye with it, and the penetration can be observed. In this case it is a little more than 4 tube diameters.

Experiments in air with a wind blowing across the top of a tube emitting hot air and smoke emitted close to the lip to reveal inflow showed that details of the velocity profile of the emerging air were important. Applications to cooling towers were also considered and the unfortunate experiences at Rugeley power station where the heat exchangers were built as a vertical wall around the base of the cooling tower. One result was that when snow was falling it tended to accumulate on the upwind side and block the flow through

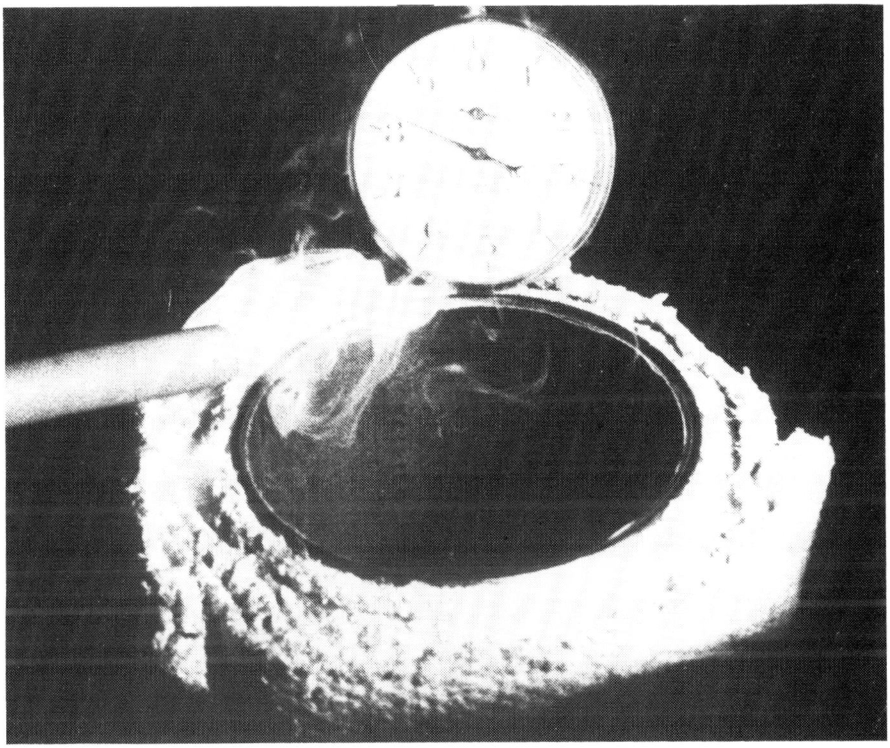

Fig. 12.6.iii This was part of the same series of experiments which gave (12.6.1). It was used to study the effect of a 'cross-wind' across the top of a tube emitting warm air, and the inflow was detected by cigarette smoke.

the exchanger with snow and the water inside became frozen so that the flow of warm water ceased.

The sensible place for the exchangers was horizontally across the base of the tower with free flow of air underneath. But because of difficulties of construction and maintenance that alternative was deliberately rejected. It was not envisaged that the exchangers would become frozen on the upwind side!

Most cooling towers exploit the principle of the natural draught, which carries the air upwards, and it was asking for trouble by planning that it should provide horizontal suction and to have no problems with driving snow.

12.7 THE SHAPE OF CHIMNEY PLUMES

Any observer soon notices that there are several different shapes of plume. Several authors have sought to list the distinct shapes by their appearance, on the following lines (see Fig. 12.7.i):

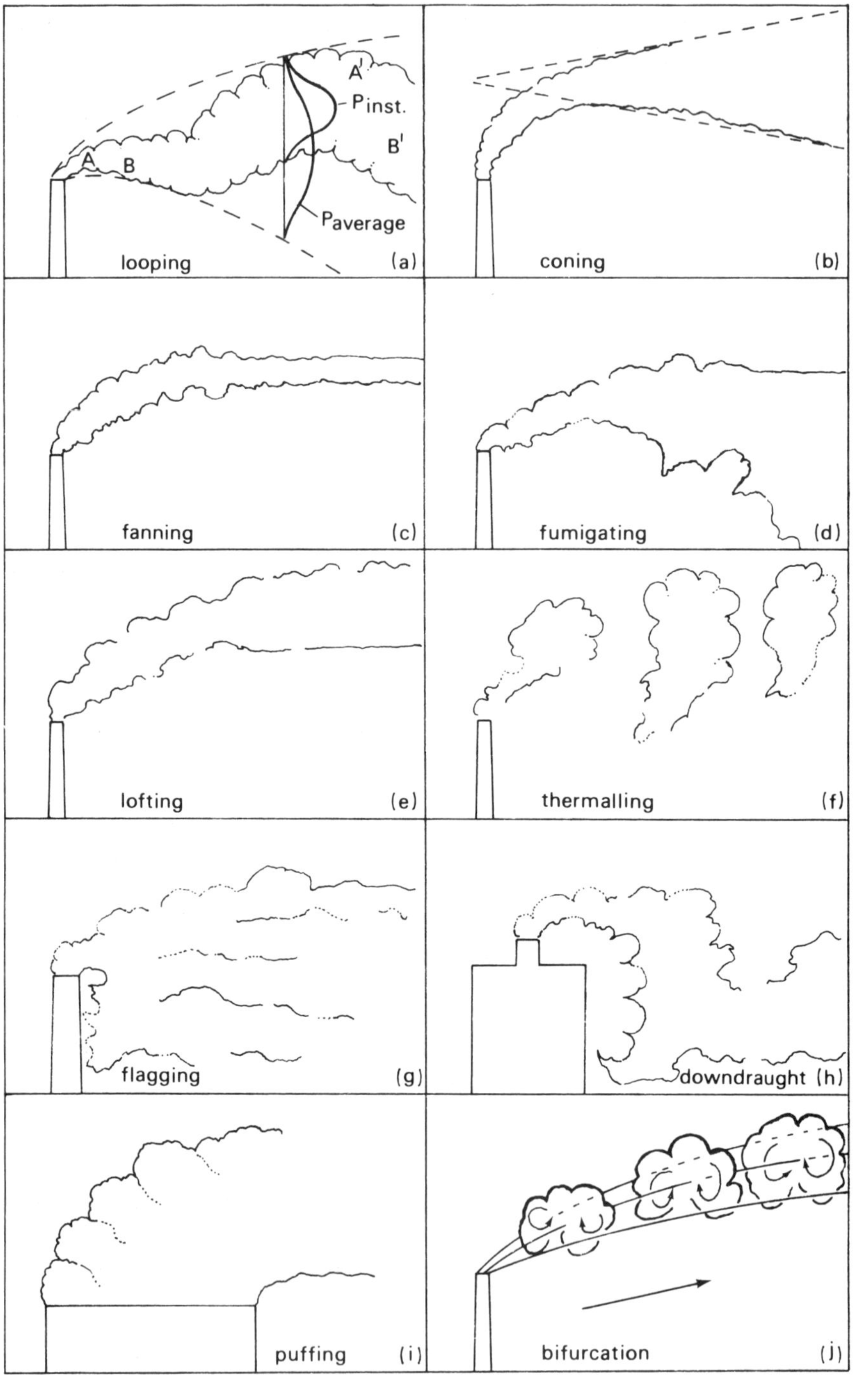

Fig. 12.7.i Names given to various plume shapes (see page 409).

(a) **looping** having large sinuosities: dispersion by eddies large compared with plume width. See 12.2.iv.

(b) **coning** having a steadily widening boundary: dispersion by eddies of size smaller than the plume width. See 12.4.iv.

(c) **fanning** dispersing horizontally more than vertically: plume achieves equilibrium level in a stably stratified airstream soon after emission. See 12.7.ii, iii, iv.

(d) **fumigating** dispersing downwards but not upwards: emission below an inversion which is not penetrated. See 12.5.iii.

(e) **lofting** dispersing upwards but not downwards: stack height sufficient to place effluent above an inversion. See 12.9.iv.

(f) **thermalling** the plume breaks up into distinct masses indicating that the buoyancy from the chimney is initiating thermals or being entrained into natural ones.

There are other features such as:

(g) **flagging (downwash)** effluent entrained into the eddies in the lee of the chimney. See 12.4.iii.

(h) **downdraught** effluent entrained from time to time in the lee of the building associated with the chimney. This is common with dwelling houses.

(i) **puffing** regular formation of cauliflower-like lumps at the upwind edge of the chimney mouth. These usually have a periodicity of 1 to 5 puffs per second.

(j) **bifurcation** this is due to the entrainment of clear air up the middle of a plume bent over by the wind so as to make it have a cross-section like a cylindrical thermal of puff. See 12.5.iv.v, and 12.4.i.

There now follow comments on these various patterns.

The large eddies which cause a fairly narrow plume to wander about within a bent-over cone may have been shed by large buildings or hills, or generated by thermal convection in sunshine. As far as any formula for dispersion is concerned such a plume is treated as if it were distributed over the **bent-over cone** which envelops it. The particles of the plume do not travel on sinuous tracks, but the particle at A travels to A′, and B to B′ (Fig. 12.7.i); thus the sinuosities move along with the wind, and the sinuous plume itself is widened more slowly by smaller eddies. There is an absence of eddy sizes in between, and this is usually because the smaller eddies are produced by the buoyancy of the plume or by small roughnesses on the ground, while the larger ones are produced by thermals or are shed from large obstacles.

A **coning plume** is usually **bifurcated** and rising under its own buoyancy, and it appears to have the shape of a bent-over cone only when seen from the side. From beneath it is usually bifurcated very clearly (Fig. 12.4.i). Indeed, a plume seen from below to be bifurcated is certainly rising because of its buoyancy, and the eddies which mix it into the surroundings are due to the buoyancy and are not present in the ambient air.

If a buoyant plume is emitted into an airstream which is strongly stratified it starts off bifurcated (and coning) but soon reaches its equilibrium level and then travels horizontally. The stable stratification is most common in rather light winds whose horizontal direction may vary, and then the plume appears to wander horizontally. This does not always happen, so that a behaviour appropriately called **fanning** is rather rare (Figs 12.7.ii and iii).

When the wind is very light, a plume may rise as a vertical hot plume and spread out horizontally on reaching its equilibrium level (Fig. 12.7.iii). The top of the plume usually overshoots a short distance, and because there is often a wind discontinuity at a very stable layer the plume is carried away as a thin layer of polluted air.

When a polluted layer is formed and thermal convection begins in the morning causing **fumigation**, the pollution is sensed at ground level over a wide area at about the same time. For this reason the source may not easily be identified, especially if the pollution has accumulated during the hours of darkness. This mechanism was first identified by Hewson when a smelter at Trail (B.C.) caused pollution to be observed about 50 miles down the valley of the Columbia river in the morning without it having been observed previously at the ground in between. The rise in pollution levels at the ground may be quite sudden because the morning convection may produce anabatic winds up the side of the valley. These spread out above the polluted layer and displace it downwards in the centre of the valley without any dilution (Fig. 12.7.iv).

When a plume is emitted above an inversion, as is often the case when there is a cold layer of air at the ground in the evening or at night, the ambient turbulent is usually very slight. Then the plume rises under its own buoyancy, with bifurcation, until it reaches the cloud base or a stable layer. **Lofting**, according to the scheme in Fig. 12.7.i, is rare because it includes situations in which the plume is dispersed upwards but not downwards by ambient turbulence. Moreover turbulence greater than that due to the buoyancy of the plume itself is rare above a very stable layer because the stability is seldom zero there.

Thermalling occurs when thermals are being generated by sunshine on the ground, the thermals being the dominant form of eddy. A plume of large buoyancy initiates its own thermals and so all the pollution is likely to be carried upwards while dispersion downwards is negligible.

Flagging is a feature mainly of plumes of low buoyancy and low efflux velocity. In the case of some chimneys, vortices with vertical axes are shed from the lee side and pollution can sometimes be seen to be entrained into the centre of these and carried away downwind. It is important only when a significant fraction is drawn down the outside of the chimney most of the time. In the case shown in Fig. 12.4.iii the chimney was

Fig. 12.7.ii The plume from the copper mine at Mt. Isa, Queensland, seen early in the morning. The mixing ceases when the plume settles at its equilibrium level. In the distance the plume can be seen just above the skyline. Fumigation occurred about an hour later at the airport.

Fig. 12.7.iii The plume of an isolated chimney on a calm day over the Humber estuary. (Photograph Ron Harrison.)

blackened down to about half its height. Some architectural designs cause flagging, but the problem of teaching architects that a chimney should be designed to emit the pollution cleanly and clear of the building has been alleviated in the last two decades.

Fig. 12.7.iv The plume from the old power station at Ironbridge. The shear in the flow above the stagnant valley air can be seen in the overshooting elements. This, like Fig. 12.7.ii, was followed by fumigation.

During these years some architects have seen their job, at last, as taking account of these phenomena, and the problem of architects has become a problem for architects.

Downdraughts can be avoided by making the chimney stand sufficiently clear of the building. Many chimneys fail to emit so as to ensure that pollution is clear of all eddies due to separation of the airflow from salient edges on the building. A special case is that of the plume from a power station employing gas washing where the descent of the effluent to the ground was clearly due to the negative buoyancy. The old rule of thumb that a chimney should be $2\frac{1}{2}$ times as high as the building on which it stands is based on nothing except its acceptance at a time when no other guidelines were offered. Sir David Brunt told me that he invented it on the basis of some observations of balloon trajectories over a hill, in which they seemed to be unaffected if they were more than $2\frac{1}{2}$ times as high as its top. That, of course, was a quite irrelevant observation, and even generalisation from it about flow over hills is incorrect. The $2\frac{1}{2}$ times rule is irrelevant for buildings which are tower shaped, and wind tunnel experiments are then recommended: but it is important that a wide variety of turbulence characteristics should be tried out in the wind tunnel airstream, otherwise possible natural events may not be produced in the model. However one should not take too seriously the problem of how the turbulence is produced because the variety of eddy forms, including cold downdraughts, is so great in nature that they cannot all be simulated by one device. It is important to experiment with a variety of obstacles, some moving, upstream of the working section in order to find out what might possibly happen if the model were built on the full scale.

Puffing is a phenomenon of little important in practice and its causes have not been studied, although guesses at its cause have been made.

12.8 SAMPLING OF POLLUTION

The weather and associated physical parameters, namely wind speed and direction, temperature and humidity, are varying enough to make it difficult to define them without reference to some standard instrument. An objective definition of any usefulness does not exist because the volume of air sampled and the duration of the sampling are finite, and anyway if small samples are taken it is not useful to know the details of the variations from one very small sample to the next in any particular case. Therefore most definitions state the manner in which the air shall be sampled, and, as we have discussed in Sections 12.1 and 12.2, there is no definition that does not to some extent, and often significantly, depend on the volume of air sampled.

Pollution introduces further variations of greater magnitude on smaller time and distance scales. This is because of our nearness to sources which increase the pollution content by several orders of magnitude. Furthermore, sinks of many forms of pollution exist both in natural and in artificial surroundings.

We therefore find it useful to define a **background level of pollution**, originating from distant sources and not varying over small distances in the air because it has been well mixed, except near sinks. (For example, the SO_2 concentration is decreased in some cases by a factor of 2 or 3 when the air passes through a hedge or well wooded patch of ground.) In addition we have **pollution due to sources near at hand** which may arrive in whiffs. For example, the pollution due to a sinuous plume may be absent at one moment and at its maximum 30 s later. Thus the average concentration over a time long enough to include many sinuosities would show a profile like P_{av} in Fig. 12.7.i(a), while if taken in a few seconds the profile would be like P_{inst}.

In practice therefore if the averaging time is to bear any relation to concentrations expected according to a dispersion formula it should be several times the time taken for a particle to travel with the wind to the point of observation. At the same time it has also to be noted that the concentration in whiffs could well be up to 10 times the average, the duration of the whiffs being determined by distance from the source. If any visible plume is observed the nature of these variations can be seen on any actual occasion. On the other hand, the average for 24 h, a month, or a year, may easily be around a fifth, a hundredth or a five-hundredth of the hourly or three-minute average. This is mainly because of the distribution around all points of the compass and the inclusion of many cases in which there is no pollution near the stack because of lofting *et cetera*. These multiples vary from case to case, and make difficult any attempt to maximise the pollution emitted into the environment. That policy is sometimes advocated on the quite spurious supposition that there exists a level up to which the pollution concentration may be raised without any harm, but beyond which harm begins. Such a level cannot, of course ever be discovered, and in any case the continual variations would be a factor determining its value. Monitoring the actual level is a very crude business unless a large number of samples is used, and then the quantity of information obtained is oppressively voluminous unless handled by a computer, in which case intelligent treatment of it is generally precluded. There is no substitute for lively and perceptive observation using a few monitoring instruments only, for visual observation of pollution incidents and behaviour of visible plumes, and for an intelligent application of simple principles of natural aerodynamics.

12.9 THE FATE OF POLLUTION

(1) Primary and secondary pollution

When air pollution is first emitted it is **primary** pollution. We may follow what happens to it by the example of smoke, which is detected visually, or in the case of CO_2 which is readily detected when mixed with compounds of sulphur by their smell; and so we may have a mental picture of what happens to other pollutants, invisible or odourless, such as CO.

They may seem to be diluted into the whole of the **mixed layer** which, it will be seen in the next chapter, usually extends from the surface up to the cloud base, when convection is occurring. But there are several mechanisms which influence the stratification of the air and the mixing of pollution into it, which are determined by the transfer of heat by radiation, including sunshine, and the associated effects of clouds.

Many pollutants retain their own physical and chemical characteristics; but some are used up in the production of **secondary** pollution. A good example is **ozone**, which arises from chemical reactions in sunshine, usually via the production of monatomic oxygen which quickly combines with diatomic oxygen to form ozone. But O_3 itself is a very reactive substance and in particular it quickly converts NO to NO_2, and both of these are included in the collective term NO_x. Sunshine plays an important part in these reactions. There are many other secondary pollutants of which a very important example is SO_3 in its various forms as sulphate or sulphuric acid.

There are many other minor components of the atmosphere which are always present and chemically active. Chief among these is OH and some other products of water vapour.

Water vapour is always available in plenty in a mixing layer, and it is appropriate to note that Brunt [7], in his famous teaching text *Physical and Dynamical Meteorology*, advised the student to regard the atmosphere as water vapour diluted to various extents by the other components of the atmosphere because the condensation to water droplets or ice particles, and the reverse process of evaporation, provide the most important source and sink of heat which dominates the motion above the condensation level (cloud base). These processes make water vapour a very variable component which commonly ranges from about 0.1% to as much as 3% by mass of a sample of air in the troposphere. CO_2, which is also universally present, shares with water vapour the most important radiative role, but varies on account of biological activity around about 0.035% by mass. Because of condensation and the fallout of rain, hail, snow, etc., air entering the stratosphere is generally less plentifully supplied with water vapour (usually less than about 0.1%) but has its usual complement of CO_2. Owing to the continuous biological production of methane (CH_4) at the surface it is carried into the stratosphere where it reacts (largely in sunshine) to produce H_2O and CO_2, and this may be responsible for some regions where the water vapour content is higher than average, which in this case is a form of secondary pollution.

(2) The mixed layer

When we consider the role of pollution on the global scale in Chapter 14 we shall be concerned with the ultimate fate of air pollution. In this chapter we are concerned with the layer in which most pollution remains, except for that which is carried higher only to be rained out fairly soon.

The mixed layer, therefore, is usually the layer up to the cloud condensation level, which is mixed mainly by convection currents over warm land (in sunshine) or sea (where the diurnal variation of temperature is very much less because the thermal capacity of the layer of water involved is much greater than of the thin skin of the land surface where the diurnal variation of temperature is usually very much larger).

Within the mixed layer is a thin layer at the bottom which becomes colder at night when there is no sunshine to counteract the continuous loss of heat by radiation to the sky. Then, pollution emitted from the low-level sources, such as cities and all that goes with them, may be confined to a shallow layer and is visible as a polluted layer, often below the tops of urban hills or the tallest buildings. During the day, even when it is cloudy, enough heat gets through so that the pollution is again spread by turbulence due to convection and the wakes of buildings etc., to fill the mixed layer with any pollution that has not yet been mixed in. The depth of this layer is typically 600 to 1000 m.

(3) Smog

This term originally meant a mixture of smoke and fog which were supposed to be respectively man-made and natural. But this convenient word has been used, particularly in the USA, to refer to the photochemical variety which is mainly due to motor vehicle exhaust acted upon by sunshine to produce a form of secondary pollution of which the most important components are $\mathbf{O_3}$ and **PAN** (peroxyacetylnitrate) in which aggregated molecules form a haze of tiny particulates (see Figs 12.9.i–iv).

The feature of smog is that it is usually trapped in a layer of stable air under an inversion, and, in the case of a fog condensed on smoke particles particularly, is prevented from dispersion upwards by the continued loss of heat from its upper surface. It is most common where the transport of air is inhibited by mountains which keep the polluted air in the same place for many hours. Smog is a very appropriate title when the top of the mixing layer is the top of a layer of cloud, for then the radiation from the cloud top causes downward convection which keeps the layer cool but under convective motion so that the cloud layer is as polluted as the air below.

An example of a 'well-cooked' smog in Mexico City is shown in Plates 2 and 3 in the colour section between pages 352 and 353.

(4) Policy

Science is seen as knowledge which enables an industrial activity to be carried on with the maximum efficiency. Clearly it is laudable to improve the efficiency in the use of resources, raw materials and, of course, labour. Efficiency may usually be defined in such contexts according to some conventional scientific principle. Thus, for example, a steam

Fig. 12.9.i A typical Los Angeles photochemical smog composed by secondary pollution products from car exhaust. (Courtesy Los Angeles Air Pollution Control District.)

Fig. 12.9.ii The Crystal Palace television mast protruding through the top of the Great London Smog of December 1952. (Courtesy Fairey Air Surveys.)

Fig. 12.9.iii The plumes of power stations penetrating above the top of the Great London Smog. (Courtesy Fairey Air Surveys.)

Fig. 12.9.iv The two plumes from Brunswick Wharf power stations penetrating the Great London Smog. One plume was blackened for ease of identification. (Courtesy Fairey Air Surveys.)

piston engine which does not make use of two or three stages in the expansion of the steam is being wasteful of the fuel used to raise the steam. But one of the necessities in industry is to get rid of the gaseous products of the combustion of fuel or of some other form of air pollution connected with manufacturing.

It is said by economists that the environment can absorb a certain amount of air pollution before it suffers unacceptable damage and that it is expensive to reduce the quantity of effluent or to process it so that it is chemically less harmful. The extent to which these damage-reducing procedures should be used depends on balancing the cost against the value of the 'benefit achieved' which to an economist is a euphemism for the reduction of the damage caused. The cost of some damage limitation is such that according to this principle it is not worth the trouble. The consequence of that is that just as some use of the world's fossil fuels is worth the ultimate result of using it all up for ever, so some permanent deliberate deterioration of the environment is economically advisable. Even cake is only work making if it is to be eaten.

That type of argument for minimising the 'unproductive' processing or reduction of waste emitted as a by-product may require aesthetic judgments which are arbitrarily measured by the vociferousness of the objections to the unprocessed emissions. But how can a scientific or technological judgment be justified in order to set a limit to unprocessed emissions by letting them increase until the objections become intolerably vociferous? It has been conventional to let environmental insult to occur until the concentration can no longer be ignored. The only way to manage such a system is to make every polluter pay according to the amount of pollution emitted according to a scale of payment which has never been properly tested. Anyway, many cleaning up procedures are allowed to be delayed on the grounds that capital has been expended earlier in the history of the industry and should be allowed to earn its expected share of profits. And to what extent should new industries be allowed to join existing ones when the planned limit of environmental insult has been almost reached? But we shall now raise purely scientific objections to the philosophy embodied in the proposals just outlined.

Given a certain level of pollution already to have become acceptable, in the sense that the authorities have not yet imposed a ban on further emissions, at what stage and by what criteria can a limit on emissions be set. If the site is close to a major source of a resource (e.g. Mt. Isa, Queensland, or Sudbury, Ontario, for non-ferrous metals; West Virginia or Silesia for sulphurous or ash-loaded coal; Biafra Nigeria, land destruction for oil) it may be argued that it is obviously more efficient to have as much production at that site as possible.

Scientists may be asked to advocate an increase in permitted levels: it maybe common that imagined episodes have not occurred in the most recent 2 or 3 decades, which would be grounds for believing that it will not happen 'in the foreseeable future'. But the complexity of reality is not made simple by refining the definitions used to formulate criteria. Real situations cannot be effectively subjected to this treatment, and the pollution problems are the result of the weather that occurs. There is no prospect of providing such detailed forforecasts as far ahead as the lifetime of a town or an industry.

Although there is some confidence that new technologies will perform as satisfactorily as envisaged in the design, circumstances may require certain design features which it may be hoped will prove never to have been necessary. Carelessness and lax maintenance are not planned. There is no certainty, and prohibitions and penalties are often delayed for economic reasons.

The above discussion may be altered if the polluting system is required to be capable of being shut down quickly if the weather expected to be unsuitable and unable of carrying away and diluting the normal effluent. This **meteorological control** has been used occasionally in several industrially advanced countries but has usually not been imposed soon enough to protect the limits agreed in advance. The relevant situation today is likely to involve a major contribution from road traffic and strict control may be regarded as worse than the inconvenience of smog. Smogs are not a good investment for public health, and even agriculture is seriously threatened by high levels of ozone.

12.10 TRANSPORT OVER LARGE DISTANCES

Even when the pollution has been confined to the mixed layer over large distances it is probable that remaining in that layer has been the major cause of it being much more widely diffused.

After a long journey the pollution may be concentrated again by the convergence into a shower cloud and subsequent deposition by being collected in the rain and absorbed by the vegetation. Such was the fate of much of the pollution from the nuclear accident at Chernobyl (Scorer 1990, Chapter 10) although there is no known forecasting method which could have forewarned the ultimate recipients of serious concentrations of radioactive material. Other occasions of similar transport have been difficult to trace back to their sources, such as the snow containing black smoke from Eastern Europe deposited on the Scottish Highlands (see Chapter 1 of Rose (1994), where the collected acid in a winter season's snow on the high mountains of Scandinavia is released in high concentrations in the spring melt, with occasional extensive kills of fish of the trout family in the rivers).

12.11 GLOBAL CONSEQUENCES

The detection of very recently emitted pollution in the air and on the ground in remote places such as Antarctica, and in the food of penguins from there, and also in the cores from Greenland showing 'industrial pollution' from Graeco-Roman times [55] all fostered the realisation that modern industry might cause sufficient change in the composition of the atmosphere which would precipitate a change in climate. The mechanism of such a change will dominate the later chapters of this book; here, we report some of the natural production of air pollution which will cause problems in the making of mathematical models of the global atmosphere.

Forest fires have been prevalent throughout all recent evolution, and naturally they have been treated as disastrous and quite probably due to human carelessness or intent. It was fashionable (or politically correct) for green movements to argue in Australia that the litter from trees should be allowed to lie on the ground and be absorbed into the soil to facilitate the growth of the younger trees.

European forests from which the trees have been harvested and new saplings planted have been found to suffer from lack of nutrients which the forest had previously obtained from the remains of fallen old trees. The spread of ages among the trees enabled a forest with a wide spectrum of ages to survive in a way that is impossible if a forest is replanted after being harvested and all the trees are of approximately the same age. This idea justified the policy of the Australian Greens, but the recent disastrous fires in Victoria and New South Wales have forced a reconsideration of the problem and the discovery that because of the prevalence of occasional drought naturally caused fires were inevitable, and the trees which survived to form the next generation were those whose seeds survived and came to require a fire to prepare them for blossoming when the rains made the ground suitable.

Of course the realisation of this characteristic was too subtle for the first generation of Green politicians, but is now a classical example of 'why things bite back' (reviewed in [56]) if we are too simple minded about the mechanisms which evolution has given to the biosphere. From the point of view of the trees, the Australian 'bush fires' were not natural disasters and the European foresters will have to develop a system equivalent to the rotation of crops to which the farmers of the industrial world will have to revert in due course when fossil fuel becomes too scarce to use up in making chemical fertilisers.

The role of haze in the evolution of the biosphere is important to understand because it is a significant factor in responding to the threat of global warming (Fig. 12.11.i). For example, the study of ship trails suggests that every continent supplies a plenitude of CCNs, and in particular these may be the Aitken Nuclei which seem to be almost

Fig. 12.11.i Haze extending the albedo of layer cloud over the Jura valleys of Western Europe.

everywhere, and are produced by every example of combustion (Fig. 12.11.ii). If this is so it is a waste of time looking for a way of increasing cloud amount by supplying CCNs artificially. Indeed it is a warning that there is unlikely to be a realistic device of that kind to alter the weather in any significant way.

Fig. 12.11.ii Straw burning by farmers provides a plethora of Aitken Nuclei (CCNs) and adds to the albedo by covering the land with a layer of haze!

13

Radiation: the climatic determinant

There is one basic idea which lies behind all our enquiries into the flow of energy through the atmosphere, because all the activities within the biosphere would appear to violate the second law of thermodynamics by their creative nature. But there is no violation because the biosphere is not an isolated system. The earth receives sunshine and emits infrared (IR) radiation outwards into space, and there is therefore an approximate energy balance because there is neither an increase nor a decrease of heat energy in the long term on the earth. Even that equality of input and outflow of energy does not seem to have been strictly adhered to in the long run at times when the climate was entering or coming out of an ice age.

The average world temperature, however it may have been defined for climatic studies, certainly seems to have fluctuated over the individual years and over the decades; and this implies that in any individual year there may be a deficit or surplus of heat energy to be got rid of to maintain the climatic state. But if the climate changes from glacial to interglacial it is reasonable to assume that one or more of the factors which determine the inflow or outflow of radiative energy has been changed. This has become a special concern of science at the present because such a factor may have been altered by the activities of humanity. We would call this an *artificial* factor.

13.1 THE DOWNFLOW OF SUNSHINE

The spectrum of radiative energy flux is determined by the temperature of the emitting body which, in the absence of any special surface characteristic, will be a **black body**. Fig. 13.1.i shows the spectrum according to Planck's theory, in which the horizontal coordinate is the product of the temperature and the wavelength. The same diagram will be appropriate whether we are discussing sunshine (solar temperature approximately 6 000 K) or the emissions from the earth (at roughly 300 K).

Sunshine has an **ultraviolet** (UV) component from about 0.2 μm to about 0.4 μm. The radiation is **visible** from 0.4 to 0.7 μm, and the maximum intensity is at about 0.5 μm. The energy in the **infrared** at wavelengths greater than 0.7 μm is roughly one third of the total. There is very little energy in the solar spectrum at IR wavelengths greater than about 0.4 μm.

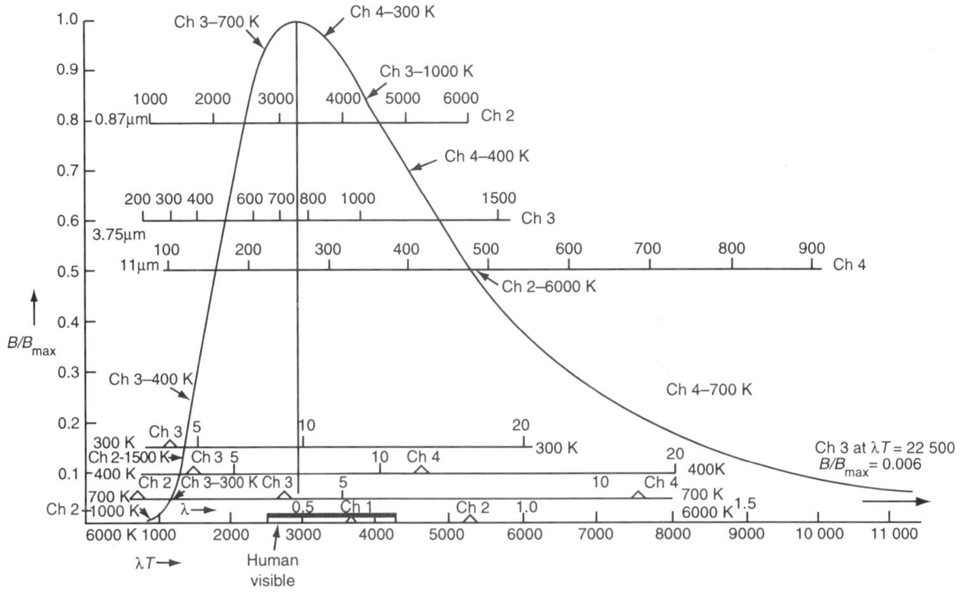

Fig. 13.1.i Planck's spectrum of black body radiation. This curve gives the intensity of the radiation from a black body as a function of the wavelength and temperature. The scale below the bottom is λT and from that is derived the scale above the bottom line for the sun's temperature, 6000 K. The next line up is for the emissions from the earth at 300 K. The three upper scales give the temperature the intensity is achieved at the wavelengths used by the meteorological satellites (see section 13.11). It is clear that for channel 4 the intensity is very sensitive to temperature, and the pictures in that channel display temperature variations.

The atmosphere above the tropopause strongly absorbs the wavelengths shorter than 0.3 μm. This is mainly due to oxygen (O_2), which becomes monatomic and combines with diatomic form to form ozone (O_3), which itself absorbs some of the UV energy and may be broken down into its components, so that there is an approximate equilibrium concentration of O_3. Thus the oxygen in the stratosphere is partly converted into O_3 which in turn absorbs almost all the sunshine in wavelengths less than 0.3 μm, which is referred to as UVB. The removal of some of the O_3 by chlorine and oxides of nitrogen (NO_x) has led to fears that the exposure of the human skin to bright sunshine would be a cause of an increased incidence of skin cancers (particularly melanomas). The amount of energy in the UVB is only a very small fraction of the total and is of no concern in that context because it is less than the inaccuracies of the treatment of the visible and IR wavelengths.

In the troposphere there is very little absorption of the solar beam, except to note that UVB is absorbed by aerosol particles, smoke pollution and ordinary clouds if it penetrates through the stratosphere. Clouds exposed to sunshine become brightly illuminated and cast strong shadows, and much of the sunshine is reflected (diffusely scattered) away from the earth and constitutes an important fraction of the Earth's emission to space. This fraction is called the **albedo**, or whiteness, and for clouds this may be up to about 85%. A water surface has an albedo of only about 20% for a beam incident vertically downwards, but it could be as much as 85% for a nearly horizontal glancing beam, at sunset for instance. For vegetation it may be over 70% when it is ripe and dry, but as low as 25% when moist. Dry soil or sand is much more reflective than wet ground. A wet road may be like a mirror, but a disturbed sea surface scatters in all directions according to the local slope of the surface. Clean, fresh snow and whitecaps on the sea and some parts of clouds close to the cloud surface may have albedos as high as 90% if we include refracted rays which have not passed through much droplet water.

Any change in the overall albedo of the earth may well be a cause of a change in climate. Equally, a change of climate might cause a significant change in the albedo, such as a change in the area covered by permanent snow, which might have an important positive or negative feedback on the climate change itself.

13.2 ABSORPTION BY CLOUDS

The question arises how much are clouds warmed by the absorption of sunshine. This has been discussed inconclusively in a total of five papers in *Nature* and *Science* very recently. The message [57] in all of them is that measurements of the radiative flux of sunshine into the top of clouds has always been measured to be about 1.2 times the flux downwards out of the cloud base. But the absorption attributed, by calculation, to the amount of water assumed in a model to be in a vertical column of the cloud is negligible by comparison with the absorption implied by the measurements. The authors 'complain' that there is something wrong with the model used for the calculation. Two recent papers have supported the suggestion made in the previous edition of this book that in so far as there is more absorption inside clouds of sunshine than in the surrounding air it is due to the increased length of the path of a ray by repetitive scattering in clouds which lengthens the path through water vapour.

It is impossible to arrive at a realistic estimate of the path of a ray of light through a cloud, or even to estimate what fraction of the rays entering unit area at the top actually emerge from the same horizontal area at the bottom. The most intense sunshine measured on a sunny day at the ground has been where, in addition to the direct solar beam, there was considerable additional intensity from bright cumulus clouds so that the warming power of the sunshine received was at least 1.25 times that of direct sunshine. That implies that much of the sunshine power may be emerging from the sides of clouds. The rays may, or may not, have travelled through cloud along many times the vertical depth of the cloud. Any sort of one-dimensional model of the cloud imagined to be transmitting the sunshine in the same straight line direction as that in which it enters the cloud at the top is utterly naive (see Fig. 13.8.i).

As a passenger looking out of the window of an aircraft on the opposite side to the sun it is easy to note than on entering a cloud the illumination becomes immediately general, with an object in the shadow of the aircraft (including the face of the passenger observer) being brightly illuminated from all directions. The physical models derived from laboratory measurements of the absorptive power of sunlight by pure water are quite unconvincing and irrelevant to the problem of absorption in clouds. Also it must be noted that the answer obtained would be determined entirely by the model chosen for the calculation (see also section 14.7).

Clouds are black bodies, which means simply that they emit and absorb radiation in the IR region of their emission spectrum very much in the same way as liquid water and water vapour. As is seen in Chapter 14, the temperature of a parcel of cloud is very significantly changed by vertical displacements, and such changes are generally sufficient to explain the cloud forms and motions observed. Thus, for example, a patch of thin altocumulus which was present at dawn, but has become evaporated during the next two hours, may have been warmed by the sunshine directly or by IR radiation from the ground which had been warmed by the sunshine since sunrise, and had become much warmer than the cloud particles. The second possibility seems more likely because liquid water is much more absorptive of deep IR than of visible radiation.

13.3 THE PHENOMENON OF DAYLIGHT

An artist knows that the natural colours of things, and of painted pictures, is best seen when illuminated by the light entering a north-facing window (in the northern hemisphere) from a particularly blue sky with some sunlit clouds, and perhaps some scattered reflections from objects on the ground. But any combination of sources of illumination from that northern scene has its own balance of blue, near white from the clouds, and browns or greens from the land. Daylight comes from all directions, and when it comes in through the window nothing in the room is hidden from it. Unless the sunshine enters the window directly, all the indoor shadows are diffuse and are mostly penumbra, not bright or dark.

The sky is bright before dawn and long after sunset because the atmosphere is loaded with particles which scatter light in all directions. In satellite pictures we can see shadows of clouds on their west sides before the sunrise could produce them. That statement refers to the satellite channels 1 and 2, which are filtered to use only visible and near IR wavelengths (up to 1.1 μm only). But we also have scenes pictured in the channel 3 waveband (3.55–3.93 μm) and the wavelength is large compared with the size of the aerosol particles which are the cause of the scattering of daylight over the skyline before sunrise or after sunset. The result is that wavelengths in channel 3 are not scattered by aerosol particles in the air, and so there is nothing corresponding to the spread in all directions by the atmosphere of daylight. The shadows in channel 3 are very dark and sharp-edged.

Channel 3 is only just within the solar spectrum, but there is enough illumination to give some good pictures. There are no shadows in pictures made in the much longer satellite wavelengths of channels 10 and 11 (10.5–11.5 and 11.5–12.5 μm respectively) because there is no general illumination and the pictures show only the thermal emissions of the components of the scene.

13.4 THE COLOUR OF CLOUDS

We consider here the appearance of clouds in daylight of various kinds. The detail of the cloud surface is much clearer if it is illuminated directly by the sun than if it is in the shadow of a higher cloud or if it is illuminated by a large area source such as a brightly illuminated cloud above it or by an area of blue sky. The detail on the top of a growing cumulus is shown best (most sharply) if it is illuminated by a sun out of the view of the observer. Some thin clouds appear very dark when seen from an aeroplane even if they are in direct sunshine. To understand how this happens we first describe their appearance when viewed from below, by an observer in their shadow.

If the sun's disc cannot be distinguished through the cloud, it appears very bright nevertheless. Many rays are getting through the cloud without a great change in direction, which implies that a single ray impacts on only a few cloud droplets or crystals. If the cloud is thin enough for some rays to penetrate the cloud without impact on any droplet or crystal, the sun's disc becomes discernible, and the cloud around the disc is very bright. If most rays pass through with only one impact on a cloud particle then special optical phenomena such as haloes of coronas are seen, and in the case of a halo the cloud is brighter just outside the ring than inside it because the number of rays which undergo random deviations and then reach the observer is greater there. The sunshine arriving at such thin clouds mostly passes through but with some deviation, but more so in the case of ice clouds because in general the cloud particles are bigger with bigger spaces in between.

When such a cloud is observed from an aircraft which is not at a lower level and receives no sun's rays which have passed through the cloud, the observer sees only rays which have come from below and have the colour of the ground or consist only of rays scattered from other clouds which are much less intense. Therefore the cloud in question appears very dark, although it is in bright sunshine. See Plates 4, 5 and 6 in the colour section between pages 352 and 353.

When the sun is low so that the sunshine is strongly reddened many clouds may appear purple. This is a mixture of the red of the sunset (or sunrise) and the blue of the sky overhead; and it is very noticeable when the clouds being observed are in the shadow of others obscuring the sun from them, so that the red sky to the west and the blue sky overhead are of similar intensity.

13.5 THE GREENHOUSE GASES, AND ABSORPTION WINDOWS

The part of the sunshine which is not scattered back to space as the albedo of the earth is absorbed, and whatever it falls on is warmed. The oceans absorb most of it in the top few metres so that a large proportion of the absorbed energy goes to warming the ocean surface. Therefore we shall see in Chapter 16 that the warm water in the Western pacific Ocean is central to understanding the climate of the **atmosphere–ocean** system, for that is the region towards which the wind drives the surface water while it is receiving this heat input.

Over the whole earth the warmed surface is radiating outwards towards space. This radiation is from a surface whose temperature averages a little below 300°K. The emission spectrum is therefore indicated by the appropriate wavelength scale on the Planck diagram in Fig. 13.1.i. The shortest detectable wavelength emitted by the earth's surface is about 3.5 μm; it can be detected by the channel 3 detector on the meteorological satellites in the absence of scattered sunshine, i.e. at night, and can then be regarded as a relative measure of temperature. The maximum amplitude of the Earth's emission is at about 10 μm wavelength. Fig. 13.5.i indicates the absorption of all wavelengths by the atmosphere. The incoming sunshine is completely absorbed by O_2 and O_3 at wavelengths less than about 0.3 μm, but then scarcely at all until we come to a series of absorption bands at wavelengths greater than 1.0 μm which become more complete as the wavelength is increased. In making satellite observations we are concerned from 1.0 μm onwards more with the bands which are not absorbed, which we call **windows**.

We have already mentioned the very short wave absorption by O_2 and O_3. At the longer wavelengths the most important absorbing gas is **water vapour**. This is the main **greenhouse gas**, so-called because it reduces the escape of radiation from the earth to outer space, and so raises the temperature at the bottom of the atmosphere. The next most important greenhouse gas is **carbon dioxide (CO_2)**. The proportion of water vapour in

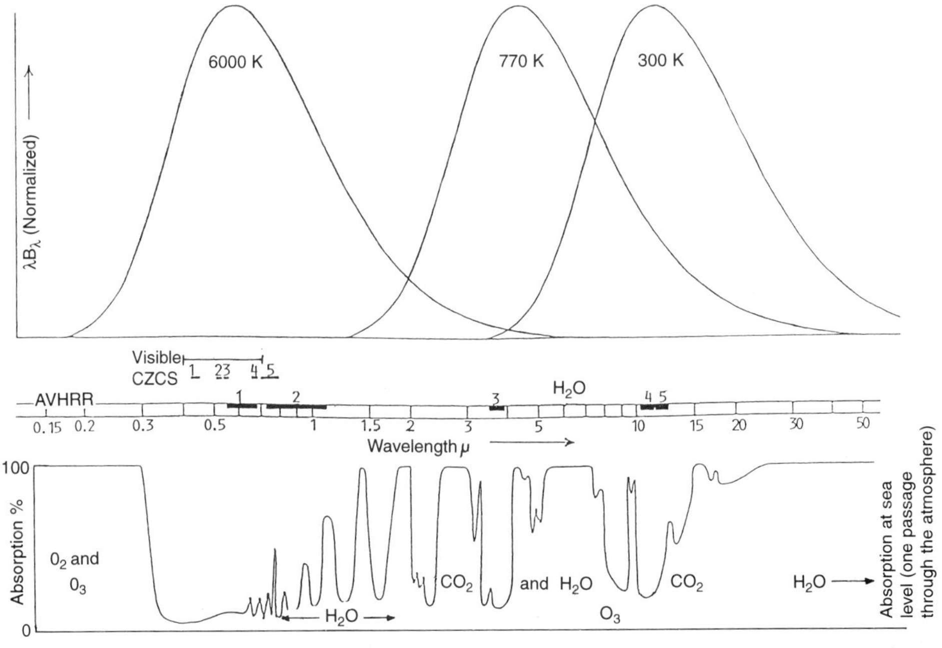

Fig. 13.5.i The Planck curve appears slightly changed in shape because the intensity is multiplied by wavelength to give an area under the curve to represent energy transmitted, so that in considering the incoming energy to the earth the emission curve for the earth, assumed to be at about 300 K (although this is probably too high an average for the emissions to space), is drawn with the same area. The curve marked 770 K, which is a dull red heat. Below the Planck curves is a wavelength scale and the range of the visible is in the middle of the sunshine range. The channels in which the satellites observe the earth are also marked. At the bottom is a representation of the amount of absorption by the atmosphere. There is seen to be very little absorption. The IR channels 3, 4 and 5 are in bands where the absorption is low, that is in windows of the spectra of water vapour and carbon dioxide. Ozone and oxygen fully absorb the UV of shorter wavelength than about 0.3 μm. Channel 3 is in the region where both the incoming and the outgoing beams are weak but makes good pictures which were of poor quality in the early days of the NOAA satellites.

the air decreases throughout the upper three quarters of the troposphere, and decreases sharply at the tropopause because it tends to be condensed and precipitated out after being condensed out as water or ice from air destined to be lifted into the stratosphere. But CO_2 by contrast becomes more or less uniformly mixed with the other gases of the atmosphere up to about 80 km above the surface. And it is not until a height of about 120 km that gravitational separation of the gases according to their density becomes important and the molecular viscosity has significant effects.

There are several other greenhouse gases, which are gases whose molecules contain more than two atoms. The most important are **methane (CH_4)** and **nitrous oxide (N_2O)** together with a variety of hydrocarbons derived from fuels and industrial solvents and finally the **chlorofluorocarbons (CFCs)** which are present in smaller quantities but have much larger molecules: they are not present naturally, but have been manufactured because they are very useful.

The question of how much attention needs to be given to the many much smaller amounts of polluting gases such as ammonia, sulphur dioxide, nitrogen peroxide and other industrial and agricultural products which have short residence times in the atmosphere because they are chemically reactive or soluble in rain must be set against the overwhelming importance of water vapour in the troposphere. Furthermore, different forms of cloud have very different greenhouse effects and we have no weather forecasting methods which can provide a forecast of the cloud amount, thickness and heights such that we can say with confidence whether any change of climate will produce a change in the occurrence of clouds which in turn will enhance or reduce the climate change.

One measure of the importance of a greenhouse gas is how many times the radiation in its absorbing wavebands will be absorbed in passing upwards through the atmosphere. The case of water vapour illustrates the intractability of formulating the parameters which would be required for a **general circulation model (GCM)**:

On an autumn morning, or late evening, when the wind is almost calm, the cooling of the ground by radiation into space through the windows in the assembly of greenhouse gases is such that the air close to the ground becomes cooled by proximity by exchanging radiation between the water vapour in it and the ground in the wavebands in which the vapour absorbs and radiates. When the air is cooled below its dew point a shallow layer of mist is formed which is often such that one can stand in it unable to see one's feet but can see over the top of it to houses and trees often miles away. That mist then becomes an absorber of all the IR radiation from the ground, and like almost any cloud layer the mist radiates like a black body to space; but in the water vapour absorption bands its emissions are again absorbed and do not penetrate more than a few metres at a time.

The observation of low lying mist formed by radiation when the sun goes down illustrates that the absorption by the water vapour in only one or two metres depth of air is enough to absorb all the radiation in the relevant waveband, and that there is enough vapour to repeat the absorption many times over before the energy finally escapes into outer space. See Plate 7 in the colour section between pages 352 and 353.

The original exchange between the ground and the water vapour illustrates the principle that all bodies emit more or less as black bodies according to Planck's formula except that gases only absorb and emit in their own particular bands. The emission from a gas or a water droplet cloud occurs in all directions—in effect equally upwards and downwards. Thus a cloud layer absorbing the IR radiation from the ground sends back to the ground an IR beam which is almost equal to its own original emission except that both beams have an intensity proportional to T^4. Usually the cloud is somewhat colder than the ground and so there is some loss of heat from the ground. But when there is a clear sky the radiation through the windows cools the ground surface so rapidly that a layer of dew is soon formed on it; while if there is a significant exchange with vapour and cloud the whole mass of air up to the top of the cloud must be cooled to produce a fog, and that represents a total mass much greater than the thin top layer on the ground. In the cooling process there will probably be convection currents initiated by the cooling of the cloud top which would be downward convection of cold thermals from the cloud to the ground.

The description just given is far from complete. In addition to the radiative exchanges between the cloud and the ground there may also be drizzle falling from the cloud which leaves the latent heat of condensation of the drizzle in the cloud, and may extract from the air enough heat to evaporate it before it reaches the ground. There is also a range of radiative exchanges between several layers of water vapour which have the effect of reducing the rate of escape of heat upwards in the wavelengths which are absorbed (and emitted) by water vapour.

A similar medley of radiative exchanges continues all the way up to the 'top' of the atmosphere, the top being thought of as the uppermost layer of each greenhouse gas above which there is not enough of that gas to absorb all the outward emissions of that top layer. The absorption by water vapour is virtually complete within the troposphere because the stratosphere is very dry, but the uppermost layer of CO_2 is well up in the stratosphere.

The different wavebands of absorption and emission of the different gases overlap to some extent. Attempts to simplify the problem have included attempts to express the effect of each gas as a CO_2 equivalent, i.e. as having the same effect as a certain amount of CO_2. But that is unsatisfactory in that if the gas absorbs in one of the CO_2 windows it produces a bigger effect than if it merely adds to the existing CO_2-absorbed wavelengths. By the same token a greenhouse gas produces a greater warming if it exists in plenty in the stratosphere where there is almost no water vapour than if it is merely and feebly, adding to what the water vapour is doing in the troposphere. To this is added the complicated fact that where there are clouds the gases have a negligible effect compared with the IR exchanges within and between cloud layers and the convection inside clouds.

13.6 COOLING BY HAZE

It has long been known that haze produces a cooling effect by radiation in IR wavelengths to space and also increases the albedo by scattering the shortwave sunshine upwards. Before the problems of chaos had been appreciated and the full complexity of the physical mechanisms had been understood it was reasonable to suppose that since every GCM has a climatology which is revealed if it is allowed to produce a forecast for a century, or perhaps even a millennium, we could see what the effect of putting greenhouse gases and haze into the model and running the long-term forecast again to see how it had changed. The idea of the greenhouse effect had been recognised, and people had even wondered why the increase in CO_2 had not already caused a warming; and they had ascribed it to the vast amounts of pollution that was having the opposite effect. Linear models of the climate had even been made and turned into computer models by which an imminent ice age had been predicted.

Later models did not use parameters for the haze because it was thought that in the modern world there was not as much haze as in the early days of the Industrial Revolution, and that it was confined to the immediate neighbourhoods of the industrial areas and was having a negligible effect. Still more recently hazes of sulphur compounds have been included in the exercise, and there was no surprise when it was revealed that the warming predicted was reduced. As with the warming the expected physical effect of haze emerged in the forecasts, and the warming was reduced when the haze was added.

13.7 THE BIOSPHERE AND CLIMATIC CHANGE

Lovelock has proposed that if there were a species of daisy which had two forms: a white flower, which increased the albedo and therefore would promote cooling, and a dark flower, which would increase the absorption of sunshine and promote warming. The plant might have a physiology which preferentially produced white or dark blossoms according to whether it would most benefit from cooling or warming.

Mathematical models of this daisy world have been made and, to the surprise of no one, performed as intended. But there is no obvious mechanism whereby natural selection might bring such a species into existence when the possibility of finding a locality with better weather would already have the well known mechanisms of seasonal migration of animals and seasonal transport of seeds of vegetation. Or would the evolution of this faculty give it any advantage in competition with other species in general. The implication is more in line with the thinking of Lysenko, which requires the species to have an imagined objective or purpose. Warm or cold weather might lead to selection for survival from among the species living, but not to selection from the environments which might put them to the test.

The original idea of Gaia, which led to the formulation of a purpose which appears to be followed, and which was to keep the earth habitable during a period in the evolution of the solar system when the earth might have fallen completely into the deadly grip of an ice age.

But that is unconvincing when it is noticed that with its large ocean and the greenhouse dominance of water vapour the equatorial belt could not have failed to provide an unfrozen home for life. Lovelock himself is not persuaded that Gaia is concerned to keep the earth warm to live on but points to the apparent preference of the algae in the ocean to increase its (the ocean's) albedo so as to extend the area of the cooler waters. In fact he suggests that Gaia may prefer a climate cooler than the present.

One of his suggestions arises from the fact that ship trails are quite commonly observed in the northeast Pacific Ocean. Unlike the trails observed in the north Atlantic, where they appear as bright white lines in very thin-layer cloud, called a **lace curtain cloud** because of the fragmentary texture, the Pacific trails often occur with no layer of other cloud, which implied at first that the air was not quite saturated for water vapour. In the Atlantic it was only just saturated. In both cases the important component of the ship's contribution was **cloud condensation nuclei (CCNs)** and not additional water vapour, which is the essential contribution in the exhaust of aircraft making condensation trails close to the tropopause. Therefore Lovelock argued that the biota in the ocean which cause condensation nuclei to be emitted could, in the appropriate circumstances, increase the amount of cloud in the **ocean deserts**, which are the areas lacking in nutrients where cloud was not formed because of this deficiency. Thus these biota would have the power to increase the albedo of the area and cool the climate.

13.8 THE DOMINANT INFLUENCE OF CLOUDS

The chief difficulty is in the idea that the biota in question would act as if conscious of a purpose. This seems to bring up again one of the difficulties which plagued the early thinkers about evolution, namely that evolution seemed to have a purpose, which implied that everything had been created for a purpose in the mind of a creator. The fact of evolution was easily accepted as the mechanism by which the great purpose would be achieved. Having thrown off the idea of a purpose-pursuing creator, Darwinism denies even the existence of a purpose imposed on life other than the purposes created by life forms themselves when they construct some sort of home for themselves. Thus there is no doubt that humanity, and many other species, mould the environment to suit that purpose. But it is a gigantic step to imagine a non-human species attempting to alter the climate—as if a bird species might seek to alter the inclination of the earth's axis instead of flying to the opposite hemisphere, as do the arctic terms, for half of each year. The present problem faced by humanity is not how to change the climate but how to avoid changing it because all the structures we have built up and can control would be put out of gear. The discussion has therefore centred on the question 'what controls the climate and its changes?'.

In discussing such problems many far too simple questions have been asked, and in particular it has become important to know what is the effect of clouds, and how will they react to any change that may occur: will they supply a positive or a negative feedback? Will clouds enhance or tend to cancel out any climatic change caused by other factors?

The background to this question is the dominance of the ocean and the presence of water vapour in the atmosphere. In his great textbook *Physical and Dynamical Meteorology*, Brunt [7] described the atmosphere as being water vapour diluted by variable amounts of other gases. Water is the only component which frequently condenses as clouds in the air and alters the temperature by the release of latent heat, and extracts it when the clouds evaporate. In addition to forming droplets the liquid falls out because droplets are coagulated into rain, hail, snow, etc. The proportion of the vapour decreases with height because of the decrease of temperature and to some extent also the decrease of pressure. Because the droplets may remain unfrozen (supercooled) at temperatures down to –40°C many clouds may be side by side with others of different composition due to differences of updraught history, coagulation, evaporation and freezing history. The vapour pressure at the surface of a liquid droplet is greater than at the surface of an ice crystal; and because condensation always begins as droplets, crystals are formed by the freezing of the droplets. The freezing may be delayed perhaps for hours at temperatures not far below 0°C, but is virtually instantaneous at and below –40°C. Freezing may be hastened by the presence of ice nuclei such as silver iodide, or by an increase in the size of the drops. Finally, surface tension increases the pressure inside small droplets so that they tend to evaporate at vapour pressures at which the larger ones may still be growing. Therefore the numbers of the smaller droplets is greatest where the updraught, and therefore the supersaturation, is greatest. As we shall see the particle size distribution in a cloud determines how a cloud processes the radiation to and from it, and so the properties just mentioned will be freely referred to.

In Fig. 13.8.i several points have been emphasised, and it is clear that the very short UV wavelengths less than 0.3 μm and the wavelengths greater than 0.7 μm are strongly absorbed compared with the visible wavelengths of sunshine. These incoming rays are refracted and reflected at the surfaces of cloud droplets and crystals; and so when the particles are very small as at the top of a growing cumulus tower or at the cloud base where condensation has only just begun the light is strongly scattered in all directions. For 90% extinction a channel 3 ray with a coefficient of about 100 must pass through about 2 mm of water, and channel 4 ray would suffer the same reduction after a passage of only 200 μm. It is not surprising therefore that clouds generally absorb the IR emissions from the ground and other clouds like black bodies. This means that they also emit like black bodies, the only exceptions being those that are very shallow.

When it was first noticed that high clouds had a similar appearance in channel 3 and channel 2 it was suggested that it was because ice was much more absorptive than water at very much the same temperature. We now understand that the absorption in channel 3 is much larger so that there is almost no scatter because the ice particles are also larger. Therefore in interpreting the appearance of the clouds in those channels all we get is a crude estimate of the dropsizes present.

Before proceeding with a large amount of pictorial evidence it is important to be as clear as possible about the scattering of sunshine by aerosol and cloud particles. Cloud droplets are very approximately spherical as are most acid droplets of similar size. Ice crystals, snow flakes and many kinds of mineral or other solid particles may be of many different shapes, and so the diagram in Fig. 13.8.ii becomes less applicable as the shape is made less spherical.

Beside coagulation the two circumstances which produce larger droplets are where the air is supersaturated for the growing particles. This could be where there are a great number of much smaller droplets present and the surface tension at the surface increases the equilibrium vapour pressure where the surface has the greater curvature: this occurs in a strong updraught where the saturation vapour pressure is most exceeded and every CCN is called into play and the number of small droplets ready to evaporate as soon as the updraught is reduced increases the size of the largest drops. The other circumstance is when some of the droplets become frozen and the ice crystals grow at the expense of the supercooled water drops around them.

Thus the similarity of the absorption coefficients of water and ice enables us to say that the appearance in channel 3 gives information about the predominant particle size. Channel 4 does not record scattered radiation whereas channel 3– will make clouds look

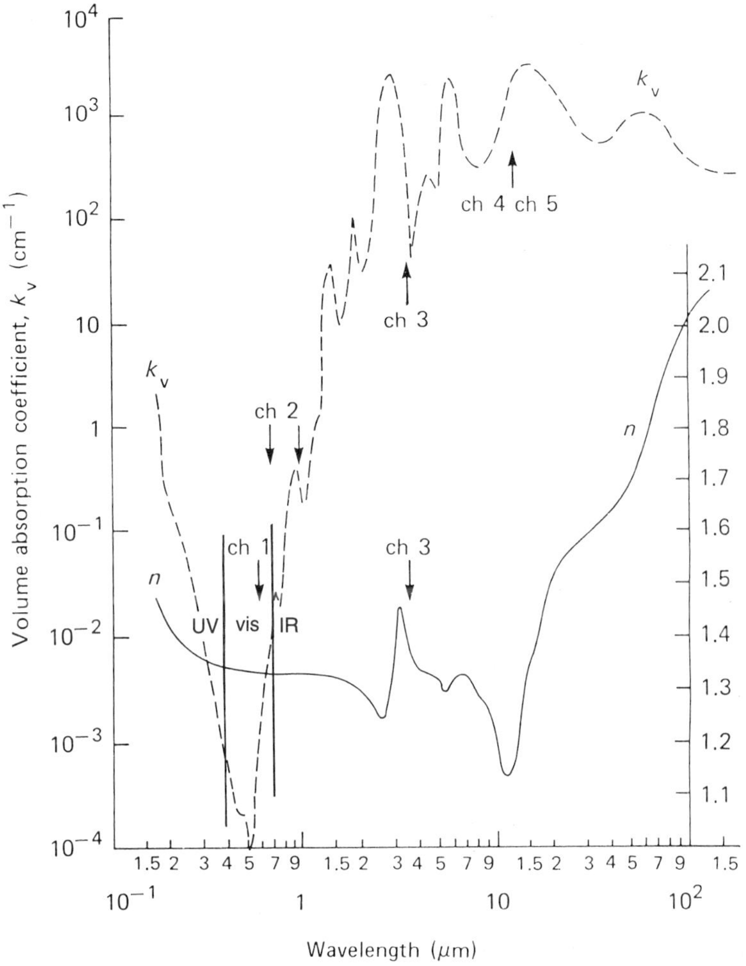

Fig. 13.8.i The absorption coefficient of radiation by water as a function of wavelength, both on logarithmic scales. The NOAA satellite channels are marked. Channel 1 absorbs very little, and the absorption increases steeply through the range of channel 2 and is multiplied 100 times in channel 3 and again increases for channel 4. The formula for **Beer's law** is $I = I_0 \times 10^{-0.4343km}$ where m is the depth of water passed through (in cm), and k is the absorption coefficient. (After Goody [58].)

black in sunshine because the scattered sunshine completely dominates the cloud's thermal emission by day (see Fig. 13.11.xii).

Where the particles are mainly large ice crystals and the cloud is shallow, much of the sunshine passes through between cloud particles, or perhaps with only one encounter so that when viewed from above one sees only the reflections from a few crystal surfaces and mostly light reflected upwards from the ground or clouds below. The cloud then looks dark from above or from the side even when it is in full sunshine, and it has a very low albedo. It may display halo phenomena when seen from below, as shown in Plates 4 and 5 of the colour section between pages 352 and 353.

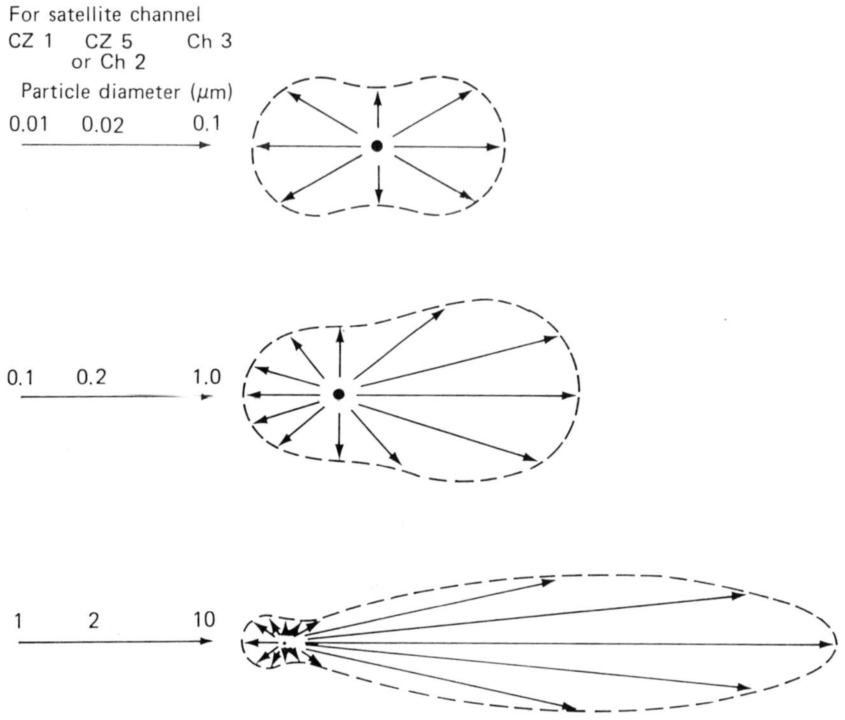

Fig. 13.8.ii The three diagrams indicate the direction and amplitude of the scattered light for three different wavelengths. Thus very small particles produce strong backward scatter and therefore give strong illumination by single scattering of the outer layer of particles. With large particles on the outside there would be very little backscatter because it would have to involve a slight change of orientation of the ray by encounter of several particles. By the time the reversal of the direction of the ray had occurred it may have been absorbed anyway, and the cloud would show no illumination. (After Minnaert [59].)

An old (meaning that the smaller droplets have been evaporated) medium-level shallow water droplet cloud has some of the same properties as the ice cloud just described, except that it may display rainbow and glory phenomena when seen from above, and possibly some iridescence. It is more likely to be thicker than the ice cloud because thin water clouds are soon evaporated and are therefore less common.

A layer of stratocumulus, which is convection cloud flattened by having a very stable layer above it, may have different parts of very different ages, and these will show different albedos according to the proportion of larger droplets. When a ship trail is formed in it a much greater proportion of smaller droplets is created in those parts, so that the trail shows up as a whiter line of cloud with a much greater albedo. (See section 13.10.)

The clouds with the greatest albedos are the rapidly growing cumulus towers; the rapid ascent produces a greater degree of supersaturation than slow ascent, and this means that all the condensation nuclei act as condensation points and are not quickly evaporated by the presence of larger drops close by. The lowest cloud albedos occur where there is a deficiency of condensation nuclei, or where no freshly condensed cloud has risen to the cloud top.

Where an area of growing cumulus has an abundance of towers casting shadows on neighbouring clouds when the sun is not high in the sky, the albedo is less than when the sun is high.

Rain is a form of cloud where the large drops are falling out and the smaller droplets may simply have been left higher up, or they may have been evaporated by a downdraught. The rain forms fine rainbows because the sun's rays make only one encounter with a raindrop passing between the other drops. If they make additional encounters that merely contributes to the brightness of the air in and around the rainbow.

From these cases it is clear that 'cloud' does not have a uniquely definable albedo, and any mathematical operations presented as manipulating a quantity called *the albedo* can only be misleading. In order to facilitate mathematical modelling using such quantities which are extremely variable, the modellers often argue that the symbol used represents an average such that the broadscale effect, namely the total proportion of sunshine reflected back into space, would be the same only if all clouds behaved identically.

Clouds also reduce the escape of IR radiation to space because they very strongly absorb it. In Fig. 13.8.i it is seen that the absorption of IR of wavelength around 10 μm is about a million times stronger than that of visible sunshine. And the only windows are then where there are no clouds. Clouds, together with water vapour, produce most of the greenhouse effect. The loss of heat to space is very largely from the tops of the highest layer of cloud; but even that depends on the IR windows in the stratosphere, and so the ultimate heat loss in addition to what escapes through the windows is from the greenhouse gases in the stratosphere. Apart from the radiation which escapes directly to space from the earth's surface through the windows left open by all the obstacles we have discussed, the loss to space is from clouds and gases which are at temperatures around 50°C colder than the surface. If there were no clouds or greenhouse gases the surface temperature would be at least 30°C colder than it actually is.

Clouds are the most important single factor in determining the world temperature. They come in a wide range of sizes which are mostly too small to be predicted by numerical forecasting models because the grid points of the models are tens or even hundreds of kilometres apart. Therefore a crude guess is made in most forecasts, using the experience gained by looking at observations, and this is turned into rules by which a forecast of cloud amount is made in numerical models.

High clouds contribute more to the greenhouse effect than they take from it by increasing the overall albedo. But it has to be remembered that thin ice clouds have a much lower albedo than thick high clouds. Low clouds contribute less to global warming because their temperature is much closer to that of the surface whose outgoing radiation they would be blocking. Daytime low clouds, caused by convection, have the highest albedo of all clouds, and there is a great diurnal variation in the amount of convection clouds, and probably there is a much smaller area covered by them at night. Thus it may be supposed that low clouds which have a great diurnal variation in amount contribute to global cooling. Therefore if global warming caused an increase in daytime convection it would be expected to produce a negative feedback to the warming. It has for a long time been supposed that warming would increase the amount of low cloud by increasing the water content of the air.

It has been said that since the cloudiness works both ways it must be a factor stabilising the climate by negative feedback, for otherwise the climate would be unstable. That feedback may take a few years to get into operation, and it may work through a mechanism such as the El Niño which might feed more warm water northwards in the Pacific Ocean and thereby take more heat away from the low-latitude ocean.

At present the various parameters in GCMs, which represent the cloud processes that cannot be represented by their actual mechanics because of problems of scale, could be varied to represent the changes in the amount of greenhouse gases, and then could be employed to produce a climatic forecast of the new state. This would be easier if the climatic forecasts, which are obtained by letting the forecast run for a century or a millennium, did not already oscillate by about the same amount as the expected climatic change (see Fig. 17.3.i), for the oscillations make it difficult to know the way the parameters work. If the models, and the real atmosphere, were not chaotic it might be possible to discover the way the mechanisms worked by comparing forecasts with accurately recorded weather sequences of the past. But the models wander off within their climatic limit in a variety of directions after about 10 days so that the precise cause of the wandering is not detectable.

The radiation processes act on a wide range of space scales and although the parametric representation of them is based in the known physics, they can only represent an average effect, as if the mechanism was everywhere acting like the average. The changes in cloud behaviour probably act on the local-scale mechanisms which produce them, and we are a long way from knowing whether these are concentrated at the coasts or far inland or far out to sea, or whether they can be represented by changes of average values over large areas, or within which latitudes they would be most active.

13.9 ON AEROSOLS

The curve showing the intensity of sunshine at the ground shows that there is some scattering away from the line of direct rays, particularly around where we would expect to find the maximum intensity (see Fig. 13.9.i) under a clear sky. This is what we would expect from our discussion of daylight, and this is most noticeable in satellite pictures which include the **terminator**, which is where we might expect sunlight to cease after sunset (or before sunrise). In the pictures made in channel 2 wavelengths the details on the ground or in low clouds remain quite clear for large distances into the dark area, which accords with our visual experience at sea level where twilight is such that lighting up time on the roads is usually half an hour *after* astronomical sunset in the latitudes of Britain and could be ignored all night in summer in Iceland. Thus the air glows in sunshine, which means that some of the sunshine is reflected back into space. This has been appreciated for well over a century, and it even led people to imagine that the haze produced by the Industrial Revolution might prevent us from ever emerging from the *Little Ice Age*, which was still causing very cold winters in the 1880s. Before the present worry about global warming it was hypothesised from time to time that additional warming would be caused by the increase in the amount of CO_2. This increase was

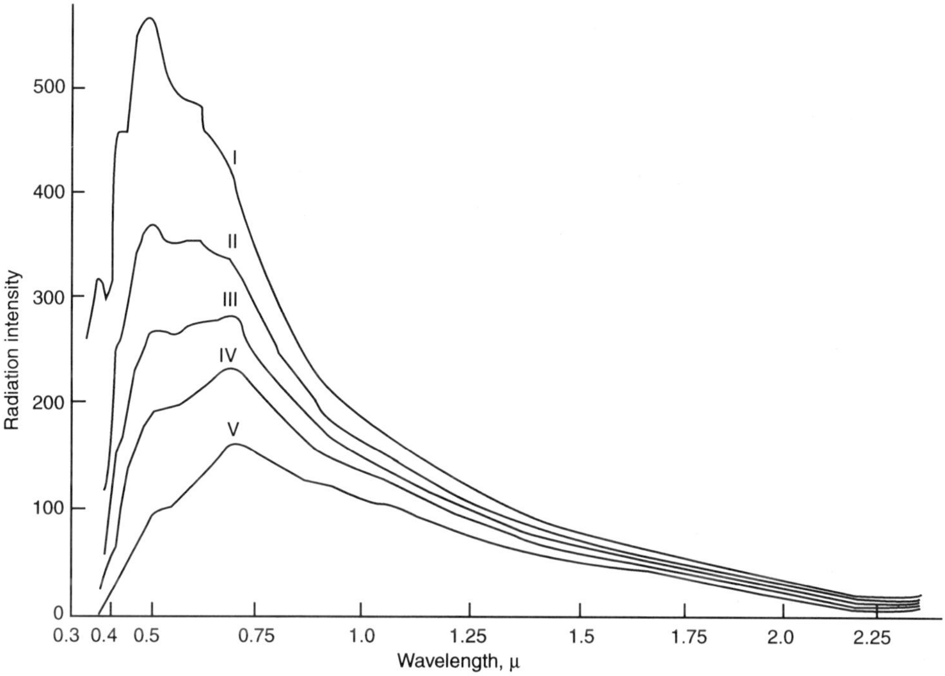

Fig. 13.9.i Curves described by Brunt [7] as the 'observed intensity distribution of solar radiation'. Curve II shows the spectral distribution of solar radiation received at Washington with zenith sun and cloudless sky, and curve I the distribution as estimated outside the atmosphere in the latitude of Washington after allowing for the absorption and scattering of the atmosphere. Curves III, IV and V give the distributions for solar altitudes 30°, 19.3° and 11.3° resp. There has long been evidence that the solar spectrum was not a clean version of the Planck curve, and this is partly due to the sun having an atmosphere from which the emissions are at a lower temperature than that from the layers beneath, of which a large proportion penetrates the atmosphere. Unfortunately, the curves reported by Brunt are not accompanied by any date or mention of the weather circumstances at the time.

discovered to be happening, and it was no longer expected that the excess would be dissolved in the sea.

Deposits of undoubted industrial origin have been observed in the Arctic and haze there has been reported on in detail by Kondratyev [60], where he tells of a pre-dominance of H_2SO_4 spherical droplets at high altitudes, which accords with early theories of mother-of-pearl clouds (shown in Plate 8 of the colour section between pages 352 and 353) which showed iridescence, due to spherical droplets, which could not have been water at temperatures around –80 K. Traces of Iron Age industrial pollution have been detected in greenland ice cores; but the predominance of sulphuric acid without the usual menu of industrial rubbish in what Kondratyev has found at high altitude makes it seem to be of volcanic origin.

The cooling effect of haze has been ignored in many GCMs which had included a parameterization of clouds because it was thought that the haze was only important in industrial areas. Although it was familiar to meteorologists that the Aitken Counter was an instrument which would measure the concentration of CCNs in the air, there was an expressed ignorance as to what these nuclei were. Thus it seemed that there always were plenty to prevent significant supersaturations occurring. The most easily measured component of industrial pollution was SO_2, and it became a belief among (indoor) chemists that somehow this became sulphate, or more particularly ammonium sulphate particles by combination with ammonia produced mainly by the presence of animals living on the land (including humanity). Therefore, far from land and industry, there was no real haze problem and so the great test for makers of GCMs was to parameterise the clouds. Thus the hit-and-miss process of getting parameters which gave good weather forecasts parameterised the effects of clouds and haze together without naming the haze. Even so that does not help to solve the problem of how much solar energy is absorbed by the haze particles, especially those of mineral origin.

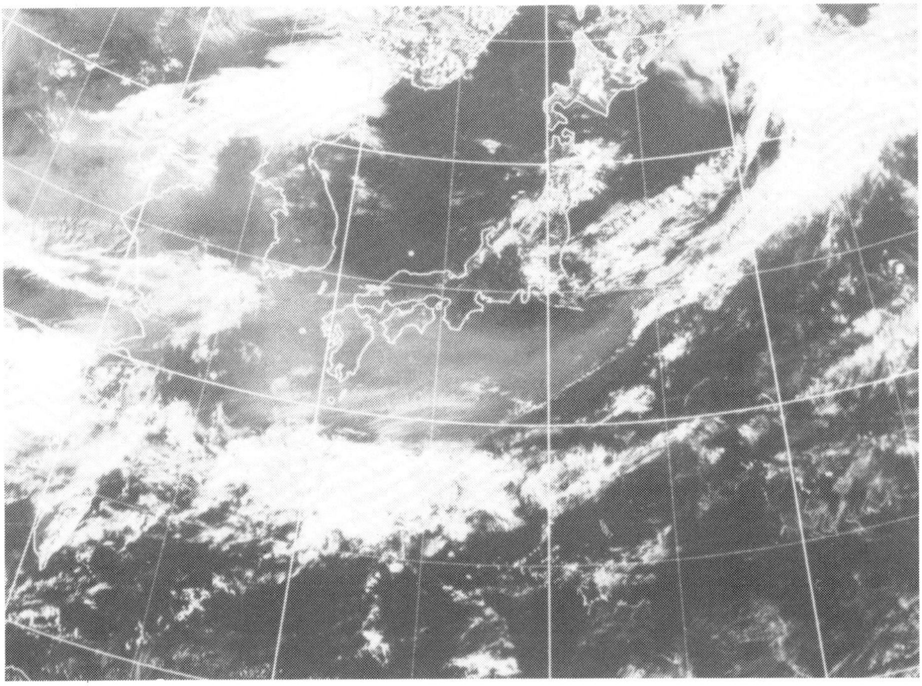

Fig. 13.9.ii 0533,15.4.79,vis (Japanese GMS I) showing dust from deserts far inland in China being carried out over the ocean behind a cold front which in the part that has crossed Japan has been reduced almost to a thin rope cloud. (Photo Japan Meteorological Agency.)

The revolution in world air transport which made it possible for scientists to go to conferences all over the world also made it possible for meteorologists to view the troposphere from above and discover (if they could not keep their eyes away from the aircraft window) that haze or other aerosols which obscured the view of the ground from eyes at the tropopause, or below, was much more widespread. Many areas were obscured by mineral dust, and in the trade wind areas the view was obscured by the sea-salt aerosol

Fig. 13.9.iii 1230,26.1.81,CZ5. It is fair to describe this picture as a typical escape of dust-laden air from NW Africa. The effect on sunshine can often be seen in the Algarve in southern Portugal. Dusts of this origin have been observed in northern Norway, having been seen over Ireland, and their arrival in the West Indies can be described as not infrequent. On this occasion the dust clearly increased the albedo of the area it occupied, but at the same time it increased the IR emission to space and in the channel 4 picture at 1958 that evening its emission showed that the dust was warmer than the land but cooler than the sea (Scorer [61]). (Courtesy of the University of Dundee.)

except vertically downwards through the tradewind inversion, which formed a sharp haze top over enormous areas of ocean.

Thus it was no cause for surprise when in recent years the modellers have announced that making a realistic representation of aerosols caused them to reduce their expectation of the amount of global warming, to be expected on account of the greenhouse gases, to about a third of their previous estimates. It is therefore possible that in some cases the effect of aerosols may have been taken into account twice, because there is no way the effect of clouds can be effectively separated from that of aerosols, in testing the parameters used to represent them.

The treatment of aerosols as if they were predominantly sulphates is unhelpful because sea-salt nuclei perform a dominant role in cloud formation. While it is possible that most of these are sulphates rather than the previously supposed chlorides, there is an argument that the chlorides are mostly larger and are observed to fall out quite close to the coast in an onshore wind; and if the nuclei over land are numerically more sulphate than chloride, this could be because more sulphates arise from the land. Further remarks on this topic arise in the discussion of ship trails. (See section 13.10.)

Because it had always seemed that there were plenty of CCNs in the air all over the world, it had been argued, during the controversies over rainmaking, that there were plenty of ice nuclei also and that there was probably nothing to be gained by attempting to supply some artificially. The most likely place to find cloud air in which the drops had not frozen would be in convection where the extra latent heat of freezing might strengthen the convection and perhaps produce a shower where there might not have been one otherwise. Whether this could be made into an economically profitable exercise was a different sort of problem not to be entered into here.

A deficiency of freezing nuclei seems to have been established in this case. The researchers had the possibility of influencing hurricanes in mind, which would be very useful, and more applicable in practice than most of the cloud seeding performed with a view to increasing rainfall as a routine. (See Fig. 13.9.iv.)

13.10 SHIP TRAILS IN THE NORTH ATLANTIC

As in the case of spike clouds (see section 8.6), ship trails were not noticed as a different kind or even special kind of cloud until they were extensively observed by satellite. At first it was thought that the contribution of ship exhaust was the water vapour resulting from the combustion of hydrogen compounds, as is well known to be the case with aircraft condensation trails. But in the North Atlantic they do not occur except where there are already some clouds present. They increase the albedo of the polluted strip of cloud. That cloud is unusually thin and has been called **lace curtain cloud** because it is a layer like a network, or sheet with many holes through which channel 4 partially detects the temperature of the sea below. The cloud is usually in a state of thermal convection on account of the cooling of the cloud top by radiation, and because of the holes the cells can be easily distinguished. (See Figs 13.10.ii–vi.)

The ship provides a very plentiful supply of CCNs. The cloud is not usually freshly formed except in the sense that cellular convection cloud is continuously circulated within the cells so that it is continuously formed and evaporated. Consequently the ship exhaust is very soon carried up into the cloud. The air mass was originally from an area where it was losing CCNs by fallout and was not being freshly supplied. The unpolluted cloud was formed with a deficiency of nuclei and therefore the droplets were larger than usual because of the small number of droplets formed at the condensation level.

The trail containing the additional nuclei was originally narrower than one convection cell, but the trail acquired a zig-zag shape because after a short time the whole cell became infected even if the trail passed through only a corner of it. This would result unless the track of the ship coincided with the orientation of the cells which usually depends on the wind direction. This result indicated that cellular convection, certainly

(a)

(b)

Fig. 13.9.iv 1401 and 1439,20.8.63. The pictures show the growth of a cloud in seeding experiments in an area where hurricanes occur. This cloud showed great growth within 38 min of being seeded. (Photos by Joanne Simpson.)

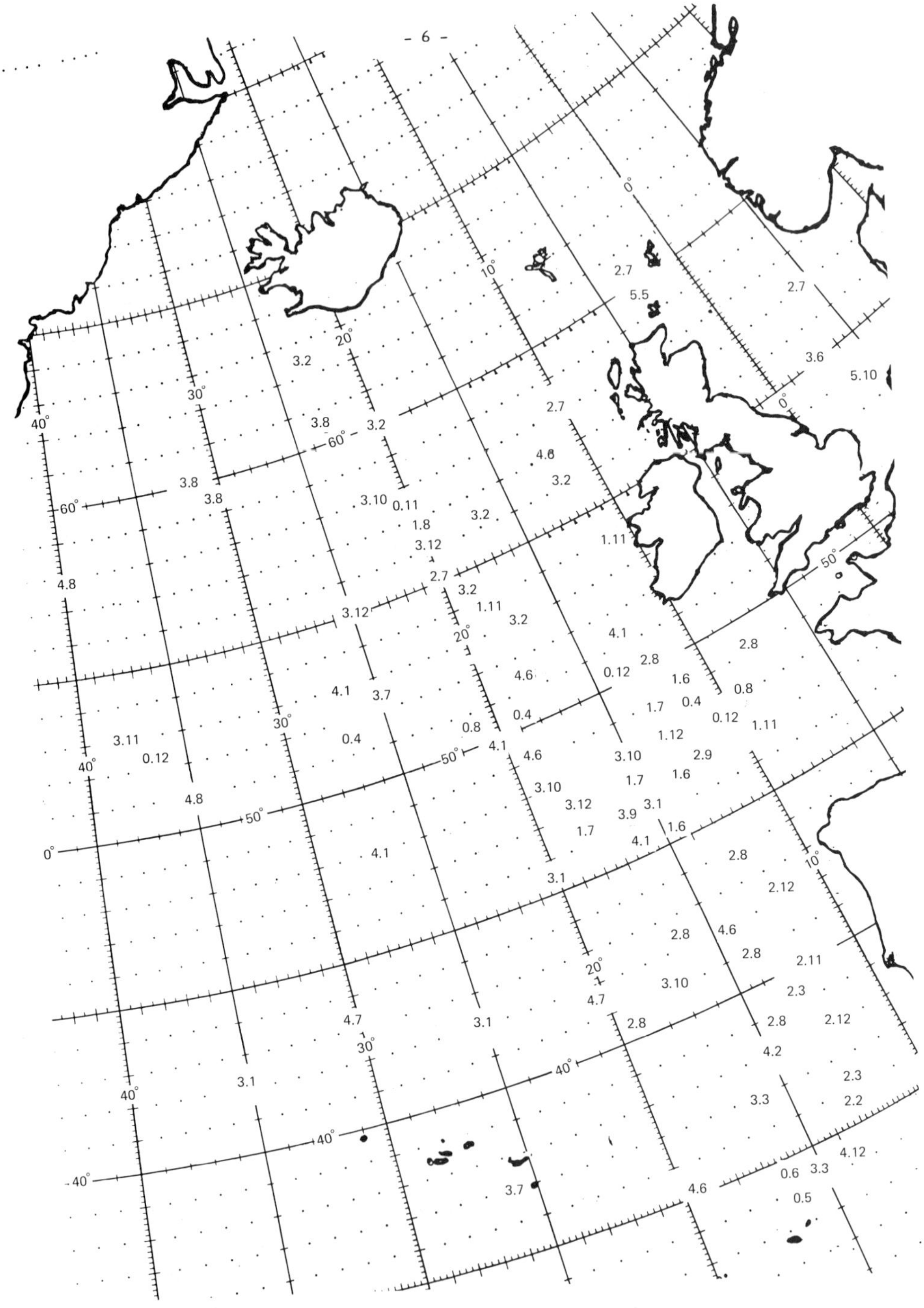

Fig. 13.10.i The positions of all cases of ship trails observed by AVHRR in the North Atlantic during 1980–85. The numbers on the map are in the middle of areas where trails were found and denote the month in the 1980s followed by the day. The pictures which follow in this section make it clear that the increased number of nuclei increases the albedo of the infected part of the lace curtain cloud, and this is not a feature readily appreciated at close quarters. Consequently they are not easily noticed from ships.

Table 13.10.i. Occurrences of ship trails during 1980–85 in the North Atlantic area. With the aid of the Dundee archive a study of the occurrence of ship trails in the northeast Atlantic during the six years 1980–1985 was made in 1986. The IR pictures of the Atlantic were examined for every day, and the channel 2 pictures also where they were available. The final year was almost barren, although it provided one of the very rare cases of trails in the North Sea

	1980	1981	1982	1983	1984	1985	Total
Jan				15–16	9–10		2
Feb			28	10–12	8		3
Mar			1	26–27			3
			11				
Apr	19						1
May	31					21	2
Jun	21	20		21	15		5
					22–24		
Jul		15–18	22–23	15	13–15		5
				26			9
Aug	21	10	7–9	10	24–25		
			18	17	31		
			22–24				
Sep			1	6			2
Oct				4		15–16	4
				19			
				23			
Nov	11	2	17–18	12			5
		30					
Dec	4–5	6	15–16	8–9	22		6
Total	7	6	10	14	8	2	47

that which is produced by cooling from the top, does not promote mixing across cell boundaries. It was found (Scorer [62]) that the air masses displaying ship trails have very light winds, and that while that state persisted the trails remained, in some cases as long as two days. The origin of the air mass was found, without exception (but every case was not tested), and the tracking was anyway not very accurate, although there was clearly an inversion at the top so that vertical mixing out of the layer did not occur. However the origin was not difficult to find because it was from the east or west side of Greenland and not anywhere else. The cases for the years 1980–85 are recorded in Fig. 13.10.i and Table 13.10.i.

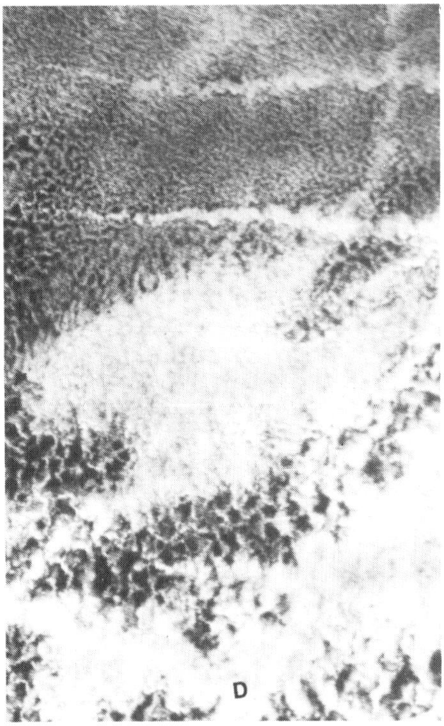

Fig. 13.10.ii 1541,15.12.82,Ch2. Ship trails in the area marked on the map as 2.12, which is near Cape Finisterre. The much recounted zigzag structure due to the cellular cloud lines having a different orientation is very evident. (Courtesy the University of Dundee.)

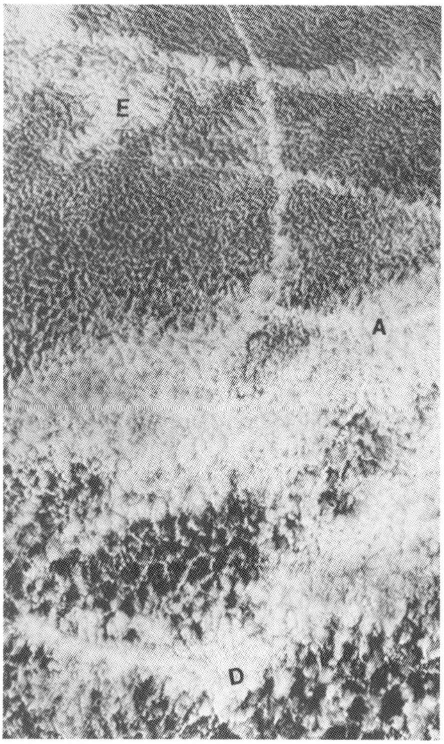

Fig. 13.10.iii 1208,15.12.82,CZ5. Here the same area is shown 3.5 h earlier and it is seen that the preliminary pollution of the convection cell the widening is very slow.

The origin of the air mass was always an area that was snow or ice covered and it is not difficult to find, but not easy to test, a mechanism which would give a crude idea of how long it takes to 'clean up' all the CCNs. The other place where they are common is the northeast Pacific Ocean, and there they are sometimes seen in air containing no other cloud. This could be that when the cloud is evaporating the trails are the last to go, and it

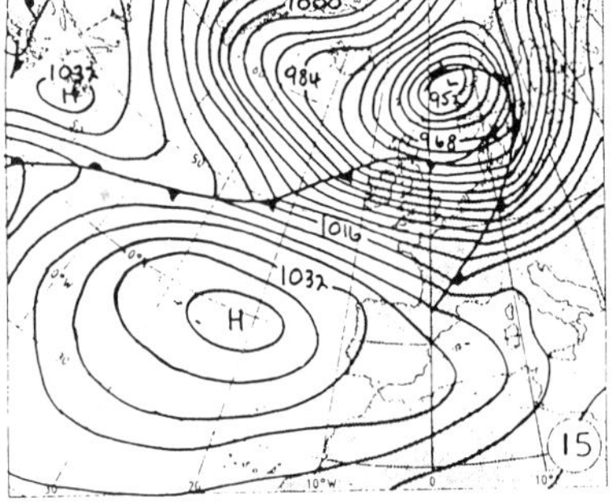

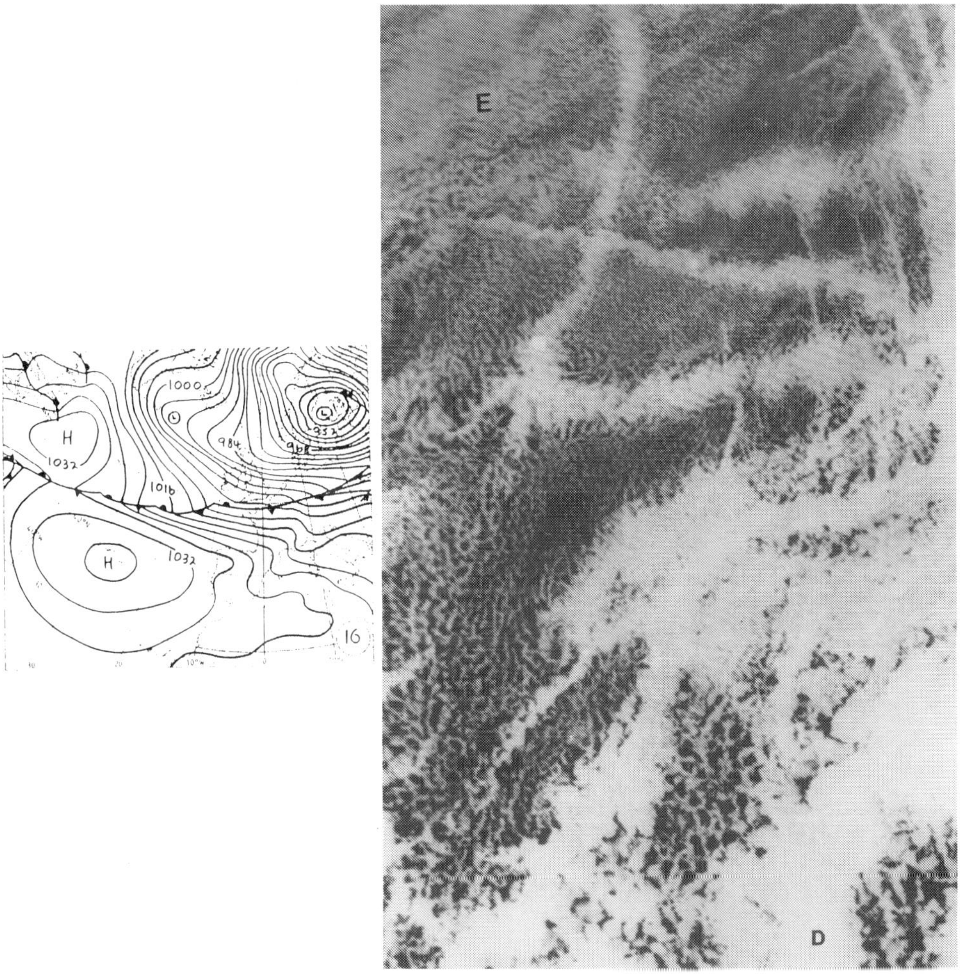

Fig. 13.10.iv 0.354,16.12.82, Ch4. Samples the same air as yesterday afternoon. (Courtesy the University of Dundee.)

is difficult to interpret the statement that they are found in otherwise clear air, the writer not being aware of the reader's possible uncertainty. The Pacific air mass in this case is not of arctic origin but was presumably cleaned of its CCNs in the anticyclonic region where there are no organic nutrients which do produce CCNs.

For a short time the suggestion that climate modification might be influenced by supplying every air mass deficient in CCN with CCNs or the wherewithal to produce them. The military wished to avoid making the trails but seemed to have no alternative but to pass through such air masses by sail and not fossil fuel power.

Ship trails are not found where the air mass comes from the land, nor where the wind is not light, nor where the convection is deep enough to mix in the air of a higher layer of different origin, such as continental. Thus there are none in the Mediterranean, and they only occur in the North Sea when the wind brings the air from the North.

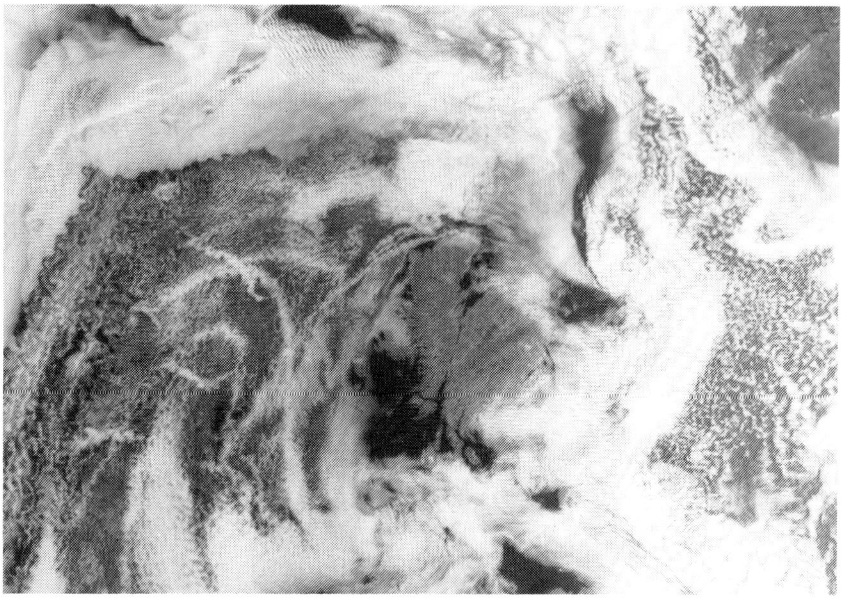

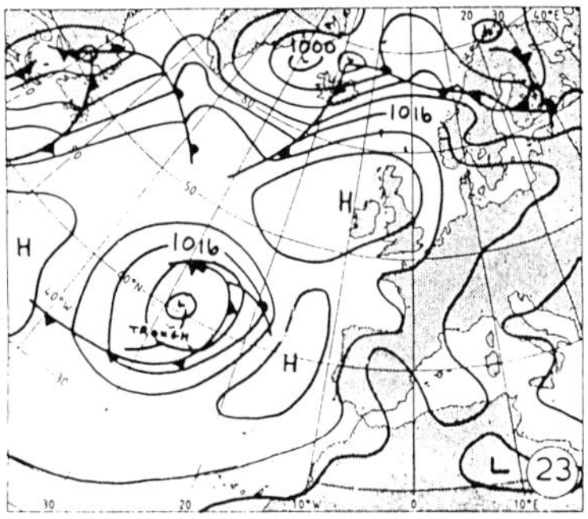

Fig. 13.10.v 0900,23.7.82,Ch2. The lightness of the wind in this case is made clear by the shortness of the lee waves of the Faroe Islands (2.7 on map 13.10.i). There are also remnants of old trails in the North Sea, which indicates that the clean air originated from the east side of Greenland. (Courtesy the University of Dundee.)

13.11 THE METEOROLOGICAL SATELLITES

(1) The choice of wavelengths

Because of the almost universal presence of haze of some sort, and because the land scatters most effectively at the red end of the visible part of the spectrum and the immediately adjacent IR it was decided to call channel 2 of the list of channels the **visible**. Thus it coincides almost with channel 5 of the Coastal Zone Colour Scanner (CSCS) of which the channels are labelled CZ. The shorter bands are in the visible and are channel 1 and CZ1–4. See page 448.

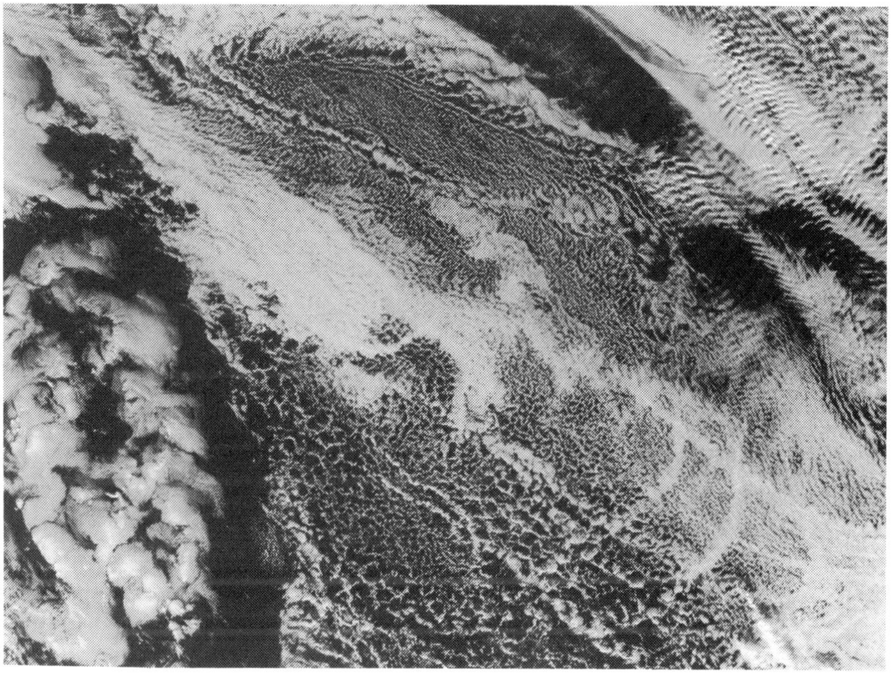

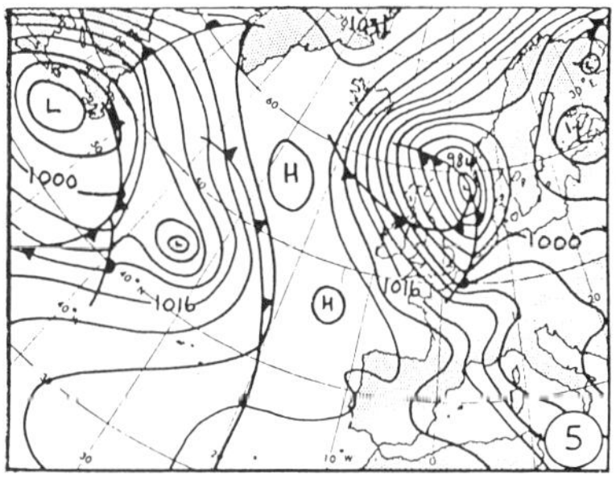

Fig. 13.10.vi 1152,5.12.80,CZ5. The position of this event can be located by the lee waves over Ireland. Map 0.12 (p. 442). Some trails are old, and others very old, judging by the fact that the lace curtain cloud has survived even though it is giving way on the edges, and in some of the waves from Ireland. (Courtesy the University of Dundee.)

The choice of the IR channels was determined by the windows in the absorption spectra of the greenhouse gases so that the emissions would reach the satellite at about 800 km. There is also a 'water vapour channel' which is not used in this work because it receives emissions from the upper layers of the water vapour present, which is of interest mainly in forecasting but scarcely ever shows cloud details below the tropopause.

Channel 3– is in that narrow band of wavelengths where the emissions from the sun overlap the emissions from the earth. For daytime pictures it is shown in the same way as channel 2 in which the sunshine which is reflected and recorded as in ordinary photography, the brightest intensities being the whitest. But at night it records the emission according to the temperature of the body emitting in the same way as the 'thermal IR' channels 4 and 5. These three IR channels are not identical in their response to ordinary earth temperatures and so they can be used to gain information about the water vapour which exists mostly in the lower layers of the troposphere. The thermal emission of channel 3 is weaker than the reflected sunshine and is usually swamped by it, and so for convenience in this book it is used that way. It is then called channel 3–, which may be confusing because when it is used in connection with channels 4 and 5 they are printed in photographic negative so that the coldest becomes the whitest. Since channel 3 was originally planned to be used in photographic negative, to use it in photographic positive has become called 3–.

Since 1986 the CZCS has not been operating, which has been a source of great disappointment to many users of its many facilities; but much information was given out while it was working and may still be available in some archives. A few of its valuable messages are becoming more relevant every day and some are included here because they refer to wavelengths within the visible range.

Table 13.11.i. Table of channels with meteorological information

Satellite	Channel code	Wavelength band (μm)
NOAA 6–10	Ch 1	0.55–0.68
	Ch 2	0.725–1.10
	Ch 3	3.55–3.93
	Ch 4	10.5–11.5
	Ch 5	11.5–12.5
CZCS	CZ 1	0.433–0.453—sky blue
	CZ 2	0.510–0.530
	CZ 3	0.540–0.560
	CZ 4	0.660–0.680
	CZ 5	0.700–0.800—v. deep red

(2) Seeing 'through' clouds

To illustrate this occasional facility we take the case of 1609,11.6.82 (Fig. 13.11.i), where the picture is taken of the southwest side of the southern tip of Greenland. In channel 2 the permanent snow in the NE, with the snow-free area down to the coast where we see fragmentary ice close to the coast and smaller filaments stretching northwards. In channel 3– the snow cap is almost black and the snow-free strip much brighter. The vegetation reflects this IR band quite strongly but the water in the fjords and the ice floating on the sea cannot be seen because they absorb channel 3– strongly. In the west, channel 3 reveals some higher clouds which can now be faintly distinguished in channel 2. We know that these clouds are higher because of the shadows cast on the sea fog below which are both darker and sharper than the same shadows in channel 2. There also appear

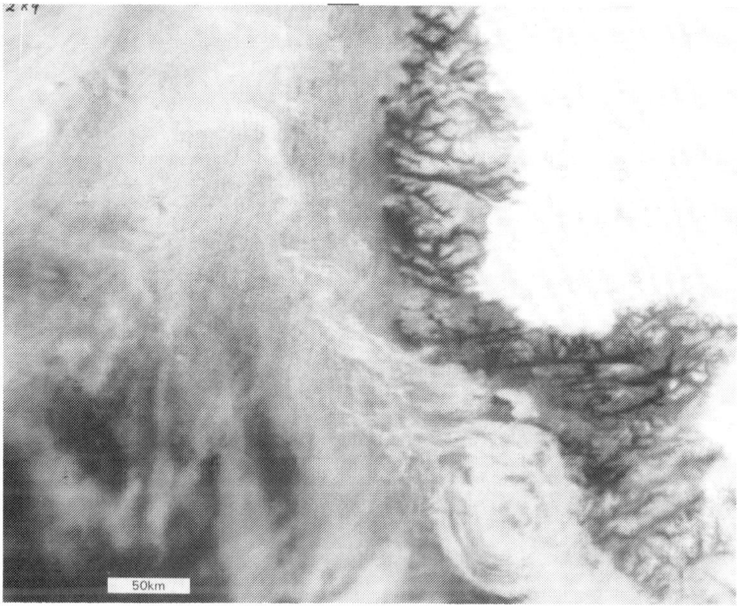

(a) (channel 2)

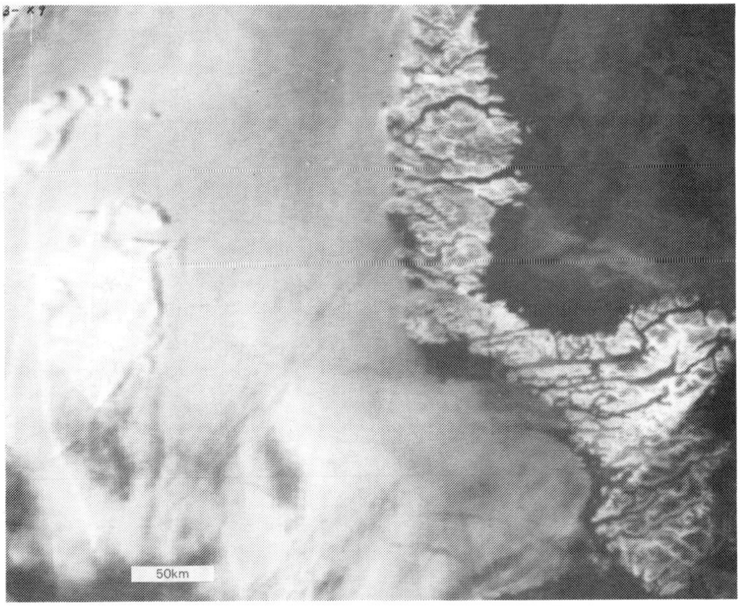

(b) (channel 3–)

Fig. 13.11.i (continues on p. 450)

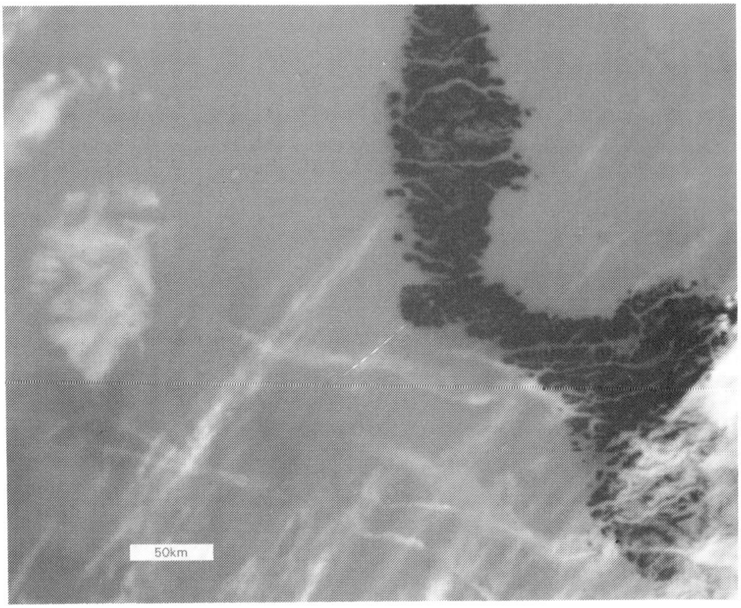

(c) (channel 4)

Fig. 13.11.i 1609,11.6.82, chs 2, 3– (see p. 449), 4. Three different channels record a scene off SW Greenland. Only channel 2 (p. 449) shows the pattern in the ice floating close to the coast. In channel 3– the ice and the inland snow are dark because the sunshine is absorbed while in 4 the floating ice is at the same temperature as the sea and the inland ice is colder. (Courtesy the University of Dundee.)

to be two converging ship trails which are bright white in channel 3–, and some dark shadows on the fog which can be identified as white streaks in channel 2. Channel 4 identifies these high cloud streaks as white (cold) cloud, while the land is black because it is warmed in the afternoon midsummer sunshine. The snow, the fog and the ship trails are all at about the same temperature.

The ice can be seen through the fog because it is bright white in the sunshine while the surrounding water is reflecting no sunshine back to the satellite. We call it fog because it has the same temperature as the water and only if it was very close to the sea could we see such detail in the pattern of the floating ice.

(3) Continental dust haze over the sea

This pair of pictures, 1150,18.9.85; C Z1 and CZ5 (Fig. 13.11.ii), gives different versions of the dust over the Bay of Biscay. CZ1 is blue/green while CZ5 is equivalent, almost, to what we call visible channel 2. Notice how the land is 'overexposed' in order to reveal the dust over the sea in CZ5. The dust is much more easily detected in the central visible colour where the sun's energy supply is greatest and therefore this CZ1 picture, of shorter wavelength, gives a much better representation of the effect of this haze on the earth's albedo. In any practical test of the validity of any parametric representation of the cloud amount this haze plays the part of cloud.

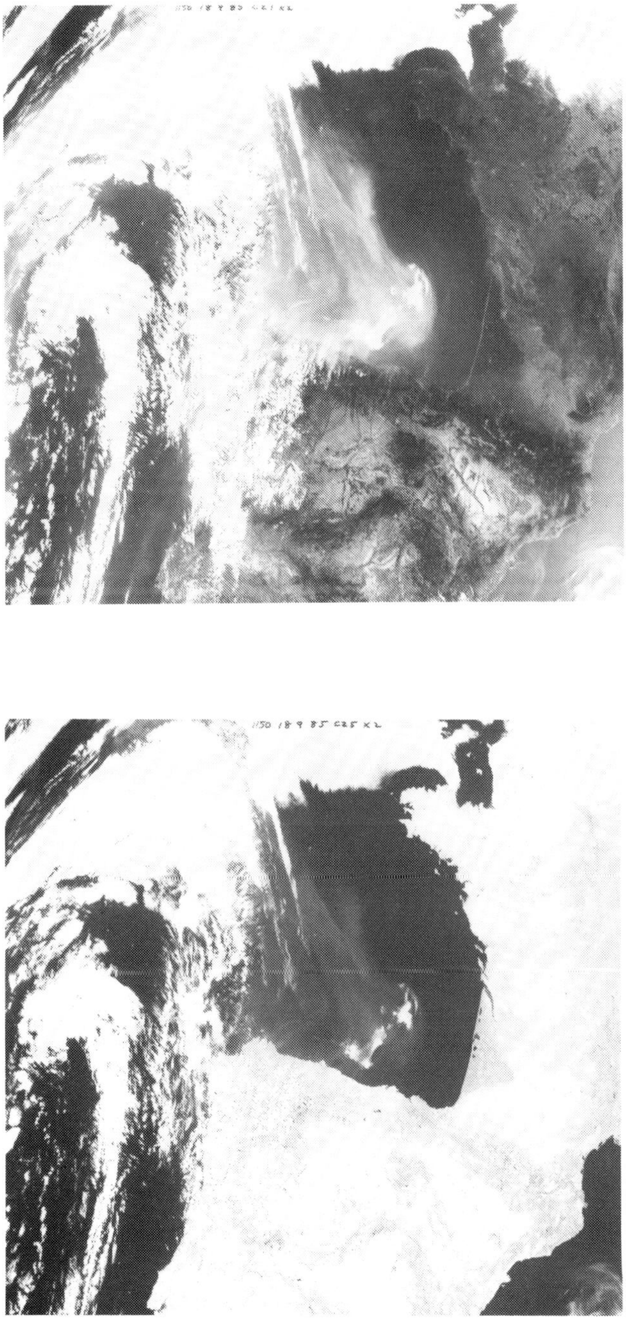

Fig. 13.11.ii 1150,18.9.85,CZ1 and CZ5. Two pictures of haze over Biscay. (Courtesy the University of Dundee.)

(4) Mediterranean haze beneath an inversion, and movement controlled by the Coriolis force

In CZ1 only the mountain tops and their small cumulus are seen clearly. CZ3 shows the coastline detail, but the land is overexposed photographically to reveal the haze over the sea. CZ5 sees through the haze (Figs 13.11.iii and 13.11.iv).

Fig. 13.11.iii (*cont*)

Fig. 13.11.iii 1013,20.8.83; CZ1, CZ3 and CZ5. A very hazy calm Mediterranean with an inversion just below the mountain tops confining the air within the area of the sea, with an accumulation of haze below. (Courtesy the University of Dundee.)

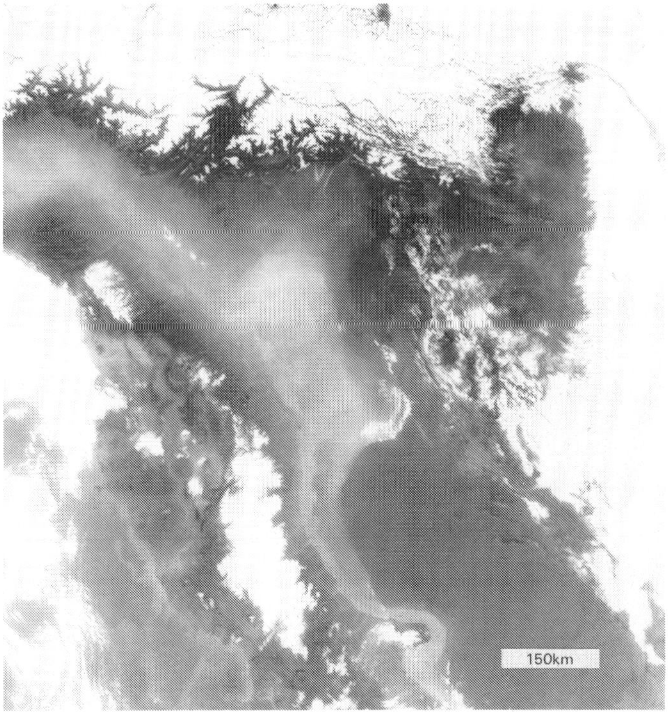

Fig. 13.11.iv 1016,21.1.83; CZ1. At this time of the year the katabatic drainage of the Po valley of all its industrial pollution, both in the air under an inversion and in the river, and this is carried down the east coast of Italy under the influence of the Coriolis force.

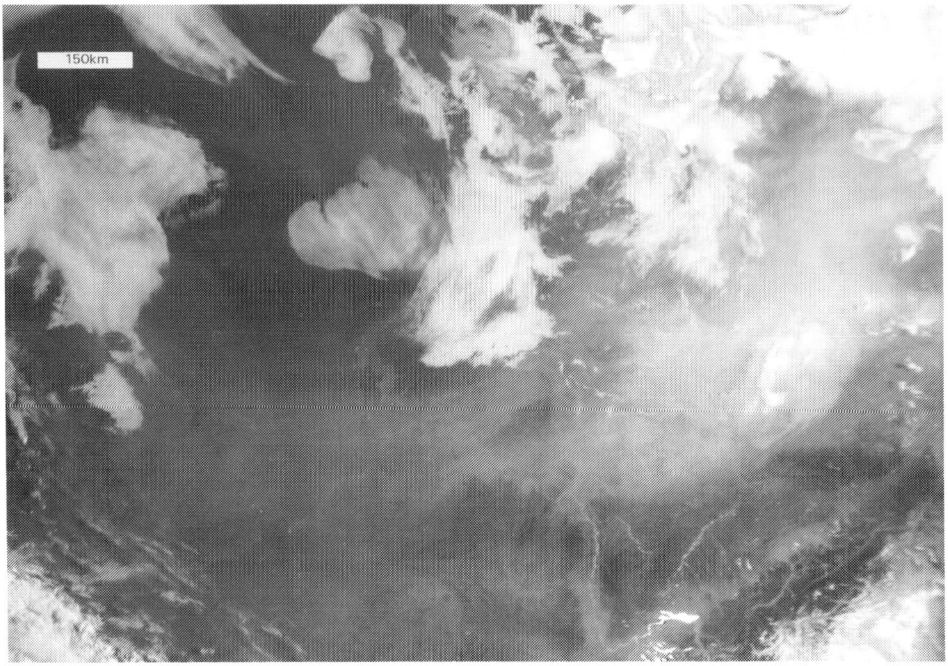

Fig. 13.11.v 0806,21.8.84, Ch1. The appearance of the aerosol haze is enhanced by viewing it towards the morning sunshine.

(5) Pure industrial haze shown up by channel 1

The smoke and sulphate haze of western German in calm conditions is quickly shown up by the most visible NOAA channel (Fig. 13.11.v).

(6) Glint and cold water in the Bise (Fig. 13.11.vi)

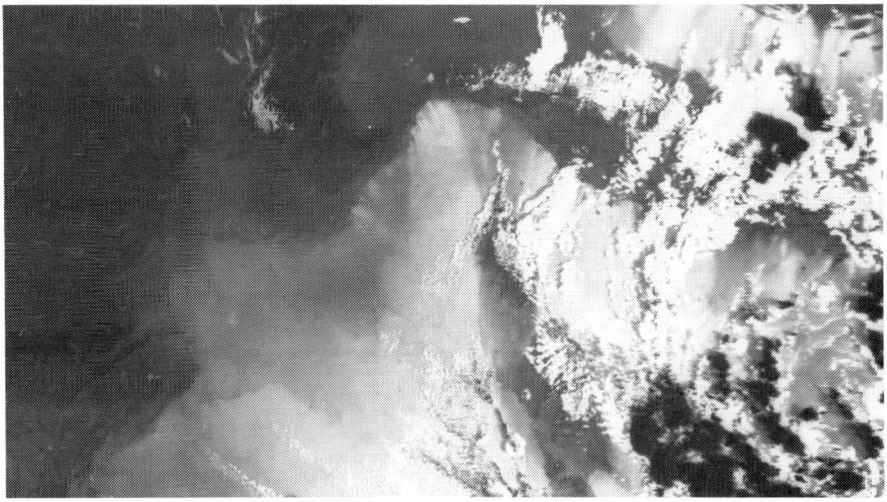

3–

Fig. 13.11.vi (*cont.*)

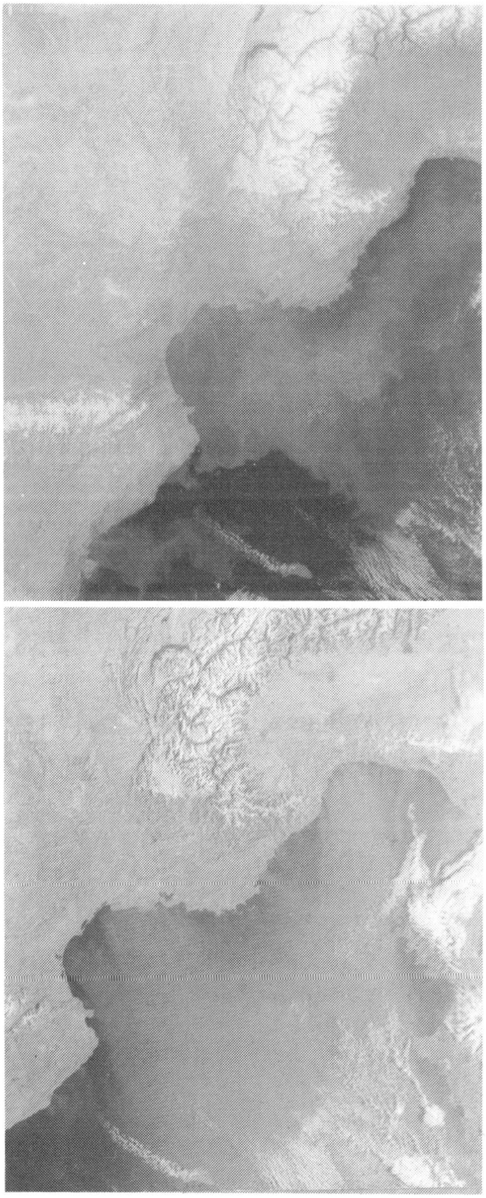

4

2

Fig. 13.11.vi 0810,20.10.80,Ch3s3–,4,2. (Fig. 13.11.vi) The first of these pictures is in channel 3 which shows glint much more brightly than visible wavelengths. This example shows the Adriatic and Tyrrhenian Seas ruffled to produce glint in a NE wind at sunrise. In the dark it measures temperature. In the twilight, which does not exist in channel 3–, channel 2 shows the sea glint caused by the Bise (a Swiss name for the cold NE wind, called Bora south of the Alps, when characteristic high clouds accompany it (see Fig. 14.5.i, Bora) blows the surface water into waves which scatter the reflection of morning sunshine in many directions and show the wind streaks across the Adriatic from the NE and from almost North in the Gulf of Genoa when the terminator prevents the glint from being seen before sunrise where the wind strikes from the Rhone Valley; in channel 3– the cold surface water drawn up from below is compared with the warmer water further out to sea which glows brighter in this wavelength.

2

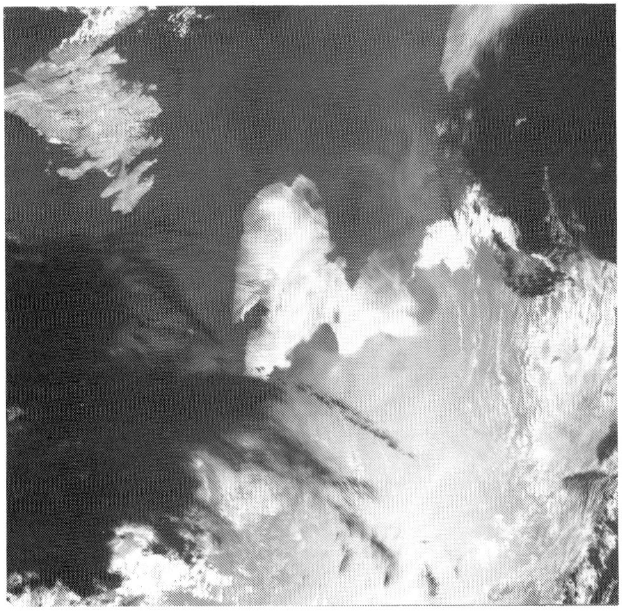

3–

Fig. 13.11.vii (*cont.*)

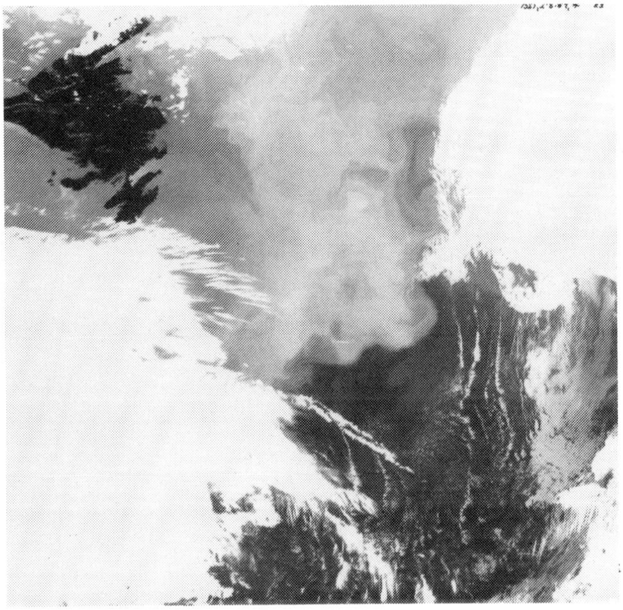

4

Fig. 13.11 vii 1551,2.8.89;Ch2,Ch3–,Ch4. The clouds which are white in channel 2 and black in channel 3–, and white again in channel 4 are high clouds, probably frozen with ice particles larger than the water particles which are bright white in channel 3–. The cloud in the centre is bright in channel 3– and shows no shadows. The sea in channel 3– is dark in the north but brighter in the south, where it is warmer and there is glint in the midday sun. In channel 4 there are no shadows and the sea cold in the north and the cloud in the centre is entirely over cold sea, which is sharp-edged. The only land visible is a bit warmer in the sunshine than the warmest part of the sea. Channel 4 shows no glint and the temperature is reversed from channel 3–. (Courtesy the University of Dundee.)

(7) Newfoundland fog bank (Fig. 13.11.vii)

The outline of Newfoundland in the NW corner of these pictures is obvious, but what about the clouds? How do we identify the famous fog banks in one of them?

(8) 'Hot' katabatic from Greenland heights (Fig. 13.11.viii)
In calm weather the tributary steep valleys generate a strong cold downslope wind which has collective momentum to carry it down to sea level with sufficient excess temperature to melt the ice in the estuary at Kangerdlugssuac on the Denmark Strait, where the ice is about at its maximum extent. There is further discussion in the frontispieces.

(9) Freezing melt water floating on warmer salt water (Fig. 13.11.ix)
In order to maintain the North Atlantic 'conveyor belt' (see sections 16.7 and 19.3) the warm Gulf Stream must sink under the melt water, which is fed into the Arctic Ocean by rain, snow and melting ice; otherwise it would not sink and return to the Atlantic as bottom water, to make way in the Arctic for more of the Gulf water.

4

2

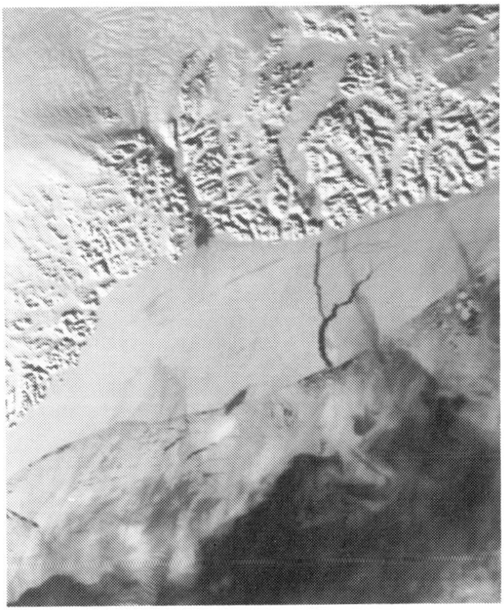

Fig. 13.11.viii 1531,20.2.90;Ch4,Ch2. Although not usually visible in channel 4 pictures, this katabatic flow must be frequent to keep the water in the estuary mostly unfrozen until late February. (Courtesy the University of Dundee.)

2

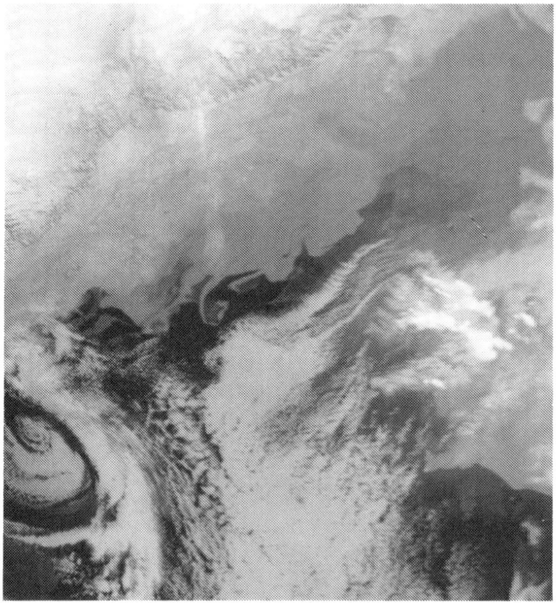

4

Fig. 13.11 ix 1637,16.3.82;Ch2,Ch4. The channel 2 picture (upper) shows the width of clear water between the coastal Greenland ice and the clouds closer to Iceland. Channel 4 shows that part of that space has a temperature close to that of the ice. In channel 2 we can see that one of the mixing whirls (which has anticyclonic rotation) still has unmelted ice, which would be expected because there the sea is seen to be colder. (Courtesy the University of Dundee.)

4

Fig. 13.11.x 1400,15.5.80;Ch4. This dark patch was in a cold area of the North Sea for a few days. Colder areas can be seen close to the coasts. (Courtesy the University of Dundee.)

(10) Warm sunshine patch in the North Sea (Fig. 13.11.x)

When the wind is calm, the sea surface is not disturbed, and so the top layer which absorbs most of the energy from sunshine and becomes perceptibly hotter than areas around it where the wind has stirred the water and mixed the absorbed heat downwards.

(11) The varied illumination of orographic cirrus (Fig. 13.11.xi)

The brightness of cloud when the sun is low and the effect of shadows is worthy of study.

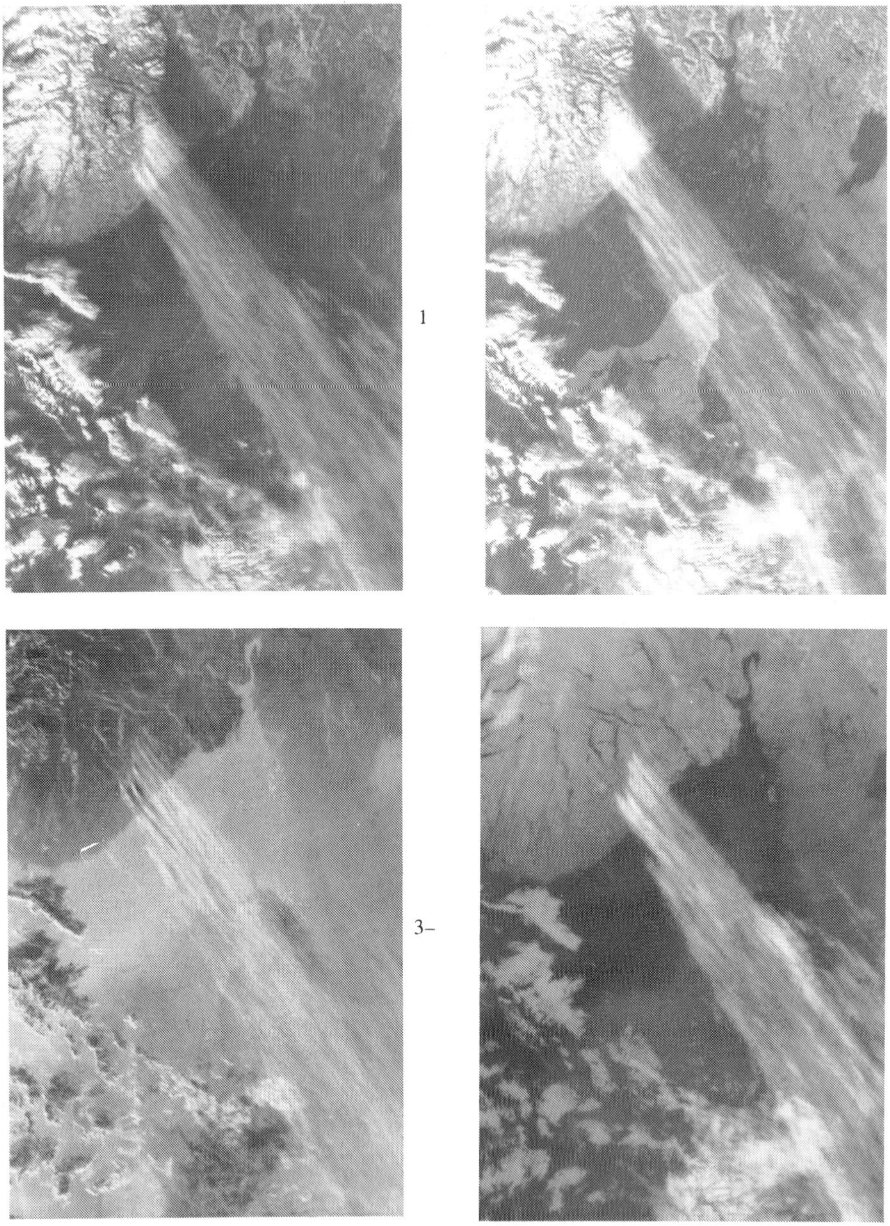

Fig. 13.11.xi 1215,9.12.88;Ch1,Ch2,Ch3–,Ch4. This cirrus is formed behind the highest peaks in southern Norway. In channel 1 the cloud is bright where there is land beneath which is bright with recent light snow. The cloud's shadow can be seen on the land to the NE. In channel 2 the land is everywhere brighter than in channel 1, and the northern tip of Jutland appears to illuminate the cloud also but not from snow but from land in sunshine scattering this channel but not channel 1. Part of the Swedish lake Vänern is seen in the NE corner, and in channel 3– it appears brighter than the land because it is warmer. There is no illumination from the land in Norway nor is there any shadow there because in channel 3– snow is dark with no scatter. In channel 4 the sea, lake and rivers are warmer than land at this midwinter date. (Courtesy the University of Dundee.)

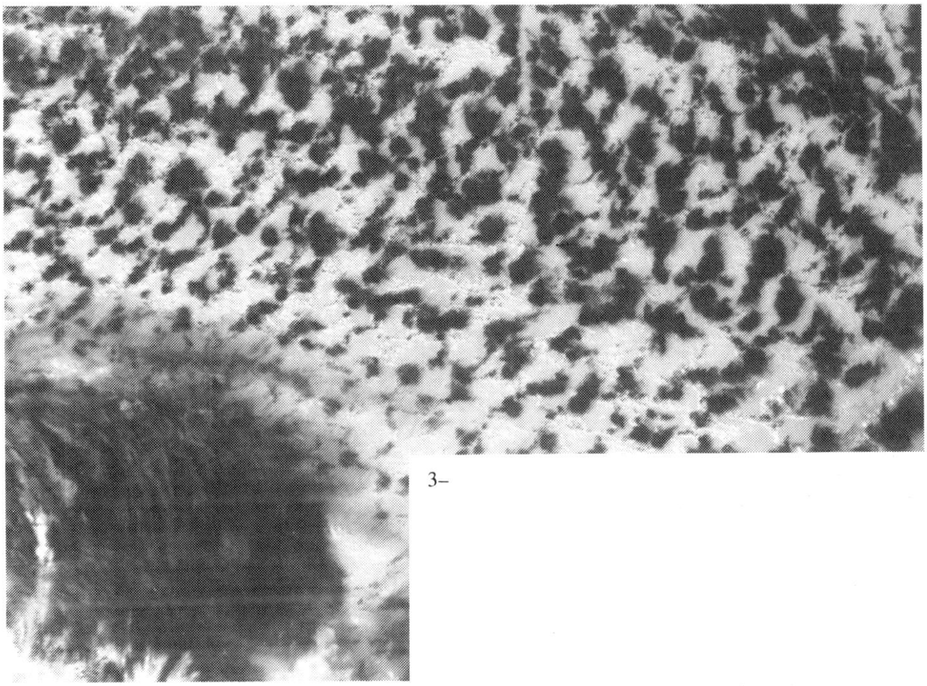

3–

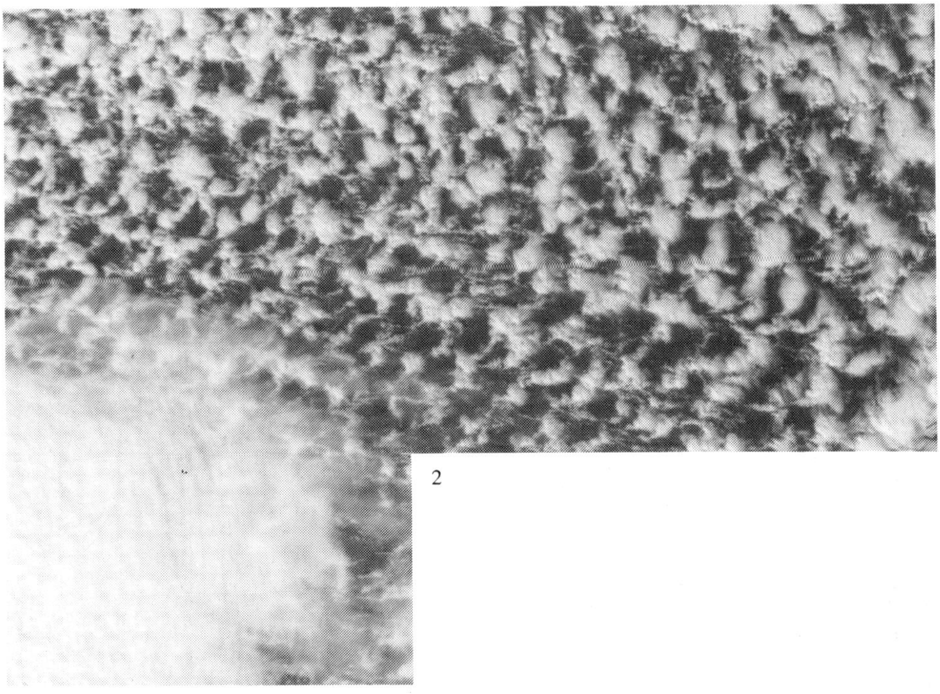

2

Fig. 13.11.xii (*cont.*)

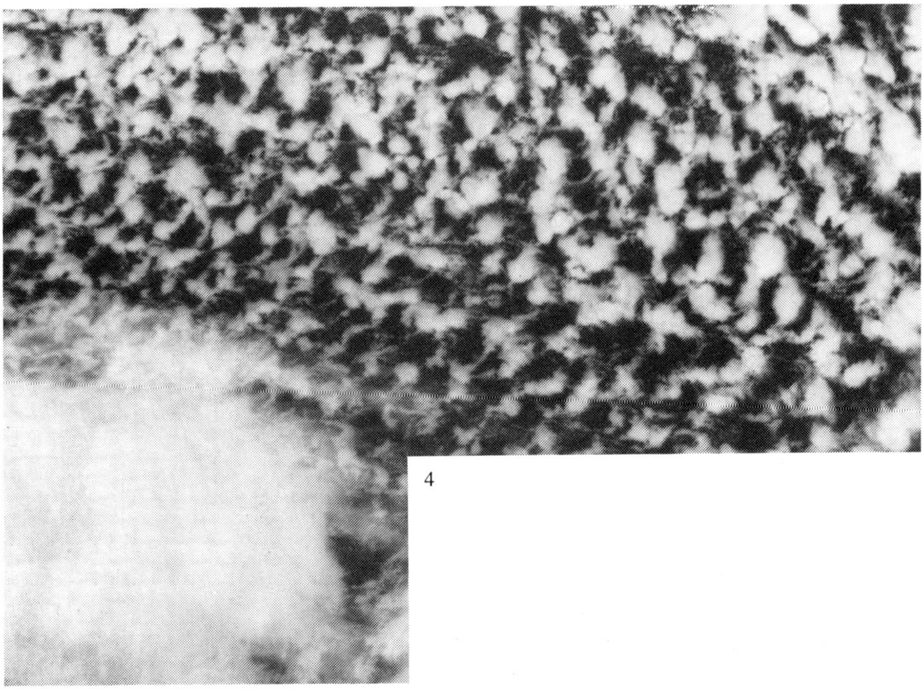

Fig. 13.11.xii 1410,11.3.82;Ch2,Ch3–,Ch4. In channel 2 we see the north coast of Ireland and a little bit of Scotland. The haziness in the west is seen in channel 4 to be thin cirrus. Channel 3– is the most interesting because the small fragments which are where the cloud growth is beginning and the droplets are all small but the older parts of the clouds where the droplets are becoming several microns in diameter, which increases their absorption and increases the scatter to a narrow forward beam which prevents significant upward scatter. In channel 4 the coldest cloud is whitest, which means the tops of the cumulus and an area in the SW of high level cirrus cloud which is composed of ice crystals which scatter sunshine well. The sea and the small cloud fragments are warm, and dark. (Courtesy the University of Dundee.)

(12) Cloud particle growth in ocean stratocumulus (Fig. 13.11.xii)

These clouds are called stratocumulus because they are confined within a stable layer, although from many points of view they are cumulus.

14

Clouds and fallout

14.1 BASIC PRINCIPLES

First, it is helpful to state certain simplifications of the actual complexities which we shall assume. We do this because any departure from these simplifications is negligible compared with the inaccuracies present in our knowledge of the more important factors. We may understand the thermodynamics of condensation physics with great precision, and we may have equations of motion which are equally accurate, but measurements are always crude and so we can never apply the equations with less crudity. There are therefore some mechanisms which can always be ignored in practice. On the other hand, some are basic:

(1) Condensation

There are always enough nuclei present in the atmosphere for condensation of water vapour to take place in the form of droplets as soon as the air is cooled below its dew point. The actual number and their chemical nature makes a difference to the sizes of the droplets, but there are always enough for a cloud to be visible. Although in some cases hygroscopic particles grow large enough to reduce the visibility when the relative humidity is between 70 and 100%, there is no effect of cloud particle size on the motion until rain or freezing occurs. Any degree of supersaturation that might occur is too small to measure in practice and has no measurable or observable effect on the buoyancy or motion except soon after glaciation (see sections 14.1(7), 14.2(6) and 14.7(4) below).

(2) Radiation

Clouds, like snow on the ground, radiate as black bodies. If a cloud is too thick to see through, its upper surface radiates to space in those long infrared wavelengths that are not absorbed by the air above. The water vapour absorbs in a very complicated way, some wavelengths being almost completely absorbed by two metres of air while at other wavelengths the air is virtually transparent. Consequently, cloud is cooling very significantly at its upper surface and is losing heat at around the same rate as the ground on a clear night when dew and frost occur.

On the other hand, clouds, like snow, are warmed only to a very slight extent by sunshine. Almost all the visible radiation and neighbouring wavelengths spanning the incoming solar energy are reflected and scattered by water droplets. In so far as clouds absorb more of this radiation than the clear air it is due to absorption not by the droplets themselves but by the water vapour in the air between them. This is because multiple scattering of the sun's rays by clouds produces a very long path length for light before it is finally directed out of the cloud. Incoming sunlight is either reflected back into space or transmitted ultimately to the ground: the number of times a ray is scattered before leaving a cloud can vary from one or two in very thin cloud to thousands in a deep cloud. The amount of heat absorbed by a deep cloud is not zero, but it is small compared with the latent heat of condensation released or absorbed as a result of up and down motions small in magnitude compared with those which are operating in a cloud anyway.

A cloud radiates in the long (infrared) wavelengths downwards. Thus, since its temperature is quite close to that of the ground below compared with the temperature of outer space (which is virtually absolute zero), a cover of even thin cloud serves to send back to earth a large proportion of the longwave radiation from the ground. Consequently, clouds very greatly reduce the incidence of frost and fog at night. They also reduce very considerably the heat received from outer space, and cloud amount is the **greatest single factor**, followed by snow cover, in reducing the amount of sunshine absorbed by the earth. At the same time they reduce the outgoing radiation by being at a lower temperature than the ground and the radiation is proportional to the 4th power of the absolute temperature.

(3) Cloud formation

Except for cases where water vapour is introduced from sources at high temperature, such as geysers, aircraft exhaust, and industrial chimneys, the only mechanisms producing cloud are adiabatic cooling due to ascent of the air or proximity to a cold surface causing fog formation. Cloud form may be modified by motion which occurs partly as a result of the radiative heat losses of the cloud, and the buoyancy forces which result. Such cloud as is formed by mixing of two unsaturated air masses of different temperature is secondary in the sense that the latent heat of condensation in that case is of no importance in producing motion. Thus billows sometimes produce cloud by such mixing, and fog is often formed by the mixing of cold air near the ground with warmer air above which has not been cooled directly by the ground. By comparison with the heat subsequently lost rather slowly by radiation, the latent heat released by the condensation of such cloud due to mixing is negligible. Compared with the impression in our knowledge of the other more important mechanisms, it is certainly irrelevant in forecasting, or even explaining, the cloud.

(4) Rain formation

Rain formation is very much dependent on the nature of the nuclei on which condensation occurs. A few large (i.e. hygroscopic) nuclei, 100 μm or larger in diameter, among a multitude of much smaller ones will generate rain quickly by collision of droplets of different sizes having different fall speeds (see section 2.10). If there is a narrow dropsize spectrum (all very nearly the same size) the collision mechanism takes very much longer, especially if the droplets are all well below 20 μm in size. See also section 14.7(4).

(5) Electrical effects
Although thunder and lightning make a great show, the amount of energy involved in such displays is very small compared with that involved in the mechanics of the storm. A hailstorm which does not generate lightning is not measurably different from one which does, and although thunderstorms are usually large that is not because of the lightning. Electrical forces do not have any effect comparable with buoyancy forces, and are due to temperature gradients within ice crystals and effects of particle collision. Electrical energy is a trivial byproduct of the buoyancy forces, and the electric fields are produced by a gravitational separation of ice crystals and hailstones. The small crystals are carried up in the updraught, acquiring a positive charge from larger hailstones falling much faster. Cloud particles are usually assumed to acquire about the same temperature as the air around them, but hailstones are warmer than the ice crystals because they are collecting supercooled droplets which release their latent heat of fusion as they freeze onto the hailstones. The ice crystals bounce off, but a charge separation takes place because of the temperature gradient.

The older parts of a glaciated cloud contain very few unfrozen particles, so that the temperature differences are reduced there. The charge separation takes place mainly in the newer, more vigorous updraughts where a large amount of supercooled water exists. The rapid generation of kinetic energy by buoyancy forces in new cloud cells complete dominates any electrical forces on the scale of the air motion. It used to be thought that when a lightning discharge took place a large electrical field was suddenly removed, and that this released a gush of rain or hail previously held in suspension by electrostatic forces. In fact the charge separation occurs because the hail has already fallen, and naturally it would be expected to reach the ground soon after a lightning flash, especially from a new cloud cell.

(6) Cloud droplet temperature
Unless otherwise stated, the temperature of water droplets and ice crystals is very nearly the same as that of the air through which they are falling. Hailstones are an exception of practical importance: they are warmer than the air when they are capturing supercooled droplets inside a cloud, and because of their great fall speed they may arrive at the ground much colder than the air into which they have descended.

Similarly, large rain drops may be cooler than the air, but that is mainly because they take up the wet bulb temperature on falling through the air (see section 14.2(5) p. 470).

(7) Glaciation and supercooling
When air is cooled to its dewpoint, condensation occurs either onto a solid surface or as cloud. On further cooling, the droplets do not freeze at 0°C but may remain liquid and supercooled down to –40°C. Below this temperature, freezing is spontaneous unless there is a high concentration of a dissolved substance such as a sulphate or chloride in the droplet.

At temperatures between 0° and –40°C the time for which a droplet may remain unfrozen depends on the droplet size and the temperature. A cloud of small droplets (less than, say, 5 μm diameter) may remain unfrozen for many hours at –30°C, while raindrops may freeze in a few seconds in a cumulus cloud at –10°C. Thus cumulus clouds glaciate at much higher temperatures than thin-layer clouds. The generation of rain is hastened by a strong updraught and a rapid condensation of liquid water. This is especially so when a few large nuclei are present to provide a wide spectrum of drop sizes, for the drops then collide and coalesce with one another more readily.

Just as the chance of the formation of an incipient crystal is greatly increased by a droplet becoming larger, so it is increased by a decrease in temperature. For practical purposes, glaciation is so rapid below –40°C that it takes place immediately even in the smallest droplets.

(8) Latent heat

The amount of heat liberated by condensation varies slightly with temperature, and is so large as to be a dominant factor in cloud dynamics. Water has a larger latent heat than most commonly occurring vapours, and condensation of only 5 to 10 per cent of the vapour is enough to warm the air by 1°C.

Glaciation of liquid produces only about one sixth as much latent heat as condensation of vapour to liquid water, and is therefore only a modifying influence. The chief, and by far the more important, effect of glaciation is to delay evaporation and so prevent the loss of latent heat of condensation from the cloud.

14.2 THE T–ϕGRAM

In order to represent the state of the atmosphere as revealed by a sounding, two coordinates are required. In the T–ϕgram these are temperature, T, and entropy. Since entropy is proportional to the logarithm of the potential temperature, θ, the vertical scale is a logarithmic one marked out with values of θ. Horizontally we have a linear scale of temperature. **A T–ϕ gram is on page 474.**

(1) Isobars

By (1.6.3) $\theta \propto p^{1/(\gamma-1)}$ at a constant temperature, and so the vertical scale is a logarithmic one for pressure, but with pressure decreasing upwards. The scales may be chosen so that the isobars are inclined at about 45° to the axes, and because we are operating far from the zero of potential temperature the logarithmic scale is not far from linear. As a consequence the isobars are only slightly curved (see Fig. 14.2.i).

(2) Dry adiabatics

If the air is neutrally stratified the potential temperature is the same at all heights, and so a sounding is represented by a line along θ = const., i.e. horizontally in the diagram of the next page. 'Upwards' is across the isobars at a lower pressure. This is called a **dry adiabatic** because it is the line followed by a particle displaced vertically to a different pressure adiabatically without condensation. Thus in Fig. 14.2.i the line AB represents the sounding of a neutral layer of the atmosphere. AC represents a stably stratified layer of the atmosphere because the potential temperature increases upwards (i.e. as pressure decreases and height increases). When a sounding is made the instruments record the pressure and temperature, and these two quantities are therefore used as the coordinates for plotting a sounding on the diagram.

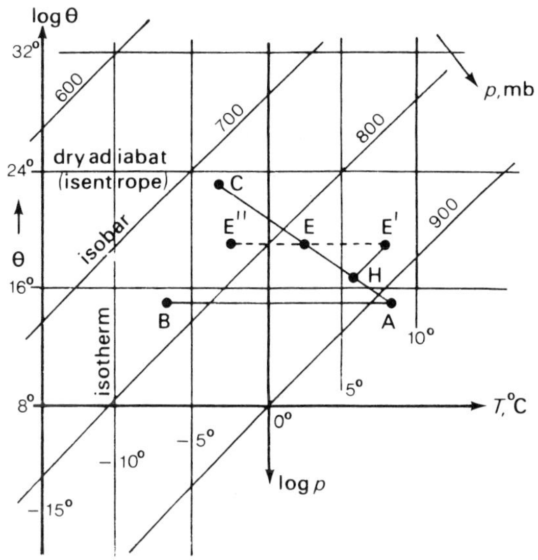

Fig. 14.2.i Part of a T–ϕgram showing isotherms at 5°C intervals and lines of constant potential temperature θ at 8°C intervals. $T = \theta$ on the 1000 mb isobar; the isobars are shown at 100 mb intervals from 900 to 600 mb. A horizontal line such as AB is a dry adiabatic, and neutral stratification on a sounding. AC represents stable stratification, or θ increasing with height (which means decreasing p).

Some meteorological services use a very similar diagram in which the coordinates are T and $\log p$. This is drawn with the axes skew so as to look like the T–ϕgram, and the only difference is that the isobars are straight and the dry adiabatics slightly curved.

A parcel of air displaced downwards adiabatically from E to E′ in the environment AC would be warmer than its new surroundings at H, and would therefore experience a restoring force towards E. If it were displaced upwards adiabatically to E″ it would be colder and denser than its new surroundings and would again suffer a restoring force. The air layer AC is therefore stable.

(3) Humidity mixing ratio: condensation level

This ratio is defined as the number of grams of water mixed with 1 kg of dry air. This is, for practical purposes, the same as the ratio of the partial vapour pressure of the water vapour to the total pressure ÷ 1000. The physical information about water vapour is placed on the diagram for saturated air in the form of lines indicating the **saturation mixing ratio**. These lines are inclined to the left of the vertical (in the diagram) as indicated in Fig. 14.2.ii. A parcel of air whose temperature and pressure placed it at A in Fig. 14.2.ii might have a humidity mixing ratio of, say, 3 g/kg. If it were lifted dry adiabatically to C, which is on the 3 g/kg line, it would become saturated. If the saturation mixing ratio at A were 5 g/kg the **relative humidity** of the air when at A would be 3/5, or 60%. When lifted to C it would be 100%, and the altitude of C is called the **condensation level** of this particular parcel.

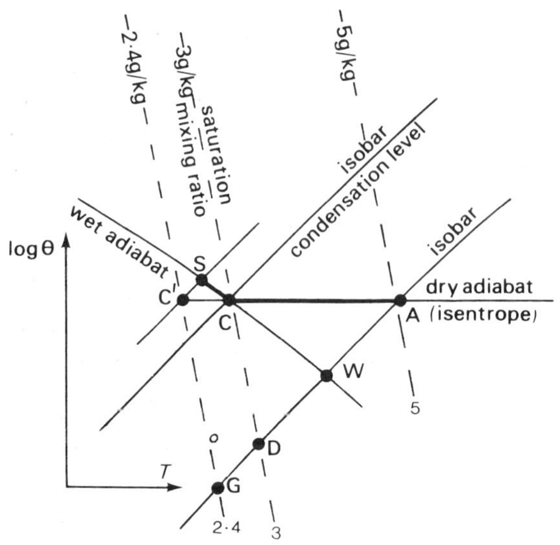

Fig. 14.2.ii The lines of constant saturation humidity mixing radio are inclined upwards towards the left on the T–ϕgram. They are marked in g per kg by means of dashed lines. If a parcel of air lifted from A adiabatically becomes saturated at C its mixing ratio is 3 g/kg. The height, or pressure, at C is called the condensation level. If lifted further, the parcel follows the wet adiabatic up to S. At the pressure of A, air having its condensation level at C has its wet bulb temperature at W and its dewpoint at D.

(4) Wet adiabatics

If the parcel of air now at C were lifted further to the isobar through C′ in Fig. 14.2.ii, some vapour would be condensed out as liquid. Thus instead of cooling dry adiabatically to the point C′ it would be warmed by the latent heat released and would be at S, where less would be condensed out.

The line CS is part of the **wet adiabatic** through C. It is calculated assuming that the liquid water is removed as it is condensed. This is done because the amount present would depend on the level at which the parcel first became saturated, but no very large error results from this. If the liquid water remains with the air, the air cools more slowly on ascent because it is bringing liquid water with it which also has to be cooled. Strictly speaking if air is brought down a wet adiabatic, the water to be evaporated into it must be supplied at the same temperature as the air. In the figure the saturation mixing ratio line through S is 2.7 g/kg, so that 0.3 g of liquid water would be condensed out in rising from C to S.

(5) Normand's theorem: wet bulb and dew point

If we now draw the isobar through A to meet the wet adiabatic through C, at W, the temperature to which the air would be cooled if water were evaporated into it in descending from C to the pressure of A, instead of descending dry to the temperature of A. A wet bulb (an ordinary mercury-in-glass thermometer covered by muslin which is kept wet by a wick) is cooled by evaporation, and because the temperature of the thermometer bulb is steady all the heat for the evaporation comes from the air, the only difference being that in the case of a wet bulb the water is evaporated at the wet bulb temperature and not at the variable temperature along CW. Nevertheless, the difference is small and for practical purposes W may be taken as the wet bulb temperature of the air at A. Errors occur owing to the conduction of heat down the thermometer stem to the wet bulb from the dry end: theoretically therefore the whole thermometer should be enclosed in a wet surface. The best way to minimise this effect is to increase the rate of evaporation by increasing the ventilation speed of the wet bulb, so that the total heat lost is greater, and insulating the rest of the thermometer in an outer casing of glass.

The constant mixing ratio line through C meets the isobar through A in D, which is the **dew point**, namely the temperature to which the air at A must be cooled at constant pressure to become saturated. **Normand's theorem** states that the dry adiabatic through the dry bulb temperature meets the wet adiabatic through the wet bulb temperature and the mixing ratio line through the dew point in a point at the condensation level.

If the air at A is cooled to G, in Fig. 14.2.ii, it can only contain 2.4 g/kg as water vapour. Thus 0.6 g must be condensed out either as dew on the cold surface producing the cooling, or in the air as fog, if the cooling is largely by radiation loss from the air (mainly, from the water vapour in it) to the cold surface. In the former case some of the air is cooled well below the dewpoint, and dew is deposited when the wind is light. If the air becomes stirred up in the morning the cooling is spread over a greater depth and fog is formed by the mechanism of mixing two parcels at different temperatures (see section 14.3).

(6) Frost point: ice evaporation level

The vapour pressure over ice is less than that over a water surface at the same temperature, which means that the mixing ratio lines for saturation over ice differ from those for water by an amount which increases as the temperature falls below 0°C. As the air is cooled on the isobar (i.e. *in situ*) through A the frost point F is reached before the dew point D (in Fig. 14.2.iii). If the air is cooled by lifting along the dry adiabatic, no condensation occurs until the water condensation level is reached at C.

If a finite amount of water were condensed out by further ascent to S and then the condensed water were frozen, first the latent heat of fusion would be liberated and then further sublimation of water vapour onto the ice crystals would take place. This would cause further warming until the air achieved the state represented by T, which is on the saturated adiabatic line representing ice saturation with the same total water content as at C, but with ice instead of liquid water.

If the air is now caused to descend from T, remaining saturated for ice by the ice crystals evaporating into it, the ice will not all be evaporated until E is reached. This is the point where the ice adiabatic through T meets the dry adiabatic through A, and it lies on the ice saturation mixing ratio line through F. The saturation mixing ratio line for water through C and D meets that for ice through E and F on the 0°C isotherm. Likewise the wet adiabatic through C (and S) meets the ice adiabatic through E (and T) on the 0°C isotherm. If at the pressure of A the water on the wet bulb is frozen, the temperature it will record is given by W′ on the isobar through A and the ice adiabatic through E.

E is not said to be at the ice condensation level because ice crystals do not form out of the vapour as the pressure falls. But if a frozen cloud is present, at T say, it will not

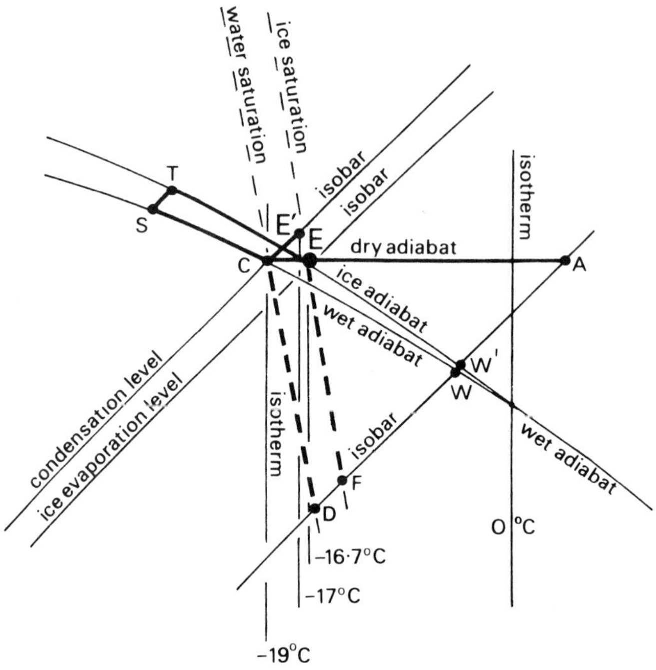

Fig. 14.2.iii The water and ice saturated adiabatic lines for an air parcel colder than 0°C meet on the 0°C isotherm. If the parcel originates at A and has condensation level at C its wet bulb (supercooled) and dewpoint temperatures are at W and D. But if the bulb is ice covered, the 'wet' bulb temperature is W′ and the frost point, at which frost begins to form on a cooled surface, is at F. Strictly F should be measured at the temperature at which frost begins to evaporate from a surface as it is warmed, because sometimes the first deposition of ice on a clean surface may occur at a temperature below that of F. If condensed water remains supercooled the parcel may be on the wet adiabat CS, but if the droplets freeze the parcel moves to TE, the ice adiabat for the same total water content. Saturation for ice at that water content is on the ice saturation line through E which meets the water saturation line through C on the 0°C isotherm, off the bottom of the diagram. Because condensation to ice does not occur out of clear air when it is cooled by ascent, the isobar through E is a pressure level called the ice evaporation level.

disappear by evaporation when the air descends until it reaches E. This is therefore called **ice evaporation level**.

The difference between the ice and water adiabatics and saturation mixing ratio lines can be expressed simply by the difference between the frost point and dew point at the same pressure for different values of the dew point. This is given in Table 14.2.i, in which T_f and T_d are the frost point and dew point respectively. Thus if the dew point is –19°C, the frost point is –17°C; air which at –19°C condensed cloud which then froze would not have it evaporated until it had descended rather more than 200 m lower than the condensation level. For this case the isobar through C meets the ice saturation line in E′, which is 2° warmer. Thus E is about 2.3° warmer than C and is 230 m lower. There are thus below the –19°C isotherm 230 m of air where an ice cloud would persist whose particles might grow, but where a water cloud would evaporate.

Table 14.2.i

T_d,	°C	0	–4.4	–9	–14	–19.4	–24.5	–32.2	–40.4
$T_f - T_d$,	°C	0	0.5	1	1.5	2	2.5	3	3.5

(7) Virtual temperature

Water vapour has a density 0.622 times that of air at the same temperature and pressure. Thus if air is 1% water vapour, i.e. has a mixing ratio of 10 g/kg, its density is reduced to 99.622% of the density of dry air. At a temperature of 270 K, a parcel of air having 1 g/kg more water vapour than another but having the same density might be 0.0388% cooler, i.e. 0.11°C cooler. Roughly, a 1 g/kg difference in water content is equivalent to 0.1°C temperature difference.

The **virtual temperature** of air is the temperature of dry air with the same pressure and density. Meteorologically this is an effect of negligible importance, but it can have important effects on the refractive index, which is influenced both by humidity and by temperature (Scorer 1961). Thus if the humidity is not uniform because of the irregular evaporation of cloud, scattering of radio waves and the refraction of light rays as seen in a powerful telescope may occur even though there are no differences in density present. Such a distribution of refractive index used to be ascribed to turbulence when it was first observed by means of radar scattering. In air of uniform composition the velocities required to produce the pressure fluctuations needed to produce the same scattering were so large as to be very unlikely. This humidity effect is a more likely explanation, and the phenomenon was given the name 'fossil turbulence' because it indicated where there had been some motion.

(8) Conservative properties of air

If a parcel of air is not mixed with others and loses no heat by radiation its potential temperature is conserved. If condensation occurs it moves along a wet adiabatic. If cloud condenses and some of the liquid water falls out and it then descends below the point where all the cloud evaporates, the same wet adiabatic remains characteristic of it. If on its descent, water which was not there before is evaporated into it, its wet bulb remains unchanged, just as it remains unchanged if water is evaporated into it at constant pressure. Thus only mixing, conduction or radiation can alter the wet adiabatic on which the parcel will move. This line can be identified by the temperature at which it crosses the 1000 mb isobar, giving the wet bulb temperature in the parcel at 1000 mb. This conservation property is called the **wet bulb potential temperature** of the air.

14.3 CONTRAIL PHYSICS: MINTRA

Two parcels of air of different density will always mix and in stably stratified surroundings will gradually find an equilibrium level (see section 14.5). But if some cloud droplets exist and the ambient air is not saturated, the evaporation of some droplets extracts some latent heat and the cooling will generate fresh mixing until all the cloud is evaporated.

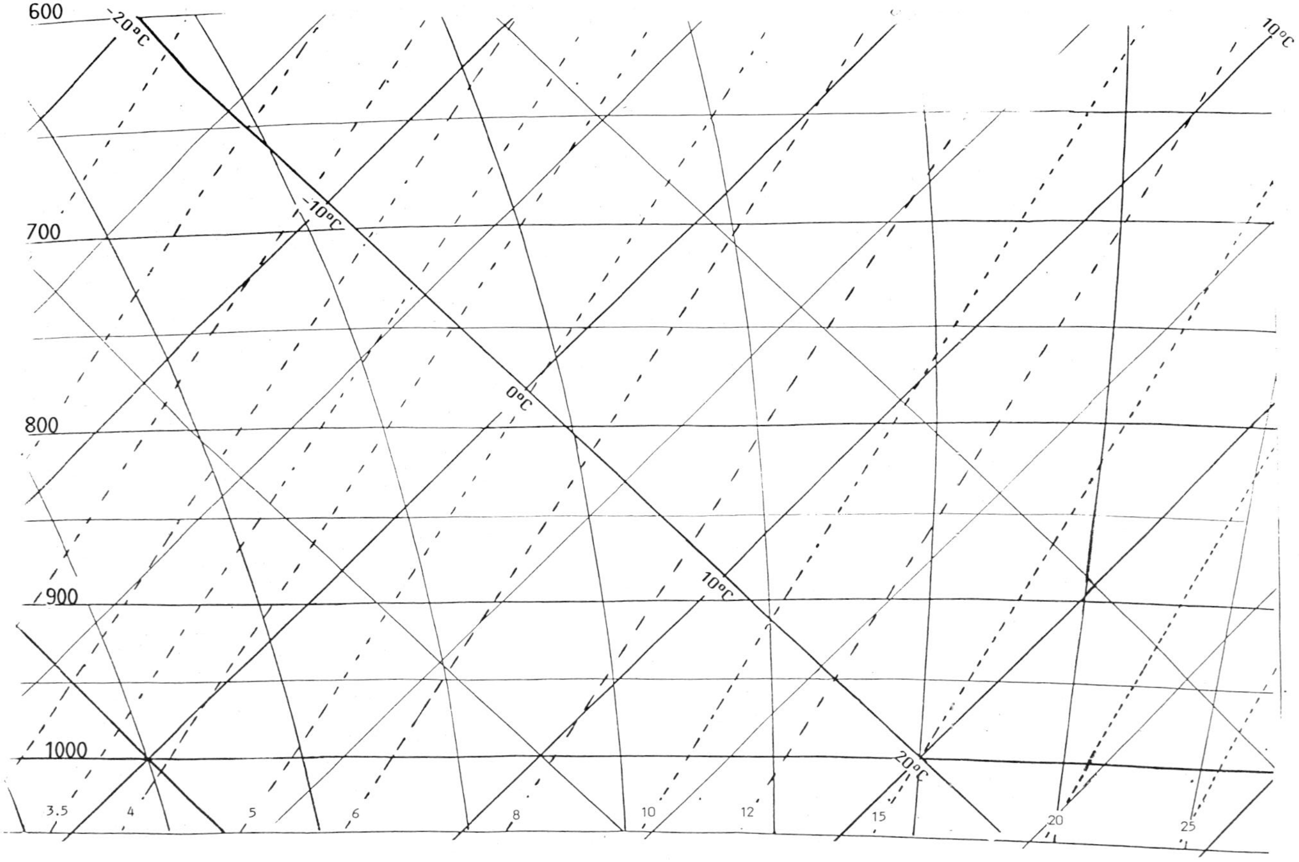

(a)

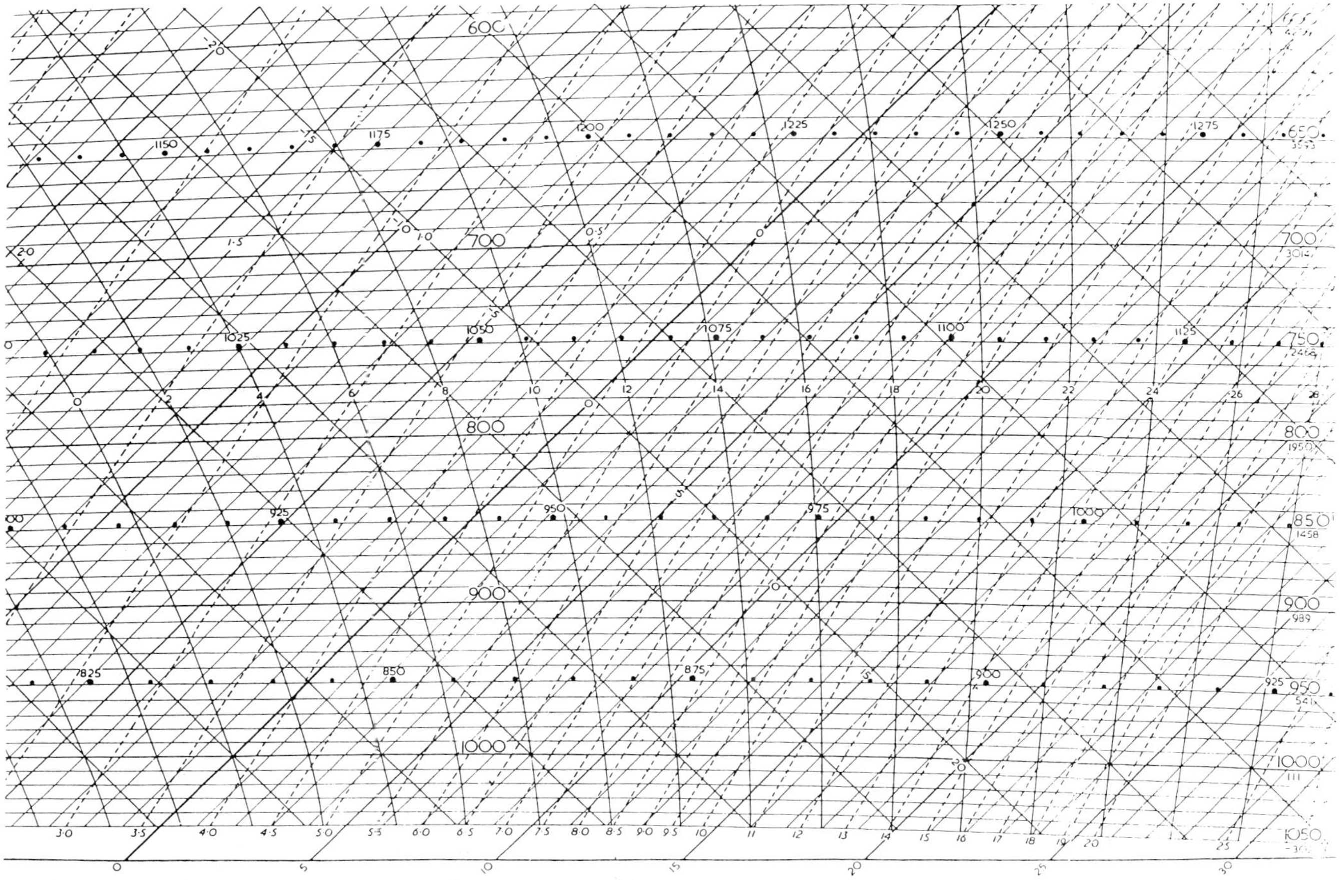

(b)

Fig. 14.2.iv (a) A T–ϕgram suitable for photocopying and experimental scribbling. (b) A T–ϕgram with layer thickness marked at mid-points of 100 mb layers at average temperature of the layer, and extended to 1050 mb.

Thus a water droplet cloud is always in a state of mixing with clear air at the same level. Fig. 14.3.i shows the saturation mixing ratio for air containing water droplets or ice particles at a pressure of 250 mb.

To use the diagram to determine the conditions for the formation of a **contrail** we find the direction E of the point representing the exhaust and draw a straight line from the point A representing the ambient air in the direction of E. As the exhaust mixes with ambient air its point moves down this line towards A. At the point M the mixture would have formed a cloud of water droplets. On passing the point D the droplets would begin to evaporate, but the fact that A is in the doubly shaded region means that if the droplets became frozen they would not evaporate and the contrail would be **persistent** because A

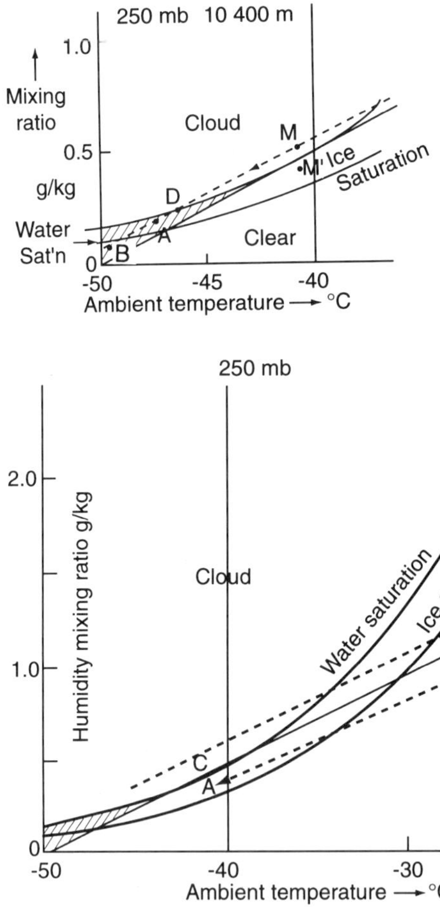

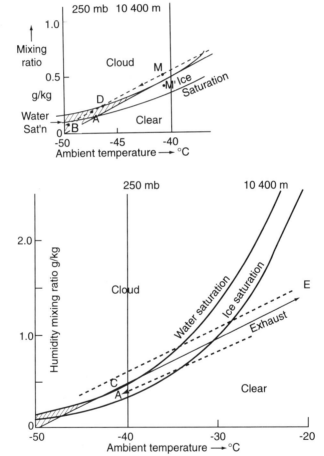

Fig. 14.3.i This diagram is drawn for a pressure of 250 mb because it will be seen that at much lower pressures the air is usually too dry and at higher pressures too warm for the formation of persistent condensation trails (**contrails**). The two main curves show the temperature at which the air is saturated with water vapour in the presence of water (upper curve) and ice (lower curve) at a range of temperatures. The coordinates are saturation mixing ratio, in g per kg of dry air and temperature. The range between is that within which persistent contrails can be formed.

is above the ice saturation line. Normally the droplets would not freeze immediately at temperatures warmer than –40° but would remain liquid, and would begin to evaporate unless the pressure was low enough for the water saturation line to be below A, which it could be in the core of one of the vortices. That happens in the core of the vortices formed by the tips of the propeller blades as illustrated in Fig. 14.4.x (p. 486) and sometimes there is condensation of cloud on the upper surface of the wing of a plane propelled by jet engines. But the conditions may be such that the air is nearly saturated already so that the pressure in the propeller tip vortices when producing maximum thrust is lower than at the condensation level of the ambient air. Condensation sometimes occurs in the flap tip vortices when the control surfaces are arranged to maximise the lift (see Fig. 14.4.ix). For a contrail to persist at high altitude the temperature has to be colder than –40°, for then the droplets freeze almost instantaneously. If the temperature is only a little warmer, only a few of the drops may freeze and then the frozen ones would absorb the vapour which comes from the others evaporating, which they must do unless the ambient air contains a water droplet cloud, which it presumably does not in the case being considered. If the ambient air were at B, in the half-shaded region, even a cloud of ice particles would evaporate.

Thus particles may remain as a cloud if the air is saturated for ice. Equally if there is no cloud in air that is supersaturated for ice but unsaturated for water there will continue to be no cloud until some ice crystals are introduced. If an aircraft deposits exhaust which, by mixing, makes the air saturated for water so that a water cloud is formed and it becomes frozen before further dilution with ambient air evaporates it, then a **persistent contrail** is formed. Fig. 4.3.ii shows the 'mintra' below which persistent trails cannot be formed.

In forming this persistent contrail the time between the freezing of the first and the last droplet to freeze determines how large the first droplets to freeze become, for when viewed by satellite in channel 3– large ice crystals will completely absorb sunshine while very small ice crystals will scatter appreciable amounts of sunshine and the trail will be seen as bright. In bromide prints of the channel 3– signal, a contrail of large crystals appears black, while one of very small crystals appears white. The trails are usually dense enough to cast sharp shadows on to lower cloud, and these look black. The larger crystals can produce the optical halo phenomena when seen from below by primary refraction (i.e. in one crystal only per light ray) and the colours would be superposed on the white-looking cloud which is the result of multiple scattering (i.e. by encounter of the rays with many crystals scattering all colours in all directions). The two coloured spectra most commonly seen in contrails are mock suns (parhelia) and bits of a circumzenithal arc, or presumably the colours of a circumhorizontal arc, all of which require prisms with vertical axes and horizontal top and bottom faces.

The temperature in aircraft exhaust, and the mixing ratio of water vapour with the other components, is represented by a point far off the diagram in the direction of the arrow. As the exhaust mixes with the ambient air, the point representing the mixture moves along a straight line from the exhaust point towards the ambient air point which is imagined to be at a temperature colder than –40° and below the water vapour saturation line. If this line joining the two points crosses into the area above the saturation line, cloud must be condensed in the mixture. Thus a cloud may be formed at temperatures

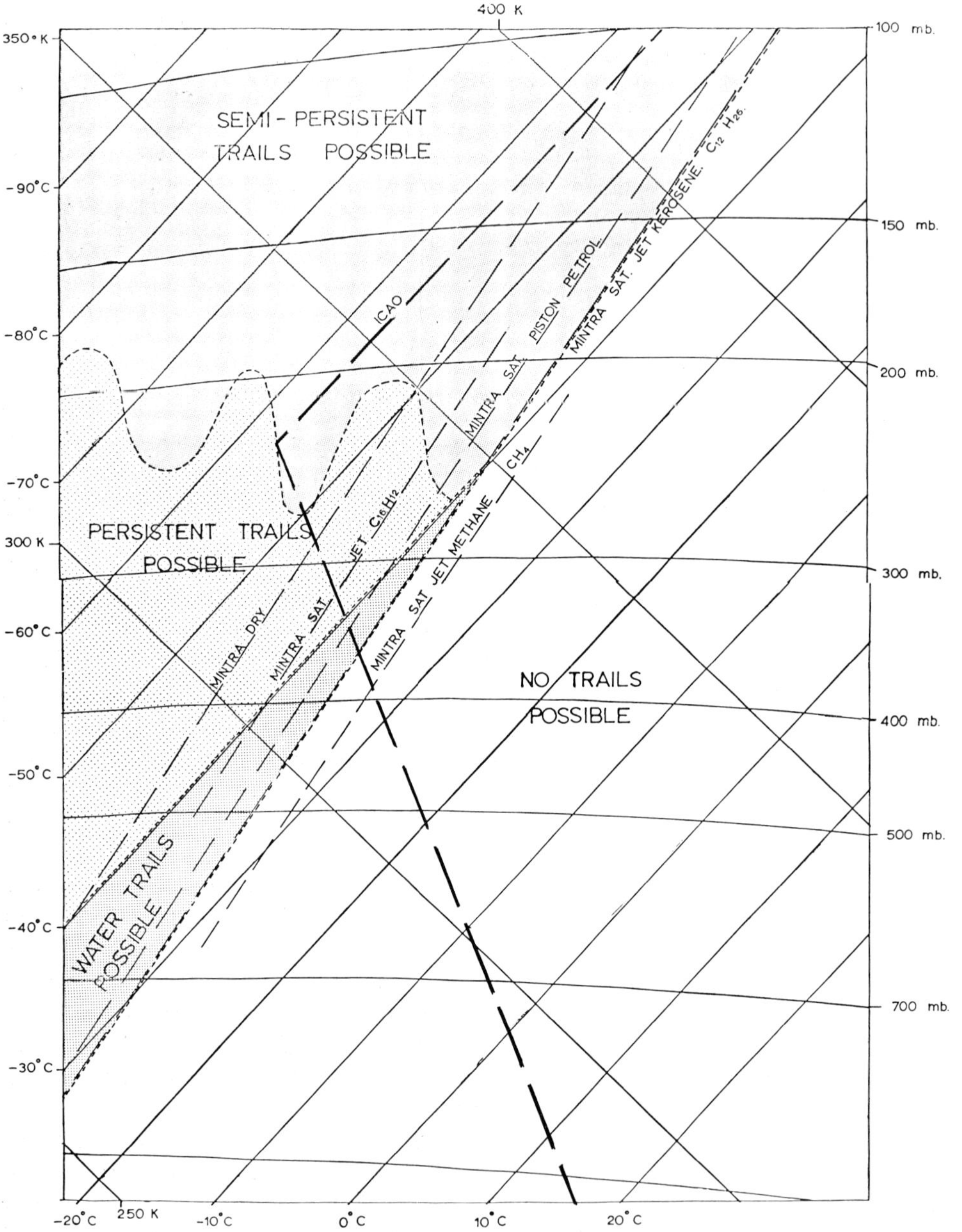

Fig. 14.3.ii A T-ϕgram placed so that the isobars are roughly horizontal. The Mintra lines are drawn assuming a known proportion of hydrogen in the fuel and knowing the temperature of the exhaust and therefore the position of the point E. Trails cannot be persistent in ambient air whose representing point is to the right of that line. The ICAO standard atmosphere from which the ambient air does not deviate by more than about 20°, other than very exceptionally, then requires the trails to be formed close to the tropopause at around 250 mb, which is why the saturation curves drawn in Fig. 14.3.i are for 250 mb.

warmer than –40° but further mixing will cause evaporation if the ambient air is not saturated. A cloud of ice particles can only be formed by freezing the droplets of a water cloud because there do not exist particles which act as CCNs for the direct formation of ice crystals out of the air. If the temperature of the (clear) ambient air and its mixing ratio place its point between the saturation lines for water and for ice, the ice crystals will grow until the point has been brought down to the ice saturation value. Thus small patches of ice cloud may persist, or even grow, over a period of several hours, but still remain thin and appear dark to an airborne observer with the sun behind him, though bright when seen from the ground.

14.4 CONTRAIL DYNAMICS

When trails form they first appear about 30 m behind the engine. The cloud is composed of water droplets (and has been seen to show iridescence which indicates that it was not frozen), but quite soon freezing begins, depending on the circumstances, and persistence of the trail indicates that it is frozen except for the case of condensation occurring in the cores of the wingtip vortices due to the low pressure. Fig. 14.4.i shows the motion streamlines inside the accompanying fluid (AF) of the trailing vortices (see section 11.10). In the centre we see a puff indicating a cross-section of the trail formed behind a single central engine. To the right we have to puffs indicating sections of the starboard trails of a four-engined aircraft. In the left half are shown three positions of a line of particles originally in line with the trailing edge of the wing. While the outer of the three particles makes more than one rotation round the vortex the two inner ones make two-thirds and one-eighth of a rotation respectively. Consequently, if viewed from below, the two inner trails from a four-engined aircraft will pass beneath the outer trails to a position outside them (Fig. 14.4.viii). All the time some diffusion and shearing takes place and they may no longer be distinguishable as four separate trails after about 20 s.

Since the trail from an outer engine is often drawn close to the centre of the vortex on each side it usually persists longer because the pressure is lower there, perhaps by as much as 50 mb. About 30 s behind the aircraft therefore the trail often has the appearance of two tubes of cloud, in which the flow appears to be non-turbulent. Much of the cloud from the inner engines, the amount depending on their position, passes close to the outside of the accompanying fluid and is detrained above the upper stagnation point as described in Fig. 11.10.iv. It is therefore seen as a vertical curtain of cloud stretching up to the level at which it was first detrained (Fig. 14.4.ii).

The vortex pair is an unstable configuration because it is accelerating downwards and any part that is slightly ahead of another part will appear as a downward and inward distortion of the trail (Fig. 14.4.iii). If there is a trail from a central engine it will appear in this motion pattern as a blob of the trail pushed downwards (Fig. 14.4.iv). However if only two trails are left in the cores of the vortices, where they are distorted downwards they quickly become unstable because of the entrainment of mixed air from the outer boundary of the accompanying fluid down the trail's centre, as described in section 11.10. If there is no contrail the passage of an aircraft may leave a **distrail** (Fig. 14.4.v), or a row of holes in a thin layer of cloud (Fig. 14.4.vi) in the place of the downward puffs. Fig. 14.4.vii shows the evolution of the pattern in the case of an artificial trail pair

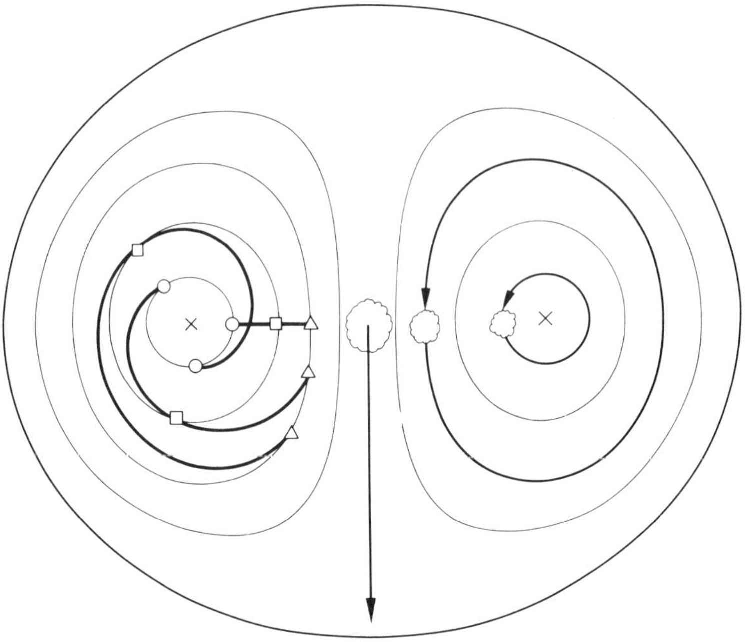

Fig. 14.4.i The streamlines of the flow inside the AF of a vortex pair which is approximately the velocity pattern in the cross-section of an aircraft wake initially. The vortices are marked X. In the right half are shown the tracks taken by the exhaust trails from outboard engines, and in the centre of the track taken by the exhaust from a central engine. On the left are shown the successive simultaneous positions of three particles initially in a horizontal line, together with successive positions of the line of particles joining them. Thus the inner engine trails rotate around the outer ones.

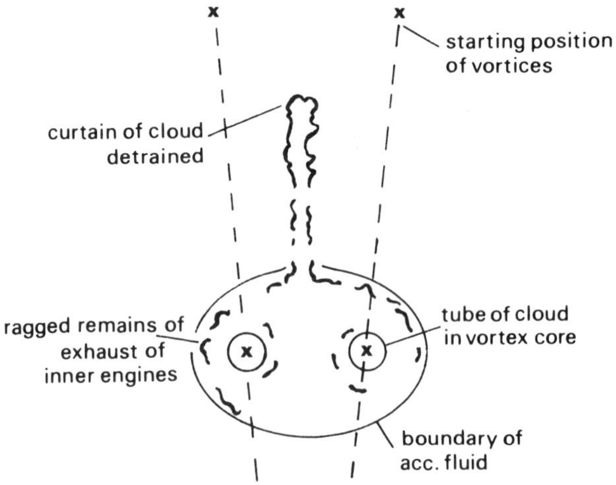

Fig. 14.4.ii The shape of a contrail cloud 30 s or more behind a four-engined aircraft. A three-engined aircraft will probably show a more distinct curtain of cloud with a sharp upper edge because the trail from a central engine is more quickly detrained at the top of the accompanying fluid than that from an inner outboard engine.

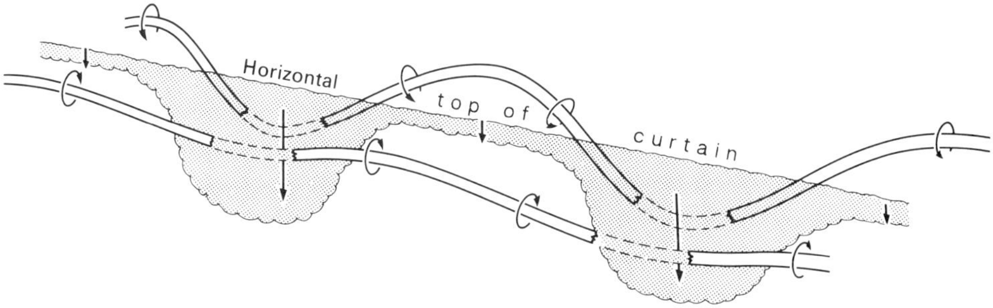

Fig. 14.4.iii The parts which are closer together have descended further than those that are further apart. This view from above and on the right.

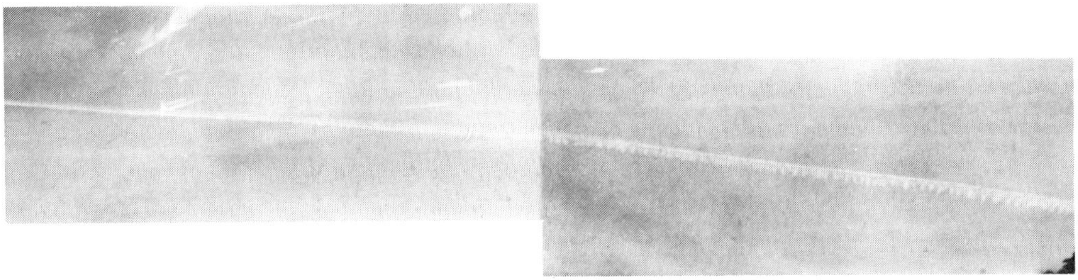

(a)

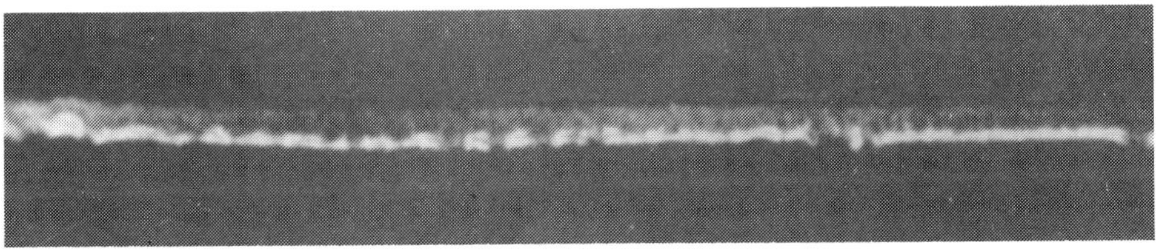

(b)

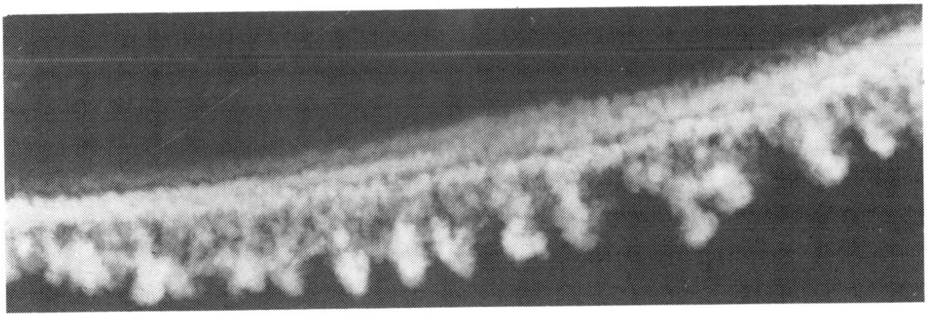

(c)

Fig. 14.4.iv (a) The development of the trail from a single (central) engined aircraft. The curtain top remains in position while the 'puffs' develop downwards. (b) Side view of a trail from a three-engined plane showing the curtain of cloud mainly from the central engine and the tube clouds in the vortex cores which are denser because of the lower pressure. The tubes are beginning to break into loops. (c) A view from below of an aircraft trail from four-engined aircraft on a turn. The downwash now has an outward component, which makes the shape visible from vertically below. The blobs are where the vortices have burst and the tube-like clouds in the vortices diluted. The blobs occur on the parts of the trails that are bent downwards (see Fig. 14.4.iii).

in the wingtip vortices. Fig. 14.4.xi shows the effect of the axial velocity in a wingtip vortex, which is a complication, as is vorticity from wing flap tips (Fig. 14.4.x).

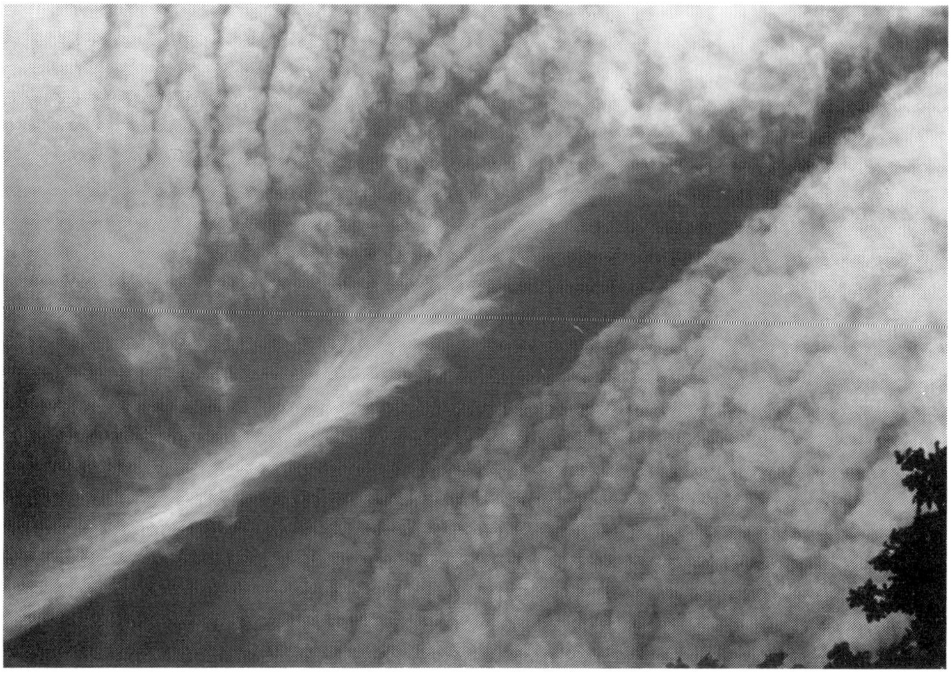

Fig. 14.4.v An aircraft flying very close to, or in, the top of a thin cloud layer of supercoded droplets generates some ice cloud but causes the water droplets to evaporate. (Courtesy of D. T. Tribble.)

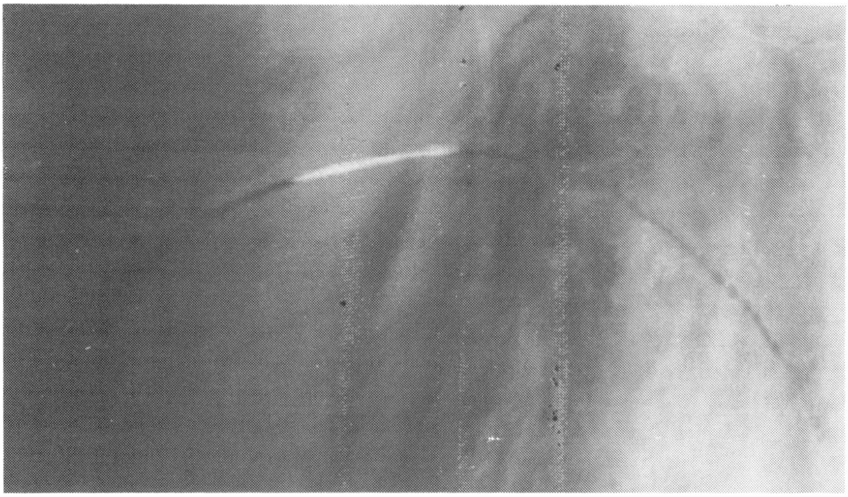

Fig. 14.4.vi The track of a plane which begins to create a distrail which becomes a frozen contrail, and then later a renewed distrail which displays as a row of holes. This photo was taken looking vertically upwards.

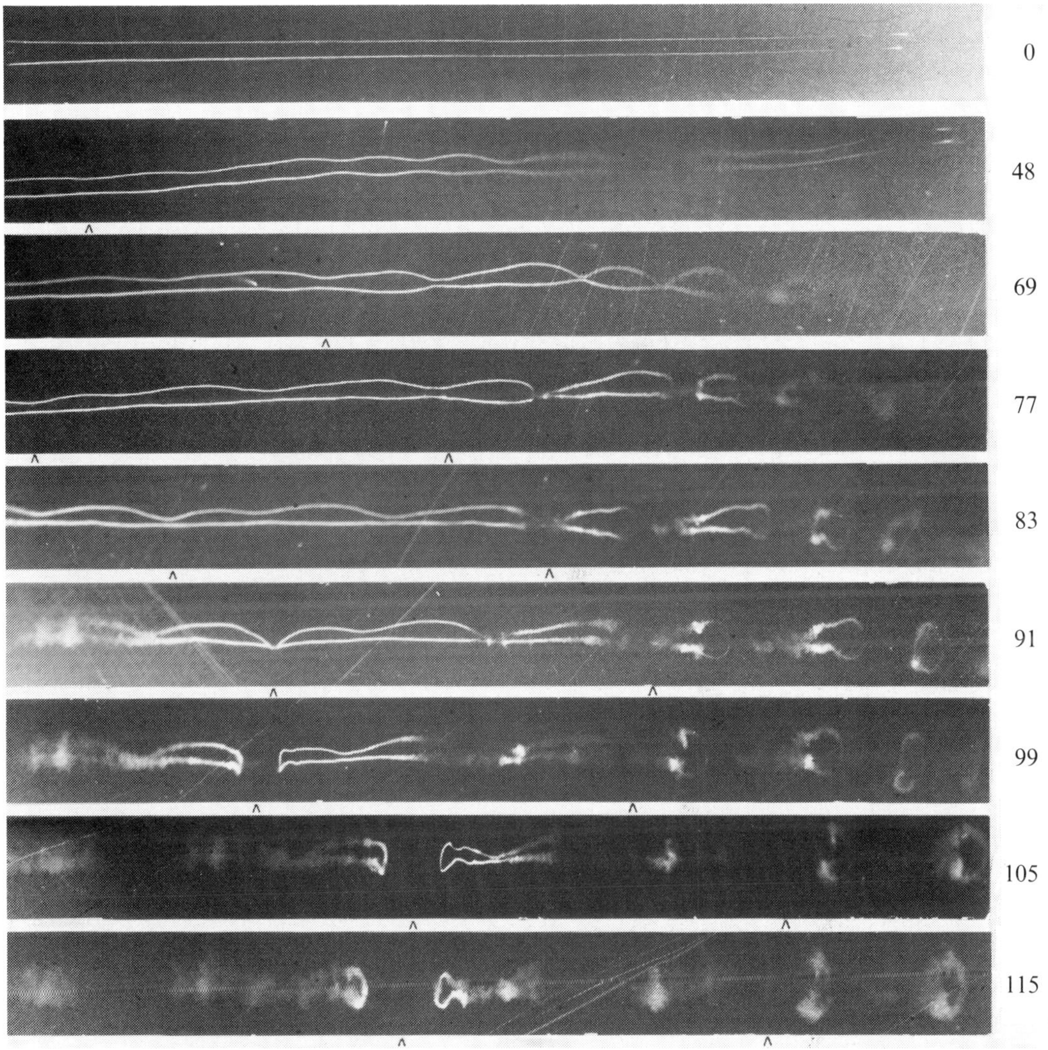

Fig. 14.4.vii Smoke trails from the wingtips of a Comet aircraft flying at 10 000 ft photographed from below at the times given in seconds after the passage of the aircraft. The arrows identify the same points of the trail, one of which is beyond the left end of the picture at 69 s. (Courtesy the Ministry of Aviation Supply.)

If an aircraft flies above a layer of cloud its downwash produces a clear lane in it (Fig. 14.4.v), and if this happens to be so placed that the blobs reach to the bottom of the layer a series of holes is made in it (Fig. 14.4.vi).

The size of the ice particles in an aircraft trail depends on the duration of the frozen droplets among supercooled unfrozen ones—as seen by satellite channel 3– (Fig. 14.4.ix(b)) and in a cloud seeded by the passage of an aircraft (Fig. 14.4.xii), which leads directly into the next section (and Plates 9 and 12 (in the colour section between pages

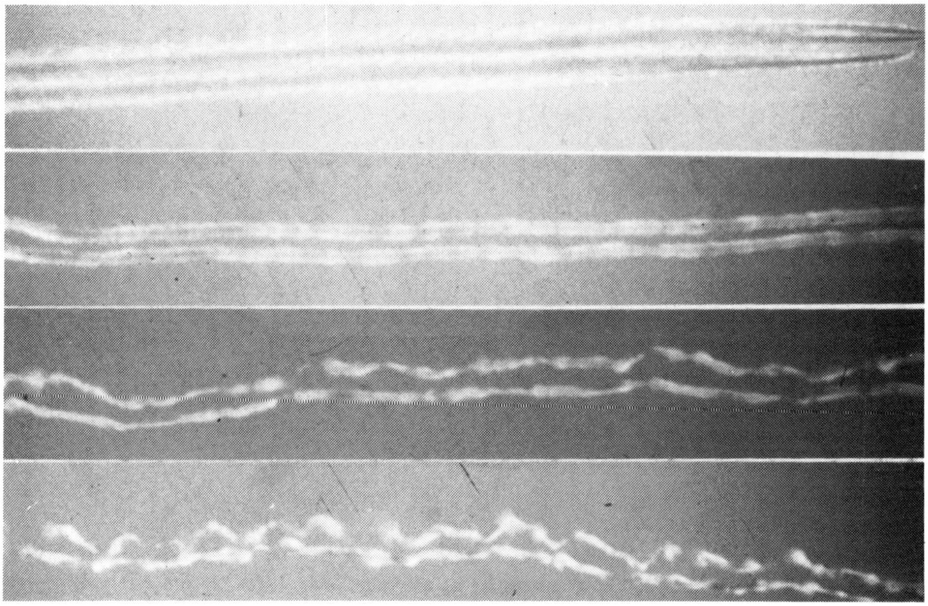

Fig. 14.4.viii Four successive pictures of an aircraft trail from a four-engined aircraft. In the first the downwash which rotates the inner engine trails around the outer can be seen to produce a different appearance of the two pairs when seen at an angle of elevation of about 30°. In the later pictures only the trails from the outer engines remain, surrounding the vortex cores. In the fourth picture the near (upper) trail appears more distorted than the far one because of the angle of view from below.

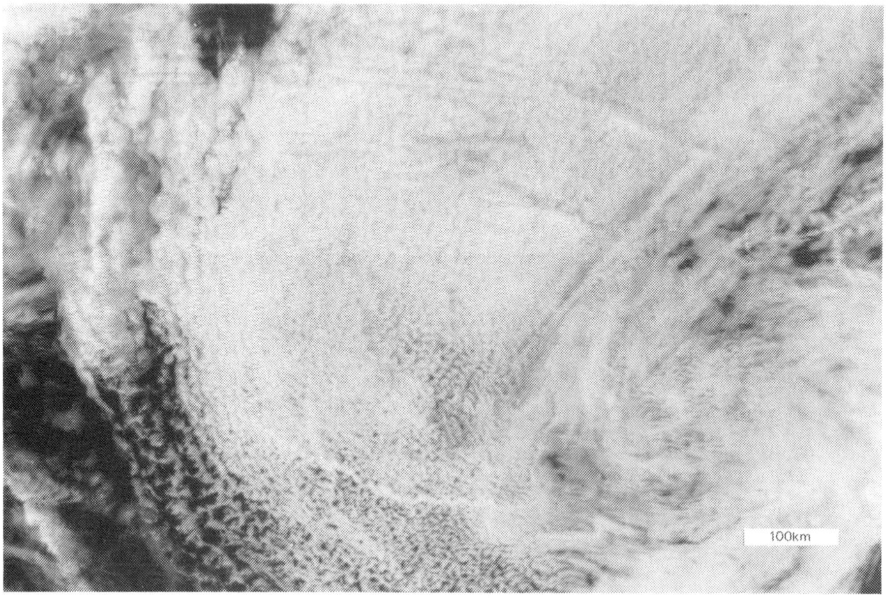

(a)

Fig. 14.4.ix (*cont.*)

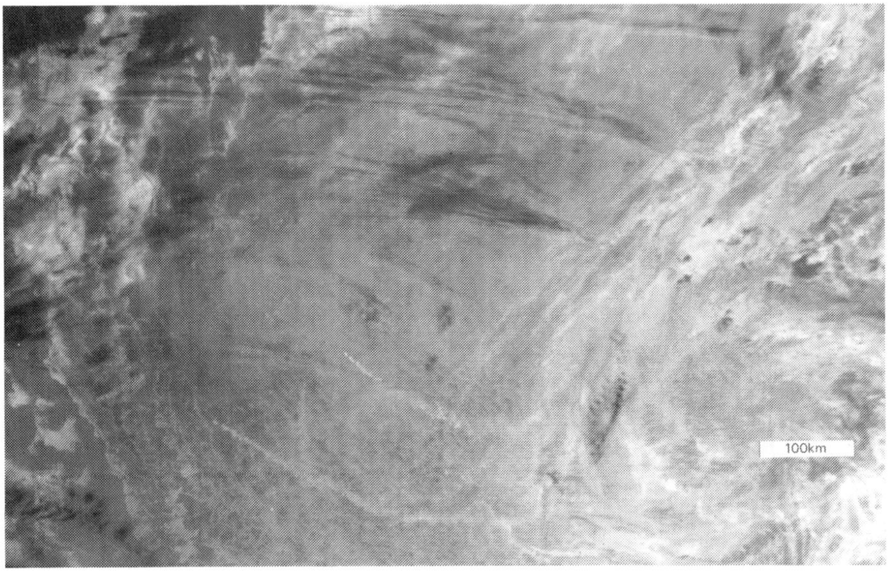

(b)

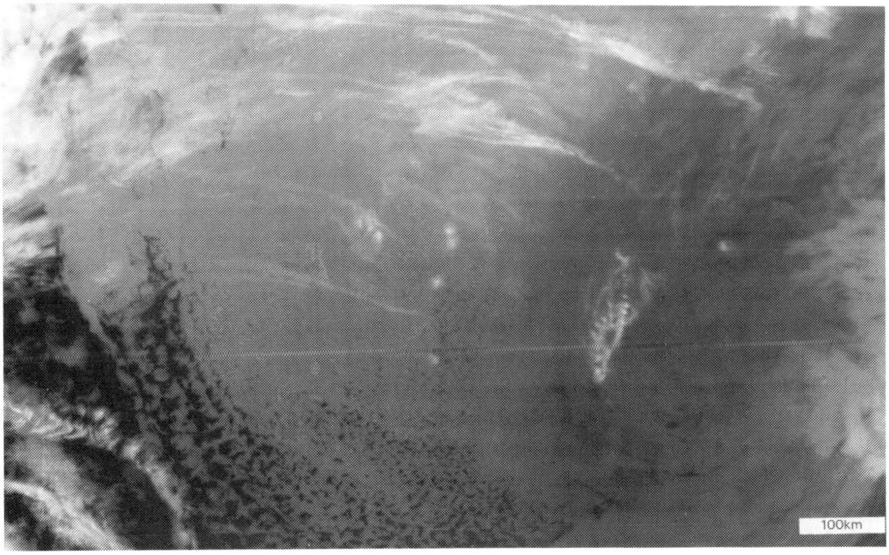

(c)

Fig. 14.4.ix (a), (b), (c) 1707,30.3.87;Ch2,Ch3–,Ch4. These pictures are of a scene with contrails above and ship trails in low cloud on the west side of the North Atlantic SE of Newfoundland. (a) The channel 2 picture shows near the bottom a clearly defined ship trail composed of whitened elements of the low cloud. Higher up there are several contrails which are most effectively discovered by their dark shadows on the white cloud below. (b) The shadows are much darker in channel 3– and two more ship trails are now clear. In this case the contrails are white in channel 3–, meaning that they are composed of very small particles and must have frozen very quickly after condensation as water cloud. (C) The ship trails are at the same temperature as the low cloud and therefore cannot be detected by channel 4 except with difficulty. The contrails are the coldest, and therefore the whitest cloud. (Courtesy the University of Dundee.)

352 and 353) showing the power of vortices due to thrust and lift of an aircraft to cause condensation and freezing).

Fig. 14.4.x The vortex trailing from the tip of a flap lowered from the trailing edge of the wing in order to increase the lift at the lower air speed on coming in to land.

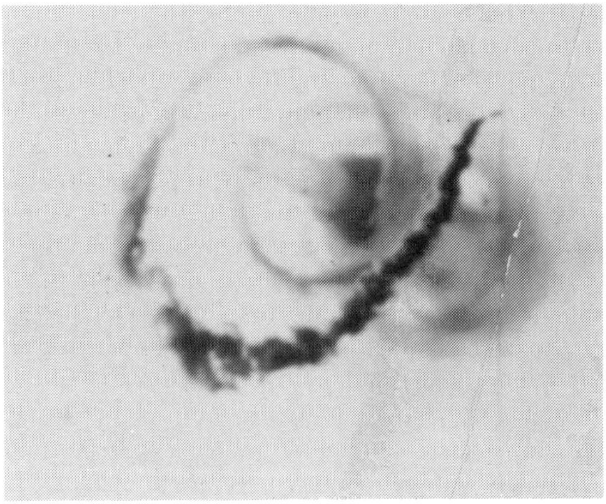

Fig. 14.4.xi An experiment in which an aircraft flew close to a smoke generator on a tall pole (during a study of the behaviour of a vortex pair on reaching the ground). For our consideration we see that there is a velocity along the vortex tube appreciably varying radially, which makes any 2-D mathematical model of the motion less relevant. (Courtesy the Ministry of Aviation Supply, Aero Flight Division 1970.)

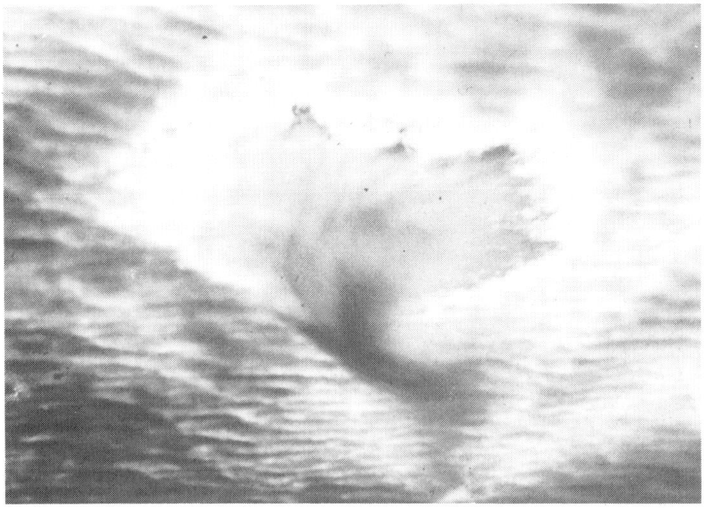

Fig. 14.4.xii The fall of ice particles from this hole in a cloud of supercooled water droplets was caused by the passage of an aeroplane taking off from Baltimore airport. The freezing of some of the droplets on contact with the aircraft without remaining stuck to the surface has the same effect as other methods of seeding the glaciation: the droplets freeze first on an enclosing shell, but the expansion of the interior causes the emission of ice spicules from cracks in the shell which seed the freezing of neighbouring droplets, so that the fallout spreads. Plate 12 in the colour section between pages 352 and 353 illustrates this same fallstreak hole when it had become large enough to let direct sunlight fall on the frozen fallout and produce a mock sun in the fallstreak.

14.5 OROGRAPHIC CIRRUS

The sequence which produces persistent contrails operates, with one major difference in mountain wave clouds (see Plates 10 and 11 in the colour section between pages 352 and 353), to produce a trail of **orographic cirrus** cloud of which Figs 14.5.i (p. 488) and 14.5.ii (pp. 490–1) show good examples. The cloud is formed when the mountain wave carries the air above its condensation level but does not take it below the ice evaporation level even when the water droplets all evaporate when it returns to below the condensation level. Wave clouds often show brightly iridescent colours which certainly indicate the presence of spherical particles, but the cloud may yet produce a cirrus trail when some of the droplets are frozen. The colours are of the same origin as the colours in a corona (see colour section) which involves single scattering, and therefore is usually seen in the edges of a cloud, but is much more common to great effect in wave clouds because all the droplets are of the same size through having had the same condensation history, particularly on the upstream side of the cloud. The upwind point of formation of the wave cloud may have been above the ice evaporation level but there may have been no ice particle clouds in the neighbourhood.

The most perplexing aspect of these wave clouds is well illustrated in Fig. 14.5.ii (pp. 490–1) where the maximum upward displacement of the air occurs where the air at the surface is **descending** from a plateau, or even has recently passed over a ridge which was

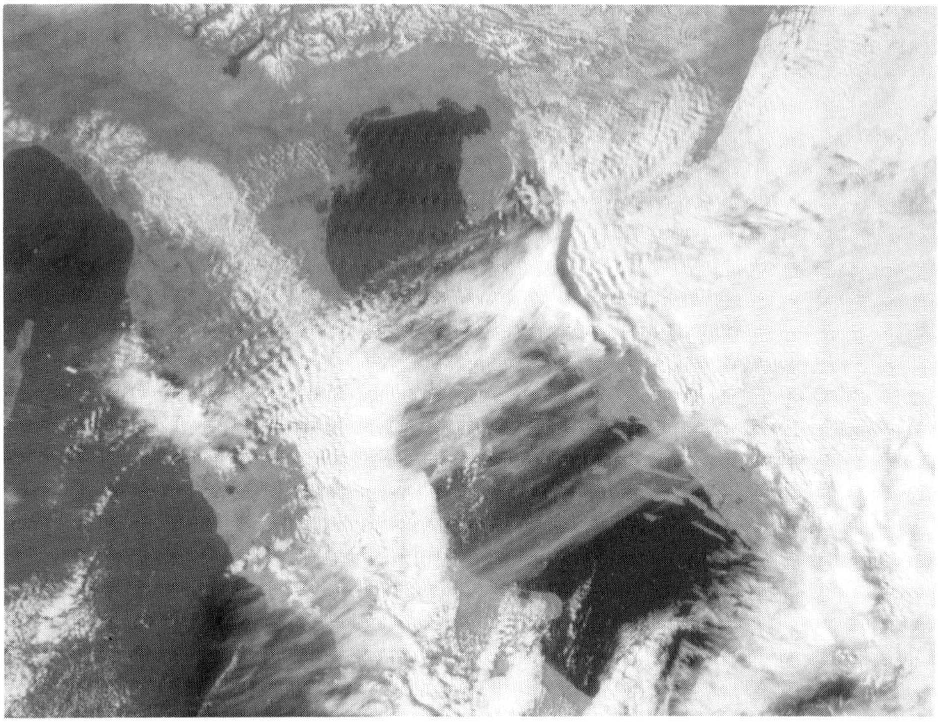

Fig. 14.5.i(a) 1356,5.3.84,Ch2. The Bora crossing the Adriatic Sea, the wind at all heights being from the ENE. The wind is very light south of Rome, which is shown by the shortness of the lee waves. The orographic cirrus is formed where the air descends from the Velebit Planina Mts at the coast of Dalmatia, and the shadow of the high wave cloud is cast on the lower-level wave clouds at 3.00 p.m. local time. (Courtesy the University of Dundee.)

not blocking the surface airflow, as shown in Fig. 14.5.i (p. 488). The pictures 14.5.ii–v illustrate various features of orographic cirrus clouds as seen by the AVHRR satellites, and the features of the different channels 2, 3– and 4.

Orographic cirrus clouds are important to know about because they cannot be used to determine the wind at their altitude. They are stationary at the upwind end where they are first formed and the cloud particles do move with the wind but are difficult to identify. Satellite pictures of clouds obtained by geostationary satellites at intervals of 20 or 30 minutes can be used to calculate wind speeds if clouds can be identified in successive pictures, and this type of cloud is usually very easy to identify but would give false results. The process cannot be fully computerised because the upwind end of orographic cirrus may move if the structure of the airstream is changing, and so zero movement cannot be a reliable criterion for exclusion from the exercise of obtaining winds. And yet they do constitute a significant proportion of all cirrus clouds, although I know of no plan to discover the magnitude of this proportion from satellite pictures of chosen mountainous areas.

In order to know a very important factor for all clouds, namely the diurnal variation in the amount, when their effect on their contribution to global warming is dependent on this because the reduction in the total solar radiation is reduced by their presence while they have a significant greenhouse effect at night by absorbing longwave IR radiation from below and emitting upwards at a lower temperature than the land or sea [65]. Most frontal cirrus would be expected to depend on factors which do not have much diurnal variation, but orographic clouds would probably be increased at night because an increase in the static stability of the bottom layers of air would probably increase the orographic effect by reducing separation of the flow from the surface.

The same considerations apply to **anvil cirrus**, which may, on the whole, be less widespread at night because it is based on thermal convection, although there is a significant persistence of big storms late into the night especially in lower latitudes. This is not useful advice about what the result of appropriate investigations is likely to be: it is merely an expression of one of the many tasks which need to be performed before a reliable parameterisation of clouds in GCMs can be framed.

Watching and reporting changes in cloud amount and types which indicate the significant mechanics of change has been a very weak part of instructions to meteorological observers which can be remedied by a proper visual study of satellite pictures. [63] An official book (1995) is a too-long-delayed, but good, start in this direction.

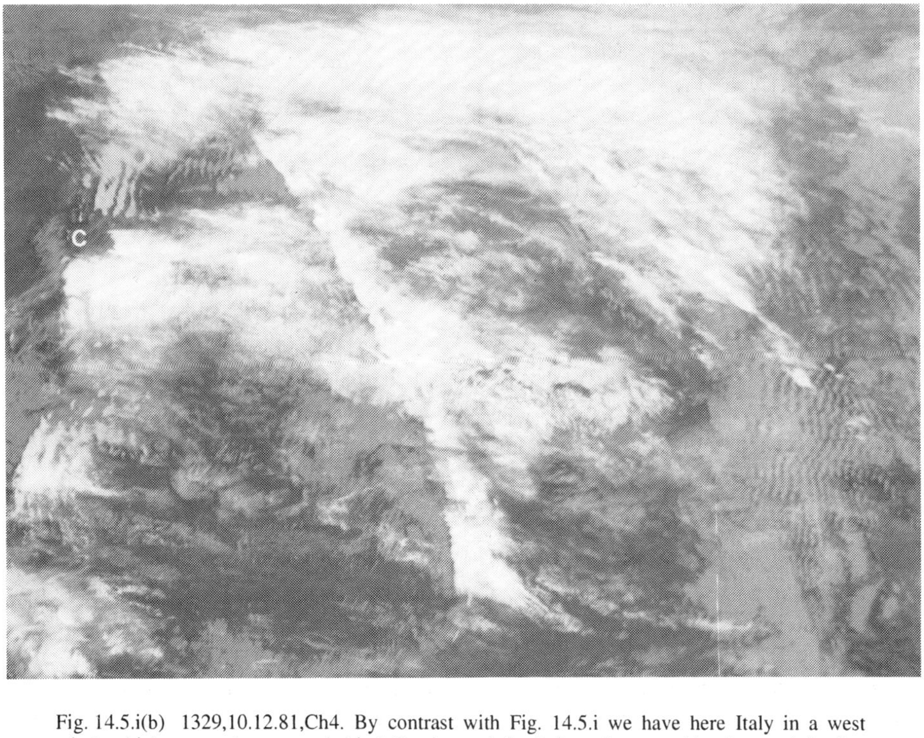

Fig. 14.5.i(b) 1329,10.12.81,Ch4. By contrast with Fig. 14.5.i we have here Italy in a west wind, which causes lee waves behind Corsica and Sardinia and orographic cirrus originating along the crest of the Apennines from the north to the south of Italy. As the air in the north descends from the Alps, a great area of orographic cloud streams away to the east. There are lee waves over Albania. (Courtesy the University of Dundee.)

(a)

Fig. 14.5.ii (a) 1235,13.4.86,CZ5. After crossing the Greenland plateau, which usually blocks the ascent of sea level air from ascending the western slopes, the air at the mountain tops descends the steep southeastern coastline and as a two-dimensional airstream widens out vertically, which requires a pressure rise to decelerate the flow. This often becomes set up by the air at higher levels rising as the lower descends and sets up a steady state of flow with an orographic cirrus cloud trailing from the crest of this wave. The last remnants of the

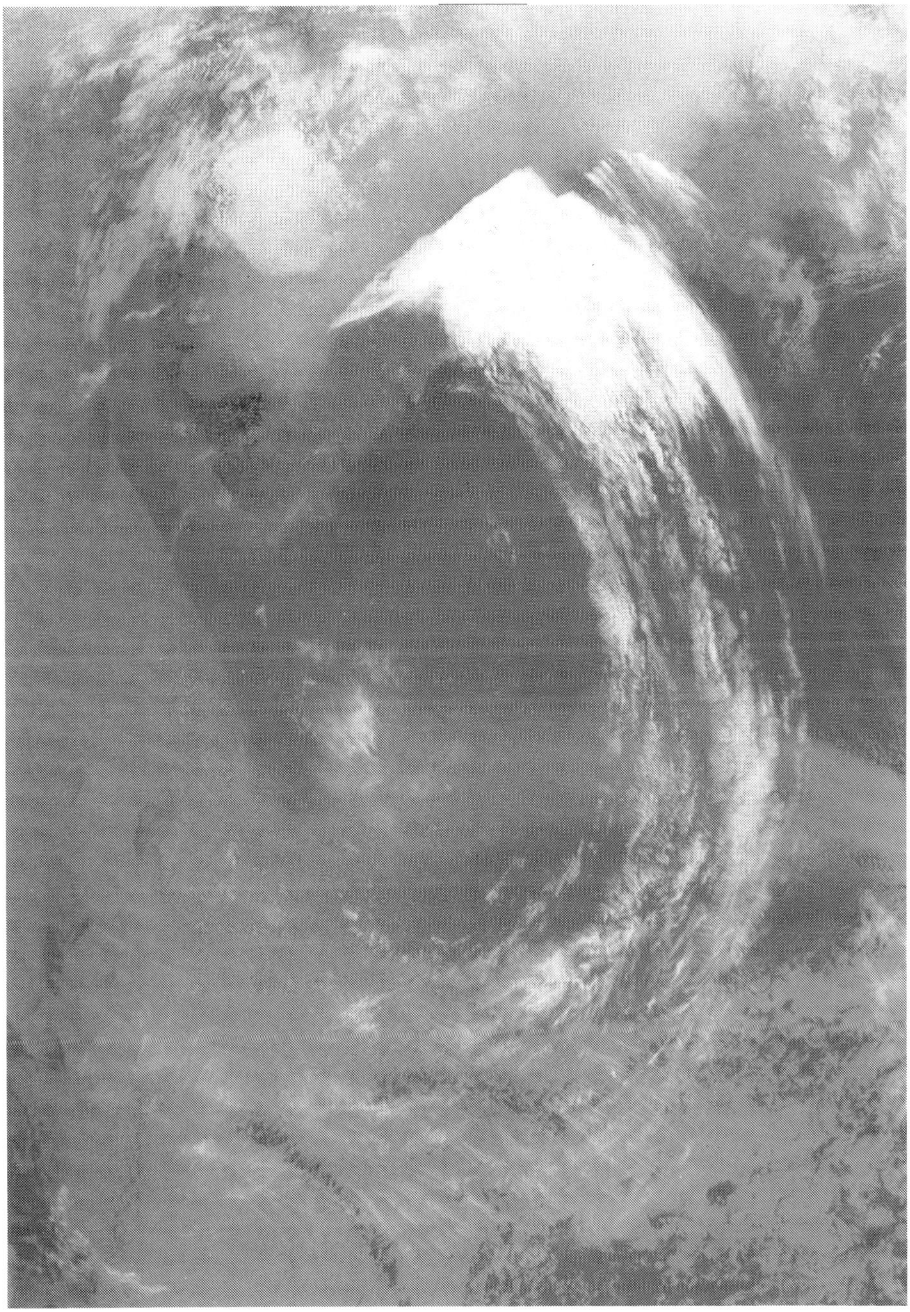

(b)

(Fig. 14.5.ii (*cont.*) downstream end of this cloud have travelled over 2500 km from the source. The curve of the flow is the effect of an anticyclone centred at 52°N 42°W, 1032 mb. (b) 1609,13.4.86,Ch4. This picture covers a greater area (not on same scale) about 90 minutes later, and shows contrails in the air supersaturated for ice at the southern end of the trail. (Courtesy the University of Dundee.)

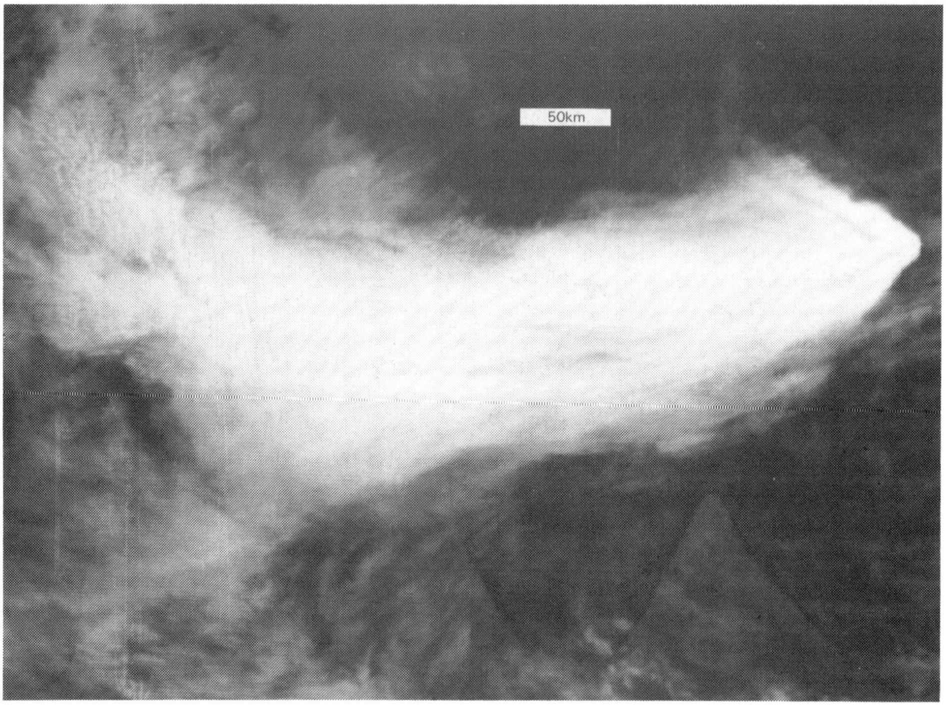

Fig. 14.5.iii 1407,14.4.84,Ch.4. The cause of this orographic cirrus is the small isolated island of Jan Mayen which has a peak, Beerenberg 2277 m, with a base diameter of around 15 km at its NE end. In this case the wind is from ENE and the island is about 55 km in length downwind from the peak. The cirrus trail is over 100 km wide and at the moment of this satellite picture was over 500 km in length, having grown from about 220 km in the previous 6 h.

The important feature of the clouds displayed in this section, with the aid of satellite pictures which show their extent, is that they stretch hundreds of kilometres from the relatively small orographic features which cause them. Every area covered by this thin cirrus cloud reflects back into space sunshine which would otherwise warm the ground and lower atmosphere. These clouds are extensive because every frontal system which lifts the air creates air that is saturated for ice but not for water. Aircraft trails produce persistent cloud in this area, and because they are more obvious to the eye, made people suggest that the weather might be altered because the contrails were produced, but did not notice the much greater areas of cirrus produced orographically.

In the Pacific area there is a scarcity of orographic detail except for small islands whose contribution to cirrus trails is from the tops of cumulonimbus formed by the sunshine on a small bit of land. But mostly the sunshine goes to heating the sea and leads to the formation of ENSO, described in section 16.8 [64].

The problem is not always seen as irrelevant to climate change, but it is important for attempts to represent the effect of clouds on the earth's albedo. So far the guessing game for the appropriate parameter in global mathematical models produces differences between models as large as the climatic changes that are predicted (see section 17.8).

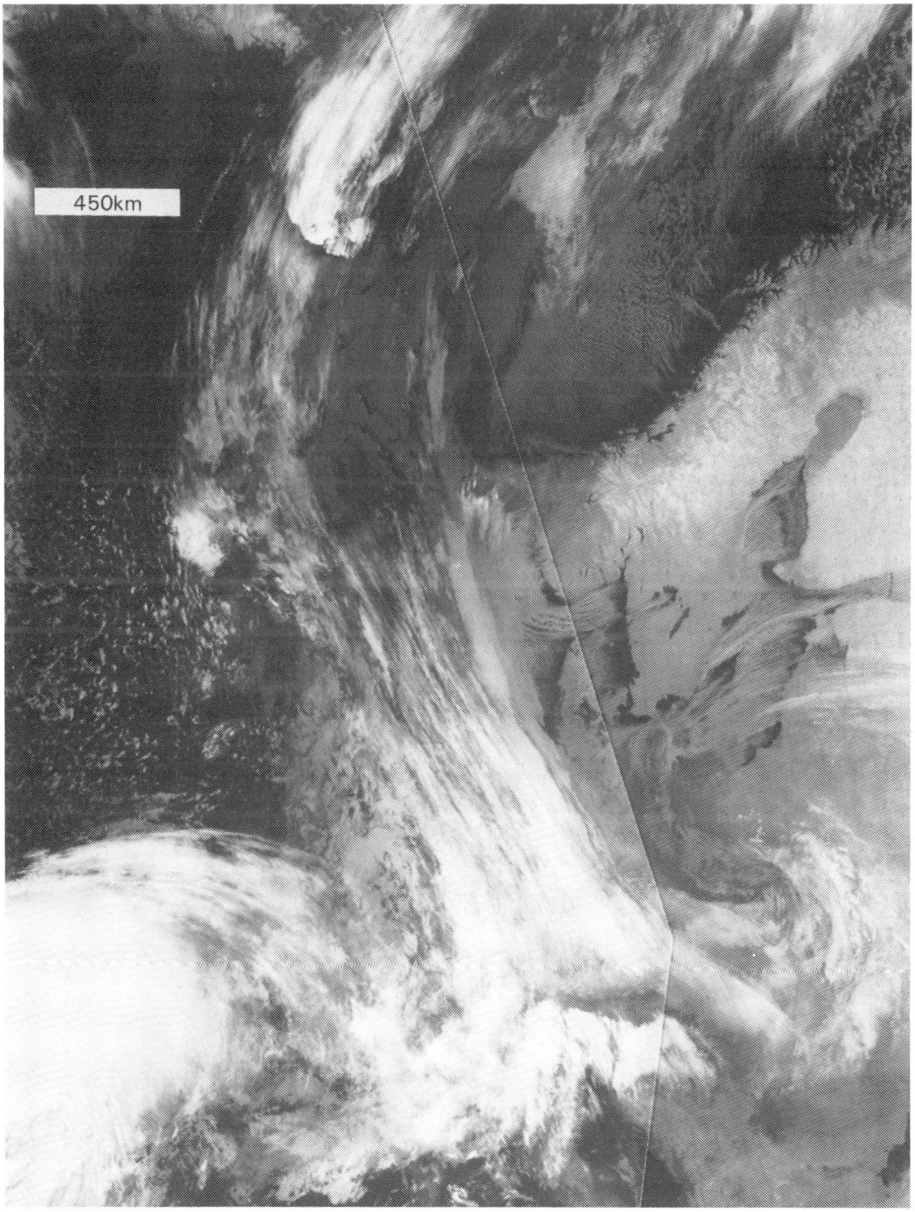

Fig. 14.5.iv 1257/1437,9.1.87,Ch4. This composite picture from two successive passes of the NOAA satellite illustrates that frontal jet streams very readily generate orographic cirrus. But in this case the line of cloud from the Alps to Iceland originated from two separate fronts which joined over north Britain. The line of cloud emerging from the Skagerak, between Norway and Denmark, produced a very heavy snowfall between Edinburgh and Newcastle, bringing traffic to a standstill. The northern half generated a cirrus trail over Iceland in a wind from the south; but in the south the high cloud can be seen to be evaporated over the Alps with orographic cirrus in the wind from the north extending over Croatia and the northern Adriatic.

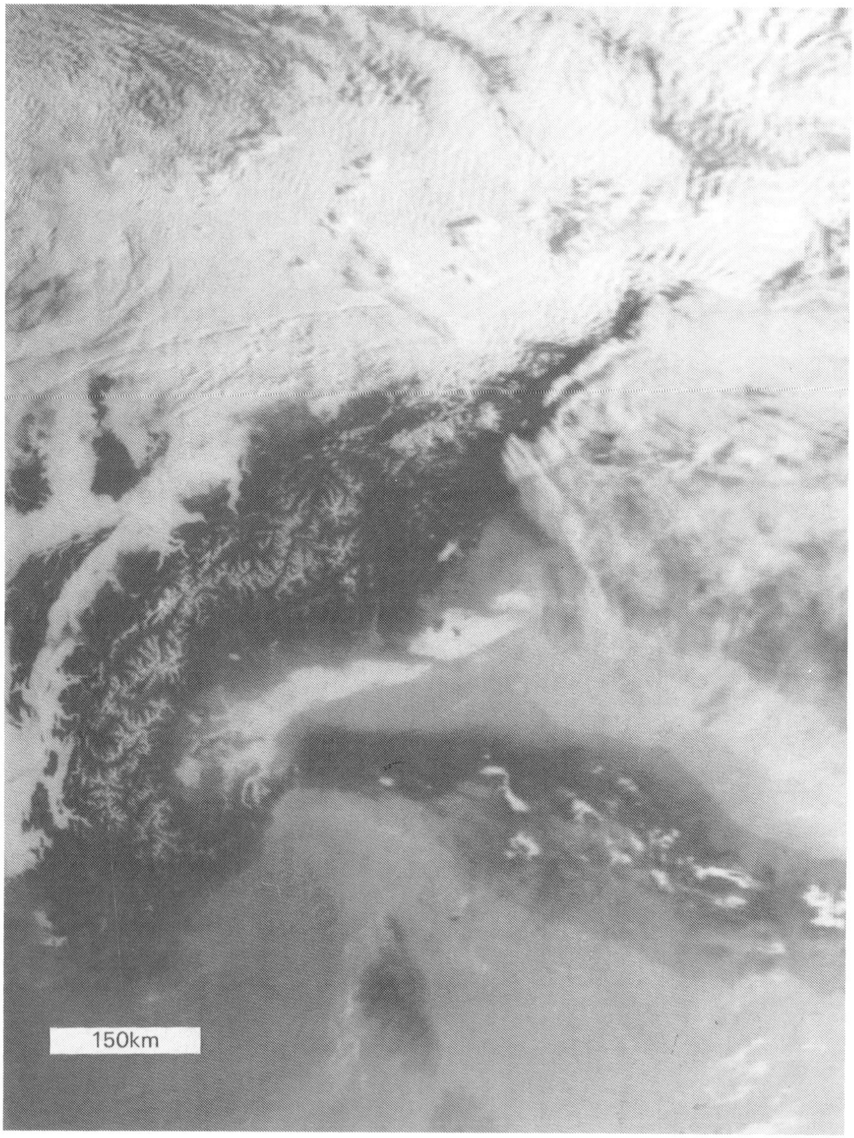

(a)

Fig. 14.5.v 0847,15.19.84,Ch.1,Ch.3–,Ch.4. (a) The channel 1 pictures shows the very hazy air at low levels over the sea and the low ground of northern Italy. It also shows the rather feeble shadows of the higher cloud on the sheet of low cloud over Austria and Bohemia and surroundings. The high cloud to the south of the Alps over Croatia does not show real distinctions in this wavelength. (Courtesy the University of Dundee.)

(b)

Fig. 14.5.v 0847,15.19.84,Ch.1,Ch.3–,Ch.4. (b) is best interpreted by its comparison with (a). The shadows on the low cloud due to the higher cloud above it are much sharper because there is no skylight in channel 3. The coastline of Italy is now clearly visible through the haze, but the cloud from Turin to Venice is clearly located. There is glint where the strongest winds blow from the land southwards over the sea from the Italian west coast. The orographic cirrus over Croatia is partly white (very small particles) and partly dark (probably larger ice particles). Corsica's snow-capped mountains are seen as black (snow absorbs channel 3 to give zero albedo to snow). The white patches on the Apennines are low cloud or fog, not snow.

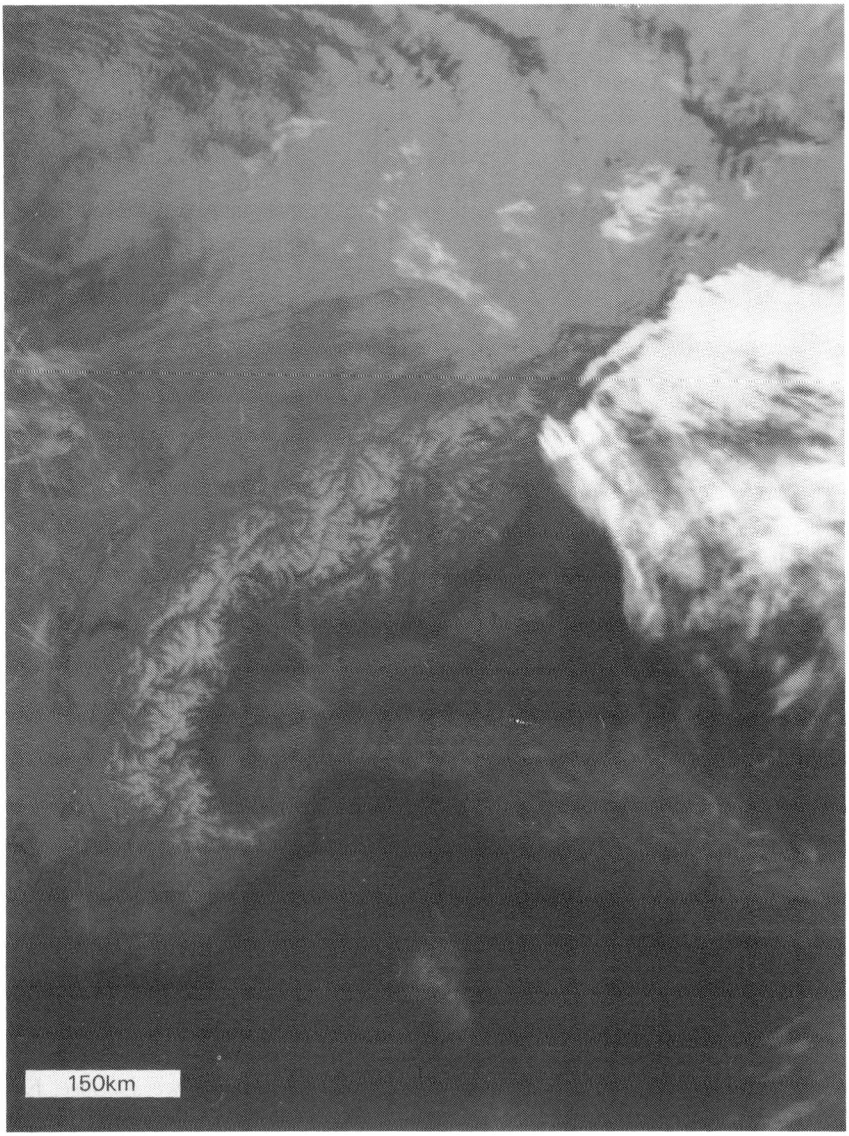

(c)

Fig. 14.5.v 0847,15.19.84,Ch.1,Ch.3–,Ch.4. (c) In the westerly part of the Croatian cirrus it looks as if there is some water droplet cloud which soon evaporates and decreases the albedo. (Courtesy the University of Dundee.)

14.6 CUMULUS DYNAMICS

Convection carries parcels of air to the condensation level, whereupon the heat liberated by condensation becomes the dominant source of buoyancy. Although some rain falls to the ground, most cumulus clouds are evaporated by mixing into the surroundings in to which they ascend, and so the heat lost when evaporation occurs is also an important feature of the mechanics. As soon as the role of water becomes important the mathematical simplicity of the situation vanishes, and vain attempts have been made by many authors to recapture the similarity love of the honeymoon period with partly turbulent flow. But the problem possesses an essential elusiveness which charts, computers and all the time in the world cannot portray better than the clouds themselves. The system is essentially very complex and is continuously reacting with everything that is going on both close at hand and many thousands of miles away. It is part of what meteorologists naively call 'the general circulation', meaning something that can only be studied with the aid of all the world's meteorological services and the world's biggest computer. However it is still elusive in the sense that nothing very useful can be done for the man in the field, that is, the field which grows the food; not the one turned into a research laboratory. The world's scientific effort has been diverted by the so-called 'rich' nations to the facilitation of their narrow purposes. Meanwhile the so-called 'poor' nations have been taught to value an airline more than an agriculture and ecology matched to their weather and longterm climate. Consequently the approach to agricultural needs has been though attempts to make rain and hurry the irrigation plans, as if some purpose were served by supplying the needs created by the previous technological effort. We, however, shall look simply at the cumulus clouds to get the feel of how they continuously circulate material through the atmosphere, creating a beautiful variety of weather in which man may evolve the variety of cultures which is the treasure of his civilisation.

(1) Erosion of thermals

A sounding of the air is depicted in Fig. 14.6.i as the line QLR on the T–ϕgram. Suppose a thermal source produces a warm parcel at A slightly warmer than the average ground temperature at Q, and that it rises to its condensation level at C along the dry adiabatic, and then up through the middle of a cumulus cloud along the wet adiabatic to D. All the way up it is warmer than the surroundings. It now emerges to the exterior of the cloud and mixes with the surroundings at R, which are unsaturated and, we suppose, have a mixing ratio of 1 g/kg. The consequences of the mixing may be discovered by imagining the parcel and the surroundings to be brought down to the condensation level, so that the parcel is back at C and the surroundings are at S, on the dry adiabatic through R.

If the two are mixed in equal quantities the mixture will be at the mid-point M, of SC, and the mixing ratio will be the mean of 9 and 1, namely 5 g/kg. If we now take the mixture back to its proper level it will go dry adiabatically to N, on the 5 g/kg saturation line, and then wet adiabatically to P. There we find that it is colder by 2°C than the surroundings, which are at R, whereas the surroundings at L were 1°C cooler than the parcel ascending from C. If this parcel is now left to find its equilibrium level this will be at L, where it has the same temperature as the sounding.

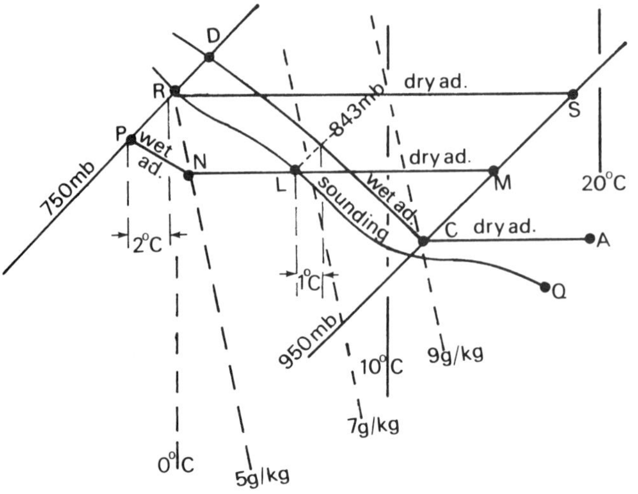

Fig. 14.6.i The ambient air is represented by the sounding QLR. A warm parcel rising without significant dilution from A would pass along ACD if it became saturated at D. If it is mixed in equal proportions with the environment at R with 20% humidity the mixture would be at P, M being the mid-point of CS. This parcel at P will find its equilibrium level at L. (See text for details.)

It is thus demonstrated by an example, and others may be multiplied at will on the T–ϕgram, that at the top of a cumulus cloud we may generate parcels which are much colder than the surroundings, especially when they are dry (this case had an environment with a 20% relative humidity). The downdraughts which occur as a consequence can easily be seen by a careful cloud watcher and are a common experience of glider pilots.

In order for a cloud to rise far into dry surroundings it must be very large so that parcels may rise to its top without being subject to evaporation, and it is most likely to do this if it is ascending along a path already moistened by previous thermals. This happens best over a good thermal source such as a mountain side where there is almost no shear or general wind to carry away the towers of air moistened by previous thermals.

The intensity of the mixing in a thermal is shown in Fig. 14.6.ii, which is a sequence showing the successive positions of a blob of surroundings which is entrained into a thermal. After the fourth position the material will be circulated around the thermal on all sides of the axis so that the blob will be completely mixed into the thermal as it rises about $1\frac{1}{2}$ diameters. When the entrained blob is exterior dry air, it is clear that the thermal cannot rise very far before all the cloud in it is evaporated, and Fig. 14.6.iii shows a sequence in the life of a cloud tower. The sharp outline on the top, which is characteristic of growing cumulus, is where fresh interior saturated air is rising to the surface. Because of the rapid rate of ascent there is plenty of water to form a droplet on every nucleus so that the cloud is very dense, and has a visibility of only a metre or less. That is why its outline is sharp. After position 4 the remains of the tower find their equilibrium level in the stably stratified surroundings and gradually evaporate, generating small motions as they do so because of the cooling by evaporation, until there is no cloud left. All that remains is a moistened tower of clear air.

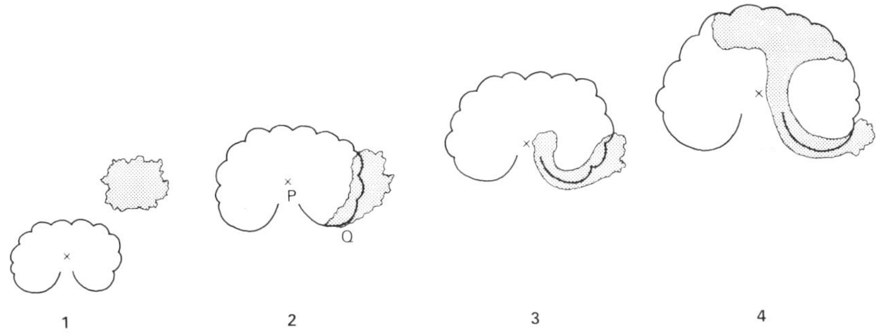

Fig. 14.6.ii Successive positions of a blob of exterior air entrained into a rising thermal. After position 4 it would be spread around all sides.

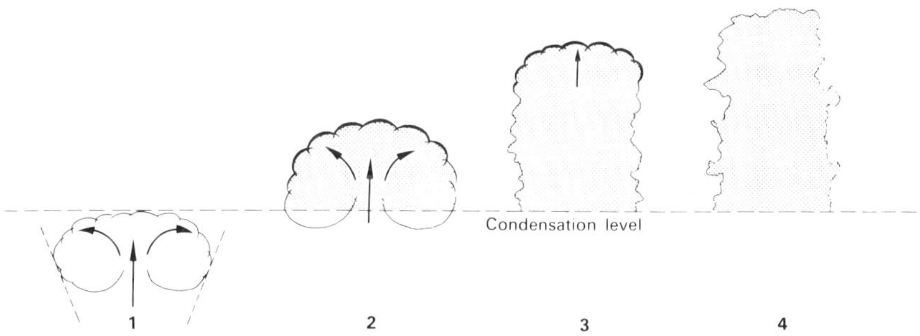

Fig. 14.6.iii Successive positions of a thermal rising above the condensation level. After the final position shown the tower would soon evaporate, sinking down as it does so.

If the thermal rises through wind shear (Fig. 14.6.iv) its axis is tilted in the opposite direction to the tilting of the tower of residual cloud. The erosion mechanism is easily demonstrated in the laboratory, by allowing a thermal to sink through a stratified mass of water. In Fig. 14.6.v the upper part of the tanks has no stratification and the thermal behaves like those in Fig. 10.2.i. As soon as it meets the stratification it leaves a trail of mixed material where each bit comes to rest at the level of its own density. The last picture shows the thermal, which has not grown since it entered the stratified layers, flattening out slightly above (in the tank) the level to which it overshot. Clouds like this tower are common in the trade winds and other oceanic areas, as shown in Fig. 14.6.vi, and see Malkus and Scorer [40].

When moistened towers of air are subjected to shear for several hours, the top may be hundreds of miles from the base and so towers become thin layers of high humidity with less humid layers in between Plate 11 in the colour section between pages 000 and 000. When these are made visible in a wave cloud we get the 'pile of plates' depicted in Fig. 14.6.vii. If a cumulus rises rather slowly in wind shear, the new growth takes place on the up-shear side because the thermals are usually rising in succession from a moving source, as depicted in Fig. 11.8.iv. The evaporation and sinking, which occurs in air mixed with a

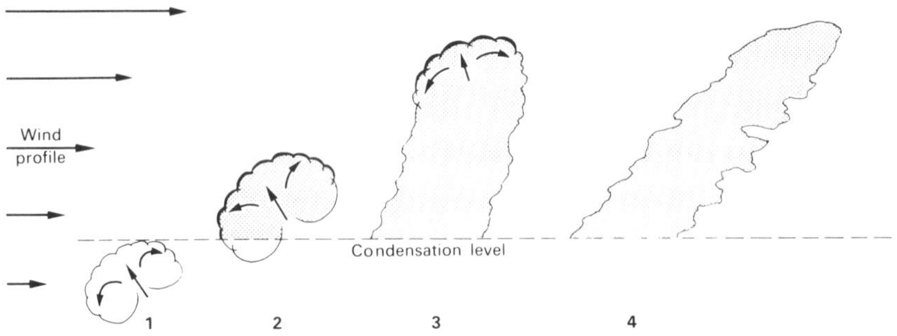

Fig. 14.6.iv Successive positions of a cloud tower rising in wind shear. Although the tower leans over in the direction of the shear, the axis of the thermal is tilted in the opposite direction while it is rising.

high proportion of surroundings and travelling nearly with the wind at that level, produces downdraughts on the down-shear side (Fig. 14.6.viii, p. 502).

(2) Cumulus and inversions: pileus; cooling and warming

When a thermal arrives at an inversion it is flattened out as at a ceiling, although it may lift it temporarily by its upward momentum. It may therefore form a moist layer beneath the inversion, and may or may not be visible as cloud. In Fig. 14.6.ix a thermal is depicted as displacing the moist layer upwards above its condensation level to form a pileus (cap) cloud which may or may not be subsequently penetrated. If the thermal is a cumulus cloud, the inversion being well above its condensation level, the pileus may appear above the top of the cumulus and may or may not be subsequently penetrated. The pileus appears temporarily because it sinks back to its own level: sometimes the thermal producing it is not visible if the pileus is at the condensation level for surface air.

Less obvious is the fact that cumulus may create inversions. We note that at the base of a cumulus cloud there is a flux of heat, and buoyancy, upwards. At the top there is evaporation of cloud at the highest level reached, and this may cause cooling if the surroundings are dry enough. Somewhere in the middle of such an average cloud is a level where the flux of heat as buoyancy is exactly equal to the flux of negative latent heat in the form of the liquid water droplets moving upwards. Somewhere a little higher up is a level at which the cooling effect predominates. We speak of the 'average' cloud because when the sky has widespread scattered cumulus each cloud passes through a life cycle of 10–45 minutes, and some are growing bigger while others are rapidly disappearing by evaporation. In Fig. 14.6.x we imagine the sounding on the T–ϕgram to be initially as indicated on page 503. As thermals rise they mix with the surroundings and therefore acquire a temperature somewhere between the wet adiabatic through their point of condensation, C, which they would follow if unmixed, and the environment. The evaporative cooling causes them to acquire the same temperature as the environment at some level M.

Some thermals rise higher than M and then fall back below that level as a result of the cooling. Others reach the level of M with less mixing and have a higher temperature than the average and therefore more liquid water. All thermals overshoot their equilibrium and

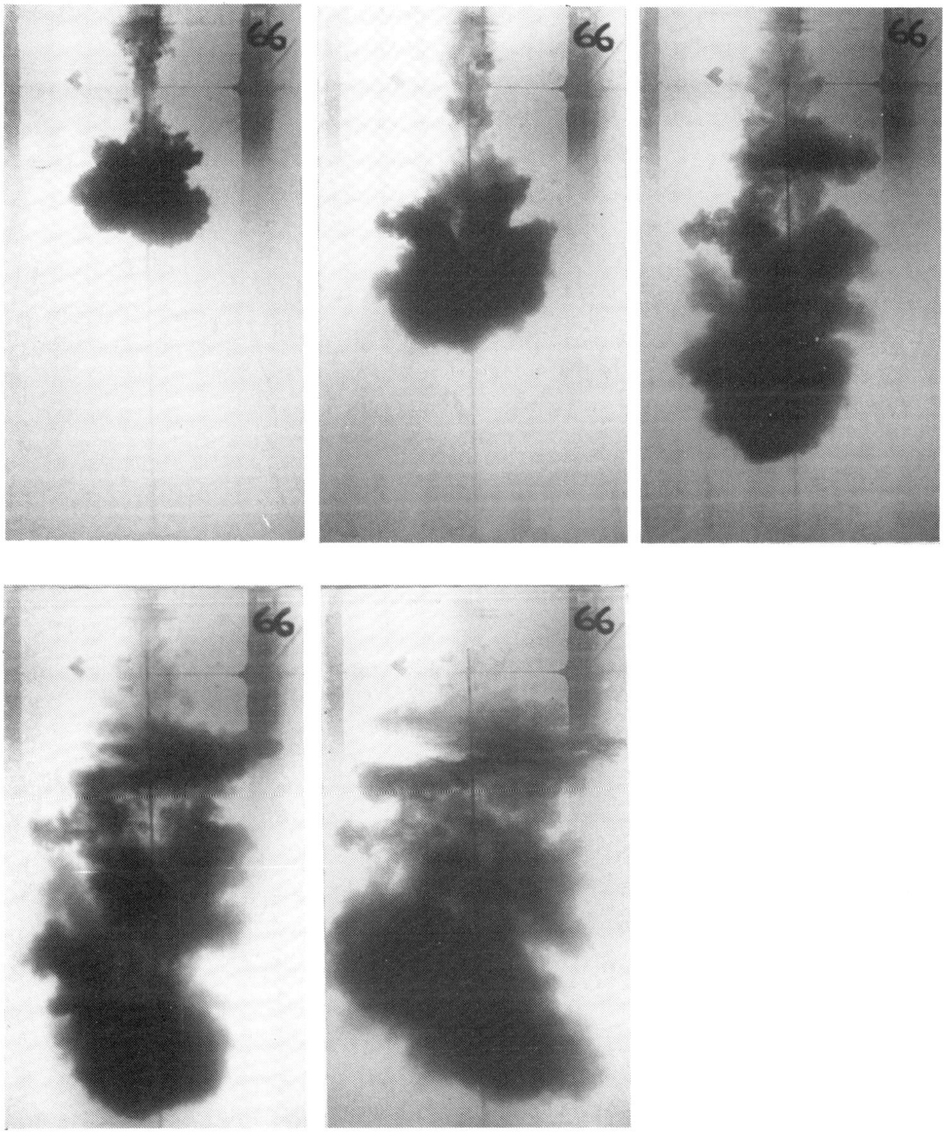

Fig. 14.6.v A model of a cloud tower rising in stratified surroundings. A thermal composed of dyed salt water is released at the top of a tank stratified by increasing the salt concentration towards the bottom. A tower of eroded material is left behind, each part at its own equilibrium level in the stratified surroundings. The front of the tower slightly overshoots its final equilibrium level.

mix with air higher up, and therefore cool it and cause it to sink; but it cannot be caused to sink below C even in an extreme case. Thermals above M cause a cooling of the air. None can possibly rise above T, where the wet adiabatic through C cuts the sounding, except by overshooting after rising unmixed. Since mixing always occurs, T is a good

Fig. 14.6.vi Trade wind cumulus of the castellatus type leaning over in wind shear. The wind is from the ENE and decreasing in strength with height above the cloud base. Often these clouds are arranged in streets. (Photo J. S. Malkus.)

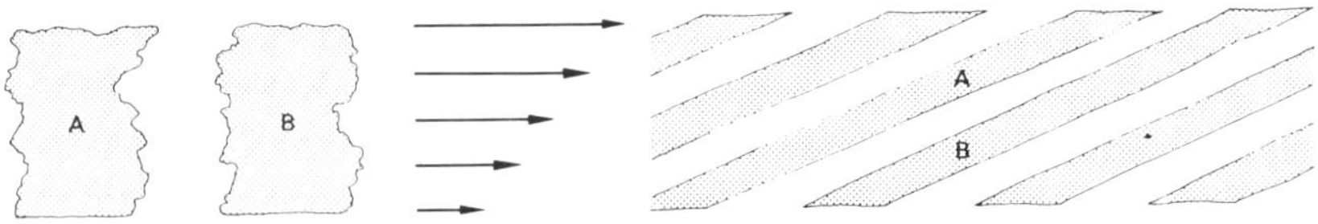

Fig. 14.6.vii When towers of moistened air are sheared over, after a few hours they take the form of layers of alternately high and low humidity. After a few days storm areas may produce the same effect, and large areas of sky with laminations of moist air are created.

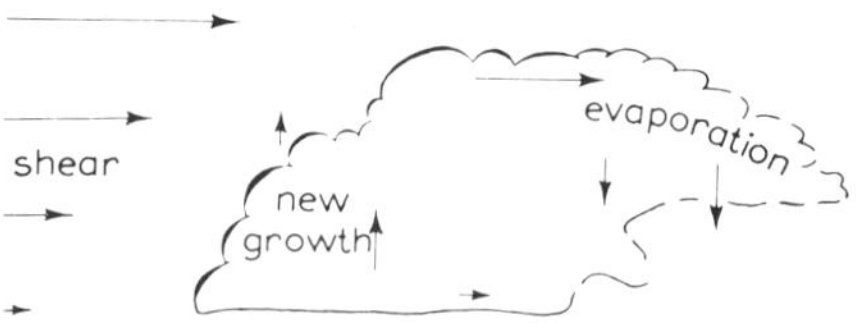

Fig. 14.6.viii In wind shear the new growth of a cumulus comes up on the upshear side from the slower-moving layers below. The evaporation and downdraughts caused by it occur on the down-shear side.

estimate of the highest level to be reached by any thermal. The environment is gradually cooled between T and M. Below M it is warmed by the subsidence necessary to compensate for the ascending thermals until the thermals can no longer rise through it, and so the

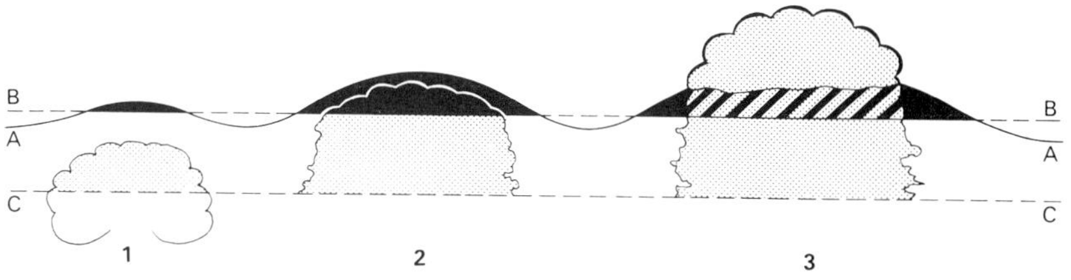

Fig. 14.6.ix Possible successive positions of a pileus (cap) cloud and the cumulus (thermal) rising beneath it. If it penetrates, the pileus becomes a 'scarf' cloud, but is soon evaporated in the downdraughts on the outside of the cumulus. C and B are the condensation levels of the cumulus- and pileus-forming air respectively. The thermal way may be invisible for a few moments if B and C coincide. A is the position to which the stable layer just below B is distorted by the thermal.

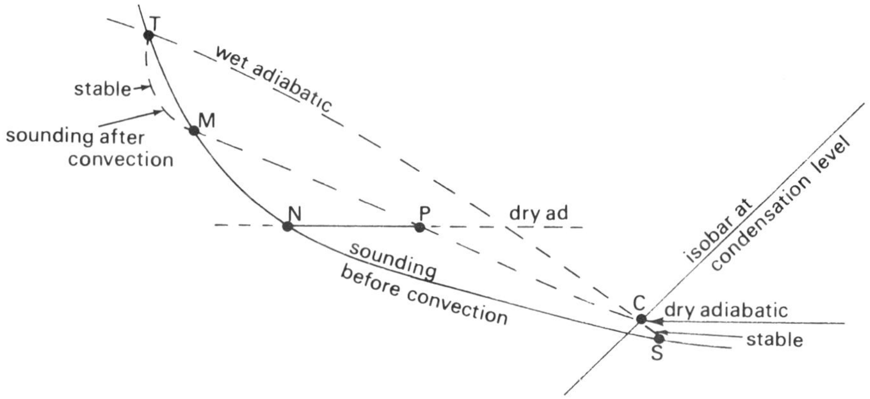

Fig. 14.6.x The initial sounding SNMT is modified by convection from below to SPMT. C is the point of condensation of thermals from below. Stable layers are generated between S and C, and between M and T by the convection.

sounding is modified to the dashed line after some time of convection. Thus an inversion is produced below T, and subsidence warming is represented by the sinking of the air at N along the dry adiabatic to P.

Below the condensation level the air is more nearly neutral, and may be slowly warmed as described in section 11.4 (end). The air just above the cloud base is stably stratified, and so the subsidence of air through the condensation level which compensates for the upward flow of air in the form of thermals generates a layer called the **sub-cloud inversion**, occupying the layer CS.

This sub-cloud inversion is intensified if there is also subsidence at that level due to large-scale motion of the air. On the other hand, if there is large-scale horizontal convergence, as over land when there are sea breezes blowing, the lifting of the whole air mass cools all the air above cloud base. This is because it is stably stratified there, which makes it possible for convection to continue even though the temperature at the ground is not increasing.

Thus the existence of convection clouds may be an indication of horizontal convergence as much as of an increasing surface temperature. Equally, warming is as often due to subsidence as to heating by thermals. This topic is discussed at length by Fraser (1968).

In a burst of cold air from the north the whole air mass is rapidly warmed by subsidence. Thus clouds which reached up to 8 or 9 km above the sea off the west of Scotland may be replaced by clouds up to only 3 or 4 km off the coast of Portugal, even though the sea is warmer there, and by a layer of stratocumulus below an intense inversion at 1 km at the Canary Islands. It then travels across the Atlantic Ocean as a Trade Wind, approaching the equator slowly and passing over ever-warmer sea. The convection penetrates higher and higher as the sea temperature rises, but the cumulus clouds are continuously evaporated into surroundings which are subsiding and becoming drier above the inversion which the thermals penetrate more and more. Gradually the inversion is lifted higher by the penetration of the still-warmer cumulus, and by the time the air arrives in the Caribbean it has reached around 3 km. All this time the air above the inversion into which it is being mixed is coming from the west.

What happens next depends on the time of year. In autumn the sea is very warm and the cumulus begin to penetrate the inversion and generate showers which may soon lead to hurricane formation (see sections 14.10 and 4.9). Otherwise, widespread showers occur from time to time. The air has been transformed from a shallow cool layer capped by warm dry air as it entered the trade winds into a warm moist layer. This is then turned northwards to become the warm sectors of the frontal cyclones of middle latitudes. Such are the intermittent and seasonally variable mechanisms of the general circulation of the atmosphere the world over. Their importance in determining how much subsidence or general convergence occurs in any area where cumulus may exist is overwhelming, and in no part of the atmosphere can the behaviour of cumulus be fully understood without reference to them. (See also El Niño [63] (16.6(3).)

By rising through a stable environment, cumulus clouds warm not only the air with which they mix but all the air in between which subsides to compensate for their updraughts as well. Thus the air over a narrow sea may be warmed by subsidence compensating for convection over the surrounding land, and this may prevent the formation of the sub-cloud inversion over the land. Because it has had the latent heat of condensation of all the cumulus clouds which appear during the day added to it, the air over the land acts as if it were all warmer on the average than the air over the sea during the daytime. It is as if each cumulus, as it rises, sends out a wave, which lowers the altitude of all the surrounding air over an ever-increasing radius, just as the level of a lake is lowered by an outward-travelling wave when a boat is lifted out of it. All the waves from an area of cumulus production add together to cause subsidence in neighbouring areas not growing cumulus. The wave travels at a speed typical of lee waves, which is typical of the wind but may be slightly faster, up to say 60 kt.

In the evening when cumulus clouds evaporate, the air mass which contained them is cooled, but not by as much as the daytime warming if some rain has fallen. This cooling sends a wave travelling outwards and causes the lifting of areas of air, over water for example, where there had been no daytime cumulus, see Scorer (1955). Sea surrounded by land may therefore appear to be producing cumulus in the evening, but they are castellatus and do not have roots in thermals generated over warm sea, and the clouds appear regardless of any land breeze which may occur.

Cumulus generated in air in which there is convergence and upward motion begin when some condensation occurs. If the sounding slopes more than the wet adiabatic, thermals may sprout spontaneously from this cloud. Such cumulus is called castellatus (turret cloud), or floccus (fleece), when it evaporates very quickly because of the dryness of the environment (see Fig. 14.6.xi). Many trade wind cumulus clouds (Fig. 14.6.vi) are castellatus. They have no roots in identifiable thermals rising through the sub-cloud layer, but derive almost all their buoyancy from condensation.

Fig. 14.6.xi Castellatus (turret) clouds sprouting from thin cloud produced by convergence and slow ascent. When they evaporate very quickly they are called floccus (fleece).

Recently, satellite pictures have shown how cloud over narrow seas such as the eastern Mediterranean comes and goes in response to the cumulus convection over the surrounding land. For example, the morning disappearance moves across the sea from the coast at about 30 kt on some occasions. Wind and wind shear together with the stability modify the way waves may travel. For example, it may be deduced from the theory in Chapters 6 and 7 that waves can more readily travel against the wind shear than with it.

14.7 THE DYNAMICS OF FALLOUT

We shall not here be concerned in much detail with the mechanics whereby a large number of cloud particles formed by condensation from vapour are collected into a smaller number of larger particles. In size these will range from those large enough to have fall speeds of the order of 1 m s^{-1} or more, to hailstones as big as grapefruit with speeds slightly greater than the largest updraughts ever occurring in hailstorms, namely 30–35 m s^{-1}.

In the early 1960s the name 'fallout' was popularly used to mean radioactive fallout because large nuclear explosions were then introducing it into the atmosphere. Even so, it was not used to refer primarily to the radioactive debris falling in the neighbourhood of the explosions, but to the much smaller total mass of very small particles carried into the stratosphere by the thermals resulting from the release of nuclear energy. Those particles were not fallout, being too small to have a fall speed comparable with the vertical motions of the air. They were nearly all removed from the stratosphere by vertical motions which brought them into the troposphere, soon after which they were brought down in rain: they were therefore fallout only parasitically. They were, however, removed from the stratosphere more quickly than was expected according to incorrect assumptions about vertical motion in the stratosphere, and the assumption that their actual speed of sink through the ambient air was not relevant.

We use the term **fallout** for rain, hail, snow and so on, namely for water, or other, particles which have a fall speed causing consequences to the air motion or to the appearance of cloud. Thus fallout leaves behind a large amount of heat that appeared when it was condensed, and perhaps was frozen. Fallout can cause downdraughts by applying its weight to air through which it falls and by cooling that air by evaporation into it; and it can alter the appearance of clouds by the falling motion.

(1) Fallout fronts

Simple experiments to reproduce the effects of the weight of the particles may be done using clouds of air bubbles in water, see section 10.4. The downdraught is the integrated effect of the wakes of the falling particles, for at their terminal velocity their weight is added to the air. The particles tend to accumulate at the bottom, i.e. the front, of the downdraught, where they are falling into slower-moving air, and the motion is rather like a thermal or new plume with the downdraught, spreading out sideways at the front. Fig. 10.4.iv show four successive positions of a cloud of air bubbles released into water. In the first they are still coming from the array of small holes; in the second the last few to emerge are rising rapidly up the axis of the motion system. By the fourth position a more or less steady state has been achieved in which they occupy a ring-shaped volume with very few near the axis, rather like a thermal. The second picture (Fig. 10.4.vi) show how, if a cloud of bubbles is allowed to rise out of a layer of cloudy water into clear water, it drags the cloudy water after it, but moving further in advance with the terminal rising speed of the individual bubbles. At this stage the rate of rise of the cloud was between two and three times the rate of rise of the individual bubbles.

With a large rate of rainfall, say 5 cm h^{-1}, which would usually be maintained for only a few minutes at most even in a heavy shower, with a rain drop fall speed of 12 m s^{-1}

(8 relative fall speed +4 for the downdraught) the amount of water in a cubic metre of air is only about 1 gram, and its weight is equivalent to cooling the air by about 0.3°C. With three times this value obtained by suitable suppositions, we may conjure up a situation in which the weight of water may be equivalent to a cooling of about 1°C. This is actually about the same order of magnitude as the weight of water condensed in a dense cloud. On the other hand, differences of 2–3°C in the temperature of the air at the same level in a vigorous cloud are likely on the grounds that the wet adiabatic for a rising parcel departs from the temperature sounding by that amount.

A more important effect of rain, particularly when it is not composed of particularly large drops, is to cool the air below cloud (or beside cloud, if rain falls out of an overhanging piece of cloud) down towards its wet bulb temperature by evaporation. This cooling increases from zero at cloud base at about 4°C 1 km below, if the air below cloud has a uniform mixing ratio. If it is drier the cooling quite possibly may be up to 10°C, and therefore the effect of evaporation is likely to be much more important than the effect of added weight of water. Furthermore, a downdraught of cold air will continue to drain outwards from a storm and to flow down sloping land far from the storm long after the rain has fallen out of it.

Such outflows from storms proceed as fronts which have much in common with sea breeze fronts (see section 11.9). They can be seen on radar because of the temperature discontinuity, and sometimes because of the birds soaring in the upcurrent ahead, for this often contains insects scooped up in the warm air ahead of the front. **Haboobs** (see Fig. 14.7.i) are such outflows, in which the cold air is filled with sand raised by the strong wind and by intense convection due to the hot desert underlying it. On the other hand, the outflow from a shower over the sea may prevent convection reaching up to the

Fig. 14.7.i A haboob is a dry, cold front, probably the outflow from a (distant) shower cloud. The air is filled with dust and/or sand particles blown up from the surface and helped by convection in the cold air as it moves over a hot surface.

condensation level for many minutes. Thus, just after a shower cloud has evaporated there may be an area surrounding its former position without any clouds at all, giving a cellular appearance in satellite views.

(2) Föhn fancies

It is almost always stated that when the air on the lee side of a mountain is found to be warmer than on the windward side, it is because of the latent heat of the rain which falls out on the mountain. This explanation is usually incorrect. It is believed on the basis of the fancy that a good-looking explanation is *the* explanation. We imagine a mountain range reaching to 700 mb (about 3000 m) above sea level, with a cloud base at 900 mb (about 1000 m) on the windward side at A′, and suppose that the cloud base on the lee side is at 700 mb. This is as high as it can possibly be in this situation (Fig. 14.7.ii). From the T–ϕgram (Fig. 14.7.iii) we deduce that if the temperature at A′ at 900 mb on the upwind side is 10°C, using the wet adiabatic to F at 700 mb and a dry adiabatic to return to B′ at 900 mb the air will be at 18.5°C at B′ on the lee side. Thus a warming of the order of 8.5°C is explained. Such an increase is more easily explained by the blocking mechanism in section 6.15 and Fig. 6.15.ii, and this is more likely because the warming is often observed when there is no cloud at the mountain top. The mixing may also be effective as described in section 11.8(6), where the effect of stirring a stably stratified layer is to warm the lower half of it and make the lapse rate dry adiabatic throughout its depth.

One must also be wary of explanations in terms of rain because before the air ascends through the cloud base on the upwind side it may have rain falling through it between A and A′ which cools it and lowers the cloud base to A'_1. The subscript 1 is used to indicate that rain has been evaporated into the air at some time. Unless there is a corresponding increase in the rain falling from the air close to the mountain surface on the upwind side,

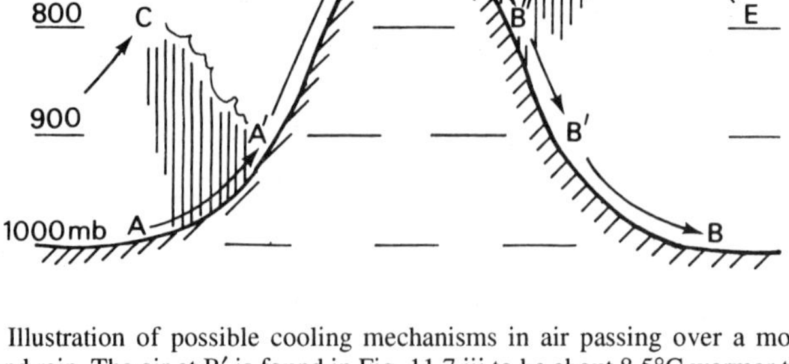

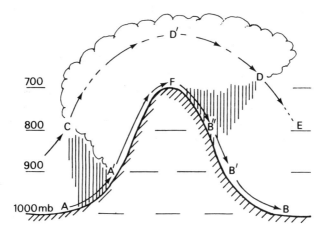

Fig. 14.7.ii Illustration of possible cooling mechanisms in air passing over a mountain range with cloud and rain. The air at B′ is found in Fig. 11.7.iii to be about 8.5°C warmer than the air at A′ if the cloud base is at 900 and 700 mb on the upwind and lee sides. Rain alters this conclusion if evaporation occurs below cloud. Other possibilities are described in the text.

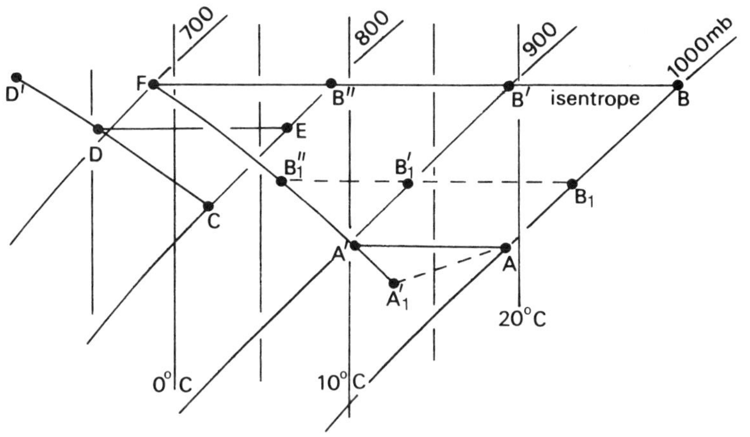

Fig. 14.7.iii A *T–φ*gram representing the possible temperature changes in air passing over a mountain range, illustrated in Fig. 14.7.ii.

the cloud base on the lee side will be lower and the parcel will return via B_1'' and B_1'. In that case the warming above the temperature at A is roughly half that previously calculated. Likewise rain falling through the air on the lee side would have the effect of bringing the air down the wet adiabatic from F, with the same consequence. Furthermore, when the cloud does engulf the top of the mountain and the air flows up the front face, it is usually observed that the cloud base on the lee side is lower close to the mountain, at B″ for example, than at D. This indicates that on the lowest streamline the air does not have as much water rained out as on higher streamlines which reached a greater altitude in the cloud.

The column of air AC is not unstable, and will normally be slightly stably stratified. In that case C li es on a warmer isentrope than A and is above AA′ on the *T–φ*gram (Fig. 14.7.iii). The air moving along the track CD′DE is represented on the *T–φ*gram, where it is seen that unless D lies above the line BF, E will be on a colder isentrope than B, and the column of air BE will be unstably stratified. This is not observed, which means that in its passage over the mountain the cloud base must be raised on the streamline CE by more than it is raised on the streamline AB, unless the column AC is very stably stratified. If it is observed that the cloud base does satisfy this condition, it is more probable that this is because the air below the mountain top does not pass over it. Moreover, because of blocking the air above the top descends on the lee side and evaporates any cloud that may be in it. This evaporation has noting to do with rain on the mountain. Furthermore, wave theory shows that the clouds do not necessarily disappear on the lee side of a hill (see Fig. 6.9.iii), and Föhn and Chinook warm winds are common with no rain on the mountains. Chinooks in their early stages take off from the ground not far from the lee of the mountain. It seems probable that, since they descend from a high level because of blocking of the lower layers, the lee wave is of large amplitude and a rotor is present which is filled with cold air not yet displaced from near the foot of the mountain. On such occasions the ground may still be snow covered at the end of the winter on the lee side.

Thus the commonly employed explanation is a facile one which depends on the cloud engulfing the mountain top with rain on the upwind side only. It may occasionally be correct, but blocking is the more common cause of the warming. Perhaps the most spectacular case of warming is provided by the Berg winds of Natal: on arrival at Durban the air sometimes has a potential temperature indicating that it has descended from higher than the mountain tops (3600 m, 150 miles away) with no cloud present. Mechanical mixing of a stably stratified layer probably contributes to the high surface temperature in this case.

(3) Dynamics of mamma

If cloud thermals spread out at a stable layer the lower surface (i.e. the base) of this 'anvil' cloud is a discontinuity of humidity mixing ratio. This surface is caused to sink, perhaps by the ascent of further thermals into the space between the inversion and the cloud base, or because the air is moving over unlevel ground. Then the air above cloud base will sink along the wet adiabatic and will quickly become colder than the air below, which sinks along the dry adiabatic. If the base sinks 100 m a temperature difference of about 0.5°C is produced, because the wet adiabatic corresponds to around 0.5°C for 100 m over a large part of the T–ϕgram. If the lapse rate in the air both above and below cloud base was the wet adiabatic (which would be typical) before the subsidence took place, a layer in each mass equal in thickness to the extent of the subsidence can be involved in the instability which follows. It takes the form of a wrinkling of the surface as illustrated in Fig. 14.7.iv, 14.7.v and 14.7.vi.

The upper surface of a cloud thermal has a typical cumulus, or cauliflower, appearance, with a sharp outline occasioned by the large number of nuclei forming droplets in the stronger upcurrent. By contrast the base of the downward thermals, or **mamma** (breast-like pendulous clouds), are smoothly and less sharply outlined. This is because the smaller droplets are quickly evaporated in the downward motion and so the visibility is greatly increased to several metres. When illuminated by the setting sun shining on the cloud base, the mamma are bright, but when the sun shines on the top of the layer the bright spaces between often show the cloud to be very thin. See also the colour section.

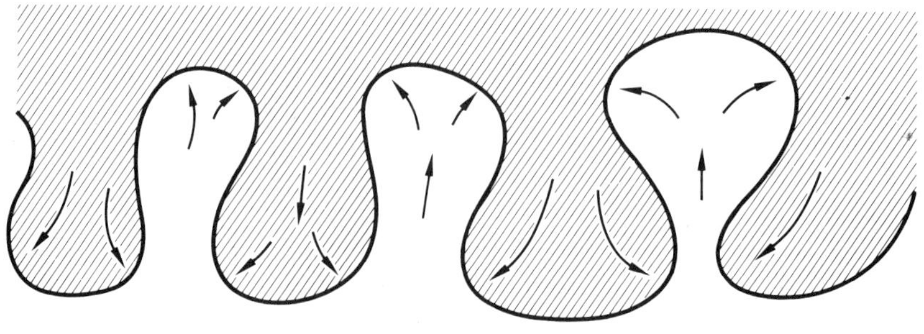

Fig. 14.7.iv The motion pattern of the base of an anvil cloud made unstable by subsidence. Such a cloud is shown in Figs 14.7.v and 14.7.vi.

Fig. 14.7.v This cloud is first formed by the spreading out of a large cumulus at a stable layer. The cumulonimbus anvil cloud is typical and spectacular, but similar motions occur when smaller cumulus spread out in the same way at their tops. As the parent cloud grows and the anvil thickens, the anvil base descends and an unstable stratification is produced there by the subsidence. The mamma (breast) clouds are the result. The instability may be enhanced by fallout which cools the air just below the mamma and increases the downdraught there. Sometimes the lower boundary of the mamma is double, the lower boundary being that of the fallout, the upper that of the cloud. See Fig. 14.7.vi. In this picture the parent cloud is beyond the lower left corner. [Photograph by M. Koldovsky.]

Fig. 14.7.vi Some mamma have a double outline, the lower outline being due to larger particles of greater fall speed than the mass of cloud particles.

When snow falls, which it does with a much smaller speed than rain, the outline of the fallout base is often quite sharp. The snow not only drags the air down with it but also cools it, and these effects cause mamma on the fallout base, for example when a snow shower or warm front snowfall begins. Many cumulonimbus anvil clouds have copious mamma, which differ from those on the stratocumulus anvil clouds just mentioned in being due to a fallout of ice crystals into air which is unsaturated and therefore cooled by the fallout.

A common form of mamma is on the trail of fallout from a high-level castellatus cloud which becomes glaciated (see section 14.7(4) below) forming dense trails. If the air into which the crystals fall is supersaturated for ice the crystals grow, and this is quite likely to happen just below castellatus where air at the base is saturated for water. If there is a slight shear the trails are carried away to one side, and then when they reach an unsaturated layer they begin to produce mamma by evaporation (see Fig. 14.7.vii).

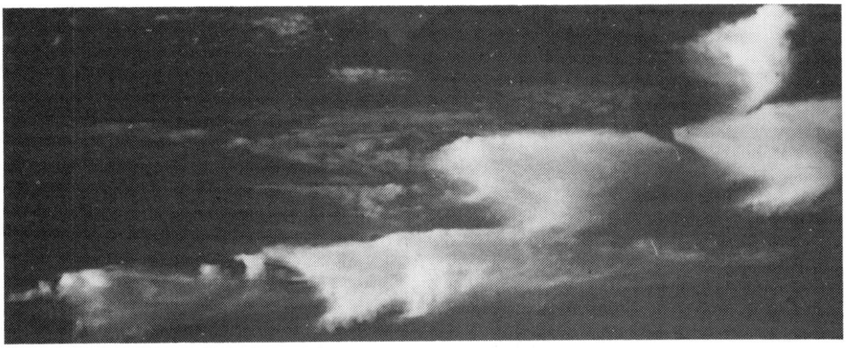

Fig. 14.7.vii The base of the ice trails formed from castellatus which became glaciated have fallout mamma. The flat tops of the clouds are near to the condensation level of the original castellatus. The water cloud has evaporated but the ice fallout from them has grown, down as far as the ice evaporation level. It then falls into air unsaturated for ice, the shear drawing the trails of falling particles to the left. The evaporation then cools the air, causing descending motion.

(4) Glaciated fallout

The **mechanism** proposed by **Bergeron** and **Findeisen**, whereby fallout particles originate as ice crystals in a cloud of supercooled droplets, is well exemplified by fallout from medium-level clouds referred to in the preceding subsection. If all the particles are glaciated, for example because the temperature is below –40°C, they will all remain small. The cirrus cloud so generated will then show some feeble effects of fallout, probably due to the slight growth of the particles in supersaturated air after they become frozen. At the other extreme we have floccus (Fig. 14.6.xi), castellatus clouds which evaporate quickly because the surrounding air is very dry and it is too warm for much glaciation (e.g. –5° to –20°C). In between about –25°C and –35°C we may have various degrees of partial glaciation where a few ice particles grow big enough to fall out. The remainder evaporate, partly because of the presence of ice among them and partly at the edges of the cloud where the surrounding air is unsaturated.

As with many theories it was assumed that *all* rain was generated by the Bergeron–Findeisen mechanism, but many 'outdoor' meteorologists knew that there could be **warm rain**, and this is illustrated in Fig. 14.7.x. But Fig. 14.7.ix illustrates the mechanism well. In both cases we can see the fallout.

More spectacular is the behaviour of a layer of supercooled cloud when glaciation is induced at one or two points and the temperature is too high (say –20°C) for any spontaneous glaciation in the tiny particles present. The glaciation may be started by ice crystals falling from some cloud frozen at a higher level, or by the flight of an aircraft. This may freeze some cloud droplets on contact and scatter crystals in its trail. Its exhaust probably does not contain freezing nuclei. Glaciation is perhaps started by deliberate artificial seeding with dry ice pellets (solid CO_2 sublimes at –72°C, so the pellets leave a trail of air temporarily cooled below –40°C) or silver iodide (see Fig. 13.9.iv). It spreads because droplets freeze first on the outside and then fragment when the inside freezes and expands accordingly, ejecting frozen spiculae. The freezing also causes the release of latent heat and thereby sets up mixing motions which spread the frozen particles among

Fig. 14.7.viii Fallout takes many different visual forms according to the lighting and the moment to make the photograph was coming as the sun was being obscured by the cloud. This was among many showers from the warm sea over S. Wales coast at 5000 m in November. The rather isolated thermal had become glaciated and the water droplets among the ice particles had evaporated and the ice grown to fallout size was all that remained of the cloud tower.

the surrounding unglaciated cloud. As a consequence the area of glaciation increases and forms a fallout trail, illustrated in Fig. 14.7.viii. The total amount of fallout in such circumstances is small, since it comes from a thin layer; nevertheless the spectacular nature of the phenomenon has often been misused in arguments in favour of rain-making programs. Fallout of this kind usually evaporates long before reaching the ground. See also Fig. 14.4.xii and Plate 12 in the colour section between pages 352 and 353.

The cirrus clouds which can be seen in the jet stream ahead of an advancing warm front often produce falling trails of great length because of the large horizontal and vertical extent of the supersaturated air in that situation. The direct circulation characteristic of a developing storm can then be seen in the sky. The orientation of the front is revealed by the long lines of fibrous cloud moving rapidly along their length with the thermal wind. The air at high levels is moving outwards and upwards from the storm; that at low levels is moving inwards and downwards, thereby decreasing the amount of low- and middle-level cloud beneath the increasing cirrus. The fall streaks lie across the lines to indicate the direct circulation, and when they do not show this feature the front may be expected to be less active, and the wind and rain reduced. This visible feature was remarked upon by Bewsey and the situation is indicated in Fig. 14.7.xiii and is seen in Fig. 14.7.xi and xii

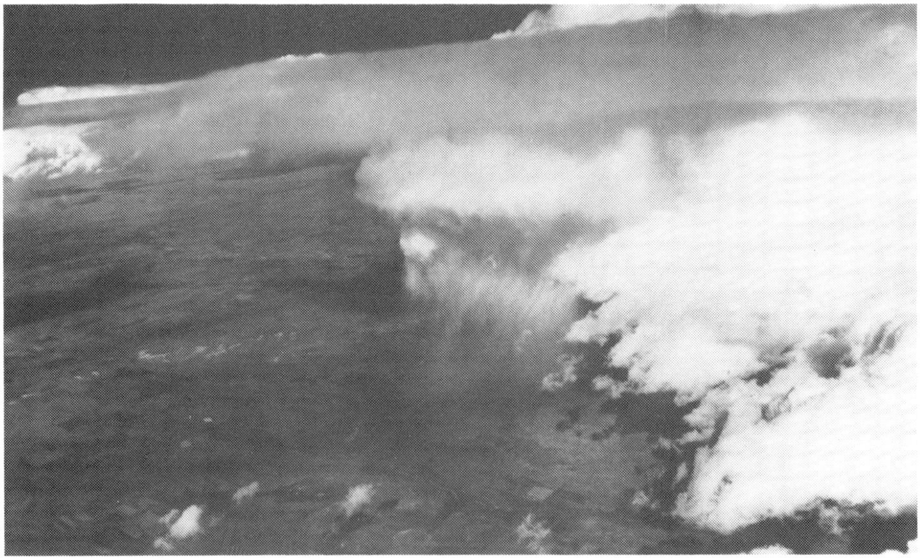

Fig. 14.7.ix The top of this cloud was well glaciated and the picture shows the fallout of a hailstorm in New Mexico. Some white patches on the ground on the left of the cloud are where hail has just fallen. [Photo by Robert Cunningham.]

Fig. 14.7.x By contrast, this cloud over the English Channel near Purbeck was all well below the freezing level, and is a good example of the effect of a few large CCNs in causing amalgamation by collision of water droplets. The large nuclei were probably sea salt. Ludlam has observed brief showers caused by the large hygroscopic nuclei mechanism inland in central Sweden. If the fallout shown here had been generated by the freezing of some droplets, the fallout, which was rather slow, would have shown a sharply defined melting level, but that was not possible because the freezing level was about 300 m above the top of the cloud. A rainbow was seen in the fallout a few minutes after the photograph.

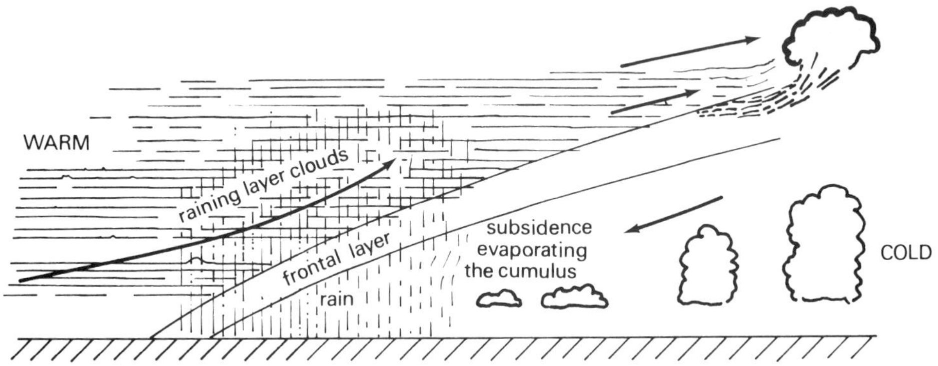

Fig. 14.7.xi A cross-section of an advancing warm front. The raining layer clouds are preceded by lines of cirrus often with fallstreaks oriented by the shear. Beneath the frontal zone the cloud is evaporated by the subsidence.

Fig. 14.7.xii An approaching cyclone. The lines of cirrus at the top are moving along their lines from the NW in the jet stream. The direct circulation (see section 4.8) is made visible by the fallstreaks pointing inwards towards the warm air. Wave clouds in the middle level are made likely by the increase of wind with height. The low-level wind is seen in the smoke to be from the SW (see Fig. 14.8.iii). [Photograph by F. H. Ludlam.]

14.8 MISCELLANEOUS CLOUD FORMS

For a fuller treatment and illustrations the reader is referred to *Spacious Skies* and *Clouds of the World*. Here we are concerned primarily with a few mechanisms which can produce an enormous variety of cloud forms according to the circumstances. **Scud** is fragmentary cloud below the main cloud base in rainy weather. It is formed when a thermal rises in air into which water has been evaporated from rain or wet ground, thereby lowering the condensation level. **Steaming fog** (Fig. 14.8.i) is produced when cold air passes over warm water, and is often seen in the autumn downstream of power stations which warm a river used for cooling. It is the same as bathroom cloud, and is produced by the mixing of air saturated at the water surface with cooler air above to produce condensation, as in a contrail.

It is unusual for layer clouds to remain formless for long. The top of a layer is losing heat by radiation to space both day and night, unless there is another layer above radiating downwards at its own temperature: consequently, downward convection from the top surface goes on all the time and both layers of cloud and fogs have a cellular structure in otherwise calm conditions. This is particularly true of sea fog which, contrary to popular supposition, is not simply due to the sea being cold. Sea fog is in fact being warmed by

Fig. 14.8.i In a warm sector near the coast at Aberystwyth the ground is made wet by drizzle. When a burst of sunshine subsequently starts to warm the ground, steaming fog is produced because the drizzle forms a very thin wet layer on soil which had been very dry and with its low thermal conductivity became warm very quickly.

the sea all the time, even in cool sea areas famous for their sea fog. In many cases the fog is first formed over the land and drifts out to sea where a steady state is reached with an upward flux of heat. Land fog shows cellular structure less often because the ground is often cooled at night to a rather low temperature before the fog is formed. Although the cells can be seen in the structure of these thin layers of cloud, the cellular motion is not easily detected because it is very slow (see Ship Trails).

Much more rapid is the formation of billow clouds which make visible the motion described in Chapter 8 without significantly affecting it. The buoyancy forces due to condensation are too small to influence this motion, which has large vorticity already, and radiation takes too long to produce significant temperature differences. The shape of clouds in billows depends on their position in the motion pattern, and the various possibilities are illustrated in Fig. 14.8.ii. Not all these forms are visible at the same time, and because the pattern is always developing, many forms are transient. If, as is very common, the billows are generated in a wave (see section 8.3) the full development of the overturning may be stopped when the air passes the wave crest and vorticity of opposite sign begins to be generated. Thus billow clouds of the kind depicted in the early stages of billow growth may actually persist and may not be transformed into the later forms (Fig. 14.8.ii).

Since the clouds represent a redistribution of humidity they may remain as a fossil pattern if evaporation by subsidence occurs as it does on the downwind side of a wave.

Fig. 14.8.ii The shaded area illustrate the various cloud shapes which may appear within the billow motion discussed in Chapter 6. The progressive development of the billows is sketched from left to right. Most of the forms occur in typical billow manner, that is several parallel elements together regularly spaced. However, subsequent billow motions may generate several crisscross patterns in the clouds they produce, because their orientations may differ (see Fig. 14.8.iii).

Fig. 14.8.iii Billows of different orientations and wavelength superimposed in a layer of cloud, together with tiny cumulus created by static instability, caused by radiative heat loss or ascent of the layer.

Thus billows or cells formed in one wave cloud may appear already formed at the upwind side of the cloud in the next wave. The fossil pattern (see section 14.2(7)) may remain for a very long time to appear a day later when cloud is reformed at a front. Very often on first appearance layer clouds at a warm front have a cell structure much larger than can be accounted for by their own thickness, and so reveal humidity patterns caused by convection many hours before.

Wave clouds may take the form of smooth lens-shaped clouds formed when the sharp top of a well mixed layer, which was the upper limit of convection somewhere upstream, is carried above its condensation level. The bases are then flat, while billows are sometimes formed on their upper surface. On the other hand, waves may be visible from below simply because of a few holes in a layer of cloud in the wave troughs. If the layer was formed by the spreading out of cumulus tops at an inversion, the base may be sharply defined by a discontinuity in humidity and may therefore be arched in waves.

Waves often produce similar effects to lifting and subsidence of a layer. Thus castellatus may be initiated by waves, and may extend in lines downstream from the point at which condensation occurs. On the other hand, the instability may be less in magnitude and merely break the first wave cloud into a cellular structure, which evaporates at the downwind edge of the wave cloud and reappears at the upwind edge of the next.

Waves may occur at very high altitudes, in which case the temperature may be below –40°C and the droplets freeze almost immediately. If the air remains above the ice evaporation level, a trail of cirrus may extend for many miles downwind as **orographic cirrus**. These trails have much in common with the trail of ice cloud from the anvil top of a cumulus situated over an oceanic island or over isolated mountain in a dry country. The source of thermals is fixed, and often the low-level wind is light, but at cirrus levels there may be a strong wind continuously carrying away the newly formed anvil cloud.

14.9 SELF-PROPAGATING STORMS

An air mass ready for shower formation often has warm moist air near the ground with much drier air aloft. Isolated thermals growing up into the dry layer are quickly evaporated. Large clouds with a long enough life to generate rain are not formed, especially if there is wind shear. Wind shear turns the humidified towers of air due to evaporating clouds into thin layers (see Fig. 14.6.vii) which scarcely reduce the evaporation of cloudy thermals rising through them.

Evaporation is greatly reduced by glaciation, and this is probably a more important consequence of glaciation than the extra 80 calories per gm released as latent heat. It also causes the onset of rain by the Bergeron-Findeisen mechanism before evaporation occurs, thereby further retaining the latent heat of condensation produced when the rain falls out as liquid. The following very much simplified description is based on work of Ludlam [66] since taken up by Newton [67] and others. It describes a simple two-dimensional flow pattern which can cause a shower to proceed through the air mass, processing the air as it goes by carrying up the warm air and releasing latent heat which is retained because of rain falling out. At the same time the shear of the wind ensures that the cloud leans over, and because the rain (R) falls out of what was the side of the cloud it falls into the dry air above cloud base and cools it. This cooling is much more than results from the evaporation of cloud because rain (R′) continues to fall into the air until it reaches the ground. Because it comes from a level of stronger wind the downdraught spreads rapidly along the ground, shovelling up the warm moist air into the cloud so as to continue the whole process. The motion is shown in Fig. 14.9.i. Many of these 2D features operate in 3D showers.

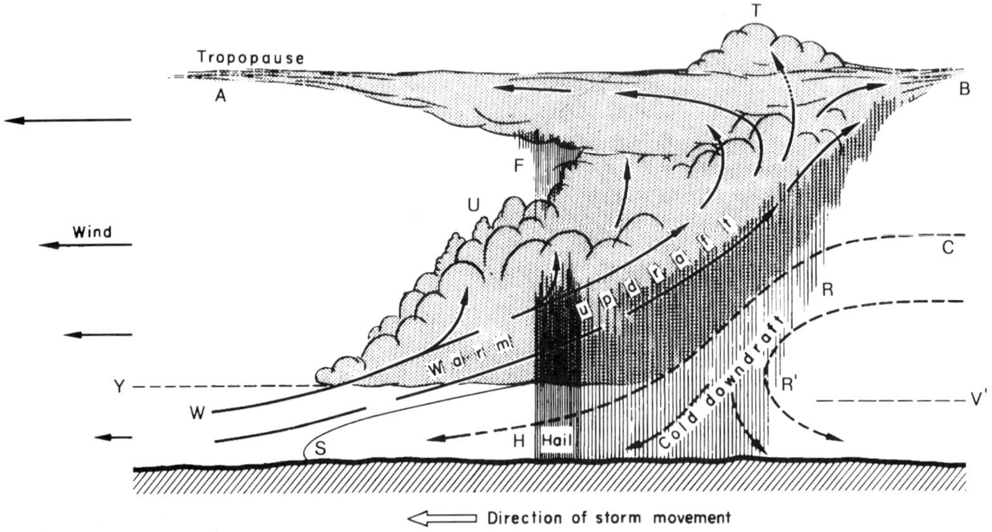

Fig. 14.9.i A cross-section of a travelling shower cloud showing features essential for self-propagation. This 'machine' processes the air, replacing the moist air ahead of it with a carpet of cooler air behind made moist by the rain. S is the front of the advancing cold air and is experienced as a sudden squall when it arrives. It may spread far ahead of the storm in some circumstances, and may support a squall cloud. (After F. H. Ludlam.)

At the highest levels, which may be just beneath the tropopause, the rising air begins to take up the speed of its surroundings and is carried forward as a great anvil cloud (A). Some of the very rapidly rising air may also spread backwards at the tropopause (B), and intermittent towers (T) penetrate several hundred metres into the stratosphere. Ice fallout from the upper forward-leaning part (F) may initiate hail (H) in the strong upcurrents (U) boiling up faster than the average in the sloping updraught.

The cold air (S) spreading forwards on the ground may move far ahead of the storm and initiate new towers of cloud separated from the main storm so that the system may proceed in jerks. Furthermore it is not a two-dimensional system, and in reality many complications produce local variations, as do features of the topography. An example of cold outflow ahead of a storm is shown in Fig. 14.9.iv, taken from Scorer [68]. The traces show the fall in temperature, rise in humidity and sudden rise in pressure.

The warm air (W) ahead of the storm may be beneath a sub-cloud inversion (V) and after the storm has passed it may leave a lower inversion (V′) topping the air that has been brought down by cooling and laid out as a carpet (C) behind the storm.

The changes in surface wind can be quite variable. The warm air may or may not be moving towards the storm relative to the ground. The cold air may or may not arrive as a squall, and if the front has a lobed structure such as is indicated in Fig. 14.9.ii, the squall will vary very much along the front in strength and direction. This is usual, and the arrival of the squall before the rain is usually accompanied by slamming of doors and windows, flapping of washing on the lines, flying of loose sheets of corrugated iron, clanging of dustbin lids, raising of dust and leaves, and so on. In the rain there may be an abundance of scud. Above the squall may be seen a variety of soaring birds reluctant to descend into the turbulent cold air but at the same time enjoying the updraught. Fig. 14.9.ii, Fig. 14.9.iii and Fig. 14.9.v are examples of such storms.

Mathematical representation of this situation is not easy, and requires that the parameters be chosen to produce the assumed pattern. This represents the essential difficulty with all complicated motion patterns involving several mechanisms which are not easily quantified: the form of the motion is never discovered by the calculation, which is, in effect, only bodged up to tell us what we already knew. No new laws of physics or dynamics are revealed by that study of these phenomena, only new combinations or complications.

Fig. 14.9.ii A storm with the anvil stretching predominantly backwards with most new growth seen on the right.

Fig. 14.9.iii A storm seen as the rain approaches, advancing from the left. [Photograph by Betsy Woodward.]

14.10 THE ROLE OF CLOUDS IN FRONTAL CYCLONES

It appears from many regional computer studies of the mechanics of cyclones that fronts may be generated in them without the intervention of condensation and the release of latent heat. On the other hand, the starting point, namely a stably stratified atmosphere with horizontal temperature gradients, could not be generated without clouds. The lapse rate up to cloud base is dry adiabatic most of the time, except for occasional shallow layers near the ground. Above the cloud base the air is always stably stratified except in cloud, and convection clouds maintain this situation. The result is that flow up frontal surfaces may remain concentrated in the frontal cloud systems, and does not extend far away from them as it otherwise would in a stably stratified air mass. This means that the direct circulations which generate winds and rain are confined mainly to the neighbourhood of fronts and last much longer that if this were not so because the general characteristics of the contrasting air masses are preserved. Computer forecasts spread the effects of fronts over a wide zone because discontinuities cannot be handled with a coarse grid of points. They are interpreted by inserting fronts in the appropriate positions and using knowledge of their properties to invent details of weather which should follow from the calculated movement of fronts and overall development.

If the earth had no clouds, cyclones would have a much shorter life history.

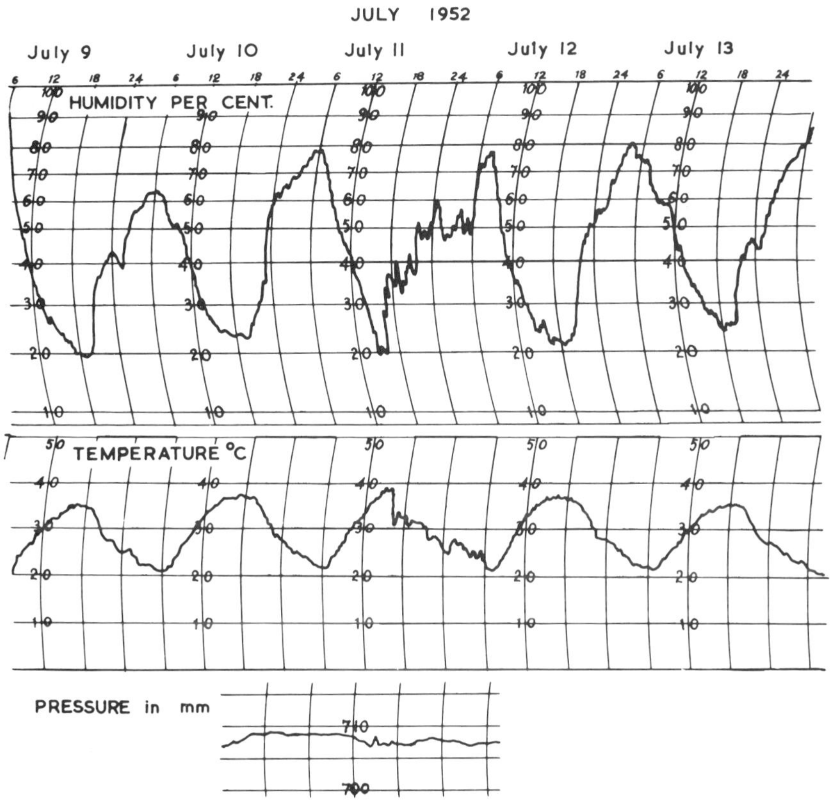

Fig. 14.9.iv Humidity, temperature and pressure traces at Madrid showing the sudden arrival of cold air from a storm at about 16 h on 11 July 1952. On the previous and following two days no showers occurred in the area so that the air mass remained unchanged.

Fig. 14.9.v The cloud at the squall front of an advancing storm showing the typical ragged base, where thermal upcurrents meet air into which rain has been evaporated. (Photo by M. Koldovsky.)

14.11 TROPICAL CYCLONES, TORNADOES AND WATERSPOUTS

Because of their small size and great intensity, particular tropical cyclones are about the most difficult phenomena to handle by computer. There is very little in the soundings over the tropical oceans which enables forecasters to predict precisely where convection will produce rain and whether it will be produced by a few large storms or several small ones. The criterion seems to be that large ones are likely if they can be sustained, and the subsidence which must take place elsewhere to compensate for the rising air in the storms warms the rest of the air mass. A new situation is created when a revolving system is set up.

When a revolving mass of air is situated over sea which has a more or less uniform temperature of, say, 28°C over a large area, the deficit of pressure at the centre may be as much as 60 mb. Air moving towards the centre may be cooled by 5°C or so by adiabatic expansion, and is therefore warmed by contact with the sea, which thus becomes a source of buoyancy as a result of rotation. The sea also becomes a source of latent heat because water is evaporated from its surface. It will be seen from the T–ϕgram that at 950 mb an increase in temperature of 5°C at 25°C raises the saturation mixing ratio by about 7 g/kg^{-1}, and that when air rises wet adiabatically to 500 mb it condenses out roughly15 g/kg^{-1}. At these warm temperatures, therefore, there is a very large supply of heat by condensation , and the turbulent nature of the air close to the sea surface due to the strong wind together with the copious spray ensure effective evaporation from the sea. Rainfall under the ring of convection cloud around the eye may commonly be of the order of 12 cm per hour, or around 3 m a day (Fig. 14.11.i).

In order to have a pressure 50 mb lower than in the surrounding air it would be necessary to warm the air up to 500 mb by 10 per cent of its absolute temperature, namely by about 30°C. This clearly cannot be achieved by the heating from the sea referred to above which causes at most about one-sixth of this temperature increase. Furthermore, if a parcel of air were to move 50 mb down the horizontal pressure gradient without retardation it would attain a speed of about 100 m s^{-1} ($\Delta p = \frac{1}{2}\rho q^2$, and so 50×10^3 dy cm^{-2} $= \frac{1}{2} \times 1.3 \times 10^3$g (100 cm)$^{-3} q^2$, giving $q = 1.14 \times 10^4$ cm s^{-1}). Consequently, air moving spirally inwards at the surface feeds the upcurrents and does not reach the centre of the storm. Instead air is sucked down dry adiabatically from the stratosphere and the tropopause is brought down to the sea surface. By descending dry from 300 mb and –50°C to 950 mb the air would attain a temperature of about 40°C and so the funnel of stratospheric air provides the low density which causes the low pressure at the centre.

Many small vortices are observed by ships and aircraft, some having winds of 30 m s^{-1} or more, but the vortices do not endure because they are not big enough to suck down the stratosphere to the surface and maintain a low enough pressure to cool the air enough to make the sea a local heat source. The initiation of a hurricane simply requires the assembly of sufficient vorticity in one small area, and this is a chance phenomenon in an area where the instability can be relieved by a large number of scattered shower clouds.

A mature hurricane is not symmetrical but has more clouds in the quadrant towards which it is advancing. It moves roughly with the wind at 3000 m, i.e. about 700 mb (see

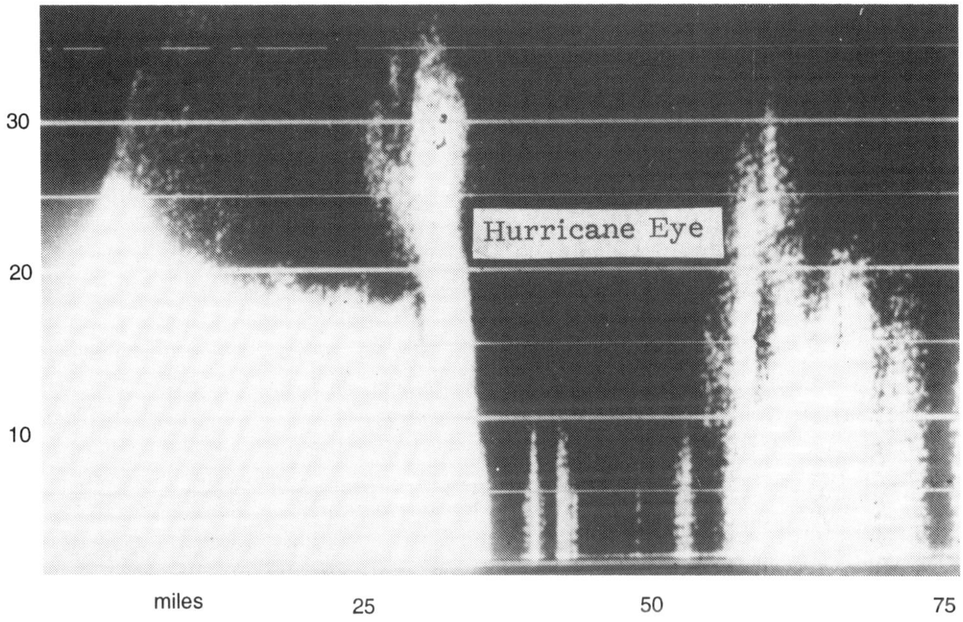

Fig. 14.11.i This is a vertical cross-section of hurricane *Donna* made by Key West Radar. The clear eye is 20 miles across and the eye wall reaches up to 35 000 ft. (Courtesy the US Weather Bureau.)

Chapter 4). It has spiral bands of cloud, shown in Fig. 14.11.i and Fig. 14.11.ii, which are radar pictures of a Caribbean hurricane. The bands move around with the wind, and are not streamlines of the flow. A vertical cross-section is drawn in Fig. 14.11.iii, in which the vertical scale is exaggerated and the clouds are therefore stretched vertically. We next draw attention to motion associated with clouds.

Descending, very dry, stratospheric air lies against the wall of cloud surrounding the eye. The liquid water content of the wall cloud is about 10 g m^{-3}. The mixing of these two utterly different masses of air would cause a cooling of 20°C if the water were evaporated at the level of mixing. Consequently, mixing causes a powerful downdraught, of the order of 30 m s^{-1} at times, which fills the bottom of the eye with much moister air than that of the stratosphere. Sometimes, but not always, there is a 'hub cloud' at about 1 km from the sea in the eye (Fig. 14.11.iv).

The drag of the sea reduces the wind at the surface so that instead of being proportional to r^{-1} as in a frictionless vortex, it is observed to be proportional to $r^{-1/2}$ roughly in a typical hurricane. In the eye the wind is calm near the centre. The drag is communicated upwards by the upcurrents, which are largely composed of retarded air, so that as it spreads out aloft (usually with a circular canopy of cirrus) it soon acquires a rotation speed slower than the earth and rotates anticyclonically. Mariners have long used this and other features to guide them as to the whereabouts of the storm centre when they are on its periphery. Storm centres move typically at 10 to 15 kt and may therefore be avoided by most ships with good navigation and reading of the sky. At a radius of 300 km the

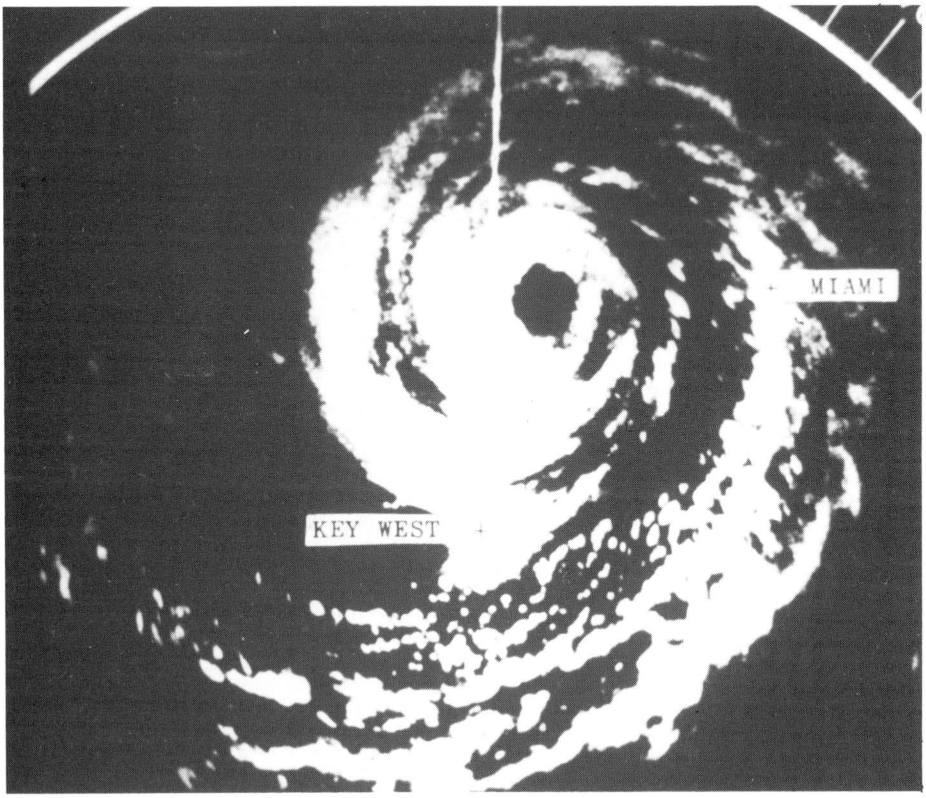

Fig. 14.11.ii Plan of hurricane *Donna* seen by Key West Radar. These may be interpreted by reference to the cloud cross-sections in Figs 14.11.i and 14.11.iii. The radar shows principally the cumulus which are raining. The largest are those surrounding the eye. The distance from Key West to Miami is about 220 km. The vertical section (Fig. 14.11.i) was made about eight hours earlier through the eye, when it was about 30 km wide. Heights are given in thousands of feet. (Courtesy the US Weather Bureau.)

vertical component of the rotation of the earth at latitude 19° (where $\cos\theta = \pi^{-1}$) causes a circumferential velocity of 25 km h^{-1} = 13.5 kt. A circumferential wind of 19 kt is therefore an actual speed of 32.5 kt in space. If the speed relative to the earth increases like $r^{-1/2}$ this leads to a wind of 60 kt near the wall of the eye at 30 km radius, which is about 61 kt in space. This decreases to 6 kt in space at a radius of 300 km, which is a speed of –7.5 kt relative to the ground at this radius, a significant anticyclonic rotation. At 600 km this circumferential velocity is increased in magnitude to $-\left(13.5\times 2 - 61 \div \frac{600}{30}\right) \approx -24$ kt.

A hurricane is subjected to more surface friction over land so that the circumferential speeds around the centre are soon significantly decreased. But, more important, the source of latent heat is removed, the rate of rainfall is decreased, and the storm quickly

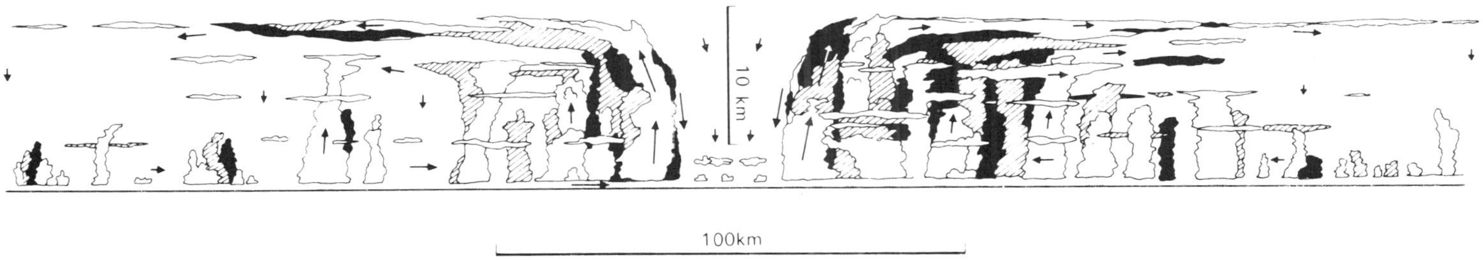

Fig. 14.11.iii A schematic cross-section of a hurricane, with typical dimensions, through the eye and the lines of shower cloud such as are illustrated in Fig. 14.11.i. The motion indicated by the arrows has the circumferential motion described in section 4.9 superimposed.

Fig. 14.11.iv 0857,31.10.81,Ch1. An Arabian Sea tropical cyclone of the kind which could drench the Hadramaut (see Plate 13 in the colour section between pages 352 and 353) with rain and start a locust plague. It could carry the insects after they begin to fly as far as Pakistan across the sea. The mountains of Iran might be a lethally cold destination.

loses its ferocity. The worst effects from the human viewpoint occur in the first few hours after it arrives near the coast.

The cold outflow from the downdraughts created in raining cumulus clouds is often responsible for very intense local updraughts. If the air lifted has vertical vorticity the rapid generation of buoyancy in the cloud creates low pressure and convergence at the cloud base similar to that into the base of a thermal. The stretching of the vortex lines then intensifies the rotation and prolongs the life of the low-pressure centre. Convergence is therefore induced in the air below, and provided that there is vorticity there also, the vortex is extended towards the ground.

The sudden growth of cumulus towers in these circumstances produces funnel clouds which gradually extend down towards the ground, and have a much prolonged life if they reach it. Over the sea these funnel clouds are called waterspouts, and they are fairly common under showers near the coast over warm seas. Many spouts last only a few minutes and do not reach down to the sea. However the vortex may reach the sea even if

Fig. 14.11.v Part of the spiral system of cloud lines which enclose a hurricane. (Courtesy National Hurricane Research Project U.S. Weather Bureau.)

Fig. 14.11.vi The hut cloud of Caribbean hurricane *Esther*, 16 Sept 1961, as seen from 20 000 ft in the eye at 1952Z. [Photo by R. W. Simpson.]

the base of the funnel cloud, which is approximately an isobaric surface, does not, and then the spray generated looks like a dust devil (see section 11.8(4)) rising upwards from the surface.

Over land, because of the much more varied nature of the surface, vorticity is greater in the lower layers. Thus when a hurricane passes over land it may produce many more

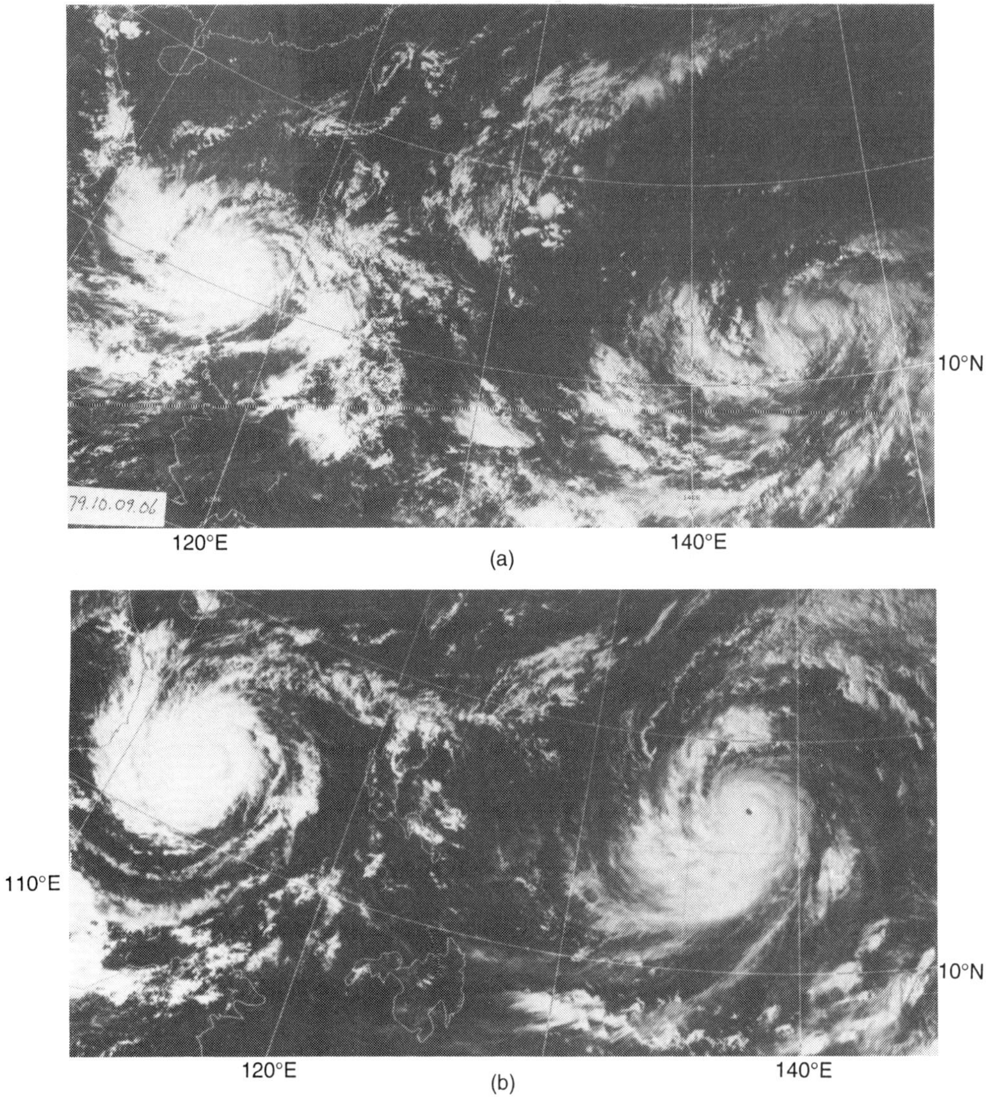

Fig. 14.11.vii (a);(b) 0600,9;12.10.79,vis. These pictures from the Japanese geostationary satellite show the most important area for the production of typhoons to the east of the Philippine Islands around 10°N, and about 140°E south of Japan. Typhoons (the name for hurricanes in the Pacific and Indian oceans) arise over the equatorial western Pacific at the end of the long equatorial stretch of sea driven westwards by the North and South Trade Winds. (a) In these two pictures we have an old typhoon on the west side of the islands at about 115E and 10°N. It has existed for a few days. At the same latitude and at 145°E we see a cloudy area just beginning to show some cyclonic vorticity. (b) 3 days later the western typhoon has moved to about 12.5°N and 113°E and does not appear to have changed much. The western area of convection cloud has become organised into a cyclone with a small eye, typical of typhoons in their most violent stage, although some typhoons display no eye at all. A similar development takes place on the southern side of the equator: the typhoons are less frequent and travel north of Australia, and some even progress across Malaysia into the Indian Ocean.

Fig. 14.11.viii The cloud, over the Gulf of Tonkin, begins to lower a 'spout' as convection stretches the vertical vorticity into a significant rotation, and this rotation extends down towards the sea surface where there happens to be an oil slick which responds to the rotation. When viewed 'end on' the slick seems to have made a hesitant response but no waves have been generated. Prof. Fujita (Univ. of Chicago) analysed this by projecting the image as if seen from above. The funnel is very narrow, and appears to be hollow. [Photo by U.S. Air Force.]

funnel clouds, called tornadoes over land, than it does over the sea. This is further enhanced by the fact that the surface air under a hurricane is now much drier than it is over the warm sea. In consequence, when a downdraught occurs it may be much more intense than one over the sea because of the greater evaporation of rain in it (see section 14.6(1)). This produces a correspondingly greater supply of air into the cloud producing the funnel and makes it possible that the downdraught front may generate a velocity discontinuity which is a much more intense supply of vorticity than is found over the sea.

Tornadoes seem to be most intense when the air above cloud base is drier than usual, because then the number of cumulus clouds is decreased and any rapidly growing one is more intense as the lapse rate above cloud base is steeper. Outside hurricanes, tornadoes are most intense at minor cold fronts produced by storms. The base of the funnel cloud is sometimes truncated by evaporation near the ground because of warming due to friction in the intense wind.

When a tornado begins to become intense it soon occupies the centre of the storm cloud because of the rapid cloud growth above it. This means that earlier tornadoes of the

Fig. 14.11.ix The base of a waterspout also appears to have a hollow surrounding its axis due to the centrifuging of the cloud droplets. Some dust devils also show this feature when the vertical vorticity is stretched, and it is often reported that there is a downcurrent in the empty middle. But this feature cannot be generalised from simple observations and models. [Photo by RAF.]

same storm have ceased to grow and their tops are sheared over so that the vortex becomes stretched and inclined. This is called the rope stage of a tornado because it becomes a long, tortuous tube of cloud.

14.12 CLOUDS AND CLIMATIC CHANGE

The mean temperature of the atmosphere is determined by the total incoming and outgoing radiation. Changes are buffered by the large thermal capacity of the ocean. Much has been written about the increased greenhouse effect due to the increase in the proportion of carbon dioxide in the air. The burning of forests and fossil fuels will continue to cause an increase, and it has been estimated that, other things being equal, the increase since 1850 has raised the world temperature by 0.5°C, and will cause a further increase of around 2°C in the coming century if the use of fuel consumption 'follows the expected trends'.

In spite of great variations in the power of the sun there has been a large part of the earth's surface at a temperature near enough to 10°C for it to be very suitable for life for 3000 million years. How the earth has avoided much larger variations in climate than have actually occurred is a much more important problem because the stabilising mechanisms might overwhelm any effect due to activities of mankind.

The most obvious factor, and probably by far the most important, is the cloud amount. This is because changes of the magnitude which are observed to occur cause changes in the earth's albedo (the fraction of sunshine which is reflected back into space) which must dominate the radiation budget. Cloud amount changes from hour to hour, and with the seasons and over the decades. If there were not a powerful stabilising mechanism the temperature would have been observed to vary in historical times by much larger amounts than are known to have taken place. A general rise in temperature increases the amount of water vapour in the air and whether this increases the global average amount of cloud accordingly and more sunshine is reflected requires a better model than exists at present.

The most important influence is nowadays thought to be the precession of the earth's axis with a period of about 45 000 years; but see [69]. At present the earth is nearest to the sun (aphelion) in the northern winter, but when it is farthest from the sun (perihelion) in that season the northern winters become more severe. This causes a decrease in the cloud amount over the greatest land masses in winter when the effect is least because sunshine is feeble. Furthermore the snow cover of the land hemisphere is increased which has the same effect as an increase of cloud amount.

This very brief reminder of the overwhelming importance of cloud amount, which is determined on a very short time scale of a few hours in stabilising and probably determining the average world temperature which varies with the seasons and over much longer time scales, is needed because of the popularity of many preposterous and fanciful theories about man's inadvertent or even deliberate influence on climate.

Unfortunately, the smallness of the details of cloud cover and its rapid changes make it very difficult to observe and quantify any measure of the global effects even with the present satellites, some of whose remarkable photographs are reproduced in this book. It will be many years before records have been obtained which give us a basis for quantitative assessment of the other natural and man-made influences which are rather small by comparison.

15

The aerodynamic environment of animal life

15.1 THE EXPECTATIONS OF THE ENQUIRY

We are in an era in which expertise in a branch of science seems to require a knowledge so detailed that it is very difficult for anyone to argue with the expert. If one takes issue with a specialist one is liable to be slapped down with statements which appear to disprove any contrary suggestion. There is a real danger that the narrowness of the specialist's imagination has excluded from his consideration ideas which are obvious to an untrained observer. Thus in Exodus X 13 it may be read, 'and the Lord brought an east wind upon the land all that day, and all the night; and when it was morning the east wind brought the locusts', and in verse 19, 'And the Lord turned an exceeding strong west wind, which took up the locusts, and drove them into the Red sea; there remained not one locust in all the border of Egypt'. However, for a long time it appeared to a trained biologist within a swarm that the locusts were in purposefully directed flight, and to them the problem was to find out where they were making for. We shall see in section 15.2 that their flight is not purposeful in direction and that they are carried by the wind on their great migrations, as Moses's 'scriptwriter' indicated; yet there was great resistance to this idea from biologists because it degraded the status of the locust, which had been thought to act purposefully. Even today many biologists have ridiculously grandiose ideas of a bird's power of migration and prefer to look to such mechanisms as a map of the heavens incorporated from birth in the bird's mind and to ability to detect the earth's magnetic field. There is, of course, no physiological evidence for either of these suggestions, and it would seem simple common sense to gain first an understanding of how species which live an important part of their lives airborne react to the air motion. Scientists appear to be so badly trained these days that their imaginations seem able to work only on concepts of laboratory origin.

We must avoid asking such questions as 'how do birds navigate?' before we are sure that they do. Life would be simpler if they had ways of travelling about with the seasons which did not involve any intellectual activity whatever (like our breathing; which we can, but don't usually, control). The mode of success requiring the minimum of biological activity is likely to be the one practised because it is the most reliable. We start, therefore, with the assumption that airborne life forms have very simple aeronautical behaviour which only involve the sense mechanisms to the minimum.

We assume that the pattern of behaviour must favour successfully the survival of the species, and that a simple pattern is more likely than a complicated one involving overt purpose.

Without any disrespect intended to that great enterprise 'Nessie', it is appropriate to use the Loch Ness Monster as a generic term (in the manner of 'boycott', 'meander', 'cardigan', etc.), indicating a marvellous being used to explain a large number of different phenomena which have come to be erroneously attributed to the same cause. Such explanations must, of necessity, be more marvellous than any of the explanations offered for the separate phenomena. Thus the Monster must be able to take the appearance of decomposing vegetation rising to the surface, motor boats, whirlwinds, fish shoals, and probably many more quite separate events. It cannot be like a motorboat because it is like the rotting weed; it cannot be like rotting weed because it moves like a whirlwind; and of course it cannot be like any other animal or it would have been caught long ago. Thus UFOs are a Loch Ness Monster, for they are like planets, like aeroplane lights, like geese or patches of clear air turbulence (to radar), like various meteorological optical phenomena (e.g. rings round a sub-moon or sub-sun due to multiple reflections on almost horizontal surfaces of ice crystals in tenuous clouds), and like mirages, together with a few other illusions and exaggerations. UFOs are required to be the single explanation of several unrelated phenomena which are only thrown together by being observed at a time when science fiction propagates certain ideas.

Such marvellous explanations are often taken seriously and the public as well as the scientists have become educated to expect instant answers. If I say that it is contrary to experimental physics to suppose that birds navigate by means of the magnetic field I receive the retort, 'well how do they do it, then?'. The situation is so complex in the great outdoors that laboratory concepts are often quite inappropriate. At the same time a long exposure to the medium can often help in finding simple explanations through experience gained living in it. In this respect the present generation, living mostly in towns and effectively protected from the weather even when they are outside the towns, is the worst equipped for many centuries to learn what is going on. The stars are effectively hidden by city lights, and we have no need to know them. A few meteorologists, astronomers and naturalists are exceptions in modern society. Most biologists seem to go out of doors with laboratory blinkers on. So, of course, do many engineers, physicists and chemists. Thus a radar engineer who obtained an ordinary photograph of the melting level which he had previously only 'seen' on the radar as a 'bright band' (it does not appear that way optically) called it a visible example of the 'bright band'.

There have been many eminent biologists in the past who have believed that a bird or a fish experiences a force equal to that on an object held fixed, relative to the ground, in a moving stream of air or water (see Ackworth 1955). This is the same as saying that an airship would feel an excess pressure on one side if it were travelling at right angles to the wind, and indeed the director of the film *Round the World in 80 Days* depicted the air as streaming past the gondola of a balloon.

In the light of these introductory remarks it will be appreciated that where the mechanisms to be described in this chapter give purely physical and mechanical explanations of phenomena involving life forms whose behaviour has been 'explained' in the past as due to biological, sensory, behaviour there is likely still to remain a residue of scepticism in some quarters previously committed to the search for such explanations.

Which is reminiscent of the entomologist who proudly claimed to have observed a particular butterfly walking on the top of the Carpathian mountains in the direction of Constantinople, which was in the direction that that species was thought to be interested in migrating. One does not have to be an engineer to see that to be airborne could be very helpful.

Thus a whole range of life forms have their own ways of reacting to the seasons, and it is not helpful to engage in moralising about the superior intelligence of the ant over that of the grasshopper *quand la Bise fut venue*, for they have survived each winter by their own brand of hibernation for thousands of years whatever the ants' pessimistic forebodings!

The way we observe members of a species may give us clues as to the mechanisms which facilitate a migratory journey, and so we begin with a study of the desert locust.

15.2 THE MIGRATION OF DESERT LOCUST SWARMS

The first comment of one observing a locust swarm in flight could be that they form a dense cloud which seems purposeful in the unanimity and determination in which they have chosen the direction of flight. They seem to have some idea that they are going somewhere interesting. Consequently a great deal of work was done to collect information about how they made their choice of direction. Cine-films taken looking vertically upwards showed that the direction of flight was not the same at all heights, but the locusts at higher levels, which appeared smaller according to their distance above the camera, were aimed in directions quite unrelated to those near the ground, although they too had a definite direction. Clearly the danger of collision would be serious if those at a particularly height did not agree on a uniform flight direction.

Under a swarm in flight there were many settled on the ground, and when they took off they would face into the wind if there was any. When observed from the air in sunshine some parts of a swarm would appear to reflect the sunshine on their wings in a particular direction determined by their direction of flight: but different parts of the swarm appeared to be going in different directions. Thus a swarm did not actually have an agreed direction of flight. When the position of a swarm was plotted on a map it was found to have been travelling in the direction of the wind but with a speed somewhat less

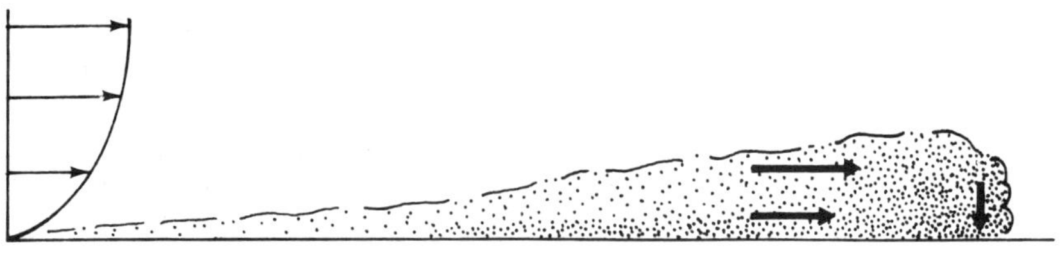

Fig. 15.2.i The distribution of desert locusts in a swarm carried along by the wind. They tend to settle at the front where the swarm is dense and take off when there are few above.

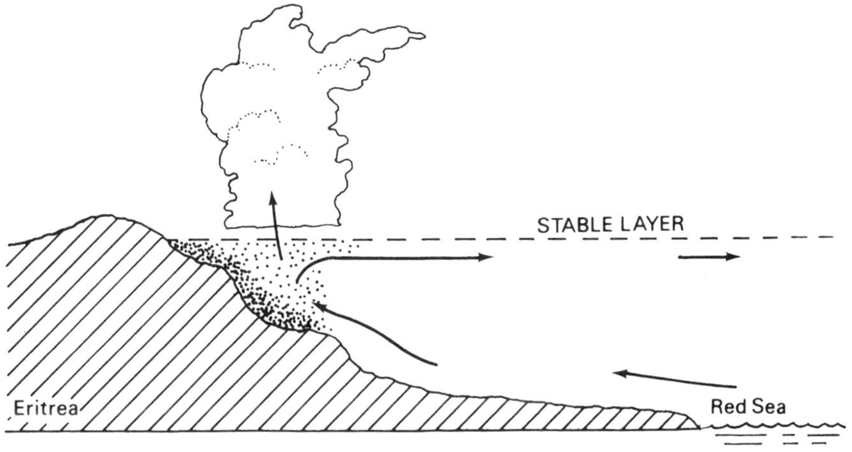

Fig. 15.2.ii Swarms are sometimes accumulated under showers on a mountain range. They avoid being carried up into the shower or away beneath the inversion.

than the wind speed. This was consistent with some locusts being on the ground at any time so that each locust spent less than all the time in the air, travelling on the average with the velocity of the wind. It appeared that as the swarm moved in the direction of the wind some of them at the front would settle on the ground, while at the back of the swarm they would take off to avoid being left behind and would positively fly towards the swarm to become part of it.

The suggestion that while flying they chose no particular direction of flight raised the question how did the swarm remain coherent and did not spread for ever more widely across the land like smoke being dispersed by the fluctuations in the wind. Swarms were observed to occupy approximately the same area from day to day, and it became necessary to suppose that those on the edge of the swarm behaved like those at the rear and were anxious to be well within the swarm. It did not require much imagination to observe those at the edge turning back into the swarm. They have been seen to become divided into two by a hill when they were confined beneath an inversion or cloud by reluctance to fly further away from the ground. Most swarms may be described as **stratiform**, but occasionally **cumuliform** swarms are seen; presumably when the upcurrents are too strong for the flying insects to avoid gaining height. The top of such a swarm may be carried away from its base in strong upper winds.

The way to destroy a swarm seemed to be to spray it with insecticide on the ground when it settled as the temperature fell in the evening, and this was found to be effective, but it could leave a noxious deposit on the ground, and when dieldrin was used because it was the most effective poison available (in the late 1950s) that could be dangerous for people. The most efficient way to apply the insecticide was to spray the swarm in the air, and because the pattern of flight was not understood it was clear that the cloud of poisonous spray should be put into the air behind the swarm, for it would be carried by the air at the speed of the wind and would move through the swarm in which each locust would pick up its dose; but no large doses would be dropped on the land such as would

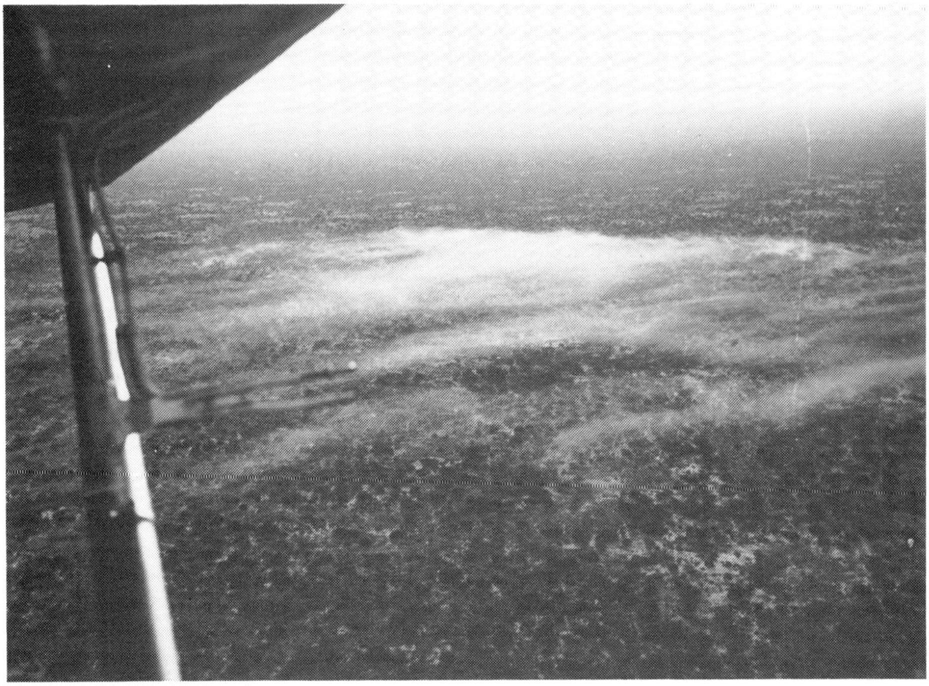

(a)

(b)

Fig. 15.2.iii (a) A small low-flying stratiform swarm about 1 km^2 near Wajir, Kenya, 31.1.53. (b) Part of a large cumuliform swarm covering about 75 km^2, seen from about 700 m above ground, but still below the level of the topmost locusts, looking up-sun. Only part of the swarm is visible in that direction because of the specular reflection required to see them. Near Hargeisa, Somalia, 1545,2.10.53. [Photographs by H. J. Sayer.]

be required if the swarm were sprayed on the ground. That which was not taken up by the swarm would be very widely dispersed.

So far the question remained why the species chose the particular flight pattern described above. The biologists who had tried hard to understand their subject could not see why they behaved that way; it was as if they were wondering what the locusts thought they were doing. The correct question to ask was, of course, what survival value the procedure had for the locust; and this case illustrates the superior merit of this Darwinian approach to the problem. It had already been observed that when there was a locust plague abroad, which means when there were several swarms turning up in the general area from Bangladesh to the west coast of North Africa, whenever there were rainstorms found in the desert there seemed to be locusts present ready to take advantage of the ideal conditions for laying their eggs in the soil (sand). When it rains on the desert the dormant seeds of grasses and perhaps other plants would blossom two or three weeks later. The locust eggs were laid in a dehydrated form so that they would only hatch out if the ground were moist enough. When they did hatch out the locust hoppers would march, eating all that was available to them until they were big enough to fly and form another swarm. Rain on the desert provided all their need in a very simple way: but how to find it?

Rain is produced in rising air, and as we saw in the last chapter, this means that the surrounding air must converge to feed the updraughts. The desert is the home of the locust because it supports none of its predators or diseases which might significantly reduce its numbers. Normally, locusts are like other grasshoppers which live in the dry semi-desert areas around the deserts where agriculture is scarcely feasible, but they have a different sort of behaviour when they are aggregated into a swarm. In this way they can breed in the deserts and go on marauding flights in the neighbouring areas, depositing fields of eggs on their travels. A few storks have often been observed following a swarm and feeding liberally on it but making little impression on its huge (millions) numbers. Swarms have been known to be carried by the wind across the Arabian Sea from southern Arabia to the coast of Pakistan where the gardens of Karachi were ravaged to destruction. It was calculated that the body of air occupied by the swarm as it set out from Oman underwent convergence so that it occupied only about 1% of its original horizontal area on arrival in Karachi, so that however many casualties the swarm suffered over the sea by thinning at the edges, the swarm style of behaviour preserved a formidable force.

The insights into the desert locusts' strategy were elucidated by Dr R. C. Rainey, who was an entomologist by profession, but had developed an interest in and acquired the skills of soaring in a glider while working in central southern Africa. During the war he trained as a meteorologist and served with the allied air force in Africa, returning to work against pests afterwards, largely to preserve the cotton crops, but was diverted on to desert locusts because of a plague of locusts in the 1950s in Ethiopia and East Africa. The vulture story was first told by Philip Wills, the World Gliding Champion of 1952 [48]. In that book he described the explanations he gave as 'blindingly obvious', but he had the advantage of having Darwin's analysis deeply embedded in his mind although he had no training as a scientist or engineer. But we may quote Moses's script-writer from Exodus X 13 'and the Lord brought an east wind upon the land all that day, and all the night; and and when it was morning the east wind brough the locusts', and in verse 19, 'And the

Lord turned an exceeding strong west wind, which took away the locusts, and cast them into the Red sea; there remained not one locust in all the borders of Egypt'.

The biologists had studied in great detail the fact previously unknown that desert locusts occur in two forms. In the **solitary** form they are not usually a serious pest to humanity but are merely another grasshopper which ensures the persistence of the species when there are no swarms. In the **gregarious** form they become physiologically different and their colour is changed somewhat by being crowded together and there is said to be a plague of locusts when there are several swarms in the general locust area. They were for a time thought to be different species. A plague may be started by the occurrence of a low-pressure centre in southern Arabia, for instance. The occurrence of more than usually copious rain in a desert area such as the Hadhramaut would produce a greater than usual solitary population which might be swept up by a cold front to form swarms which could be carried by the wind either into East Africa, Iran or Baluchistan. Since rain on the deserts is a rarity, the species has developed a life style which can support a much larger population while the opportunity lasts (see Plate 13 in the colour section between pages 352 and 353).

When the eggs laid by a swarm hatch out, the insects, called **hoppers**, are small and wingless. They form **hopper bands** which **march** forward across the ground with a circulatory motion bringing fresh hoppers to the front, especially when those at the front halt to feed. Almost everything edible in the way of the marching bands is eaten.

The most effective method of destroying a hopper band by poisonous spray has been to use **drift spraying**. In the case of closely grown crops the widely used method has been to spray the crop; but the efficiency, defined as the fraction of the insecticide which actually kills the pest, may then be very low: large drops are used so as to deposit the spray by gravity onto the crop. The upper leaves protect the lower ones, and spray falls on the bare ground which is not covered by crop. Calm weather has to be used to achieve the deposition. If the evening is chosen, the air is likely to be stably stratified close to the ground, in which case very small droplets are used and these are captured by vegetation in drift spraying when there is a wind. Then the hoppers eat the vegetation until they have taken the required dose. The technique also deposits spray on the lower parts of the vegetation.

By contrast, the method favoured by the Indian Government has been to dig a trench ahead of the marching hoppers and bury them when they march into it. This uses human labour, which is in plentiful supply in India, so that it is preferred to other more expensive methods.

A large swarm may cover several square kilometres and have as big an appetite as 100 000 people. It has long been recognised that the swarms can be a very formidable enemy, and unlike bees and ants they have no leader, yet 'go they forth all of them by bands' (Proverbs 30, line 27). Nevertheless the assumption that they have some unknown decision-making process seems almost universal (Nahum 3.17). Thus they were assumed to fly only when it was warm enough; but where to?

The problem of control has not been fully solved partly because during periods of no plague, governments are unable to maintain the international organisation required to pounce on any new swarms. Nor has an efficient use of pesticide been devised. The primitive idea is that a crop is protected if it is sprayed with insecticide; but this is pure spite because the insect may damage the crop before it becomes poisoned and the main reservoir of the pest may be unaffected. It would therefore be necessary to spray all the crops in a large area together and so action under some supervisory authority is required. If 60% of a crop is covered with spray, that is a measure of the spraying process. It could be said that 40% of the spray becomes environmental pollution, while only one part in 10^8 is all that does the killing. We do not know this figure within a factor of perhaps 10^2, but in any case the efficiency of use of the poison is very low. But why spray the crop? We do not spray people to kill malarial mosquitoes; least of all we do not try to cover people effectively all over. Yet this is often the tactic of the sprayer—to cover the crop all over with a uniform layer of spray of minimum depth. The solution is to spray the pest in the air, and Rainey calculated that in that case as much as 20 per cent of the pesticide was employed in destroying locusts. The only way to fight a pest is to grow a crop next to the crop to be protected which will harbour the pests' most dangerous predator.

The concentration of locusts under convection clouds with converging horizontal winds supplying the upcurrents in the clouds may be likened to a raingauge carried along beneath a raining cloud. As we now from some occurrences of floods, the records of rain showers in fixed raingauges do not usually record the cause of floods. Sometimes the falling rain brings pollution down to the ground that had been captured by water in the air when it condensed from the vapour, the best example being the radioactive particles from the explosion at Chernobyl which had been carried over long distances (several hundreds of kilometres) without detection, but which were deposited on the ground by rain showers in Britain and Ireland. Locusts which refused to be deposited on the ground but determined to remain below the cloud base would remain under the shower cloud. They would be aggregated under the upcurrents of any converging system of winds, as mentioned in the case of those which arrived in Karachi, as a very dense swarm (see above).

15.3 MARGINAL TERRITORY

Climatic variations may be like the El Niño, showing what seem to us as major fluctuations of the weather; they may extend into a decade or more of drought, which makes it very difficult to establish a viable town in what we call semi-desert. But we are not the only form of life which has concerned itself with such regions. They are not like the retreat of an ice age, which opens up new evolutionary possibilities; nor are they like the return of the ice which may destroy most of what the interstadial had created. Ice ages put what has evolved to severe test, and retreat of the ice opens up the land for quite new creations of evolution. The short term, decadal or even centennial changes in weather style leave remains of villages and cultivation to which people hope to return with only the most viable of oases staying in business. Such is the fate of many attempts by people to live in the semi-deserts which stretch from the west coast of Africa through the Middle East, Afghanistan, and beyond to the Thar Desert, and are to be found in many other continents. Donald Campbell planned to set up a land speed record on the salt flats of Lake Eyre in central Australia, only to find that it had become a lake with metre-high windblown waves by the time he was ready to go.

People can have a go at colonising semi-desert areas by an input of machinery-based irrigation and agriculture, but all they achieve is temporary success until the investment is frustrated by drought. In the case of the Aral Sea, the irrigation caused greater evaporation than the flow into the inland sea could sustain, and the initial success has become a dusty desert poisoned by the insecticides used to protect an alien agriculture. Outside the edges of the artificially irrigated areas of the western Nile delta, an airline passenger can see channels, walls, villages and all the remains of quite recently attempted cultivation abandoned in the face of advancing sand drifts.

Vegetable species have developed seeds which can survive a few years of drought and can therefore blossom when the rains come. And in southern Africa, where modern attempts to feed the explosively growing populations by introducing maize and ground nuts fail from time to time because the use of fertilisers has made soil care unnecessary so that without regular rains, complete failure of crops is occasionally inevitable, and then surplusses have to be brought in from outside.

Any wild species attempting to exploit the fruitful years must have methods of very rapid population growth with efficient means of occupation of the territory. Such a species is the **quelea**, a weaver bird which builds its nests obsessively if the materials can be found. Also called the 'locust bird' because it appears in gigantic flocks which darken the sky like the densest of desert locust swarms. Any human attempt to cultivate by irrigation and thereby produce a food supply typical of an occasional rainy year is liable to be destroyed by one visit of the quelea. There is a true story that in the Sahel area of northern Niger, where the river has been exploited for irrigation, the grain harvest had been taken by quelea, and maize was tried the following year in the expectation that it would be harvested while the grain was still covered by the leaves. But as soon as one quelea had discovered how to steal the grain, within two days all the quelea of North Africa knew how to get it. It could be said that the farms were merely making easier for the quelea to breed without having to wait for the occasional rains. If one farm was 'protected' by a poison which killed the birds the flocks would breed nearby: if there was any crop available where mankind had invaded the desert the quelea would find it. None of the bird's predators can support themselves in the lean years; nor can they multiply rapidly enough to make any impression on the quelea's explosive growth rate. We have evolved so that the only way we can exploit marginal territory is by continual outside support, and frequent reconstruction, or by cultivating appropriate vegetation such as is found in the outback of Australia.

15.4 VULTURES AND FEEDERS ON CARRION

Philip Wills [48] noticed in 1936 in South Africa that the glider pilots only practised slope soaring, whereas he expected to get good thermal updraughts in the sunny weather. He had seen many soaring birds but was told that the thermals would not support a glider in those conditions because, with the ground already at 5000 ft, the air density was not enough. However, having been launched he found himself sharing the experience of the birds, with their apparent appreciation of his discovery of a good thermal which they were glad to join.

When he reached a height of 2500 ft (half a mile) above the ground, not only did they seem reluctant to go higher but they tended to leave him and spread out over the countryside. Their purpose was to survey the ground for carrion, and from their behaviour he deduced that they spread out at a network of points about half a mile apart at half a mile above the ground. At a higher altitude their eyesight would not be good enough to see any opportunity of carrion; at that height and with that spacing between them all the neighbours would get the message if one flew down on seeing a chance of a meal. In that way any discovery of carrion would be surrounded by several members of the network in a few moments.

A most interesting study of the flight of vultures and similar birds in India was made by Hankin [69]. He was of the opinion that soaring birds gained energy from sunshine on their wings, by a mechanism unknown, and that they were not using upcurrents except in obvious cases where the air flowed upwards over an obstacle such as a hill, a building or a ship. In retrospect it is difficult to understand why we had to wait for Wolf Hirth to demonstrate in 1928 that thermals were quite adequate to raise the height of a glider, for nowadays the imagination of someone watching soaring birds circling immediately fills in the picture with a rising mass of warm air. Indeed Hankin's writing suggests that even in his time it was widely believed that birds soared in upcurrents. The fact that the obvious was not accepted as the correct explanation suggests that scientific thinking can produce a very obtuse wit. But then, Hankin was a chemist, who believed in the paranormal—which he expected would be made normal.

Much of the detailed behaviour of soaring birds confirms the opinion that thermal convection consists of thermals, sometimes in complicated relationship with one another, and shaped more like new plumes in their early stages near the ground (see Chapter 11). A very brief summary of Hankin's observations was given by Scorer [70], at a time before the significance of the early experiments on thermals was fully appreciated and the concept of bubbles was still prevalent; nevertheless those notes may be stimulating to readers interested in intelligent observation of soaring birds.

The most important point which emerges from the phenomenon of bird soaring is that certain species, such as vultures in desert areas, are completely dependent on soaring. Like buzzards and eagles they may live in mountainous areas where slope soaring is readily available on most days. Similarly hawks, kestrels, ravens, crows and, of course, many coastal birds also exploit lift in upslope airstreams to remain airborne without wing flapping. Hankin noted that near Agra, vultures were only seen flapping extensively when returning home to roost in the evening when thermals were failing. In the morning, vultures were seen sampling the air from time to time to see how good the thermals were, and only becoming airborne in large numbers when some of them had established that it was worthwhile. Indeed, some glider pilots even suggested that the vultures, just noted as sampling the air, had the ability to 'trigger off' thermals in air that was becoming unstable. Some of the larger soaring birds emigrate from areas in which thermals are not reliable in the winter, while the others which remain scarcely take to the air during the season, except in slope soaring. The fish eagle, or black kite, does not depend on carrion.

There are no vultures in Australia, which is at first sight surprising because it is a continent of varying degrees of desert. However, it suffers from long periods of drought as a whole and so does not contain enough variety of climate to provide a reliable base for vultures to exist in when most of the continent is arid. There are two factors militating against the establishment of vultures there. The first is that soaring birds of prey and feeders on varrion do not have the wing power which would enable them to cross the ocean to Australia, and so a colony cannot, except artificially, be established from an outside population. The winds to not carry birds towards Australia from Asia. The second factor is the rapidity with which the well established fly population of Australia can consume any carrion available. Flies can survive a period of drought in the form of dehydrated eggs and can multiply very rapidly in a few days if the opportunity arises. In order to reduce the number of flies it might be worth considering the introduction of vultures or quelea–awful thought! However, they would have to be fed artificially through periods of drought and the numbers kept up in order to be effective in stealing the flies' food if a situation arose favourable for an increase in flies. There is, by contrast, a viable but small population of herons and shags dependent on fish in the few permanent saline lakes among the mountain ranges of the Centre.

The wing form of birds dependent on thermal soaring, or slope soaring in small spaces, is shown in Fig. 15.4.i. This shape has several advantages. First, the curvature of the section containing the secondaries is increased when the primaries are extended fully

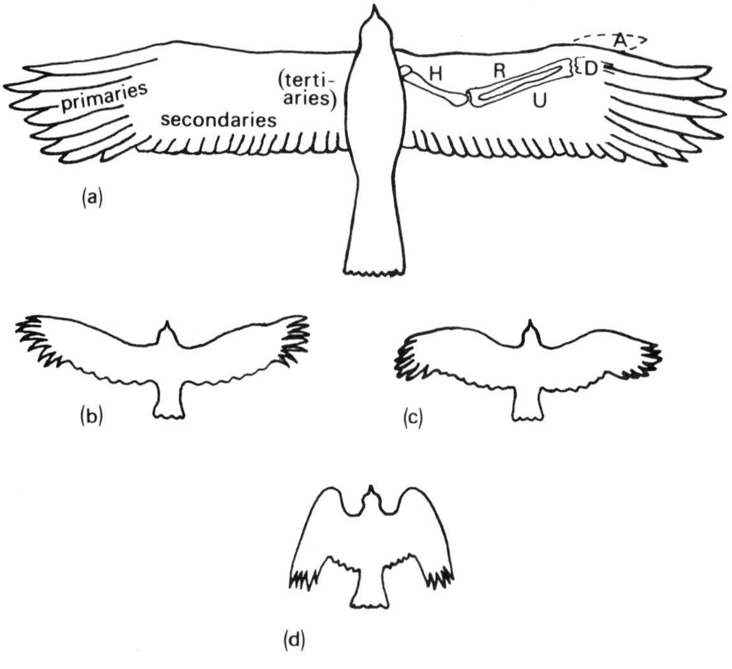

Fig. 15.4.i The wings of birds dependent on thermal or slope soaring. The span/chord ratio is small, and the primaries separated at the tips so as to improve aerodynamic control at low air speeds, minimum sinking speed and small turning radius. The bones are indicated thus: H = humerus, U = ulna, R = radius, D = fingers. The alula (A), or thumb, carries a large feather in some species which when held off the wing acts as a slot to prevent stalling at low speeds. The wing positions are (a) normal soaring, (b) for minimum sink, (c) for high-speed gliding, (d) for fast descent.

forwards. This increases the wing area and reduces the best flying speed (i.e. the speed at which the rate of descent relative to the air is a minimum). Secondly, primary feathers at the tip are spaced apart and provide boundary layer control as in a slotted wing, and the alula (thumb) appears to serve the same purpose in some large species. These effects reduce the stalling speed, especially on the inside wing during circling in small thermals or turning in confined upslope winds when a larger wing span would be an embarrassment. When the primaries are closed together and turned backwards, thereby reducing the curvature of the primary section, very much higher air speeds are possible. This is especially so in diving downwards without loss of control, or in flying fast upwind between thermals, which may be necessary to avoid being carried progressively downwind while thermal soaring.

Hankin's description of soaring flight as he observed it is spoiled to some extent by his conviction that soaring was not due to the air having an upward component of velocity. He says that he offers no explanations, which is true, but he was determined to disprove the orthodox, and very sensible, explanation of the day. He makes a Loch Ness monster out of soaring. The upward motion of the air can be of several kinds due to different causes. By making a single phenomenon of it he proves, by instances, that each of the suggested origins of the upcurrents is incorrect because he can find cases to which they are each in turn inapplicable. In some cases he even provides good evidence from observation of clouds that the soaring was due to upcurrents, but does not appreciate that he does so. This is because in these cases he is seeking to prove that the upcurrents cannot be of small dimension, as suggested by one contemporary.

15.5 SEASONAL MIGRATION: SWALLOWS AND CUCKOOS

These two species are characteristic of migrant birds and are very easily recognised by their manner of flight and call respectively. Swallows can be observed soaring in thermal upcurrents, these being the parts of the air which carry up the insects, such as aphids, on which they feed. It has been objected that they are not 'soaring birds', but this does not mean that they do not gain considerable advantage from the upward motion of the air in which they find their food. Indeed, it is probable that they could not sustain such prolonged flight otherwise. Some species, such as swifts, travel hundreds of miles in upcurrents in cold fronts, feeding on the insects rising in warm air shovelled up by the advancing cold air. They have been observed also in sea breeze fronts and in outflow fronts from rainstorms (Simpson, Berson and Simpson, Harper [71]).

Since significant numbers of swallows and cuckoos do not simply travel through Britain *en route* to countries beyond, their arrival and departure here is easily observed to accompany winds in their direction of travel. Therefore it is likely that they perform most of their migrations by moving with the wind. Other species are known to fly in large flocks by night on migration flights of considerable distance. The effect of wind and weather on these is described by Williamson [72]. The discovery that birds were responsible for strong echoes at cloudless cold fronts was made by Harper by telescope mounted on radar receiver .

The philosophy behind these studies is that the birds' behaviour is more likely to be made up of instinctive patterns of flight which exploit the air motion than of purposeful activities which are required to operate from time to time in order to be successful, like finding food.

15.6 BIRD NAVIGATION AND PASSIVE FLIGHT

The regular migration of birds, and particularly stories about ringed birds making exactly the same journeys in successive years, having wintered in Africa and returned to the same nesting point in Britain as in the previous summer, had caused people to ask 'How do birds navigate?'. It has been answered by those who do not understand what is involved in navigation by mariners, and they suppose that because a compass is useful it would be very useful for a bird to be able to detect the earth's magnetic field. Columbus almost certainly had a compass, but because he had neither map nor a clock he not unreasonably thought he had reached India or China when he arrived in America. The declination of the magnetic field has also been suggested as a clue which birds might use. This unprofitable discussion will not be pursued except to note that the suggested navigational aids might not actually be good for the survival of the species.

The number of birds crossing from Europe into Africa each year may be of the order of 5 000 000 000 (five billion), but in any case far more than might be necessary to sustain the numbers of all of the species. We have seen in other chapters that the weather is not identical every year and so the same old locations for wintering or for raising a family may not be the best. Anyway they will be contested and predators, including mankind, may be more deadly this year. This means that other possibilities in the wintering or breeding territories may prove profitable. The world could not support population growths corresponding to the usual number of offspring every year.

The usual incentives, or triggers, to migration are straightforwardly indicative of what is required. The cold north winds and the sinking of the sun lower in the south each day and the shortening of the days suggest to any species that it would be good to go south. The fattening up for a long-distance flight, the moults, and other annual occurrences are likely to produce rather similar results each year. But without making a special study of migration it can be said that it makes better use of the environmental possibilities than staying put all the year round. The transhumance practised in alpine regions indicates a human readiness to be constrained by the annual changes of weather for the benefit of their flocks.

15.7 APHIDS

Greenfly, blackfly, greyfly, and a few other insects are covered by this name. During the summer they live on growing green vegetation in colonies consisting only of females which reproduce numerous female young without sex. The production continues as long as the vegetation survives. During this time, when the air is warm, members of a colony mature enough to fly away do so each morning. Each afternoon there is another flight, this time of those which have become mature enough since sunrise. They are carried away by the air motion, and may ascend to heights of several kilometres according to what natural convection currents they may enter by chance. They are more likely to be carried upwards if they begin to fly only when the temperature reaches a certain minimum.

Aphids are thereby dispersed over a wide area and on descending to ground level they fly around, with an air speed usually less than the average wind speed which has transported them at higher altitudes, until they find a suitable host plant on which to start a new colony. During the day they have been found by traps carried on aircraft and balloons to decrease in numbers logarithmically with height, as would be expected on the assumption that they are carried up by thermal convection. However at night and in the early morning they are often found in layers in the air (Johnson, Johnson *et al.* [73]). These are presumed to be at temperature inversions: as the aphids descend, the air becomes warmer, except at an inversion. Here it is supposed that they fly upwards into warmer air as soon as they sense a significantly colder mass of air; causing a rapid change of around 2°C or more. Equally, when in the warm air near the ground they fly about but then stop propelling themselves in order to descend once carried to much colder levels by convection currents.

As the days lengthen in the autumn, males are produced and copulation leads to the production of eggs which survive the winter and hatch out the following spring.

Dragonflies have also been observed to make use of upcurrents for soaring. They find the aerial plankton on which they feed much more numerous in ascending air (see Scorer 1954) and soar only over fixed objects such as walls from which the sunshine causes continuous currents of rising air (Hankin 1913).

15.8 ALBATROSSES AND DYNAMIC SOARING

If for the moment we ignore rate of sink of a gliding bird on the grounds that any method of gaining height can be used to overcome this, we may think of the problem as finding a way of increasing the bird's air speed. If there were a thin layer of strong shear in the wind, a bird could fly across it repeatedly, always in the direction which would cause an increase in air speed each time. In Fig. 15.8.i we imagine two air currents, side by side, in opposite directions. A bird crossing the interface in the directions shown by the heavy arrows in the presence of the wind discontinuity given on the left experiences an increase

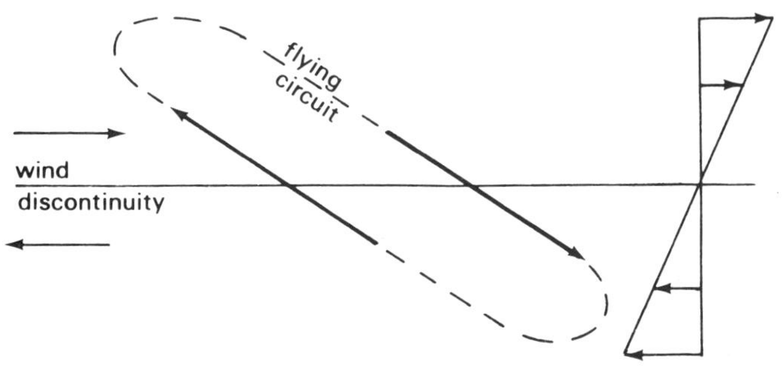

Fig. 15.8.i By flying the circuit shown across a wind discontinuity a bird repeatedly increases its air speed where the arrows are thickened. A wind gradient as indicated on the right will support unpowered flight in the same way.

in air speed which can be used up to gain height. It is only necessary to complete the circuit along the dashed part of the track to continue to ascend up the discontinuity.

If we now imagine the diagram to be a vertical section of one airstream superposed on another, the bird will convert as much gravitational potential energy into kinetic energy on the descending arm as it converts the other way on the ascending arm, and so flight can be supported indefinitely if the discontinuity is sufficient. If the discontinuity is replaced by a wind gradient, as shown on the right of the diagram, the same circuit will, with sufficient wind gradient, support flight without power indefinitely.

If there is a wind gradient $\mathrm{d}v/\mathrm{d}z$, and the vertical component of the bird's velocity is w, on ascending the upward arm the bird's air speed, u, will be increased at the rate $w\,\mathrm{d}v/\mathrm{d}z$. Its kinetic energy will therefore increase at the rate

$$\frac{\mathrm{d}}{\mathrm{d}t}\frac{1}{2}u^2 = u\frac{\mathrm{d}u}{\mathrm{d}t} = u\,w\frac{\mathrm{d}v}{\mathrm{d}z}, \tag{15.8.1}$$

while potential energy is stored at the rage gw, and must therefore be subtracted from the kinetic energy. The bird will be able to ascend this arm without loss of air speed (neglecting the aerodynamic drag) if

$$gw < uw\frac{\mathrm{d}v}{\mathrm{d}z},$$

i.e., if

$$\frac{\mathrm{d}v}{\mathrm{d}z} > \frac{g}{u}. \tag{15.8.2}$$

Clearly, a bird with a large air speed, u, will be most likely to satisfy this criterion. On the downward arm of the circuit the potential energy will be converted back into kinetic energy and the loss to aerodynamic drag made up. Consequently the downslope arm of the circuit will be the most prolonged so as to minimise the loss due to drag while gaining height.

A tactic often observed to be employed by birds is to fly into a region of increasing air speed, gaining height rapidly, then to turn quickly and descend near to the surface with a greatly enhanced air speed. This is used up as required, for instance, to search the sea surface for edible material. The sudden upward leap into the wind increase and rapid descent is then repeated.

The rapid ascent is often made by passing over the crest of a wave from the slow-moving air on the lee side into the larger velocity at the wave crest. This is particularly advantageous if the flow separates at the wave crest, often made visible by the motion of spray blown off at a cusped crest (Fig. 15.8.ii). Birds following a ship may use the wave pattern of the ship's wake to proceed upwind with the ship. This is more effective than using waves raised by the wind because the greater speed of the waves relative to the air makes separation at their cusped crests more likely to occur. Birds undoubtedly develop an unconscious skill from experience in performing these soaring feats, and skill is required since the waves are moving relative to the sea. Birds have the advantage that

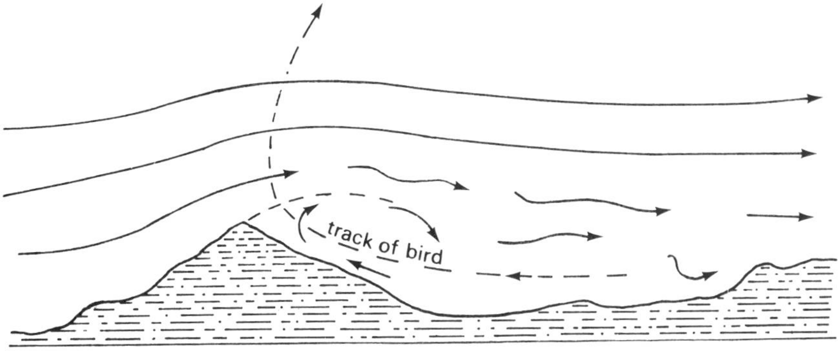

Fig. 15.8.ii Birds exploit the large wind gradient behind the crest of a cusped wave to gain air speed and height. This is most effective when the waves move against the wind, as may be the case when the waves are produced by a ship.

mistakes can be rectified by wing flapping or settling on the water; dynamic soaring is unlikely to be practical in fixed-wing gliders because of the difficulty of learning the art without frequent accidents.

If the airflow over wave crests is not used, the wind gradient required can be guessed from (15.8.2). Thus if the bird's air speed is 30 m s^{-1} (about 60 kt), since $g = 9.81$ m s^{-2} a wind gradient of about $\frac{1}{3}$ m s^{-1} per metre is required, and this is only likely to be found in the lowest 10 m of the air over the sea.

Albatrosses have a long (almost two dimensional) wings which minimise the aerodynamic drag at high speeds (Fig. 15.8.iii) and have much higher air speeds than birds which soar in upslope winds or thermals. The largest birds have difficulty in flapping flight close to the sea surface and cannot take off from it unless the wind is strong, i.e. comparable with their best air speed. The great southern albatross breeds on remote oceanic islands where it can get airborne by running down a steep wind-facing slope. These birds do not inhabit the northern hemisphere because, as was pointed out above, the non-ascending part of the flight circuit in dynamic soaring will carry the bird

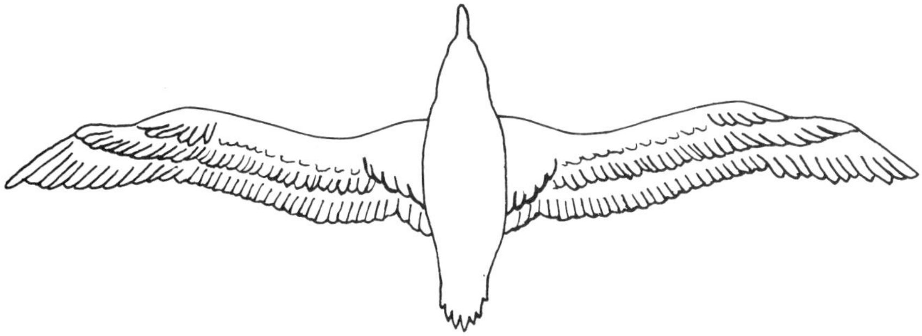

Fig. 15.8.iii Albatross wings have a small chord/span ratio and a shape more suitable for high speed flight at minimum sink, which is required for dynamic soaring.

downwind unless a special effort is made to direct it upwind. I have observed smaller albatrosses and similar sea birds travelling with the ocean liners *Georgic* and *Fairstar* at speeds of 15–20 kt in still air or against the wind, but they do not seem able to keep level with the Queens travelling at around 29 kt by any means other than exploiting upsloping air motion produced by the ship. The consequences of this limitation are that birds living by dynamic soaring alone must drift downwind, and that means having to travel round the earth in the direction of the wind without crossing land. This can only be done in the southern hemisphere westerlies. Large albatrosses cannot survive on a continental coast, nor in an ocean subject to frequent or prolonged periods of light wind in which they could not soar. The northern ocean westerlies have anticyclonic periods in which dynamic soaring would fail. The land obstacles in the southern oceans, notably New Zealand, South Africa and the tip of South America and South Georgia, are easily circumvented.

Flying fishes, which remain airborne for up to half a mile, may use dynamic soaring, although this seems unlikely because their main object is to escape momentarily from predators. In such cases unidirectional travel would be preferred. They could however use the dynamic soaring routine while heading nearly across the wind all the time, and I have seen this technique used by gannets. On the other hand, it is said that flying fishes dip their tails into the water surface while airborne and propel themselves by means of sideways oscillations of the fin (Hankin, *loc. cit.*).

15.9 RECAPITULATION: TACTICS THROUGH UNDERSTANDING

Desert locusts occur in two forms, **solitary** and **gregarious**. In the solitary form they are not usually a serious pest. The gregarious form is physiologically different and individuals can be changed into the gregarious form by being crowded together. The existence of the solitary form ensures the persistence of the species during periods when there are no swarms of the gregarious locust. When there are numerous swarms travelling about over a large area there is said to be a plague of desert locusts.

A plague begins when there are rainstorms over a desert area, such as Southern Arabia, provided that there are enough locusts there to start with. The swarms occur only as a result of rain on the desert, which is a rare phenomenon, and they are of interest because they have thrown light on how a species which spends an important part of its life flying can exploit the air motions in a simple manner.

In order to produce rain the air must ascend and that means that there must be horizontal convergence. If locusts flew at a constant height in a random manner they would be diffused by their random motion but aggregated by the converging air motion. At cold fronts the convergence is very large because the warm air is lifted bodily off the ground by the cold air and so locusts in the warm air mass may be collected from a very large area as in a shovel.

Swarms lay fields of eggs at a depth of around 15 cm in sandy ground after rain and provided they remain moist the eggs hatch out in just under three weeks. If the rains are not sufficient, the eggs, which are dehydrated, will not hatch out but will remain buried until sufficient rain occurs. When the rain is sufficient it produces a luscious growth of vegetation which provides the hatching swarm with food, for which there are usually no competitors.

The insects which emerge are small and wingless, and are called **hoppers**., They aggregate together in **hopper bands** which **march** forward with a circulatory motion continually bringing fresh hoppers to the front, especially when those at the front halt to feed. Almost everything edible in the way of the marching bands is eaten. In some areas where sufficient human labour is available, trenches are dug in which the hoppers are buried as soon as they march in. Another technique is to put down poisoned bait in front of them.

The most effective method of destroying hopper bands so far used has been to cover the vegetation with pesticide by **drift spraying**. The technique with a closely growing crop has usually been to spray insecticide from the air in calm weather so that the spray settles on to the crop by gravity. This is not fully satisfactory because of the sheltering of the lower parts of the crops by the upper. It is a very wasteful technique in desert areas because most of the insecticide falls on the ground and it has the serious limitation that it can only be carried out in suitable wind conditions, which do not obtain most of the time. In drift spraying a moderate wind is used to carry the spray horizontally across the ground, and the vegetation is covered by a cloud of spray droplets which are kept airborne for a few kilometres by the turbulence but are caught on parts of the vegetation with small geometry by impact. Even if the hoppers have to march a mile or two they will eventually consume a lethal dose of insecticide. Since insecticides which are very persistent can be used, a barrier to their progress can be set down in the desert.

The locust swarms reach the flying stage in about 6 weeks. They then fly by day, provided the temperature is warm enough, and settle at night. A large flying swarm may cover a few square miles and have as big an appetite as 100 000 people. The destructive power of a swarm has been recognised since civilisation began, and the Old Testament recognised several elementary attributes. For instance, unlike ants and bees they have no leader although they look like a well ordered army (Proverbs 30.27), and they remain settled in cold weather, flying to an unknown destination when it is warm (Nahum 3.17).

It used to be thought by ecologists that the swarms were aggregated by topographical features or vegetation; thus the early arrivals would stay around while others arrived to concentrate the swarm. There is no evidence of this sort of effect, quite the contrary, for they depart from an area of food to one of desert if the wind carries them that way.

Rainey demonstrated that swarms move with the wind, any apparent departure from this behaviour being of the order of the uncertainty in knowledge of the wind vector at the swarm. The time at which swarms take off in the morning and settle in the evening is not usually known accurately, and uncertainty about the flying time is increased because under a swarm there is usually a large number of settle locusts. The swarm therefore travels somewhat slower than the wind except in very warm conditions. There is circulatory motion (Fig. 15.2.i), with those at higher altitude travelling faster both because of the greater wind strength and also because locusts tend to fly upwind when near the ground, especially when taking off or landing. There tends to be a greater accumulation of locusts at the front of the swarm (Fig. 15.2.iii), which can add to the mass of the air to such an extent as to produce a significant downdraught causing those nearest the ground to settle due to congestion of numbers in flight. This retards the advance of the front: the stragglers at the back take off when the sky clears of locusts above them and they can catch up with the swarm, probably by dispersing in the warm sunshine to a greater altitude than the majority.

At the edge of the swarm it appears, from considerable observational evidence, that when a group of locusts finds itself flying outwards into unoccupied air it turns round and flies towards the greater concentration. This skin effect is not like surface tension, which tends always to reduce the surface area. If a downdraught produced a hole in the middle of a big swarm those on the boundary of the hole would fly away from it so as to increase its size and eventually, perhaps, divide the swarm. It does, however, have the effect of preventing the random flight of the locusts from dispersing the swarm by slowly decreasing its concentration at the edge. A swarm is considered dense if it has a concentration of more than one locust, per cubic metre, or has more than 100 locusts above a square metre of ground.

In situations in which a swarm remains compact, smoke or any other airborne material carried more or less passively by the air (e.g. water vapour, odour, pollen, aphids) is dispersed. Even when the dispersion is confined to the air between the ground and a stable layer, the air motion spreads smoke out horizontally. Locusts differ in that they tend to remain in the air either at a more or less constant height or in the lowest layers. When convection is strong over a considerable depth, locusts are carried up in columns, but, as explained in section 10.5, even in that case they may often fall back to a lower level in high concentrations. Locusts are not observed above the levels reached by thermal convection except that when there is no convection they may nevertheless fly, in which case they remain close to the ground. If they were merely to partake of the horizontal component of the air motion and remain at a constant height they would be concentrated at the base of the upcurrents. If a downcurrent were then to occur, unless it scored a direct hit on the centre of an aggregation found in the base of a previous upcurrent, the concentration of locusts would simply be transported bodily into the base of another upcurrent. Dispersion like smoke would therefore not occur.

A large swarm viewed from the air appears more like a network, with high concentrations in the threads and low concentrations between. This structure is to some extent illusory because the locusts are most easily seen when sunlight is reflected to the observer from their shiny wings. In a swarm, groups of locust are seen flying in random directions, the local unanimity of direction serving to avoid collision. However the direction of flight relative to the air often changes with height in a deep swarm. Consequently some parts of the swarm are more easily seen than others, and become visible or invisible quite quickly to an observer circling around the swarm in an aircraft.

The small-scale air motion and random flight are thus prevented by simple mechanisms from dispersing the swarm. A swarm may be divided by airflow around a mountain when it does not pass over the top, and may even reform beyond an isolated mountain which divides it. Swarms do not fly over mountains except when the air flows over the top. If there is a convergence on one side (as on the eastern slopes of the mountains of Eritrea) the swarms are concentrated against the mountain by their reluctance to be carried away from the ground (Fig. 15.2.ii).

Often there are showers in the airflow from the Red Sea and then locusts would avoid, if possible, being carried up into the clouds, which would be located above the upslope wind. There is often a layer of cloud beneath the stable layer, which is less pronounced if there are showers. The storms may be above the mountain tops, in which case the airflow would carry locusts there.

It is not certain what mechanism causes locusts to resist being carried upwards but there are several obvious possibilities—sight of the ground or sense of movement relative to it, or avoidance of the lower temperature at greater altitude. This could be achieved either by deliberate powered flight to lower levels in an upcurrent, or by an automatic cessation of wing beating when a temperature is reached at which the locusts would not have taken off from the ground. These mechanisms can be represented mathematically by imagining particles to be carried with the horizontal air velocity but having a downward velocity relative to the air which is a simple function of height above the ground (or above a level of desired flight) or a function of the upward velocity of the air. Cold fronts can concentrate widely dispersed insects as they lift warm air off the ground. Very large concentrations of locusts have been observed near Hargeisha in Somalia when air currents from the north and southeast converge at a front.

The rate of change of insect concentration σ following the insects which have a velocity **v** relative to the air which has a velocity **u** is

$$\frac{\mathrm{D}\sigma}{\mathrm{D}t} = \frac{\partial \sigma}{\partial t} + (\mathbf{v} + \mathbf{u}) . \mathbf{grad}\ \sigma = \frac{\partial \sigma}{\partial t} + \mathrm{div}\ \sigma(\mathbf{v} + \mathbf{v}) - \sigma\, \mathrm{div}\ \mathbf{v} \qquad (15.9.1)$$

if div $\mathbf{u} = 0$. If the concentration is steady in a region, the first two terms on the right are zero and the equation tells us the rate of change of σ experienced by an insect. It is positive if div **v** is negative. If the horizontal air velocity of the insects is small or random the only component which can be effective in bringing the insects towards higher concentrations is the vertical velocity v_z, and so if $\partial v_z / \partial z$ is negative the regions of high concentration are maintained even when the wind is from a direction where the concentration is lower. In practical terms, if the downward velocity of the insects relative to the air increases upwards, conditions favour aggregation of swarms in a region towards which the wind blows.

Small-scale details such as fronts may thus produce and maintain very large concentrations of locusts. Cochemé [74] performed some simple calculations and found that on the larger scale also particles following only the horizontal component of air velocity could increase their concentration by a factor of 10^3 in a few days. Furthermore, he showed that in particular cases air from a widely dispersed area of Southern Arabia and East Africa, where gregarious locusts had been observed, would be concentrated in areas of Baluchistan, where large swarms were observed 2 or 3 weeks later. This converging motion produces the rain which the locusts exploit to perpetuate the species.

Even the long distance 'migrations' of desert locusts serve the same purpose of carrying swarms towards the convergence zones which produce rain in the desert. Such rain is very unreliable, and does not support human settlement except at oases or other places where the water is stored underground; but desert locusts have evolved a flight behaviour to exploit the opportunities provided by the 'random' and sparse rainfall.

Many other species such as the swarming weaver bird (*Quelea quelea*) or locust bird, and many varieties of moth (e.g. army worm) and grasshopper, probably have a similar survival mechanism. Indeed the understanding of the lifestyle of locusts has opened up quite new possibilities in the control of insect pests. The development of this theme is clear in the works of Rainey (1951, 1963, 1969, 1973).

15.10 CONTROL OF PESTS THROUGH SWARMS

It is conventional to think that the best way to protect a crop from a pest is to spray the crop with pesticide. This may often be done too late, and since we know that insects are carried over large distances by air motion, to spray the crop certainly has a negligible effect on the population reserves of the pest elsewhere. Joyce *et al.* (1970) found that when farmers in a large part of Java sprayed at the same time as one another at intervals of 2 to 3 weeks, the effect was greater than if they sprayed at different times (which had made much more economic sense to them in terms of utilisation of aerial- or ground-based spray teams). This was because when the pest population is reduced over a large area it takes longer to recover, particularly if the territory is an island with only infrequent communication with other population reserves. On the other hand, a small area can be quickly reinfested from outside.

In the case of crop spraying, the proportion of spray which actually enters the pest is about one part in $10^{8\pm2}$. This is very wasteful and means that almost all the pesticide is harmful environmental pollution. Efficiency is usually measured in terms of the proportion of the crop surface covered by insecticide, or the proportion of the total insecticide that is deposited either on the crop or within the field containing the crop and not carried across its boundary: in such terms efficiencies between 0.1 and 0.6 can be claimed, but this is a very misleading measure of efficiency. One of the most successful uses of insecticide was a case in which about half of the insecticide sprayed into a swarm of desert locusts was taken up by the flying locusts: the rest was deposited on the ground over a wide area at very low concentrations. Since many of the locusts received a dose considerably more than the minimum lethal one, the proportion of insecticide which actually killed insects in this case where the swarm was totally destroyed was about one part in five. According to this measure the case mentioned is about the best ever achieved with any pest having an airborne phase.

It is therefore important to discover whether the swarm-forming mechanisms operate for many insects. Clearly they are not important for aphids, but for any species which propagates entirely as a result of copulation with egg and grub phases, aggregation into swarms is important for mating. Using an aircraft equipped with doppler radar navigation and insect-trapping devices in the Sudan, it has been found that the concentration of insects is greatest by far, in small-scale regions of horizontal wind convergence. It is probable that most of the infestations take place when a swarm created at convergence zone passes over the crop. There is a great deal of folklore about the arrival of dense aggregations of moths, locusts or other insects such as ladybirds, some of which assumes that the swarm was hatched out locally from eggs buried 7 years previously. The lore suggests that the storm which accompanied their appearance caused the outbreak. It is much more likely that the storm and the concentration of insects were both the result of the horizontal convergence of air.

Locust swarms have no natural enemies other than starvation, cold or disease. Predators such as storks can make no impression on their vast numbers. No grounded animal can concern itself with them because of their unpredictable and unreliable movements. Although they succumb to certain viruses in high humidities, and do not survive in well vegetated tropical areas for that reason, the concentration for breeding into a region which cannot be predicted by predators has great survival value for a species. All that is required for this to happen is that the insect should be inclined to fly at a constant height above the ground or at a constant temperature, and wait for its mates to accumulate around it at a suitable breeding ground.

The fact that desert locusts have two forms and that these were thought for some time to be different species obscured the relationship between solitary locusts and the creation of populations by swarms. Many insects found in swarms may depend on swarms almost entirely for their reproduction. A species which could propagate that way would have many survival advantages against both predators and mankind. Small insects are not easy to observe in swarms, and moreover many moths probably only swarm at night or at dusk, so it requires radar to seek them out. Birds which feed on airborne insects (e.g. swallows, swifts) certainly benefit from insect swarms in cold fronts, including sea breezes and other microfronts.

G. W. Schaefer [75] has shown how wing beat and breathing frequencies can be used to identify insect species by radar, and thus find swarms of particular species. Many swarms seem to occupy thin layers in the atmosphere, and this apparently horizontal arrangement of the high concentrations could be the result of the shearing over of more compact volumes containing the insects. If the layers are bounded below by colder air, the insects' reaction could well be to fly upwards if they find themselves sinking downward into the cold air, thus preserving the high concentration at the bottom of the warm air. Already successful operations against the spruce budworm moth have used knowledge of these mechanisms.

Early studies of bird migration caused it to be supposed that birds had wonderful powers of navigation. Insects were presumed, on the other hand, to be dispersed by air motion, their reproduction rate being so high as to make this an advantage. Such simple hypotheses created problems of how a moth found its mate and how birds navigated, and quite miraculous powers have been ascribed to them as a result. Actually we see that no intelligence is required. It is not even an advantage for a bird species to be very successful at navigation, because that would prevent it from exploiting the possibility of colonising new territory made suitable by long-term (even decades) changes in climate. In a successful year, parent birds would not wish their larger than usual broods to accompany them on migration but would more likely instruct them to 'get lost'. Through the work of Williamson [72] it is now known that birds too are subject to wind trends in a very important way: when making an ocean crossing they appear to choose to fly downwind if their stellar means of navigation are obscured by cloud. This would be the best tactic for making the quickest land fall, but equally it would often lead the birds hundreds or thousands of miles from their normally expected destination. Such almost random flights are necessary for the preservation of a species existing in a continually changing climate and in competition with other species of continually changing numbers.

At the time of writing our knowledge of the movements of low-intensity lines of insect convergence, of the kind found by Vernon Joyce, is small because no meteorological observing system has been designed to reveal them. The best prospects are certainly through the use of radar, which can see the insects, and it is probable that improved understanding of microfronts will come from the study of insects. That will probably provide the economic incentive for more detailed work. Meanwhile, knowledge of the mechanics of sea breeze fronts is the best guide.

Part 3

Forecasting and climatic change

(NOAA11 1529,20.2.90,3–) The first bits of information about events that had only been guessed at are treated initially as if they were typical of a wide range of events. The human imagination tests out the new information for the implications that would follow if it actually were typical. In the frontispiece to Part 2 on page 274 it was implied that the katabatic wind which melted the ice in the Kangerdlugssuaq estuary made it a likely place for early human occupation; but a quick look at other occasions [76] showed that this one was not typical: and so we immediately wonder whether its unrepresentative nature made it an indication of change, particularly as a second fjord in the SW corner of this picture showed it also to be ice-free, with no hinterland funnelling warm air to it. So perhaps there has been more warm salty water in 1990 than usual, with a very warm spot in the big crack in the fast ice which we see here. This channel is 3–, and presents several differences from the pictures on pages 274 and 676, which are of the same scene.

16

Forecasting: general circulation models (GCMs)

16.1 POSSIBLE ERRORS IN OBSERVATIONS AND STARTING VALUES

A mathematical model must first be made to represent the known laws of physics and mechanics as far as possible. Since these laws are to be applied to a continuum (a gas or a liquid) we have to make an approximation by representing its state (its temperature, pressure, velocity, humidity, etc. as necessary) at a system of grid points, with the assumption that at points in between the grid points the state changes smoothly from one grid point value to the next. For small areas such as Western Europe with a bit of the NE Atlantic there could be a rectangular system of grid points at equal intervals imposed on a map with a conventional conical, or polar-stereographic, projection; or the points could be where the lines of latitude and longitude intersect. Whatever system is chosen, the mathematical model must contain appropriate parameters for the calculation at each point because the actual distance between points will vary from place to place on the earth's surface because it is part of a sphere and cannot be represented on the same scale everywhere on a flat plane surface.

But more important is the fact that the points where measurements of the state of the air are actually made are not the same as the grid points. Therefore values have to be estimated at the grid points. This is a big task: for over the land, the observation points may be 15 km, or even less, apart but as much as several hundred kilometres over the ocean, while the grid points may, for example, be at a regular 50 km apart. To make this conversion, especially where the observations are far apart, estimates are helped by using a weather chart which will indicate approximately how things change in the present situation from one observation or grid point to the next. Nevertheless this estimation must be programmed to be performed automatically.

The observations themselves are measurements inevitably made with some degree of approximation. The measurements will already have been transmitted to some central point so that errors may have been introduced by imperfections which will gradually be found by regular use of the system.

The average forecaster may be able to recognise an error on the chart at sight, but rules have to be invented whereby the computer will recognise errors and provide for their omission or substitution. In order to minimise the time required to make a forecast it is obviously desirable to feed the observational material into the part of the program which calculates the forecast as soon as possible. In a system using all information in digital form, 'noise' can be removed in the same way as scratches and other unwanted 'information' are easily distinguished and removed from music. Thus any number which represents a gross departure from its neighbours is easily replaced by an interpolated value.

It may sound very arbitrary to decide that inconvenient information is 'obviously wrong', but a forecaster scanning the chart with a critical eye tries to fit every observation into his ideas of how it all works. At this point satellite pictures can be very helpful by revealing the presence of a small polar air cyclone, Fig. 16.1.i, which has developed over the ocean or a large cumulonimbus over the mountains, either of which might produce a very strong wind in an unexpected direction. The profession is awash with anecdotes about fore-casters who 'hadn't looked out of the window' and to whom an unexpected observation was not immediately credible. The direction of strong winds is unexpected when it goes against the surrounding flow—the rapid cyclonic rotation of a convection outburst or the strong katabatic outflow from a shower cloud which suddenly produces a downdraught in heavy rain.

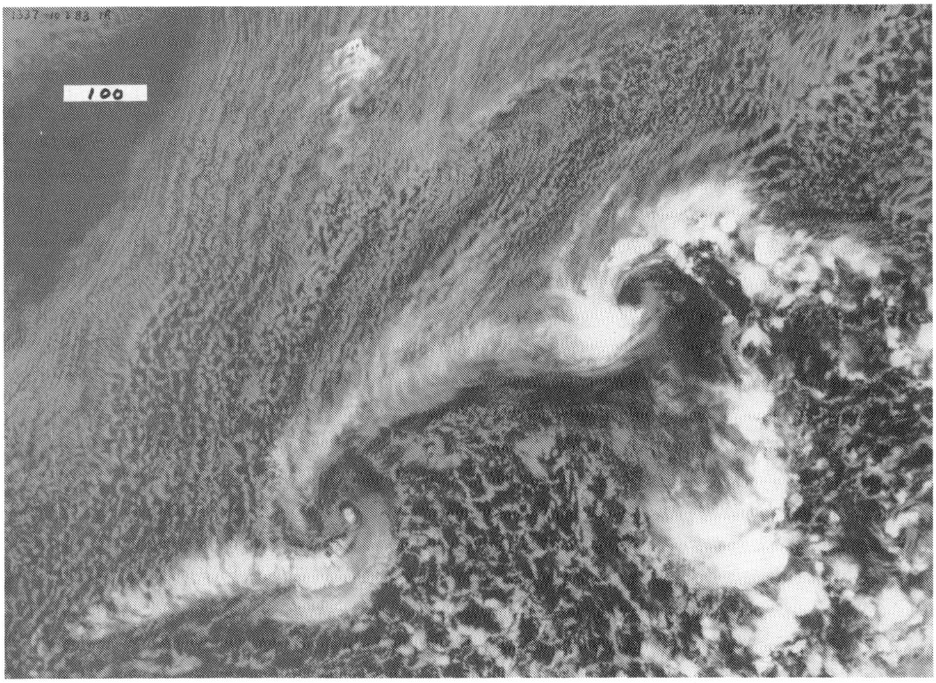

Fig. 16.1.i 1337.10.3.83,Ch4. Small polar air lows which might be sampled by a reporting ship whose observation of winds in unexpected directions might be subject to disbelief in the absence of this picture. These lows might be too small to arise in the model because they would not involve the required number of grid points to represent the circulation. The wake of Jan Mayen can be seen near the top of the picture.

The task of setting a mathematical model calculating the weather situation in one hour's time according to the program, which is a statement of the rate of change of pressure etc. in terms of the present state of the system, may seem to be a straightforward expression of the laws of mechanics and physics in terms of finite differences. These laws are well known and are believed to be a correct representation of reality, certainly within the accuracy of the measurements. But there is one very important task which must be carefully carried out first. The laws relate the accelerations of the fluid elements to the pressure forces acting on them. The pressure at the surface is one of the most accurately measurable quantities, and there are many ways of measuring the horizontal wind speed and direction both by balloon and from aircraft whose navigators are continually measuring it at many different heights. They also report temperatures, and these are measured by radiosonde balloons as well as at the ground and on ships at sea from which the sea temperature is also measured. But the vertical component of wind velocity and the acceleration of the air are basically too difficult to measure directly. Therefore we do not know in advance whether the wind (acceleration) and the pressure force field are in balance. When **in balance** the wind would not be accelerating and the pressure force would balance the Coriolis force which arises because the air is moving in a curved path in space, and therefore has some acceleration (see Chapter 4). In the balanced state the wind is equal to the **geostrophic wind**; but, as explained in Chapter 4, *all* development of anything interesting in the weather requires some acceleration and ageostrophic flow and vertical movement. The observations give no information on these points, see Figs. 16.1.ii and iii. The most reliable information comes from the previous states of the air, and if the model has been giving reasonable forecasts during the last 24 h or more then it will have already given a forecast for the moment at which we set it on the next forecast. In principle, therefore, the forecaster will wish to understand, and correct if necessary, any differences between the previous forecast and the starting conditions of the next one.

If there are any parts of the system which are not 'in balance' this could mean that there were gravity waves being propagated through the air. There may be special reasons why there are such waves, but they must not be confused with the out of balance forces that go with development. It is not obvious how to distinguish between these two kinds of out of balance force, but if they are not eliminated the waves might grow and put the model on an erroneous forecasting track. The essential difference between waves and development is that waves are propagated stably, while development is the realisation of an instability, and we shall have more to say about that later when discussing chaos. It may be remarked here that erroneous observations would start the propagation of a gravity wave, and so they are not a trivial matter. Since the waves are stable they could be propagated backwards, but the model as a whole cannot be run backwards because there are irreversible processes such as transmission of heat by radiation, and in the model by parametric conduction, and this is not a reversible process. It is a part of the technology of running a model which is now a specialised part of the profession and quite outside the scope of this book.

Besides getting the model started there is one final bit of running it to be worked out. This does not worry the forecaster once the forecast has begun, but it can be very important if the forecast is run for three days but is open to correction one or two days

Fig. 16.1.ii 0900,8.4.84,vis. A tropical cyclone in the Indian Ocean close to the northern end of Madagascar. This typhoon has strong winds circulating clockwise in a central area which is about 300 km in diameter. A ship in this area might report almost any wind direction and strength, and rather low pressures are possible. The circulation creates a wall of cloud on the steep east slopes of Madagascar.

Fig. 16.1.iii The **anvil** of a rapidly growing storm is carried away by strong upper winds. This New Mexico cumulonimbus would produce strong downdraughts when its hail begins to fall causing sudden rises of pressure and wind changes. [Photo by R. M. Cunningham.]

later. Indeed it can be important if the forecast has been run for only 3 h. This is the problem of incorporating genuine observational material which comes in 'late'; for example, aircraft reports of winds which have been found to differ from the forecast or a sea temperature measured from a ship which may come in when the ship happens to steam into colder, or warmer, water. The navigating techniques and instrumental accuracy of aircraft equipment are steadily being raised and so the measurement of changes of temperature and wind at a front in the middle of the ocean where the intensity is continuously changing and must be incorporated in the starting values for the next forecasting period.

16.2 THE PARAMETERS OF A GCM

The spacing of the grid points in a **general circulation model (GCM)** of the whole globe is at the present time about 5° of latitude or longitude but may be up to three or four times this or as small as 1° of either according to the size of the computer available and the purpose of the exercise. Thus near the poles the spacing of longitude could be up to 60° and in areas where special detail was being studied it could be reduced to a fraction of 1° for a mesoscale model. Appropriate instructions in the program would have to be written to join areas with different spacing.

For a GCM this means that coastlines and high mountains have to be very crudely represented and any mechanism which actually had a much smaller scale would have to be represented by a parameter which was designed to produce the same overall effect as the actual phenomenon on the smallest scale of the model. The most obvious case for parametric representation is clouds, for cloud-spacing is actually anything from a few metres to 100 or more kilometres. In some models even a thing as large as a tropical cyclone could not be identifiable by the two or three grid points which were located in it. Even the change of values from one side of a front to the other could not place the position of the front with better accuracy than the grid point spacing and would be carried on from one chart to the next by moving its position with the local wind speed: but some GCMs do not note the positions of fronts. Certain radiation properties also have to be parameterised.

Thus the heat transferred from one place to another is imagined to be transferred by a conductivity which would depend on the cloud thickness, and the cloud amount would depend on the humidity calculated at each point. It will be obvious that at every grid point the values of temperature, wind and humidity (and one or two cloud factors) must be specified at several heights. The model would be described as having a certain number of layers in the troposphere and one or two more in the stratosphere with a necessarily arbitrary boundary condition at the 'top' of the model, where for the short periods usually required would be regarded as rigid, or having a fixed pressure. On the other hand, a model could have about 30 layers in the troposphere and have layers in the stratosphere up to a height where the pressure is only 5 mb, or a two-hundredth of an atmosphere! But that can be accommodated only in a very large computer.

It is widely recognised that the parameters representing clouds and their radiative properties are, to say the least, very crude, and are justified only by being a reasonable guess from our physical knowledge and the fact that they have been tested and found to lead to fairly good forecasts. The main difficulty in the present state of the subject is that we do not know how these parameters are determined by changes in the basic values of the measured quantities. Of course, people have been ready to make guesses in their groupings to see what the models tell us about things that might happen, like the transport of large quantities of Sahara dust out of Africa, or the unusual diversion of the Gulf Stream or the Kuroshio current.

It is clear from the behaviour of different models of the atmosphere alone for which the sea temperature must be specified that for application to the problem of climate change the model must be extended to include the ocean. The grid point spacing may be different, and it could be seen as valid to use a different (longer) time interval because the changes are much slower in the ocean. But the two systems would have to define the boundary conditions of each other at every forward time step. Many atmospheric GCMs are worked using climatological mean sea temperature tables, but that is not good enough for the study of climate change.

16.3 RECAPITULATION

This chapter began by looking at the task of correctly starting a model-based calculation first of future weather and then of future climate. Observations made over the oceans are very far apart and it is necessary to use earlier forecasts to get a new one going. Any initial errors lead developments into wrong directions and models have to be designed not to be driven into errors and rare extremes, and gravity waves which are very difficult to detect on a weather chart. Getting a model correctly into action is the most arduous part of the whole task, for without all the detailed preparation its output could be very misleading, and the beauties of the model calculation would be frustrated. To keep it on the rails we require a technique of continuous input of what the observations tell us.

Anyone who doubts the overwhelming importance and difficulties of simply getting a numerical forecast properly started would be well advised to bury himself or herself in a book like that by Roger Daley [77] for a month or two.

In the next section we will have in mind investigations into how far ahead it is theoretically possible to forecast the weather.

16.4 CHAOS: setting limits

(1) Newton's legacy

By comparison with previous eras the advance of applied mathematics since the time of Newton has been both remarkable and exceptional.

Newton put a new orderliness into the scientific thinking of the time by his statement of laws of mechanics. He stated the very simple relationship between forces and the motion they give rise to. In addition he stated what the forces were which caused the motion of the planets to be in orbit round the sun and included their motion in the same theory as the motion of much smaller objects on the earth, namely the familiar theory of gravity. Since that time the law of gravity has been assumed to be of universal application to include the forces within and between the galaxies throughout all of space, as well as between the earth and small objects held by gravity on its surface.

(2) The calculus and the solar system

In order to exploit the laws of mechanics, Newton invented the calculus, which is a branch of mathematics dealing with the way in which things change as a result of motion or any other circumstances which result in change. These changes may result from the progress of time, movement from place to place, or movement of some other circumstance such as money or ideas. The calculus has become fundamental in the mathematical treatment of any problem in physics or any other fields of thought such as economics, sociology, psychology, demography or anything in which numerical measurements are the main expression of information.

Thus Newton stated that the rate of change of velocity is proportional to the force producing the change. When applied to the orbits of the planets, the equations expressing the law of dynamics can be applied to the future of the orbits, or to the past by simply going backwards in time. This reversal of time is possible because there is only a very tiny dissipation of energy. Thus the loss of energy due to the winds in the atmosphere and the tides and other currents in the oceans, seismic stresses in the solid earth, and tides in the sun, is very small compared with the energy of motion which is already present in the solar system. The planets do not need any energy to be continually fed into their motion, and the energy lost by the types of friction just mentioned is actually obtained by the earth's rotation slowing down very slightly, or by the earth's velocity in its orbit being slightly reduced.

As it happens there is a very slow wastage of energy in the orbits of the planets in the solar system. A very long time ago the moon was closer to the earth than it is today. But its field of gravity raised tides in the earth's ocean. The tide was retarded by the land obstacles and lagged behind the gravity field with the result that it slowed the earth's rotation and transferred its angular momentum to the moon in its orbit. That meant that the moon moved away from the earth. This loss by the earth of angular momentum is today being made up to an unknown extent by the sun, which raises a tide in the atmosphere by its daily heating. That tide is in such a position that the gravitational field of the sun provides forces which are increasing the earth's rotation. At the present time therefore the moon is slowing down the earth's rotation and tending to increase the length of the day while the sun is speeding it up and tending to keep the length of the day constant. No other planet has an important thermally raised tide in the atmosphere.

The earth raised tides in the moon as it rotated on its axis, which take the form of seismic stresses which had the result of wasting energy and slowing down the moon's rotation until the present state was reached in which the moon presents the same face to the earth and the tide remains in the same place [27]. By a similar mechanism, Mercury has been forced to present the same face always towards the sun.

The west winds contribute positively to the earth's momentum of rotation, but they vary from day to day and so the solid earth speeds up or slows down according to the amount of angular momentum in the atmosphere at each moment. The exact position of the axis of rotation varies very slightly at the same time and these variations make it more difficult for astronomers to aim their telescopes with the very high degree of accuracy required to make long-exposure photographs of distant galaxies. Therefore the European Centre issues predictions for tomorrow indicating any changes in the west wind angular momentum of the atmosphere, which is equal in amount to the negative change of the angular momentum of the solid earth, and is therefore a prediction of the change in the length of the day, tomorrow. This estimate of the length of tomorrow is given with an accuracy of a thousandth of a second.

Matter is held to a particular shape when it is solid, to a particular volume when it is liquid, but as a gas it tends to occupy any volume available although on the larger scale it is held by gravity to the surface of the earth, or other heavenly bodies.

Atomic and nuclear forces produce attractions and repulsions, and tend therefore to cancel out at ordinary and large distances, but electric and magnetic forces become appreciable at ordinary distances when the particles of a material are organised to add their magnetic fields together, or when the material is charged electrically and produces appreciable forces on other charges. Electromagnetic fields of force are produced by moving magnetised material or moving electric charges, but the forces are only appreciable close to the moving charges or magnets.

The laws of behaviour of gases and electromagnetic fields of force have been developed to place physics at the centre of scientific advance since Newton's day because the theory of gravity gave people the confidence to believe in and treat mathematically the fields of force acting at a distance through empty space. Newton had been ridiculed by some of his contemporaries for even suggesting that forces could act through empty space, because their experience taught them that real forces were of a kind which the body could feel when they were applied.

(3) Atmosphere not isolated

For the whole period of human history the dissipation of energy in all the planetary orbits is only a tiny fraction in a thousand years, so that it is possible to calculate the precise date of events such as eclipses and previous appearances of comets which are mentioned in historical records. The point relevant to this book is that although we use the laws of motion in calculating future weather by means of mathematical models of the weather, we cannot use them like the equations for the astronomical movements by reversing time to go back even to the weather of last week because the dissipation of the motion of the atmosphere each day is a very significant fraction of the total. It would be slowed down to almost zero in perhaps three or four weeks in a model with no heat (and gravitational) energy continuously fed into it by sunshine and removed from other parts of it by radiation into space.

It is nowadays fairly well known that Newton's laws are not absolutely and exactly accurate, and they have been slightly modified by Einstein's laws of relativity which make corrections when velocities reach a speed at which the modifications can be measured, the ultimate speed being the speed of light, which cannot be exceeded. But these improvements are ignored by those who make the mathematical models of the weather because they represent modifications which are extremely tiny compared with the approximations we have to make in the models in order to enable even the world's biggest computers to contain all the instructions and information involved. Thus even the trajectories of the spacecraft we send to the other planets are guided by mathematical equations based on Newton's version of the laws. The speed of light is between 30 and 40 thousand times the speed of travel of a spacecraft, and so the navigation errors which have to be corrected anyway are very much larger than any errors of Newton's laws.

Very simply, we cannot use the equations backwards in meteorology because they represent the dissipation of energy, and we would have to make the process which converts motion into heat go backwards. The process which describes the dissipation of energy of motion is a very crude one using parameters, and it is simply not possible to describe properly the reverse of the actual energy flows. The actual energy flows destroy details of information about how the energy was previously distributed and nature has no way of recreating that information.

The second law of thermodynamics indicates that all temperature differences in the universe are gradually being evened out. New temperature differences are created by the conversion of other forms of energy into heat, and when that heat is spread out into colder places it cannot be recalled, but is ultimately radiated into space.

Thus mathematical logic is in accordance with our view of, and assumption of, causation. Since it would take a few weeks to make the atmosphere and ocean run down to a standstill we could in fact go backwards one day, or perhaps two days; for during that period the energy loss might be only a small fraction of the total. Some of the early numerical forecasts (using what was called a **barotropic** model, meaning one without horizontal temperature gradients or vertical velocities) did in fact go forwards for one day, or perhaps two days, by ignoring dissipation and creation of energy of motion and concentrating on the conservation of angular momentum; but it was absolutely necessary to include some description of the input and outward radiation of heat to have any hope of getting further into the future. The second law of thermodynamics tells us in which direction time must flow in a model. In view of this it is rather remarkable that we can infer very much about the history of the earth and the solar system: much information becomes lost beyond recall, and the future cannot be determined for certain in detail for reasons about to be discussed in this chapter.

To sum up this discussion of energy dissipation: it is desirable to distinguish the dissipation of the wind energy in the atmosphere which occurs relatively rapidly from the dissipation of tidal energy which is due to the action of the gravitation fields of each of the earth, moon and sun on the tides of the other two which redistributes the momentum of rotation within the whole system, while the dissipation of energy of motion (kinetic) is very small. The ultimate result of this redistribution would be to move the planets and moons so that each body in an orbit around another would always face the same side towards the bigger body at the centre of its orbit. That state has been reached in the case of the moon orbiting the earth, and the planet Mercury which faces the same side always towards the sun. Venus has a very long day which is not far from a year in length, but for causes not at present understood parts of its atmosphere are said to be rotating round its axis in as little as four (earth) days. Mars has no ocean and a very tenuous atmosphere and is still rotating at about the same rate as the earth.

Although the rotational momentum of the bodies in the solar system is very gradually being redistributed between them there is no decrease of the total of that momentum because the communication with other heavenly bodies outside the solar system is virtually zero. Nevertheless kinetic energy of motion is necessarily lost when the gravitational forces cause that redistribution. The solar system has evolved quite a long way from arrangements it must have had earlier on with quite different distributions of the matter that it is made of, and some very special arrangements have developed. For instance, Saturn has rings, and the orbits of Pluto and Neptune move inside and outside each other, with Pluto usually being further from the sun but sometimes nearer to it. There is an asteroid belt and occasional comets with very elongated orbits. The axis of Uranus is almost in the plane of the ecliptic while the axes of all the other planets are not far from perpendicular to that plane.

The smaller bodies, such as our moon, and some of the moons of Jupiter and Saturn have lost any atmosphere they may have had because the escape velocities from their gravitation have been less than the velocity of thermal agitation of the gas molecules, which have therefore drifted out of reach. And these smaller bodies still retain craters clearly visible on their surfaces, where smaller fragments have collided with them. Such craters are easily and clearly visible on the moon even with no more magnification than is obtained with good binoculars.

Among the moons of the larger outer planets are cases where the moon has ceased to rotate relative to the planet and all the craters due to collisions with asteroids or comet material are on the side which is at the front in its orbit round the planet. From this it is possible to put the period bombardment and the creation of the craters into a solar system chronology, which implies that it was all more or less set up as now before the era of bombardment.

The earth is unique in having a clearly visible ocean with continents raised above its surface. No other planet has an atmosphere which is always only partly cloudy, an ocean which is only partly frozen, and an abundance of life everywhere close to the surface in the atmosphere and ocean. There is also a variety of life forms deep in the darkness of the oceans, and recently the discovery of Lake Vostok beneath the Antarctic ice at a temperature only a very few degrees warmer than 0°C implies the possibility of life having been isolated there, away from the rest of the biosphere for between half a million and a million years.

There have been many changes on earth since it became established in its present orbit, and the varied sequence of changes seems likely to continue for many millions of years into the future. It is most probable that life appeared spontaneously on earth and was not planted here from outside in any form recognisable as life as we usually define it or its precursors. This background is relevant to the discussion towards the end of this book about the prospects of climatic change and whether we, as a unique species have now, or may have in the future, any power deliberately to influence the sequence of events.

Our species appeared only very recently and there has not been much time for other species to change very much since we became dominant. A great deal of adjustment and adaptation by the others has occurred in response to our occupation of our extremely advantageous position. But we have cheated in the sense that our power has been derived from our use of special weapons and the building of effective fortresses which the others have difficulty in penetrating. Above all, we have harnessed mechanical power. The control of the muscle power of horses, dogs and the domestication of others and the control of their breeding for our advantage was undoubtedly a civilising influence, but the control of wind and water power and then the power of heat engines through the control of fire has taken us into a way of life and fuel consumption that cannot possibly be sustained on a time scale comparable with even our own past of the mere two million years since we spread out of Africa, or with our historical time scale of about 8000 years before which almost no coherent records or relics were made which gave a clear indication of people's thoughts.

(4) Linear differential equations

The remarks in this particular section about equations refer not to ordinary equations but only to differential equations, which describe the rate of change of the quantities in them such as velocity etc., and the forces causing changes. For this, the calculus is essential. Mathematics has become the main vehicle of thought in mechanics and physics.

The remarkable progress since the time of Newton in understanding the mechanics of nature through applied mathematics has been largely due to the simple fact that the equations in most cases have been linear. This means that the equations for the velocity (represented by the symbol v) contain only first-order terms like 'v', and none like 'v^2' or 'v^3', or products with their derivatives, which we call non-linear terms.

A linear equation has the same form however large the quantity v may become, so that it is controlled only by the coefficients and parameters which represent the physics. But if there are non-linear terms they take on different relationships when the quantities get very large or very small. Thus a term in v is the same magnitude as one in v^2 when v is equal to 1. But when it is doubled, the term in v^2 becomes four times as large. On the other hand, when v is equal to 0.01 the terms in v^2 or any other power are reduced by factors like 0.0001 and can be neglected. Equations then become more nearly 'linear', meaning that if we plot the relationship described by the equation it is represented by a straight line.

Many physical situations are discussed mathematically be supposing that they depart only slightly from their initial state. If we then discuss the developments in terms of the departure from the initial state, we call it a **small perturbation**, and the equation remains virtually linear as long as the perturbation is small. The method of 'small perturbations' has been developed with quite brilliant success, particularly in cases where a system oscillates (e.g. radio or sound waves) or when a system is becoming unstable (e.g. when the atmosphere is heated at the bottom and convection currents are set up) and then only the early stages of the instability can be studied by a linear equation because the quantity studied becomes large and the equation becomes non-linear.

As a general rule, linear differential equations have solutions which either increase or decrease exponentially (the quantity studied becomes indefinitely large or infinitesimally small) or it oscillates (its changes are described by the phase and amplitude of the oscillations) with some sort of regularity.

Another, and most important, factor about linear equations is that any two different solutions of them can be added together and the sum of them is still a solution. This is not true for non-linear equations. The result of this is quite startling. Fourier (Jean-Baptiste Joseph 1768–1830), Bessel (Friedrich Wilhelm 1784–1846), Legendre (Adrien Marie 1752–1833), Laplace (Pierre Simon 1749–1827) and Poisson (Simeon Denis 1781–1840) were among the leaders in the solution of the second-order (but still linear) differential equations which gave answers to many natural problems in applied mathematics. It was suggested (and widely believed) according to the mechanics initiated by Newton a century earlier that it would be possible to predict the future behaviour of any mechanical system if the initial state could be described. The further work of applied mathematicians such as Airy (George Biddell 1801–1892), Kelvin (Ld. William Thompson 1824–1907) and Rayleigh (Ld. John William Strutt 1842–1919), and pure mathematical investigations by Poincaré (Jules Henri 1854–1912) and many others, including J. E. Littlewood and M. L. Cartwright more recently at Cambridge, have been frustrated in their attempts to find ways of describing the solutions of non-linear differential equations in terms of the mathematical functions which their predecessors had made into conventional applied methods. That difficulty was compounded by the impossibility of describing any general solutions of non-linear differential equations in terms of mathematical functions such as sine, cosine, logarithm, exponential and many others named after the scientist (Airy, Bessel, Fourier and others) who proposed them as solutions of physical problems.

In many problems it had been possible to use a whole series of the solutions of linear equations to describe some very complicated shapes, Fourier analysis being probably the best known example, and to follow their evolution and dispersion in cases where the shapes changed but the differential equation remained linear. The transmission of sound, seismic or electromagnetic waves has been subjected to much revealing analysis, especially in dispersive media (in which the wave speed varies with the wavelength). (See Chapters 5–7.)

Much of the art in applied mathematics has been in deciding what terms might be neglected in order to keep the equations linear, and soluble. As the non-linear terms became important it was only possible to obtain an indication of the general way in which changes would proceed by using numerical methods like those used to calculate books of tables of the functions defined. That procedure became progressively less and less reliable as time went on, and it became evident that the basic equations of fluid mechanics which govern the motion of the atmosphere and the ocean were unavoidably non-linear. But they can still be solved in any particular case by numerical methods, and the big computers were brought into service of weather forecasters by using greater and greater accuracy. But it had not been realised how widely divergent two different solutions of the same non-linear differential equation could be when they started from very close points. It was not like two exponentially increasing solutions getting further apart: it was like two solutions, both of which were oscillatory, being close together for a time and then rather suddenly going in different directions. The non-linear terms acted rather like the parameters and coefficients in linear differential equations changing their values to make the solutions exponential.

(5) Non-linear differential equations: bifurcations

At first sight it might seem that attempts to solve non-linear differential equations might only occasionally be successful. After all it is easy to make up non-linear cases which do have solutions in terms of the familiar mathematical functions. The leading pure mathematicians attempted to introduce methods of finding solutions, but they had no hope of being able to add lots of different solutions together to get more or less any form of solution, as had been done with linear differential equations with enormous success. The real difficulty arises in non-linear equations when two slightly different starting points actually produce very different results with the same equation after only a fairly short calculation. And sometimes, if two situations look very similar and have only small differences in the coefficients in their equations, the solutions may rather suddenly part company and go in quite different ways.

That this difficulty was going to be very important was dimly foreseen after it had been noticed that the flow of a fluid could often be found to stay in one pattern, yet it would only require a very small disturbance to make it assume a very different pattern and stay in that. This was particularly true of flow guided by boundary shapes and when there was a significant amount of viscosity (friction) in the fluid leading to separation (see Chapter 2). A well known example is when an aeroplane wing which has been giving a good and controllable response to the elevators (the 'control surfaces') suddenly becomes stalled, and produces much less lift for the plane, when a further slight change is made. This could lead to disaster because the control surfaces cease to control.

A stage may be reached when the choice between the two possible types of flow depends on minute differences at the start, there is said to be a 'bifurcation'. This is very important because it used to be thought that if the equations were deterministic, and therefore represented the physical mechanisms which determine what will actually happen, then even in the case of instability, which is a very common and familiar feature of solutions of many linear differential equations, the solution could be obtained on a computer if the calculation was carried out with sufficient accuracy. So with the arrival of modern giant computers capable of working to 20 or more significant figures (such as used to be employed using mathematical tables in the astronomical calculations two centuries ago) it seemed to be merely a case of using enough accuracy to solve non-linear differential equations.

It therefore came as a surprise to find that solutions of some non-linear equations became bifurcated rather quickly, and seemed to go off in unexpected directions rather soon. This was at first simply seen as a kind of instability not physically different from the well known unstable solutions of linear cases. But that was not all. It soon emerged that some cases went on from the first bifurcation to a second, and a third, and a seemingly endless sequence of bifurcations. What is more, the solution that seemed to have been chosen at the first bifurcation was sometimes abandoned and the system went into the alternative one. And even this might be abandoned for the first one again. The different solutions seemed to have their own domains and in any actual case the system wandered around sometimes in one domain and sometimes in another.

(6) Thoughts of God: determinism

As each successive bifurcation would require a rather large increase in the accuracy of the calculation, to make it deterministic from the start, the exact detail of numerical solutions soon becomes unpredictable in every practical case. As the realities concerning serious computations aimed at predicting the future became better known, but not at the same time correspondingly better understood, many opportunistic 'philosophers' thought they were seeing the relationship of cause and effect, which is the basis of all the physical sciences, disintegrating before their eyes. Many had already thought that Heisenberg's principle of uncertainty had opened the door to some sort of spiritual determinism (i.e. miracles and the like). The revelation that commonplace differential equations also revealed that computational determinism was necessarily of very limited scope as far as the future was concerned gave these people the chance to talk as if they had gained some insight into the mind of God, which makes one suspect that such 'insights' are probably nonsense anyway.

The presumption of determinism made some people think that predictions could be relied upon in any case in which a correct differential equation was used. But that had depended upon the creation of a correct mathematical model, a process which became all the rage in the 1960s and led many people astray into trusting predictions based on models which soon went wrong.

To be quite clear we should say that determinism has not been in doubt because of a theory of chaos. There is no theoretical objection to solving a non-linear differential equation as accurately as may be required by numerical methods (given the required computer power). But all we get is one particular solution which is different from any other solution which starts slightly differently but which might become very different indeed after going not very far from the start. What has not been devised is a general formula for the solution. Such formulae have been devised for linear differential equations so that we *can* calculate the value at any distant point (e.g. in the future) from a formula and a book of tables of the formal solutions. Astronomers can calculate the dates for future eclipses without having to calculate all the details of how we shall get there because the answer has been expressed in terms of mathematical functions which have been tabulated in books waiting to be used. As it becomes useful to know the answers, they can be worked out for the particular occasion of interest. In principle, all that is necessary is to put the starting values, based on astronomical observations, into their places in the formulae which are the solutions of the equations, and calculate the values. The equations have to represent the mechanics of correct models which are correctly linear in the particular case.

If the equations had been non-linear we would not have any formulae for the solutions except in a few very special cases which are not usually the ones we are interested in. The great dream of many philosophers and mathematicians since the time of Newton has been that if the positions and velocities of all the particles in the universe were known at one time they could be calculated for any time in the future using the mathematical equations which represent the laws of mechanics and physics, gaily called collectively *the laws of Nature*. The futility, indeed the stupidity, of such an all-inclusive thesis can be appreciated by noting that however small a piece of material we might measure in order to set up a calculation of the future behaviour it would be impossible to specify all the numbers required for the calculation, and to set everything up for this calculation would take longer than waiting for it all to happen and letting Nature do its own calculation!

To express Nature's law in mathematics gives them no additional legitimacy, and always knowing the outcome in advance would be of no interest because there would always be infinitely more potential information remaining unknown than we could claim it would be useful to know or understand.

I am not in the least decrying the achievements of our science, which I am proud to have been involved in. I am laughing at the philosophical pretensions of those who contemplate a sort of ultimate power of scientific theory or alternatively claim to have discovered a way in which causation may be cheated and God explained thereby.

(7) Strange attractors

When the solution of a differential equation may go off in either of two very different directions the mathematicians studying the problem invented a phrase to describe the situation. Each of the two patterns of behaviour was said to take the solution into the territory of an 'attractor'. If the behaviour was represented on a graph of some sort so that at any moment a point was made to represent the situation, the evolution of the system's behaviour was represented on the graph by progression of the representative point along a track. It seemed as though the moving point was being drawn into the influence of something at a particular spot, where there could be imagined to be something attracting it. The moving point seemed to go into orbit around the attractor, and then at a moment which seemed quite unpredictable (and was usually unpredicted) it might become attracted to another point which may have represented an alternative pattern which it had avoided at the previous bifurcation, or it might represent a choice made at a second bifurcation.

As long as the precise nature of the attracting point remained unknown and how to calculate its exact position was not immediately evident, it was promoted to the status of *strange attractor*, especially when it seemed that the system might suddenly abandon it and begin to encircle another different attractor. Behaviour of this kind is still seen as containing some unpredictable elements. For example small changes in the starting conditions of the calculation would be followed after a not very long time by the system going in quite a different direction, and perhaps moving in orbit round a different attractor. Thus it was said that the system was essentially unstable, not following any specifiable pattern, and correspondingly unpredictable. Its behaviour was not just unstable; it was *chaotic*. The orbits around an attractor are not closed, repeated orbits, but vary from one orbit to the next, usually only slightly.

(8) Chaos takes charge

The word **chaos** originally meant a void, a chasm, in which there was nothing. The use of the word chaos to mean a muddle which cannot be sorted out or a state of extreme complexity is quite modern. It has become a state in which tiny alterations can soon lead to big changes—the flap of a butterfly's wing in Brazil would alter the course of a full-blooded Atlantic cyclone. The message from that particular popularised example was that the future was essentially unpredictable and undeterministic, and actual events could be said to be due to infinitesimal causes, or possibly that nothing had fully determinable causes or outcomes.

If we accept anything like such an unhelpful conclusion, which strikes at the very roots of scientific enquiry and technological practice it will mean abandoning all that appears to have been the very successful justification for the technology developed during the most recent two or three centuries of the Industrial Revolution. After all, science has enabled us to design and plan and turn our hopes into beneficial consequences by developing technologies. Actually, of course, all our architectural and engineering designs are very simple by comparison with most natural situations and are backed up by a great deal of preliminary experiment and investigation in which it is hoped to avoid the possibility of our machines coming to a 'bifurcation' at which it would be impossible to predict the consequences. A bifurcation could cause the breakage of a component of our machine followed by disaster. The failure of a component might be caused by a sudden increase in stresses or oscillations resulting from a bifurcation of the system's correct equation; but that something had been ignored and had caused unjustified confidence in the design. Such dangerous situations are avoided by a great deal of testing, and by what the engineers call research and development (R&D). This particularly important when we are concerned with engines of great power which depend on the chemistry of combustion (about which neither Newton nor his 19th century followers had even a pretence of a precise theory).

The atmosphere is not a system which *becomes* chaotic. It has been fully chaotic for billions of years. Attractors are rife on all size and time scales, most of them quite strange. Most solutions based on models are probably not realistic until they have become chaotic.

The problem posed by the possibility that the tiniest of disturbances might be responsible for big changes soon afterwards is not relevant except when the behaviour of the model is believed as if it were absolutely correct. Yet no one could believe that the tiniest disturbance could have a significant effect when there are millions of other comparable disturbances dissipating the energy of the system. Thus a butterfly's wing makes less disturbance than a leaf on a tree in a breeze, and there are billions of those, and none of them could be described as the cause of any particular measured happening in the atmosphere.

16.5 MODEL CLIMATOLOGIES OF GCMS

A GCM is not created overnight but takes many months of trials, making of forecasts and continual modification of the parameters, all of which takes several years. If the model is then run continuously to cover a year, or several years, of weather based on one set of starting values and including the usual diurnal and seasonal variations, the statistics of the weather covered by that forecasting exercise amount to a climate, which is characteristic of the model.

By choice of the parameters during the long trials with real situations the climate may be altered so that it possesses certain characteristics. One very difficult feature to manage is the frequency of extremes, particularly rare extremes. In routine forecasting, these can arise as a result of the model, in routine use, being kept in line with the weather as it happens every day and thereby causing it to possess a degree of reality which it would not generate in its own synthetic climate. In routine use, the failures of a model can be detected. Trials with some of the parameters modified to make it perform better on the basis of **physical reasoning** can result in a reduction of the frequency of the failures. But this can be a dangerous procedure in the sense that if the failures refer to cases when extremes were not forecast, to modify a parameter so that the model would have forecast the extreme (which actually occurred but was not forecast) may cause it to be predicted too often in future. This is not a feature only of using a model to make a forecast; for example, on one day at Gibraltar a strong wind from the southeast blew up, which made the approach to the runway very turbulent in the lee of the Rock; also the wind was loaded with sand from North Africa, which caused other inconveniences. This occurrence was of considerable interest to the forecasters because it 'explained' how a large amount of sand came to be piled high up against the steep east face of the Rock. For the next few days the forecasts included warnings that strong and dangerous southeast winds might occur. As it happened, the phenomenon was not repeated and the warnings were soon dropped out of the forecasts.

The older generation of forecasters knows from experience that there used to be skill gained from experience which made forecasting to some extent an art, which was very much supported by physical reasoning. Today that reasoning and art have become associated with learning to exploit the advantages of the predictions of a GCM. When, and when not, to accept the details of a numerical prediction—that is the question.

Physical reasoning has told three generations of forecasters that the current increase in the greenhouse gas CO_2 will cause global warming, and also that the industrial increase in air pollution has increased the earth's albedo, which will cause global cooling (and might even drive the climate into an ice age). The models are presumed to encapsulate physical reasoning and therefore may be of help in generating a more confident expectation. Any guidance of this kind must be based on the inclusion of the role of the oceans, now to be discussed.

16.6 THE OCEANS

Any warming of the oceans caused by the greenhouse gases has to be communicated through the ocean surface and carried to the lower layers by downward motion, which will be compensated somewhere by upward motion of colder water from below. This seems at first sight to be a mechanism which will take a very long time because warming a fluid at the top will merely produce a warm layer floating on the top of the colder layers. The warming of the atmosphere starts at the bottom and warmed air floats upwards to warm the upper layers by mixing into them. In general, to warm a fluid which expands when it is warmed, the heat must be applied at the base.

The heat capacity of the oceans is much greater than that of the atmosphere and so the reaction to the **ocean–atmosphere system** will take much longer to react to a warming process due to any increase of the greenhouse gases in the air. The warming may have been going on for several decades already. On the other hand, we seem to have been coming out of the Little Ice Age, which began in the 14th century and was still quite severe during parts of the 19th century when, for example, the river Thames was frozen over for weeks on end so that occasionally it was possible to light a fire and roast an ox on the ice. Since about 1880 there appears (in London) to have been a gradual global warming of about 0.5°C, although the measurement used to obtain the representative world temperature showed that it varied up and down with a range of about 0.4°C in any decade (see section 17.3).

There has also been a gradual rise in the sea level of about 1 mm per decade during this last century, which can suggestively be explained as being due to the decrease of water stored on land. This has been partly due to the melting at the snouts of glaciers, which are known to have receded by comparing old pictures of the scenes with modern photographs. In addition many inland lakes and seas have been exploited to irrigate land for farming. The most notorious case is that of the Aral Sea, which has been so reduced that ships there have been stranded on dry land and are now far from the receding coast. The level has fallen so far that the irrigation has had to be abandoned, and the land surrounding it is now desert on which the dust is highly polluted with pesticide poisons left over from the intense agriculture. In the southwest of the USA the same process has its victims, notably Pyramid Lake in Nevada where no more water can now be taken, and where an island which had been a breeding centre for pelicans has been jointed to land and has been invaded by rats, which makes it useless for safe nesting.

In addition to the lake water taken for irrigation and evaporated into the air, which reduces the evaporation from the sea overall, the demands of cities for water supplies have been met by extraction from underground aquifers. The lowering of the water table under the land has raised the sea level because the drainage of cities using underground water sources has been fed into rivers and thence into the sea. An extreme case is in northern Algeria close to the Mediterranean coast where in the areas of growing population the first buildings could be supplied with ground water at a depth of about 1 m. But now the water has to be pumped from a depth of 100 m, and so it is in danger of becoming economically inaccessible.

So far the changes in sea level are not significant because they could have been caused by rising or sinking of the land or sea bed. For example, marine deposits on high ground on Oahu [78], whose age has been estimated by thorium-230 analysis, have a uniform distribution of ages dating them between 131 000 and 114 000 years ago, indicating a high sea level lasting around 17 000 years. This contrast with 8000 years derived from oxygen isotope analysis of ocean bed cores on the assumption that the isotope composition of sea water is controlled by the global ice volume and this is recorded in the isotope variations in foraminifera. The chronology is derived by assuming that this in turn was controlled by orbital changes according to the Milankovitch theory. However, it was inferred from the Devil's Hole calcite that the interglacial (Eemian, warmer period) began 140 000 years ago and lasted about 20 000 years, according to chronology based on the same theory [79]. This suggests that that theory of the warming mechanism is thrown into doubt as the major cause of the climatic change. But at the same time the actual elevation of the sea level derived from the present altitude of the Hawaiian deposits has had to be modified by assuming that the land has been uplifted at a rate of about 0.05 m per thousand years (i.e. about 7 m higher now than when the coral was laid down over 130 000 years ago). Furthermore, the sea-level high occurred 4000 years before the maximum heating according to the Milankovitch mechanism.

The chief fear has always been the possible melting of the great mountains of ice resting on land in Greenland and Antarctica. If the ice in the Arctic Ocean (the North Polar Ice Cap) were all melted it would not raise the sea level because that ice is already floating in the ocean. There is a large area of Antarctica (the Ross Ice Shelf) where it is not known how much of it is resting on the sea bed. The ice stored on firm land may possibly be augmented as a result of global warming, for the air will then hold more water vapour so that the rainfall and snowfall everywhere on land may be increased, although at present there cannot be said to be any certainty about this. Equally possible is that an increase in rainfall might raise the level of the Caspian Sea and of Lake Chad both of which have shrunk in recent years. Also the underground water in the Sahara might be replenished and increased. If there is any increase in the snow and rain deposited on land it must be remembered that in any area occupied by the human race the water available will be used either for irrigation, which means more irrigation and therefore more evaporation, or to supply the needs of the population, which means that it will be used and ultimately drained into the sea.

In this connection a discussion has developed over the importance of artificial reservoirs accommodating more water on land. Some very big dams exemplified by the Boulder dam on the Colorado river and the Asswan dam on the Nile have been built in this century as well as many smaller ones; and the argument is about whether the growth of land-stored water will continue, and so about whether possible explanations of the slight rise during the last century will remain, or indeed are, valid as proof of an expansion of the ocean's volume.

Measurements were made of sea temperature in the North Atlantic at 125 m depth, a level at which the rapid seasonal variations do not occur. There is a suggestion in the record at 125 m at Ocean Weather Ship 'C' from 1947 to 1985 [80] of a periodicity of around a decade varying between 5°C and 7°C, which is about twice the range of the annual cycle. But there is no evidence of a trend over that whole period. But there was a cooling north of about 45°N and a warming south of that latitude centred on about 35°N. That may seem to be a very crude statement of what the measurements reveal; but the period of measurements is so short that we can have no confidence in a trend which is everywhere small compared with the variations. Still less can we attribute any trend to a particular cause. The analysis by a sophisticated statistical method known as the empirical orthogonal function technique on a 5° latitude by 15° longitude grid certainly cannot justify extrapolation into the future.

An attempt to measure average ocean temperature changes in the Pacific Ocean by observing the speed of sound over long distances from two sources, Point Sur (160 km south of San Francisco) and the north coast of Kauai (the northernmost of the Hawaiin Islands), has been halted because of a public outcry that it would be desperately harmful to whales and dophins in particular because they communicate acoustically. A preliminary test of the technique was made in 1991 from Heard Island (appropriately named?), and now it has been proposed to emit a 20-minute signal six times a day for 30 months and the argument rests on strong opinions based on little more than anecdotal information obtained during the Heard Island test. The outcome is likely to include an intense study of whale and dolphin language and related behaviour, which will be of great interest.

Many of the alleged rises and falls of mean sea level recorded on tide gauges throughout the world may be seen in a new light now that satellite measurements of sea level have reached an accuracy of very few centimetres and there are slow-moving Kelvin and Rossby waves in the Pacific and probably also the Atlantic and Indian oceans to be reckoned with.

We know by experience that floods occur because the rain occurs in far greater amount in one area rather than because the total amount falling on the whole earth has been increased. Thus they are caused by storms remaining stationary and not distributing the rain over large areas. The total amount of water in the atmosphere in the form of water vapour or as cloud droplets is equal to a layer of water about 2.5 cm thick over the whole earth, and this is very roughly equivalent to about 2 weeks rainfall. The mechanisms which produce the clouds operate on a small time scale compared with the seasonal and climatic changes, and at the average wind speed the air circulates round the world in about a month so that the pattern of winds ensures that on the average each parcel of air is refilled with vapour equivalent to the rain it has dropped out at least twice as it travels round the world. The temperature of the sea is of fundamental importance not so much in getting the water molecules out through the sea surface as in supplying the latent heat required to convert that water into vapour.

Water vapour is lighter than air, and so evaporating water into the air might often increase the buoyancy as much as if it had been warmed by about 1°C. Thus warm sea supports thermal convection very strongly.

16.7 THE NORTH ATLANTIC CONVEYOR BELT

It is not possible for any global warming to be restricted to the atmosphere alone. The circulation is not capable of leaving a thin top layer of the ocean warmer than before without also warming much of the depths beneath. How this happens depends on the topography of the ocean floor, just as the motion of the air is guided by the presence of the large mountain ranges. The Gulf Stream'conveyor belt' illustrates this point well, and reminds us that until we have a model of the ocean circulation as good as that of the global atmosphere, we shall not be able to get insight into the problem of future climate by study of a global atmospheric model alone.

The ocean circulation is driven by the wind, but it does not simply follow the wind. For example, where the wind blows continuously from the land to the sea it carries the surface water away from the coast and causes upwelling of water from below at the coast. This is well illustrated by the Mistral (see Fig. 13.11.vi), and many other coastal upwellings. Nor does the water simply follow the wind direction but tends to veer off to the right (in the northern hemisphere) so that a wind from the north will generate an ocean current from the east, other things being equal and steady (neither of which they usually are). According to the same principle, which is usually associated with the name of the oceanographer Walfrid Ekman (1874–1954), a wind blowing along a coast would produce a flow of surface water either towards or away from the coast, in a direction to the right of the wind direction (in the northern hemisphere). The Gulf Stream flows from the Gulf of Mexico up the coast of Florida and northeastwards from New England across the North Atlantic south of Iceland and then up the coast of Norway. It does not follow a steady route but deviates very considerably and at any one moment occupies a very wiggly configuration. The warm water illustrates the dominance of the ocean as a cause of climate by keeping the whole of northwest Europe mild as long as the wind blows from it to the land, and has to cross the warm water.

Fig. 16.7.i 1651,13.4.83,Ch2;Ch3–. (a) The west coast of Greenland is kept clear of ice by the katabatic wind which produces a northward flow of the sea water and holds off the coast as far north as Disco Island. The ice of Baffin Land and Labrador is gradually being eroded and fragmented. The Greenland southeast coast has ice which has been driven by the wind from the Greenland Sea (see 16.7.v). (b) The Ch3– picture shows the clouds which have small enough droplets to scatter sunshine, while the sea and ice, and probably cloud ice particles absorb it and appear black. These two pictures show the identical scene in different channels.

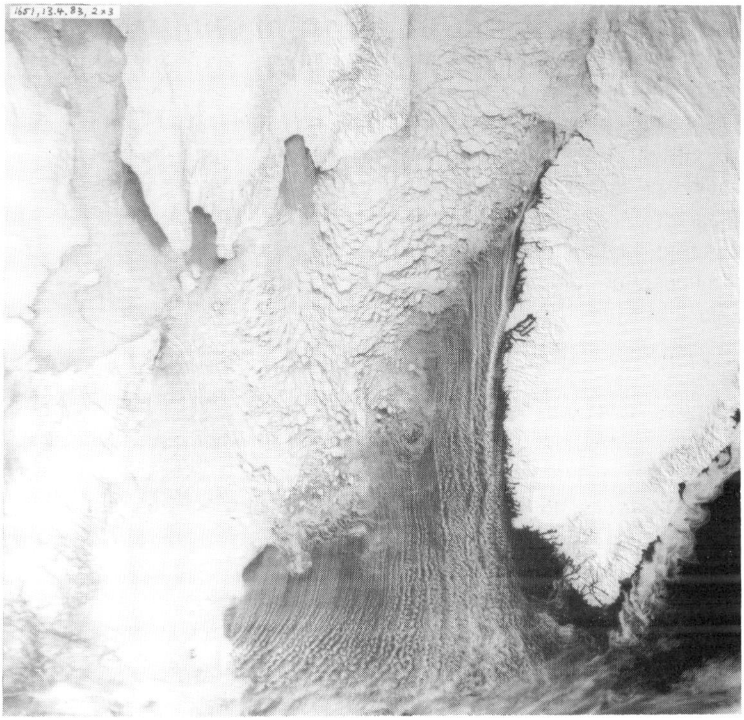

(a)

(b)

The North Sea has remained ice-free every winter in the 20th century except for appearance of coastal ice in Holland and southeast England in very occasional winters. But it is in the same latitude as the Sea of Okhotsk, which is tucked in behind the Kurile Islands to the northeast of Japan, and which is completely frozen over every winter (Fig. 16.6.ii). The virtually ice-free winters of Norway extend round the North Cape and the Kola peninsula to the north Russian ports. The **warm water is salty sea water**, and salt may easily add 10% to the density of it. But most of the surface water in the Arctic Ocean is almost salt-free, having originated as melted snow or rain and floating ice, and so even though it is much colder it begins to flow over the top of the warm salty water where they meet. The black body radiation to the sky begins to equalise the temperature. The water of the Arctic Ocean under the sea is at a temperature between 0°C and 4°C, the latter being the temperature at which the water has a maximum density; but it is still not as dense as the salty Gulf water.

There is at the same time a prevailing ocean current from the north in the Greenland Sea, which, with the Norwegian Sea to its east, makes up the ocean between Greenland and Norway. But this current does not make up for the inflow from the south of the warm surface water. Fortunately for Europe there are gaps in the ridges of the ocean bed, of which the Faroe Islands between Scotland and Iceland are evidence, so that cold bottom water can flow into the Atlantic Ocean, and this makes possible the persistence of the supply of warm Gulf Stream water, which constitutes the conveyor belt of heat into the Arctic which is continuously emitting black body radiation to the sky, cooling and sinking beneath the surface of less salty water.

At the same time the atmosphere is doing its bit and operates a conveyor belt of its own along the polar front carrying moist tropical air into the Arctic and the north of the Eurasian continent. The air is not restricted by the continents, and so in as far as there is a return flow it consists of bursts of arctic air from the north of America into the North Atlantic through the Davis Strait, and by the time that cold air has crossed the wiggles of the Gulf Stream and reached Western Europe it has been warmed to what we call acceptable European temperatures. These outbursts of polar air often produce hail and snow showers in Scotland, where it is nevertheless much warmer than New York in winter, although Scotland is in the same latitude as the much-longer-frozen wastes of Labrador. These fairly frequent bursts of arctic air may be regarded as exporting the cold of the Arctic towards the equator, a sort of sea-level conveyor belt via the Eastern Atlantic and the Canary Islands into the northeast trade winds which carry it to the Caribbean.

The **return flow** of the ocean conveyor belt is an important part of the ocean circulation. There are pools of very deep water in the oceans which may remain almost undisturbed for a thousand years, but it is easy to appreciate that it will be a long time before we have an account of the whole of the ocean circulation and its daily and weekly variations comparable with our present observing of the global atmosphere.

Fig. 16.7.iii 1623,24.5.82,Ch4. This picture is centred at 44°N 47°W on the edge of the Grand Banks of Newfoundland where the cold melt water (light) of Labrador and the warm salty water (dark) of the Gulf are found providing strong temperature gradients and flow patterns contrasting with those of the air.

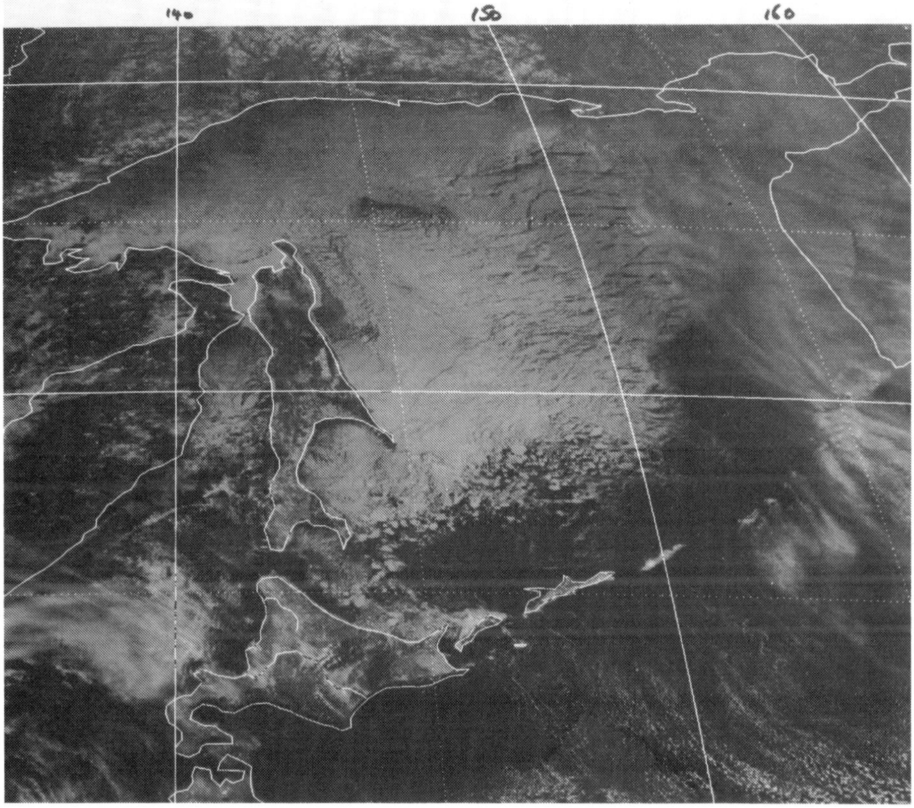

Fig. 16.7.ii 0000,6.4.80,vis. The Sea of Okhotsk as seen by the Japanese geostationary satellite GMS1, after the spring warming had begun. This sea is almost covered with ice every winter although it is in the same latitudes as the North Sea, which has little or no ice in winter. The coast of Hokkaido and Sakhalin and of the Kurile Islands leading to the southern end of the Kamchatka Peninsula are outlined. The cloud of a frontal system covers Japan.

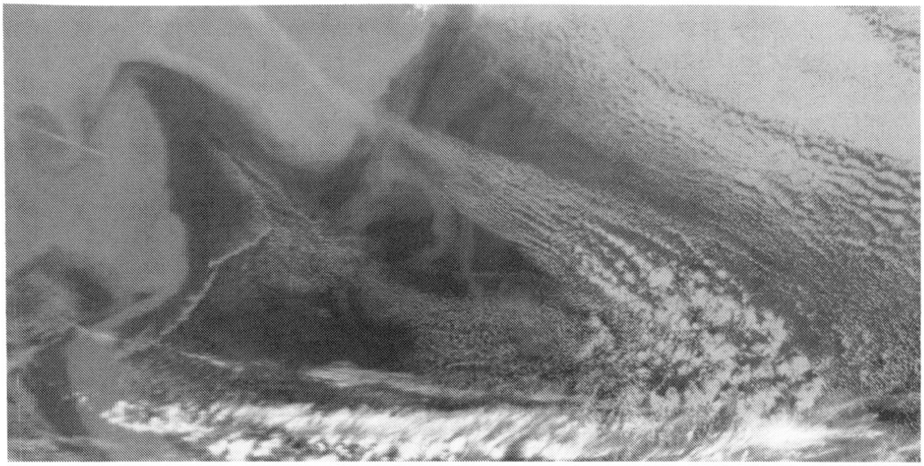

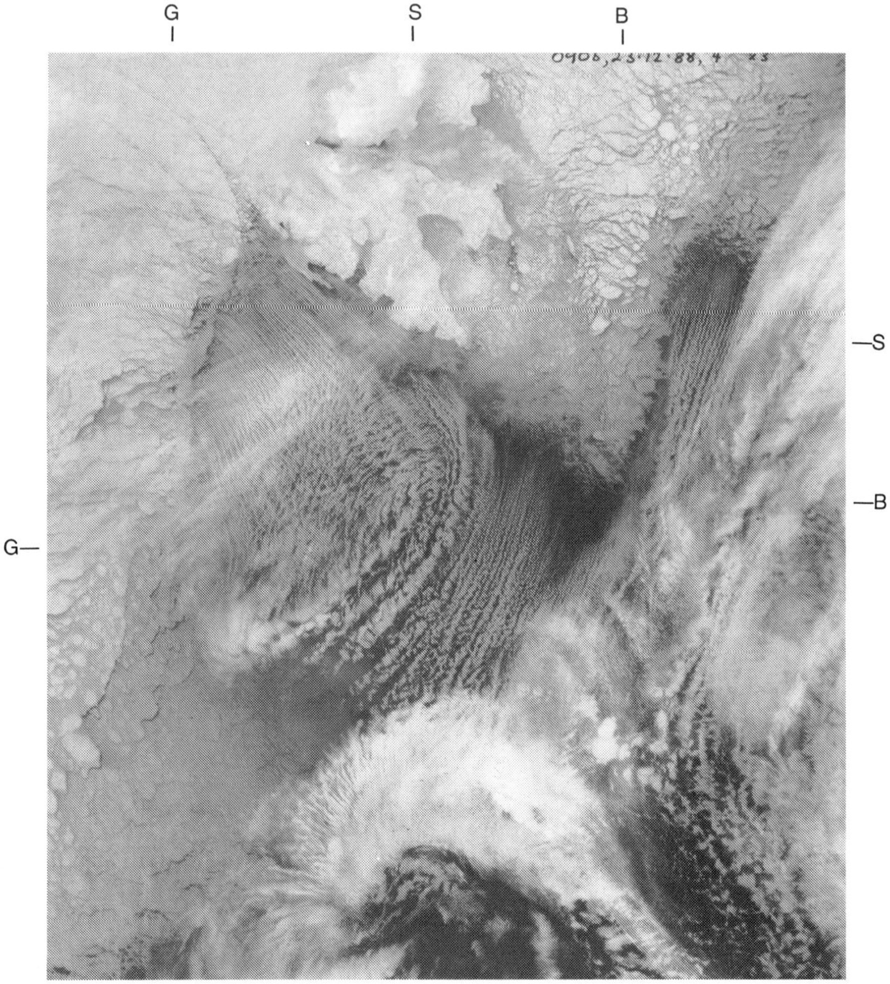

Fig. 16.7.iv 0905,12.12.88,Ch4. In the North at the end of the year the west coast of Svalbard (S) still has a few ice-free bits of coast while the fragmented ice melts as it is carried southwards. Bear Island (Bjorneya (B)) is just surrounded by ice and Norway is under the cloud halfway up the east side of the picture. In the south there is some high convection cloud over the warm salty water brought by the 'conveyor belt'. There is extensive ice over the Greenland Sea (G) in the west. The sides of the picture are aligned NNE.

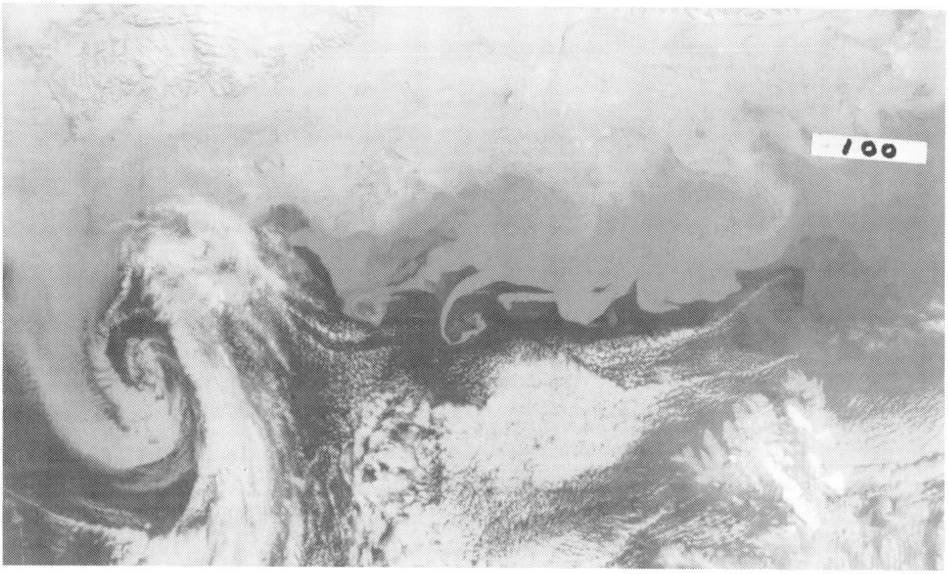

(a)

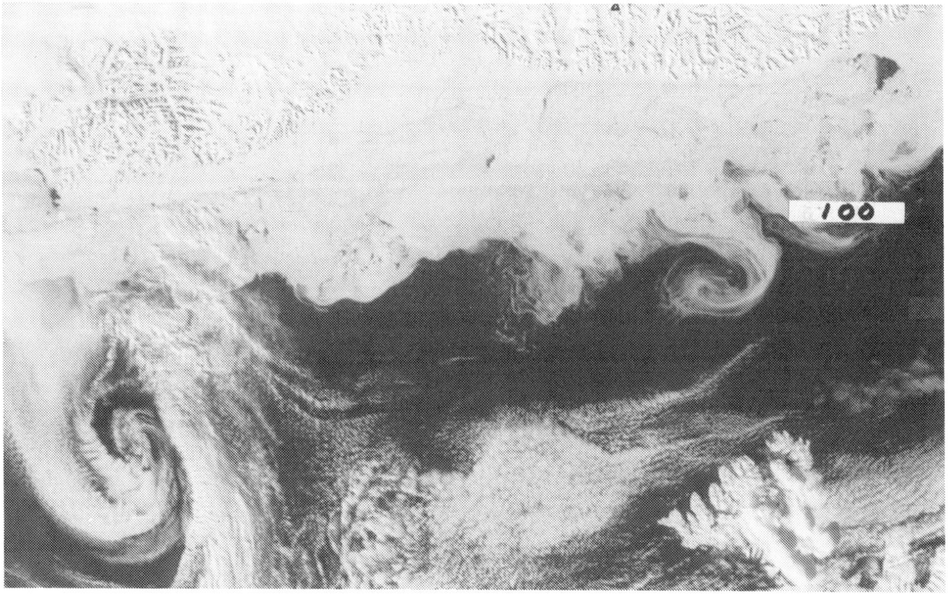

(b)

Fig. 16.7.v 1452,16.3.82,Ch4;Ch2. (a) On the north side of the Denmark Strait between Iceland and Greenland there is a continual transport of ice from the Greenland Sea towards Cape Farewell, melting as it goes. The IR picture shows a narrow belt of open sea between the cold meltwater on the north side and the snow-covered peninsula of Iceland. (b) The VIS picture shows a much wider space from Iceland to the northern ice edge, showing that the meltwater still floats on the warm salt water of southern origin. (Courtesy the University of Dundee.)

6.8 ENSO

This is the first phenomenon to be discovered to have relevance to other events worldwide. To express that another way: the state of ENSO, which by its name states that El Niño is of more than local importance and is an essential part of the Southern Oscillation. It can be said to be the first phenomenon to be recognised as a world index which is nowhere irrelevant.

Another ocean current, as famous now as the Gulf Stream, is called El Niño, named by Peruvian fishermen after the Christ child because it brought warmer water to their Pacific coast, most usually at Christmas time. More usually the wind from the southeast up the Peruvian coast causes the surface water to begin to flow northwards but it soon turns westwards away from the coast under the influence of the earth's rotation (Coriolis force) with the result that there is upwelling of colder water at the coast. The east wind (i.e. *from* the east) at the equator drives the surface water as a *westerly* current into the West Pacific where the surface temperature is usually warmer than in the east because the water is exposed to equatorial sunshine on its journey to the Philippines. Thus the Western Pacific is where the sea temperature is the highest anywhere on earth (about 30°C) and the cumulonimbus clouds become their largest. They frequently congregate into tropical cyclones (called typhoons in that region (see Fig. 14.11.vii, p. 526) and the large spreading tops of the storm clouds are composed of cirrus thick enough to have a greater than usual albedo which cuts off the direct sunshine and sets a limit to the high sea temperature. The high vapour pressure of the water at that high sea temperature is contributory to the buoyancy and humidity at cirrus level, but there is still dispute about the increased albedo of some of the cirrus. (See Plate 6 in the colour section between pp. 352 and 353.)

The Bay of Bengal, where tropical cyclones also occur, has a much more restricted time of year for them, and the sea temperature does not rise as high because for half the year the wind comes from between north and west, whereas it may be from the east for the whole year in the equatorial Western Pacific (Fig. 16.8.ii).

This equatorial current under the stress of the east wind has only a very slight deviating force (which is, of course, zero on the equator) and so it piles up the water in the west with the result that from time to time there is a surge back towards the coast of South America bringing the warmer El Niño. The extent of this surge of the equatorial water eastwards is held together because the deviating force tends to push the north and south sides of the eastward streams towards each other. That mechanism has the name Kelvin wave, and is more common as a surge travelling along a coastline or boundary of shallow water like the polluted water of the River Po travelling down the east coast of Italy without spreading out across the Adriatic Sea, or the flow round the Greenland coast.

The area of warm surface water spread into the Eastern Pacific has a big influence on the weather of the world and of the southern hemisphere in particular, and the resulting changes were named the **southern oscillation** by Sir Gilbert Walker and his students at Imperial College during the 1920s and 1930s. It has required extensive research since 1950 to discover that it was linked to major movements of Pacific surface water. The natural period of oscillation of the ocean does not coincide with that of the atmosphere

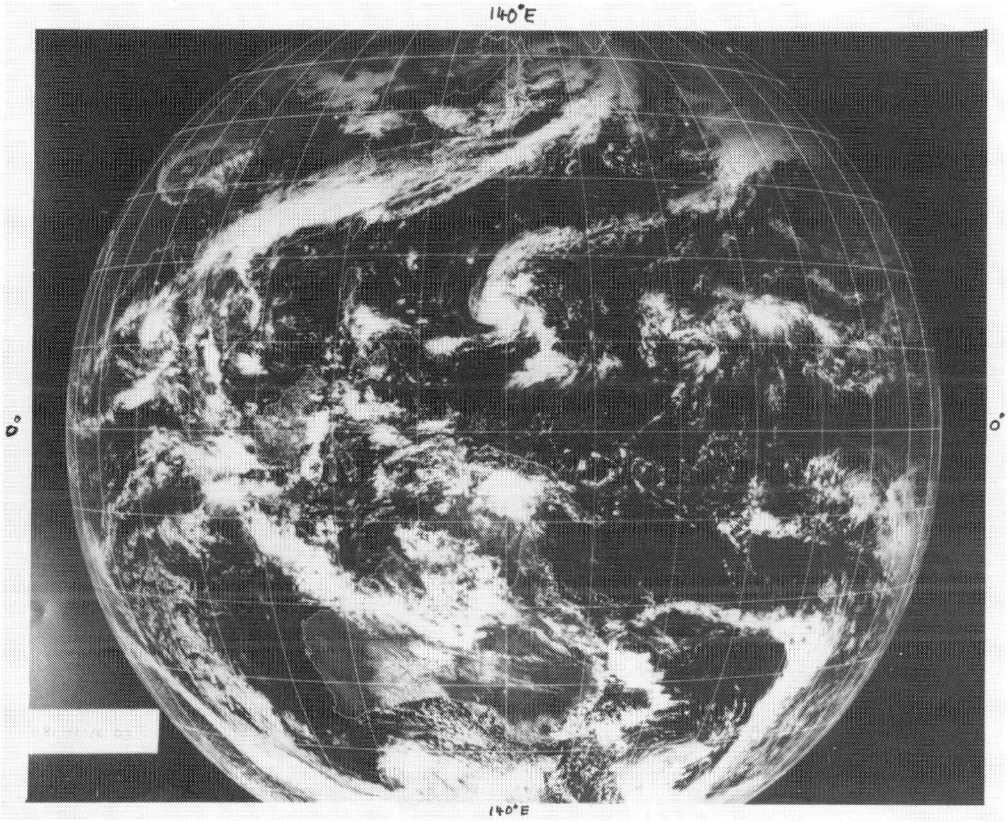

Fig. 16.8.i 0300,16.11.81,vis. This GMS1 picture shows the meeting of the NE and SE trade winds driving westwards along the equator to the West Pacific area where typhoons are created as the air approaches the Philippines in the North and New Guinea in the South. Typically the fronts lie across China having passed across Japan with similar orientation in the south. The cloud lines at 10°N and S of the equator stretch from South America most of the year. The sunshine heats the sea all across the Pacific and the great typhoon breeding area begins around 150°E. In the north they cannot cross into continental Asia, but in the south some cross into the Indian Ocean. (Courtesy the Japan Meteorological Agency.)

nor with the annual forcing by the seasons of the ocean–atmosphere system. This has had important consequences for the planning of research into the combined system and there has been much sharing of understanding between meteorologists and oceanographers in recent years. The El Niño forecasts have been found useful in forecasting drought in Zimbabwe, which has become a more important issue than in the past because of the introduction of more mechanised farming in the production of maize [81].

It is a case of the atmosphere being the cause of a change in the ocean current system and the ocean currents causing a change in the weather system, and it illustrates well the dependence of climate on the combined atmosphere–ocean system, and is, even alone, a sufficient reason for seeking to establish a good model of that double system. The combined example just described has been given the code name ENSO.

(a)

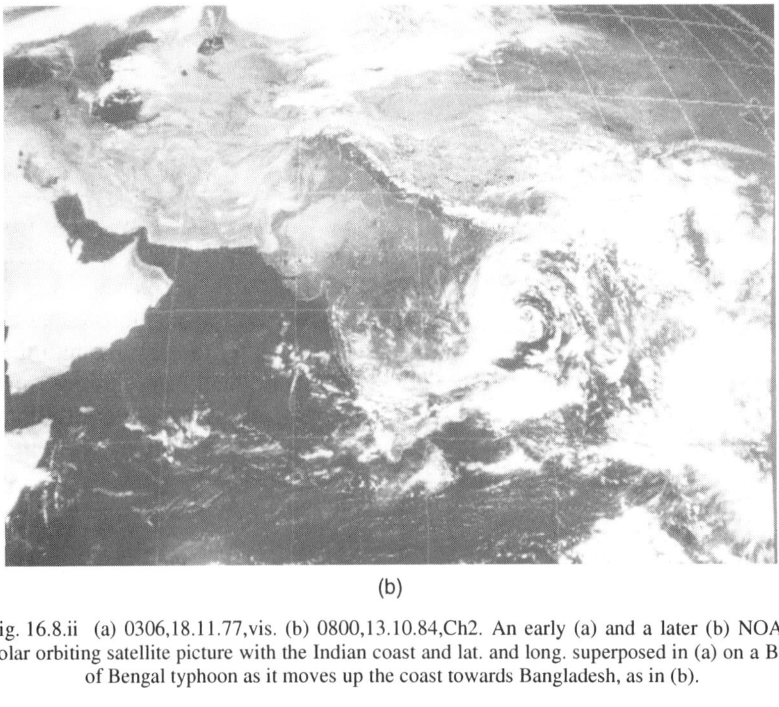

(b)

Fig. 16.8.ii (a) 0306,18.11.77,vis. (b) 0800,13.10.84,Ch2. An early (a) and a later (b) NOAA polar orbiting satellite picture with the Indian coast and lat. and long. superposed in (a) on a Bay of Bengal typhoon as it moves up the coast towards Bangladesh, as in (b).

Fig. 16.8.iii 1800,18.10.79,vis. The early morning sun describes the cloud top with shadows as this powerful typhoon gathers strength by incorporating a front at 30°N into its circulation when it is centred over Okinawa. This picture was a warning as it approached Japan.

The outstanding feature of ENSO is that it appears at **irregular intervals** of between about 3 and 6 years. Or we could say that its behaviour seems to keep to one period, usually 3 years, for only a very few repetitions, whereupon it switches to a longer period. Such attempts as have so far been made to produce a mathematical model which represents crudely the mechanics of the connection between the two components have, without becoming more complicated than the observations of the whole area would justify, represented it by non-linear differential equations. The generalities of this topic which lead to the new mathematical concept of chaos are treated earlier in Chapter 8. It has been the habit of climatologists to seek any periodicities in the weather and climate that appear to be reliable enough for use in forecasting for a season or up to a decade into the future. The obvious periodicities are the daily and annual cycles, and there has been much talk accompanied by less convection of what is called the **quasi-biennial oscillation**, which refers to changes in the dominant high-level winds which have a period of about 26 months. The mechanism causing such a periodicity to occur has received much attention, but the essential difficulty arises because traces of any of the normal mechanisms which drive the atmosphere seem to die out in a few days or weeks at most if they rely on the inertia of the atmosphere to produce the next cycle, which in this case is over two years away. But the geographical and oceanic influences to remain in place long enough to push the next cycle into action. The quasi-biennial oscillation seems to originate high in the stratosphere and so there has been no search for a very direct cause in the ocean.

It is not intended here to describe details of the research into the ENSO phenomeno except to note that the periodicity is a phenomenon without any obvious origin like the daily and annual cycles. The early study of the southern oscillation illustrates the strong wish to categorise it as probably periodic. Much of the success of solving linear differential equations stems from the fact that any motion can be analysed into components, all of which are solutions of the same equation. That technique can be used in several global atmospheric models because the earth is circular and so if we describe the pattern of weather in the atmosphere starting 'here' and going eastwards we 'come round again' after one encirclement of the globe, and this gives us a basic periodicity. The ENSO case is certainly not one for linear differential equations and so we must expect chaos.

Indeed the appearance of periodicity in the records of the southern oscillation and the appearance of the warm El Niño has been misleading in that it suggested a sort of resonance. Thus the more or less repetition of the sequence of events every 3, or sometimes 4 or more, years was actually an interaction between the ocean and the atmosphere, each of which would probably have had a precise repetition every 3 years or perhaps every 1 year if the conditions in the other had remained constant. This would mean that models which reproduced resonant periodicities like what was observed would seem to have faithfully imitated the reality. A good combined model would produce the appearance of a 3-year period which did 3 or 4 cycles and then repeated with a different frequency a few times with some reduced oscillations in between. If we think in terms of resonant frequencies, which are essentially a linear phenomenon, we would say that these two systems each had natural frequencies which did not have a rational relationship (i.e. did not have a ratio described by the ratio of two integers). This produces what the scientists Jin, Neelin and Ghil describe [82] in a recent article (perhaps with not-unintended irreverence for El Niño) as the Devil's Staircase, for when the system is represented graphically it would jump erratically from one near resonant state to another like a drunken attempt to ascend a staircase of uneven step height. The irrational relationship is a recipe for chaos (see also Tziperman *et al.* [83]). We shall see, in Chapter 17, that this is indeed a new approach to the reality which faces model makers.

The joint atmosphere–ocean thinking does have repercussions in climatology. This is well illustrated by investigations which use information about sea temperature obtained from satellites which have become available only in the last decade or so. The IR emission intensities from the sea surface in the long wavelength bands do suffer from some unavoidable errors due to the emissions from the variable amounts of water vapour above the water and from the fact that clouds prevent regular observations from being made. Even so it is possible to use the occasions when there is a gap in the clouds, which occur in a random manner and can be selected by suitable programming which enables a decade of observations made at the rate of four a day to be processed to give monthly averages of sea temperature variations. Recent work by Dr A. Nykaer [84] has revealed that the predominant wind from the northeast across the Atlantic coast of northwest Africa causes an upwelling of ocean cool water along the coast which varies with the season in different ways on different parts of the coast from the coast of Portugal (latitude 42°) down to cape Verde (latitude 15°N) as the trade winds in those latitudes vary with the season. Most interestingly he found that the upwelling was not the same every year, but appears to have varied in the same way as the upwelling off the Pacific coast of South America, having a weak year at the time of the unusually strong El Niño during 1982–1983. This links the wind-driven upwelling with the two elements of ENSO. They thus appear to be in near resonance as a single system with a period of about three years, but slip from time to time on to another step of the Devil's Staircase into a different period. The fishermen of Peru expect the El Niño will occur no more often than every third year, and can pray for it too, knowing that the prayers will certainly be answered in due course, when the Almighty gets round to it. It is even possible that a non-linear model may be invented to give better forecasts of this phenomenon which we believe to be physically determined although in a chaotic manner.

16.9 RETURNING FLOW

The idea of a return flow in the form of an internal gravity wave such as we have just described is not a new discovery. I now describe three examples.

First we note that the complementary mechanism to the making of sea water fresher in the Arctic Sea is the making of the sea saltier in warm oceans by evaporation at the surface, which is caused by the strong sunshine. The process is even greater in the Mediterranean because it too has plenty of sunshine, and evaporation is enhanced by the flow of very dry Sirocco from North Africa and Föhn winds from Europe, so that it becomes saltier than the Atlantic water to the west of Gibraltar. This result is enhanced by the fact that the Mediterranean is much shallower than the Atlantic.

During World War II the Royal Air Force patrolled the western and eastern approaches of the Straits of Gibraltar in order to prevent German U-boats from entering the Mediterranean. The U-boats could be detected from aircraft if they moved close to the surface using their schnorkels, and if they attempted to traverse the straits at greater depth using their engines they would be detected by the acoustic waves produced in the water by their noisy engines. Therefore they had to find a way of drifting with the ocean current from the Atlantic into the Mediterranean close to the bottom, without engines. Classical oceanography taught that the Mediterranean was more salty than the Atlantic because of the more intense evaporation so that very salty Mediterranean water flowed out at the bottom while lighter Atlantic water flowed into the enclosed sea, from the west. However, actuality was more complicated. The weather at Gibraltar is characterised by intermittent periods of wind blowing from the west and then from the east alternately through straits, and if one of these spells lasted rather longer than usual, a west wind would tend to overfill the Mediterranean, and a prolonged period of east winds would empty it to a slightly lower level than usual. After such an easterly spell the water would tend to move into the Mediterranean at all levels right down to the bottom for a day or two to fill the enclosed sea up to its usual level, and that would provide an opportunity for a U-boat to drift through without engine power. One day in 1943 a U-boat was successful in this enterprise but was forced to surface for air close to the coast of (then) Spanish Morocco and was beached there, on what was neutral territory, after being attacked. There were diplomatic flutterings when a British navy crew went ashore and acquired documents from the stranded submarine, some of which revealed that they were privy to the occasional departures of the bottom currents from the classical oceanographic model.

That case was a good illustration of the power of the wind to drive ocean currents and to build up a higher water level so as to produce a return flow as an internal gravity wave, rather like the east wind which drives the westward surface current across the equatorial Pacific Ocean. In the case of the Pacific, the current is driven in the direction of the slope of the water surface because there is no Coriolis force at the equator; in the case of the Straits of Gibraltar the water has to flow towards the lower surface level because it is constrained by the coasts on both sides, but as soon as the current reached the wider Alboran Channel east of the straits the current was forced to the right and hugged the Moroccan coast, which is where the damaged submarine was beached.

The second example is taken from Lake Victoria in Central Africa. That lake has internal stratification with a relatively thin layer of warm water on the top and cooler water in which most of the fish live underneath. Gravity waves are set up by the gravitational force of the moon, as they are in all the oceans on earth. This sets up a wave on the interface between the cold and the warm water layers, and because of the small temperature difference these waves move very slowly. It takes about a month for the wave to travel the length of the lake and back again and so the wave is amplified by resonance with the period of the moon. At one point on the coast there is an underwater shelf, at about the same height as the interface, where regular fishing takes place; and the fishermen know that the fish come with a particular phase of the moon because only then do the fish in the lower layer come up on to the underwater shelf. One can imagine the stories about how the fish come in response to the light of the moon.

The primary message from this chapter is that the most important factor in oceanography is the wind which drives the surface currents; the most important factor in meteorology (climate in particular) is the sea surface temperature. Thus the ocean and the atmosphere form one single combined system in which the sea temperature, which was taken as the starting point for weather forecasts, and the wind stress at the surface, which is taken by oceanographers as the starting point for their ocean current analysis, and the whole system must be included in any attempt to predict climate.

Surface wind stress and temperature have recently begun to be measured regularly by buoys attached by cable to the ocean bottom, and sea surface temperature is now being measured more reliably by satellite, but we have a long way to go before a routine reporting of all the oceans becomes possible.

The frequency of serious floods has varied through the ages, even during the present interglacial. James C. Knox [85] investigated the gravels left by floods in the central states of North America; the larger stones indicate greater floods. He was able to produce a flood chronology back 7000 years in which the period from 5000 to 3300 years ago was relatively free from the more serious floods. After that came a period of greater frequency. More recently, the Middle Ages up to about 1250 AD was warm and relatively free from the worst floods, but with the transition to the Little Ice Age up to 1450 experienced larger floods than subsequently during the Little Ice Age Itself. The more serious flooding seems to be associated with climatic change not only in the Mississippi–Missouri basin but also in China and Northern Europe. This casts doubt on some of the statistical predictions of serious flooding frequency which specify the expected intervals between floods of various severity. Knox also suggests that the present predictions of climatic change due to greenhouse warming should render flood frequency warning less reliable. This paper was accepted on 21 December 1992.

Since that acceptance the extensive floods in the Mississippi–Missouri basin have occurred, and they have been attributed by G. A. Jacobs *et al.* [86] to the long-term effect of the 1982–3 El Niño in redistributing the temperature of the eastern Pacific Ocean. Before briefly describing those conclusions it must be remarked that the perennial problem of assigning (first) causes to particular events in world weather is that many antecedent global factors may be involved. To attribute to El Niño the fact that the rain-producing storms tended to occur over a particular area rather than travel over large distances and distribute their rain, thereby causing serious floods, may be followed by the question 'what caused that El Niño to be exceptional in its legacy of ocean temperature distribution?' and so on. There is no beginning and no end, but important change nevertheless, dominated by physical happenings.

This emphasises the need for global-evolving models of weather and climate rather than documentation of incidents, as if they were a sequence of separate events which could be labelled as causes or as effects. Weather forecasting is not the same as climate analysis. Weather depends on local causes, and recent local events. Specific weather is not forecastable beyond a very few days: this is not only because events outside the immediate region begin to affect what comes into the area but because the regime may shift so that the locality of interest begins to occupy a different part of the global system.

16.10 KELVIN AND ROSSBY WAVES

In the case now to be described, the long-term influence of an ocean incident is seen to be, perhaps, much greater than the periods between incidents [87]. Because the El Niño of 1982–3 was more than usually powerful—and we remember that that was caused by Trade Winds of previous years—so that the return El Niño current reached a wider expanse of the western shore of the Americas than usual. This made the El Niño event itself unusually memorable. It piled up water on the west coasts which began to flow northwards up the coasts of Mexico, California and northwards as a Kelvin wave hugging the coast. A similar current travelled down the west coast of South America, with consequences which will soon be studied because the total incident becomes daily more interesting.

The Kelvin wave travelled progressively more slowly, reaching the Gulf of Alaska about 6 months later. The elevation of the eastward-flowing water at the equator gradually subsided, but the elevated water along the whole west coast began to generate a wave travelling westwards. At their highest the waves we are discussing reached a maximum height or depth of the order of 15 cm up or down and over distances of hundreds of miles so that the slopes needed the new technique of satellite radio-altimetry to reveal them, but they must have affected the judgments made from tide gauges that the global sea level has gone up by 10 cm in the last century.

The westward-travelling waves depend on the variation of the earth's rotational effect from the pole to the equator (where it is zero and becomes reversed in the opposite hemisphere), and are called Rossby waves which are better known in the atmosphere in the North Atlantic. In the present case they travel westwards more rapidly at their equatorial (i.e. southern) end. This very slow wave encountered the Kuroshio warm current, which flows all the time up the east Asian coast and turns to the right as it approaches Japan to cross the open ocean. The Rossby wave caused this current to be pushed further north than usual by the faster-moving southern end of the wave.

In the years following, the northern part of the Pacific was made warmer than usual with significant effect on the frontal cyclones (which are thought of climatologically as *the Aleutian Low*, corresponding to what the Europeans think of as the Iceland Low which feeds frontal cyclones into Europe). The significant detail of the result was that the travelling lows tended to become virtually stationary over the central United States, causing the exceptional flooding of 1993.

Two factors made this analysis possible. The first was the employment of a Pacific Ocean mathematical model which demonstrated the wave-travelling mechanism at work. The southern end of the Rossby wave had reached Taiwan, while the northern end was in the Gulf of Alaska 10 years after the Kelvin wave which brought the warm water to the west coasts of America. That this Rossby wave existed, and was observed by satellite to be travelling superposed on all the other oceanographic turmoil, is a triumph of mathematical analysis which told us what to look for. The height of the wave was less than that of many of the windblown waves which rode on top of it, ruffling its surface. The Kuroshio too was located in detail by satellite temperature measurements, and the whole episode demonstrates the enormous inertia of the ocean, which every oceanographer first learns is essentially driven by the wind, which every meteorologist learns is given its detailed marching orders by the temperature of the sea.

The message seems clear: changes in climate may cause changes in the frequency of occurrence of extremes of weather which do not necessarily represent the climate towards which the change is moving. In the case of the effects of the enhanced El Niño of 1982–3 we cannot take this as evidence of a move to a warmer climate, although it may be, because the transition into the Little Ice Age was also accompanied by more severe floods for about two centuries, so that if we are now definitely in a transition to a warmer climate it could be that the floods of 1993 were a symptom of that change.

The next two decades will see the invention of mathematical models of the **ocean–troposphere–stratosphere system**, which is a chaotic system, but may be said to possess a characteristic climate at which the mind can learn not to boggle. Even Gaia may be involved because one of the major processes involved concerns the absorption of CO_2 into the sea in which the role of algae and coral are gradually being understood. There is plenty of exciting history of the evolution of this planet to be elucidated and lots of wonder waiting to replace the awful nonsense of dogmatic religions which pretend that the world's future history and purpose have already been revealed. The legacy of religions is that they have left in humanity a belief that we will have some sort of control or ability to forecast detail and take avoiding action if we don't like the look of what is forecast. It is important to realise that an increase of temperature in one area of a continent might cause change in the winds which would change the ocean currents which in turn might cause a cooling in another area. Global warming is not likely to produce an equal increase in temperature everywhere.

16.11 INTERNAL MIXING

The ocean cannot be warmed effectively from the top because the stable stratification prevents mixing. This is quite different from the atmosphere, which can very readily be warmed from the bottom or cooled from the top, and undergoes these changes every day. It has been said that it would take a thousand years to change the ocean temperature significantly, and since the weather depends very fundamentally on the sea surface temperature and the ocean has a much greater heat capacity than the atmosphere, any change in climate could only be very slow.

The wind blows surface water from high to low latitudes where it is placed above warmer water so that convection would take place. The water also has its density altered by the salt dissolved in it. The water in the polar seas is reduced in salinity by the water being frozen when it falls as rain or when it snows onto the floating ice, which is not salty. When the ice melts, it floats away on top of the warmer salty water (see cover picture).

Heat is transferred from warm to cold water more speedily than salt is transferred to less salty water, and so we can make a mental model of the matter by imagining a vertical tube with its top in very hot salty water which is lying above much colder fresh water into which the bottom of the tube is open. Heat can flow through the tube wall so that the water in the tube will begin to move upwards as the fresh water in it gets warmed a little. From then on, a fountain will be generated as the fresh water rises up the tube and is warmed by being surrounded by hotter salt water. Alternatively, if the salt water penetrated down the tube to start with, it would be cooled by the surrounding fresh water and become denser than it and generate a downward fountain in the tube. This is called **double diffusive convection** (see Turner [51]) because heat and salinity have different diffusion rates. The mechanism is reminiscent of the dynamics of mamma (p. 509).

The exchange of fluids would be quite likely to take place in the form of fingers or towers of salty and warm water sinking and rising in among each other. The shapes may be like towers of castellanus cloud, which is caused by the complication of latent heat release, which makes its motion different from that of dry air.

If hot salty water were beneath cold fresh water, bubbles of each would move vertically into the other mass because of the buoyancy due to the temperature difference; but as that difference became smoothed out we would end up with less salty water at the top and with saltier water at the bottom, but with smaller temperature and salinity differences than at the start.

The idea of static stability, in which a light fluid rests above a heavier one and does not mix with it, has been a powerful model in the minds of many who like to think of scientific theories in terms of homely or atmospheric analogies. Likewise, plain parables have been widely used by preachers to teach their moral principles. But they often need only a small variation to render them quite misleading. Thus, if an interface between an upper layer of water and a lower layer of denser water is tilted, the lighter fluid begins to run up the slope and the heavier fluid down the slope. In this way a strong shearing motion is generated which fairly soon becomes unstable and is turned into a series of rolls, or **billows**, which are unstable waves. This is usually called Kelvin–Helmholtz (K–H) instability. The billows each roll up, which causes the interior fluid to become well mixed and end up with an intermediate density. Thus the system settles down to the two original layers with an intermediate density layer between them. This process is described with appropriate theory in section 8.1. The stable layers are tilted by tidal waves travelling in shallow coastal waters particularly.

Thus we now have two interfaces with lighter and heavier fluid above and below each of them; and when they are tilted we get two rows of billows, and so on until what was originally a single interface becomes a pile of several interfaces which, unless very carefully measured, would look like a continuously graded density, but would actually be a pile of very thin layers, each of intermediate density, in the place of the original single discontinuity.

Such billows have been observed in the shallow coastal waters of the Mediterranean and are likely to occur by the same mechanism in the deep ocean. Billows are very commonly seen in clouds in the atmosphere when a very stable interface at the top of a cloud layer (which has been caused by the radiative loss of heat from the cloud top). They have been observed with a wide range of wavelengths from a few tens to a few thousand metres.

This means that systems which look very stably stratified can become much more mixed. The phenomenon is very frequently seen in the sky, and the rolls appear as billow clouds (see Fig. 16.11.i). There is a very interesting theoretical point about this phenomenon: the shearing motion generated by the tilting of the interface away from the

Fig. 16.11.i A large wave cloud in sunset illumination over Aberdeen with a shallow wave cloud below fragmented into billows in two directions. (Courtesy the Clarke Collection; Royal Meteorological Society.)

horizontal is proportional to the density difference across it; the instability of the shear layer which makes it roll up into billows is proportional to the square of the velocity difference. The interfaces which are initially the most stable in a stratified air mass are therefore the ones which are most quickly transformed into billows (see section 8.5). The stabilising forces are proportional to the density differences, while the destabilising forces are inertia forces and are proportional to the square of the velocities which are proportional to the density differences. If that argument is followed, there should be no surprise that billows are most likely to be found in the stable layers at the base of the stratosphere. This is just where clear air turbulence was most often discovered when jet planes first started to fly above the weather of the troposphere, but it has been a surprise for those with only intuitive ideas about static stability.

Mechanisms like the salt fingers and billows operate in the ocean on a small time scale such as a few days at most, and in order to detect their effect, measurements must be made at that sort of frequency. Therefore if we wish to make observations of changes in the ocean temperature, which might be due to global warming, it is not helpful to make the measurements once every few years—which is exactly what has been done very recently. The present comments do not detract from the value of the observations which were made along the traditional route of Columbus, in 1957, 1981 and 1992 [88]. Each survey was made along the latitude 24°N from Africa to Florida at intervals of 160 km, 70 km and 60 km respectively, down to the ocean bed, and maps of their average temperature contours were compared. The changes near the surface were very much more varied than down below, which is to be expected because they varied by 1°C or 2°C and were on the whole cooler in the later years. At a depth of 1000 m the changes were much less but more consistent and showed a rise of around 0.25°C to 0.5°C. The changes near the surface were not very meaningful because they could have been characteristic of the month of the particular year of the measurements and are within the normal variations, and 'could have happened anyway', meaning that it cannot be regarded as evidence that it is due to global warming which would have been deduced to be taking place.

17

Prediction and proof of climatic change

With all the difficulties of predicting the weather even a week ahead we now examine the possibility that we may have good physical reasons for predicting changes in the climate, without being able to predict either the rapidity or the precise nature of the change. It has been noted that there is, and always has been, a greenhouse effect caused mainly by the presence of water vapour and carbon dioxide, and some other less important greenhouse gases. Rough calculations indicate that without this effect the world would have been about 30°C colder than it actually is: there would have been a permanent ice age, and advanced forms of life could not have evolved.

17.1 PHYSICAL REASONING

It is therefore good physical reasoning to expect that if the atmosphere were made to contain more carbon dioxide the greenhouse effect would be increased and the temperature would be raised. Not unreasonably with all the modern computer assistance we might hope to estimate the magnitude of this increase; and with the aid of modern modelling techniques already used to predict tomorrow's weather, a fairly accurate idea of the effect on the weather of such an increase would follow. We could then plan for the future with a good idea of the details of the changes of weather different places in the world might experience. So far the biggest element of doubt is in predicting the detail of the changes: to the absence of a good model with realistic feedbacks is now added the spectre of chaos, which throws doubt on the details even when a very good model is used.

Models are very crude. But they do obey the laws of physics and mechanics even though they are in a chaotic mode all the time. The grid of points at which values of the weather parameters are calculated is typically spaced at 4.5° (500 km) intervals of latitude and 7.5° intervals of longitude (which goes from 833 km along the equator to zero at the poles, close to which obviously all the points are not used). By the standards of accuracy used in ordinary forecasting such a grid is very crude and much more of the ordinary detail of the weather has to be represented by something parametric which has roughly the same effect on the broad scale as the smaller weather phenomena. Thus the actual lifetime of individual clouds, their thickness and the amount of rain they produce are not represented, but the effect they would have on the emission and absorption of radiation is represented, it may be hoped, by numbers deduced from the calculated humidity. Likewise the radiative effects of the greenhouse gases are represented by parameters thought to have the same effect on the large scale as the real gases actually have. The accuracy of these and other representations of physical effects by guesses thought to be about right are thoroughly tested by comparing actual weather with the forecasts produced by a model calculation.

17.2 DEPENDENCE ON PARAMETERS

The parameterisation is considered good if the forecasts are usually quite close to the mark. If they are not, parameters can be altered until they are: and by doing this, forecasters and model researchers can acquire some skill and knowledge of the effects of the changes. It is then theoretically possible to change a parameter so as to represent any different circumstance we may wish to investigate. For instance, the effect of having, from tomorrow onwards, twice as much carbon dioxide in the atmosphere as there actually is today can be tested on the model. However, the result of that forecast cannot yet be tested for accuracy, but the numerical experiment can be carried out on several different models and the results, which will not all be the same, will represent a range within which the actual future outcome may be expected. Such a procedure involves a great deal of clear-headed research with the software of the models and may consume expensive computer time. We know that the details of the models' predictions cease to be good weather forecasts after only a few days; but nevertheless the average weather predicted by a model when the forecast is carried on for the next year or so has been managed by the choice of the physical parameters to be very similar to the real average weather as recorded in the past. Thus each forecasting model has a climate which can be examined for any important climatic detail, such as the average temperature, and so the effect on the temperature of having twice as much carbon dioxide can be calculated for the model system.

Thus it may be possible to obtain a numerical estimate of the magnitude of a physical effect which we may feel fairly certain will occur because we have a good understanding of the physical mechanisms and have not misrepresented them in the models: it alerts us to the 'probable effect'. But there are so many different effects that might be happening that we need a new kind of study to tell us whether what we have come to expect is actually taking place.

We know that there have been variations of climate which will not take place in our models because the models have been constructed to avoid them. Thus most models are bad at forecasting the extremes of weather. It is found that if they are allowed to predict extremes they would probably do so far too often. If they could wobble between extremes they would be far more wobbly than reality; and so they are likely to be made less wobbly than reality. This may seem to be cheating, and avoiding the interesting part of the soothsayer's art. But this smoothing of the weather's temper is quite justified when we are trying to make forecasts for only the next few days. There are influences which cause changes in the climate and thereby widen the range of wobbling that actually occurs. Thus we usually make our models so that they do not, and probably cannot, forecast changes of climate and we have to study these by feeding into them influences which we think are causes of climate change.

The climate of the past is known to have varied under one or more of the influences described in this book. To test any of the predictions of such changes we need to have accurate records of past changes, just as we need them to test our daily weather forecasts. At the present time, therefore, serious attempts to accumulate the climatic records required are under way. Without records of actual events we cannot know whether our parameterisations of the physical mechanisms are satisfactory. The parameters chosen for our models may happen to be fairly good for tomorrow's forecast if we restrict the wobbliness of the model, but may turn out to be rather poor if the model is allowed to wander into new territory. In particular many of the parameters are made to work more or less correctly when many of the surrounding values such as the temperature of the sea are prescribed in advance on the basis of averages from the past. In that way many feedback mechanisms are ruled out because these values are fixed, although they would vary if a real climatic change were started.

A recent example of what may be done to make progress in predicting the magnitude of changes which are forced on the atmosphere will now be described to illustrate the difficulties that have to be overcome. First we need to have a 'coupled ocean–atmosphere model'. We know that the ocean will change too if the atmosphere is forced into a new regime. Very large changes in the climate of some localities would occur if the ocean currents carrying water that is warmer or colder than average were diverted to different areas. In arctic regions particularly, the density of water is determined not only by its temperature but also by its salinity. Rain and melting snow and ice are much less salty than the average ocean, and so when the much more salty Gulf Stream is driven by the wind into the far north of the Norwegian Sea, as soon as it has given up its warmth by emitting more radiative energy than the polar sunshine provides, it slides under the less salty arctic water and eventually flows back at a deeper level into the North Atlantic (see the Atlantic conveyor belt in Chapter 16). This is an important mechanism in generating exchanges of water between the surface and the deeper layers of the oceans. Things might be very different if the ocean bed between Norway and Scotland on one side and Greenland and Iceland on the other side were higher, making the ocean very shallow, for the climate of Norway and much of North West Europe might then be more like that of Greenland. If the return flow at the deeper levels were prevented, the flow of the warm water to the far North would be inhibited as it is in the Pacific Ocean. The climate in previous eras, when the distribution of the continents and the coastlines which restrained the ocean currents were different, could have been very different from today.

To make an ocean–atmosphere model which would represent the part played by the enormous inertia of the ocean is a very ambitious and expensive undertaking. The attempt by a group at Princeton University illustrates well what can or cannot be done [89]. The horizontal spacing of grid points for the atmosphere was as described above, and this grid was used at nine different altitudes in the atmosphere. The grid for the ocean was spaced at twelve depths in the vertical and at intervals of 4.5° degrees of latitude and 3.7° of longitude. The mixing process in the ocean was represented by a diffusion parameter for horizontal eddies, which is arguably misleading in some respects, but is probably about as good at representing the effects of the eddies in the highly stratified ocean as any other could be in this model, which is necessarily very crude anyway. For the ocean coastline and depths together with the mountain heights and the way in which the ocean and atmosphere flow over and around them is still no more than a plausible caricature of reality. For instance, to parameterise the sinking of salty warm water below the almost fresh Arctic water it would have to note that water has a maximum density at 4°C, and therefore it is not easy to see how the bottom water might be colder than that except by conduction and radiation. But salinity is a more important determinant of density in the likely temperature range, but the same parameters cannot be used to represent the transfer of both heat and salinity because heat is more effectively transferred than salt. So bottom water is more probably salty than at a temperature of about 4°C. To state these problems is not the same as discussing how they may be parameterised, which must be an arduously long job to complete.

17.3 MEASURED FACTS

A number which has been given by climatologists the title of, and defined as, 'global temperature' has been derived from actual weather records from about 1800 to the present. Cores of ice in Greenland and Antarctica, and a few cores from ocean beds, have been used to give an indication of the temperature and some of the ancient life forms deposited in those particular places, and they give very useful information going back nearly 200 000 years; but without knowing much more about how the weather was distributed in those far-back eras long before accurate measurements began to be made, a good guess at the value of a single temperature to represent the whole globe is scarcely possible if based only on these widely spaced core samples. So we depend on at most about a century of acceptable measurements or a series of numbers obtained from the computer's figures, which is easy—given a clear definition. Even so, useful information about floods, failures of normal crops and records of severe frost or even wine-growing is contained in reliable historical records, and significant deviations from normals or averages can be crudely deduced.

During the century since 1880 the measurements indicate, according to a particular definition by climatologists with whom a mere dynamicist would hesitate to disagree, that the atmosphere has become about 0.5°C warmer, but has oscillated up and down by almost that much from one decade to the next. The Princeton coupled model [89] was run forward for 1000 years to see what sort of fluctuations of global climate might be expected starting from now, and assuming that no outside forces changed the circumstances. Several statistical tests were applied to this 1000-year prediction and these showed that the fluctuations were very similar in scale and frequency to the variations in the actual records of the last 100 years. Thus the model can be taken as not unrepresentative of the present climate. The 10-year measured fluctuations can be compared visually with the computer's millennium in Fig. 17.3.i.

The warming trend in one century of historical record of average world temperature (+0.5°C in 100 years) is reasonably thought to have been due to the greenhouse effect, or some combination of causes which most probably included the greenhouse effect. Anyway, whatever the cause, the comparison is thought to show that this trend is not a normal fluctuation but is the result of a **contemporary cause** of warming.

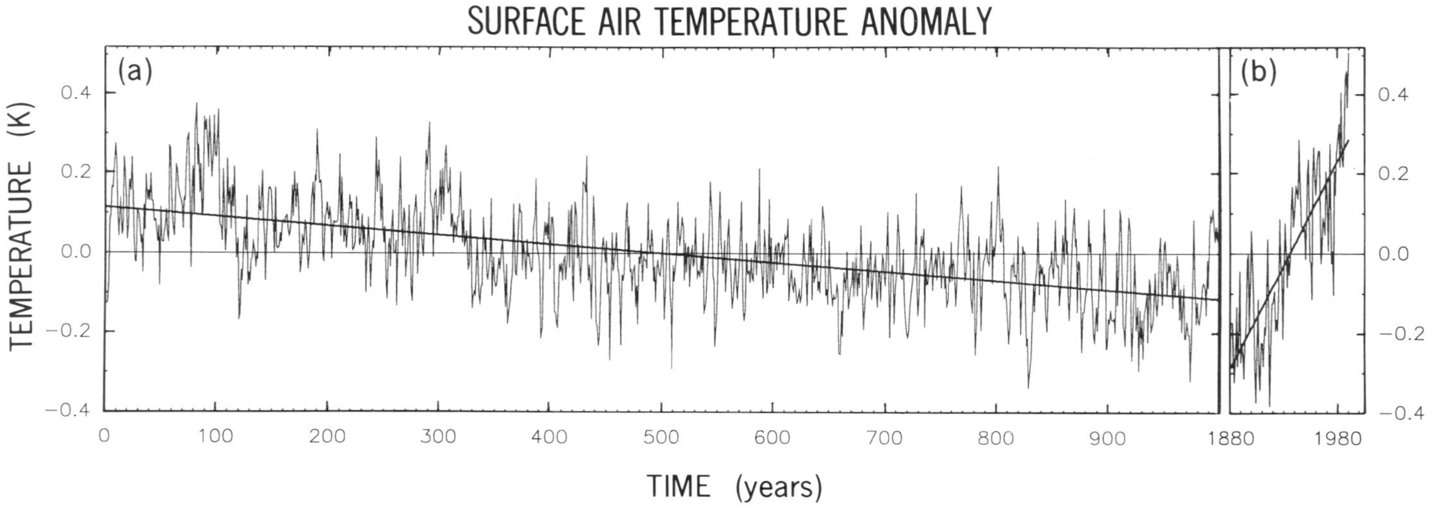

Fig. 17.3.i Surface air temperature anomaly (a) Globally averaged annual mean surface air temperature during 1000 years predicted by the Princeton combined ocean–atmosphere numerical model. This, of course, is not an actual prediction because we know that chaos destroys its value as a precise forecast after a few days; but over periods of a few years the model records changes which are very similar to the real combined system on the earth. What we see is the computer model's climate. So this is a 'time series' in which the number called mean temperature is generated under influences which were designed to be like the mechanisms which actually determine that number in the real world. Its oscillations are statistically similar to the real thing as actually measured with thermometers and so the parameters which make it generate each year's number are thought to produce the same result (statistically) as the mechanisms of which they are an analogy. (b) The graph of global temperature derived from actual temperature measurements made between 1880 and 1991. The statistics of this time series are similar to those of the model except for the trends shown by the straight lines through the middles of the series. The cooling in (a) shown by the trend (−0.2°C in 1000 years) is thought to be due to the imbalance of the ocean and atmosphere at the beginning of the series, at which time (i.e. now) the ocean lagged behind the atmosphere, and was more slowly becoming warmer, and therefore continued to extract heat from the atmosphere. [Reproduced with permission from Prof. Manabe and *Nature*, **367**, p. 635, 17 Feb. (1994).]

17.4 INTERPRETATION

In order to make use of this result it was noted that while the records showed a rise of world air temperature of about 0.5° in the last century, the model showed a slight downward trend of around 0.25° in 1000 years. It is thus credible to ascribe the rise of the last 100 years to the greenhouse effect because the amount of carbon dioxide has been steadily increasing for all of that time. Actually, the carbon dioxide began to increase many decades before 1880, and so the atmosphere must already have been warming, and would have been changed ahead of the ocean which has greater inertia and is warmed by the atmosphere. Thus the model ocean was lagging behind the model atmosphere, which must have been in the process of being cooled by the ocean during all of the 1000 computed years as a result of starting with the real conditions as they are at present. Thus the calculation started with the model slightly out of balance.

All that can be said with confidence from this exercise is that the present rise at the rate of 0.5°C per century in the atmosphere is not incompatible with a climatic warming due to an unspecified cause of climatic change. We ought not to be surprised, and might even feel very clever, if that cause turns out to have been greenhouse warming, provided that we can be sure there was no other possible cause. That is not a very firm conclusion but we should note that it is well within the range of warming rates suggested by other calculations, and will probably be taken by journalists as a very precise statement of the magnitude of the warming that is actually taking place due to the greenhouse effect, and that it confirms the forecasts made earlier.

This helpful result merely states that the temperature rise of the last hundred years is more than would be expected to occur within the fluctuations that a steady climate would undergo, and so, presumably, some particular cause must have been operating. Anyone well briefed in the psychology of public reactions to specific predictions about the environment will understand that the public see it as confirming what was already believed to be a confident forecast made several years previously. Thus the public tend not to notice the detailed qualifications made at the time of the previous 'forecasts'. The US vice-president, Al Gore, will be able to justify the policy advocated in his book *Earth in the Balance* [90], in which he wrote as if he believed that the forecast of global warming should be treated as a very confident prediction, in spite of the expressed uncertainty about the probable warming to be expected.

Perhaps politicians have to be unequivocal in order to be effective, but that does not turn their naivety into reliable prophecy. It is too easy to believe that when an exercise in soothsaying turns out to have been followed by events it indicates that the prophet is usually to be believed. When they are wrong—forget it, or think of another reason why things went the way they did. When they seem to be going in the expected direction, that is the important time to think of all possible alternative causes!

If we follow the most reputable philosophers of science, we could say that this investigation of the forecast of climate change failed to prove that the expectation was an error, but did not prove that it was correct. What is needed now is more, but different, tests. Any test which shows that the recent rise could not have been due to the greenhouse effect will represent an advance in human knowledge. Equally, any test which shows that any of the other possible causes we may think of were not operating and therefore could not have been the cause will also represent an advance. But first it is necessary to show that the property of the model, which prevents it from predicting periods of 100 years in which the atmospheric temperature can rise by as much as 0.5°, is actually a property of the real atmosphere, for otherwise the deductions from this numerical experiment are not valid. We should, therefore, remind ourselves that the best daily forecasting models also have the property that they have been deliberately made incapable of forecasting some extremes, in order to avoid forecasting them too often. They may have been made incapable of executing long-term fluctuations of a few hundred years caused by natural events.

While possible climate change is a major concern of today, and is being used as a reason for quite specific action to limit the burning of carbon fuel, it is important to investigate the theoretical limit of computer model forecasting, especially because the grid used in the combined ocean–atmosphere model is very crude and allocated only two observation points to the whole of Great Britain. Everyone who has lived in that island for a few years is aware that the climates of London, the Isles of Scilly and Inverness are very different! It is important to note that there are periods up to about 80 years in length in the 1000 years displayed during which the model climate did appear to be warming but which did not halt the slow decline of the straight-line mean value drawn in the diagram.

We must form a valid opinion about how much improvement in short-term forecasting is likely to be obtained by having bigger and more powerful computers than we have at present. Unless we know more about what has caused climatic changes in the past we shall not be able to test the evidence which we are now collecting as an indication of what may be causing any change that may emerge in the near future. Other chapters have dealt with the handling of the enormous amount of information that is collected from all over the world in order to improve the accuracy of our forecasts for the next few days. And we must understand the implications of chaos theory for the validity of all forecasting, and ask whether we are really facing the most important feature of our rapid exploitation of the environment. The suggestion of this experiment is that in the actual record of the period 1880–1980 we have evidence of a trend indicating that the most recent century has been warmed by more than the natural fluctuations likely within an atmosphere which was subject to no environmental changes. To put it another way, the warming of the last century was caused by some changing outside influence; it is reasonable to say that the increase of greenhouse gases, which we know to have taken place, may (but possibly may not) have been that cause.

Should we be treating other matters as more urgent? Is there any effect of the warming trend which the international community ought to take effective steps to avoid? A firm answer must await the discussion of this and other questions posed in this book.

An argument for caution, and action now to halt undesired change, is that if the change produces disaster we may discover its reality too late to halt it, for an effect that takes a long time (by human chronology) is likely to take a long time to put into reverse.

17.5 MOONSHINE

It is generally agreed that the albedo of the earth is about 30%, and that clouds cover about 60% of the surface. The clouds reduce the global average rate of absorption of sunshine by about 50 W m^{-2}, but at the same time the clouds have a greenhouse effect equivalent to about 30 W m^{-2}. Therefore the net effect of clouds is a cooling of about 20 W m^{-2}. These figures have large seasonal and regional variations [91].

A doubling of the CO_2 content of the atmosphere would add about 4 W m^{-2} to the energy budget of the climate system so that any feedback from the clouds which already have a budget equal to five times this amount may be the most important feedback. At present it is not even known whether any change in the effect of clouds is positive or negative.

Several models designed to compute the global warming by doubling the CO_2 are very similar in all respects except in their treatment of clouds; and they predict a global warming of between 4°C and 5°C as the highest and 1°C and 2°C as the lowest estimate, according to the different cloud parameters.

It is generally agreed that the warming by the increased CO_2 during the most recent century has amounted to about 2 W m^{-2}. Any recent talk about cloud amount has been obscured by talk about the particulate aerosol generated by industry, which is said to have caused about the same amount of cooling. When a bit of cooling due to dust etc. in the atmosphere has been included in the models, the magnitude of the warming predicted has been reduced. Not surprisingly. Even at the present stage the possibility of the cooling being of greater magnitude than the warming cannot be ruled out because of the contribution of several volcanoes to the dust load in the atmosphere; also it might happen that a warming altered the general circulation in such a way that it carried the heat to the poles more efficiently and caused the climate to be colder in some places. Not long ago, volcanoes were [92] considered to have been a major cause of climatic change. A consultation with the reference given may be found to be very persuasive!

Satellite remote sensing has not yet been shown capable of providing a useful estimate of the earth's 'normal' albedo from which its changes might be deduced. A 70-year-old idea that the magnitude of earthshine might be deduced from the intensity of the illumination of the dark part of the moon, which is clearly visible to the naked eye on a clear new moon, was worked on by the French astronomer Andre Danjon in 1925 and his student, J. E. Dubois, through the 1950s. Gary Taubes (*News Report*, [93]) writes, 'Decrease of the albedo by as little as 1%, for example, and global temperature will rise by 2°F. Conversely, change global temperatures, and the albedo might change in response, resulting in a feedback that could reduce or amplify a climatic shift. For this reason, trustworthy long term measurements of the albedo are vital to understanding greenhouse warming.'

Donald Huffman of the University of Arizona has taken up the challenge to develop the earlier work. By modelling the physics more carefully than was earlier attempted, Huffman has announced that the measurements of Dubois 'show years-long variations in the albedo that are much larger than can be explained by noise,—large enough to have left their mark on the global temperature record'.

If the project is successful it could provide at the very least a clue as to the feedback of global albedo. We have seen in the record of global temperature used in the exercise just described that there have been variations with a much greater rate of change than those predicted for global warming over several individual decades. The above argument offers an acceptable explanation of the observed variations. The very similar variations in the

1000-year climatology displayed by Manabe's computer model could be caused by the same mechanism, if there is a mechanism, in the model which makes the cloud amount vary, because that would provide the right kind of feedback.

The fluctuations in the climatology produced by the model are not explained by calling them 'random fluctuations' or naming them 'chaos', because the physical system embodied in the model produces chaos anyway, and we must remember that a chaotic pattern is still in accordance with the physical laws expressed in the computer model program. In this case we may have a model which offers a good mechanism for causing the fluctuations. But I am not confident that such a crisp conclusion is valid unless the maker of the model can say that there is no other factor producing such a very effective negative feedback.

The above idea seems to offer the believers in Gaia an explanation as to why the climate has been so stable since the last ice age, but it will not satisfy them because it is not a mechanism centred on the biosphere. There is an irresistible attraction for them in proposing that the changes in clouds which provide this stabilising negative feedback are due to changes in the supply of cloud condensation nuclei. It will be very interesting to know whether the fluctuations recorded by Dubois show variations which mimic recorded variations of global temperature used in the Princeton experiment described above.

17.6 ALBEDO OF CLOUDS AND HAZE

A serious criticism of investigations into the radiational properties of clouds is that they use the far out-dated names, such as cirrus and stratocumulus, as if all clouds, for which that name seems to be the appropriate one, had the same properties. Thus they seem to assume that they all have the same drop or particle size distribution. Of course the thickness of the cloud is sometimes presumed to make a difference, but always the same difference. In reality every cloud passes through an evolutionary history with its size and strength of updraughts and duration, and even the time of day and its temperature range, making a difference to its radiational properties. The reason for the complexity of the situation is that the scattering and absorption properties depend on the ratio of particle dimension to wavelength. There is no single, effective, representative particle size which produces the same overall effect as the actual particle size spread. A given particle scatters a given wavelength mainly according to the ratio of its dimension to the wavelength. That scattering is indicated in Fig. 13.8.ii on page 434. This is a purely scaling effect; the absorption depends on the wavelength and the distance through the particles traversed by the ray. Thus a light ray is scattered to produce part of a rainbow system if it traverses only one water drop between the sun and the viewing eye, whereas an IR or near-IR ray may be completely absorbed in only one or two larger drops. Light rays can cause a dense cloud (dense meaning that a straight line must pass through very many drop(let)s whatever its direction) to appear bright white; while IR rays are completely absorbed. Thin cirrus or medium-level clouds may look bright white when seen from the opposite side to the sun but as dark brown when the illumination scattered towards the observer (who is typically in an aeroplane above the cloud) originated from the ground while direct sunshine on the clouds is scattered in downwards directions.

These complexities are important particularly when interpreting satellite observations of clouds, and if we consider the heat budget of the layer of air below a cloud layer it is necessary to include the absorption of IR in the 1 μm to 2 μm range by the water vapour present, the latent heat of drizzle particles falling into the clear air below cloud and evaporating, and, one may say without being frivolous, any other similar mechanisms which affect the air temperature and density. Among these is the albedo of haze, which is very widespread, and has been made into a more serious problem which is not connected with the albedo: forest fires cause a lot of haze, and are generally thought of as in need of extinction and prevention. Tree species in areas where naturally occurring fires are caused by droughts have evolved to survive these conditions by requiring a fire to make the seeds fertile. This prevents the forest from becoming very thick and preserves the species. In recent years there have been fires, in builtup areas of both Victoria and New South Wales, which have been more serious than if they had been allowed to occur naturally. Enthusiastic environmentalists had advised against clearing away the litter annually (to tidy the place up) because it was thought that the trees needed the nutrients provided thereby. But the result was the storage of an enormous pile of fuel for the fire.

There is now a serious problem in estimating the albedo of haze and there has been an inclination to assume that most haze is man-made and quick to disperse, and the albedo thus not allowed for is treated in parameterisations for a model as if it were due to clouds.

The effects of clouds and particulates cannot be separated in any test for the parameters used to represent them. The difficulty is illustrated in the comparison between natural and artificial haze in Figs 17.6.i and ii.

Fig. 17.6.i 0930,16.12.78. The forest of E Burma seen from the air.

Fig. 17.6.ii A haze, presumably caused by human activity, seen from the air in the Jura area of Western Europe. If we try to find new parameters for a model to represent the changed radiation fluxes no measurements of the accuracy of forecasts can distinguish, on a global scale between new artificial haze and cloud changes in a form of feedback in a climate change.

This may appear to be a counsel of abdication from model making, but of course it is not. However, the idea of getting information from the earthshine on the new moon or any other method of measuring the variations in time of one of the radiational exchanges is bound to help in detailed understanding of what affects what, in this complex scene. Enthusiasts for the Gaia idea are well advised that cloud condensation nuclei may be important in a few special cases such as those which produce ship trails, but that these are so predominantly in the ocean deserts of the Pacific that it would be a very expensive and possibly ineffective exercise to attempt to provide nuclei from an artificial source, with a view to affecting the earth's albedo.

Forests may be a source of haze produced either by the terpenes which form by aggregation of the molecules into a blue haze, or smoke from forest fires. In either case the haze becomes more dense under an inversion formed at night in the valleys and the haze obscures vision when seen at a low angle, as in Figs 17.6.i and ii.

Perhaps the most exciting possibility is to provide the ocean biota with whatever they need to make them accelerate the absorption of CO_2 into the ocean. An interesting example is the supplying of iron (described by Dr Andrew Watson of the Plymouth Marine Laboratory, on the American research ship *Columbus Iselin* in the *Guardian* in November 1993). It is thought that the iron is needed if the algae are to absorb CO_2 from the atmosphere. What is often called *the carbon cycle* is really a one-way process whereby the carbon exuding from the rocks, largely in the form of methane (CH_4), which has always been a problem in deep coal mines, eventually becomes coral or limestone deposited on the sea bed [94].

In all these problems which concern any alteration of the atmosphere caused by human activity it is difficult to forget that an atmosphere composed mostly of oxygen and nitrogen, rather than a mixture of compounds containing these two gases and others and particularly hydrogen and sulphur, has been transformed entirely by life in such a way that evolution has now filled the biosphere with animals and fishes, most of which require dilute oxygen to stay alive. For that reason the next chapter discusses why ozone has appeared to take so much of people's attention in the last half century.

17.7 THE ROLE OF THE FORECASTER WITH BIG MODELS

The mechanisms described in section 4.7 illustrate well the kind of reasoning used by forecasters before numerical modelling had been developed. It has often been said that L. F. Richardson in the 1920s had envisaged that such thinking would be replaced by more precise calculations, but that the mechanisms would not be far from the mind of the forecaster. But the models have become so complex that first of all it has become necessary to feed the observational material directly into the computer, and from them the calculation first of all produces a set of synthetic observations at the grid points of the model.

The observations are plotted mechanically onto a chart which the forecaster begins to analyse. The isobars and fronts which are contained in his analysis will never be exactly the same as another forecaster would draw on the same set of observations, but the analysis is no longer used to create a forecast chart although it is used to help him express his forecast in terms of the same weather elements—wind, temperature, cloud amount and heights, visibility, rain, etc. The calculating machine uses the laws of mechanics and physics to produce a forecast of the position of the isobars expected in the next chart, and at several 6- or 12-hourly intervals from then onwards. The forecaster will not have gone through a reasoning process to make his own prediction of the same series of future charts, and would have to have some quite exceptionally good reason for not accepting the machine's forecast: but he will have to clothe that 'cold' calculation with the 'warmth' of details about the weather that will be produced by the conditions of motion and temperature described by the machine.

The latest attempt to use a computer to discover whether the conditions imposed by the action of mankind are, or are not, causing a change of climate involves the authors of the previous exercise, described above, together with those of several other inter-nationally reputed institutions which produced three different GCMs and compared their results with a selected range of more recent years (1963–1987) of recorded weather with the same general criteria to see how close the models are to the actual trends that can be discerned.

Their paper [89] is not for anyone not already well versed in the jargon and objectives of computer modelling of the ocean–atmosphere system, and with the statistical techniques which have to be used to make sure that the variations described by the models are typical of the real system.

In this case the effects of ozone and its depletion were carefully included in order to find out how important ozone is as a greenhouse gas. In the UVB range, O_3 is a strong absorber of sunshine but the amount of energy in that range is a very small fraction of the total. But in the IR emissions from the earth there are some very significant ozone absorption bands which are not strongly absorbed by CO_2 or water vapour, the most notable being at 9.8 μm.

Their conclusions are expressed in very guarded language because, while they may have lighted on the true explanation of the recent trends, there may be other important causes of change not represented in their models. In particular they used the same parameters to represent the clouds, which assumes that the climate change processes have not changed the way clouds are generated and evaporated, and it is widely recognised that all the **effects of the greenhouse gases amount to less than the uncertainties in the representation of clouds**.

They also do not allow the weather to alter the sea temperature, which is therefore assumed to vary in the same way as before through the seasons. The importance of this is that they assume that the weather types have remained much the same as before. Thus they do not claim to have discovered where the chief warming has begun to take place. It is quite possible that while some places become significantly warmer, others may become colder although there is warming on the average. One 'fear' that workers in this field have is that the warm 'Gulf Stream' might cease to go as far north as at present: it only goes beyond the north cape of Norway because there is a return southward flow at the bottom of the North Atlantic which is described in the next chapter. Without this return flow, the warm water could not go as far north, but would be diverted down the coast of Portugal and North Africa, and Western Europe would be much colder all the year round.

The final verdict is that probably a significant part of the warming trend particularly in the upper troposphere and the cooling in the lower stratosphere IS due to the greenhouse gases and the changes in the amount of ozone should not be ignored as of little importance. Crudely, they say that the similarity between the model predictions and the observed trends is unlikely not to have had a cause and effect relationship.

17.8 THE CHIEF OUTSTANDING DIFFICULTY: PARAMETERISATIONS

As stated in the opening part of Chapter 9 the subject of this section has become a rather special aspect of turbulence in the context of this chapter. The scale of clouds is smaller than any distribution of a model's grid points or of observing stations is likely to be able to describe. But we have not merely cited a basic meteorological example of turbulent motions, we have seen that particulate haze plays a role of the same kind, in that it is an important factor in the radiative exchanges which ultimately determine almost everything else connected with the weather and climate; but that role is not influenced by condensation and evaporation of water which are dominant in clouds.

Haze is dominant in the processing of sunshine, but like clouds it is not definable from the behaviour of one or two cases. Its chief role in the global warming equation is in its power to direct sunshine back into space. But the effect of a haze layer depends on the elevation of the sun and therefore has a basic diurnal variation. Any model of it needs to describe the particle size and vertical depth of the layer and this has not been done in any model other than the crude assumption that volcanic dust is in the stratosphere.

Beside volcanoes there are forest fire smoke and desert dust raised as in Fig. 13.9.iii, and on Mars and industrial haze mainly located in centres of large population. Yet the only kind of haze included on the global scale by modellers has been sulphate particulates, which mainly originate from the ocean [*15]. These few words describe an almost unexplored research field because most research into aerosols has been oriented towards their effect on clouds (and visibility), and not as a global factor interfering in the greenhouse effects. The state of this branch of the subject is well summarised by Harshvardham in Chapter 3 of the book by Hobbs [91].

As already mentioned sulphate haze has only recently been included as a global factor, but it has been widely noted that what was suspected before World War I is fact; that particulates do affect the global heat exchange mechanisms: that is why volcanoes were seen as a major cause of climate change.

The outstanding difficulty is that no theory for use in a global model can be based on validated parameters because the effect is crudely global and cannot be disentangled from the still largely undefined effect of clouds. This is explained by Somerville [91] in his review of Hobbs's book.

18

Ozone

18.1 STARTING POINTS

Concern about ozone (O_3) is quite a recent phenomenon. It used to be thought that since the body needs to inhale oxygen (O_2) it would be even better to inhale ozone. It has long been said that fresh air is good for you, and the so-called fresh-air factor, which kills germs and improves health, remained unknown, although it had been associated with countryside and more especially with seaside air. In the later 1930s it was alleged that ozone was the factor, and at exhibitions a machine was demonstrated which made sparks and imparted a smell to the air to make it reminiscent of the seaside. It was alleged that the odour was ozone, although it was more probably NO_x (which is a shorthand for a variable mixture of NO and NO_2), which may be poisonous but is stimulating to breathe in very small quantities occasionally.

More recently, the fresh air property has been attributed to OH, the radical which is found in the air, particularly in sunshine, and is found to be involved in many photochemical reactions.

Smog, a mixture of smoke and fog, was plaguing many cities in the late 1940s. There was also serious pollution in the air of Los Angeles, which the inhabitants called smog although its make-up involved neither smoke nor fog. It increased as a layer which could be viewed from above on tall buildings as the morning rush-hour developed and was gone again by the next dawn but soon reappeared. It was obviously associated mainly with motor traffic and was found to contain a very fine particulate haze, peroxyacetyl nitrate (PAN), together with various smaller proportions of acrolein, formaldehyde and sometimes even peroxybenzoyl nitrate (PBN), which is carcinogenic. These were all **secondary pollutants**; and the components necessary for their formation were found mainly in car exhaust, and were oxides of nitrogen (NO_x) and hydrocarbons which were emitted by cars because about 4% of the fuel remained unburnt in the very rich fuel/air mixture. Hydrocarbons were also released into the air at filling stations and many other activities in a city of which dry-cleaning solvents were the most obvious example.

The chemical reactions needed sunshine to produce these secondary pollutants, which are therefore called photochemical. They are very complicated and the relatively new subject of **photochemistry** has developed rapidly. The NO_x is produced in any combustion of fuel in air, which is about 80% nitrogen, but is worse when the air/fuel ratio mixture is variable within the mixture and at a high temperature and pressure. On cars the catalytic converters complete the oxidation of the fuel and partially reduce the NO_x. In gas turbine aero engines it is reduced by combustion chambers with better mixing and the continuous flow leading to the elimination of the hot high-pressure spots which are harder to remove in spark-ignited piston engines.

18.2 GROWTH OF PHOTOCHEMISTRY

Photochemistry is concerned with incomplete reactions which depend on the concentrations of the reactants and the rate-constants, which have to be discovered experimentally in the laboratory. Consequently, many new compounds are created and take part in the reactions which are many and simultaneous. It was a surprise therefore to be told that, although the presence of NO_x generated ozone in the Los Angeles photochemical smog, its presence in the stratosphere would, through a series of photochemical reactions, destroy ozone. Therefore it would be very dangerous to fly aircraft extensively in the stratosphere because it would remove part of the 'ozone layer', whose presence caused the absorption of the UV component of sunshine and particularly the wavelengths shorter than 0.3 μm (UVB); and if this radiation penetrated to sea level it would cause an increase in human skin cancers. The damage to other life forms was scarcely known, but it had to be presumed that it might be harmful because all life forms had possibly evolved without being exposed to UVB. Also DNA seemed to have a structure sensitive to UVB.

The increase in skin cancers was expected to be around 10%, but when the medical profession was asked about the seriousness of this they pointed out that since the 1930s there had already been about a twelve-fold (1200%) increase caused by a great increase in the exposure of the skin to sunshine: people had taken little notice of the medical recommendation to regulate sun-bathing, particularly in mid-winter when the skin is unprotected by the development of a tan. However, the use of the word 'cancer' by the opposition to the creation of fleets of supersonic airliners was deliberate, and the campaign was carried on with a 'better to be safe than sorry' approach. It was also argued that it would be a triumph to persuade governments to take action essentially on environmental grounds.

The objections to stratospheric aviation were mainly directed against proposals to build supersonic passenger airliners, but on grounds of the noise and shock-wave bangs. There was also an objection to the supersonic aircraft on the grounds that it had about twice the fuel consumption per passenger mile of the subsonic aircraft. But this latter was still only about one-tenth of the fuel used by the big ocean liners per passenger crossing of the Atlantic! Interest in getting research grants for photochemistry was also a concern. The concentrations of chlorofluorocarbons (CFCs, commercially named 'freons') measured around the world were such that all the freons ever manufactured seemed still to be present in the atmosphere. Thus they were indestructible in the troposphere, but would in due course be carried up into the stratosphere where they would be decomposed by UV radiation and their chemical elements would become the raw material for new photochemical reactions. These would include chlorine, which could play a role similar to NO by taking on oxygen atoms and using them to convert O_3 into O_2 in sunshine.

There are two reminders that must be made at this point: first, the decision not to build large fleets of supersonic airliners was made mainly on commercial and political grounds and noise problems. There was undoubtedly a nationalistic feeling in the USA against a European supersonic aircraft being the only one flying in their country, a feeling supported by the American aircraft makers. Secondly, only 15 Concordes were ever built by the Anglo-French makers and the environmentalists claimed that this low figure was as a result of their campaign. The ozone-destroying effect would be catalytic, meaning that the Cl and NO_x would not be used up and would continue to cause destruction of O_3.

The forecasts of the photochemical reactions were based on computer models with assumptions about the mixing processes and reaction rates at stratospheric pressures and temperatures. As each newly proposed chemical reaction was included, the atmosphere was found to be more robust than previously thought. The presence of the ozone in the stratosphere in the first place had been explained on the very simple theory that the UV which did not reach the troposphere would cause the disassociation of the two atoms of the oxygen molecules to make atomic oxygen, and that some of these would combine with other O_2 to form O_3. Thus there would be both odd and even oxygen molecules always present: and the molecules containing three atoms are more absorptive of both solar UV and earth IR radiation. This absorption causes a warming of the stratosphere at about 50 km altitude to temperatures about the same as at the surface and it also means that ozone is a greenhouse gas which absorbs long wavelength IR. But the UV of sunshine would warm the bottom of the atmosphere a very little if there were less ozone; thus the destruction of ozone is not a very significant factor in global warming.

With more complex photochemical models the effect of NO_x in destroying O_3 turned out to be less than previously thought, and it became obvious that there had always been some NO_x in the stratosphere. N_2O, which is generated in the soil when vegetable waste decays, is decomposed by UV and is a natural source of NO_x in the stratosphere. It seemed obvious that the widespread use of nitrogenous fertiliser (NPK) on farms would increase the production of N_2O and add to the O_3 destruction as much as the NO_x from aircraft, cars and industry. The use of this fertiliser in excess was already being criticised because it was causing the presence of unhealthy amounts of nitrates in rivers used for water supplies, and that had raised the question whether it was worth removing the nitrate from public water supplies, most of which was used for washing where the nitrate did not matter. The suppliers of bottled mineral drinking water gladly joined in the campaign to emphasise the unhealthy aspects of nitrate in drinking water.

18.3 DIFFUSION MODELS

The model making used to calculate the photochemical reactions in the stratosphere raised several contentious points. It had been supposed that any chemical substance, particulate or gaseous, released at the earth's surface into the air would, in due course, be diffused up into the stratosphere unless it was rained out from the troposphere. Thus rain not only prevented almost all the water vapour which came from the sea or wet ground from reaching the tropopause, but the same was true of most water-soluble gases such as

SO_2, NH_3 and any salt or smoke particles which acted as condensation nuclei for cloud droplets. Radiosonde balloon measurements regularly showed the dryness of the stratosphere, and at the same time measurements of CO_2 showed that it constituted the same proportion of the air above and below the tropopause. Methane (CH_4), another very important greenhouse gas, was also carried through the tropopause and was therefore, like the others which got through, gradually well mixed into the whole atmosphere up to at least 100 km altitude. Like all other compounds, methane would be decomposed by UV at high altitudes in the stratosphere and the hydrogen thereby released became the origin of much of the water vapour in the upper stratosphere.

Hydrocarbons, like methane, refrigerants and solvents such as the chlorofluorocarbons and halo-hydrocarbons (CFCs) and carbon tetrachloride (CCl_4, a favourite solvent) would also be carried up into the stratosphere where the strong UV component of sunshine would decompose them so that with the presence of NO_x they would all contribute to the destruction of ozone. The main problem in model making was to include the rate at which they would be diffused upwards, and it was suggested that the diffusion rate could be calculated from measurements of other gases or vapours which were there anyway. This was a brilliant suggestion because there were other gases, and in particular HCl, and the models had to predict the concentrations of such gases so as to agree with the measurements of those which were much easier to sample because they were more plentiful and less reactive.

At this stage many 'one-dimensional' models were made. They assumed that the photochemical effects would be at least represented with the correct order of magnitude if the atmosphere was represented in a model by a single column of air. It was, of course, known that the mechanisms which carried all these gases into the stratosphere involved large eddies and that it was not really a matter of them being diffused up by the random motion of the molecules as in an aerodynamic boundary layer. But the same mechanisms worked for the gases for which fairly good measurements of concentrations in the stratosphere had been made, and it was naively assumed that the same diffusion coefficient would give the right answers for the new chemicals now being investigated.

Nature is not quite so simple. As more and more measurements were made, it became widely recognised that as they were made more sensitive and accurate, all components were found to exist in many surprisingly thin layers with thicker layers of lower concentration in between. This pattern of unmixed layers is caused by the fact that large volumes of typical tropospheric air are carried in bulk into the stratosphere where they are converted into very extensive thin layers stretched horizontally by the winds which were not the same at all heights. The large blobs of polluted air were sheared over like a blob of lubricating oil between two metal surfaces in relative motion. If the wind at the top of a blob was 10 m s^{-1} and 40 m s^{-1} at the bottom, which could be quite typical, after two days the air samples at the bottom and top would be 5000 km apart instead of one beneath the other. It is not surprising that a radar looking upwards in the Hawaiian Islands frequently observed thin but concentrated layers of SO_2 or smoke-like particles many kilometres above the tropopause. The details of these thin layers have been more reliably observed using laser lidar. The significant fact thereby observed was that these layers were not diffused into the air above and below them in the manner supposed in the models. Until this pattern of spreading of pollution in the stratosphere was understood and the observations believed, serious attempts were not made to obtain accurate measurements of the profile of concentrations at these high altitudes which are measurable only with difficulty.

The stretching of compact masses of gas could equally by caused by horizontal gradients of wind so that the vertically thin sheets would become narrow and shaped like tape.

It had been said that since several 'one-dimensional' models made by different people had given similar results there was therefore a **consensus** within the scientific community, and the results ought to be believed. The agreement arose because all the scientists had used the same (erroneous) model, and got the same (erroneous) answer unless their calculations were wrong. It is a criticism of the model-maker's ignorance that they did not see the obvious error that in the model no pollutant could rise up the one-dimensional column without having to react with any molecule of any other component of the air which happened to be diffusing its way down. This was obviously not true of parcels of gas either going up or coming down in the stratosphere in different geographical locations, and therefore all the models were likely to be equally wrong and the answers incorrect and not to be believed. The community of workers in aerodynamic research is to be criticised for not making the crudities of their diffusion theories better known. It is to the shame of scientists in general that when an author uses a (diffusion) technique in a published paper in a well recognised journal, that technique is not recognised as unacceptable in quite different circumstances. Thus a crude approximation to a physical phenomenon does not become a law of nature simply because no one has got a better approximation into publication.

The idea that there is special value in a 'scientific consensus' has been used by campaigning groups in many environmental controversies, and the idea that there is such a consensus is often argued by politicians and campaigners with very honourable intentions. They desperately need definitive answers: Vice-President Al Gore [90], whose book was widely welcomed by people who have genuine concerns and serious worries about the way our industrial society is moving, has no doubt that global warming is under way, and that the detailed predictions of what will happen are to be believed. Such anxious people may be unable to find an authoritative scientist who will confidently state the opposite of the alleged consensus. They feel entitled to assume that the supposed consensus really exists and is correct. But they are on a par with the cranks who have not been able to get an idea of theirs accepted for publication in any good scientific journal and then go on to argue either that there is a conspiracy of the establishment to suppress their idea; or that because they can find cases of a good idea actually unable to achieve immediate acceptance, but accepted a generation later, their own similar early experience proves that they, themselves, are also right.

Al Gore's book is written on the assumption that it has been indisputably proved that global warming is about to happen, with all the detailed awful consequences that have been bandied about in sociological conferences. My own comment is that he borders on telling himself that population growth and exhaustion of fossil fuels will, with much greater certainty than any global warming, actually happen; but it will happen much later on. Yet that will present much more serious problems for our not very distant descendants than any of the floods, droughts and sea level rises hinted at in the reliable literature. The simple unqualified AlGorithms ought not to be accepted as a basis of decisions.

18.4 GETTING INTERNATIONAL ACTION

Preaching the relatively mild inconveniences expected of global warming in Gore's book helps to make people realise what inadvertent harm we can do to the environment on which we depend; and the reduction of dependence on fossil fuels will help to ameliorate both fuel scarcity and population problems. A recent paper showed [95] that life in some arctic ponds has been changed significantly from previous millennia, beginning in the 19th century. These ponds are in an area (Cape Herschel, Ellesmere Island) relatively free from modern air pollution and water circulation and are probably associated with general atmospheric warming, but beginning before there was significant industrialisation and increased output of CO_2 and chlorine compounds.

The common view of a 'clean' pre-industrial environment is challenged by early deposition of lead in Swedish lakes, where levels of lead prior to its exploitation in Greece (700–400 BC) during the refinement of silver and in the Roman period in Europe and Asia Minor (300 BC–AD 350 approx), when smelting (producing about 800 000 tons/yr of lead) was so crude that 5% of the lead was emitted to the atmosphere [55]. There was an increase beginning in about 1000 AD in southern Sweden, but in northern Sweden the lakes remained as clean as before that time right until the late 1980s. Even so, the lead accumulated in southern Sweden in what we think of as pre-industrial times 'was as large as, or even larger than, the deposition during the comparatively short industrial time period'.

People may be ready to accept the assurances of economists that there will be plenty of time to deal with problems for the foreseeable future for, of course, we will not jog along with no concerns until the petrol pumps all simultaneously become dry on the same Monday morning. But pollution has grown 100 times in the last 2 or 3 millennia. The desperation of Rwanda and Bosnia are but tiny examples of the panic and ungovernability of a fearful crowd. Saint Augustine pointed out that 'government without justice is robbery', meaning that the boundary between government and a protection racket is very fuzzy. An unfed fearful crowd is without power and is a danger to itself, but indignation with power can be indiscriminately destructive. Injustice has been exacerbated terribly by a six-times increase in population in half a century in some areas, as in Rwanda, and in a century and a half in the world as a whole.

18.5 SCIENCE-BASED THREATS

To discuss our scientific problems without these terrible consequences being regarded as possible, even probable in the next generation, is like discussing a scientific problem with a computer using a faulty model, ignorant of the other branches of science and politics on which the model is based.

The ozone problem hit the Western world in the form of newspaper headlines predicting that we were soon to be 'fried up' by the UVB rays which would reach the earth's surface when the 'ozone layer' had been destroyed. This was probably a journalist's chosen paraphrase of what Dr Molina probably said with different words, and it raises some interesting issues. Without doubt Prof. Rowland and his colleague, Molina, were deeply worried about what might happen, and in a recent paper Rowland and others describe how in this case international action to reduce the manufacture of CFCs is showing commendable signs of success. They call for future problems to be tackled in a similarly effective way; but the short-term vision of economists and politicians is not helpful.

The ozone hole may be expected to shrink in the coming few decades [96].

It could be said that the effect of volcanoes throughout the geological ages had been to try out almost every photochemical excess, but the destruction of ozone has never been tried out as an explanation of the demise of the dinosaurs about 65 million years ago. There is always a temptation to suppose that environmental extremes will result from the effects about which one's own research gives one special insight. Scientists are certainly tempted to frighten those who have control of the research purse strings, and many scientists have said to me that it was obviously good tactics to use such fears to get financial support. In the case of ozone destruction and the ignorant use of one-dimensional models by people who were not well briefed about the mechanisms of mixing in the atmosphere, they could have noted the several decades during which a very prolonged effort was needed by people with good experience in forecasting gradually to produce models which made forecasts for the next very few days consistently reliable. Clearly such good understanding of the mechanisms in the stratosphere will require a similar period supplied with regular detailed measurements within the stratosphere before the photochemists will be in possession of good parameterisations of the mixing of the various chemical components injected into the air. For example, almost no work has been done on the spreading of aircraft contrails and tubes of high humidity left behind when the visible trails commonly produced have evaporated.

Suppose that a scientist does have a theory and has used the best available model put onto a computer, and out comes a startling result about which ordinary forecasters are at least sceptical or very surprised. Should he or she announce the new result in order to wake the world up to the danger and get headlines like the one just mentioned in the case of Dr Molina? It might be effective in getting the world of science thinking about a new problem or it could possibly get the science a bad reputation for crying 'wolf'. My own instinct is to examine the assumptions in the model used in the computer, perhaps to reduce the credulity with which people are inclined to believe 'what the computer says', knowing that a computer has done some calculations which even the brainiest mathematician could not have done in a lifetime. 'Rubbish in—rubbish out' is a better logical sequence with which to approach a computer.

18.6 THE OZONE HOLE

In the ozone case the first measurement of a significant reduction of ozone in the stratosphere was the discovery of the so-called *ozone hole*. This started a frantic search for its cause because it had not occurred in the region where ozone destruction had been predicted. The explanation offered was that the reactions had taken place where the spring sunshine began to shine on the **polar stratospheric clouds** (PSCs) in whose particles there were rather high concentrations of sulphuric and nitric acid. The clouds had been known for several decades to exist, and it was known that because they are wave clouds the air passes with the wind through them so that the volume of air exposed to the potential photochemical reactions is much greater than in most clouds, which remain in the same piece of air in which they are first formed. Also, there has been found much more chlorine and its compounds in the clouds and their immediate neighbourhoods, which had not been expected to have anything to do with the formation of the clouds. Sulphuric acid had been invoked to explain how the cloud showed iridescence characteristic of clouds of spherical particles (see colour section). Ice clouds do not produce iridescence, and the temperature of the clouds has been found to be at least as low as –80°C. The unexpected chlorine was explained as having originated from CFCs which had been decomposed by UV sunlight in the stratosphere.

Experience has not demonstrated that the photochemical reactions outside the clouds have caused as much ozone depletion as predicted, probably because the computer models did not have a model of the mixing mechanisms in the stratosphere which resembled reality. It is interesting that Professor Rowland, who first drew attention to the ozone-destroying properties of chlorine, was more attentive to criticisms than were others who had become campaigners, seeing themselves as his supporters. He found that atomic chlorine, and chlorine in the form of chlorine oxide, was playing a role similar to NO and NO_2 in the first described ozone-destruction process. He also discovered that they formed a compound which is described as **chlorine nitrate** ($ClONO_2$), and this removed both components from the destruction process. However, he was also the first to discover that $ClONO_2$ is itself decomposed by UV so that the two components were being put back again into their catalytic ozone-destroying role. But, being essentially a photochemist, he naturally accepted the expectations of the modellers' one-dimensional computer calculations of the destruction process, and it seemed to him that the world of politics needed to be warned about what chemical changes were taking place.

The PSCs, being standing wave clouds, have air passing through them; but they are not like filters, which would capture and collect the chlorine. The cloud condenses at its upwind side and the chlorine is collected onto the cloud particles, and as they pass out through the downwind side of the cloud they become attached to the cloud condensation nuclei ready to be at the centre of the droplets formed at the upwind side of the next stationary cloud through which the air passes. Seeing that the ozone destruction takes place in the clouds, the air in the layer passing through the clouds becomes cleared of ozone. Thus the ozone hole is not such that if you look upwards through it here is no ozone on your line of sight, because there are layers above and below from which no ozone has been removed but in which the concentration of ozone is less anyway. What the clouds achieve is to collect chlorine onto cloud nuclei so that whenever cloud is formed in sunshine on those nuclei, ozone destruction may take place as long as those nuclei remain in the stratosphere at those low temperatures colder than –80°C and in sunshine. The air from which ozone has been depleted may travel to other regions which in the past have been supplied with ozone-laden air which has previously occupied much of the air which occupies the darkness of the polar stratospheric night.

The clouds, by radiation into outer space, become the coldest regions of the atmosphere, and the very low temperatures sometimes as low as –90°C are required for the ozone-destruction processes which have produced the ozone hole. The surface of Antarctica is high mountains covered with snow which radiates as a black body in some wavelengths which are unabsorbed by the atmosphere. The surface, having a low-temperature start because of its high altitude, becomes the coldest place on earth (except perhaps for the famous frost hollow at Verkhoyansk) during the polar night, and when the PSCs are formed they receive less IR radiation from below than PSCs formed in the northern hemisphere.

Measurements have been made possible by the fear of catastrophic results from rashly planned industrialisation. The measurements include some from satellites which make possible spectroscopic analysis of light coming through the high atmosphere tangentially around sunset as seen by sun-synchronous satellites. The technique of unscrambling observations of radiation which has traversed great depths or lengths of the atmosphere have been well developed recently. They have made possible vertical soundings by analysis of the radiation at several different wavelengths which come predominantly from layers at various altitudes [98].

In [97], G. L. Manney *et al.*, an international team, have analysed the winter and early spring concentrations of several trace components of the atmosphere in order to distinguish the apparent changes due to the vertical motion of the air from those due to photochemical reactions between them. The effects of the motion are studied mainly by following N_2O, which does not take part in the reactions under study. It must be remembered that N_2O which is produced at the ground is largely responsible for the appearance of NO_x in the stratosphere, but that is because it is dissociated at much higher altitudes than the layers under study in this research. The reactions concerned mainly involve ClO and O_3 and take place in (on) the particles of polar stratospheric clouds (PSCs) at very low temperatures [99] which often occur over Antarctica but are rather rare in the Arctic. The southern hemisphere circumpolar vortex, which results from these low temperatures, is therefore more intense and is better structured and steady than the northern one because there the coldest surface is on the mountains of Greenland, which lies between 60°N and 83°N, and the ice at the pole is at sea level, where it is much warmer than in Antarctica.

This means that the PSCs are cold enough to support the photochemical reaction in which ClO destroys O_3 for a period which varies between about 2 weeks and possibly as much as 2 months after the spring sunshine begins to shine on the clouds in the north. In the south, however, this period is more nearly about 4 months and so the O_3 destruction is much more intense.

From this brilliant work, which involves scientists from California, Colorado, Edinburgh, Reading and Bracknell, it is to be expected that there will be increased O_3 destruction in the next few years as more chlorine is carried up into the stratosphere, although the change in the south may be difficult to detect because more chlorine compounds are generated in the north. There is more exchange in the north with the higher levels of the stratosphere where most of the ozone is generated, and the less regular motion patterns make it more difficult to measure the total loss. As in many cases of this kind where good observations were begun only recently, our expectations for ozone loss will be more reliable when 2 or 3 more decades of measurements have been accumulated.

We need many years more of observing the stratosphere before computer models can tell us much more about it. However, a recent internationally cooperative article on the effect of including ozone depletion in climate change calculations [*6] has made some interesting results available to those prepared to read the very involved description of what the calculations have achieved.

Theories that ozone will be greatly reduced must, of course, be held in mind, but the predicted consequences have been greatly exaggerated and have not so far been observed (people have become much more conscious of the incidence of common skin cancers and the deadly melanoma, but the earlier increase in their occurrence, before ozone depletion could have been caused by human activity, has become more widely known and appreciated, and some people are more careful about sun-bathing). However, there is encouragement in the agreement of governments to do something about an environmental issue. The ozone issue is trivial by comparison with the global warming and fossil fuel depletion problems. The increase of CO_2 in the stratosphere is a cause of cooling there, and this tends to increase O_3 production.

18.7 OZONE AT THE SURFACE

So far we have discussed ozone as a component of the air which achieves an equilibrium concentration in the stratosphere through a multiplicity of photochemical reactions in sunshine. Reference is customarily made to the **ozone layer**, but ozone exists throughout the whole depth of the stratosphere and the maximum proportional concentration is between 25 and 50 km, and the warming due to the absorption of sunshine and IR from below is a maximum at about 50 km. The maximum concentration per unit volume is in the lower stratosphere at between 17 and 25 km, and this contains most of the ozone, and is borne in mind when the ozone 'layer' is mentioned. This is the layer in which most of the ozone destruction occurs because the UV of sunshine penetrates down to this layer and, by decomposing various pollutants, such as the CFCs, enables the photochemical reactions to take place.

The exchange of air between the troposphere and stratosphere brings ozone down to the surface and it becomes used up as an oxidant on a variety of substances such as rubber and does some damage to growing vegetation. Most particularly it reacts with NO to form NO_2 in regions where there is plentiful car exhaust. Its typical concentration is 20 to 30 parts per billion, (ppb) but only becomes a seriously damaging pollutant at over 50 ppb for long periods.

When all the NO has been oxidised, catalytic reactions begin using NO_x and hydrocarbons in sunshine to generate more ozone which becomes a damaging secondary pollutant, and is an important component of new photochemical smog. This production of ozone in urban pollution, which is mainly car exhaust at present, can only be eliminated by reducing the emission of the primary pollution. But it is quite a different problem from the destruction of the (stratospheric) ozone layer which, we are warned, will permit the arrival of UVB radiation at the surface and increase the incidence of cancers due to excessive sun-bathing (which can be, and has been, caused even when there has been no destruction of natural stratospheric ozone). In order to avoid misunderstanding it needs to be stated that the UVB which ozone destruction allows to reach sea level is strongly absorbed by clouds and aerosols, and is therefore not a widespread danger to light skins except in prolonged sun-bathing, particularly in winter.

One of the odd results of the ozone at the surface coming in the first place from the stratosphere and gradually being neutralised as it descends towards the ground, particularly over land, is that there appears to be more of it over higher ground (around 30 ppb) whereas it is less concentrated over towns (in the morning because the O_3 is used up in converting the NO to NO_x). But later on as the pollution concentration rises, more O_3 is produced in the sunshine. The same phenomenon occurs around the mouths of Alpine tunnels: close to the exit of a tunnel, car exhaust fumes containing NO destroy the O_3 and so it appears that the fumes are doing a useful job. But further away, fresh O_3 is synthesised in sunshine to generate urban-type photochemical smog, especially in light winds.

19

The climate of the past: limits to modelling

A thousand ages in thy sight are but an evening gone. (Popular hymn)

19.1 SOURCES OF INFORMATION: CORE ANALYSIS

There are many opportunities to gather information about the past by making a bore hole where there has been continuous deposition from the atmosphere for a long time. The core is the material extracted from the hole, and it may be examined in several different ways, as follows.

(1) Isotope analysis

In nature, the isotopes of the common elements are found in the ocean, the rocks and in the atmosphere in proportions which have been measured with great accuracy. The most reactive and common are hydrogen, oxygen and carbon. Because these elements are found in almost all cores they are the most used for this purpose. The results are expressed as a departure from the common proportions, for example, the proportion of ^{18}O in the mixture with ^{16}O is expressed $\delta^{18}O$. The departure from the common value occurs where diffusion is a major mechanism in the incorporation of the element into the core material because the lighter isotope is more rapidly diffused and appears in greater proportion. This variation is then interpreted as a temperature variation because that is the most important factor in determining the rate of diffusion. In the case of condensation of water vapour into ice crystals there is considerable uncertainty as to what temperature is represented, but nevertheless it is assumed to represent the same temperature throughout the time represented by the core material. There is therefore uncertainty about the altitude at which snow was condensed and deposited at different times at a site in Greenland where the depth of snow is as much as 2 or 3 km.

(2) Ocean bed cores

The material in these cores may have been created close to the sea surface by the diffusion of carbon and oxygen into carbonate shells of foraminifera, which have a dimension of 1 or 2 mm. Their shells fall to the sea bed along with many other sorts of material after it ceases to be formed. During a warming period when there were many more icebergs released from the terminal snouts of glaciers the moraine-like material may be released from the base of the icebergs over a very long distance, so that the track of the ocean currents which carried the melting ice may be revealed. The age of the carbonate may be revealed roughly by carbon dating. Alternatively, organic material may have been created on the ocean bed.

(3) Pollen and mineral dust
This material may be used to indicate the weather conditions where the pollen or dust originated; and if the place of origin can be discovered with confidence we have clues as to the wind of the prevailing climate.

(4) Electrical conductivity measurements (ECMs)
It has been discovered that this quantity correlates well with other factors and may even be used to discriminate more detail than them.

(5) Analysis of trapped gas
This has been used particularly to study the concentration of CO_2 throughout the ages, and in this case it is assumed that the sample represents the concentration throughout the well mixed atmosphere. A similar study has been applied to methane (CH_4).

(6) Chemical analysis
This obvious method has been used to sample for chlorine or sulphur, which come largely from the sea, although some of this material may have been deposited because of volcanic activity.

(7) Location
The further into the past that information refers to, the fewer are the sampling points. In particular, the ice cores from Greenland and Antarctica have been very prolific and it has been important to discover how much they refer to global conditions and how much only to local weather.

(8) Accumulation
The rate of accumulation of core material both in ice and on the ocean bed has been found to be well correlated with other indicators of temperature change.

(9) Peat cores
Sites have been found where peat has accumulated over thousands of years, for example in places undisturbed by mankind in South America.

(10) Dust
The accumulation of dust may indicate the conditions (desert, volcanic, etc.) in places, probably upwind, at least in the same hemisphere unless the same is found all over. The dust may be carbonate which, however, may be coral or foraminifer shell.

19.2 HISTORICAL OUTLINE

(1) Geological

The earth has been in its present orbit for over 4 billion (10^9) years. Life has been traced back to over half a billion years ago (fossils in the **Cambrian**) and since that time evolution has produced biological species of improving quality and diversity. The era of the **dinosaurs** lasted from about 250 million until about 65 million years (Ma) before the present (BP) (i.e. **Triassic** until **Cretaceous**) the 'present' usually meaning the year 1950.

The **Phanerozoic** period (beginning with the Cambrian 550 Ma BP) begins with the **Paleozoic** (up to 150 Ma BP) in which life flourished in the sea, and the early air-breathing animals appeared. Then followed the **Mesozoic** (Triassic—Jurassic—Cretaceous, up to 65 Ma BP) dominated by the dinosaurs, followed by the present era called **Cenozoic** (**Paleogene—Neogene—Pleistocene**, the last of which together with the **Holocene** makes up the Quaternary). This includes the **Tertiary**, the period of great advance of birds and mammals (warm blood) and the building of the great mountain ranges. This saw a cooling of the climate up to about 1.5 Ma BP and was followed by the **Quaternary** period up to the present.

The Quaternary consists of the icy Pleistocene which has been studied in great geological detail, leading into the present warm **Holocene**, which is the name of the 11.7-thousand-year period since the last ice age. The knowledge of this more recent geological history is now understood mainly through mapping of the rocks near the surface and their fossil contents. The Quaternary is roughly the last million-and-a-half years during which mankind has come to be the dominant species and the environmental scene has been changed by erosive weather, coastal storms, meandering, gorge-cutting and delta-forming rivers. The evolution of animals, including mammals in the sea, many new birds and fishes, trees and insects, also moulded a more powerful biosphere.

The cores have supplied information only about the last quarter of a million years, as we shall see. The most likely extension beyond that will probably come from deltas which are still in the process of formation, such as those on the Nile and Mississippi, and they may take us back to the formation of stratified rocks.

The most interesting features revealed by analysis of the cores are those which enable us to develop a single chronology for places which are distant such as Greenland, the Indian Ocean and Antarctica. This is of special importance for what has been called the 'end of the last ice age', *about* 10 000 years ago, because the Greenland cores give a more accurate date as more like 11 700 years. Before that, a cold period, which has been given the name 'Younger Dryas', which began about 12 900 ago and before which there had been several interstadial (warm) events going back to almost 140 000 y BP. The information about the years back to that time is more detailed and reliable than that beyond the Eemian, and in order to work out the mechanism of change we need to know whether the variations were local or world-wide.

(2) Continental drift

Continental drift is now generally accepted as having taken place, and contemporary speculation begins with a large land mass in early Triassic times, at the end of the

Carboniferous period, which had been very productive of life in the sea, and swamp land which probably favoured the dinosaurs. (See Figs 19.2.i, ii, and iii.)

This land mass included Antarctica, India and Australia, joined at the southern end to Africa and South America with traces of the Northern continents variously detectable. By cenozoic times the Atlantic was half formed and Europe, Britain, Greenland and North America had begun to become separate and the two Americas were showing signs of

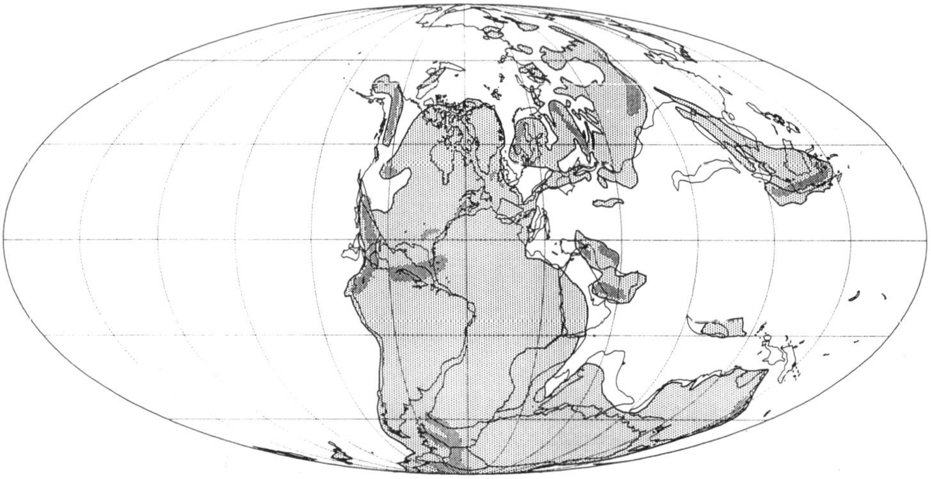

Fig. 19.2.i Map of coastlines about 250 Ma BP. The beginning of the Triassic. [Reproduced with permission from Atlas of Mesozoic and Cenozoic Coastlines, by A. G. Smith, D. G. Smith and B. M. Funnell, Cambridge University Press, 1994.]

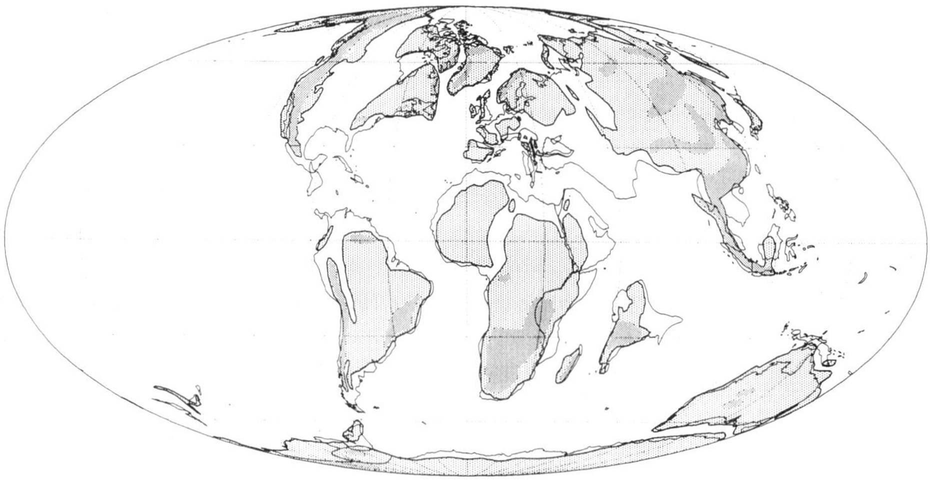

Fig. 19.2.ii Map of coastlines about the end of Mesozoic 70 Ma BP, about the time of the demise of the dinosaurs. The beginning of the Tertiary. [Reproduced with permission from Atlas of Mesozoic and Cenozoic Coastlines, by A. G. Smith, D. G. Smith and B. M. Funnell, Cambridge University Press, 1994.]

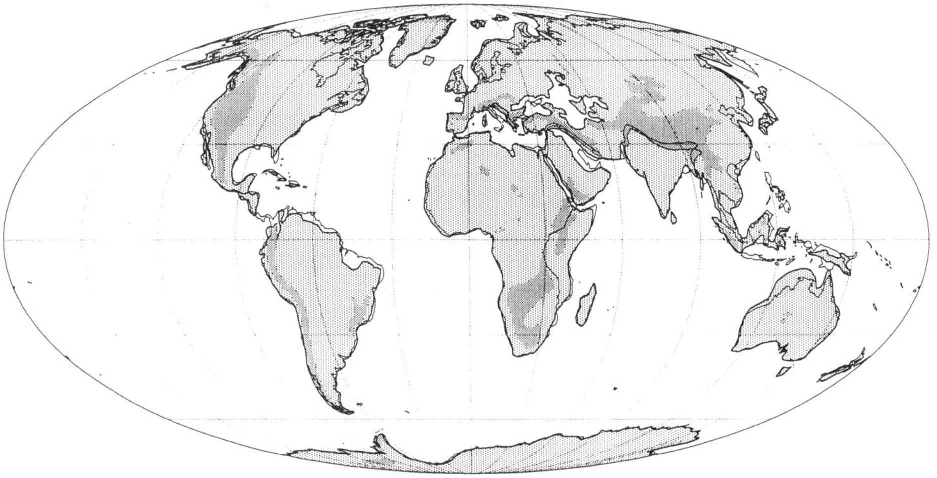

Fig. 19.2.iii The Pliocene coastline 5 Ma BP. Mammals growing larger and more numerous. Middle of great mountain growing, and cooling of climate. [Reproduced with permission from Atlas of Mesozoic and Cenozoic Coastlines, by A. G. Smith, D. G. Smith and B. M. Funnell, Cambridge University Press, 1994.]

coming together. India was half way to joining South Asia: Australia, Africa and South America had long been separated from Antarctica.

By 10 Ma BP the Caspian/Black Sea complex was still as large as the Mediterranean. The certainties of continental drift remain hidden by the convention of using the coastline to define the events, which is useful except in the details of the ups and downs of sea level. The present outline of the land was established well before the beginning of the Quarternary a million-and-a-half years ago.

Throughout all the continental movements there were up and down movements of large areas of land which altered the sea level and therefore also the coastline. But perhaps the greatest cause of the ups and downs of sea level was the accumulation of great quantities of ice on land, and the subsequent melting of it.

(3) Climatological

During the Pleistocene, i.e. the Quaternary until the Holocene, the climate seems to have been based on ice ages over the polar caps down to around 50° of latitude, and only one period as warm as today has been the **Eemian**, which occurred between about 135 and 115 thousand years ago (135–115 kyr BP) (see Fig. 19.2.iv). It must be remarked that we have almost no certain observations, direct or indirect, about the climate, and certainly none about the weather, before about 200 kyr BP. Thus as far as we know the predominant weather since animal life began has been ice ages beyond 50° of latitude, with gradually warmer weather as the equator is approached. There have been relatively short periods of warmth, and sea levels up to 100 m above the present surface. It is therefore important to understand the changes which have taken place since the Eemian

(140–115 kyr BP), much better than we do at present, before we can presume to predict the probable long-term (centuries or millennia) climate from now onwards.

The original papers with these diagrams give many interesting details about the correlations, and among other things suggest that the glacial periods were dustier, i.e. drier and more desert as far as the continents were concerned. The effect of the orbital factor is very evident throughout the last 200 kyr, but the weather of the Eemian is exceptional for causes which are not clear in any of the cores. However, the end of the Eemian has been

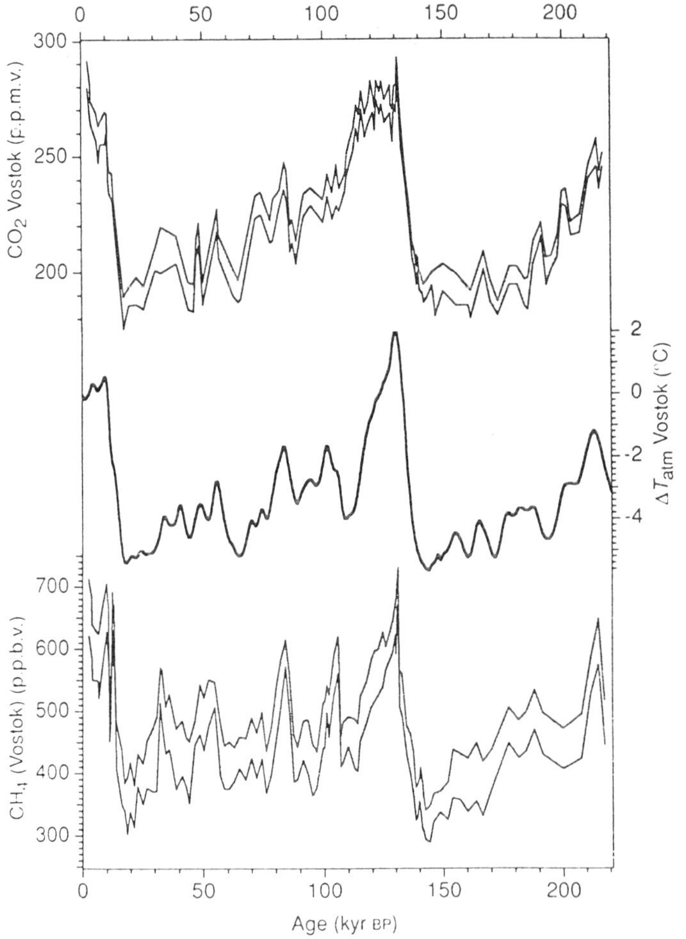

Fig. 19.2.iv The Vostok core analysis. The record deduced from the Vostok core for the concentration of CO_2 and CH_4, with probable error limits indicated together with temperature deduced from other well correlated measurements. The startling fact revealed by this profile, and first made known in 1985 presented here with a few later adjustments [100], is that the temperature and CO_2 concentration increased together at the beginning of the Holocene with no cause-and-effect relationship discernible. The Eemian is clearly shown to have begun with a similar steep rise, only to have been whittled away by a series of step coolings and subsequent warmings referred to as Heinrich events. [Reproduced with permission from *Nature*, **364**, 411 29 July (1993).]

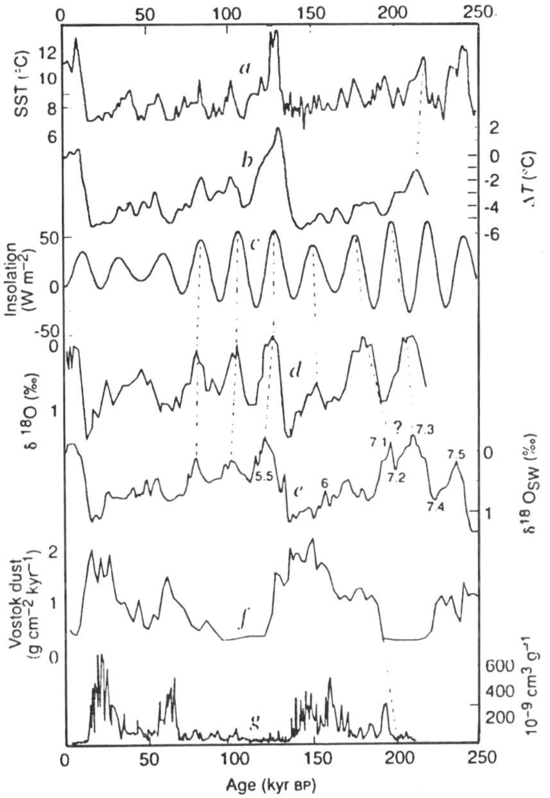

Fig. 19.2.v(a) Different information sources: a. Summer sea surface temperature (SST) at an Indian Ocean site at 16°37N 59°52E; b. Atmos. tempt at Vostok derived from deuterium; c. Atmos. temp. variation at 20°N referred to as the orbital variation on extended geological timescale (EGT); d. Oxygen isotope measurements referred by EGT to temperature; e. Foraminifera-derived ^{18}O estimate of sea water temperature with average Holocene value of 0%; f. Mass accumulation rate at Indian Ocean site RC27-61; g. Dust concentration in Vostok ice. The correlations indicated between c, d and e seem to confirm the importance of the orbital effects as expressed by Milankovitch. [Reproduced with permission from *Nature*, **364**, p. 409, 29 July (1993).]

eloquently explained as being due to the replacement of boreal forest by tundra. Every winter the forest protruded through the snow and made the melting of it more rapid because the tree tops absorbed the spring sunshine. But when the forest was replaced by tundra the surface remained snow-covered with a high albedo and the spring was delayed, and possibly the snow lasted through the summer. Gallimore and Kutzbach [101] applied a global atmospheric model to this problem to show that the albedo in high latitudes can be very important in climate changes of this kind.

Since the Eemian there have been several periods of steady warming, but each time there was a more rapid but intermittent reversion to ice. The output of icebergs during each of these warmings (called Heinrich events [*19], see Fig. 19.2.(a) and (b)), carried

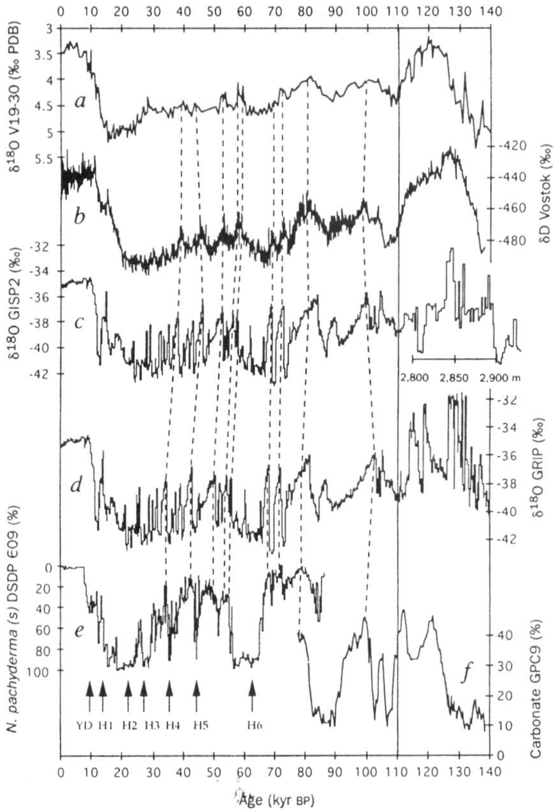

Fig. 19.2.v(b) Different sources (Ref. 0000) emphasising the disagreement between GRIP and GISP2 before 110 kyr BP due to a bulk displacement of the ice and giving more detail of the Heinrich events which were essentially a North Atlantic phenomenon: a. Benthic core at 3°21S 83°21W (referred to as Vostok 19-30, ^{18}O; b. Vostok deuterium (adjusted chronology); c. ^{18}O at GISP2; d. ^{18}O at GRIP to show difference from c; e. Foraminifera (proxy for SST) from DSDP-609 (50°N, 27°W); f. Carbonate on core. Heinrich events are numbered and highlighted in chronology by dashed lines. [Reproduced with permission from *Nature*, **372** p. 663, 15 December (1994).]

material across the sea (see [102]). The difficulties arise at the present stage of knowledge when the arrival of new evidence begins to correlate the climate of different parts of the world only to smother us with detail which makes the simplicity of the earlier story more difficult to maintain.

The Heinrich events can be associated with **orbital** causes. These are summarised by the Milankovitch theory [79], which describes how the increase of sunshine when earth is at apogee in summer receives a positive feedback from the area of the earth's surface frozen throughout the year being at a minimum see Fig. 19.2.vi(d). At the present time we are at apogee in midwinter and so we can expect this effect arising from the precession of the earth's eccentric orbit to take us into a warming period during the coming 20 000 years. These variations in the strength of sunshine are shown in Fig. 19.2.v(a) curve labelled c, and also in Fig. 19.2.vi d.

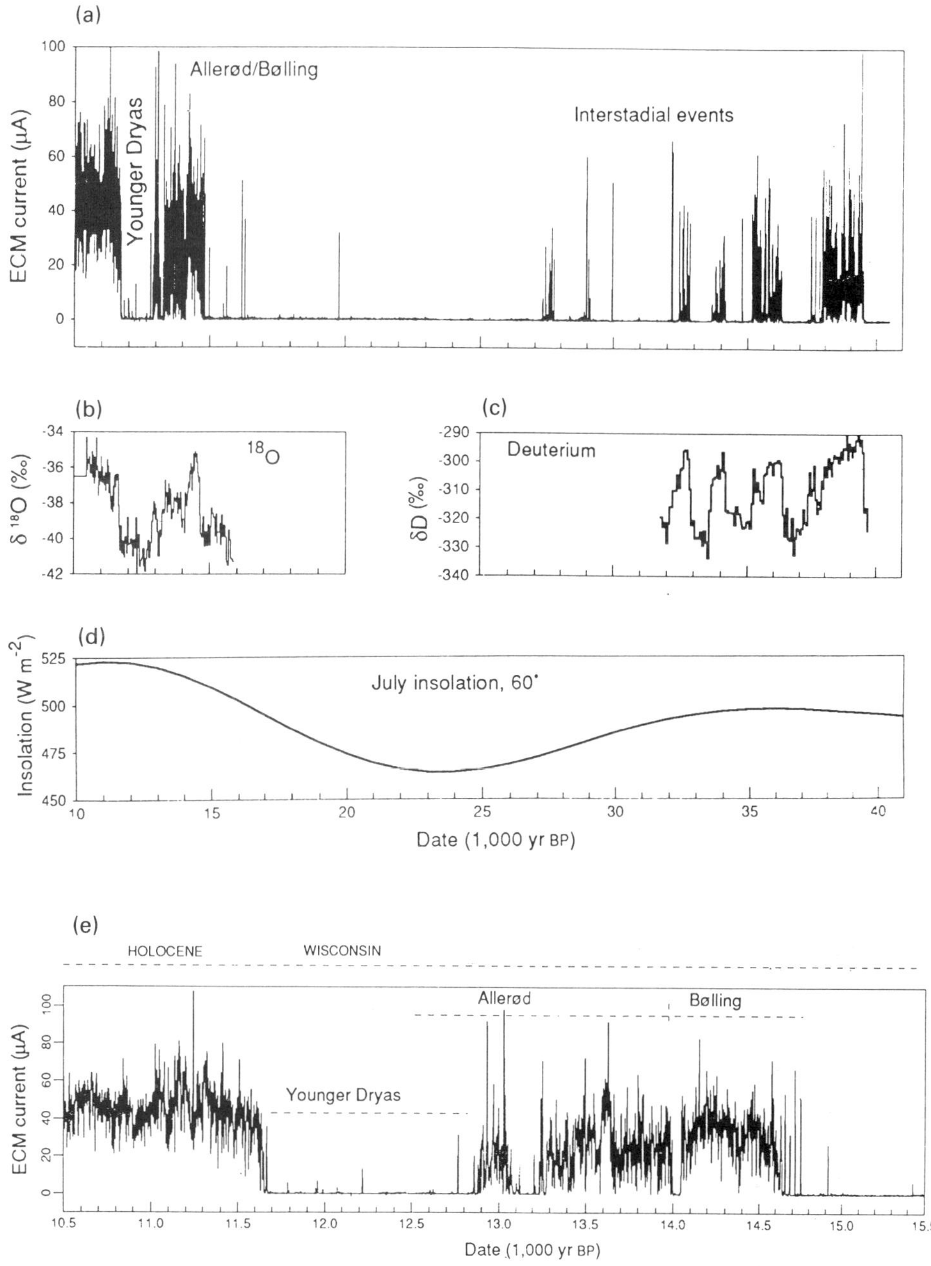

Fig. 19.2.vi GISP record emphasising detail from the last 40 kyr BP: a. shows the ECM record, and the detail surrounding the Younger Dryas appears in e. d shows the orbital effect; b. and c. show the kind of record obtained by the oxygen isotope and deuterium analysis. The time scale along the bottom is the same for all four diagrams. [Reproduced with permission from *Nature*, **361**, p. 432, 4 February (1993) [103].]

The Vostok core analysis (Fig. 19.2.iv) showed that the two greenhouse gases, CO_2 and CH_4, were greatly increased in the atmosphere at the time of the Eemian, but the onset of the warmer climate was so rapid that the sequence of events did not make it clear whether the increase of the gases should be regarded as cause or effect of the warming. Thus the core record of that distant time would not tell us whether the increase was instant or took a few hundred years to occur. We tend to think of the 250 years of the increase of CO_2 due to the Industrial Revolution as a long time and yet the cores do not indicate whether there was a series of rapid increases or one slow increase lasting ten thousand years. When traces of the warming which we now enjoy first arrived about 15 000 years ago, all the indicators did not immediately and in parallel tell the same story. The same indicators tracing the Eemian changes do not give the same agreement with each other during the Younger Dryas event. See Figs 19.2.vii and viii [104].

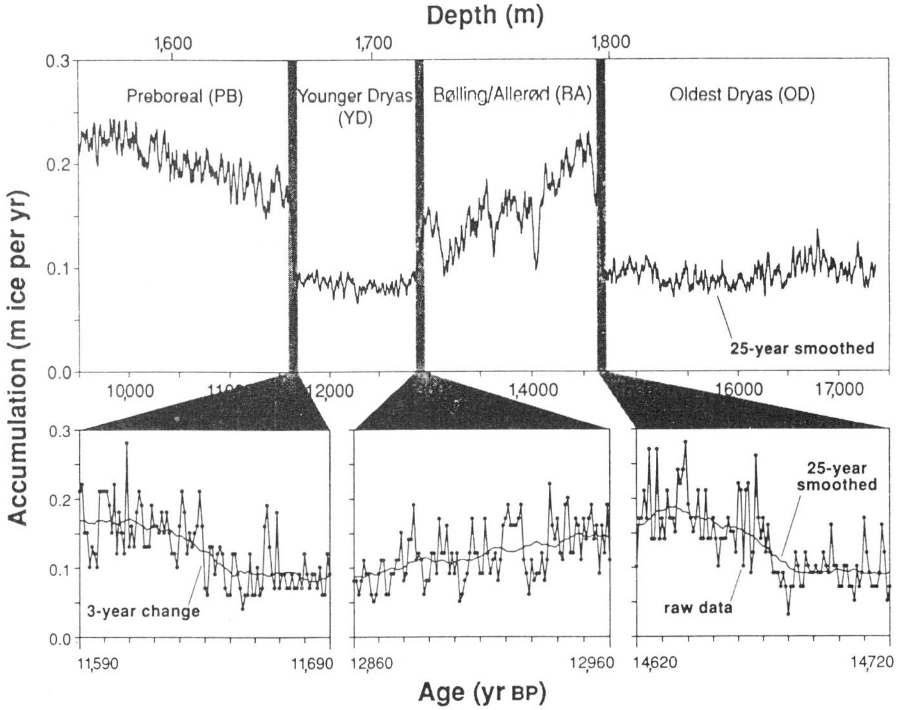

Fig. 19.2.vii In this diagram the **transition periods** from the Oldest Dryas to the Bølling/Allerød interstadials, then to the Younger Dryas, and thirdly into the Holocene. The top curve shows the accumulation rate at GISP2, a core in which individual years could be identified by a shiny ice layer presumably caused by a slight summer melt. The transition between the distinct periods marked in (a) are shown as expanded transition periods between stadials and interstadials during the passage into the Holocene (here marked as Preboreal). It is to be noted that the transitions OD–BA and more particularly YD–PB were accompanied by very sudden rises in accumulation rate in individual years. The accumulated snow in a very few years can say nothing about the weather experienced during the single years. [Reproduced with permission from *Nature*, **362**, p. 528, 8 April (1993).]

It is suggested that there was also a temperature rise of about 6°C, which for a meteorologist means that the core site was put on the opposite side of a front quite frequently instead of scarcely at all. We can imagine it happening for the first time one afternoon. In terminology more familiar, the North Atlantic conveyor belt came into action to bring the YD to an end when warm salt water was able to flow into the extreme north of the Atlantic because cold salty bottom water began to flow southwards across the gap between Greenland and the British Isles. For geologists it is less of a problem how that process took 40 kyr to get going.

The most recent warm period began about 39 kyr ago with a series of interstadial (warm period) events which in turn gave way intermittently to the ice from about 32 kyr to about 15 kyr ago. At that time the Allerød and Bølling interstadials lasted with small breaks until the **Younger Dryas**, which was the last intense spell of ice age between 12.9 kyr and 11.7 kyr ago. That was ushered out, so it might seem, by a rise in the orbital warming. Much of the mechanics of the Younger Dryas has been associated with the melting of the North American ice sheet and the action of the North Atlantic conveyor belt, which threw doubt on it being a global phenomenon, and this aspect was studied by Mayewski *et al.* [105]. They showed that the Taylor Dome core, which was the closest to the sea of the Antarctic cores and therefore was most likely to resemble the Greenland cores, did show mutual resemblances only in some respects (see Fig. 19.2.viii).

Many of the factors measured in the deep cores are still in the exploratory stage in which correlations are being established with the ocean bed cores in order to discover which features are global and which represent only local or regional variations of climate. Such standards as **SMOW** (standard mean ocean water) are found to make useful comparisons with the deep ice cores of Greenland and Antarctica in respect of oxygen isotopes in particular. The atmosphere and the ocean both circulate with sufficient rapidity to give useful correlations between the Southern and Indian Oceans and the North Atlantic (see Fig. 19.2.v). It has become possible to find traces of the Heinrich events, which are essentially North Atlantic phenomena, in Vostok isotope records. This may mean that the release of icebergs (and therefore fresh water) from the North American ice might have the same cause as the release of greater numbers from Antarctica, or it might be that the happening in one place had repercussions in the other through the global ocean–atmosphere system.

For these reasons, comparisons of each of the oxygen isotope, calcium, and chloride found in the cores at GISP2 in central Greenland and at the Taylor Dome close to the Ross Shelf in Antarctica during the period 16 to 10 kyr BP were made to see how different they might be in their record of the Younger and Older Dryas, for it is important to discover causes of these recent climatic events if we are to make any impression on foreknowing beyond the present. Calcium makes very little progress from the sea into Antarctica, although it reaches the Greenland central sites and so the Taylor Dome was selected for this comparison. One diagram shows the GISP2 records only faintly so that the traces may be disentangled, and the sample lengths are written close to the appropriate parts of the curves.

The Antarctic cold reversal (ACR) appears to have lasted longer than its counterpart—the Younger Drayas—but only some of this difference might be due to uncertain synchronisation. The amplitude scales for the two stations are not identical and are quite

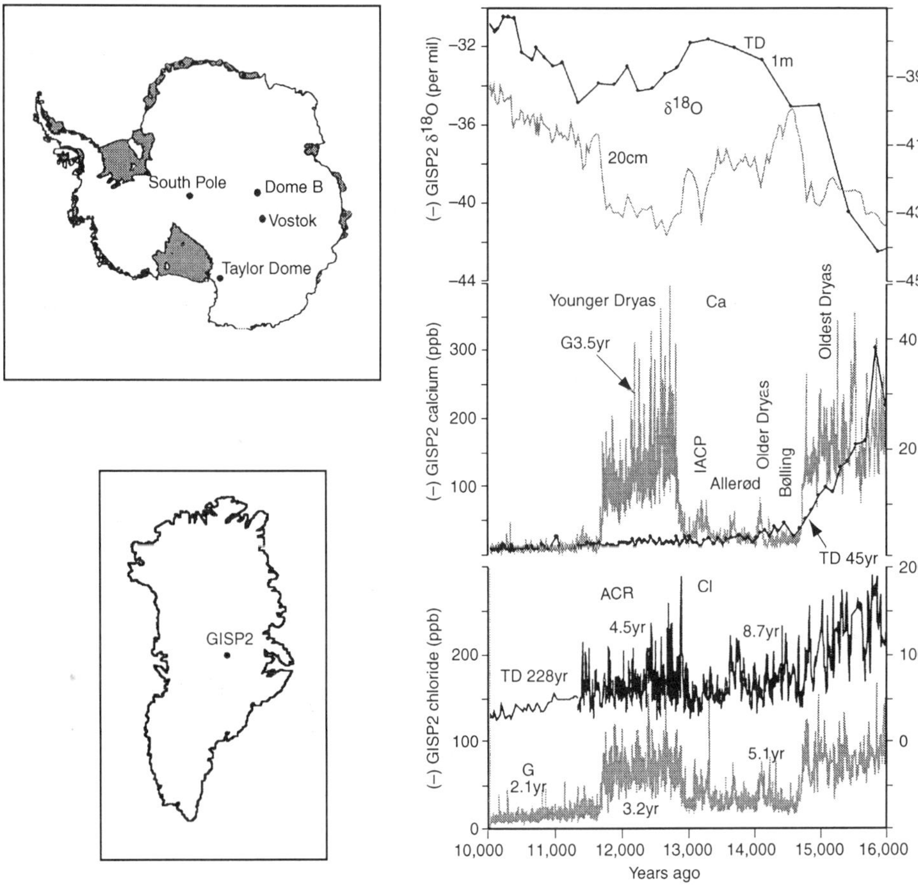

Fig. 19.2.viii Comparisons are shown between samples of $\delta^{18}O$, calcium and chloride, each having its own scale, but on the same chronology. The sample intervals are shown against each of the curves (in three sections for chloride). The classic North Atlantic sequence is as follows: oldest Dryas, Bølling, Older Dryas, Allerød, Intra-Allerød cold period (ICAP), Younger Dryas, all of which is given according to GISP2 chronology. In Antarctica the oldest Dryas appears not to have ended as abruptly as in Greenland and the Antarctic cold reversal to have lasted longer than the Younger Dryas. [From *Science*, **272**, p. 1637, 14 June (1996) [106].]

well separated for chlorine to avoid confusion. Evidently, calcium does not easily reach Antarctica, while North American dust easily gets to central Greenland.

19.3 THE NORTH ATLANTIC CONVEYOR BELT

The temptation to break up the climate, which is produced by a global system, into a number of distinct phenomena is irresistible, especially to those of a geographical frame of mind which keeps reminding itself of the differences in the experience of the climate in different places. One of the most outstanding is the climate of the North Atlantic, where the sea remains unfrozen up to 78°N on the west coast of Svalbard throughout the year and the coast of Norway stays open to ships. This is due to the warm water of the Gulf Stream, which continues to flow northwards thereby warming all the air which comes into Europe from the west and causing the climate of West Europe to be very mild compared with all other places in as high latitudes in both hemispheres.

The very salty warm water of the Gulf Stream heats the air of the Norwegian Sea (see Fig. 19.3.i) but eventually sinks under the colder, much less salty, water of the Arctic Ocean and the Greenland Sea. This is made possible by this salty, and now quite cold, water returning on the ocean bed between Iceland and Scotland to replenish the 'bottom water' of the North Atlantic. Without this return flow, the Arctic end of the Atlantic Ocean would become filled to a greater sea level height so that the Gulf Stream would not flow into the Norwegian Sea but would pass southwards towards the Canary Islands. The North Sea would almost certainly freeze over every winter like the Sea of Okhotsk, which is in the same latitude to the North East of Japan (see Fig. 6.7.ii) hidden behind the Kurile Islands.

Fig. 19.3.i 0355,28.1.85,Ch4. Strong convection in the Norwegian Sea, enough to generate significant cyclonic rotation, contrasted with the small cloud sheets coming from the Greenland ice. The coast of Iceland is seen in the SW corner. (Courtesy the University of Dundee.)

Fig. 19.3.ii 0834/1028,25.5.789,Ch2. A typical component of the North Atlantic conveyor belt: an old cyclone, which has drawn warm water up the coast of Norway by a wind from the SE, and has also driven the cold current down the east coast of Greenland. At this time of year the ice is well broken and is carried to the southern tip at Cape Farewell and even some distance up the coast. (Courtesy the University of Dundee.)

Thus far we have the ocean conveyor belt. But this area between the mountains of Greenland and the Scandinavian peninsula typically guides cyclonic developments off the east coast of North America up into the Greenland and Barents seas. The surface wind becomes typically southerly or southeasterly, driving the warm surface water towards the northeast. The sequence of low-pressure centres travelling this route constitutes the atmospheric conveyor belt which brings warm air into Western Europe and much of the far North, where the bird life is profuse, and from where migration takes off to half the globe, or more. See map on p. 637—Fig. 19.3.v, from [76].

19.4 THE LAST 1500 YEARS

A detailed account of this most recent period is given by Meese *et al.* [107]. The warmth of the period from AD 600 to 1300 is illustrated by thin layers of ice formed by the melting of the snow surface which were readily identifiable in the GISP2 core from

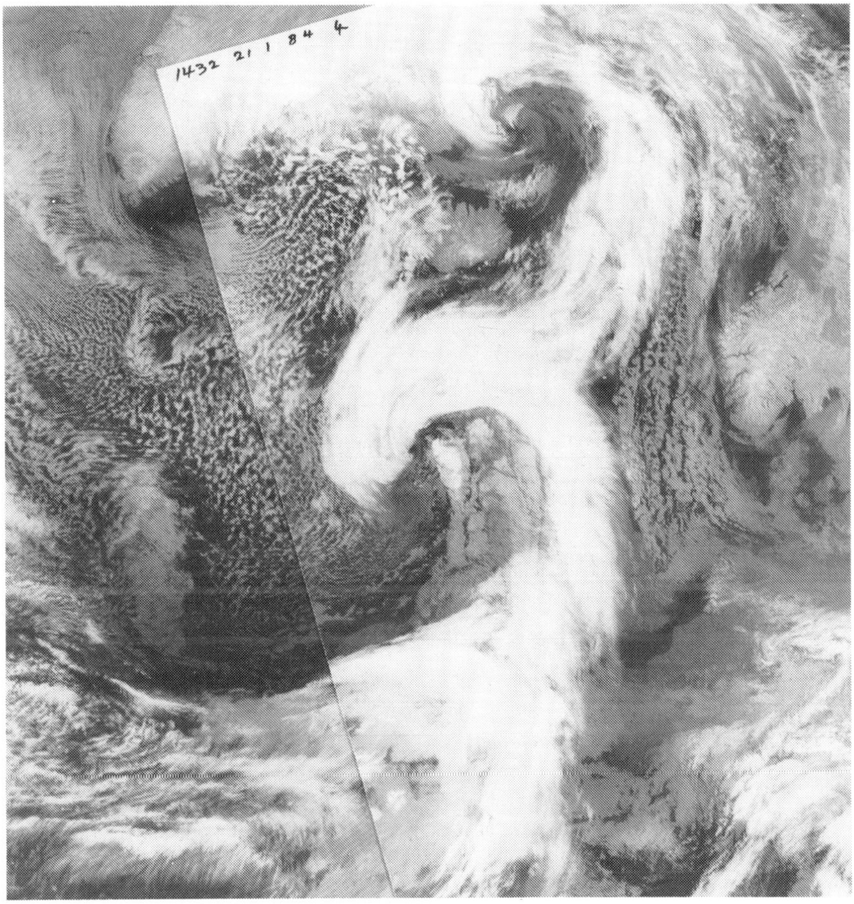

Fig. 19.3.iii 1432/1614,21.1.84,Ch4. Because of the great gradient of temperature caused by the cold southerly flow west of Greenland and the northward transport of air on the east (except close to the coasts) there is often a meridional flow established which drives cyclones northwards between Iceland and Norway. Thus the Gulf Stream flows northeastwards into the Barents Sea in the summer, although it may be submerged earlier on its way in winter. (Courtesy the University of Dundee.)

central Greenland. During that period we can romanticise about the growing of wine in Northern England, but we cannot claim to know how that warmth was caused. The orbital factor was in favour of it getting cooler, and indeed it did begin a colder period in which according to historical, but not yet instrumental, records the climate of western Europe became significantly cooler during the 14th century [108]. On the evidence presented there was good reason for expecting a reappearance of the ice, and when climate change was modelled before the understanding of chaos theory, it was common for the climate system to be described as unstable, and only requiring a slight push to bring on the ice.

The cloudy summers and cold winters of the gloomy 14th century were amplified in misery by the poor harvests and the appearance of the plague centred around the Black

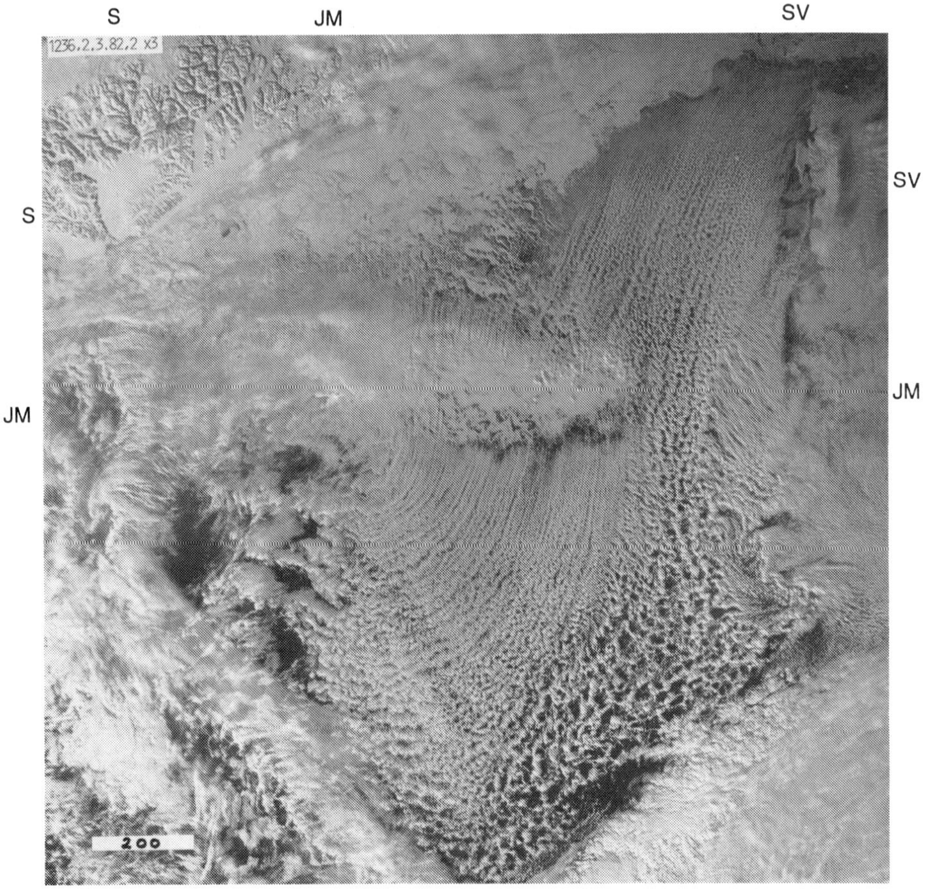

Fig. 19.3.iv 1236,2.3.82,Ch2. A Greenland Sea scene as the demolition of the Greenland sea ice carried southwards. The white peak of Jan Mayen (JM) is about the southerly limit of the ice in the present regime. Note the clear coast of Svalbard (SV) in the NE. Scoresbysund (S) indicated in the NE, where the ice is flat among the mountains. The convection calls disappear over the big ice floe lying to the east of Jan Mayen. That ice was driven by the wind, but it had all melted and left no trace two days later.

Death of 1348, and recurring, particularly in big cities until the middle of the 18th century.

This miserable period during which the intellectual advances were very considerable has been called the **Little Ice Age**, and its end has been placed variously between the 1880s and the present when the global warming due to the industrial increase in carbon dioxide is said, at last, to be in evidence.

19.5 LIMITS TO MODELLING; USE OF ENSEMBLES

There is now a great deal of evidence about the climate of the past, and much more detail will be discovered in the next few years as the analyses of different cores becomes

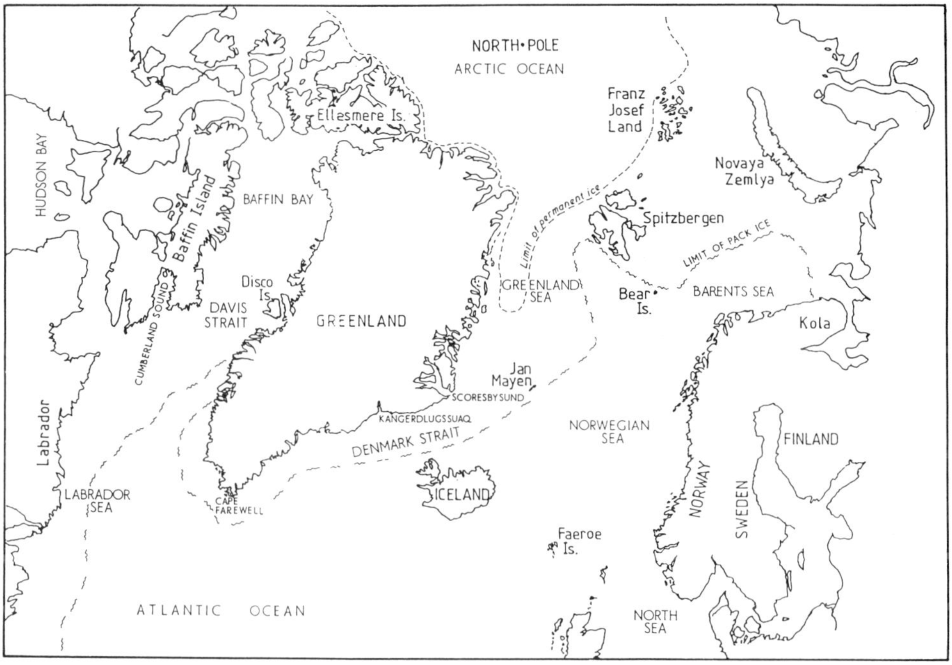

Fig. 19.3.v The Northern seas.

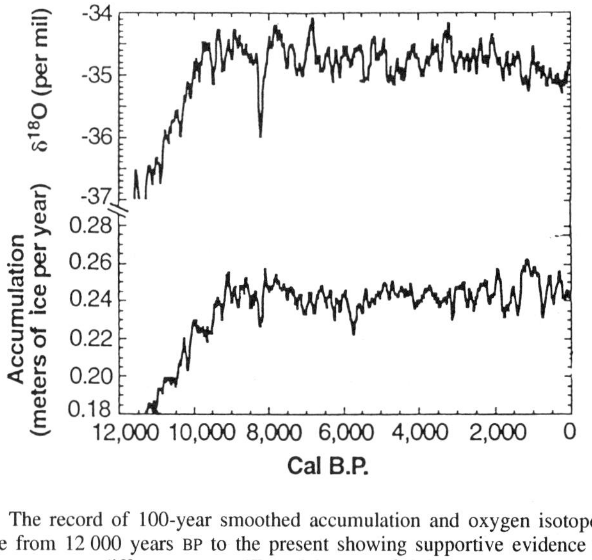

Fig. 19.4.i The record of 100-year smoothed accumulation and oxygen isotope profiles from GISP2 core from 12 000 years BP to the present showing supportive evidence from two very different tracers. [From Meese *et al.* (1994) [107].]

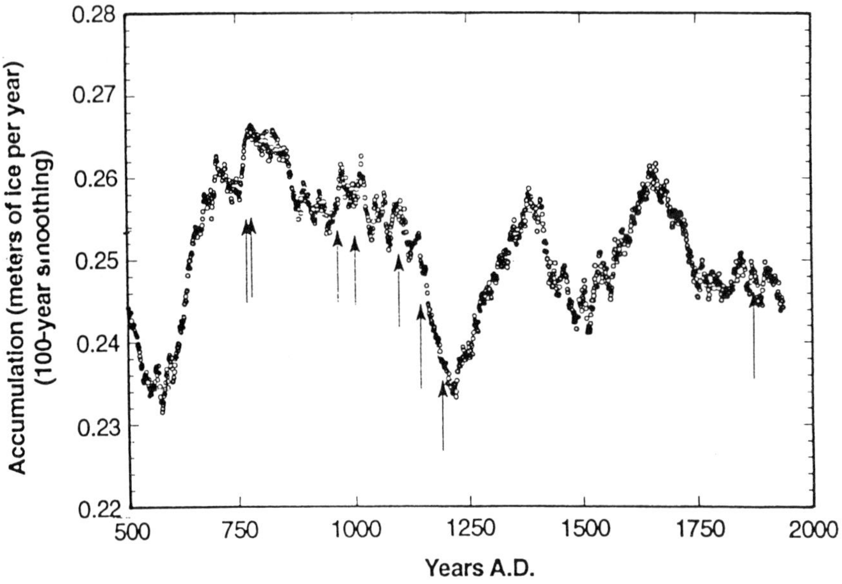

Fig. 19.4.ii The 100-year smoothed accumulation record from the GISP2 core for the period AD 500 to the present. The arrows show locations of visually identified melt layers in the ice core. The fluctuations evident in this record may be of interest in connection with the historical events in the **Medieval Warm Period** in Europe and their correlation with the snowfall in Central Greenland. [From Meese *et al.* (1994).]

reliably synchronised and the variations in different places become understood in terms of the weather in different regions. As this text is written the anchored buoys in the Eastern Pacific are collecting routine observations which will become absorbed into the collective knowledge which will include more detail about the stratosphere and the deeper parts of the oceans. The ocean–atmosphere models will become progressively more accurate and thereby able to play the role of test-laboratory for improved parameters representing those features such as clouds and haze for which at present there is no prospect of using the basic physical and mechanical equations directly because the detail is small compared with the spacing of the grid points.

The present limitation to the application of the fundamental equations, in the same direct way as they are applied to the air and will soon be applied to the oceans to describe the cause-and-effect results of deterministic laws, arises because the equations are not linear and therefore the solution of them is chaotic, as described in Chapter 16.

The demonstration of this feature of reality was devised by Dr Tim Palmer of the European Centre for Medium Range Weather Forecasting (ECMWF) where the calculations were performed on one of the world's largest computer systems. A weather factor is chosen, in this case described as the midday temperature in London. It must be remembered that all the observations made close to one of the eight 3-hour observation times every day are used to determine the observations assigned to the grid points of the model. The grid points are not the exact points where the observations were made, but in order to include as much real weather observations in the calculations as possible each

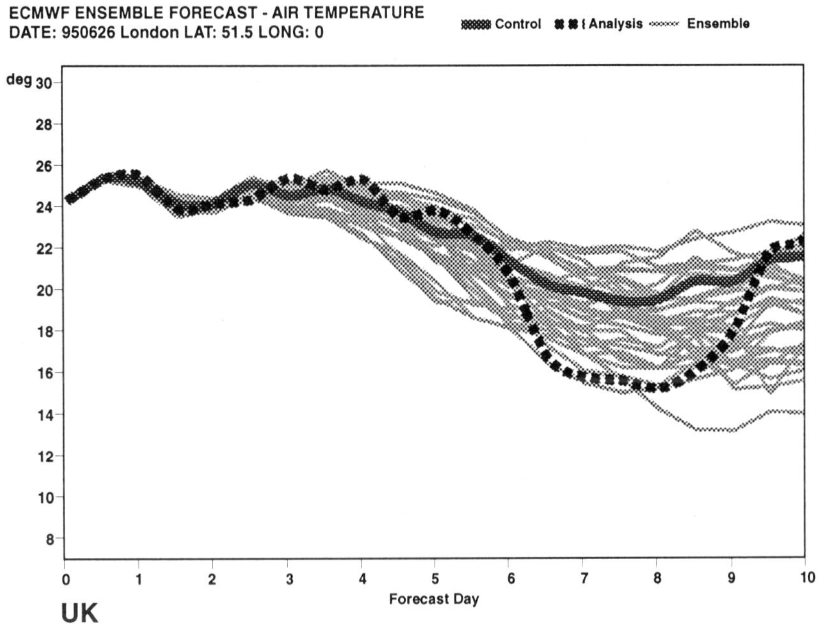

Fig. 19.5.i Ten-day forecast (by Dr Tim Palmer and ECMWF) of the notional midday air temperature in London beginning on 26 June 1995, with an ensemble of forecasts with slightly changed starting states.

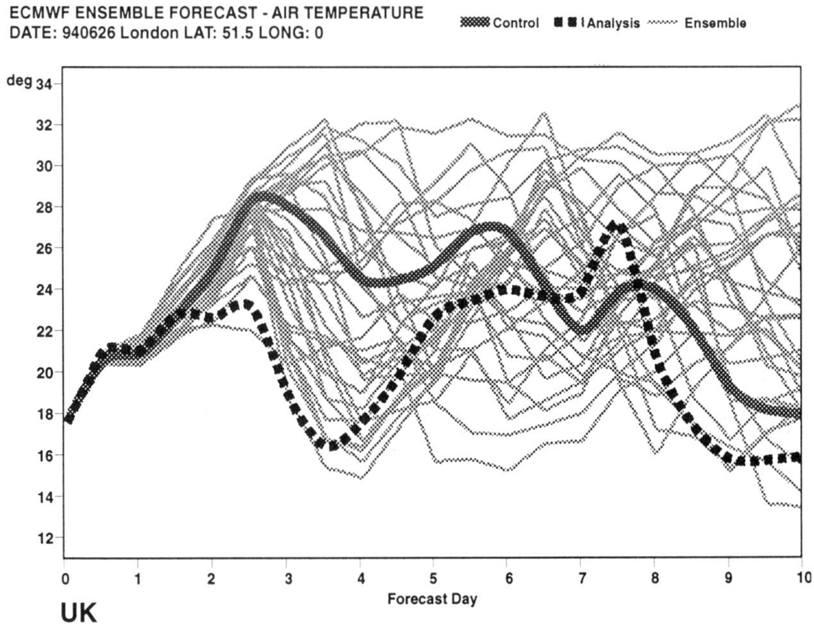

Fig. 19.5.ii The same as (i) but beginning on 26 June 1994.

has some influence on the values assigned to the grid points by using interpolation formulae. The exercise starts with the latest weather pattern arrived at and it is known which areas are likely to develop new growing disturbances, and that is where the forecast calculated by the model is most likely to go wrong. See Figs 19.5.i and 19.5.ii.

At this point of greatest uncertainty, several small disturbances are introduced one at a time and the forecast generated by the model is calculated starting with the original system and again starting with one of the small disturbances and so an ensemble of forecasts is generated. The diagram shows the best forecast, shown as a bolder thick line, together with the slightly disturbed forecasts, which are all shown as feinter thin lines. This calculation was done for 26 June 1994 and again with the real weather for 26 June 1995.

On both occasions the various forecasts stayed close together for about 2 days, but within very few more days the ensemble had spread out, rather more quickly for the 1994 case than for 1995. The bolder, thick line remains within the general spread of the feinter, thin lines and, on the assumption that the spread represents the region in which the forecast made for the whole of the 10 days will be found, the extent to which the feinter, thin lines cluster around the bolder,thick line is an expression of the confidence with which the forecast is made. It is fairly obvious on the basis of this example that there is no obvious rule about picking out the most probable series of values for the temperature for the next ten days. The black dashed line delineates the temperature obtained from the final chart drawn for each day, or to be more correct it shows the temperature assigned to the grid point from which the values were taken at the time the ensemble of forecasts was made, and so it is the nearest thing to the actual value which ever gets into the computer in the course of making the forecasts during the next ten days shown in the diagram. In 1994 the forecast would have been good for about two days and then went rather badly wrong; but in 1995 the forecast was very good up to the sixth day. It is understandable that the modellers claim to give a good forecast up to two or three days only with confidence, but have to have a very good reason to dispute the model's forecast for the next three days.

To summarise: the discovery that non-linear differential equations have the property of becoming chaotic does not, in the least, alter the view that the mechanisms described by them are deterministic. The best way to deal with the problem which is presented to us when the solution may proceed in time in directions which are determined by inaccuracies which inevitably arise in any numerical finite step method of solution. In the case of linear equations we can express the solution in terms of mathematical functions (sine, logarithm, Bessel functions, etc.) which are themselves defined in terms of solutions of such equations. But it is not possible to proceed in the same way with non-liner ones. The **ensemble** of solutions may provide an indication of the way the uncertainties may develop. If there is a crowding of the members of the ensemble around one direction, that may be taken as expressing a probability that events will tend to go that way too. The uncertainties are inevitable when the outcome is dependent on parameterisation of some of the mechanisms, and determinism must fail when the calculated outcome is dependent on such uncertain methods of approximation.

20

The study of climate change in evolution

20.1 WHENCE WE CAME

Of the four and a half billion years since the earth took up its orbit as the mildest, the wettest and the bluest of the planets, all but the last half billion were spent in violent eruptive and erosive stormy sorting out of the tectonic plates with life taking hold in the shifting tidal shallows where stratified rocks were laid down with increasing residues of new experimental biological entities, cells which have remained as the sole witness of the beginnings of our most distance ancestors.

In subsequent due course, only described as 'due', implying a search for some sort of later attainment, life took a mere half-billion revolutions around the sun to evolve air-breathing animals as well as vegetation. By carboniferous times (300 Ma BP) land areas had increased and even winged insects accompanied the ammonites which gradually (by later standards) generated reptiles in the ill-defined territory available. A quarter of a million years ago, the dinosaurs had become the chief appreciators of the swampy warm environment in which they found themselves. Their supremacy endured until the Cretaceous (100 Ma BP), and during their reign more modern flowering plants and the insects which served them flourished. Volcanic activity was rife through the Triassic, Jurassic and Cretaceous, by the end of which the world was ready for the mammals of the Tertiary era around 65 million years ago.

During the Tertiary, plate tectonics was a main agent of environmental and climate change; modern mountains soon (? i.e. after only about 50 Ma) began to accumulate glaciers, and mammals had learned to exploit the opportunities of the night; the large dinosaurs found impossible the pursuit of their lazy lifestyle among thetorrential rivers (or, if you prefer, the unlikely alternative of climatic catastrophe: a comet landed in Yucatan with great force and covered the world with darkness whichselectively exterminated the kings of the reptile world). Environmental change established new continents in North America and Asia, where the mammals learnt to run and the grasses and forests provided new homes and sources of food. By 50 million years ago (50 Ma BP) the birds and mammals had perfected the arts of evolutionary living and growing. The increasing number of different life forms had 'perfected' the evolutionary machinery and there was scarcely a location which did not have its seasonal visitors. They may have plagued its permanent residents, who were learning to build their homes and live by recycling each other's waste. Any species which became too dominant and populous in its area fell prey to new pests and disease which themselves stimulated new herbs and blood components which saved those who could make allies of them.

The world was now a place of continually changing environments providing new opportunities for inventive responses, and a thousand years of one fluctuation appears in our account like a rapid alteration of the norm with an overnight response. But while a new insect or virus seems to appear with unrestrained immediacy, we have begun to learn the lesson of the advance by natural selection: overpopulation of a region by a single species soon attracts a multitude of parasites, many of them having been in virtual obscurity because of the biologically various enemies who discovered them to be a source of food, or who in turn may support their own varieties of more minute and intimate parasite.

Our own mammalian kind of consciousness facilitated the exploitation of sources of energy which our manual skills could learn to control, and it is now a cliché to say that we are endangering the support systems of our own way of life by flourishing to excess. It has become unnecessary for us each to be conscious of the second law of thermodynamics, that to create order the degradation of energy is required. While the animals and birds use their own muscle power to make their homes, and have to consume food to do this, the food is created by the degradation of sunshine by chlorophyll. It passes up the food chain to us and keeps the mammals warm. But from the principle that the energy coming into our world as UV and visible sunshine is equalled by the energy leaving the earth as IR radiation to space, it follows that if we impede the escape of the IR, the earth must become warmer. And we know that if the climate becomes too warm and enervating, the qualities of active life are not readily enjoyed. Some mammals' coats are seasonal: ours are diurnal, and some of our artificial societies in warm climates are dependent on refrigeration as a routine.

20.2 THE HUMAN PREDICAMENT

If we are to believe our demographers, the world's human population will reach a total of 10 billion in 50 years from now, and our politicians have accepted that they are powerless to avoid this. In so far as they have responded to the situation which became clear to a few two centuries ago and to the advanced and well informed populations during the last few decades, the effective response has been somehow as far as possible to facilitate the feeding of whatever human numbers appear at any time. The consequence has been to heighten the predicament: supporting the numbers by technologies which depend on the expenditure of finite resources which were laid down in the carboniferous swamps. The time scale for the aggravation or resolution of the ultimately unsustainable conflict is 'not geological', nor are the resources 'astronomical', and the only rational approach is to search for more evidence and understanding of what is happening around us. The next generation grows up accustomed to what would have been regarded as a disastrous challenge if it had all happened in a year, or even a decade instead of taking a few human generations.

The international community reacted to the threat of ozone depletion in a way which appears to be having success. But it seems that it had been necessary to scarethe politicians. To what extent is it the duty of the scientific community to scare the world? That has been attempted to some extent in the case of the threatened greenhouse warming, but because the warming seems to be taking more than a human lifetime to be in indisputable evidence, there are economically strong lobbies growing up arguing that the warming is not happening and the exploitation of the 'widow's megacruse' (see I Kings 17.16) can continue unabated, until the warming begins (Elijah said 'until it rains').

The threatened rise in sea level which could inundate the living space of several million people is not taken seriously even by the populations most threatened because many have survived inundations caused by Bengal typhoons and have returned to the flooded territories to have more babies when the typhoon season was over. On the West Pacific seaboard they know that the season is the whole year round.

Disbelief in the seriousness of the threat overcomes the rational explanation of why it will become real. The comforting nature of the disbelief is exploited by those whose economic interest is threatened: they argue that chaos theory makes a reliable forecast impossible. It becomes important therefore to understand reliably the climate changes of the past. *T. rex* does not threaten us, but cardiovascular wrecks won't go away!

There are some scenarios which start with assumptions about the probable future world fuel consumption. The one called 'business as usual' (BAU) comes in the middle because until the 1970s the expectation of eternal exponential growth was the BAU economic advice given to politicians as an expectation, as well as a mere hope. BAU is least bother!

We have looked at the problems of chaos with the hope that probabilities may be extracted from the alternatives. But even this is not seen as a way of understanding the limitation while retaining a genuine belief in causation. If it is thought that it provides an argument against paying for improved and more powerful computers it can be obvious that in due course we will need them to handle the ocean models when we have enough routine observations to make monthly forecasts. Because the ocean has a larger scale of its smallest detail the models of it will encounter the chaos barrier some time after the atmosphere has met it, and this will benefit the weather forecasts when the ocean model is running properly.

But we cannot envisage a GCM which does not depend on at least a few parameterisations because clouds, haze, fronts and shear layers have such small time and space detail. It is not possible to test the acceptability of more than one parameterisation at a time, and that must depend on assuming that all the others in the model are up to the required standard. This is not possible if the effect of one overlaps, but does not coincide with, the effect of the other. This is best understood by the case of clouds and haze. Low clouds which appear by day and disappear by night have a similar cooling effect to haze which scatters some sunshine back into space but lets through IR from the earth all the time. Layer clouds, especially any which tend to disappear by day, reduce the albedo but additionally produce a nocturnal warming effect. The calculation of the escape of radiation to space is difficult enough in the case of the extent to which water vapour, with its complicated windows and variable distribution in the upper troposphere, contributes to warming and probably to a positive feedback. Clouds produce a cooling by their

overwhelming contribution to the albedo, but a very uncertain feedback. Haze, meaning everything from Aitken nuclei to the dust of the Sahara and everything in between produced by human activity and natural
combustion, produces a cooling.

To test the parameters used to represent clouds, the effects of haze and water vapour must also be present, and their parameters cannot be tested separately. The parameters are therefore not really any use in determining the feedback of all three categories together or separately. It may be possible to measure the water vapour effect using the w.v. satellite. There is no similar satellite for clouds; and do clouds above a snow-covered surface, which may be warmer or cooler, have a warming or cooling effect?

These problems are eased a bit in mesoscale modelling, and so a GCM with similar grid spacing and reduced dependence on parameterisation would push the chaos barrier a little further away and give us better probabilities in weather forecasts, but it would need

Fig. 20.i This picture shows four typhoons alive simultaneously at 20°N 113°E near Hong Kong; 28°N 126°E between Shanghai and S. Japan; SE of Japan 23°N 148°E; 33°N 182°E in Central Pacific. A GCM could not contain a proper record of all these typhoons and could make an analysis that did not represent reality. Each is a separate mesoscale phenomenon. (Courtesy Japan Meteorological Agency.)

a much enlarged computer. In testing parameterisations one learns a little more about causes of observed behaviour of the atmosphere. There is still much to be learned about how heat and water vapour get out of the sea into the air. It was seriously thought in the 1950s that if the molecular coefficients of viscosity and diffusion were very different, the weather would also be very different. Radiation and splashing in strong winds make all the difference, however, and the effect of surface tension on small droplets (a parameterisation on a different scale) got over some of the colloid problems in clouds, but we have got into some serious problems with the chemistry and physics at –90°C on PSCs which have themselves become a laboratory for halogen behaviour and the fate of ozone e.g. [109].

We have argued that parameters, and perhaps even some coefficients, need to be improved upon if the chaos barrier is to be perceptibly rolled back, and that means that mesoscale phenomena will need to be studied in more detail, and it is hoped that because that is what this book is about it will be helpful in that work.

Whether we prove that human activity is causing climate change or not, to answer this question we certainly need to know what changes there have been in the past during and before the Holocene, and how they were caused. Otherwise we cannot feel that we have understood the changes which may happen to us. To wait until the change is certainly happening before we prepare to meet the consequences may be too late!

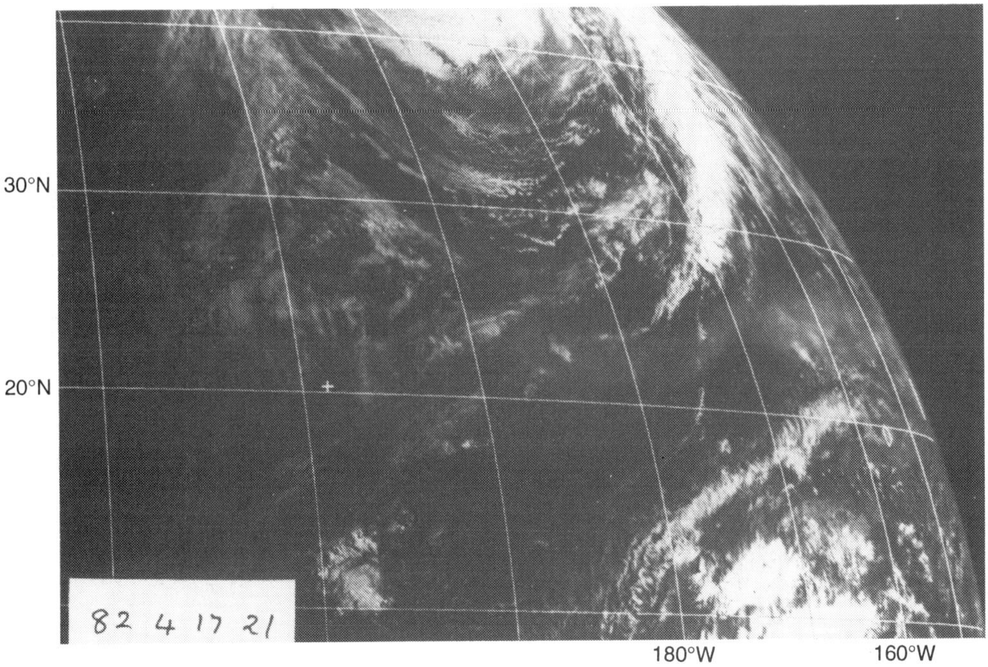

Fig. 20.ii 2100,17.4.82 GMS2, Vis. The dust in the central Pacific, mainly around 25°N 160°W and extending to 20°N 165°E, from the volcano Chichon in Lower California at 30° 40N, 115°W. (Photo from GMS I and Japan Meteorological Agency.)

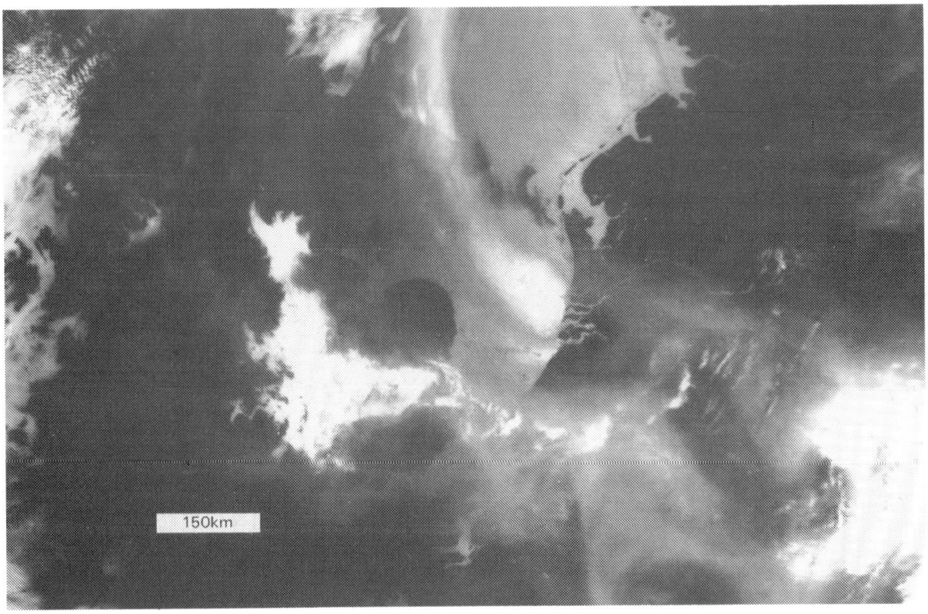

Fig. 20.iii 0852,27.4.87,Ch1. The visible wave channel is used to identify smoke haze from western Germany and Belgium crossing from the mouths of the Schelde and curving round towards Jutland, and from the Pas de Calais into SE England. These whirls of pollution may falsify measurements made on the coast as an indication of the condition upwind of London during experiments to find the contribution of London by making further measurements downwind in central England.

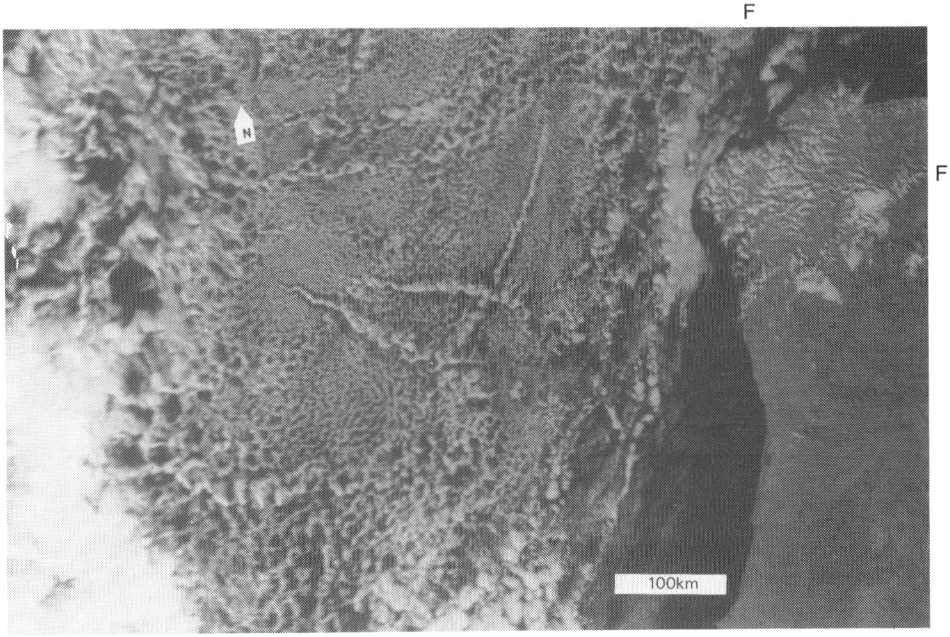

Fig. 20.iv 1436,17.11.82,Ch2. Ship trails not far west of Cape Finisterre (F) which are unlike the majority of ship trails in that there appears to be cloud-free gap between the trail and the rest of the lace curtain cloud. This is a mesoscale phenomenon.

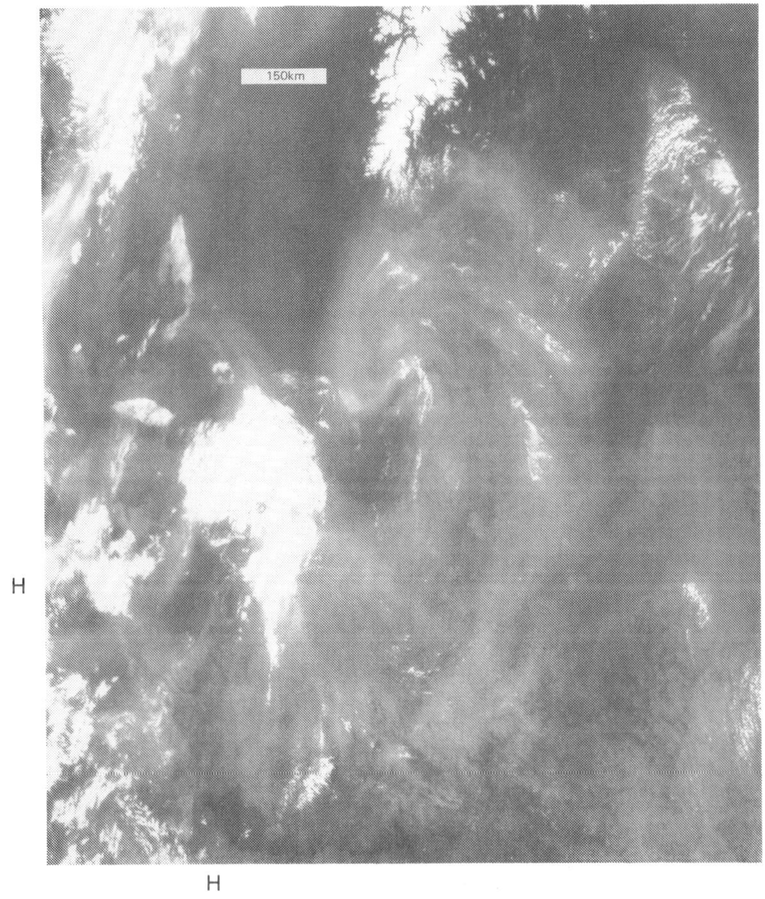

(a)

Fig. 20.v(a) 1046,2.6.82,CZ1. In the top of the picture we see the snow on the Norwegian mountains. The south of Norway we see dense smoke pollution carried up from W. Europe. The coast of Holland (H) extends southwards from the cloud patch.

20.3 VARIETY UNLIMITED

We are so used to having mass-produced machines and products at our service that it is easy to see in these artificial surroundings evidence of our power over the natural world. Although some events like the appearance, growth, and ultimate vanishing of a cumulus cloud seam to be a rather simple repetition, the natural scene never seems identically the same as on a previous occasion. In these last pictures are assembled events which it would have been reasonable to expect, but not in detail; meaning that we can understand or interpret how it happended, but we would not have been able to predict the event which characterises the occasion—or could we?

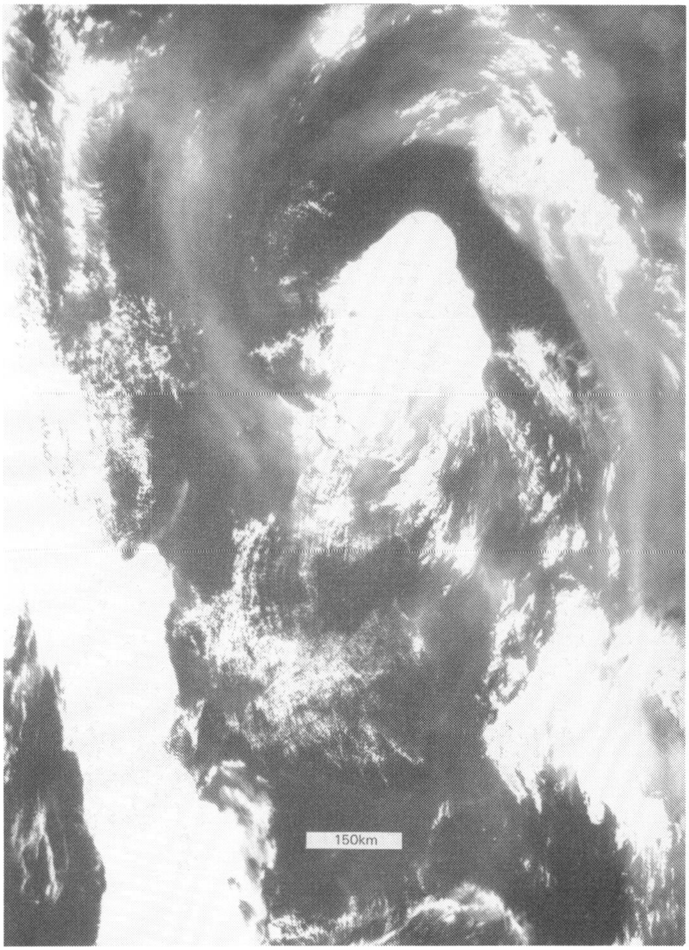

(b)

Fig. 20.v(b) 1122,4.6.82,CZ1. A similar scene two days later with the haze wound up around the cloud now situated in the middle of the North Sea. This blue channel was not used for the purposes of locating haze!

In Fig. 20.i we have four typhoons active at the same time. We like to say that the tropic cyclones occur during their season—which in the Pacific is from January to December inclusive.

In Fig. 20.ii illustrates the contribution of volcanoes to possible climate change; but all who try have great difficulty in predicting the great dust emission that occurs. We have all heard of Etna, but where is Chichon?

Fig. 20.iii is difficult enough to interpret, let alone explain or predict. Why does a plume of pollution from the area of the Schelde produce a bright patch which is isolated?

And why does a ship trail only sometimes clear a lane of cloud on each side in the lace-curtain cloud?

There is no research to use satellites to tell us where our industrial haze goes, and like desert dust, it is very varied in its behaviour. (See 20.iv and v(a) and (b).) Spike clouds become common when you look for them; but not much effort has gone into predicting them or finding a use for observing them.

In Fig. 20.vii we see grass fires in Zimbabwe, which are common, so common that the vegetation may be damaged by preventing them. They may be used to fires to alert their seeds.

Fig. 20.viii. Arabian Sea cyclones are not well known because the surroundings are not densely populated. But they can be very damaging if they support desert locust swarms. See Fig. 14.11.vi.

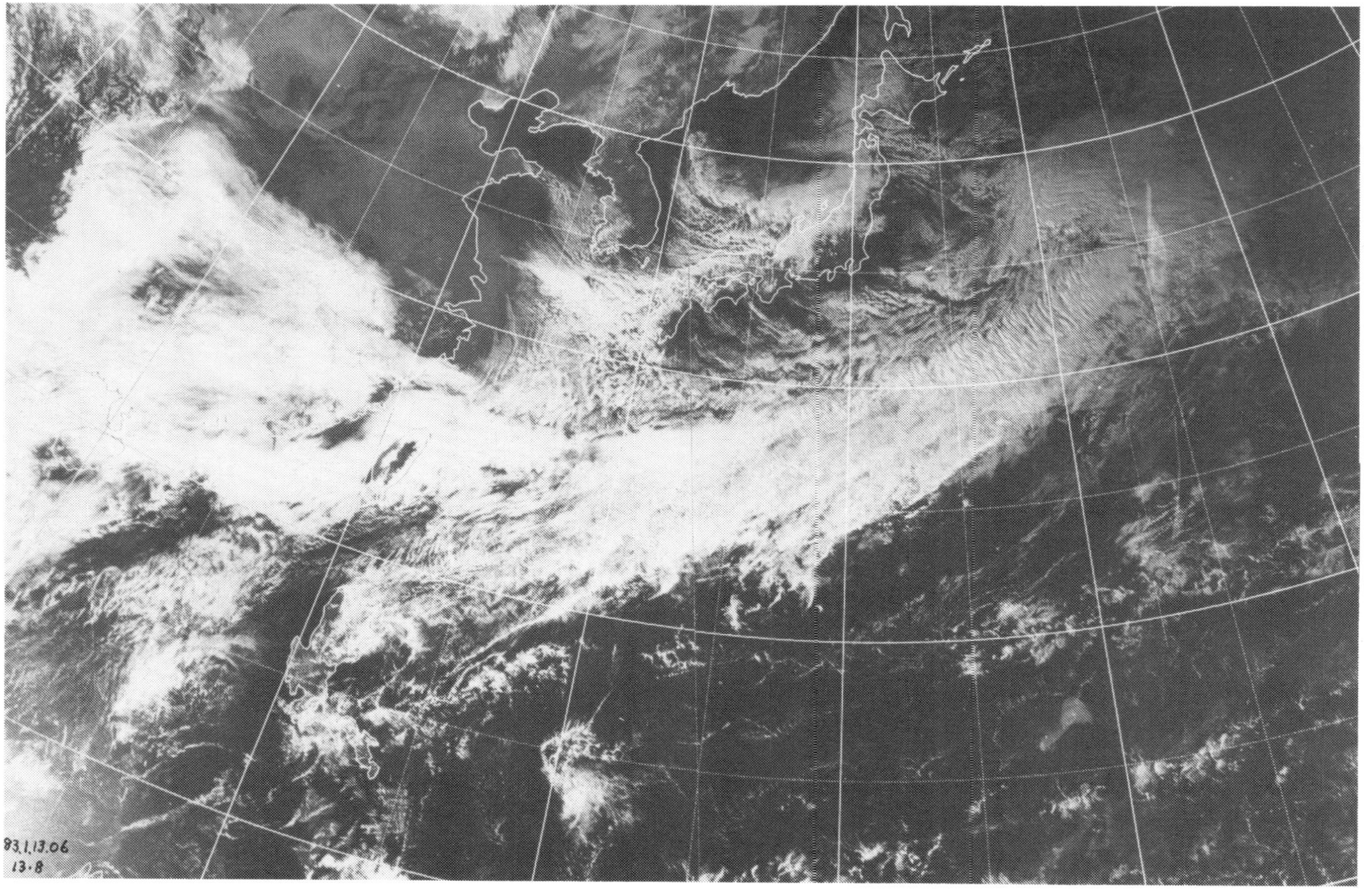

Fig. 20.vi 0600,13.1.83,Vis. A cold front has crossed Japan: a rope cloud is at the advancing front, and at the top of the high cloud there are some spike cloud formations. (GMS1; courtesy Japan Meteorological Agency.)

Fig. 20.vii Typical grass fire in central southern Africa. The smoke reaches up to the sub-cloud stable layer (inversion) and a thermal penetrates through the inversion temporarily and the contrasting very white cumulus appears. (Photo D. M. Cookson.)

Fig. 20.viii 0837,31.10.81,Ch1. A cyclone in the Arabian Sea such as may well have carried a swarm of desert locusts from the Hadhramaut of South Yemen to Pakistan across the open sea. (Courtesy Indian Meteorological Service.)

References

SECTION 1

The two weekly journals *Nature* and *Science* are essential reading to remain abreast of climate study and so, as they are the most commonly referred to, they are abbreviated as N and S respectively. The *Quarterly Journal of the Royal Meteorological Society* is abbreviated to Q.J.

The full titles are given (in italics) only for books or papers of a special nature.

[1] H. B. Squire and K. G. Winter 1951 J. Aero Sci. **18**, 271

[2] H.-H. Stolum *River meandering as a self-organising process* S**271**, 1710 22/3/96

[3] W. R. Hawthorne 1951 Proc. Roy. Soc. A **206**, 374
W. R. Hawthorne and M. E. Martin 1951 Proc. Roy. Soc. A **232**, 1843

[4] W. R. Dean 1927 Phil. Mag. **4**, 208

[5] H. Görtler Ingenieur Archiv **28**, 71

[6] J. Eustice 1925 Engineering, Nov. 13

[7] D. Brunt 1939 *Physical and Dynamical Meteorology*, 2nd edn (C.U.P.)

[8] P. Queney 1947 Misc. Rep. Dep. Met. Univ. Chicago No. 23
1948 Bull. Amer. Met. Soc. **29**, 16
P. Queney Ed. et 4al 1960 Tech. Note No. 34 WMO, No. 98, TP43

[9] R. R. Long 1953 Tellus **5**,42
1955 Ibid **7**, 341
C.-S. Yih 1960 J. Fluid Mech. **9**, 161
1967 Ibid **29**, 539
1969 "*Mechanics of non-homogeneous fluids*", Chap. 1–3 (Macmillan)

[10] G. Lyra 1943 Z. Angew. Math. u. Mech. (Berlin) **23** No. 1

[11] R. S. Scorer 1949 Q.J. **75**, 41

[12] 1953 Q.J. **79**, 70
1953(b) Gliding **4**, 119

[13] N. Gerbier and M. Berenger Q.J. **87**, 13

[14] R. S. Scorer and H. Klieforth 1959 Q.J. **85**, 131

[15] F. W. G. Warren 1961 J. Fluid Mech. **10**, 584
1950(a) Quart. J. Math. Appl. Mech. **3** 107

[16] R. S. Scorer 1950 Proc. Roy. Soc. A.**201**, 138
1951(a) Q.J. **77**, 76
F. J. W. Whipple 1930 Q.J. **56**, 287
1935 Ibid **61**, 285
I. S. Astapowitsch 1934 Ibid **60**, 493
[17] F. H. Berkshire 1975 Pageoph. **113**, 561
[18] R. S. Scorer 1956 Q.J. **82**, 75
[19] R. S. Scorer and M. Wilkinson Q.J. **82**, 419
[20] R. D. Roper 1952 Q.J. **78**, 415
[21] G. Manley 1945 Q.J. **71**, 197
[22] P. Queney 1955 Tellus **3**, 367
[23] R. S. Scorer 1953 Gliding **4**, 61
[24] G. I. Taylor 1936 Proc. Roy. Soc. A**156**, 318
[25] C. L. Pekeris 1948 Phys. Rev. **73** No. 2, 145
[26] M. V. Wilkes and K. Weekes 1947 Proc. Roy. Soc. A**192**, 80
[27] C. P. Sonett et 4al *Late Proterozoic and Palaeozoic tides, retreat of the Moon, and rotation of the Earth* S**273**, 100 5.7.96
D. A. Salstein and R. D. Rosen *Topographic forcing of the atmosphere and a rapid change in the length of the day* S**264**, 407 15.4.94
R. D. Ray et 3al *Diurnal and semi diurnal variations in the Earth's rotation rate induced by oceanic tides* S**264**, 830 6.5.94
[28] P. G. Drazin 1957 J. Fluid Mech. **4**, 215
1961 Ibid **10**, 571
1970 Ibid **42**, 321
1974 Ibid **65**, 781
[29] L. N. Howard 1961 Ibid **10**, 509
J. W. Miles 1961 Ibid **4**, 538
1959 Ibid **6**, 583
F. W. G. Warren 1968 Q.J. Mech. & Appl. Math. **21**, 433
1974 J. Fluid Mech. **68**, 413
[30] S. A. Thorpe 1968 J. Fluid Mech. **32**, 683
[31] J. D. Woods and R. L. Wiley 1972 Deep Sea Res. **19**, 87
[32] J. D. Woods 1968 J. Fluid Mech. **32**, 791
[33] R. S. Scorer 1961 Chapter in "Atmospheric Turbulence in relation to aircraft operation", R.A.E. Symposium (HMSO, London)
[34] H. Görtler 1959 Ingenieur Archiv, **28**, 71
[35] R. S. Scorer and S. D. R. Wilson 1963 Q.J. **89**, 532
R. S. Scorer 1967 J. Inst. Maths. Applic. **3**, 250
[36] J. O. Hinze 1959 "*Turbulence*" (McGraw-Hill)
[37] R. M. Davies and G. I. Taylor 1950 Proc. Roy. Soc. A **200**, 375
[38] R. S. Scorer 1957 J. Fluid Mech. **2**, 583
[39] J. M. Richards 1961 J. Fluid Mech. **11**, 369
1963(a) Int. J. Air & Wat. Poll. **7**, 17
1963(b) Q.J. **89**, 254
1965 J. Fluid Mech. **21**, 97

J. M. Richards 1968 J. Fluid Mech. **32**, 681
1970 Q.J. **96**, 702
[40] J. S. Malkus and R. S. Scorer 1955 J. Met. **12**, 43
[41] B. Woodward 1959 Q.J. **85**, 144
[42] H. C. N. Goodhart 1958 Swiss Aero. Revue, OSTIV Section **6**, 324
[43] Spackman, 1964 DIC Thesis Imp. Coll. Sci. Tech. Dept. Math.
[44] W. Schmidt 1941 Z.A.M.M. **21**, 265
[45] H. Rouse, C.-S. Yih and H. W. Humphreys 1952 Tellus **4**, 201
[46] Youxue Zhang *Dynamics of CO_2-driven lake eruptions* N**379**, 57 4.1.96
[47] T. G. Konrad and R. A. Kropfli 1968 Proc. 13th Radar Meteor. Conf. Amer Met. Soc., 262
[48] P. A. Wills 1977 "*On being a bird*" (David & Charles)
[49] J. E. Simpson 1969 Q.J. **95**, 758
1972 J. Fluid Mech. **53**, 759
1987 "*Gravity Currents*" (Ellis Horwood)
1994 "*Sea Breeze and Local Winds*" (C.U.P.)
[50] R. S. Scorer and J. L. Davenport 1970 J. Fluid Mech. **43**, 451
[51] J. S. Turner 1973 "*Buoyancy Effects in Fluids*" (C.U.P.) 43
[52] R. S. Scorer 1959 Int. J. Air Poll. **1**, 198
J. Rose Ed. 1994 "*Acid Rain*" (Gordon & Breach)
R. S. Scorer 1994 *Long distance transport*, Chapter 1 in Rose (above)
[53] D. H. Lucas et 2al 1963 Int. J. Air & Water Poll. **7**, 473
[54] O. Jorg and R. S. Scorer 1967 Atmos. Envir. **1**, 64554
[55] Hong Sungmin et 3al *Greenland ice evidence of 1hemispheric lead pollution two millennia ago by Greek and Roman civilisations* S**265**, 1841 23.9.94
[56] L. Winner *Bad surprises*, Review of "*Why things bite back*" by E. Tenner (Knopf NY) S**273**, 1052 23.8.96
[57] R. S. Scorer *The cloud absorption anomaly* Weather **51** No. 8 Aug. 1996
[58] R. M. Goody 1964 "*Atmospheric Radiation: Theoretical Basis* (O.U.P.)
[59] M. Minnaert "*Light and Colour in the Open Air*" (Dover)
[60] K. Y. Kondratyev et 2al 1996 "*High Latitude Climate and Remote Sensing*" (Wiley–Praxis)
[61] R. S. Scorer 1986 "*Cloud Investigation by Satellite*" (Ellis Horwood)
[62] R. S. Scorer 1990 "*Satellite as Microscope*" (Ellis Horwood)
[63] M. J. Bader and A. J. Waters et 3al 1995 "*Satellite and Radar Images in Weather Forecasting*" (C.U.P.)
[64] T. J. DeVries and L. Ortlieb and discussion by 10al *Determining the early history of El Nino* S**276**, 965 9.5.97
Masood, E. *El Nino forecast fails to convince sceptics* N**388**, 108 10.7.97
[65] Fu Rong et 3al *Cirrus-cloud thermostat for tropical sea surface temperatures tested using satellite data* N**358**, 394 30.7.92 and comment by:
G. Stephens and T. Sligo *An air-conditioned greenhouse* N**358**, 369 30.7.92
Further discussion N**361**, 410 4.2.93
[66] F. H. Ludlam 1963 *Severe local storms: a review* Met. Monographs **5**, 1
1966 Tellus **18**, 687

[67] C. W. Newton Tellus **18**, 699
[68] R. S. Scorer 1952 Weather **8** No. 7
[69] E. H. Hankin 1913 *"Animal Flight: a Record of Observation"* (Iliffe)
[70] R. S. Scorer 1954 *The nature of convection as revealed by soaring birds and dragonflies* Q.J. **80**
[71] J. E. Simpson 1967 British Birds **60**, 225
[72] F. A. Berson and K. G. Simpson 1971 Weather **26**, 23
W. G. Harper 1957 N**180**, 847
1958 Proc. Roy. Soc. B **149**, 484
1959 Ibid **101**, 201
[73] C. G. Johnson 1957 Q.J. **83**194
C. G. Johnson et 2al 1962 J. Anim. Ecol. **31**, 373
[74] J. Cochemé and R. S. Scorer inter al 1965 *Meteorology and the desert locust* WHO/FAO seminar—WMO Tech. Note No. 69
[75] G. W. Schaefer 1976 in *"Insect Flight"* ed. R. C. Rainey (Blackwell) Ch. 2- "Radar observations of insect flight"
[76] R. S. Scorer *"Sunny Greenland"* Q.J. **114**, 3 1.1.88
[77] Roger Daley 1991 *"Atmospheric Data Analysis"* (C.U.P.)
[78] B. J. Szabo et 3al *Thorium-230 ages of corals and duration of the last interglacial sea level high stand on Oahu Hawaii* S**266**, 93 7.10.94
[79] Orbital—Milankovitch
H.-S. Liu *Frequency variations of the Earth's obliquity and the 100 kyr ice age cycles* N**358**, 387 30.7.92
N. J. Shackleton *Last interglacial in Devil's Hole* N**362**, 596 15.4.93
J. Imbrie et 2al *Milankovitch theory viewed from Devil's Hole* N**363**, 531 10.6.93
C. Emiliani *Milankovitch theory verified* N**364**, 583 12.8.93
J. M. Landwehr et 2al *No verification for Milankovitch* N**368**, 594 14.4.94
B. Saltzman and M. Verbitsky *CO_2 and glacial cycles* N**267**, 419 3.2.94
T. J. Crawley and Kim Kuang-Yul *Milankovitch forcing of the last interglacial sea level* S**265**, 1566 7.9.94
Orbital—Inclination
R. A. Muller and G. J. MacDonald *Glacial cycles and orbital inclination* N**377**, 107 14.9.95
R. A. Muller and G. J. MacDonald *Glacial cycles and astronomical forcing* S**277**, 215 11.7.97
[80] S. Levitus et 2al *Interannual variability of temperature at a depth of 125 m in the North Atlantic Ocean* S**266**, 96 7.10.94
[81] C. Rosenzweig *Maize suffers a sea change* N**370**, 175 21.7.94
M. A. Cane et 2al *Forecasting Zimbabwean maize yield using eastern equatorial Pacific sea surface temperature* N**370**, 204 21.7.94
[82] F.-F. Jin, J. D. Neelin and M. Ghil *El Nino on the devil's staircase: Annual subharmonic steps to chaos* S**264**, 70 1.4.94
[83] E. Tziperman et al *El Nino chaos: Overlapping of resonances betweeen the seasonal cycle and the Pacific Ocean–Atmosphere oscillator* S**264**, 72 1.4.94

[84] L. Nykjaer and L. VanCamp 1997 *Seasonal and annual variability of coastal upwelling along Northwest Africa and Portugal from 1981 to 1991* recently accepted by Journal of Geophysical Research

[85] J. C. Knox *Large increases in flood magnitude in response to modest changes in climate* N**361**, 430 4.2.93

[86] G. A. Jacobs et 6al *Decade-scale trans-Pacific propagation and warming effects of an El Nino anomaly* N**370**, 360 4.8.94

A. Shemesh et 2al *Meltwater input to the Southern ocean during the last glacial maximum* S**266**, 1542 2.12.94

[87] G. A. Meehl *Coupled land–ocean–atmosphere processes and south Asian monsoon variability* S**266**, 263 14.10.94

[88] G. Parilla et 4al *Rising temperatures in the subtropical North Atlantic Ocean over the past 35 years* N**369**, 48 (see also p. 20) 5.3.94

T. Fichefet et 2al *A model study of the Atlantic thermohaline circulation during the last glacial maximum* N**372**, 252 17.11.94

[89] S. Manabe and R. J. Stouffer *Century-scale effects of increased atmospheric* CO_2 *on the ocean–atmosphere system* N**364**, 215 15.7.93

R. J. Stouffer et 2al *Model assessment of the role of natural variability in recent global warming* N**367**, 215 15.7.93

[90] Al Gore 1992 *"Earth in the Balance"* (Houghton Mifflin)

[91] P. V. Hobbs, Ed. 1993 *Aerosol–Cloud–Climate Interactions* (Academic Press)

C. J. Somerville Review of Hobbs (above) S**264**, 115 1.4.94

[92] W. J. Humphreys 1964 *"Physics of the Air"* Part V—Factors of climatic control (Dover 1964; Franklin Inst. 1920)

[93] Gary Taubes *Physicist watch global change mirrored on the Moon* S**264**, 1529 10.6.94

[94] A. Watson *Iron in the Oceans* The Guardian 4.11.93

M. L. Wells *Pumping iron into the pacific* N**368**, 295 24.3.94

J. H. Martin et 43al *Testing the iron hypothesis in ecosystems of the Equatorial Pacific Ocean* N**371**, 123 8.9.94

A. J. Watson et 9al *Minimal effect of iron fertilisation on sea surface carbon dioxide concentrations* N**371**, 143 8.9.94

Z. S. Koller et 7al *Iron limitation of phytoplankton photosynthesis in the Equatorial Pacific Ocean* N**371**, 145 8.9.94

[95] I. Renberg et 2al, *Pre-industrial atmospheric lead contamination detected in Swedish lake sediments.* N**368**, 323 24.3.94

[96] D. J. Hofman *The 1996 Antarctic ozone hole* N**383**, 129 and *Recovery of Antarctic ozone hole* N**384**, 222 21.11.96

[97] G. L. Manney et 2al *Chemical depletion of ozone in the Arctic lower stratosphere during winter 1992–93* N**370**, 429 11.8.94

[98] J. E. Harries 1994 *"Earthwatch: The Climate from Space"* (Wiley–Praxis)

[99] M. P. McCormick et 2al 1985 *Characteristics of polar stratospheric clouds as observed by SAM II, SAGE, and lidar* J. Meteor. Soc. of Japan **63** No. 2. 267

[100] Mark Chandler *Trees retreat and ice advances* N**381**, 477 6.6.96

[101] R. G. Gallimore and J. E. Kutzbach *Role of orbitally induced changes in tundra area in the onset of glaciation* N**381**, 503 6.6.96
[102] E. Jansen *Following iceberg footprints* N**360**, 212 19.11.92
[103] K. C. Taylor et 7al *The flickering switch of late Pleistocene climate change* N**361**, 432 4.2.93
[104] R. B. Alley et 10al *Abrupt increase in Greenland snow accumulation at the end of the Younger Dryas event* N**362**, 527 8.4.93
[105] P. A. Mayewski et 13al *Changes in atmospheric circulation and ocean ice cover over the North Atlantic during the last 41,000 years* S**263**, 1747 25.3.94
[106] P. A. Mayewski *Climate change during the last deglaciation in Antarctica* S**272**, 1636 14.6.96
[107] D. A. Meese et 8al *Accumulation record from the GISP2 core as an indicator of climate change throughout the Holocene* S**266**, 1680 9.12.94
[108] Greenland Ice Core Project (GRIP) Members *Climatic instability during the last interglacial period recorded in the GRIP ice core* N**364**, 203 15.7.93
[109] P. O. Wennberg et 18al *Removal of stratospheric O_3 by radicals: in situ measurements of OH, HO_2, NO, NO_2, ClO and BrO* S**266**, 398 21.10.94

SECTION 2

The following references were not specifically mentioned in the text but are important reading on climate change mainly from the two current journals *Nature* and *Science* which are abbreviated to N and S and the titles of the works are given to indicate their specific relevance. Understanding the problems of climate change has become very exciting in recent years.

*1 Ice cores

S. J. Johnson et 9al *Irregular glacial interstadials recorded in a new Greenland ice core* N**359**, 311 24.9.92
J. W. C. White *Don't touch that dial* N**364**, 186 15.7.93
G. S. Boulton *Two cores are better than one* N**366**, 507 9.12.93
W. Dansgaard et 10al *Evidence for general instability of past climate from a 250-kyr ice-core record* N**364**, 218 15.7.93
R. B. Alley et 5al *Comparison of deep ice cores* N**373**, 393 2.2.95
C. Lorius et 4al *The ice core record: climate sensitivity and future greenhouse warming* N**347**, 139 13.9.90
J. R. Petit et 11al *Four climate cycles in the Vostok ice core (to 450 Kyr BP)* N**387**, 359 22.5.97
P. Newton *Hot air, big chill* (how to get a tropical mountain snow core) N**387**, 864 26.6.97
L. G. Thompson et 9al *Tropical climate instability: the last glacial cycle from Qinghai-Tibetan ice core* S**276**, 1821 20.6.97
W. Dansgaard *Ice cores and human history* S**276**, 1013 16.5.97
J. C. Stager and P. A. Mayewski *Abrupt early to mid-holocene climatic transition registered at the equator and the poles* S**276**, 1834 20.6.97

*2 Chemistry, and ocean sediment

R. L. Edwards et 3al *Protoactinium-231 dating of carbonates by thermal ionisation mass spectrometry: implications for quaternary climate change* S**276**, 782 2.5.97

R. A. Kerr *Second clock supports orbital pacing of the ice ages* (Devil's Hole chronology) S**276**, 680 2.5.97

M. A. Altabet et 3al *Climate related variations in denitrification in the Arabian Sea from sediment* $^{15}N/^{14}N$ *ratios* N**373**, 506 9.2.95

*3 Land sediment

S. L. O'Hara et 2al *Accelerated soil erosion around a Mexican highland lake caused by prehistoric agriculture* N**362**, 48 1.3.93 and discussion by:

C. Vita-Finzi *Climate and soil erosion* N**364**, 197 15.7.93

Rainer Zahn *Fast flickers in the tropics* N**372**, 621 15.12.94

J. Quade et 3al *The late Neogene* $^{87}Sr/^{86}Sr$ *record of lowland Himalayan rivers* S**276**, 1828 20.6.97

*4 Peat cores

D. J. Bearling *Clues to the laws of ecosystems* (review of *"Climate Change and its Biological Consequences"* by D. M. Gates and of *"Biotic Interactions and Global Consequences"* Ed. P. M. Kareiva) N**364**, 24 1.7.93

R. Sukumar et 3al *A* ^{13}C *record of late quaternary climate change from tropical peats in Southern India* N**364**, 703 19.8.93

*5 Trees and tundra

Mark Chandler *Trees retreat and ice advances* N**381**, 477 6.6.96

R. G. Gallimore and J. E. Kutzbach *Role of orbitally induced changes in tundra area in the onset of glaciation* N**381**, 503 6.6.96.

I. D. Campbell and J. H. McAndrews *Forest disequilibrium caused by rapid Little Ice Age cooling* N**366**, 336 25.11.93

G. M. MacDonald et 4al *Rapid response of treeline vegetation and lakes to past climatic warming* N**361**, 243 21.1.93

X. Feng and S. Epstein *Climatic implicaations of an 8,000 yr hydrogen isotope time series from bristlecone pine trees* S**265**, 1079 19.8.94

J. Pastor *Northward march of spruce* N**361**, 208 21.1.93

H. J. Kronzucker et 2al *Conifer root discrimination against soil nitrate and the ecology of forest succession* N**385**, 59 2.1.97

P. K. Van de Water et 2al *Trends in stomatal density and* $^{13}C/^{12}C$ *ratios of Pinus flexilis needles during last glacial–interglacial cycle* S**264**, 239 8.4.94

J. A. Foley et 3al *Feedback between climate and boreal forests during the holocene epoch* N**371**, 52 1.9.94

P. C. Tzedakis *Long term tree populations in northwest Greece through multiple quaternary climatic cycles* N**364**, 437 29.7.93

P. C. Tzedakis et 2al *Climate and the pollen record* N**370**, 513 18.8.94

***6 Ozone**

J. Mahlman *A looming Arctic ozone hole* **N360**, 209 14.11.92

R. Toumi et 2al *Stratospheric ozone depletion by $ClONO_2$ photolysis* **N365**, 37 2.9.93

J. Austin et 2al *Possibility of an Arctic ozone hole in a doubled-CO_2 climate* **N360**, 221 19.11.92

A. McMinn et 2al *Minimal effects of UVB radiation on Antarctic diatoms over the past 20 years* **N370**, 547 18.8.94

C. Holden *Quicker ozone recovery forecast* **S265**, 1806 23.9.94

R. A. Kerr *Arctic ozone hole fails to recover* **S266**, 217 14.10.94

S. Mano & M. O. Andreae *Emission of methyl bromide from biomass burning* **S263**, 1255 4.3.94

R. S. Scorer *A commentary on ozone depletion theories* Atmos. Envir. **C10** 177 1976

R. S. Scorer *The stability of the stratosphere and its importance* Atmos. Envir. **11** 277 1976

R. S. Scorer *Haze in the stratosphere* **N258**, 134

P. Goldsmith et al *Nitrogen oxides, nuclear weapon testing, Concorde and stratospheric ozone* Nature 1973

M. A. Tolbert *Sulfate aerosols and polar stratospheric cloud formation* **S264**, 527 22.4.94

R. L. Miller et 5al *The "Ozone Deficit" problem: ... from 226 nm ozone photodissociation* **S265**, 1831 23.9.94

R. Toumi et 2al *Indirect influence of ozone depletion on climate forcing by clouds* **N372**, 348 24.11.94 (see also p. 322)

P. A. Newman *Antarctic total ozone in 1958* **S264**, 543 22.4.94

***7 Pollution transported**

P. V. Hobbs et 4al *Direct radiative forcing by smoke from biomass burning* **S275**, 1776 21.3.97

G. Danabasoglu et 2al *The role of mesoscale tracer transports in the global ocean circulation* **S264**, 1123 20.5.94

S. M. Stanley and X. Yang *A double mass extinction at the end of the Paleozoic era* **S266**, 1340 25.11.94

N. Roberts et 5al *Timing of the younger Dryas event in East Africa from lake level changes* **N366**, 146 11.11.93

***8 Sunshine**

R. Dawkins *The eye in a twinkling* **N368**, 690 21.4.94

R. S. Scorer *Haze in the stratosphere* **N258**, 13

P. Fourkal *Stellar luminosity variations and global warming* **S264**, 238 8.4.94

R. A. Radick *Stellar variability and global warming* **S266**, 1072 11.11.94

A. Jones et 2al *A climate model study of indirect radiative forcing by anthropogenic sulphate aerosols* **N370**, 450 (see also p. 420) 11.8.94

***9 Clouds**

R. S. Scorer *Ship trails* Atmos. Envir. **21** No. 6 417 1987

F. H. Ludlam and R. S. Scorer *Convection in the atmosphere* Quart. J. Roy. Met. Soc. **79** 317 1953

J. H. Conover J. Atmos. Sci. **23** 778 1966

T. Novakov and J. E. Penner *Large contribution of organic aerosols to cloud condensation nuclei concentrations* N**365**, 823 28.10.93

R. S. Scorer *"Cloud Investigation by Satellite"* (Ellis Horwood 1986)

R. S. Scorer *"Satellite as Microscope"* (Ellis Horwood 1990)

R. S. Scorer *Spreading out of downdraughts* Weather **8** No. 7, July 1953

J. M. Wallace *Dynamics in the balance* (review of *"Physics of Climate"* by J. P. Piexoto and A. H. Oort) N**360**, 220 1992

V. Ramanathan and W. Collins N**351**, 27 1991

T. Novakov and J. E. Penner *Large contribution of organic aerosols to cloud-condensation-nuclei concentrations* N**365**, 823 28.10.93

Aigno Dai et 2al *Clouds, precipitation and temperature range* N**386**, 665 17.4.97

D. R. Easterling et 10al *Maximum and minimum temperature trends for the globe* S**277**, 364 18.7.97

G. Malin *Sulphur, climate and the microbial maze* N**387**, 857 26.6.97

*10 Ocean temperature and algal blooms

P. J. Valero et 4al *Direct radiometric observations of the water vapour greenhouse effect over the equatorial Pacific Ocean* S**275**, 1773 21.3.97

D. Paillard and L. Labeyrie *Role of the thermohaline circulation in the abrupt warming after Heinrich events* N**372**, 162 10.11.94

J. W. Beck et 4al *Abrupt changes in early Holocene tropical sea surface temperature derived from coral records* N**385**, 705 20.2.97

U. Send et 9al *Acoustic observations of heat content across the* (Western) *Mediterranean sea* N**385**, 615 13.2.97

R. S. Webb et 4al *Influence of ocean heat transport on the climate of the last glacial maximum* N**385**, 695 20.2.97

Ed Boyle and A. Weaver *Conveying past climates* N**372**, 41 3.11.94

P. C. Sereno et 4al *Early Cretaceous dinosaurs from the Sahara* N**266**, 267 14.10.94 (see also p. 219)

W. R. Hammer and W. J. Hickerson *A crested theropod dinosaur from Antarctica* S**264**, 828 6.5.94

K. R. Arrigo et 4al *Primary production in Antarctic sea ice* S**276**, 394 18.4.97

K. S. Brown *Taking the measure of life in the ice* S**276**, 353 18.4.97

C. H. Fritzen et 3al *Autumn bloom of the Arctic pack ice algae* S**266**, 782 4.11.94

M. S. V. Douglas et 2al *Marked post-18th century environmental change in high-arctic ecosystems* S**266**, 416 21.10.94

K. R. Arrigo and C. R. McClain *Spring phytoplankton production in the western Ross Sea* S**266** 14.10.94

K. E. Clifton *Mass spawning by green algae on coral reefs* S**275**, 1116 21.2.97

J. E. Smith et 3al *Rapid climate change in the North Atlantic during the Younger Dryas recorded by deep sea corals* N**386**, 818 17.4.97

J. D. Kahl et 4al *Absence of evidence for Greenhouse warming over the Arctic Ocean in the past 40 years* N**361**, 355 28.1.93

E. Cortijo et 5al *Eemian cooling in the Norwegian sea and North Atlantic Ocean preceding continental ice sheet growth* N**372**, 446 1.12.94

Guoqui Gao *The temperatures and oxygen isotope composition of early Devonian oceans* N**361**, 712 25.3.93

H. F. Lamb et 7al *Relation between century-scale Holocene arid intervals in tropical and temperature zones* N**373**, 134 12.1.95

*11 Ocean mixing

J. M. Toole et 2al *Estimates of diapycnal mixing in the abyssal ocean* S**264**, 1120 20.5.94

B. W. Sellwood et 2al *Cooler estimates of cretaceous temperatures* N**370**, 453 11.8.94

A. Kumar et 2al *Simulations of atmospheric variability induced by sea surface temperatures and implications for global warming* S**266**, 632 28.10.94

K. Schmidt *ATOC delayed as report laments research gaps* S**264**, 339 15.4.94

A. Sanyal et 3al *Evidence for a higher pH in the glacial ocean from boron isotopes in foraminifera* N**373** 234 19.1.95

P. L. Woodworth *Ocean topography measured by satellite altimetry* Hydrographic Society Technical Seminar, London, Oct. 1987

A. Sy et 6al *Surprisingly rapid spreading of newly formed intermediate waters across the North Atlantic Ocean* N**386**, 675 17.4.97

T. Rice *Review of "Deep Atlantic: Life, Death and Exploration in the Abyss" by Richard Ellis* N**386**, 238 20.3.97

R. G. Lueck and T. D. Mudge *Topographically induced mixing around a shallow sea mount* S**276**, 1831 20.6.97

*12 Iron and CO_2 in the oceans

Z. S. Kolber et 7al *Iron limitation of phytoplankton photosynthesis in the equatorial Pacific Ocean* N**371**, 145 8.8.94

J. C. Williams et 2al *Anisotropy and coherent vortex structures in planetary turbulence* S**264**, 410 15.4.94

A. F. Michaels et 4al *Carbon-cycle imbalances in the Sargasso sea* N**372**, 537 8.12.94

H. J. de Baar et 6al *Importance of iron for plankton blooms and carbon dioxide drawdown in the southern ocean* N**373**, 412 2.2.95

P. A. del Giorgio et 2al *Respiration rates in bacteria exceed phytoplankton production in unproductive aquatic systems* N**385**, 148 9.1.97

J. W. Murray et 4al *Physical and biological controls on carbon cycling in the Equatorial Pacific* S**266**, 58 7.10.94

P. G. Falkowski *Evolution of the nitrogen cycle and its influence on the biological sequestration of* CO_2 *in the ocean* N**387**, 272 15.5.97

D. A. Hansell et 2al *Predominance of vertical loss of carbon from surface waters of the equatorial Pacific Ocean* N**386**, 59 6.3.97

***13 Chaos**

R. V. Sole and J. Bascompte *Egological chaos* and reply by L. Stone N**367**, 418 3.2.94

J. N. Perry et 3al *Estimating chaos in an insect population* S**276**, 1881 20.6.97

***14 Proof of change**

K. Hasselman *Are we seeing global warming?* N**226**, 914 9.5.97

R. N. Harris and D. S. Chapman *Borehole temperatures and a baseline for 20th century warming estimates* S**275**, 1618 14.3.97

Christopher Charles *Cool tropical punch of the ice ages* S**385**, 681 20.2.97

P. C. Frisch *Morphology and ionisation of the interstellar cloud surrounding the solar system* S**265**, 1423 2.9.94

R. K. Kaufmann and D. I. Stern *Evidence for human influence on climate from hemispheric temperature relations* N**383**, 674 24.10.96

J. A. Carton *See saw sea* N**385**, 487 6.2.97

***15 Making models**

J. D. Neelin and J. Marotzke *Representing ocean eddies in climate models* S**264**, 1099 20.5.94

S. Rahmstorf *Rapid climate transition in a couple ocean–atmosphere model* N**372**, 82 3.11.94

M. Latif and T. P. Barnett *Causes of decadal climate variability over the North Pacific and North America* S**266**, 634 (see also p. 544) 28.10.94

J. E. Lovelock and L. R. Kump *Failure of climate regulation in a geophysical model* N**369**, 732 30.6.94

R. A. Kerr *Climate modelling's fudge factor comes under fire* S**265**, 1528 9.9.94

R.A. Kerr *Greenhouse forecasting still cloudy* S**276**, 1040 followed by group of articles on atmospheric chemistry 16.5.97

R. S. Lindzen *Paleoclimate sensitivity* N**363**, 25 6.5.93

K. E. Trenberth *The use and abuse of climate models* N**386**, 131 13.3.97

J. W. Hurrell and K. E. Trenberth *Spurious trends in satellite MSU temperatures from merging different satellite records* N**386**, 165 13.3.97

U. Mikolajewicz et 3al *Modelling teleconnections between the North Atlantic and North Pacific during the Younger Dryas* N**387**, 384 22.5.97

S. Edouard et 3al *Effect of small scale inhomogeneities on* (model) of *ozone depletion in the Arctic* N**384**, 444 5.12.96

***16 Icebergs**

R. Stein et al *The last deglaciation event in the eastern central Arctic Ocean* S**264**, 692 29.4.94

T. Brey et 9al *Antarctic benthic diversity* N**368**, 297 24.3.94

***17 North Sea**

H. Schmidt et H. Von Storch *German Bight storms analysed* N**365**, 791 28.10.93

R. Koenig *Black spots blot German coastal flats* S**273**, 25 5.7.96

*18 Sea level

D. J. Jacobs et 2al *Climate induced fluctuations in sea level during non-glacial times* N**361**, 710 8 25.2.93

D. J. Stanley and A. G. Warne *Sea level and initiation of predynastic culture in the Nile Delta* N**363**, 435 3.6.93

D. L. Holmes *Rise of the Nile Delta* N**363**, 437 3.6.93

J. B. Laronne and I. Reid *Very high rates of bedload sediment transport by ephemeral desert rivers* N**366**, 113 and 148 7.10.93

W. R. Peltier *Paleotopography* S**265**, 195 8.7.94

R. L. Edwards *Paleotopography of glacial age ice sheets* (and reply by W. R. Peltier) S**267**, 536 27.1.95

V. Gorlitz, C. Rosenzweig and D. Hillel *Is sea level rising or falling?* N**371**, 481 6.10.94

D. L. Sahagian et 2al *Direct anthropogenic contributions to sea level rise in the 20th century* N**367**, 54 6.1.94

B. Fong Chao and E. Rodenburg *Discussion and reply* by D. L. Sahagian et 2al N**370**, 258 28.7.94

C. G. A. Harrison *Rises and falls* (review of *"Phanerozoic Sea Level Changes"* by A. Hallam) N**365**, 220 16.9.93

D. Bromwich *Ice sheets and sea level* N**373**, 18 5.1.95

*19 Heinrich (Dansgaard–Oeschger) events

D. Paillard and L. Labeyrie *Role of the thermohaline circulation in the abrupt warming after Heinrich events* N**372**, 162 10.11.94

G. C. Bond and R. Lotti *Iceberg discharges into the North Atlantic on Millennial time scale during the last glaciation* S**267**, 1005 17.2.95

G. C. Bond et 6al *Correlations between climate records from North Atlantic sediments and Greenland ice* N**365**, 143 9.9.93

S. Lehman *Ice sheets, wayward winds and sea change* N**365**, 108 9.9.93

J. C. Jezek *Drought in Greenland* N**366**, 17 4.11.93

J. Adams vs J. E. Lovelock and L. R. Kump (correspondence) *Weathering and glacial cycles* N**373**, 110 12.1.95

*20 Glaciers

N. R. Iverson et 4al *Flow mechanism of glaciers on soft beds* S**267**, 80 6.1.95

J. Oerlemans *Quantifying global warming from the retreat of glaciers* S**264**, 243 8.4.94

J. L. Best and P. J. Ashworth *Scour in large braided rivers and the recognition of sequence stratigraphic boundaries* N**387**, 275 15.5.97

*21 Food, and ENSO

C. Rosenzweig and M. L. Parry *Potential impact of climate change on world food supply* N**367**, 133 and 118 13.1.94

Per Gloersen *Modulation of hemispheric sea-ice cover by ENSO events* N**373**, 503 9.2.95

W. R. Kapsner et 4al *Dominant influence of atmospheric circulation in Greenland over the past 18,000 years* N**373**, 52 5.1.95

D. M. Karl et 6al *Ecosystem changes in the North Pacific subtropical gyre attributed to the 1991–92 El Nino* N**373**, 230 19.1.95

R. C. Stone et 2al *Prediction of global rainfall probabilities using phases of the Southern Oscillation Index* N**384**, 252 21.11.96

D. H. Sandweiss et 4al *Geoarchaeological evidence from Peru for a 5,000 years BP onset of El Nino* S**273**, 1531 13.9.96

*22 Carbon dioxide

K. Caldiera and J. F. Kasting *Insensitivity of global warming potentials to carbon dioxide emission scenarios* N**366**, 251 18.11.93

R. A. Berner *The rise of plants and their effect on weathering and atmospheric* CO_2 S**276**, 544 25.4.97

R. S. Webb and J. T. Overpeck *Carbon reserves released? (source in Canadian tundra)* N**361**, 497 11.2.93

W. C. Oeckel et 5al *Recent change in arctic tundra ecosystems from a net carbon dioxide sink to a source* N**361**, 520 11.2.93

T. M. Smith and H. H. Shugart *Transient response of terrestrial carbon storage to a perturbed climate* N**361**, 523 11.2.93

L. J. Rothschild *Carbon dioxide and diatom mats (High levels of* CO_2 *20 Ma BP)* N**368**, 817 28.4.94

J. W. C. White et 4al *A high resolution record of atmospheric* CO_2 *content from carbon isotopes in peat* N**367**, 153 1.3.93

R. J. Francey et 5al *Changes in oceanic and terrestrial carbon uptake since 1982* N**373**, 326 26.1.95

*23 Volcanos, earth movements

G. A. Zielinski et 8al *Record of volcanism since 7,000 B.C. from the GISP2 Greenland ice core and implications for the volcano–climate system* S**264**, 948 13.5.94

T. Simkin *Distant effects of volcanism—how big and how often?* S**264**, 913 13.5.94

P. Allard et 4al *Sulphur output and magma degassing budget of Stromboli volcano* N**368**, 326 24.3.94 W. J. Humphreys *"Physics of the Air"* (Dover 1964) *Part V "Factors of Climatic Control".*

L. H. Roydon et 6al *Surface deformation and lower crustal flow in Eastern Tibet* S**276**, 788 2.5.97

L. J. Wolfe et 3al *Seismic structure of the Iceland mantle plume* N**385**, 245 16.1.97

G. H. Miller et 2al *Low latitude glacial cooling in the southern hemisphere from amino acid racemization in emu eggshells* N**385**, 241 16.1.97

A. J. Kaufman *An ice age in the tropics* N**386**, 227 20.3.97

D. R. Hilton et 2al *Effect of shallow-level contamination in the helium isotope systematics of ocean-island lavas* N**373**, 330 26.1.95

R. A. Beck et 13al *Stratigraphic evidence for an early collision between Northwest India and Asia* N**373**, 55 5.1. 95

J. W. Head III *Venus after the flood* N**372**, 729 22.12.94

M. P. McCormick et 2al *Atmospheric effects of the Mt Pinatubo eruption* N**373**, 399 2.2.95

G. Ramstein et 3al *Effect of orogeny, plate motion and land–sea distribution on Eurasian climate change over the last 30 million years* N**386**, 788 17.4.97

*24 Ice

K. S. Brown *Taking the measure of life in the ice* S**276**, 353 18.4.97
K. R. Arrigo et 4al *Primary production in Antarctic sea ice* S**276**, 394 18.4.97
D. A. Evans et 2al *Low-latitude glaciation in the Palaeoproterozoic era* N**386**, 262 20.3.97

*25 Vegetation

J. Kutzbach et 3al *Vegetation and soil feedbacks on the response of the African monsoon to orbital forcing in the early to middle Holocene* N**384**, 623 19.12.96
D. Jolly and A. Haxeltine *Effect of low glacial atmospheric CO_2 on tropical African montane vegetation* S**276**, 786 1.5.97
C. Whitlock and P. J. Bartlein *Vegetation and climate change in Northwest America during the past 125 kyr* N**388**, 57 3.7.97
R. B. Myneni et 4al *Increased plant growth in the northern high latitudes from 1981 to 1991* N**386**, 698 17.4.97
N. Nichols *Increased Australian wheat yield due to recent climate trends* N**387**, 484 29.5.97
B. L. Otto-Bliesner and G. R. Upchurch *Vegetation induced warming of high latitude regions during the late cretaceous period* N**385**, 805 27.2.97
J. Guiot *Back at the last interglacial* N**388**, 25 3.7.97
S. Stokes et 2al *Multiple episodes of aridity in Southern Africa since the last interglacial period* N**388**, 154 10.7.97
M. S. V. Douglas et 2al *Marked post-18th century environmental change in high-arctic ecosystems* S**266**, 416 21.10.94

*26 Orbital

T. J. Crowley and Kwang Yul Kim *Milankovitch forcing of the last interglacial sea level* N**265**, 1566 9.9.94
R. A. Kerr *Upstart ice age theory gets attentive but chilly hearing* S**277**, 183 11.7.97
Han-Shou Liu *Frequency variations of the Earth's obliquity and the 100 kyr ice age cycles* N**358**, 397 30.7.92
B. Steinberger and R. J. O'Connell *Changes of the earth's rotation axis owing to advection of mantle density heterogeneities* N**387**, 169 8.5.97

*27 Disasters

A. Cooper and D. Penny *Mass survival of birds across the Cretaceous–Tertiary boundary: molecular evidence* S**275**, 1109 21.2.97
S. Rojstaczer et 2al *Permeability enhancement in the shallow crust as a cause of earthquake-induced hydrological changes* N**373**, 237 19.1.95
K. Zahle *Leaving no stone unburned* (Tunguska; see [16]) N**383**, 674 24.10.96

LIA Little Ice Age, lasting in NW Europe from roughly 1350 to 1880, since when the warming may have been due to the Industrial Revolution.

LIG Last Interglacial.

Ma or Myr Million years.

MAR Mass accumulation rate.

Meltwater Rain, snow or melting ice produce fresh water but the original freezing of salty water leaves the salt unfrozen: surface ice generates saltier water below. Arctic winters therefore generate fresh top water and salty bottom water. Evaporation in low latitudes generates saltier water (see Fig. 16.7.v).

Micron 10^{-6} m. A unit of length used as a measure for wavelength and droplet sizes particularly. 10^{-3} = 1 mm.

Milankovitch The originator of the 'orbital' theory of ice-age change, in which it depends on the variations of sunshine depending on the ellipticity of the earth's orbit and the precession of the equinoxes and variations in the inclination of the earth's axis to the ecliptic.

Mintra The lowest altitude at which persistent contrails can be formed. It depends on the hydrogen content of the fuel (see section 14.3).

Mistral A cold north wind blowing down the Rhone valley to the Mediterranean coast.

MWP Medieval Warm Period, roughly 500–1300 AD.

NADW North Atlantic Deep Water.

NO_x A mixture of the gaseous oxides of nitrogen usually meaning only NO and NO_2 from combustion.

NPK Fertiliser containing nitrogen, phosphorus and potassium.

PAN Peroxyacetyl nitrate (see p. 610).

Parameter A number used to represent some physical feature of a situation; specifically used to replace complex features such as clouds by a number giving roughly the same effect in a GCM or computer model used in forecasting.

Particle In describing clouds this word is taken to include droplets as well as ice crystals and even hail.

PBN Peroxybenzoyl nitrate (see p. 610).

ppb(v) Parts per billion (by volume).

PSC Polar stratospheric cloud (see section 18.6).

Puff The motion pattern of a body of fluid injected into an environment of similar fluid at rest, but not with concentrated vorticity as in the case of a vortex ring, and mixing into it with a motion pattern very similar to that of a thermal.

Rough A crude approximation correct to order of magnitude; far from exact.

SMOW Standard Mean Ocean Water, defining a standard for the proportion of $^{18}O/^{16}O$.

Spicula A small spike-like protuberance ejected from a supercooled water drop which freezes on the outside to form a shell which is shattered when the interior also freezes, the result being that the freezing is initiated in neighbouring drops.

Spectrum The distribution of a factor defining the quality of a phenomenon. Used specially of the strength of the wavelength distribution in a radiation beam.

SST Sea Surface Temperature.

Stadial A substage of the Ice Age era of the World's weather evolution. A period dominated by ice.

Strake A helical salient edge built around a metal chimney to spoil the creation of a vortex street downwind and the periodic oscillations which it would create.

Stratosphere The part of the atmosphere immediately above the troposphere and below the mesophere, i.e. roughly from 10 km up to 30 km, stably stratified but having large wind shears, and heights greater in equatorial but less in polar latitudes.

Stratus Literally a layer of cloud or fog, usually a thin layer in the case of cirrus, but may be thick at alto level.

Super-cooled Sometimes called under-cooled, and meaning having a temperature colder than would be needed to change its state, but not actually changed, e.g. water at a temperature below 0°C but not frozen.

Surface tension The effect whereby a liquid behaves as if it had a membrane on its surface in a state of tension.

Terminator The line of the edge of direct sunshine at the earth's surface. This is not sharp in Channel 1 and 2 images because of the twilight scatter due to cloud and haze, but is sharp in Channel 3; although sea glint and forward scatter by the very small droplets at cloud base may form a 'wave guide' into the darkness.

Thermal Adjectivally meaning to do with temperature. Thus 'thermal wind' refers to the vertical wind gradient caused by a horizontal gradient of temperature. As a noun, a 'thermal' is a mass of buoyant fluid rising through its surroundings, its buoyancy having been the result of thermal expansion.

Tropopause The surface between the stratosphere and the troposphere in which the mixing due to weather mechanisms which end at the tropopause.

Troposphere The lowest layer of the atmosphere in which the weather causes a mixing of the constituents.

Vortex Mathematically a fairly good approximation to the motion pattern left behind by an aircraft (see sections 11.10 and sections 14.3). A rough model for flapping birds.

Water cloud Cloud composed of water droplets only, no ice crystals; notable when its temperature is all below freezing point.

Window A gap in an absorption spectrum through which radiation passes if it is of an appropriate wavelength.

Younger Dryas The most recent climatic period dominated by ice, lasting in Greenland from 12.9 to 11.7 kyr BP.

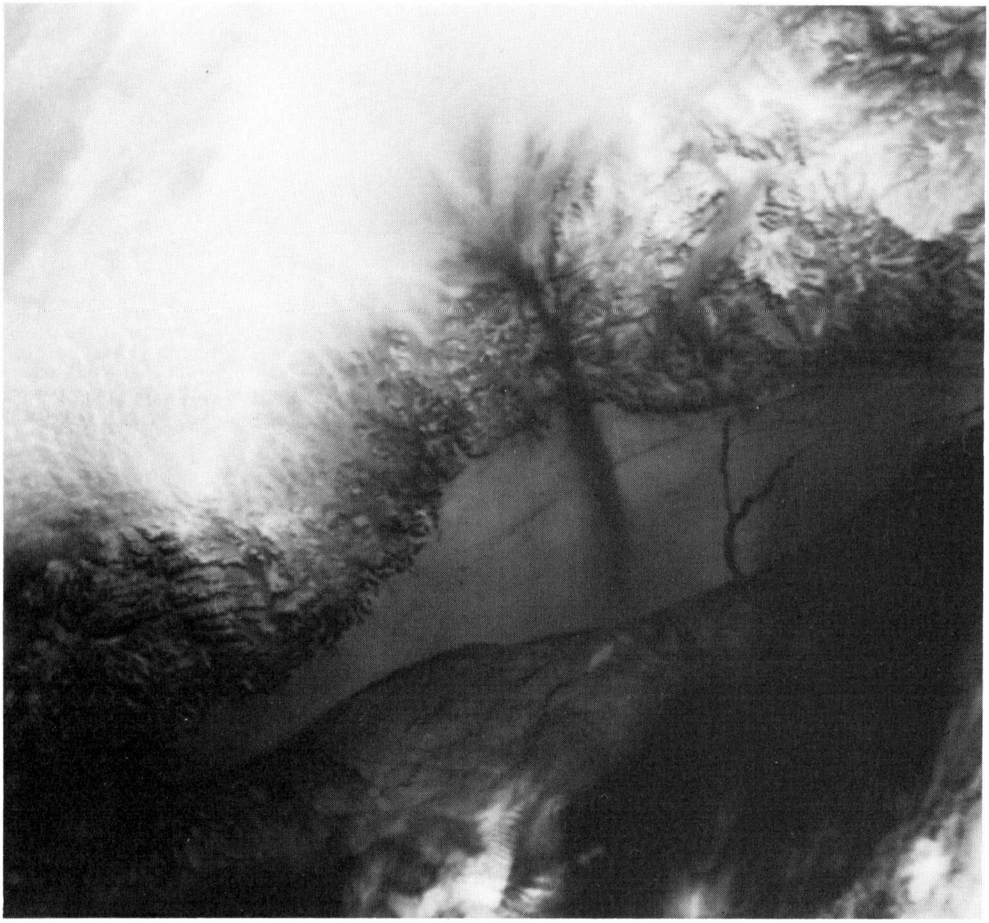

This is a normal enlargement of a small part of the frontispiece to Part 1 (page 2). It shows very well how the infra-red (Channel 4) image, designed to show temperature variations over the ground and cloud tops, picks out the warm and relatively moist air in the katabatic wind from the Kangerdslugssuaq hinterland and can easily show temperature effects which cannot be seen in Channel 2 although they are partially seen in Channel 3 frontispieces to Parts 2 and 3 (pages 274 and 558). To detect this intensity is a matter of chance, and probably depends on having a few hours only of very light wind which allows the katabatic effect to be fully developed at the moment that the NOAA satellite passes overhead. We are seeing the water vapour in the wind when its temperature is well above that of the background scene. (Courtesy the University of Dundee.)

Mathematical symbols

A few symbols which occur only in one section have not been included: their meaning is clear from the context. References to equations are in the form (2.3.4), to figures, 2.3.iv, to sections, 2.3, and to pages, 234.

a	radius of sphere (2.4.10); chimney radius; particle dimension (2.10.2); tangential acceleration (3.1.12)
$\mathbf{a}$	non-divergent vector (1.4.5)
b	horizontal obstacle dimension (5.13.i); plume width (12.2.i)
$\mathbf{b}$	unit binormal (3.1.6)
B	buoyancy number (2.14.1) (10.2.1) (10.2.7) (10.13.5)
$\mathbf{B}$	Bernoulli vector $\mathbf{t} \times \mathbf{w}$
C	universal dimensionless constant (10.2.1), (11.9.1); dimensional constant (10.11.12) (10.12.14)
C_D	drag coefficient (2.12.1)
c	velocity of sound (1.8.8); transition height (10.11.14) (10.11.17)
d	body dimension (2.10.1); density current depth (11.9.1)
D	drag force (2.10.2)
E	turbulent energy (9.1.1)
e_{ij}	rate of strain (1.3.1); $e_{r\theta}$ (9.9.17)
f	Coriolis parameter (5.7.1); f_3 Ch. 4
$\mathbf{f}$	fluid acceleration = $\mathrm{D}\mathbf{v}/\mathrm{D}t$ (1.4.1) (3.1.9); earth's rotation (4.1.1)
$\mathbf{F}$, F_i	body or total force
F	Froude number (2.14.3) (10.10.6), Rossby number (5.8.14)
F	buoyancy or pollution flux (3.4.6) (11.4.3) (12.4.3)
$\mathbf{g}$	gravity, magnitude g
$\mathbf{g}^*$	apparent gravity (4.1.6)
h	depth (2.7.1); representative height (2.14.2); isobaric contour height (4.4.1); layer depth Chs 5, 6; chimney height (12.2.iii); h' = thermal rise (12.2.vi)
H	Bernoulli constant gH (1.7.4) or (1.7.9); layer depth (6.2.i); effective chimney height (12.2.vi)
$\mathbf{j}$	unit vector across flow (5.1.6)

k line vortex strength (2.7.4) (9.8.2); wave number Chs 5, 6; k^* lee wave number; dimensional constant (10.2.9); momentum factor (10.7.1) (10.11.2)
k_0 isentropic constant (1.8.1)
K eddy transfer coefficient (4.12.1); vortex strength (9.10.ii)
K_M momentum transfer coefficient (11.6.5)
K_H heat transfer coefficient (11.6.1)
1 unit vector (1.1.1)
ℓ Scorer parameter 153; mixing length
L $= \ln(\frac{1}{2}\rho q^2)$ (3.2.6)
m source strength (2.4.7); volume factor 311, (10.2.9)
M Mach number $= U/c$
$\dot{M}$ momentum flux (10.9.20)
n widening factor (10.2.6)
n unit normal (3.1.3)
N plume dilution factor (12.2.1)
p pressure, p_m mean pressure (1.3.18); lee wave parameter (6.4.19)
p_{ij} stress 14
P pollution concentration (12.2.1)
q fluid speed (1.2.5)
Q volume or pollution flux (10.13.2), (12.2.1)
r radial coordinate
r position vector (4.1)
R gas constant (1.6.3); radius of earth Ch. 4; radius or width of thermal, puff, jet, plume Ch. 10; vortex pair (11.10.1)
Ra Rayleigh number (11.2.1)
Re Reynolds number (2.10.1) (9.3.11)
Ri Richardson number (2.14.4), 236, (9.5.3)
R $= \frac{1}{\rho}\,\mathbf{grad}\,\rho$ or $\frac{1}{\tau}\,\mathbf{grad}\,\tau$ (1.4.1)
s distance along trajectory
S Strouhal number (2.13.1); stability number (10.12.7)
S stream stratification (5.4.5); stratification factor 319
t time
t unit tangent (3.1.1) $= \mathbf{v}/q$
T absolute temperature (1.6.1); kinetic energy 6
u x- or r-velocity component, velocity perturbation 5.8
u velocity vector, magnitude u
U inflow velocity 336; stream velocity or wind speed
v y- or θ-velocity component
v velocity $= (u, v, w)$ in Cartesian or cylindrical polar coordinates
$\mathbf{v}_A$ ageostrophic component of velocity (4.3.5)
$\mathbf{v}_G$ geostrophic component of velocity (4.3.4)
$\mathbf{v}_i$ isallobaric wind (4.3.12)
V transverse wind 5.8; speed of front (11.9.1)

w	vertical velocity component
w_c	vertical velocity of cap of thermal
W	vertical velocity (8.11.1)
x	horizontal streamwise Cartesian coordinate
$\mathbf{x}$	position vector
y	horizontal transverse Cartesian coordinate
Y	yield factor (10.4.1)
z	vertical or axial coordinate
α	angle of deflection of flow (2.6.1), 3.6.i; inclination of front (4.8.1); velocity stratification (shear) (5.4.4); entrainment ratio (10.11.3); plume cone angle 12.2.i
β	generally static stability (1.8.2) (5.2.4) (10.12.1); angle of helix 8.6.ii; β-plane parameter (7.4.2)
γ	ratio of specific heats (1.6.6), helical angle of vortex line 8.6.iii
Γ	adiabatic lapse rate (1.8.3)
δ	boundary layer thickness (2.9.1)
ϵ	energy dissipation rate (9.3.12)
ζ	vertical displacement (1.8.5), 5.1.1; vertical or z-component of vorticity (1.4.7), (5.1.1)
η	y-component of vorticity; = vorticity in 2D motion (Chs 5, 6, 7); magnitude of shear
H	vorticity density (11.10.4)
θ	potential temperature (1.6.10); angle of bend (Ch. 3); colatitude (8.6.4), 4.1.i and spherical polar coordinates; inclination of thermal (11.5.1)
κ	Hele–Shaw porosity (2.8.6); trajectory curvature (3.1.3); von Karman's constant (9.3.7)
λ	dynamic viscosity (1.3.7) and (1.3.23); stratification (3.6.2); turbulent intensity (12.4.7)
μ	dynamic viscosity (1.3.7) and (1.3.23); doublet strength (2.4.8); stability coefficient for curved flow (8.6.14); vertical decay exponent (6.4.12); bending-over factor, 12.4.ii
ν	kinematic viscosity (1.3.28); vertical wave number (5.10.3)
Π	impulsive pressure (1.1.11)
ρ	density
σ	D'Arcy porosity (2.8.7); concentration of pollution (10.9.11), of insencts (15.9.1), frequency (2.13.1), (7.3.19), (9.11.17); convection fraction (9.4.1); dimensionless constant (10.12.13)
τ	$1/\theta$ (potential temperature)$^{-1}$ (1.6.10); trajectory torsion (3.1.7); shear stress (9.3.7); mixing time (9.3.8)
τ_{ij}	stress tensor (4.12.7)
ϕ	potential (1.1.6) (2.4.2) (8.1), angular displacement (1.8.9), (3.2.ii) (3.5.ii), ridge inclination 184; entropy 14.2
Φ	buoyancy flux (10.11.8)
ψ	stream function (8.3.1); streamline inclination (8.4.2)
ξ	x-component of vorticity

$\boldsymbol{\omega}$	vorticity = **curl v** (1.2.3), = (ξ, η, ζ) (1.3.2)
ω_s	secondary vorticity = $\boldsymbol{\omega}\,.\,\mathbf{t}$ (3.1.18)
ω_n	normal component vorticity = $\boldsymbol{\omega}\,.\,\mathbf{n}$ (3.3.1)
Ω	angular velocity of rotation (1.3.3) (2.7.2) (9.8.2) (7.11.1)
$\boldsymbol{\Omega}$	earth's angular velocity (Ch. 4)
ϖ	modified pressure (1.6.11)

Index

References to equations are in the form (2.3.4), to figures, 2.3.iv, to sections, 2.3, and to pages, 234.

***28 God, animals**
E. J. Larson and L. Witham *Scientists are still keeping the faith* N**386**, 435 3.4.97
B. Appleyard *"Understanding the Present"* (Picador 1992)
P. Davies *"The Mind of God"* (Simon & Schuster 1992)
W. R. Hammer and W. J. Hickerson *A crested theropod dinosaur from Antarctica* S**264**, 828 6.5.94
P. C. Sereno et 4al *Early Cretaceous dinosaurs from the Sahara* S**266**, 267 14.10.94
C. R. Chapman *Review of "T. Rex and the Crater of Doom" by Walter Alvarez* N**387**, 33 1.5.97

Glossary

AADW	Antarctic deep water.
Absorption	Usually refers to Beer's law for the rate of reduction of a ray or signal on passage through a medium such as a gas or liquid which is partly transparent.
ACR	Arctic cold reversal, the southern hemisphere's equivalent to the Younger Dryas.
AF	Accompanying fluid—the fluid which travels with a vortex system and is not part of the fluid which passes by or through the vortex (system) with the main stream.
Albedo	The whiteness of a surface, the proportion of sunshine which is reflected (scattered) back into space and not absorbed by the earth.
Allerød–Bølling	Warm periods immediately preceding the Younger Dryas.
Ambient air	The surrounding atmospheric environment.
Anthropogenic	A word not used in this book. Used elsewhere to imply 'man-made' but is used because that word is thought not to be politically correct, although it would be as sexist in Greek. Etymologically it means 'man-making', as carcinogenic means cancer-making, anthropophagi means cannibal (*phagein* = to eat, Gk), anthropomorphic means changed to be like man.
Anvil	The flat-topped cloud formed when thermal convection reaches a stable layer and spreads horizontally at its equilibrium level.
Approximate	A close approach to … being the best substitute for exact equality but not rough (q.v.).
ATOC	Acoustic thermometry of ocean climates.
Baroclinic	Having horizontal gradients of density.
Barotropic	Of uniform density; well stirred.

Benthic	Ocean bed.
Bise	A cold N/NE wind of central Europe overlapping both the Mistral and Bora.
Billows	Unstable waves in a statically stable layer, usually caused by large shear, the waves being of a particular wavelength.
Black body	A material body or substance emitting and absorbing a range of wavelengths and emitting a radiation spectrum according to Planck's law for temperature and wavelength. Effectively non-reflecting.
Bølling	The period between the Older and the Oldest Dryas.
Boussinesq approximation	Neglect terms in a density anomaly except when multiplied by g(ravity) the assumption being that the fluid acceleration is small comared with g, having been caused by buoyancy (Archimedes) forces.
BP	Before the present (=1950)—referring to a number of years.
CCN	Cloud condensation nucleus.
Cirrus	A cloud form with the appearance of a curl of hair, composed of ice particles, most often hexagonal crystals although other forms occur.
CFC	Chlorofluorocarbon (*Indus*: 'Freon').
Chaos, chaotic	Of a state in which future states are not exactly predictable by means of a formula although they may be determined according to a specified manner of change (e.g. by a differential equation describing rate of change, or according to tabulated mathematical functions).
Channel	Referring to a wavelength band of NOAA meteorological satellites (see p. 446ff.)
Coefficient	A number having dimensions, which gives the magnitude of the essential feature of a mechanism or a multiplier in a (differential) equation.
Colloid	A suspension of particles all of the same size in a fluid with such similar fall speeds that coagulation does not occur.
Condensation level	The altitude at which ascending air becomes saturated and begins to form cloud.
Continuous	Having no breaks or points where a finite change in value takes place in an infinitesimal distance of time.
Convection	Transport of material, heat, pollution, etc. by fluid flow. Commonly used as an abbreviation for thermal (or buoyant) convection.
Conveyor belt	A descriptive name for a meteorological or oceanic current which carries heat from low to high latitudes in a particular region.
Contrail	Condensation trail, specifically of high level aircraft.

Coriolis The name of a philosopher who identified the apparent force found to accompany motion on a rotating body expressed mathematically with reference to a system of axes which is rotating in space. Thus, unlike linear velocity, rotation has an absolute value, i.e. a zero at which no Coriolis forces (e.g. centrifugal) are present.

Cumulonimbus A cumulus cloud producing rain, hail or snow.

Cumulus Literally a heap (Latin); applied to clouds shaped by thermal convection, like a cauliflower.

DE Differential equation.

Diffluent Of a fluid current—widening in the direction of flow, usually meaning sideways only.

Diffusion A spreading, usually in a turbulent manner, of a substance or particles, heat, pollution, etc. into a greater volume causing dilution.

Dispersion Of pollution to mean diffusion or scattering; a property of wave groups or packets in which different wavelengths have different speeds of transmission—velocity is a variable function of wavelength.

Divergent Of motion in which distance between any two nearby particles is increasing. Divergence = rate of expansion/per unit volume.

Doubly diffusive Of a system or medium in which two different qualities (e.g. heat, salinity) are diffused at different rates in the same medium (e.g. water).

DSDP Deep Sea Drilling Project.

Ecliptic The plane of the earth's orbit round the sun.

ECM Electrical conductivity measurement.

ECMWF European Centre for Medium-range Weather Forecasts (Reading, UK).

Eemian The last long interstadial before the Holocene, lasting from 135 to 115 kyr BP.

EGT Extended Glacialogical Timescale. Used to guide early core datings.

Ekman An oceanographer whose name is given to the spiral of the current vector below a wind-driven surface current.

Emulsion Densely packed fluid droplets held stably in a surrounding liquid.

Feedback When a system is subject to an outside influence a component of the system may be affected so that it either reduces or enhances the outside influence and the feedback is described as negative or positive accordingly.

Foraminifera Millimetre or smaller carbonate shells sampled in benthic cores.

GCM General circulation model. A mathematical description of the global atmosphere designed to predict its future state by computer.

GISP2 The Greenland Ice-Sheet Project 2. A bore hole in the ice in central Greenland at 3200 m at 72.6°N 38.5°W 30 km west of GRIP a previous bore in which the lowest layers cannot be interpreted to agree, probably because of ice flow in those layers due to the weight above.

Glint Reflection of sunshine from the sea surface seen in satellite images, the scatter depending on the sea roughness, prominent in Channels 1 and 2 but much more prominent in Channel 3.

GRIP The Greenland Icecore Project (1992) providing corroborative data about the abruptness of the changes during the last 40 kyr.

Heinrich event One of several relatively short-lived warm periods between the Eemian and the most recent warming. They refer to periods when icebergs thought to originate from North America carried deposits into the North Atlantic which have been detected in ocean bed boreholes.

Holocene The most recent (present) interstadial which began about 11.7 kyr BP.

IACP Inter Allerød Cold Period well indicated in GISP2, and enduring for about 250 y from 31.3 kyr BP. A brief foretaste of the Younger Dryas.

Ice cloud Cloud composed of particles all frozen.

Ice evaporation level The altitude at which an ice cloud in descending air begins to evaporate. It is not the condensation level in ascending air because ice crystals do not form when the air becomes saturated for ice (reaches the frost point) but has first to form supercooled water droplets at the dew point (see p. 471).

Interstadial A geologically warm period interrupting the stadials which are substages of the glacial stage which is the dominant feature.

Inversion Where the potential temperature increases upwards, a term used when the increase amounts almost to a discontinuity.

kyr Kiloyear(s) = 1000 yrs.

Lace-curtain cloud A very thin type of ocean stratocumulus, thin because of the unusually few CCN and larger than normal droplet size and low albedo in which ship trails are prominent with high albedo. The cloud is in a state of cellular convection.

Lapse rate Rate of decrease upwards (in the atmosphere) most commonly used in reference to temperature, potential temperature and humidity.

LGM Last Glacial Maximum.